W. Greiner

RELATIVISTIC QUANTUM MECHANICS

D1524608

Springer
Berlin
Heidelberg
New York
Barcelona
Budapest
Hong Kong
London
Milan
Paris
Santa Clara
Singapore
Tokyo

Greiner
Quantum Mechanics
An Introduction 3rd Edition

Greiner
Quantum Mechanics
Special Chapters

Greiner · Müller
Quantum Mechanics
Symmetries 2nd Edition

Greiner
Relativistic Quantum Mechanics
Wave Equations 2nd Edition

Greiner · Reinhardt
Field Quantization

Greiner · Reinhardt
Quantum Electrodynamics
2nd Edition

Greiner · Schäfer
Quantum Chromodynamics

Greiner · Maruhn
Nuclear Models

Greiner · Müller
Gauge Theory of Weak Interactions
2nd Edition

Greiner
Mechanics I
(in preparation)

Greiner
Mechanics II
(in preparation)

Greiner
Electrodynamics
(in preparation)

Greiner · Neise · Stöcker
Thermodynamics
and Statistical Mechanics

Walter Greiner

RELATIVISTIC QUANTUM MECHANICS

WAVE EQUATIONS

With a Foreword by
D. A. Bromley

Second Revised Edition
With 62 Figures
and 89 Worked Examples and Problems

Springer

Professor Dr. Walter Greiner
Institut für Theoretische Physik der
Johann Wolfgang Goethe-Universität Frankfurt
Postfach 111932
D-60054 Frankfurt am Main
Germany

Street address:

Robert-Mayer-Strasse 8–10
D-60325 Frankfurt am Main
Germany

email: greiner@th.physik.uni-frankfurt.de

Title of the original German edition: *Theoretische Physik,* Ein Lehr- und Übungsbuch,
Band 6: Relativistische Quantenmechanik, Wellengleichungen

Library of Congress Cataloging-in-Publication Data.

Greiner, Walter: Relativistic quantum mechanics: wave equations; with 89 worked examples and problems / Walter Greiner. With a foreword by D. A. Bromley. – 2., rev. ed. – Berlin; Heidelberg; New York; Barcelona; Budapest; Hong Kong; London; Milan; Paris; Santa Clara; Singapore; Tokyo: Springer, 1997
Einheitssacht.: Relativistische Quantenmechanik <engl.>
ISBN 3-540-61621-7

ISBN 3-540-61621-7 2nd Edition Springer-Verlag Berlin Heidelberg New York

ISBN 3-540-50986-0 1st Edition Springer-Verlag Berlin Heidelberg New York

Typesetting: A. Leinz, Karlsruhe
Cover design; Design Concept, Emil Smejkal, Heidelberg
Copy Editor: V. Wicks
Production Editor: P. Treiber
SPIN 10547305 56/3144 - 5 4 3 2 1 0 - Printed on acid-free paper

Preface to the Second Edition

For its second edition the book *Relativistic Quantum Mechanics – Wave Equations* has undergone only minor revisions. A number of misprints and errors in a few equations have been corrected. Also the typographical appearance and layout of the book has been improved. I hope that the book will continue to be useful to students and teachers alike.

I thank Markus Bleicher for his help and acknowledge the agreeable collaboration with Dr. H. J. Kölsch and his team at Springer-Verlag, Heidelberg.

Frankfurt am Main, *Walter Greiner*
March 1997

Preface to the First Edition

Theoretical physics has become a many-faceted science. For the young student it is difficult enough to cope with the overwhelming amount of new scientific material that has to be learned, let alone obtain an overview of the entire field, which ranges from mechanics through electrodynamics, quantum mechanics, field theory, nuclear and heavy-ion science, statistical mechanics, thermodynamics, and solid-state theory to elementary-particle physics. And this knowledge should be acquired in just 8–10 semesters, during which, in addition, a Diploma or Master's thesis has to be worked on or examinations prepared for. All this can be achieved only if the university teachers help to introduce the student to the new disciplines as early on as possible, in order to create interest and excitement that in turn set free essential new energy. Naturally, all inessential material must simply be eliminated.

At the Johann Wolfgang Goethe University in Frankfurt we therefore confront the student with theoretical physics immediately, in the first semester. Theoretical Mechanics I and II, Electrodynamics, and Quantum Mechanics I – An Introduction are the basic courses during the first two years. These lectures are supplemented with many mathematical explanations and much support material. After the fourth semester of studies, graduate work begins, and Quantum Mechanics II – Symmetries, Statistical Mechanics and Thermodynamics, Relativistic Quantum Mechanics, Quantum Electrodynamics, the Gauge Theory of Weak Interactions, and Quantum Chromodynamics are obligatory. Apart from these a number of supplementary courses on special topics are offered, such as Hydrodynamics, Classical Field Theory, Special and General Relativity, Many-Body Theories, Nuclear Models, Models of Elementary Particles, and Solid-State Theory. Some of them, for example the two-semester courses Theoretical Nuclear Physics and Theoretical Solid-State Physics, are also obligatory.

The form of the lectures that comprise *Relativistic Quantum Mechanics – Wave Equations* follows that of all the others: together with a broad presentation of the necessary mathematical tools, many examples and exercises are worked through. We try to offer science in as interesting a way as possible. With relativistic quantum mechanics we are dealing with a broad, yet beautiful, theme. Therefore we have had to restrict ourselves to relativistic wave equations. The selected material is perhaps unconventional, but corresponds, in our opinion, to the importance of this field in modern physics:

The Klein–Gordon equation (for spin-0 particles) and the Dirac equation (for spin-$\frac{1}{2}$ particles) and their applications constitute the backbone of these lectures. Wave equations for particles with higher spin (the Rarita–Schwinger, spin-$\frac{3}{2}$, Kemmer and Proca, spin-1, and general Bargmann–Wigner equations) are confined to the last chapters.

After introducing the Klein–Gordon equation we discuss its properties and difficulties (especially with respect to the single-particle interpretation); the Feshbach–Villars representation is given. In many worked-out exercises and examples its practical applications can be found: pionic atoms as a modern field of research and the particularly challenging examples on the effective pion–nucleus potential (the Kisslinger potential) and its improvement by Ericson and Ericson stand in the foreground.

Most of these lectures deal with Dirac's theory. The covariance properties of the Dirac equation are discussed in detail. So, for example, its free solutions are on the one hand determined directly and on the other hand through Lorentz transformations from the simple solutions in the rest frame. Here the methodical issue is emphasized: the same physical phenomenon is illuminated from different angles. We proceed in a similar manner in the discussion of single-particle operators (the odd and even parts of an operator) and the so-called *Zitterbewegung*, which is also derived from the consideration of wave packets of plane Dirac waves. In many worked-out problems and examples the new tools are exercised. Thus the whole of Chap. 9 is dedicated to the motion of Dirac particles in external potentials. It contains simple potential problems, extensively the case of the electron in a Coulomb potential (the fine-structure formula), and muonic atoms. In Chap. 10 we present the two-centre Dirac equation, which is of importance in the modern field of heavy-ion atomic physics. The fundamental problem of overcritical fields and the decay of the electron–positron vacuum is only touched upon. A full treatment is reserved for *Quantum Electrodynamics* (Vol. 4 of this series). However, we give an extended discussion of hole theory and also of Klein's paradox. The Weyl equation for the neutrino (Chap. 14) and relativistic wave equations for particles with arbitrary spin (Chap. 15) follow. Starting with the Bargmann–Wigner equations the general frame for these equations is set, and in numerous worked-out examples and exercises special cases (spin-1 particles with and without mass, and spin-$\frac{3}{2}$ particles according to Rarita and Schwinger) are considered in greater detail. In the last chapter we give an overview of relativistic symmetry principles, which we enjoy from a superior point of view, since by now we have studied *Quantum Mechanics – Symmetries* (Vol. 2 of this series).

We hope that in this way the lectures will become ever more complete and may lead to new insights.

Biographical notes help to obtain an impression, however short, of the life and work of outstanding physicists and mathematicians. We gratefully acknowledge the publishers Harri Deutsch and F.A. Brockhaus (*Brockhaus Enzyklopädie*, F.A. Brockhaus – Wiesbaden indicated by BR) for giving permission to use relevant information from their publications.

Special thanks go to Prof. Dr. Gerhard Soff, Dr. Joachim Reinhardt, and Dr. David Vasak for their critical reading of the original draft of these lectures. Many students and collaborators have helped during the years to work out examples and exercises. For this first English edition we enjoyed the help of Maria Berenguer, Christian Borchert, Snježana Butorac, Christian Derreth, Carsten Greiner, Kordt Griepenkerl, Christian Hofmann, Raffele Mattiello, Dieter Neubauer, Jochen Rau, Wolfgang Renner, Dirk Rischke, Alexander Scherdin, Thomas Schönfeld, and

Dr. Stefan Schramm. Miss Astrid Steidl drew the graphs and prepared the figures. To all of them we express our sincere thanks.

We would especially like to thank Mr. Béla Waldhauser, Dipl.-Phys., for his overall assistance. His organizational talent and his advice in technical matters are very much appreciated.

Finally, we wish to thank Springer-Verlag; in particular, Dr. H.-U. Daniel, for his encouragement and patience, Mr. Michael Edmeades for expertly copy-editing the English edition, and Mr. R. Michels and his team for the excellent layout.

Frankfurt am Main, *Walter Greiner*
May 1990

Foreword to Earlier Series Editions

More than a generation of German-speaking students around the world have worked their way to an understanding and appreciation of the power and beauty of modern theoretical physics – with mathematics, the most fundamental of sciences – using Walter Greiner's textbooks as their guide.

The idea of developing a coherent, complete presentation of an entire field of science in a series of closely related textbooks is not a new one. Many older physicists remember with real pleasure their sense of adventure and discovery as they worked their ways through the classic series by Sommerfeld, by Planck and by Landau and Lifshitz. From the students' viewpoint, there are a great many obvious advantages to be gained through use of consistent notation, logical ordering of topics and coherence of presentation; beyond this, the complete coverage of the science provides a unique opportunity for the author to convey his personal enthusiasm and love for his subject.

The present five-volume set, *Theoretical Physics*, is in fact only that part of the complete set of textbooks developed by Greiner and his students that presents the quantum theory. I have long urged him to make the remaining volumes on classical mechanics and dynamics, on electromagnetism, on nuclear and particle physics, and on special topics available to an English-speaking audience as well, and we can hope for these companion volumes covering all of theoretical physics some time in the future.

What makes Greiner's volumes of particular value to the student and professor alike is their completeness. Greiner avoids the all too common "it follows that ..." which conceals several pages of mathematical manipulation and confounds the student. He does not hesitate to include experimental data to illuminate or illustrate a theoretical point and these data, like the theoretical content, have been kept up to date and topical through frequent revision and expansion of the lecture notes upon which these volumes are based.

Moreover, Greiner greatly increases the value of his presentation by including something like one hundred completely worked examples in each volume. Nothing is of greater importance to the student than seeing, in detail, how the theoretical concepts and tools under study are applied to actual problems of interest to a working physicist. And, finally, Greiner adds brief biographical sketches to each chapter covering the people responsible for the development of the theoretical ideas and/or the experimental data presented. It was Auguste Comte (1798–1857) in his *Positive Philosophy* who noted, "To understand a science it is necessary to know its history". This is all too often forgotten in modern physics teaching and the bridges that Greiner builds to the pioneering figures of our science upon whose work we build are welcome ones.

Greiner's lectures, which underlie these volumes, are internationally noted for their clarity, their completeness and for the effort that he has devoted to making physics an integral whole; his enthusiasm for his science is contagious and shines through almost every page.

These volumes represent only a part of a unique and Herculean effort to make all of theoretical physics accessible to the interested student. Beyond that, they are of enormous value to the professional physicist and to all others working with quantum phenomena. Again and again the reader will find that, after dipping into a particular volume to review a specific topic, he will end up browsing, caught up by often fascinating new insights and developments with which he had not previously been familiar.

Having used a number of Greiner's volumes in their original German in my teaching and research at Yale, I welcome these new and revised English translations and would recommend them enthusiastically to anyone searching for a coherent overview of physics.

Yale University
New Haven, CT, USA
1989

D. Allan Bromley
Henry Ford II Professor of Physics

Contents

Contents of Examples and Exercises

1. Relativistic Wave Equation for Spin-0 Particles: The Klein–Gordon Equation and Its Applications

The description of phenomena at high energies requires the investigation of relativistic wave equations. This means equations which are invariant under Lorentz transformations. The transition from a nonrelativistic to a relativistic description implies that several concepts of the nonrelativistic theory have to be reinvestigated, in particular:

(1) Spatial and temporal coordinates have to be treated equally within the theory.

(2) Since

$$\Delta x \sim \frac{\hbar}{\Delta p} \sim \frac{\hbar}{m_0 c} \quad ,$$

a relativistic particle cannot be localized more accurately than $\approx \hbar/m_0 c$; otherwise pair creation occurs for $E > 2m_0c^2$. Thus, the idea of a free particle only makes sense if the particle is not confined by external constraints to a volume which is smaller than approximately the Compton wavelength $\lambda_c = \hbar/m_0 c$. Otherwise the particle automatically has companions due to particle–antiparticle creation.

(3) If the position of the particle is uncertain, i.e. if

$$\Delta x > \frac{\hbar}{m_0 c} \quad ,$$

then the time is also uncertain, because

$$\Delta t \sim \frac{\Delta x}{c} > \frac{\hbar}{m_0 c^2} \quad .$$

In a nonrelativistic theory Δt can become arbitrarily small, because $c \to \infty$. Thereby, we recognize the necessity to reconsider the concept of probability density

$$\varrho(x, y, z, t) \quad ,$$

which describes the probability of finding a particle at a definite place $\boldsymbol{r}$ at fixed time t.

(4) At high (relativistic) energies pair creation and annihilation processes occur, usually in the form of creating particle–antiparticle pairs. Thus, at relativistic energies particle conservation is no longer a valid assumption. A relativistic theory must be able to describe pair creation, vacuum polarization, particle conversion, etc.

1.1 The Notation

First we shall remark on the notation used. Until now we have expressed four-vectors by Minkowski's notation, with an imaginary fourth component, as for example

$$
\begin{aligned}
x &= \{x, y, z, \mathrm{i}ct\} && \text{(world vector)} \quad , \\
p &= \{p_x, p_y, p_z, \mathrm{i}E/c\} && \text{(four-momentum)} \quad , \\
A &= \{A_x, A_y, A_z, \mathrm{i}A_0\} && \text{(four-potential)} \quad , \\
\nabla &= \left\{\frac{\partial}{\partial x}, \frac{\partial}{\partial y}, \frac{\partial}{\partial z}, \frac{\partial}{\mathrm{i}\partial(ct)}\right\} && \text{(four-gradient), etc.} \quad .
\end{aligned} \tag{1.1}
$$

The letters x, p, A, ∇ abbreviate the full four-vector. Sometimes we shall also denote them by $\overset{\Rightarrow}{x}$, $\overset{\Rightarrow}{p}$, $\overset{\Rightarrow}{A}$, $\overset{\Rightarrow}{\nabla}$, etc., i.e. with a double arrow. As long as there is no confusion arising, we prefer the former notation. For the following it is useful to introduce the metric tensor (*covariant compontents*)

$$
g_{\mu\nu} = \begin{pmatrix} g_{00} & g_{01} & g_{02} & g_{03} \\ g_{10} & g_{11} & g_{12} & g_{13} \\ g_{20} & g_{21} & g_{22} & g_{23} \\ g_{30} & g_{31} & g_{32} & g_{33} \end{pmatrix} = \begin{pmatrix} 1 & 0 & 0 & 0 \\ 0 & -1 & 0 & 0 \\ 0 & 0 & -1 & 0 \\ 0 & 0 & 0 & -1 \end{pmatrix} \quad . \tag{1.2}
$$

Thereby, one can denote the length of the vector $\mathrm{d}x = \{\mathrm{d}x^\mu\}$ as $\mathrm{d}s^2 = \mathrm{d}x\,\mathrm{d}x = g_{\mu\nu}\,\mathrm{d}x^\mu\,\mathrm{d}x^\nu$. This relation is often taken as the defining relation of the metric tensor.[1] The *contravariant form* $g^{\mu\nu}$ of the metric tensor follows from the condition

$$
g^{\mu\sigma} g_{\sigma\nu} = \delta^\mu_\nu \overset{\text{def}}{\equiv} \begin{pmatrix} 1 & 0 & 0 & 0 \\ 0 & 1 & 0 & 0 \\ 0 & 0 & 1 & 0 \\ 0 & 0 & 0 & 1 \end{pmatrix} \quad , \tag{1.3}
$$

$$
g^{\mu\sigma} = \left(g^{-1}\right)_{\mu\sigma} = \frac{\Delta_{\mu\sigma}}{g} = \begin{pmatrix} 1 & 0 & 0 & 0 \\ 0 & -1 & 0 & 0 \\ 0 & 0 & -1 & 0 \\ 0 & 0 & 0 & -1 \end{pmatrix} \quad . \tag{1.4}
$$

Here $\Delta_{\mu\sigma}$ is the cofactor of $g_{\mu\sigma}$ [i.e. the subdeterminant, obtained by crossing out the μth row and the σth column and multiplying it with the phase $(-1)^{\mu+\sigma}$] and g is given by $g = \det(g_{\mu\nu}) = -1$. For the special Lorentz metric the contravariant and covariant metric tensor are identical:

$$
g^{\mu\nu} = g_{\mu\nu} \quad \text{[for Lorentz metric!]} \quad .
$$

From now on we will use the contravariant four-vector

$$
x^\mu = \left\{x^0, x^1, x^2, x^3\right\} \equiv \{ct, x, y, z\} \tag{1.5}
$$

for the description of the space-time coordinates, where the time-like component is denoted as zero component. We get the covariant form of the four-vector by "lowering" the index μ with the help of the metric tensor, i.e.

---

[1] We adopt the same notation as J.D. Bjorken, S.D. Drell: *Relativistic Quantum Mechanics* (McGraw Hill, New York 1964).

$$x_\mu = g_{\mu\nu}x^\nu = \{ct, -x, -y, -z\} = \{x_0, x_1, x_2, x_3\} \quad . \tag{1.6}$$

Similarly the indices can be "raised" to give

$$x^\mu = g^{\mu\nu}x_\nu = \{x^0, x^1, x^2, x^3\} \quad .$$

This means that one can easily transform the covariant into the contravariant form of a vector (respectively of a tensor) and vice versa. Except in special cases, where we denote it explicitly, we use *Einstein's summation convention*: We automatically add from 0 to 3 over indices occuring doubly (one upper and one lower index). So we have, for example,

$$\begin{aligned} x \cdot x = x^\mu x_\mu \equiv \sum_{\mu=0}^{3} x^\mu x_\mu &= x^0 x_0 + x^1 x_1 + x^2 x_2 + x^3 x_3 \\ &= c^2t^2 - x^2 - y^2 - z^2 \\ &= c^2t^2 - \boldsymbol{x}^2 \quad . \end{aligned} \tag{1.7}$$

The definition of the four-momentum vector is analogous,

$$p^\mu = \{E/c, p_x, p_y, p_z\} \quad , \tag{1.8}$$

and we write the scalar product in four dimensions (space-time) as

$$p_1 \cdot p_2 = p_1^\mu p_{2\mu} = \frac{E_1}{c}\frac{E_2}{c} - \boldsymbol{p}_1 \cdot \boldsymbol{p}_2 \quad , \tag{1.9}$$

or equally

$$x \cdot p = x^\mu p_\mu = x_\mu p^\mu = Et - \boldsymbol{x} \cdot \boldsymbol{p} \quad . \tag{1.10}$$

We identify the four-vectors by a common letter. Thus, for instance,

$$a = \{a_0, a_1, a_2, a_3\} \quad .$$

In contrast to this we denote three-vectors by bold type as in

$$\boldsymbol{a} = \{a_1, a_2, a_3\} \quad .$$

Often we write only the components. Hence,

$$a^\mu = \{a^0, a^1, a^2, a^3\}$$

means a four-vector with contravariant components. Greek indices, such as μ, always run from 0 to 3. Latin indices, as for example i, imply values from 1 to 3. A three-vector can thus also be written in contravariant form as

$$a^i = \{a^1, a^2, a^3\} \quad \text{or as} \quad a_i = \{a_1, a_2, a_3\}$$

in covariant form. So the *four-momentum operator* is therefore denoted by

$$\begin{aligned} \hat{p}^\mu = \mathrm{i}\hbar\frac{\partial}{\partial x_\mu} &= \left\{\mathrm{i}\hbar\frac{\partial}{\partial(ct)}, +\mathrm{i}\hbar\frac{\partial}{\partial x_1}, +\mathrm{i}\hbar\frac{\partial}{\partial x_2}, +\mathrm{i}\hbar\frac{\partial}{\partial x_3}\right\} \\ \equiv \mathrm{i}\hbar\boldsymbol{\nabla}^\mu &= \left\{\frac{\partial}{\partial(ct)}, -\mathrm{i}\hbar\frac{\partial}{\partial x}, -\mathrm{i}\hbar\frac{\partial}{\partial y}, -\mathrm{i}\hbar\frac{\partial}{\partial z}\right\} \\ &= \mathrm{i}\hbar\left\{\frac{\partial}{\partial(ct)}, -\boldsymbol{\nabla}\right\} \quad . \end{aligned} \tag{1.11}$$

It transforms as a contravariant four-vector, so that

$$\begin{aligned}\hat{p}^{\mu}\hat{p}_{\mu} &= -\hbar^2 \frac{\partial}{\partial x_\mu}\frac{\partial}{\partial x^\mu} = -\hbar^2\left(\frac{1}{c^2}\frac{\partial^2}{\partial t^2} - \left(\frac{\partial^2}{\partial x^2}+\frac{\partial^2}{\partial y^2}+\frac{\partial^2}{\partial z^2}\right)\right)\\ &\equiv -\hbar^2\Box = -\hbar^2\left(\frac{1}{c^2}\frac{\partial^2}{\partial t^2} - \Delta\right) \quad . \end{aligned} \tag{1.12}$$

This equation defines both the three-dimensional delta operator ($\Delta = \nabla^2$) and the four-dimensional d'Alembertian ($\Box = \partial^2/(c^2\partial t^2) - \Delta$). Finally we check the commutation relations of momentum and position by means of (1.11 and 1.5), obtaining

$$\begin{aligned}\left[\hat{p}^{\mu}, x^{\nu}\right]_- &= \mathrm{i}\hbar\left[\frac{\partial}{\partial x_\mu}, g^{\nu\sigma}x_\sigma\right]_- = \mathrm{i}\hbar g^{\nu\sigma}\frac{\partial x_\sigma}{\partial x_\mu}\\ &= \mathrm{i}\hbar g^{\nu\sigma}\delta^{\mu}_{\sigma} = \mathrm{i}\hbar g^{\nu\mu} = \mathrm{i}\hbar g^{\mu\nu} \quad . \end{aligned} \tag{1.13}$$

On the right-hand side (rhs), the metric tensor $g^{\mu\nu}$ appears expressing the *covariant form of the commutation relation.*

The four-potential of the electromagnetic field is given by

$$A^{\mu} = \{A_0, \boldsymbol{A}\} = \{A_0, A_x, A_y, A_z\} = g^{\mu\nu}A_\nu \quad . \tag{1.14}$$

Here A^μ are the contravariant, and $A_\mu = \{A_0, -A_x, -A_y, -A_z\}$ the covariant components. From A^μ the electromagnetic field tensor follows in the well-known way:

$$F^{\mu\nu} = \frac{\partial A^\mu}{\partial x_\nu} - \frac{\partial A^\nu}{\partial x_\mu} = \begin{pmatrix} 0 & E_x & E_y & E_z \\ -E_x & 0 & B_z & -B_y \\ -E_y & -B_z & 0 & B_x \\ -E_z & B_y & -B_x & 0 \end{pmatrix} \quad . \tag{1.15}$$

1.2 The Klein–Gordon Equation

From elementary quantum mechanics[2] we know the Schrödinger equation

$$\mathrm{i}\hbar\frac{\partial\psi}{\partial t} = \left[-\frac{\hbar^2}{2m_0}\boldsymbol{\nabla}^2 + V(\boldsymbol{x})\right]\psi(\boldsymbol{x}, t) \tag{1.16}$$

corresponds to the nonrelativistic energy relation in operator form,

$$\hat{E} = \frac{\hat{\boldsymbol{p}}^2}{2m_0} + V(\boldsymbol{x}) \quad , \quad \text{where} \tag{1.17}$$

$$\hat{E} = \mathrm{i}\hbar\frac{\partial}{\partial t} \quad , \quad \hat{\boldsymbol{p}} = -\mathrm{i}\hbar\boldsymbol{\nabla} \tag{1.18}$$

are the operators of energy and momentum, respectively. In order to obtain a relativistic wave equation we start by considering free particles with the relativistic relation

[2] See W. Greiner: *Quantum Mechanics – An Introduction*, 3rd ed. (Springer, Berlin, Heidelberg 1994).

$$p^\mu p_\mu = \frac{E^2}{c^2} - \boldsymbol{p}\cdot\boldsymbol{p} = m_0^2 c^2 \quad . \tag{1.19}$$

We now replace the four-momentum p^μ by the four-momentum operator

$$\begin{aligned} \hat{p}^\mu = \mathrm{i}\hbar\frac{\partial}{\partial x_\mu} &= \mathrm{i}\hbar\left\{\frac{\partial}{\partial(ct)}, -\frac{\partial}{\partial x}, -\frac{\partial}{\partial y}, -\frac{\partial}{\partial z}\right\} \\ &= \mathrm{i}\hbar\left\{\frac{\partial}{\partial(ct)}, -\boldsymbol{\nabla}\right\} = \{\hat{p}_0, \hat{\boldsymbol{p}}\} \quad . \end{aligned} \tag{1.20}$$

Following (1.6) and (1.11), the result is in accordance with (1.18). Thus, we obtain the **Klein–Gordon** equation for free particles,

$$\hat{p}^\mu \hat{p}_\mu \psi = m_0^2 c^2 \psi \quad . \tag{1.21}$$

Here m_0 is the rest mass of the particle and c the velocity of light in vacuum. With the help of (1.12) we can write (1.21) in the form

$$\left(\Box + \frac{m_0^2c^2}{\hbar^2}\right)\psi = \left(\frac{\partial^2}{c^2\partial t^2} - \frac{\partial^2}{\partial x^2} - \frac{\partial^2}{\partial y^2} - \frac{\partial^2}{\partial z^2} + \frac{m_0^2c^2}{\hbar^2}\right)\psi = 0 \quad . \tag{1.22}$$

We can immediately verify the Lorentz covariance of the Klein–Gordon equation, as $\hat{p}^\mu \hat{p}_\mu$ is Lorentz invariant. We also recognize (1.22) as the classical wave equation including the *mass term* $m_0^2c^2/\hbar^2$. Free solutions are of the form

$$\begin{aligned} \psi = \exp\left(-\frac{\mathrm{i}}{\hbar}p_\mu x^\mu\right) &= \exp\left[-\frac{\mathrm{i}}{\hbar}\left(p_0 x^0 - \boldsymbol{p}\cdot\boldsymbol{x}\right)\right] \\ = \exp\left[+\frac{\mathrm{i}}{\hbar}(\boldsymbol{p}\cdot\boldsymbol{x} - Et)\right] & \quad . \end{aligned} \tag{1.23}$$

Indeed, insertion of (1.23) into (1.21) leads to the condition

$$\begin{aligned} \hat{p}_\mu \hat{p}^\mu \psi = m_0^2c^2\psi \to p^\mu p_\mu \exp\left(-\frac{\mathrm{i}}{\hbar}p_\mu x^\mu\right) &= m_0^2c^2 \exp\left(-\frac{\mathrm{i}}{\hbar}p_\mu x^\mu\right) \\ \to p^\mu p_\mu = m_0^2c^2 \quad \text{or} \quad \frac{E^2}{c^2} - \boldsymbol{p}\cdot\boldsymbol{p} &= m_0^2c^2 \quad , \end{aligned}$$

which results in

$$E = \pm\sqrt{m_0^2c^2 + \boldsymbol{p}^2} \quad . \tag{1.24}$$

Thus, there exist solutions both for positive $E = +c(m_0^2c^2 + \boldsymbol{p}^2)^{1/2}$ as well as for negative $E = -c(m_0^2c^2 + \boldsymbol{p}^2)^{1/2}$ energies respectively (see Fig. 1.1). We shall see later that the solutions yielding negative energy are physically connected with antiparticles. Since antiparticles can indeed be observed in nature, we have already obtained an indication of the value of extending the nonrelativistic theory.

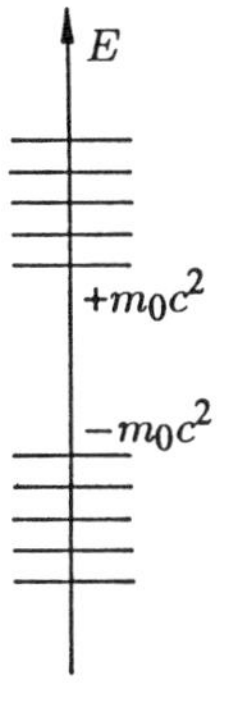

Fig. 1.1. Energy spectrum of the free Klein–Gordon equation

Next we construct the four-current j_μ connected with (1.21). In analogy to our considerations concerning the Schrödinger equation, we expect a conservation law for the j_μ. We start from (1.22), in the form

$$\left(\hat{p}_\mu \hat{p}^\mu - m_0^2c^2\right)\psi = 0 \quad ,$$

and take the complex conjugate of this equation, i.e.

$$\left(\hat{p}_\mu \hat{p}^\mu - m_0^2 c^2\right) \psi^* = 0 \quad .$$

Multiplying both equations from the left, the first by ψ^* and the second by ψ, and calculating the difference of the resulting two equations yields

$$\psi^* \left(\hat{p}_\mu \hat{p}^\mu - m_0^2 c^2\right) \psi - \psi \left(\hat{p}_\mu \hat{p}^\mu - m_0^2 c^2\right) \psi^* = 0$$

or

$$\begin{aligned} & -\psi^* \left(\hbar^2 \nabla_\mu \nabla^\mu + m_0^2 c^2\right) \psi + \psi \left(\hbar^2 \nabla_\mu \nabla^\mu + m_0^2 c^2\right) \psi^* = 0 \\ & \quad \Rightarrow \nabla_\mu \left(\psi^* \nabla^\mu \psi - \psi \nabla^\mu \psi^*\right) \equiv \nabla_\mu j^\mu = 0 \quad . \end{aligned} \tag{1.25}$$

The four-current density is therefore

$$j_\mu = \frac{\mathrm{i}\hbar}{2m_0} \left(\psi^* \nabla_\mu \psi - \psi \nabla_\mu \psi^*\right) \quad . \tag{1.26}$$

Here we have multiplied by $\mathrm{i}\hbar/2m_0$, so that the zero component j_0 has the dimension of a probability density (that is 1/cm^3). Furthermore this ensures that we obtain the correct nonrelativistic limit [cf. (1.30–31)] below. In detail, (1.25) reads

$$\begin{aligned} & \frac{\partial}{\partial t} \left[\frac{\mathrm{i}\hbar}{2m_0 c^2} \left(\psi^* \frac{\partial \psi}{\partial t} - \psi \frac{\partial \psi^*}{\partial t}\right)\right] \\ & \quad + \operatorname{div} \left(\frac{-\mathrm{i}\hbar}{2m_0}\right) \left[\psi^* (\nabla \psi) - \psi (\nabla \psi^*)\right] = 0 \quad . \end{aligned} \tag{1.27}$$

This expression possesses the form of a continuity equation

$$\frac{\partial \varrho}{\partial t} + \operatorname{div} \boldsymbol{j} = 0 \quad . \tag{1.28}$$

As usual integration over the entire configuration space yields

$$\int_V \frac{\partial \varrho}{\partial t} \mathrm{d}^3 x = \frac{\partial}{\partial t} \int_V \varrho \, \mathrm{d}^3 x = -\int_V \operatorname{div} \boldsymbol{j} \, \mathrm{d}^3 x = -\int_F \boldsymbol{j} \cdot \mathrm{d}\boldsymbol{F} = 0 \quad .$$

Hence,

$$\int_V \varrho \, \mathrm{d}^3 x = \text{const.} \quad ,$$

i.e. $\int_V \varrho \, \mathrm{d}^3 x$ is constant in time. It would be a natural guess to interpret

$$\varrho = \frac{\mathrm{i}\hbar}{2m_0 c^2} \left(\psi^* \frac{\partial \psi}{\partial t} - \psi \frac{\partial \psi^*}{\partial t}\right) \tag{1.29}$$

as a probability density. However, there is a problem with such an interpretation: At a given time t both ψ and $\partial\psi/\partial t$ may have arbitrary values; therefore, $\varrho(\boldsymbol{x}, t)$ in (1.29) may be either positive or negative. Hence, $\varrho(\boldsymbol{x}, t)$ is not positive definite and thus not a probability density. The deeper reason for this is that the Klein–Gordon equation is of second order in time, so that we must know both $\psi(\boldsymbol{x}, t)$ and $\partial\psi(\boldsymbol{x}, t)/\partial t$ for a given t. Furthermore there exist solutions for negative energy [see

(12.4) and (1.38) below]. This and the difficulty with the probability interpretation was the reason that, for a long time, the Klein–Gordon equation was regarded to be physically senseless. One therefore looked for a relativistic wave equation of first order in time with positive definite probability, which was finally derived by Dirac (cf. Chap. 2). However, it turns out that this equation has negative energy solutions too. As we have previously remarked and as we shall discuss in greater detail later, in Chap. 2, these solutions are connected with the existence of antiparticles.

1.3 The Nonrelativistic Limit

We can study the nonrelativistic limit of the Klein–Gordon equation (1.21). In order to do this we make the ansatz

$$\psi(\boldsymbol{r},t) = \varphi(\boldsymbol{r},t)\exp\left(-\frac{\mathrm{i}}{\hbar}m_0c^2t\right) \quad , \tag{1.30}$$

i.e. we split the time dependence of ψ into two terms, one containing the rest mass. In the nonrelativistic limit the difference of total energy E of the particle and the rest mass m_0c^2 is small. Therefore we define

$$E' = E - m_0c^2$$

and remark that the kinetic energy E' is nonrelativistic, which means $E' \ll m_0c^2$. Hence,

$$\left|\mathrm{i}\hbar\frac{\partial\varphi}{\partial t}\right| \approx E'\varphi \ll m_0c^2\varphi \tag{1.31}$$

holds also and with (1.30) we have

$$\frac{\partial\psi}{\partial t} = \left(\frac{\partial\varphi}{\partial t} - \mathrm{i}\frac{m_0c^2}{\hbar}\varphi\right)\exp\left(-\frac{\mathrm{i}}{\hbar}m_0c^2t\right) \approx -\mathrm{i}\frac{m_0c^2}{\hbar}\varphi\exp\left(-\frac{\mathrm{i}}{\hbar}m_0c^2t\right)$$

$$\begin{aligned}\frac{\partial^2\psi}{\partial t^2} &= \frac{\partial}{\partial t}\left(\frac{\partial\varphi}{\partial t} - \mathrm{i}\frac{m_0c^2}{\hbar}\varphi\right)\exp\left(-\frac{\mathrm{i}}{\hbar}m_0c^2t\right)\\ &\approx \left[-\mathrm{i}\frac{m_0c^2}{\hbar}\frac{\partial\varphi}{\partial t} - \mathrm{i}\frac{m_0c^2}{\hbar}\frac{\partial\varphi}{\partial t} - \frac{m_0^2c^4}{\hbar^2}\varphi\right]\exp\left(-\frac{\mathrm{i}}{\hbar}m_0c^2t\right)\\ &= -\left[\mathrm{i}\frac{2m_0c^2}{\hbar}\frac{\partial\varphi}{\partial t} + \frac{m_0^2c^4}{\hbar^2}\varphi\right]\exp\left(-\frac{\mathrm{i}}{\hbar}m_0c^2t\right) \quad .\end{aligned}$$

Inserting this result into (1.21) yields

$$\begin{aligned}&-\frac{1}{c^2}\left[\mathrm{i}\frac{2m_0c^2}{\hbar}\frac{\partial\varphi}{\partial t} + \frac{m_0^2c^4}{\hbar^2}\varphi\right]\exp\left(-\frac{\mathrm{i}}{\hbar}m_0c^2t\right)\\ &\quad = \left(\frac{\partial^2}{\partial x^2} + \frac{\partial^2}{\partial y^2} + \frac{\partial^2}{\partial z^2} - \frac{m_0^2c^2}{\hbar^2}\right)\varphi\exp\left(-\frac{\mathrm{i}}{\hbar}m_0c^2t\right)\end{aligned}$$

or

$$\mathrm{i}\hbar\frac{\partial\varphi}{\partial t} = -\frac{\hbar^2}{2m_0}\left(\frac{\partial^2}{\partial x^2} + \frac{\partial^2}{\partial y^2} + \frac{\partial^2}{\partial z^2}\right)\varphi = -\frac{\hbar^2}{2m_0}\Delta\varphi \quad . \tag{1.32}$$

This is the free Schrödinger equation for spinless particles. As the type of particle which is described by a wave equation does not depend upon whether the particle is relativistic or nonrelativistic, we infer that *the Klein–Gordon equation describes spin-zero particles.* Later on we will obtain this important result in a totally different manner by making use of the transformation properties of the Klein–Gordon field ψ.

1.4 Free Spin-0 Particles

Previously we have remarked that in a relativistic theory *the concept of a free particle is an idealization.* Furthermore spin-zero particles, like pions or kaons, interact strongly with other particles and fields. Nevertheless we can discover some of the practical methods for dealing with these problems by studying the free solutions of (1.21). We return to the interpretation of the current density (1.26) that we discarded due to ϱ in (1.29) not being positive definite. As usual integrating the continuity equation (1.28) yields

$$\int_V \frac{\partial \varrho}{\partial t} \mathrm{d}^3x = \frac{\partial}{\partial t} \int_V \varrho(\boldsymbol{x}, t) \mathrm{d}^3x = -\int_V \operatorname{div} \boldsymbol{j} \, \mathrm{d}^3x = -\int_F \boldsymbol{j} \cdot \mathrm{d}\boldsymbol{F} = 0 \quad ,$$

which means that

$$\int_V \varrho \, \mathrm{d}^3x = \text{const.} \quad , \tag{1.33}$$

i.e. the constancy in time of the normalization (which is a reasonable result). The question remains of how to interpret ϱ and $\boldsymbol{j}$. The probability interpretation is not applicable, as we have just seen in context with (1.29). However, we do have the following alternative: We obtain the *four-current density of charge* by multiplication of the current density (1.26) with the elementary charge e to give

$$j'_\mu = \frac{\mathrm{i}e\hbar}{2m_0} \left(\psi^* \boldsymbol{\nabla}_\mu \psi - \psi \boldsymbol{\nabla}_\mu \psi^* \right) = \left\{ c\varrho', -\boldsymbol{j}' \right\} \quad , \tag{1.34}$$

where

$$\varrho' = \frac{\mathrm{i}\hbar e}{2m_0c^2} \left(\psi^* \frac{\partial \psi}{\partial t} - \psi \frac{\partial \psi^*}{\partial t} \right) \tag{1.35}$$

signifies the *charge density*, and

$$\boldsymbol{j}' = \frac{\mathrm{i}e\hbar}{2m_0} \left(\psi^* \boldsymbol{\nabla} \psi - \psi \boldsymbol{\nabla} \psi^* \right) \tag{1.36}$$

denotes the *charge-current density*. The charge density (1.35) is allowed to be positive, negative or zero. This equates with the existence of particles and antiparticles in the theory. By calculating the *solutions for free particles*, we may understand this still better. Starting from (1.22), written in the form

$$\left(\hat{p}^\mu \hat{p}_\mu - m_0^2 c^2 \right) \psi = 0 \quad ,$$

and from the ansatz (1.23) for free waves

$$\psi = A\exp\left(-\frac{\mathrm{i}}{\hbar}p_\mu x^\mu\right) = A\exp\left[\frac{\mathrm{i}}{\hbar}(\boldsymbol{p}\cdot\boldsymbol{x} - Et)\right] \quad ,$$

we obtain the necessary condition that

$$p^\mu p_\mu - m_0^2c^2 = 0 = p_0^2 - \boldsymbol{p}^2 - m_0^2c^2$$

or, because of $p_0 = E/c$,

$$E^2 = c^2\left(\boldsymbol{p}^2 + m_0^2c^2\right) \quad . \tag{1.37}$$

Consequently, there exist two possible solutions for a given momentum $\boldsymbol{p}$: one with positive, the other with negative energy,

$$E_p = \pm c\sqrt{\boldsymbol{p}^2 + m_0^2c^2} \quad , \quad \psi_{(\pm)} = A_{(\pm)}\exp\left[\frac{\mathrm{i}}{\hbar}(\boldsymbol{p}\cdot\boldsymbol{x} \mp |E_p|t)\right] \quad . \tag{1.38}$$

$A_{(\pm)}$ are normalization constants, to be determined later. Inserting this into the density formula (1.35), we find

$$\varrho_{(\pm)} = \pm\frac{e|E_p|}{m_0c^2}\psi^*_{(\pm)}\psi_{(\pm)} \quad . \tag{1.39}$$

This suggests the following interpretation: $\psi_{(+)}$ specifies particles with charge $+e$; $\psi_{(-)}$ specifies particles with the same mass, but with charge $-e$. The general solution of the wave equation is always a linear combination of both types of functions. This point may be further clarified by discretizing the continuous plane waves (1.23). For that purpose we confine the waves to a large cubic box (*normalization box*) with edge length L (see Fig. 1.2) and, as usual, we demand periodic boundary conditions at the box walls. This yields in the well-known manner

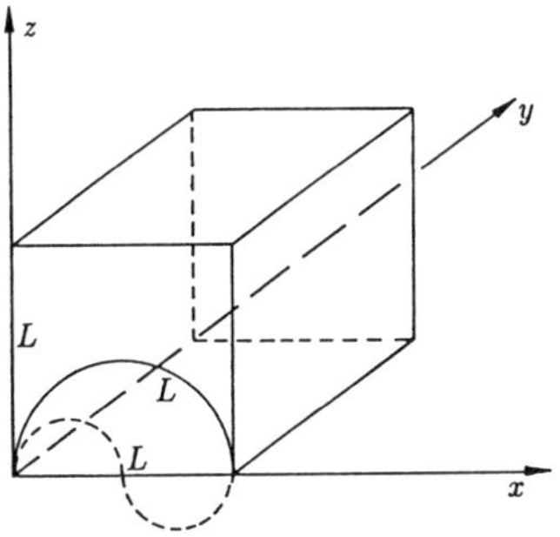

Fig. 1.2. The figure shows a normalization box. Two standing waves along the x axis are indicated

$$\psi_{n(\pm)} = A_{n(\pm)}\exp\left[\frac{\mathrm{i}}{\hbar}\left(\boldsymbol{p}_n\cdot\boldsymbol{x} \mp E_{p_n}t\right)\right] \quad , \tag{1.40}$$

where

$$\boldsymbol{p}_n = \frac{2\pi}{L}\boldsymbol{n} \quad , \quad \boldsymbol{n} = \{n_x, n_y, n_z\} \quad ; \quad n_i \in \mathbb{N}$$

and

$$E_{p_n} = c\sqrt{p_n^2 + m_0^2c^2} \equiv E_n \quad . \tag{1.41}$$

Here $\boldsymbol{n}$ is a (discrete) vector in the lattice space with axes n_x, n_y, n_z. Using (1.39), the normalization factors $A_{(\pm)}$ are determined by the requirement that

$$\pm e = \int_{L^3} \mathrm{d}^3x\,\varrho_{(\pm)}(\boldsymbol{x}) = \pm\frac{\mathrm{e}E_{p_n}}{m_0c^2}\left|A_{n(\pm)}\right|^2 L^3 \quad .$$

Choosing the phases in such a way that the amplitudes are real, we get

$$A_{n(\pm)} = \sqrt{\frac{m_0c^2}{L^3E_{p_n}}} \tag{1.42}$$

and thereby

$$\psi_{n(\pm)} = \sqrt{\frac{m_0c^2}{L^3E_{p_n}}} \exp\left[\frac{\mathrm{i}}{\hbar}\left(\boldsymbol{p}_n \cdot \boldsymbol{x} \mp E_{p_n}t\right)\right] \quad . \tag{1.43}$$

Notice that the normalization of either type of solution (corresponding to positive and negative charge) is the same. The only difference is due to the time factor $\exp(\pm\mathrm{i}E_{p_n}t/\hbar)$. The most general solutions of the Klein–Gordon equation for positive and negative spin-0 particles read then as

$$\psi_{(+)} = \sum_n a_n \psi_{n(+)} = \sum_n a_n \sqrt{\frac{m_0c^2}{L^3E_{p_n}}} \exp\left[\frac{\mathrm{i}}{\hbar}\left(\boldsymbol{p}_n \cdot \boldsymbol{x} - E_{p_n}t\right)\right]$$

and

$$\psi_{(-)} = \sum_n b_n \psi_{n(-)} = \sum_n b_n \sqrt{\frac{m_0c^2}{L^3E_{p_n}}} \exp\left[\frac{\mathrm{i}}{\hbar}\left(\boldsymbol{p}_n \cdot \boldsymbol{x} + E_{p_n}t\right)\right] \quad , \tag{1.44}$$

respectively. Solutions for spin-0 particles with zero charge can be constructed too. One immediately recognizes from the form of the expression for the charge density (1.35) that *the Klein–Gordon field* ψ *has no be real for neutral particles*, in which case

$$\psi^* = \psi \quad . \tag{1.45}$$

By means of (1.43) we can easily describe a wave front for a neutral particle:

$$\begin{aligned}
\psi_{n(0)} &= \frac{1}{\sqrt{2}}\left(\psi^n_{(+)}(\boldsymbol{p}_n) + \psi^n_{(-)}(-\boldsymbol{p}_n)\right) \\
&= \sqrt{\frac{m_0c^2}{2L^3E_{p_n}}}\left\{\exp\left[\frac{\mathrm{i}}{\hbar}\left(\boldsymbol{p}_n \cdot \boldsymbol{x} - E_{p_n}t\right)\right] + \exp\left[\frac{\mathrm{i}}{\hbar}\left(-\boldsymbol{p}_n \cdot \boldsymbol{x} + E_{p_n}t\right)\right]\right\} \\
&= \sqrt{\frac{m_0c^2}{2L^3E_{p_n}}}\left\{\exp\left[\frac{\mathrm{i}}{\hbar}\left(\boldsymbol{p}_n \cdot \boldsymbol{x} - E_{p_n}t\right)\right] + \exp\left[-\frac{\mathrm{i}}{\hbar}\left(\boldsymbol{p}_n \cdot \boldsymbol{x} - E_{p_n}t\right)\right]\right\} \\
&= \sqrt{\frac{m_0c^2}{2L^3E_{p_n}}}\, 2\cos\left(\frac{\boldsymbol{p}_n \cdot \boldsymbol{x} - E_{p_n}t}{\hbar}\right) \quad .
\end{aligned} \tag{1.46}$$

Thus $\psi^n_{(0)} = \psi^{n*}_{(0)}$ holds and, therefore, according to (1.35)

$$\varrho' = \frac{\mathrm{i}\hbar e}{2m_0c^2}\left(\psi^*_{n(0)}\frac{\partial\psi_{n(0)}}{\partial t} - \psi_{n(0)}\frac{\partial\psi^*_{n(0)}}{\partial t}\right) = 0 \quad .$$

Furthermore, we realize that the current density $\boldsymbol{j}'(\boldsymbol{x}, t)$ of neutral particles (1.36) vanishes too. Consequently in this case there is no conservation law. Obviously the relativistic quantum theory necessarily leads to novel degrees of freedom, that is so say the *charge degrees of freedom* of particles. In a nonrelativistic theory free, spinless particles can propagate freely with a well-defined momentum $\boldsymbol{p}$. In the relativistic case of free, spinless particle, *three solutions, which correspond to the electric charge* $(+, -, 0)$ *of the particles, exist for every momentum* $\boldsymbol{p}$.

EXAMPLE

1.1 The Charged Klein–Gordon Field

Hitherto we have inspected the Klein–Gordon equation both for a real, i.e. uncharged, and for a complex, i.e. charged, scalar field. In the case of the complex field we specified a current

$$j^\mu = \frac{\mathrm{i}e\hbar}{2m_0}\left(\varphi^*\nabla^\mu\varphi - \varphi\nabla^\mu\varphi^*\right) \quad , \tag{1}$$

with $\partial j^\mu/\partial x^\mu = 0$ and a charge

$$Q = \left(\mathrm{i}e\hbar/2m_0c^2\right)\int \mathrm{d}^3x(\varphi^*\dot{\varphi} - \varphi\dot{\varphi}^*) \quad . \tag{2}$$

Now we want to examine charged fields in a little bit more detail. To that end we decompose the complex wave function $\varphi(x)$ into real and imaginary components as

$$\varphi(x) = \frac{1}{\sqrt{2}}\left[\varphi_1(x) + \mathrm{i}\varphi_2(x)\right] \quad , \tag{3}$$

where $\varphi_1(x)$ and $\varphi_2(x)$ are real. If $\varphi(x)$ fulfills the Klein–Gordon equation

$$\left(\Box + \frac{m_0^2c^2}{\hbar^2}\right)\varphi(x) = 0 \quad , \tag{4}$$

then it immediately follows that φ_1 and φ_2 also obey the Klein–Gordon equation, i.e.

$$\left(\Box + \frac{m_0^2c^2}{\hbar^2}\right)\varphi_1(x) = 0 \quad \text{and} \quad \left(\Box + \frac{m_0^2c^2}{\hbar^2}\right)\varphi_2(x) = 0 \quad . \tag{5}$$

Conversely the following holds: If two fields $\varphi_1(x)$ and $\varphi_2(x)$ separately fulfill a Klein–Gordon equation with the same mass $m = m_1 = m_2$, then the equations can be replaced by *one* equation for a complex field, i.e.

$$\varphi = \frac{1}{\sqrt{2}}(\varphi_1 + \mathrm{i}\varphi_2) \quad \text{and} \quad \varphi^* = \frac{1}{\sqrt{2}}(\varphi_1 - \mathrm{i}\varphi_2) \tag{6}$$

fulfill

$$\left(\Box + \frac{m_0^2c^2}{\hbar^2}\right)\varphi = 0 \quad \text{and} \quad \left(\Box + \frac{m_0^2c^2}{\hbar^2}\right)\varphi^* = 0 \quad . \tag{7}$$

By interchanging φ and φ^* in

$$Q = \frac{\mathrm{i}e\hbar}{2m_0c^2}\int \mathrm{d}^3x(\varphi^*\dot{\varphi} - \varphi\dot{\varphi}^*) \tag{8}$$

we obtain the opposite charge. Hence φ and φ^* *characterize opposite charges*. These studies can, for example, be applied to the pion triplet (π^+, π^-, π^0): The π^0, being a neutral particle, is characterized by a real wave function, whereas π^+ and

Example 1.1.

π^-, being charged fields, have to be represented by complex wave functions. π^+ and π^- have the same mass and opposite charges, i.e. we can define

$$\varphi_{\pi^+} = \varphi^* = \frac{1}{\sqrt{2}}(\varphi_1 - \mathrm{i}\varphi_2) \quad ,$$

$$\varphi_{\pi^-} = \varphi = \frac{1}{\sqrt{2}}(\varphi_1 + \mathrm{i}\varphi_2) \tag{9}$$

and

$$\varphi_{\pi^0} = \varphi_0 = \varphi_0^* \quad . \tag{10}$$

Before we start to discuss the degrees of freedom, with respect to charge in detail, let us consider the energy. Is the energy positive (for positive particles), negative (for negative particles) or even equal to zero (for neutral particles)? In order to answer this question, we have to discuss the energy of the Klein–Gordon field within the scope of the general canonical formalism.

1.5 Energy-Momentum Tensor of the Klein–Gordon Field

In classical mechanics the energy is always given by the Hamilton function H which depends on the **Lagrange** function by means of the relation

$$H = \sum_i \pi_i \dot{q}_i - L \quad . \tag{1.47}$$

The classical equations of motion are obtained from the variational principle

$$\delta \int L\,\mathrm{d}t = 0 \quad , \tag{1.48}$$

which leads to the Lagrange equations in the familiar way,[3] i.e.

$$\frac{\partial}{\partial t}\frac{\partial L}{\partial \dot{q}_i} - \frac{\partial L}{\partial q_i} = 0 \quad . \tag{1.49}$$

The field theory is based on a similar concept[4] where the starting point is the *Lagrange density*

$$\mathcal{L}\left(\psi_\sigma, \frac{\partial \psi_\sigma}{\partial x_\mu}\right) \quad , \tag{1.50}$$

from which one obtains the *Lagrange function* by integration over the three-dimensional volume V

[3] See H. Goldstein: *Classical Mechanics*, 2nd ed. (Addison-Wesley, Reading, MA 1980) or W. Greiner: *Theoretische Physik II: Mechanik II* (Harri Deutsch, Frankfurt a.M. 1989).
[4] See J.D. Jackson: *Classical Electrodynamics*, 2nd ed. (Wiley, New York 1975).

$$L = \int_V \mathcal{L}\left(\psi_\sigma, \frac{\partial \psi_\sigma}{\partial x_\mu}\right) \mathrm{d}^3x \quad . \tag{1.51}$$

In general, the Lagrange density $\mathcal{L}$ depends on the various wave fields ψ_σ and all their derivatives $\partial\psi_\sigma/\partial x_\mu$. Derivatives of higher order are not considered because they would lead to nonlocal theories. So the variational principle (1.48) reads

$$\begin{aligned} \delta \int L\,\mathrm{d}t &= \delta \int \mathcal{L}\left(\psi_\sigma, \frac{\partial \psi_\sigma}{\partial x_\mu}\right) \mathrm{d}^3x\,\mathrm{d}t \\ &= \delta \int \mathcal{L}\left(\psi_\sigma, \frac{\partial \psi_\sigma}{\partial x_\mu}\right) \mathrm{d}^4x = 0 \end{aligned} \tag{1.52}$$

and yields in the well-known way (see the following Exercise 1.2) the field equations

$$\frac{\partial}{\partial x^\mu}\frac{\partial \mathcal{L}}{\partial(\partial\psi_\sigma/\partial x^\mu)} - \frac{\partial \mathcal{L}}{\partial \psi_\sigma} = 0 \quad . \tag{1.53}$$

Within the theory of wave fields (1.53) represent the equations of motion and are analogous to the classical equations of motion (1.49). Note the similarities and differences between (1.49) and (1.53): the variable t (time) distinguished in (1.49) is of the same significance as all other coordinates of x^μ in (1.53). In (1.49) one considers the Lagrange function L, whereas in the field equations (1.53) L is replaced by the Lagrange density $\mathcal{L}$.

EXERCISE

1.2 Derivation of the Field Equations for Wavefields

Problem. Derive the equations of motion (field equations) from the variation principle for the fields $\psi_\sigma(x)$

$$\delta \int_{V_4} \mathcal{L}\left(\psi_\sigma, \frac{\partial \psi_\sigma}{\partial x_\mu}\right) \mathrm{d}^4x = 0 \quad .$$

Solution. The variation is defined by

$$\begin{aligned} &\delta \int \mathcal{L}\left(\psi_\sigma, \frac{\partial \psi_\sigma}{\partial x^\mu}\right) \mathrm{d}^4x \\ &= \int \left[\frac{\partial \mathcal{L}}{\partial \psi_\sigma}\delta\psi_\sigma + \frac{\partial \mathcal{L}}{\partial(\partial\psi_\sigma/\partial x^\mu)}\delta\left(\frac{\partial\psi_\sigma}{\partial x^\mu}\right)\right] \mathrm{d}^4x = 0 \quad . \end{aligned} \tag{1}$$

We make use of the possibility of interchanging variation and partial differentiation,

$$\delta\frac{\partial\psi_\sigma}{\partial x^\mu} = \frac{\partial}{\partial x^\mu}(\psi_\sigma + \delta\psi_\sigma) - \frac{\partial}{\partial x^\mu}\psi_\sigma = \frac{\partial}{\partial x^\mu}(\delta\psi_\sigma) \quad . \tag{2}$$

The second term in (1) is integrated by parts, taking into account that the variation of ψ_σ vanishes at the integration boundaries. This leads to

Exercise 1.2.

$$\int \delta\psi_\sigma \left[\frac{\partial \mathcal{L}}{\partial \psi_\sigma} - \frac{\partial}{\partial x^\mu} \frac{\partial \mathcal{L}}{\partial(\partial\psi_\sigma/\partial x^\mu)} \right] \mathrm{d}^4x = 0 \quad . \tag{3}$$

As these equations are valid for arbitrary variations $\delta\psi_\sigma$, one obtains the field equations (*Euler–Lagrange equations*)

$$\frac{\partial \mathcal{L}}{\partial \psi_\sigma} - \frac{\partial}{\partial x^\mu} \frac{\partial \mathcal{L}}{\partial(\partial\psi_\sigma/\partial x^\mu)} = 0 \quad . \tag{4}$$

We will now demonstrate the procedure of the Lagrange formalism for the Klein–Gordon field. In this case we deal with two wave fields ψ and ψ^* (for charged particles). We could also choose the real and imaginary part of ψ as independent fields, but prefer the choice of ψ and ψ^*. Now, the *Lagrange density for the Klein–Gordon field* is of the form

$$\mathcal{L}\left(\psi, \psi^*, \frac{\partial\psi}{\partial x^\mu}, \frac{\partial\psi^*}{\partial x^\nu}\right) = \frac{\hbar^2}{2m_0}\left(g^{\mu\nu}\frac{\partial\psi^*}{\partial x^\mu}\frac{\partial\psi}{\partial x^\nu} - \frac{m_0^2c^2}{\hbar^2}\psi^*\psi\right) \quad . \tag{1.54}$$

The constant $\hbar^2/2m_0$ is chosen in such a way that $\int \mathcal{L}\,\mathrm{d}^3x$ has the dimension of energy, so that in particular the plane waves (1.43) carry the energy E_{p_n} [see (1.61) below]. The proof that this expression is the correct Lagrange density for the Klein–Gordon field is given by showing that, with the help of (1.53), one obtains the Klein–Gordon equations for ψ and ψ^* as resulting field equations. We note that one can immediately recognize from (1.54) that $\mathcal{L}$ is a Lorentz scalar if ψ and ψ^* are scalar fields, which we assume. Because of the covariance of (1.53), the resulting field equations (in our case the Klein–Gordon equation) are also Lorentz invariant. Indeed, by inserting (1.54) into (1.53) one immediately obtains

$$\frac{\partial}{\partial x^\nu}\frac{\partial\mathcal{L}}{\partial(\partial\psi/\partial x^\nu)} - \frac{\partial\mathcal{L}}{\partial\psi} = 0 = \frac{1}{2}\frac{\hbar^2}{m_0}\left(\frac{\partial}{\partial x^\nu}g^{\mu\nu}\frac{\partial\psi^*}{\partial x^\mu} + \frac{m_0^2c^2}{\hbar^2}\psi^*\right)$$

or

$$g^{\mu\nu}\frac{\partial}{\partial x^\mu}\frac{\partial}{\partial x^\nu}\psi^* + \frac{m_0^2c^2}{\hbar^2}\psi^* = 0 \quad , \tag{1.55}$$

and, analogously for the ψ^* field,

$$\frac{\partial}{\partial x^\nu}\frac{\partial\mathcal{L}}{\partial(\partial\psi/\partial x^\nu)} - \frac{\partial\mathcal{L}}{\partial\psi} = 0 = \frac{1}{2}\frac{\hbar^2}{m_0}\left(\frac{\partial}{\partial x^\nu}g^{\mu\nu}\frac{\partial\psi^*}{\partial x^\mu} + \frac{m_0^2c^2}{\hbar^2}\psi^*\right)$$

or

$$g^{\mu\nu}\frac{\partial}{\partial x^\mu}\frac{\partial}{\partial x^\nu}\psi^* + \frac{m_0^2c^2}{\hbar^2}\psi^* = 0 \quad . \tag{1.56}$$

Hence, the Lagrange density (1.54) yields precisely the Klein–Gordon equation for the ψ field as well as for the ψ^* field.

The energy and momentum of the wave field are described by the *energy-momentum tensor*,[5] defined as

$$T_\mu{}^\nu = \sum_\sigma \frac{\partial \psi_\sigma}{\partial x^\mu} \frac{\partial \mathcal{L}}{\partial(\partial\psi_\sigma/\partial x^\nu)} - \mathcal{L} g_\mu{}^\nu \quad . \tag{1.57}$$

The energy density $\mathcal{H}(x)$ of the field is identical with T_0^0,

$$\mathcal{H}(x) = T_0^0(x) \quad . \tag{1.58}$$

In our case one obtains from (1.54) that

$$\begin{aligned} T_\mu{}^\nu &= \frac{\partial \psi}{\partial x^\mu} \frac{\partial \mathcal{L}}{\partial(\partial\psi/\partial x^\nu)} + \frac{\partial \psi^*}{\partial x^\mu} \frac{\partial \mathcal{L}}{\partial(\partial\psi^*/\partial x^\nu)} - \mathcal{L} g_\mu{}^\nu \\ &= \frac{\hbar^2}{2m_0}\left[g^{\sigma\nu} \frac{\partial\psi^*}{\partial x^\sigma} \frac{\partial\psi}{\partial x^\mu} + g^{\sigma\nu} \frac{\partial\psi}{\partial x^\sigma} \frac{\partial\psi^*}{\partial x^\mu} \right. \\ &\quad \left. - \left(g^{\sigma\varrho} \frac{\partial\psi^*}{\partial x^\sigma} \frac{\partial\psi}{\partial x^\varrho} - \frac{m_0^2 c^2}{\hbar} \psi^*\psi \right) g_\mu{}^\nu \right] \end{aligned} \tag{1.59}$$

and, hence

$$\begin{aligned} \mathcal{H}(x) &\equiv T_0^0 \\ &= \frac{\hbar^2}{2m_0}\left[g^{\sigma 0} \frac{\partial\psi^*}{\partial x^\sigma} \frac{\partial\psi}{\partial x^0} + g^{\sigma 0} \frac{\partial\psi}{\partial x^\sigma} \frac{\partial\psi^*}{\partial x^0} \right. \\ &\quad \left. - \left(g^{\sigma\varrho} \frac{\partial\psi^*}{\partial x^\sigma} \frac{\partial\psi}{x^\varrho} - \frac{m_0^2 c^2}{\hbar^2} \psi^*\psi \right) g_0{}^0 \right] \\ &= \frac{\hbar^2}{2m_0}\left[\frac{\partial\psi^*}{\partial x^0} \frac{\partial\psi}{\partial x^0} + \frac{\partial\psi}{\partial x^0} \frac{\partial\psi^*}{\partial x^0} \right. \\ &\quad \left. - \left(\frac{\partial\psi^*}{\partial x^0} \frac{\partial\psi}{x^0} - \sum_{i=1}^{3} \frac{\partial\psi^*}{\partial x^i} \frac{\partial\psi}{\partial x^i} - \frac{m_0^2 c^2}{\hbar^2} \psi^*\psi \right) \right] \\ &= \frac{\hbar^2}{2m_0}\left[\frac{1}{c^2} \frac{\partial\psi^*}{\partial t} \frac{\partial\psi}{\partial t} + (\boldsymbol{\nabla}\psi^*)\cdot(\boldsymbol{\nabla}\psi) + \frac{m_0^2 c^2}{\hbar^2} \psi^*\psi \right] \quad . \end{aligned} \tag{1.60}$$

The energy H, which belongs to the plane wave solutions (1.43), is given by the integral, over the energy density in the volume L^3,

$$\begin{aligned} H_{n(\pm)} &= \int_{L^3} T_0^0(n,\pm)\, \mathrm{d}^3x \\ &= \int_{L^3} \frac{\hbar^2}{2m_0} \left[\frac{m_0c^2}{L^3 E_{p_n}} \frac{(\mp E_{p_n})^2}{\hbar^2 c^2} + \frac{m_0 c^2}{L^3 E_{p_n}} \frac{\boldsymbol{p}_n\cdot\boldsymbol{p}_n}{\hbar^2} + \frac{m_0^2c^2}{\hbar^2} \frac{m_0c^2}{L^3 E_{p_n}} \right] \mathrm{d}^3x \\ &= \frac{\hbar^2}{2m_0} \frac{m_0c^2}{L^3 E_{p_n}\hbar^2} \left[\frac{E_{p_n}^2}{c^2} + \boldsymbol{p}_n^2 + m_0^2c^2 \right] L^3 \\ &= \frac{\hbar^2}{2m_0} \frac{m_0c^2}{L^3 E_{p_n}\hbar^2} \frac{2E_{p_n}^2}{c^2} L^3 = E_{p_n} \quad . \end{aligned} \tag{1.61}$$

[5] See Example 1.3 and, for a detailed discussion, J.D. Jackson: *Classical Electrodynamics*, 2nd ed. (Wiley, New York 1975) or W. Greiner: *Theoretische Physik III, Klassische Elektrodynamik* (Harri Deutsch, Frankfurt a.M. 1985).

Although we chose the constant $\hbar^2/2m_0$ of the Lagrange density (1.54) in such a way that the wave $\psi_{n(+)}$ carried the energy $+E_{p_n}$, the result (1.61) shows that the wave $\psi_{n(-)}$ also carries energy $+E_{p_n}$. Thus, we have found the interesting result that plane waves $\psi_{n(\pm)}$ from (1.43) describe particles with positive and negative charge, respectively [see (1.39) and (1.42)], but that both waves carry positive energy $+E_{p_n} = +c(p_n^2 + m_0^2c^2)^{1/2}$. Hence, energy plays two roles: On one hand $+E_p$ characterizes particles with positive charge according to (1.39) – the plane wave in this case reading as $\psi_{(+)} \sim \exp[\mathrm{i}(\boldsymbol{p}\cdot\boldsymbol{x} - E_pt)/\hbar]$ – and $-E_p$, particles of negative charge – the plane wave in this case reads as $\psi_{(-)} \sim \exp[\mathrm{i}(\boldsymbol{p}\cdot\boldsymbol{x}+E_pt)/\hbar]$. On the other hand $|E_p|$ always gives the energy of the particles.

EXAMPLE

1.3 Determination of the Energy-Momentum Tensor for a General Lagrange Density $\mathcal{L}(\psi_\sigma, \partial\psi_\sigma/\partial x^\mu)$

In order to calculate the energy-momentum tensor we start with the Noether theorem: Each continuous symmetry transformation which leaves the Lagrange density invariant, corresponds to a conservation law and, hence, to a constant of motion. Specifically we shall discuss the conservation laws, which follow from the translational invariance of classical field theory.

Consider the infinitesimal displacement

$$x'_\mu = x_\mu + \varepsilon_\mu \quad ,$$

$$\delta x_\mu = x'_\mu - x_\mu = \varepsilon_\mu \quad . \tag{1}$$

The corresponding variation of the Lagrangian is given by

$$\delta\mathcal{L} = \frac{\partial\mathcal{L}}{\partial x_\mu}\delta x_\mu = \varepsilon_\mu\frac{\partial\mathcal{L}}{\partial x_\mu} \quad . \tag{2}$$

If, on the other hand, $\mathcal{L}$ is translationally invariant then $\mathcal{L}$ does not explicitly depend on the coordinates. Thus, we can write $\mathcal{L} = \mathcal{L}(\psi_\sigma, \partial\psi_\sigma/\partial x_\mu)$. From this follows the variation

$$\delta\mathcal{L} = \sum_\sigma\left[\frac{\partial\mathcal{L}}{\partial\psi_\sigma}\delta\psi_\sigma + \frac{\partial\mathcal{L}}{\partial(\partial\psi_\sigma/\partial x_\mu)}\delta\left(\frac{\partial\psi_\sigma}{\partial x_\mu}\right)\right] \quad . \tag{3}$$

The variation of ψ_σ is obtained as

$$\delta\psi_\sigma = \frac{\partial\psi_\sigma(x)}{\partial x_\nu}\delta x_\nu = \frac{\partial\psi_\sigma}{\partial x_\nu}\varepsilon_\nu \quad . \tag{4}$$

In the following we will also use the Euler–Lagrange equations (field equations)

$$\frac{\partial\mathcal{L}}{\partial\psi_\sigma} - \frac{\partial}{\partial x^\mu}\frac{\partial\mathcal{L}}{\partial(\partial\psi_\sigma/\partial x^\mu)} = 0 \quad . \tag{5}$$

Example 1.3.

Equating (2) and (3) yields

$$\begin{aligned}\varepsilon_\mu \frac{\partial \mathcal{L}}{\partial x_\mu} &= \sum_\sigma \left[\frac{\partial \mathcal{L}}{\partial \psi_\sigma} \delta\psi_\sigma + \frac{\partial \mathcal{L}}{\partial(\partial\psi_\sigma/\partial x_\mu)} \delta\left(\frac{\partial \psi_\sigma}{\partial x_\mu}\right)\right] \\ &= \sum_\sigma \left[\left\{ \frac{\partial}{\partial x^\mu} \frac{\partial \mathcal{L}}{\partial(\partial\psi_\sigma/\partial x^\mu)} \right\} \varepsilon_\nu \frac{\partial \psi_\sigma}{\partial x_\nu} + \frac{\partial \mathcal{L}}{\partial(\partial\psi_\sigma/\partial x_\mu)} \frac{\partial}{\partial x^\nu}\left(\frac{\partial \psi_\sigma}{\partial x_\mu}\right)\varepsilon_\nu \right] , \end{aligned} \tag{6}$$

where we have used

$$\delta\left(\frac{\partial \psi_\sigma}{\partial x_\mu}\right) = \frac{\partial}{\partial x_\nu}\left(\frac{\partial \psi_\sigma}{\partial x_\mu}\right)\delta x_\nu = \frac{\partial}{\partial x_\nu}\left(\frac{\partial \psi_\sigma}{\partial x_\mu}\right)\varepsilon_\nu \quad . \tag{7}$$

The rhs of (6) can be expressed as

$$\varepsilon_\mu \frac{\partial \mathcal{L}}{\partial x_\mu} = \frac{\partial}{\partial x_\mu}\left[\sum_\sigma \frac{\partial \mathcal{L}}{\partial(\partial\psi_\sigma/\partial x_\mu)} \varepsilon_\nu \frac{\partial \psi_\sigma}{\partial x_\nu}\right] \quad . \tag{8}$$

Since the result holds for arbitrary translations ε_μ we can write

$$\frac{\partial}{\partial x_\mu} T_{\mu\nu} = 0 \quad . \tag{9}$$

These are four continuity equations (one for every ν) and, thus, four conservation laws. $T_{\mu\nu}$ is the *energy-momentum tensor* (stress tensor), given by

$$T_{\mu\nu} = -g_{\mu\nu}\mathcal{L} + \sum_\sigma \frac{\partial \mathcal{L}}{\partial(\partial\psi_\sigma/\partial x_\mu)} \frac{\partial \psi_\sigma}{\partial x^\nu} \quad . \tag{10}$$

For a better understanding of its physical content we will illustrate the meaning of the T_{00} component. For that reason, in analogy to classical mechanics, where the generalized momentum is given by $p_n = \partial L/\partial \dot{q}_n$, we define the momentum-density conjugate to $\psi_\sigma(\boldsymbol{x}, t)$ by

$$\pi_\sigma(\boldsymbol{x}, t) = \frac{\partial \mathcal{L}(\psi_\sigma, \partial\psi_\sigma/\partial x_\nu)}{\partial \dot{\psi}_\sigma(\boldsymbol{x}, t)} \quad . \tag{11}$$

Here the short-hand notation $\dot{\psi}_\sigma \equiv \partial\psi_\sigma/\partial(ct)$ has been used. The classical Hamiltonian reads

$$H(p, q) = p\dot{q} - L(q, \dot{q}) \quad . \tag{12}$$

Analogously, we express the Hamiltonian as a volume integral over a *Hamiltonian density* $\mathcal{H}(\pi, \psi)$ with the definition

$$H = \int \mathrm{d}^3x\, \mathcal{H}\big(\pi(\boldsymbol{x}, t), \psi(\boldsymbol{x}, t)\big) \quad , \tag{13}$$

and

$$\mathcal{H} = \sum_\sigma \pi_\sigma \dot{\psi}_\sigma - \mathcal{L} \quad . \tag{14}$$

Example 1.3.

Using (14) and (11), we find that (10) yields

$$p_\nu = \int \mathrm{d}^3x\, T_{0\nu} = \int \mathrm{d}^3x \left[\sum_\sigma \pi_\sigma \frac{\partial \psi_\sigma}{\partial x^\nu} - g_{0\nu}\mathcal{L} \right] \quad , \tag{15}$$

with $\partial p_\nu / \partial t = 0$. The quantities p_ν are therefore constant in time, which follows from (9), because $\int \mathrm{d}^3x\, \partial T_{0\nu}/c\partial t = -\int \mathrm{d}^3x\, \partial T_{i\nu}/\partial x_i = 0$ according to Gauß's theorem. For the T_{00} component of the stress tensor it clearly follows that

$$T_{00} = \sum_\sigma \pi_\sigma \dot{\psi}_\sigma - \mathcal{L} = \mathcal{H} \quad , \tag{16}$$

and, therefore, from (13)

$$\int \mathrm{d}^3x\, T_{00} = H \quad . \tag{17}$$

Hence, we can identify p_ν with the energy-momentum four-vector. Its time component is clearly the energy, and also each component of p_ν is conserved in time, i.e. it is a constant of motion.

EXERCISE

1.4 Lagrange Density and Energy-Momentum Tensor of the Schrödinger Equation

Problem. Derive the Lagrangian for the Schrödinger equation and discuss the corresponding energy-momentum tensor.

Solution. The Lagrangian is given by

$$\mathcal{L} = -\frac{\hbar^2}{2m_0}(\boldsymbol{\nabla}\psi^*)\cdot(\boldsymbol{\nabla}\psi) - \frac{\hbar}{2\mathrm{i}}\left(\psi^*\frac{\partial\psi}{\partial t} - \frac{\partial\psi^*}{\partial t}\psi\right) - \psi^* V \psi \quad . \tag{1}$$

Using the Euler–Lagrange differential equation we will show that the variation of this Lagrangian yields the Schrödinger equation. ψ and ψ^* have to be varied independently.

The Euler–Lagrange equation, split up into space and time components, reads

$$\frac{\partial \mathcal{L}}{\partial \psi_\sigma} - \frac{\partial}{\partial x^i}\frac{\partial \mathcal{L}}{\partial(\partial\psi_\sigma/\partial x^i)} - \frac{\partial}{\partial t}\frac{\partial \mathcal{L}}{\partial \dot{\psi}_\sigma} = 0 \quad , \tag{2}$$

where the summation over i runs through $i = 1, 2, 3$ or

$$\frac{\partial}{\partial t}\frac{\partial \mathcal{L}}{\partial \dot{\psi}_\sigma} = \frac{\partial \mathcal{L}}{\partial \psi_\sigma} - \boldsymbol{\nabla}\cdot\frac{\partial \mathcal{L}}{\partial(\boldsymbol{\nabla}\psi_\sigma)} \quad . \tag{3}$$

First we vary with respect to ψ^* and obtain

$$-V\psi + \frac{\hbar^2}{2m_0}\boldsymbol{\nabla}^2\psi - \frac{\hbar}{\mathrm{i}}\dot{\psi} = 0 \quad . \tag{4}$$

Analogously, variation with respect to ψ yields *Exercise 1.4.*

$$-V\psi^* + \frac{\hbar^2}{2m_0}\nabla^2\psi^* + \frac{\hbar}{\mathrm{i}}\dot{\psi}^* = 0 \quad . \tag{5}$$

Rewriting (4) and (5) yields the Schrödinger equation for ψ and ψ^* in the familiar form,

$$\begin{aligned} \mathrm{i}\hbar\dot{\psi} &= -\frac{\hbar^2}{2m_0}\nabla^2\psi + V\psi \equiv \hat{H}\psi \quad , \\ -\mathrm{i}\hbar\dot{\psi}^* &= -\frac{\hbar^2}{2m_0}\nabla^2\psi^* + V\psi^* \equiv \hat{H}^\dagger\psi^* \quad , \end{aligned} \tag{6}$$

where

$$\hat{H} = -\frac{\hbar^2}{2m_0}\Delta + V(\boldsymbol{x})$$

is the Hamiltonian. The conjugate momenta to ψ are

$$\pi = \frac{\partial\mathcal{L}}{\partial\dot{\psi}} = \frac{\mathrm{i}\hbar}{2}\psi^* \quad , \tag{7}$$

and

$$\pi^* = \frac{\partial\mathcal{L}}{\partial\dot{\psi}^*} = -\frac{\mathrm{i}\hbar}{2}\psi \quad . \tag{8}$$

Hence, the Hamiltonian density reads

$$\mathcal{H} = \sum_\sigma \pi_\sigma\dot{\psi}_\sigma - \mathcal{L} = -\frac{\mathrm{i}\hbar}{m_0}\nabla\pi\cdot\nabla\psi - \frac{2\mathrm{i}}{\hbar}V\pi\psi \quad . \tag{9}$$

We also determine the stress tensor belonging to the Lagrange density (1),

$$T_\mu{}^\nu = \frac{\partial\psi}{\partial x^\mu}\frac{\partial\mathcal{L}}{\partial(\partial\psi/\partial x^\nu)} + \frac{\partial\psi^*}{\partial x^\mu}\frac{\partial\mathcal{L}}{\partial(\partial\psi^*/\partial x^\nu)} - \mathcal{L}\delta_\mu^\nu \quad . \tag{10}$$

In particular the $T_0{}^0$ component, split up into space and time components, is

$$\begin{aligned} T_0{}^0 &= \dot{\psi}\frac{\partial\mathcal{L}}{\partial\dot{\psi}} + \dot{\psi}^*\frac{\partial\mathcal{L}}{\partial\dot{\psi}^*} - \mathcal{L} \\ &= -\frac{\hbar}{2\mathrm{i}}\dot{\psi}\psi^* + \dot{\psi}^*\frac{\hbar}{2\mathrm{i}}\psi + \frac{\hbar^2}{2m_0}\nabla\psi^*\cdot\nabla\psi + \frac{\hbar}{2\mathrm{i}}(\psi^*\dot{\psi} - \dot{\psi}^*\psi) + \psi^*V\psi \\ &= \frac{\hbar^2}{2m_0}\nabla\psi^*\cdot\nabla\psi + \psi^*V\psi \equiv \mathcal{H} \quad , \end{aligned} \tag{11}$$

that is, in accordance with (9). $T_0{}^0$ gives the energy density of a given system. The total energy of the Schrödinger field is then

$$\begin{aligned} H &= \int T_0{}^0\,\mathrm{d}^3x = \int\left(\frac{\hbar^2}{2m_0}\nabla\psi^*\cdot\nabla\psi + \psi^*V\psi\right)\mathrm{d}^3x \\ &= \int\psi^*\left(-\frac{\hbar^2}{2m_0}\nabla^2 + V\right)\psi\,\mathrm{d}^3x = \int\psi^*\hat{H}\psi\,\mathrm{d}^3x \quad , \end{aligned}$$

i.e. the expectation value of the Hamiltonian.

Exercise 1.4.

Calculation of the energy flux $\boldsymbol{S}$: In analogy to the Poynting vector in electrodynamics, $\boldsymbol{S} = \boldsymbol{E} \times \boldsymbol{B}$, the canonical formulation generally follows as

$$\boldsymbol{S} = e_1 T_0{}^1 + e_2 T_0{}^2 + e_3 T_0{}^3 \quad , \tag{12}$$

where e_i are the Cartesian unit vectors. In the case of the Schrödinger field we get

$$\boldsymbol{S} = \dot{\psi} \frac{\partial \mathcal{L}}{\partial(\boldsymbol{\nabla}\psi)} + \dot{\psi}^* \frac{\partial \mathcal{L}}{\partial(\boldsymbol{\nabla}\psi^*)} = -\frac{\hbar^2}{2m_0}(\dot{\psi}^* \boldsymbol{\nabla}\psi + \dot{\psi}\boldsymbol{\nabla}\psi^*) \quad . \tag{13}$$

Calculation of the momentum density $\boldsymbol{p}$:

$$\begin{aligned} \boldsymbol{p} &= e_1 T_1{}^0 + e_2 T_2{}^0 + e_3 T_3{}^0 \\ &= (\boldsymbol{\nabla}\psi)\frac{\partial \mathcal{L}}{\partial \dot{\psi}} + (\boldsymbol{\nabla}\psi^*)\frac{\partial \mathcal{L}}{\partial \dot{\psi}^*} \\ &= -\frac{\hbar}{2\mathrm{i}}(\psi^* \boldsymbol{\nabla}\psi - \psi\boldsymbol{\nabla}\psi^*) \quad . \end{aligned} \tag{14}$$

This is the expression known to us from elementary quantum mechanics.[6] The remaining components of the tensor are called momentum fluxes.

EXERCISE

1.5 Lorentz Invariance of the Klein–Gordon Equation

Problem. Show the Lorentz invariance of the Klein–Gordon equation.

Solution. The Lorentz invariance of the Klein–Gordon equation is a direct consequence of the invariance of the underlying energy-momentum relation. In four-vector notation it reads

$$\sum_{\mu=0}^{3} p_\mu p^\mu = p_\mu p^\mu = E^2/c^2 - \boldsymbol{p}\cdot\boldsymbol{p} = m_0^2 c^2 \quad .$$

Now it is helpful to express the Klein–Gordon equation in four-vector notation, i.e.

$$-\hbar^2 \frac{\partial^2}{\partial t^2}\psi = \left(-\hbar^2 c^2 \boldsymbol{\nabla}^2 + m_0^2 c^4\right)\psi$$

changes to

$$\left[\Box + \left(\frac{m_0 c}{\hbar}\right)^2\right]\psi(x_\mu) = 0 \quad , \quad \text{where}$$

$$\Box = \frac{\partial}{\partial x_\mu}\frac{\partial}{\partial x^\mu} \quad .$$

[6] See W. Greiner: *Quantum Mechanics – An Introduction*, 3rd ed. (Springer, Berlin, Heidelberg 1994).

Hence, in the transformed system the Klein–Gordon equation should read

Exercise 1.5.

$$\left[\frac{\partial}{\partial x'_\mu}\frac{\partial}{\partial x'^\mu}+\kappa^2\right]\psi'(x'_\mu)=0 \quad \text{with} \quad \kappa=\frac{m_0 c}{\hbar} \quad ,$$

thus, it has to be shown that the operator $(\partial/\partial x_\mu)(\partial/\partial x^\mu)$ is invariant under Lorentz transformations. This may easily be seen, because

$$\hat{p}_\mu=+\mathrm{i}\hbar\frac{\partial}{\partial x^\mu} \quad \text{and, therefore,}$$

$$\frac{\partial}{\partial x_\mu}\frac{\partial}{\partial x^\mu}\sim\hat{p}_\mu\hat{p}^\mu \quad .$$

The length of the vector p_μ is, as stated before, Lorentz invariant and, therefore, the operator $\Box$ is too.

The transformation properties of the wave function can easily be deduced by looking at the plane-wave solutions of the Klein–Gordon equation,

$$\psi(x_\mu)=\mathrm{e}^{\mathrm{i}k_\mu x^\mu} \quad , \quad \text{with} \quad k_0=\frac{w}{c}=\sqrt{\boldsymbol{k}^2+\kappa^2} \quad .$$

The difference between $\psi'(x'_\mu)$ and $\psi'(x_\mu)$ should be noticed: $\psi(x_\mu)$ and $\psi'(x'_\mu)$ refer to the *same* space-time point, i.e., $x'_\mu=\sum_{\nu=0}^{3}a_\mu{}^\nu x_\nu$, where $a_\mu{}^\nu$ is a Lorentz transformation, whereas $\psi(x_\mu)$ and $\psi'(x_\mu)$ refer to two different points with coordinates x_μ in the old and new system, respectively.

1.6 The Klein–Gordon Equation in Schrödinger Form

To demonstrate the new degrees of freedom of the charge in a more distinct way, it is advantageous to transform the Klein–Gordon equation (1.22) – which is of second order in the time coordinate – into a system of two coupled differential equations that are of first order in time. This is achieved by the ansatz

$$\psi=\varphi+\chi \quad , \quad \mathrm{i}\hbar\frac{\partial\psi}{\partial t}=m_0c^2(\varphi-\chi) \quad , \tag{1.62}$$

in which ψ and the time derivative $\partial\psi/\partial t$ are expressed by the two functions φ and χ. According to (1.22), the Klein–Gordon field fulfills

$$\frac{1}{c^2}\frac{\partial^2}{\partial t^2}\psi=\left(\frac{\partial^2}{\partial x^2}+\frac{\partial^2}{\partial y^2}+\frac{\partial^2}{\partial z^2}-\frac{m_0^2c^2}{\hbar^2}\right)\psi \quad .$$

It is easily proved that the two coupled differential equations,

$$\mathrm{i}\hbar\frac{\partial\varphi}{\partial t}=-\frac{\hbar^2}{2m_0}\Delta(\varphi+\chi)+m_0c^2\varphi \quad , \tag{1.63a}$$

$$\mathrm{i}\hbar\frac{\partial\chi}{\partial t}=\frac{\hbar^2}{2m_0}\Delta(\varphi+\chi)-m_0c^2\chi \quad , \tag{1.63b}$$

are equivalent to the Klein–Gordon equation (1.22). By adding and subtracting the equations (1.63), the following arguments can be made:

(a) *addition yields*

$$\mathrm{i}\hbar\frac{\partial}{\partial t}(\varphi+\chi) = m_0c^2(\varphi-\chi) \quad .$$

This is the second equation of (1.62); it leads to the trivial equation $\partial\psi/\partial t = \partial\psi/\partial t$.

(b) *subtraction yields*

$$\mathrm{i}\hbar\frac{\partial}{\partial t}(\varphi-\chi) = -\frac{\hbar^2}{m_0}\Delta(\varphi-\chi) + m_0c^2(\varphi+\chi) \quad ,$$

or, using (1.62),

$$\begin{aligned}
\mathrm{i}\hbar\frac{\partial}{\partial t}\left(\frac{\mathrm{i}\hbar}{m_0c^2}\frac{\partial\psi}{\partial t}\right) &= -\frac{\hbar^2}{m_0}\Delta\psi + m_0c^2\psi \\
\Leftrightarrow\ -\frac{\hbar^2}{m_0}\frac{1}{c^2}\frac{\partial^2\psi}{\partial t^2} &= -\frac{\hbar^2}{m_0}\Delta\psi + m_0c^2\psi \\
\Leftrightarrow\ \frac{1}{c^2}\frac{\partial^2\psi}{\partial t^2} &= \Delta\psi - \frac{m_0^2c^2}{\hbar^2}\psi \quad ,
\end{aligned}$$

which is just the Klein–Gordon equation (1.22).

The coupled equations (1.63) may be combined to form one equation. For this purpose we introduce the column vector

$$\Psi = \begin{pmatrix}\varphi \\ \chi\end{pmatrix} \tag{1.64}$$

and make use of the four 2×2 matrices

$$\hat{\tau}_1 = \begin{pmatrix}0 & 1\\ 1 & 0\end{pmatrix}, \quad \hat{\tau}_2 = \begin{pmatrix}0 & -\mathrm{i}\\ \mathrm{i} & 0\end{pmatrix}, \quad \hat{\tau}_3 = \begin{pmatrix}1 & 0\\ 0 & -1\end{pmatrix}, \quad \mathbb{1} = \begin{pmatrix}1 & 0\\ 0 & 1\end{pmatrix} . \tag{1.65}$$

These are identical to the Pauli matrices,[7] with the significant difference that the matrices (1.65) do not act in spin space, but in the vector space defined by (1.64). The Pauli matrices fulfill the algebraic relations

$$\hat{\tau}_k^2 = \mathbb{1} \quad , \quad \hat{\tau}_k\hat{\tau}_l = -\hat{\tau}_l\hat{\tau}_k = \mathrm{i}\hat{\tau}_m \quad \{k,l,m = 1,2,3 - \text{cyclic}\} \quad . \tag{1.66}$$

Using (1.64–66) we can combine the coupled equations (1.63) to form a Schrödinger-type equation, namely,

$$\mathrm{i}\hbar\frac{\partial}{\partial t}\Psi = \hat{H}_\mathrm{f}\Psi \quad , \quad \text{or}$$

$$\left(\mathrm{i}\hbar\frac{\partial}{\partial t} - \hat{H}_\mathrm{f}\right)\Psi = 0 \quad , \tag{1.67}$$

[7] These were introduced in W. Greiner: *Quantum Mechanics – An Introduction*, 3rd ed. (Springer, Berlin, Heidelberg 1994).

where the Hamiltonian $\hat{H}_{\mathrm{f}}$ for free particles is given by

$$\hat{H}_{\mathrm{f}} = (\hat{\tau}_3 + \mathrm{i}\hat{\tau}_2)\frac{\hat{\boldsymbol{p}}^2}{2m_0} + \hat{\tau}_{3m_0c^2} = \begin{pmatrix} 1 & 1 \\ -1 & -1 \end{pmatrix}\frac{\hat{\boldsymbol{p}}^2}{2m_0} + \begin{pmatrix} 1 & 0 \\ 0 & -1 \end{pmatrix} m_0c^2 \,. \quad (1.68)$$

Hence, in (1.67) we have found a *Schrödinger formulation* of the Klein–Gordon equation. Starting from (1.67) and using the relation

$$\hat{H}_{\mathrm{f}}^2 = c^2\hat{\boldsymbol{p}}^2 + m_0^2c^4 \quad , \quad (1.69)$$

it is easily proved that *each component* of the vector Ψ of (1.64) *individually satisfies the Klein–Gordon equation.* This is most elegantly shown by applying $(\mathrm{i}\partial/\partial t + \hat{H}_{\mathrm{f}})$ to the lhs of (1.67), yielding

$$\begin{aligned}
\left(\mathrm{i}\hbar\frac{\partial}{\partial t} + \hat{H}_{\mathrm{f}}\right)\left(\mathrm{i}\hbar\frac{\partial}{\partial t} - \hat{H}_{\mathrm{f}}\right)\Psi &= 0 \\
\left(-\hbar^2\frac{\partial^2}{\partial t^2} - \hat{H}_{\mathrm{f}}^2\right)\Psi &= 0 \\
\left(-\hbar^2\frac{\partial^2}{\partial t^2} + \hbar^2c^2\Delta - m_0^2c^4\right)\Psi &= 0 \quad ,
\end{aligned}$$

or

$$\left(+\frac{1}{c^2}\frac{\partial^2}{\partial t^2} - \Delta + \frac{m_0^2c^2}{\hbar^2}\right)\Psi = 0 \quad . \quad (1.70)$$

In this representation the expression for the density (1.35) becomes especially simple. With (1.62) we find

$$\begin{aligned}
\varrho' &= \frac{\mathrm{i}e\hbar}{2m_0c^2}\left(\psi^*\frac{\partial\psi}{\partial t} - \psi\frac{\partial\psi^*}{\partial t}\right) \\
&= \frac{em_0c^2}{2m_0c^2}\left(\psi^*(\varphi - \chi) + \psi(\varphi^* - \chi^*)\right) \\
&= \frac{e}{2}\left[(\varphi^* + \chi^*)(\varphi - \chi) + (\varphi + \chi)(\varphi^* - \chi^*)\right] \\
&= \frac{e}{2}\left[\varphi^*\varphi - \chi^*\chi - \cancel{\varphi^*\chi} + \cancel{\varphi\chi^*} + \varphi\varphi^* - \chi\chi^* - \cancel{\varphi\chi^*} + \cancel{\varphi^*\chi}\right] \\
&= e\left(\varphi^*\varphi - \chi^*\chi\right) \\
&= e\Psi^\dagger\hat{\tau}_3\Psi \quad . \qquad (1.71)
\end{aligned}$$

Similarly, from (1.36) we can infer the current vector in Schrödinger representation

$$\boldsymbol{j}' = \frac{e\hbar}{2m_0\mathrm{i}}\left[\Psi^\dagger\hat{\tau}_3(\hat{\tau}_3 + \mathrm{i}\hat{\tau}_2)\boldsymbol{\nabla}\Psi - (\boldsymbol{\nabla}\Psi^\dagger)\hat{\tau}_3(\hat{\tau}_3 + \mathrm{i}\hat{\tau}_2)\Psi\right] \quad . \quad (1.72)$$

For the normalization of charge it follows that

$$\int \varrho'(\boldsymbol{x})\,\mathrm{d}^3x = \pm e \quad , \quad \text{or}$$

$$\int \Psi^\dagger\hat{\tau}_3\Psi\,\mathrm{d}^3x = \pm 1 = \int (\varphi\varphi^* - \chi\chi^*)\,\mathrm{d}^3x \quad . \quad (1.73)$$

Let us once again consider free particles in this new representation. If we write

$$\Psi = \begin{pmatrix} \varphi \\ \chi \end{pmatrix} = A \begin{pmatrix} \varphi_0 \\ \chi_0 \end{pmatrix} \exp\left[\frac{\mathrm{i}}{\hbar}(\boldsymbol{p}\cdot\boldsymbol{x} - Et)\right] \tag{1.74}$$

and substitute this ansatz into (1.67) by means of (1.68), we find

$$E \begin{pmatrix} \varphi \\ \chi \end{pmatrix} = \begin{pmatrix} 1 & 1 \\ -1 & -1 \end{pmatrix} \frac{p^2}{2m_0} \begin{pmatrix} \varphi \\ \chi \end{pmatrix} + \begin{pmatrix} 1 & 0 \\ 0 & -1 \end{pmatrix} m_0c^2 \begin{pmatrix} \varphi \\ \chi \end{pmatrix} \tag{1.75}$$

or, alternatively,

$$E\varphi = \frac{p^2}{2m_0}(\varphi + \chi) + m_0c^2\varphi \quad , \quad E\chi = -\frac{p^2}{2m_0}(\varphi + \chi) - m_0c^2\chi \quad . \tag{1.76}$$

φ_0 and χ_0 are therefore readily determined by the solution of the coupled equations

$$\left(E - \frac{p^2}{2m_0} - m_0c^2\right)\varphi_0 - \frac{p^2}{2m_0}\chi_0 = 0 \quad ,$$

$$\left(\frac{p^2}{2m_0}\right)\varphi_0 + \left(E + \frac{p^2}{2m_0} + m_0c^2\right)\chi_0 = 0 \quad , \tag{1.77}$$

and, since the determinant necessarily needs to vanish,

$$\begin{vmatrix} E - \dfrac{p^2}{2m_0} - m_0c^2 & -\dfrac{p^2}{2m_0} \\ +\dfrac{p^2}{2m_0} & E + \dfrac{p^2}{2m_0} + m_0c^2 \end{vmatrix} = 0 \quad .$$

It follows that

$$E^2 - \left(\frac{p^2}{2m_0} + m_0c^2\right)^2 + \left(\frac{p^2}{2m_0}\right)^2 = 0 \quad .$$

We thus recover the relativistic energy-momentum relation

$$E^2 = p^2c^2 + m_0^2c^4 \quad \text{or} \quad E = \pm c\sqrt{p^2 + m_0^2c^2} \equiv \pm E_p \quad ,$$

the corresponding solutions following from (1.77) and (1.74). We shall discuss the positive and negative energy solutions separately:

(1) $E = +E_p$:

$$\Psi^{(+)}(\boldsymbol{p}) = A_{(+)} \begin{pmatrix} \varphi_0^{(+)} \\ \chi_0^{(+)} \end{pmatrix} \exp\left[\mathrm{i}(\boldsymbol{p}\cdot\boldsymbol{x} - E_pt)/\hbar\right] \equiv \begin{pmatrix} \varphi^{(+)}(\boldsymbol{p}) \\ \chi^{(+)}(\boldsymbol{p}) \end{pmatrix} \quad , \tag{1.78a}$$

where

$$\begin{pmatrix} \varphi_0^{(+)} \\ \chi_0^{(+)} \end{pmatrix} = \begin{pmatrix} m_0c^2 + E_p \\ m_0c^2 - E_p \end{pmatrix} \tag{1.78b}$$

and

$$A_{(+)} = \frac{1}{\sqrt{E_p L^3}} \frac{1}{\sqrt{4m_0c^2}} \quad . \tag{1.78c}$$

Equations (1.78) are readily understood if we refer to (1.76), from which we can deduce that

$$\left(E_p - m_0c^2\right) \varphi_0^{(+)} = -\left(E_p + m_0c^2\right) \chi_0^{(+)} \quad \text{or} \quad \varphi_0^{(+)} = \frac{m_0c^2 + E_p}{m_0c^2 - E_p} \chi_0^{(+)} \quad .$$

Choosing

$$\chi_0^{(+)} = m_0c^2 - E_p \quad ,$$

it consequently follows that

$$\varphi_0^{(+)} = m_0c^2 + E_p \quad .$$

Equation (1.73) allows us to calculate the normalization constant $A_{(+)}$ from

$$\begin{aligned} &|A_{(+)}|^2 \int \left(\varphi_0^{(+)*}\varphi_0^{(-)} - \chi_0^{(+)*}\varphi_0^{(+)}\right) \mathrm{d}^3x \\ &\quad = |A_{(+)}|^2 L^3 \left[\left(m_0c^2 + E_p\right)^2 - \left(m_0c^2 - E_p\right)^2\right] = 1 \quad . \end{aligned}$$

If we chose the phase to be real, the result is

$$A_{(+)} = \frac{1}{\sqrt{L^3(4m_0c^2E_p)}} = \frac{1}{\sqrt{4m_0c^2}\sqrt{L^3E_p}} \quad .$$

We proceed similarly in the other case.

(2) $E = -E_p$:

$$\psi^{(-)}(\boldsymbol{p}) = A_{(-)} \begin{pmatrix} \varphi_0^{(-)} \\ \chi_0^{(-)} \end{pmatrix} \exp\left[\mathrm{i}\left(\boldsymbol{p}\cdot\boldsymbol{x} + E_p t\right)/\hbar\right] \equiv \begin{pmatrix} \varphi_0^{(-)}(\boldsymbol{p}) \\ \chi_0^{(-)}(\boldsymbol{p}) \end{pmatrix} \quad , \tag{1.79a}$$

where

$$\begin{pmatrix} \varphi_0^{(-)} \\ \chi_0^{(-)} \end{pmatrix} = \begin{pmatrix} m_0c^2 - E_p \\ m_0c^2 + E_p \end{pmatrix} \quad \text{and} \tag{1.79b}$$

$$A_{(-)} = A_{(+)} = \frac{1}{\sqrt{4m_0c^2L^3E_p}} = \frac{1}{\sqrt{4m_0c^2}} \frac{1}{\sqrt{L^3E_p}} \quad . \tag{1.79c}$$

In the nonrelativistic limit we obtain

$$\begin{aligned} E_p &= c\sqrt{\boldsymbol{p}^2 + m_0^2c^2} = m_0c^2\sqrt{1 + \frac{\boldsymbol{p}^2}{m_0^2c^2}} \\ &\approx m_0c^2\left(1 + \frac{1}{2}\frac{\boldsymbol{p}^2}{m_0^2c^2}\right) = m_0c^2 + \frac{\boldsymbol{p}^2}{2m_0} \end{aligned} \tag{1.80}$$

and, hence,

$$\begin{pmatrix} A_{(+)}\varphi_0^{(+)} \\ A_{(+)}\chi_0^{(+)} \end{pmatrix} = \frac{1}{\sqrt{L^3}} \begin{pmatrix} (m_0c^2 + E_p)/\sqrt{E_p 4m_0c^2} \\ (m_0c^2 - E_p)/\sqrt{E_p 4m_0c^2} \end{pmatrix}$$
$$\approx \frac{1}{\sqrt{L^3}} \begin{pmatrix} 2m_0c^2/2m_0c^2 \\ [-\boldsymbol{p}^2/(2m_0)]/2m_0c^2 \end{pmatrix}$$
$$= \frac{1}{\sqrt{L^3}} \begin{pmatrix} 1 \\ -\frac{1}{4}\left(\frac{v}{c}\right)^2 \end{pmatrix} \underset{v/c\to 0}{\Rightarrow} \frac{1}{\sqrt{L^3}} \begin{pmatrix} 1 \\ 0 \end{pmatrix} \tag{1.81}$$

and

$$\begin{pmatrix} A_{(-)}\varphi_0^{(-)} \\ A_{(-)}\chi_0^{(-)} \end{pmatrix} = \frac{1}{\sqrt{L^3}} \begin{pmatrix} (m_0c^2 - E_p)/\sqrt{E_p 4m_0c^2} \\ (m_0c^2 + E_p)/\sqrt{E_p 4m_0c^2} \end{pmatrix}$$
$$\approx \frac{1}{\sqrt{L^3}} \begin{pmatrix} -\frac{1}{4}\left(\frac{v}{c}\right)^2 \\ 1 \end{pmatrix} \underset{v/c\to 0}{\Rightarrow} \frac{1}{\sqrt{L^3}} \begin{pmatrix} 0 \\ 1 \end{pmatrix} \quad . \tag{1.82}$$

Thus, we can see that in the nonrelativistic limit for states with positive charge, the upper component is large and the lower one is small and vice versa for states with negative charge.

1.7 Charge Conjugation

By comparing (1.78a) with (1.79a), we may write down the relation

$$\Psi^{(-)}(-\boldsymbol{p}) = \begin{pmatrix} \varphi^{(-)}(-\boldsymbol{p}) \\ \chi^{(-)}(-\boldsymbol{p}) \end{pmatrix} = \begin{pmatrix} \chi^{(+)*}(+\boldsymbol{p}) \\ \varphi^{(+)*}(+\boldsymbol{p}) \end{pmatrix}$$
$$= \begin{pmatrix} \chi^{(+)*}(+\boldsymbol{p}) \\ \varphi^{(+)*}(+\boldsymbol{p}) \end{pmatrix} = \hat{\tau}_1 \Psi^{(+)*}(+\boldsymbol{p}) \quad . \tag{1.83}$$

This can be interpreted in the following way: If the state

$$\Psi = \begin{pmatrix} \varphi \\ \chi \end{pmatrix} \tag{1.84}$$

represents a positive charge, then the state

$$\hat{C}\Psi\hat{C}^{-1} \equiv \Psi_c = \tau_1\Psi^* = \begin{pmatrix} \chi^* \\ \varphi^* \end{pmatrix} \tag{1.85}$$

describes a particle with negative charge. We call Ψ_c the *charge-conjugated state of* Ψ. Similarly, Ψ is the charge-conjugated state of Ψ_c because it obeys

$$(\Psi_c)_c = \hat{\tau}_1(\hat{\tau}_1\Psi^*)^* = \Psi \quad . \tag{1.86}$$

Explicitly, the charge conjugation implies the following transformations according to (1.83):

$$\begin{aligned} \varphi_0^{(+)} &\longrightarrow \chi_0^{(-)} \quad , \\ \chi_0^{(+)} &\longrightarrow \varphi_0^{(-)} \quad , \\ \boldsymbol{p} &\longrightarrow -\boldsymbol{p} \quad , \\ +E_p &\longrightarrow -E_p \quad . \end{aligned} \tag{1.87}$$

If we (arbitrarily) call the particle described by Ψ as *the particle*, then we call the particle described by Ψ_c *antiparticles*. If we call, for example, the π^- mesons particles, then the π^+ mesons are antiparticles. Neutral particles fit into this picture too, in that for these the charge-conjugated state is the state itself. In other words, neutral particles are their own antiparticles. So we have

$$\Psi_c = \hat{\tau}_1 \Psi^* = \alpha \Psi \quad . \tag{1.88}$$

The factor α has to be real. This important point can be understood if we imagine that, for a neutral particle, the Klein–Gordon wave function $\Psi = \varphi + \chi$ (1.62) has to be real; therefore,

$$\operatorname{Im} \varphi = -\operatorname{Im} \chi \tag{1.89}$$

must always hold. Since ψ_c in (1.88) is describing neutral particles, then similarly,

$$\operatorname{Im}(\alpha\varphi) = -\operatorname{Im}(\alpha\chi) \tag{1.90}$$

must hold: Both conditions, (1.89) and (1.90), necessarily lead to

$$\alpha \quad \text{real} \quad .$$

This can also be deduced from (1.88), where for neutral particles both ψ and $\alpha\psi$ are real. From

$$(\Psi_c)_c = \Psi \quad ,$$

it follows that

$$(\alpha\Psi)_c = \hat{\tau}_1 (\alpha\Psi)^* = \alpha\hat{\tau}_1 \Psi^* = \alpha\alpha\Psi = \Psi \quad ,$$

so that

$$a^2 = 1 \quad , \quad \alpha = \pm 1 \quad . \tag{1.91}$$

Accordingly there exist two different kinds of neutral particles, namely

(a) neutral particles with *positive charge parity*, i.e. $\alpha = +1$

$$\Psi_c \equiv \hat{\tau}_1 \Psi^* = \Psi \quad (\text{or} \ \ \varphi^* = \chi) \quad ; \tag{1.92}$$

(b) neutral particles with *negative charge parity*, i.e. $\alpha = -1$

$$\Psi_c \equiv \hat{\tau}_1 \Psi^* = -\Psi \quad (\text{or} \ \ \varphi^* = -\chi) \quad . \tag{1.93}$$

EXAMPLE

1.6 *C* Parity

C parity stands for *charge-conjugation parity* or, less accurately, particle–antiparticle conjugation symmetry. To obtain a general definition, we choose to characterize a particle state in the following way:

$$|\psi\rangle = |M, \boldsymbol{p}, J, \lambda;\ B, Q, L, N_\mu\rangle \quad . \tag{1}$$

This characterization is based on a set of quantum numbers with the following definitions:

M = mass ≡ energy of the system,
$\boldsymbol{p}$ = momentum of the system,
J = angular momentum quantum number,
λ = helicity = eigenvalue of the helicity operator,
(λ characterizes the spin projection onto the momentum direction),
B = baryon number,
Q = charge,
L = lepton number,
N_μ = muon number, compiled in the following table.

Equivalently we can use the hypercharge Y, the strangeness S and the isospin T_3, which are connected via

$$Q = \tfrac{1}{2}Y + T_3 \quad , \quad B = Y - S$$

to the charge and baryon number.[8]

Value of N_μ for different particles

Particle	e^+, e^-, $\bar{\nu}_e\nu_e$	μ^-, ν_μ	μ^+, $\bar{\nu}_\mu$	All other particles
N_μ	0	+1	−1	0

The following processes are forbidden due to muon number conservation:

$$\left.\begin{array}{l}\pi^- \rightarrow \mu^- + \bar{\nu}_e \\ \bar{\nu}_\mu + p \rightarrow e^+ + n\end{array}\right\}$$

On the other hand reactions like

$$\left.\begin{array}{l}\bar{\nu}_e + p \rightarrow e^+ + n \\ \bar{\nu}_\mu + p \rightarrow \mu^+ + n\end{array}\right\}$$

are allowed.

---

[8] See W. Greiner, B. Müller: *Quantum Mechanics – Symmetries*, 2nd ed. (Springer, Berlin, Heidelberg 1994).

Example 1.6.

The muonic charge N_μ and the lepton number L may be replaced by the alternative lepton numbers L_e and L_μ, which are listed, for convenience, in the following table.

Value of L_e and L_μ for different particles

Particle	e^+, $\bar{\nu}_e$	e^-, ν_e	μ^+, $\bar{\nu}_\mu$	μ^-, ν_μ	All other particles
L_e	-1	$+1$	0	0	0
L_μ	0	0	-1	$+1$	0

The introduction of L_e and L_μ has the advantage that the set of quantum numbers is symmetric with respect to electrons and muons. Clearly one has

$$L = L_e + L_\mu \tag{2}$$

and state (1) can be written as

$$|\psi\rangle = |M, \boldsymbol{p}, J, \lambda; B, Q, L_e, L_\mu\rangle \quad . \tag{3}$$

The charge conjugation is now defined by the equation

$$\begin{aligned}\hat{C}\,&|M, \boldsymbol{p}, J, \lambda; B, Q, L_e, L_\mu\rangle \\ &\equiv |M, \boldsymbol{p}, J, \lambda; B, Q, L_e, L_\mu\rangle_c \\ &= \eta_c|M, \boldsymbol{p}, J, \lambda; -B, -Q, -L_e, -L_\mu\rangle \quad ,\end{aligned} \tag{4}$$

which is simply saying that the $\hat{C}$ operator reflects (i.e. changes the sign of) charge-like quantum numbers, whereas other properties such as M, $\boldsymbol{p}$, J, λ remain unchanged. The former quantities are called *intrinsic*, while the latter are named *external*. The state $|M, \boldsymbol{p}, J, \lambda; B, Q, L_e, L_\mu\rangle$ is called a *particle state*, the state $|M, \boldsymbol{p}, J, \lambda; B, -Q, -L_e, -L_\mu\rangle$ is called an *antiparticle state*. Since $\hat{C}\,|\rangle = \psi_c$ should be normalized, e.g.

$$\langle\psi_c|\psi_c\rangle = 1 \quad ,$$

and because the states (3) are also normalized, it follows that

$$|\eta_c|^2 = 1 \quad \text{or} \quad \eta_c = e^{i\alpha} \quad , \quad \alpha \text{ real} \quad . \tag{5}$$

This formal property does not yet guarantee that the operator $\hat{C}$ corresponds to a physical symmetry. For this, the states on the rhs of (4) have to be physically realized, which is the case for $\hat{C}$ in nearly all known theories; however, an important exception occurs for neutrinos and antineutrinos.

Although neutrinos, in particular, are of great importance in the theory of weak interactions, the $\hat{C}$ conjugation cannot be chosen in this case as a symmetry operation.[9] On the other hand careful investigations have shown that the Hamiltonians of the strong and the electromagnetic interactions have a vanishing commutator with $\hat{C}$, i.e.

[9] This is discussed in more detail in W. Greiner, B. Müller: *Gauge Theory of Weak Interactions*, 2nd ed. (Springer, Berlin, Heidelberg 1996).

Example 1.6.

$$\begin{aligned}\hat{C}\hat{H}_{\text{strong}} &= \hat{H}_{\text{strong}}\hat{C} \quad ,\\ \hat{C}\hat{H}_{\text{elm}} &= \hat{H}_{\text{elm}}\hat{C} \quad .\end{aligned} \tag{6}$$

Consequently, $\hat{C}$ is, as a matter of fact, a symmetry operation for all strong and electromagnetic processes.

Let us now return to (4). If one of the charge-like quantum numbers is not equal to zero, then η_c has no physical meaning. Since, in this case, one is free to choose a relative phase factor between the states $|M, \boldsymbol{p}, J, \lambda; B, Q, L_e, L_\mu\rangle$ and $|M, \boldsymbol{p}, J, \lambda; -B, -Q, -L_e, -L_\mu\rangle$ (which are assumed to be different), then we can choose $\eta_c = 1$. *But if particles and antiparticles are identical*, then (4) turns to be an eigenvalue equation

$$|M, \boldsymbol{p}, J, \lambda; 0,0,0,0\rangle_c = \eta_c|M, \boldsymbol{p}, J, \lambda; 0,0,0,0\rangle \quad . \tag{7}$$

The eigenvalue η_c is named *C parity*. It is natural to postulate that the double application of the charge conjugation $\hat{C}$ leads back to the original state. Thus,

$$\begin{aligned}\left(|M, \boldsymbol{p}, J, \lambda; 0,0,0,0\rangle_c\right)_c &= |M, \boldsymbol{p}, J, \lambda; 0,0,0,0\rangle\\ &= \eta_c^2|M, \boldsymbol{p}, J, \lambda; 0,0,0,0\rangle \quad .\end{aligned} \tag{8}$$

Therefore,

$$\eta_c^2 = 1 \quad \text{or} \quad \eta_c = \pm 1 \quad . \tag{9}$$

If one considers a many-particle state, e.g. a two-particle state like

$$|\phi_1, \phi_2\rangle = |\phi_1\rangle|\phi_2\rangle \quad , \tag{10}$$

where each particle state $|\phi_i\rangle$ has the C parity $\eta_c^{(i)}$, then it holds that

$$\begin{aligned}\hat{C}|\phi_1, \phi_2\rangle &= \eta_c|\phi_1, \phi_2\rangle = |\phi_1\rangle_c|\phi_1\rangle_c\\ &= \eta_c^{(1)}|\phi_1\rangle\eta_c^{(2)}|\phi_2\rangle = \eta_c^{(1)}\eta_c^{(2)}|\phi_1\rangle|\phi_2\rangle\\ &= \eta_c^{(1)}\eta_c^{(2)}|\phi_1, \phi_2\rangle \quad .\end{aligned} \tag{11}$$

Thus,

$$\eta_c = \eta_c^{(1)}\eta_c^{(2)} \quad . \tag{12}$$

Since, from (6), $\hat{C}$ should be a symmetry operation, then C parity is a conserved quantum number for all strong and electromagnetic reactions. To illustrate this we consider the following three neutral particles:

(a) The η meson, for which $m_\eta c^2 = 548.8\,\text{MeV}$, $\Gamma_\eta = \hbar/\tau_\eta = 2.3\,\text{keV}$ [$1\,\text{MeV} \mathrel{\hat{=}} (3/2)\times 10^{31}\,\text{s}^{-1}$], charge = 0, spin/parity = 0^+;
(b) the pion π^0, with $m_{\pi^0}c^2 = 134.97\,\text{MeV}$, $\Gamma_{\pi^0} = \hbar/\tau_{\pi^0} = 7.9\,\text{eV}$, charge = 0, spin/parity = 0^-;
(c) the photon, having $m_\gamma c^2 = 0$, $\Gamma_\gamma = \hbar/\tau_\gamma = 0$, charge = 0, spin/parity = 1^-.

For these particles, the decay processes

$$\begin{aligned}\eta &\to \gamma + \gamma \quad ,\\ \eta &\to 3\pi^0 \quad ,\\ \eta &\to \pi^0 + \gamma + \gamma \quad ,\\ \pi^0 &\to \gamma + \gamma \quad ,\end{aligned} \tag{13}$$

Example 1.6.

may be observed, but the following reactions do not occur:

$$\eta \not\to \pi^0 + \gamma$$
$$\eta \not\to 3\gamma$$
$$\pi^0 \not\to 3\gamma \quad . \tag{14}$$

We understand both the allowed as well as the forbidden processes by assuming the *conservation of C parity* and assigning the C parities to the particles as shown in the accompanying table.

Particle	γ	π_0	η
C parity	-1	1	$+1$

Of course, within a given process the C parity is only conserved if there occur solely particles with a well-defined C parity. This is true for reactions (13) and (14). The *negative C parity of the photon*, as shown in the table, can be understood more precisely: The photons are coupled to all other particles by the electric current j_μ. The interaction part of the Lagrangian is

$$\mathcal{L}_{\mathrm{elm}} = j_\mu(x)A^\mu(x) \tag{15}$$

where $A^\mu(x)$ is the four-potential of the photon. Obviously, the current $j_\mu(x)$ changes its sign under $\hat{C}$ transformation,

$$\hat{C}j_\mu(x)\hat{C}^{-1} = -j_\mu(x) \quad . \tag{16}$$

Therefore,

$$\hat{C}A^\mu(x)\hat{C}^{-1} = -A^\mu(x) \tag{17}$$

must hold to save the invariance of the Lagrangian (15) under $\hat{C}$ transformation, i.e.

$$\hat{C}\mathcal{L}_{\mathrm{elm}}\hat{C}^{-1} = \mathcal{L}_{\mathrm{elm}} \quad . \tag{18}$$

The positive C parities of π^0 and η follow from the existence of the decays (13) using (12).

1.8 Free Spin-0 Particles in the Feshbach–Villars Representation[10]

We saw in (1.81) and (1.82) that positively charged particles possess a large upper component in the nonrelativistic limit ($|\varphi^{(+)}| \gg |\chi^{(+)}|$) while negatively charged particles have a large lower component ($|\chi^{(-)}| \gg |\varphi^{(-)}|$). Now there exists a

[10] H. Feshbach, F. Villars: Rev. Mod. Phys. **30**, 24 (1958).

representation – the so-called Φ *representation* – in which the positive and negative solutions are always of the following form

$$\Phi^{(+)}(\boldsymbol{p}) \sim \begin{pmatrix}1\\0\end{pmatrix} \quad , \quad \Phi^{(-)}(\boldsymbol{p}) \sim \begin{pmatrix}0\\1\end{pmatrix} \quad . \tag{1.94}$$

This representation is established by the transformation

$$\Phi = \hat{U}\Phi \quad , \quad \Phi^\dagger = \Phi^\dagger \hat{U}^\dagger \quad , \tag{1.95}$$

where $\hat{U}$ is the operator

$$\hat{U} = 1\!\!1 \cdot \frac{(m_0c^2 + E_p) - \hat{\tau}_1(m_0c^2 - E_p)}{\sqrt{4m_0c^2E_p}} \tag{1.96}$$

with $E_p = c(\boldsymbol{p}^2 + m_0^2c^2)^{1/2}$ and $\phi^\dagger$ denotes the Hermitian conjugate of ϕ; similarly $\hat{U}^\dagger$. The 2×2 matrix $\hat{U}$ is not unitary in the usual sense ($\hat{U}^{-1} \neq U^\dagger$) since

$$\hat{U}^{-1} = \hat{\tau}_3\hat{U}\hat{\tau}_3 = 1\!\!1 \cdot \frac{(m_0c^2 + E_p) + \hat{\tau}_1(m_0c^2 - E_p)}{\sqrt{4m_0c^2E_p}} \quad . \tag{1.97}$$

This can be seen directly with the aid of (1.66) and by the relation,

$$\hat{U}\hat{U}^{-1} = 1\!\!1 \cdot \frac{(m_0c^2 + E_p)^2 - (m_0c^2 - E_p)^2}{4m_0c^2E_p} = 1\!\!1 \cdot \frac{4m_0c^2E_p}{4m_0c^2E_p} = 1\!\!1 \quad . \tag{1.98}$$

Besides this, according to (1.78a) and (1.79a), plane waves are transformed by

$$\begin{aligned}
\phi^{(+)}(\boldsymbol{p}) &= \hat{U}\Psi^{(+)}(\boldsymbol{p}) \\
&= \hat{U}\frac{1}{\sqrt{L^3}}\frac{1}{\sqrt{4m_0c^2E_p}}\begin{pmatrix}m_0c^2 + E_p\\ m_0c^2 - E_p\end{pmatrix}\exp[\mathrm{i}(\boldsymbol{p}\cdot\boldsymbol{x} - E_pt)/\hbar] \\
&= \frac{1}{\sqrt{L^3}}\frac{\begin{pmatrix}(m_0c^2 + E_p)^2\\ (m_0c^2 + E_p)(m_0c^2 - E_p)\end{pmatrix} - \begin{pmatrix}(m_0c^2 - E_p)^2\\ (m_0c^2 - E_p)(m_0c^2 + E_p)\end{pmatrix}}{4m_0c^2E_p} \\
&\quad \times \exp[\mathrm{i}(\boldsymbol{p}\cdot\boldsymbol{x} - E_pt)/\hbar] \\
&= \frac{1}{\sqrt{L^3}}\begin{pmatrix}1\\0\end{pmatrix}\exp[\mathrm{i}(\boldsymbol{p}\cdot\boldsymbol{x} - E_pt)/\hbar] \quad .
\end{aligned} \tag{1.99}$$

Similarly, we get

$$\begin{aligned}
\phi^{(-)}(\boldsymbol{p}) &= \hat{U}\Psi^{(-)}(\boldsymbol{p}) \\
&= \hat{U}\frac{1}{\sqrt{L^3}}\frac{1}{\sqrt{4m_0c^2E_p}}\begin{pmatrix}m_0c^2 - E_p\\ m_0c^2 + E_p\end{pmatrix}\exp[\mathrm{i}(\boldsymbol{p}\cdot\boldsymbol{x} + E_pt)/\hbar] \\
&= \frac{1}{\sqrt{L^3}}\frac{\begin{pmatrix}(m_0c^2 + E_p)(m_0c^2 - E_p)\\ (m_0c^2 + E_p)^2\end{pmatrix} - \begin{pmatrix}(m_0c^2 + E_p)(m_0c^2 - E_p)\\ (m_0c^2 + E_p)^2\end{pmatrix}}{4m_0c^2E_p} \\
&\quad \times \exp[\mathrm{i}(\boldsymbol{p}\cdot\boldsymbol{x} + E_pt)/\hbar] \\
&= \frac{1}{\sqrt{L^3}}\begin{pmatrix}0\\1\end{pmatrix}\exp[\mathrm{i}(\boldsymbol{p}\cdot\boldsymbol{x} + E_pt)/\hbar] \quad .
\end{aligned} \tag{1.100}$$

This is, in fact, the result required by (1.94). The normalization of the Φ representation follows from that of Ψ (1.73):

$$\pm 1 = \int \Psi^\dagger \hat{\tau}_3 \Psi \, \mathrm{d}^3x = \int (\hat{U}^{-1}\Phi)^\dagger \hat{\tau}_3 \hat{U}^{-1} \Phi \, \mathrm{d}^3x$$

$$\int \Phi^\dagger (\hat{U}^{-1})^\dagger \hat{\tau}_3 \hat{U}^{-1} \Phi \, \mathrm{d}^3x = \int \Phi^\dagger \hat{\tau}_3 \Phi \, \mathrm{d}^3x \quad (1.101)$$

since, due to (1.97), $\hat{\tau}_3 \hat{U}^{-1} = \hat{U} \hat{\tau}_3$ and therefore

$$(\hat{U}^{-1})^\dagger \hat{\tau}_3 \hat{U}^{-1} = (\hat{U}^{-1})^\dagger \hat{U} \hat{\tau}_3 = \hat{U}^{-1} \hat{U} \hat{\tau}_3 = \hat{\tau}_3 \quad . \quad (1.102)$$

Here we have made use of $(\hat{U}^{-1})^\dagger = \hat{U}^{-1}$ which also follows from (1.97). In analogy to (1.101) we define a *generalized scalar product* or *Φ product*

$$\langle \Psi | \Psi' \rangle_\Phi \stackrel{\text{def}}{\equiv} \int \Psi^\dagger \hat{\tau}_3 \Psi \, \mathrm{d}^3x \quad . \quad (1.103)$$

One recognizes immediately that, as in (1.101),

$$\langle \Psi | \Psi' \rangle_\Phi = \langle \Phi | \Phi' \rangle_\Phi \quad , \quad (1.104)$$

i.e. the generalized scalar product is invariant under the transformation (1.95). It seems natural to call an operator $\hat{A}$, with the property

$$\langle \Psi | \Psi' \rangle_\Phi = \langle \hat{A}\Phi | \hat{A}\Phi' \rangle_\Phi \quad , \quad (1.105)$$

Φ unitary. Such an operator has to fulfill the condition

$$\hat{A}^H \stackrel{\text{def}}{\equiv} \hat{\tau}_3 \hat{A}^\dagger \hat{\tau}_3 = \hat{A}^{-1} \quad , \quad (1.106)$$

since $\int \Psi^\dagger \hat{\tau}_3 \Psi' \, \mathrm{d}^3x = \int \Phi^\dagger \hat{A}^\dagger \hat{\tau}_3 \hat{A} \Phi' \, \mathrm{d}^3x$ and, thus, $\hat{A}^\dagger \hat{\tau}_3 \hat{A} = \hat{\tau}_3$ or $\hat{\tau}_3 \hat{A}^\dagger \hat{\tau}_3 = \hat{A}^{-1}$. The $\hat{U}$ operator (1.96) is a member of this class. If $\hat{A}$ and $\hat{\tau}_3$ commute, then the relation $\hat{A}^\dagger = \hat{A}^{-1}$ follows from (1.106), i.e. the usual unitary relation.

The charge Q of a state Ψ is given by the integral

$$Q = e \int \Psi^\dagger \hat{\tau}_3 \Psi \, \mathrm{d}^3x \quad . \quad (1.107)$$

In the following exercise we will show that the average energy of a state Ψ is determined by

$$E = \int \Psi^\dagger \hat{\tau}_3 \hat{H}_\mathrm{f} \Psi \, \mathrm{d}^3x \quad . \quad (1.108)$$

EXERCISE

1.7 Lagrange Density and Energy-Momentum Tensor of the Free Klein–Gordon Equation in the Feshbach–Villars Representation

Problem. Determine the Lagrange density of the free Klein–Gordon equation in the Schrödinger representation (Feshbach–Villars representation). Subsequently calculate the energy-momentum tensor and show that the energy is given by the expression $E = \int \Psi^\dagger \hat{\tau}_3 \hat{H}_\mathrm{f} \Psi \, \mathrm{d}^3x$.

Solution. In the Schrödinger representation the equation of motion for the Klein–Gordon field reads

$$\mathrm{i}\hbar\partial_t \Psi = \hat{H}_\mathrm{f}\Psi \quad \text{with}$$

$$\hat{H}_\mathrm{f} = (\hat{\tau}_3 + \mathrm{i}\hat{\tau}_2)\frac{\hat{\boldsymbol{p}}^2}{2m_0} + \hat{\tau}_3 m_0 c^2 \quad . \tag{1}$$

The vector Ψ has two components, $\Psi = \binom{\phi}{\chi} \Rightarrow \Psi^\dagger = (\phi^*, \chi^*)$, and we define $\overline{\Psi} \equiv \Psi^\dagger \hat{\tau}_3$. To prove that the Lagrange density

$$\mathcal{L} = \mathrm{i}\hbar\overline{\Psi}\partial_t\Psi - \frac{\hbar^2}{2m_0}\boldsymbol{\nabla}\overline{\Psi}\,(\hat{\tau}_3 + \mathrm{i}\hat{\tau}_2)\,\boldsymbol{\nabla}\Psi - m_0c^2\overline{\Psi}\hat{\tau}_3\Psi \tag{2}$$

yields the correct equation of motion, we vary the action integral $I = \int \mathcal{L}\,\mathrm{d}^4x$ using the standard method. This variation with respect to the components of $\overline{\Psi}$ results in equation (1),

$$\begin{aligned}
&\frac{\partial I}{\partial \overline{\Psi}_\alpha} = 0 \\
&\quad\Rightarrow \partial_\nu \frac{\partial \mathcal{L}}{\partial(\partial_\nu \overline{\Psi}_\alpha)} - \frac{\partial \mathcal{L}}{\partial \overline{\Psi}_\alpha} = 0 \\
&\quad\Rightarrow \partial_t \frac{\partial \mathcal{L}}{\partial \dot{\overline{\Psi}}_\alpha} + \boldsymbol{\nabla}\cdot\frac{\partial \mathcal{L}}{\partial(\boldsymbol{\nabla}\overline{\Psi}_\alpha)} - \frac{\partial \mathcal{L}}{\partial \overline{\Psi}_\alpha} = 0 \\
&\quad\Rightarrow -\frac{\hbar^2}{2m_0}(\hat{\tau}_3 + \mathrm{i}\hat{\tau}_2)\,\nabla^2\Psi_\alpha - \left(\mathrm{i}\hbar\partial_t\Psi_\alpha - m_0c^2\hat{\tau}\Psi_\alpha\right) = 0
\end{aligned}$$

with $\alpha = 1, 2$.

Similarly, variation of Ψ results in the corresponding equation of motion for $\overline{\Psi}$, i.e.

$$\begin{aligned}
\frac{\partial I}{\partial \Psi_\alpha} = 0 &\Rightarrow \partial_t \frac{\partial \mathcal{L}}{\partial \dot{\Psi}_\alpha} + \boldsymbol{\nabla}\cdot\frac{\partial \mathcal{L}}{\partial(\boldsymbol{\nabla}\Psi_\alpha)} - \frac{\partial \mathcal{L}}{\partial \Psi_\alpha} = 0 \quad , \quad (\alpha = 1, 2) \\
&\Rightarrow -\mathrm{i}\hbar\partial_t\overline{\Psi}_\alpha = -\frac{\hbar^2}{2m_0}\boldsymbol{\nabla}^2\left(\overline{\Psi}(\hat{\tau}_3 + \mathrm{i}\hat{\tau}_2)\right)_\alpha + m_0c^2\left(\overline{\Psi}\hat{\tau}_3\right)_\alpha \quad .
\end{aligned}$$

If we had defined the Lagrange density with $\Psi^\dagger$ instead of $\overline{\Psi}$, the same equation of motion (1) would have resulted from the variation with respect to $\Psi^\dagger_\alpha$. However, we demand the action $I = \int \mathrm{d}^3x\,\mathrm{d}t\,\mathcal{L}$ to be real which results in the condition that

Exercise 1.7.

$$\int \left[\overline{\Psi}(+\mathrm{i}\hbar\partial_t)\Psi - \nabla\overline{\Psi}\frac{\hbar^2(\hat{\tau}_3+\mathrm{i}\hat{\tau}_2)}{2m_0}\cdot\nabla\Psi - m_0c^2\overline{\Psi}\hat{\tau}_3\Psi\right]\mathrm{d}^3x\,\mathrm{d}t$$

(partial integration)

$$= \int \left[\overline{\Psi}\left(\mathrm{i}\hbar\partial_t + \frac{\hbar^2\nabla^2}{2m_0}(\hat{\tau}_3+\mathrm{i}\hat{\tau}_2) - m_0c^2\hat{\tau}_3\right)\Psi\right]\mathrm{d}^3x\,\mathrm{d}t$$

must be real. This is the case if each of the operators

$$\mathrm{i}\hbar\partial_t \quad , \qquad -\frac{\hbar^2\nabla^2}{2m_0}(\hat{\tau}_3+\mathrm{i}\hat{\tau}_2) + m_0c^2\hat{\tau}_3 = \hat{H}_\mathrm{f}$$

fulfills the generalized hermiticity condition

$$\hat{O}^\mathrm{H} = \hat{\tau}_3\hat{O}^\dagger\hat{\tau}_3 = \hat{O} \quad .$$

This has already been proven [see (1.109) and (1.110)] to be true. The integral I would not be real if, instead of $\overline{\Psi}$, the spinor $\Psi^\dagger$ had been used.

We now calculate the energy-momentum tensor from $\mathcal{L}$, i.e.

$$T_{\mu\nu} = \frac{\partial\mathcal{L}}{\partial(\partial_\mu\Psi)}\partial_\nu\Psi + \frac{\partial\mathcal{L}}{\partial(\partial_\mu\overline{\Psi})}\partial_\nu\overline{\Psi} - \mathcal{L}g_{\mu\nu} \quad , \tag{3}$$

to obtain

$$T_{00} = \mathrm{i}\hbar\overline{\Psi}\partial_t\Psi - \mathrm{i}\hbar\overline{\Psi}\partial_t\Psi + \frac{\hbar^2}{2m_0}\nabla\overline{\Psi}(\hat{\tau}_3+\mathrm{i}\hat{\tau}_2)\cdot\nabla\Psi + m_0c^2\overline{\Psi}\hat{\tau}_3\Psi \quad , \tag{4}$$

and

$$\begin{aligned} E &= \int T^0_0\,\mathrm{d}^3x = \int T_{00}\,\mathrm{d}^3x \\ &= \int\left(\frac{\hbar^2}{2m_0}\nabla\overline{\Psi}\cdot(\hat{\tau}_3+\mathrm{i}\hat{\tau}_2)\,\nabla\Psi + m_0c^2\overline{\Psi}\hat{\tau}_3\Psi\right)\mathrm{d}^3x \\ &= \int\overline{\Psi}\left(-\frac{\hbar^2}{2m_0}(\hat{\tau}_3+\mathrm{i}\hat{\tau}_2)\,\nabla^2 + m_0c^2\hat{\tau}_3\right)\Psi\,\mathrm{d}^3x \\ &= \int\Psi^\dagger\hat{\tau}_3\hat{H}_\mathrm{f}\Psi\,\mathrm{d}^3x \end{aligned} \tag{5}$$

which is just (1.108).

From (1.107) and (1.108) we can now guess the generalization of the expectation values of arbitrary operators $\hat{L}$ and define the mean value (expectation value) $\langle L\rangle$ by

$$\langle L\rangle = \int\Psi^\dagger\hat{\tau}_3\hat{L}\Psi\,\mathrm{d}^3x \quad , \tag{1.109}$$

where $\langle L\rangle$ must be real. This results from the condition that

$$\int \Psi^\dagger \hat{\tau}_3 \hat{L} \Psi \, d^3x = \left(\int \Psi^\dagger \hat{\tau}_3 \hat{L} \Psi \, d^3x \right)^\dagger = \int \Psi^\dagger \hat{L}^\dagger \hat{\tau}_3^\dagger \Psi \, d^3x \quad .$$

Therefore,

$$\hat{\tau}_3 \hat{L} = \hat{L}^\dagger \hat{\tau}_3 \quad \text{or} \quad L^{\mathrm{H}} \stackrel{\text{def}}{\equiv} \hat{\tau}_3 \hat{L}^\dagger \hat{\tau}_3 = \hat{L} \quad , \tag{1.110}$$

which is the *generalized hermiticity* for an operator $\hat{L}$. We immediately see that the hermiticity of the Hamiltonian $\hat{H}_{\mathrm{f}}$ is *general*, because

$$\begin{aligned} \hat{H}_{\mathrm{f}}^{\mathrm{H}} &= \hat{\tau}_3 \hat{H}_{\mathrm{f}}^\dagger \hat{\tau}_3 = \hat{\tau}_3 \left[(\hat{\tau}_3 - \mathrm{i}\hat{\tau}_2) \frac{\hat{\boldsymbol{p}}^2}{2m_0} + \hat{\tau}_3 m_0 c^2 \right] \hat{\tau}_3 \\ &= \left[(\hat{\tau}_3 + \mathrm{i}\hat{\tau}_2) \frac{\hat{\boldsymbol{p}}^2}{2m_0} + \hat{\tau}_3 m_0 c^2 \right] \hat{\tau}_3 \hat{\tau}_3 \equiv \hat{H}_{\mathrm{f}} \quad . \end{aligned} \tag{1.111}$$

Nevertheless, the operator $\hat{U}$ (1.96) is *ordinary hermitian* (i.e. hermitian in the standard sense), i.e. $\hat{U}^\dagger = \hat{U}$.

The generalized scalar product (1.103), (1.109) necessarily leads to a transformation law for the operators $\hat{L}$ when changing the states according to (1.95). We simply calculate

$$\begin{aligned} \int \Psi^\dagger \hat{\tau}_3 \hat{L} \Psi' \, d^3x &= \int \Phi^\dagger (\hat{U}^{-1})^\dagger \hat{\tau}_3 \hat{L} \hat{U}^{-1} \Phi' \, d^3x \\ &= \int \Phi^\dagger \hat{\tau}_3 \hat{U} \hat{L} \hat{U}^{-1} \Phi' \, d^3x \\ &\equiv \int \Phi^\dagger \hat{\tau}_3 \hat{L}_\Phi \Phi' \, d^3x \quad , \end{aligned}$$

using (1.102). Thus,

$$\hat{L}_\Phi = \hat{U} \hat{L} \hat{U}^{-1} \quad . \tag{1.112}$$

This is the transformation law for operators covering the transition from Ψ *representation* to Φ *representation* (1.95).

Remark: Notice that the transformation law for operators depends on the definition of the scalar product. If, instead of the Φ product (1.103, 109). we had used the *ordinary scalar product*

$$\langle \Psi | \Psi' \rangle = \int \Psi^\dagger \Psi' \, d^3x \quad ,$$

with the matrix elements for operators $\hat{L}$ being

$$\langle \Psi | \hat{L} | \Psi' \rangle = \int \Psi^\dagger \hat{L} \Psi' \, d^3x \quad ,$$

then the operator $\hat{L}$ in Φ representation (we denote it by $\hat{L}'_\Phi$) would have been

$$\int \Psi^\dagger \hat{L} \Psi \, d^3x = \int \Phi^\dagger (\hat{U}^{-1})^\dagger \hat{L} \hat{U}^{-1} \Phi' \, d^3x \equiv \int \Phi^\dagger \hat{L}'_\Phi \Phi' \, d^3x \quad .$$

Thus,

$$\hat{L}'_\Phi = (\hat{U}^{-1})^\dagger \hat{L}\hat{U}^{-1} = \hat{U}^{-1}\hat{L}\hat{U}^{-1} \quad ,$$

i.e. a different law from (1.112). It is important that, because of physical reasons (charge interpretation), only the generalized scalar product (Φ product) makes sense to us and thus the transformation law (1.112) will be used in the following.

EXERCISE

1.8 The Hamiltonian in the Feshbach–Villars Representation

Problem. Show that, in the Feshbach–Villars representation, the Hamiltonian for all momentum eigenstates is given by $\hat{H}_\Phi = \hat{U}\hat{H}_\mathrm{f}\hat{U}^{-1} = \hat{\tau}_3 E_p$.

Solution. According to (1.112) the Hamiltonian $\hat{H}_\mathrm{f}$ of (1.68) for a free Klein–Gordon field transforms, under the transition to the Feshbach–Villars representation (1.95), according to

$$\hat{H}_\Phi = \hat{U}\hat{H}_\mathrm{f}\hat{U}^{-1} = \left[\frac{(E_p + m_0c^2)\mathbb{1} - (m_0c^2 - E_p)\hat{\tau}_1}{\sqrt{4m_0c^2E_p}}\right]$$
$$\times \left[(\hat{\tau}_3 + \mathrm{i}\hat{\tau}_2)\frac{\hat{\boldsymbol{p}}^2}{2m_0} + \hat{\tau}_3 m_0c^2\right]\left[\frac{(m_0c^2 + E_p)\mathbb{1} + (m_0c^2 - E_p)\hat{\tau}_1}{\sqrt{4m_0c^2E_p}}\right] \quad .$$

If we define

$$a_+ = m_0c^2 + E_p \quad , \quad a_- = m_0c^2 - E_p \quad , \quad a = m_0c^2 \quad ,$$

then it follows that

$$\begin{aligned}
\hat{H}_\Phi &= \frac{1}{4m_0c^2E_p}\left[a_+ - a_-\hat{\tau}_1\right]\left[\frac{\hat{\boldsymbol{p}}^2}{2m_0}(\hat{\tau}_3 + \mathrm{i}\hat{\tau}_2) + a\hat{\tau}_3\right](a_+ + a_-\hat{\tau}_1) \\
&= \frac{1}{4m_0c^2E_p}\left[a_+ - a_-\hat{\tau}_1\right]\left[a_+\frac{\hat{\boldsymbol{p}}^2}{2m_0}(\hat{\tau}_3 + \mathrm{i}\hat{\tau}_2)\right. \\
&\quad \left. + a_-\frac{\hat{\boldsymbol{p}}^2}{2m_0}(\hat{\tau}_3\hat{\tau}_1 + \mathrm{i}\hat{\tau}_2\hat{\tau}_1) + aa_+\hat{\tau}_3 + aa_-\hat{\tau}_3\hat{\tau}_1\right] \\
&= \frac{1}{4m_0c^2E_p}\left[a_+^2\frac{\hat{\boldsymbol{p}}^2}{2m_0}(\hat{\tau}_3 + \mathrm{i}\hat{\tau}_2) + a_+a_-\frac{\hat{\boldsymbol{p}}^2}{2m_0}(\hat{\tau}_3\hat{\tau}_1 + \mathrm{i}\hat{\tau}_2\hat{\tau}_1)\right. \\
&\quad + aa_+^2\hat{\tau}_3 + aa_-a_+\hat{\tau}_3\hat{\tau}_1 - a_-a_+\frac{\hat{\boldsymbol{p}}^2}{2m_0}(\hat{\tau}_1\hat{\tau}_3 + \mathrm{i}\hat{\tau}_1\hat{\tau}_2) \\
&\quad \left. - a_-^2\frac{\hat{\boldsymbol{p}}^2}{2m_0}(\hat{\tau}_1\hat{\tau}_3\hat{\tau}_1 + \mathrm{i}\hat{\tau}_1\hat{\tau}_2\hat{\tau}_1) - aa_+a_-\hat{\tau}_1\hat{\tau}_3 - aa_-^2\hat{\tau}_1\hat{\tau}_3\hat{\tau}_1\right] \quad .
\end{aligned}$$

Exercise 1.8.

For momentum eigenstates the operator $\hat{\boldsymbol{p}}$ can be substituted by its eigenvalue. Therefore, with $\hat{\tau}_i\hat{\tau}_j = \mathrm{i}\varepsilon_{ijk}\hat{\tau}_k$ $(i \neq j = 1,2,3)$ one has

$$\hat{\tau}_3 + \mathrm{i}\hat{\tau}_2 = +\mathrm{i}\hat{\tau}_2\hat{\tau}_1 + \hat{\tau}_3\hat{\tau}_1 = -\mathrm{i}\hat{\tau}_1\hat{\tau}_2 - \hat{\tau}_1\hat{\tau}_3$$
$$= -\hat{\tau}_1\hat{\tau}_3\hat{\tau}_1 - \mathrm{i}\hat{\tau}_1\hat{\tau}_2\hat{\tau}_1 = \begin{pmatrix} 1 & 1 \\ -1 & -1 \end{pmatrix} \quad ,$$

and

$$\hat{\tau}_1\hat{\tau}_3\hat{\tau}_1 = -\hat{\tau}_3 = \begin{pmatrix} -1 & 0 \\ 0 & 1 \end{pmatrix} \quad ,$$

as well as

$$-\hat{\tau}_1\hat{\tau}_3 = \begin{pmatrix} 0 & 1 \\ -1 & 0 \end{pmatrix} \quad ,$$

and we get

$$\hat{H}_\Phi = \frac{1}{4m_0c^2E_p}\left[\begin{pmatrix} 1 & 1 \\ -1 & -1 \end{pmatrix}\left\{a_+^2\frac{\boldsymbol{p}^2}{2m_0} + a_+a_-\frac{\boldsymbol{p}^2}{2m_0} + a_-a_+\frac{\boldsymbol{p}^2}{2m_0} + a_-^2\frac{\boldsymbol{p}^2}{2m_0}\right\}\right.$$
$$\left. + \begin{pmatrix} 1 & 0 \\ 0 & -1 \end{pmatrix}\left\{a_+^2a + a_-^2a\right\} + \begin{pmatrix} 0 & 1 \\ -1 & 0 \end{pmatrix}\left\{2aa_-a_+\right\}\right]$$
$$= \frac{1}{4m_0c^2E_p}\left[\hat{\tau}_3\left\{(a_+ + a_-)^2\frac{\boldsymbol{p}^2}{m_0} + (a_+^2 + a_-^2)a\right\}\right.$$
$$\left. + \mathrm{i}\hat{\tau}_2\left\{(a_+ + a_-)^2\frac{\boldsymbol{p}^2}{2m_0} + 2aa_-a_+\right\}\right] \quad .$$

Using the relation

$$\begin{aligned}
(a_+ + a_-)^2\frac{\boldsymbol{p}^2}{2m_0} &= 4m_0^2c^4\frac{\boldsymbol{p}^2}{2m_0} = 2m_0c^2\boldsymbol{p}^2c^2 \quad , \\
\left(a_+^2 + a_-^2\right)a &= 2m_0c^2 = \left(m_0^2c^4 + E_p^2\right) = 2m_0c^2\left(2m_0^2c^4 + \boldsymbol{p}^2c^2\right) \quad , \\
2aa_-a_+ &= 2m_0c^2\left(m_0^2c^4 - E_p^2\right) = -2m_0c^2\boldsymbol{p}^2c^2 \quad ,
\end{aligned}$$

we obtain

$$\hat{H}_\Phi = \frac{1}{4m_0c^2E_p}\left[\hat{\tau}_3 \cdot 4m_0c^2E_p^2 + \mathrm{i}\hat{\tau}_2 \cdot 0\right] = \hat{\tau}_3E_p \quad .$$

From the free Klein–Gordon equation in Schrödinger representation (1.67)

$$\mathrm{i}\hbar\frac{\partial\Psi}{\partial t} = \hat{H}_\mathrm{f}\Psi$$

and by the application of $\hat{U}$ from the left it follows that

$$\mathrm{i}\hbar\frac{\partial\hat{U}\Psi}{\partial t} = \hat{U}\hat{H}_\mathrm{f}\hat{U}^{-1}\hat{U}\Psi \quad .$$

Using the results of Exercise 1.8 and (1.95), we obtain the free Klein–Gordon equation in the *Feshbach–Villars representation*, i.e.

$$i\hbar\frac{\partial\Phi}{\partial t} = \hat{\tau}_3 E_p \Phi \quad . \tag{1.113}$$

This equation yields two different solutions for any given momentum $\boldsymbol{p}$, one with positive ($+E_p$) and one with negative ($-E_p$) time factor. They are precisely the solutions (1.99) and (1.100), obtained by the direct transformation, and can be interpreted, according to the time factor, as belonging to positive charge or negative charge, respectively (see Exercise 1.9 for further details).

EXERCISE

1.9 Solution of the Free Klein–Gordon Equation in the Feshbach–Villars Representation

Problem. Solve the free Klein–Gordon equation in the Feshbach–Villars representation (1.113) directly and show that the solutions are identical to (1.99) and (1.100).

Solution. The Feshbach–Villars representation of the Schrödinger equation reads

$$i\hbar\frac{\partial\Phi}{\partial t} = \hat{\tau}_3 E_p \Phi \quad ,$$

where the ϕ's are eigenstates of the momentum operator

$$\Phi = \exp(i\boldsymbol{p}\cdot\boldsymbol{r}/\hbar)\theta \quad , \quad \theta = \begin{pmatrix}\vartheta_1\\ \vartheta_2\end{pmatrix} \quad ,$$

$$E_p = \sqrt{m_0^2c^4 + \boldsymbol{p}^2c^2} \quad .$$

By inserting the matrix $\hat{\tau}_3$ we get

$$i\hbar\frac{\partial}{\partial t}\begin{pmatrix}\vartheta_1\\ \vartheta_2\end{pmatrix} = E_p\begin{pmatrix}\vartheta_1\\ -\vartheta_2\end{pmatrix}$$

and, thus,

$$i\hbar\dot{\vartheta}_1 = E_p\vartheta_1 \quad , \quad i\hbar\dot{\vartheta}_2 = -E_p\vartheta_2 \quad .$$

Integration yields θ_1 and θ_2 is the form of

$$\vartheta_1 = N_1\exp\left(-\frac{i}{\hbar}E_p t\right)$$

$$\vartheta_2 = N_2\exp\left(+\frac{i}{\hbar}E_p t\right) \quad .$$

N_1 and N_2 are normalization constants which are determined by the normalization condition (1.101),

Exercise 1.9.

$$\int \Phi^\dagger \hat{\tau}_3 \Phi \, d^3r = \int \theta^\dagger \hat{\tau}_3 \theta \, d^3r = \pm 1 \quad ,$$

yielding

$$|N_1|^2 - |N_2|^2 = \pm \frac{1}{V} \quad .$$

Hence, we obtain two independent solutions [see (1.99) and (1.100)].

$$\Phi^{(+)} = \frac{1}{\sqrt{V}} \begin{pmatrix} 1 \\ 0 \end{pmatrix} \exp\left[\frac{\mathrm{i}}{\hbar}(\boldsymbol{p}\cdot\boldsymbol{r} - E_p t)\right] \quad , \quad \text{charge } +1$$

$$\Phi^{(-)} = \frac{1}{\sqrt{V}} \begin{pmatrix} 0 \\ 1 \end{pmatrix} \exp\left[\frac{\mathrm{i}}{\hbar}(\boldsymbol{p}\cdot\boldsymbol{r} + E_p t)\right] \quad , \quad \text{charge } -1$$

Again each linear combination

$$n_1 \Phi^{(+)} + n_2 \Phi^{(-)} \quad , \quad \text{with} \quad |n_1|^2 - |n_2|^2 = 1 \quad ,$$

is a normalized eigenfunction of the momentum $\boldsymbol{p}$ with charge $+1$, and each linear combination with $|n_1|^2 - |n_2|^2 = -1$ is a normalized solution with charge -1.

We denote the solution (1.99) and (1.100) of the Klein–Gordon equation (1.113) for fixed momentum $\boldsymbol{p}$ (see Exercise 1.9) by

$$\Phi_{\boldsymbol{p},\lambda} \quad ,$$

$$\boldsymbol{p} = \{p_i\} = \left\{\frac{2\pi\hbar}{L} n_i\right\} \quad n_i = 0, \pm 1, \pm 2, \ldots ; \quad \lambda = 1, -1 \quad . \tag{1.114}$$

From now on we drop the index i on p_i, which has its origin in the box normalization, to simplify the notation, e.g. $\sum_{\boldsymbol{p}}$ means $\sum_{p_i}$. The index $\lambda = +1$ denotes a positive charge state whereas $\lambda = -1$ denotes a negative one. The states $\Phi_{\boldsymbol{p},\lambda}$ in (1.114) form a complete orthonormalized system. Hence, the following relation holds:

$$\int \Phi^\dagger_{\boldsymbol{p},\lambda} \hat{\tau}_3 \Phi_{\boldsymbol{p}',\lambda'} \, d^3x = \lambda \delta_{\boldsymbol{p},\boldsymbol{p}'} \delta_{\lambda\lambda'} \quad , \tag{1.115}$$

where $\lambda, \lambda' = \pm 1$ and $\boldsymbol{p}$, $\boldsymbol{p}'$ take the values given in (1.114). Let us now consider a general state Φ containing particles with positive as well as negative charge. We can recognize and distinguish between the charged particles by expanding Φ in terms of $\phi_{\boldsymbol{p},\lambda}$ to give

$$\Phi = \sum_{\boldsymbol{p},\lambda} a_{\boldsymbol{p},\lambda} \Phi_{\boldsymbol{p},\lambda} = \sum_{\boldsymbol{p}} \left(a_{\boldsymbol{p},+1} \Phi_{\boldsymbol{p},+1} + a_{\boldsymbol{p},-1} \Phi_{\boldsymbol{p},-1}\right) \quad , \tag{1.116}$$

where the expansion coefficient $a_{\boldsymbol{p},\lambda}$ is given by

$$a_{\boldsymbol{p},\lambda} = \int \Phi^\dagger_{\boldsymbol{p},\lambda} \hat{\tau}_3 \Phi \, d^3x \quad . \tag{1.117}$$

For the total charge we obtain

$$e \int \Phi^\dagger \hat{\tau}_3 \Phi \, \mathrm{d}^3x = e \sum_p \left(|a_{p,+1}|^2 - |a_{p,-1}|^2 \right) = \pm Ne \quad . \tag{1.118}$$

The total number N of elementary charges which are contained in the state Φ can be positive, negative or zero. $\sum_p |a_{p,+1}|^2$ is equal to the total number of positive elementary charges and $\sum_p |a_{p,-1}|^2$ equal to the total number of negative elementary charges in Φ.

1.9 The Interaction of a Spin-0 Particle with an Electromagnetic Field

The electromagnetic field is described by the four-vector [defined in (1.14)]

$$\begin{aligned} A^\mu &= \{A_0, \boldsymbol{A}\} = \{A_0, A_x, A_y, A_z\} = g^{\mu\nu} A_\nu \quad , \\ A_\mu &= g_{\mu\nu} A^\nu = \{A_0, -\boldsymbol{A}\} \quad . \end{aligned}$$

In the case of nonrelativistic quantum mechanics we specified minimal coupling of the electromagnetic field,

$$\hat{E} \Rightarrow \mathrm{i}\hbar \frac{\partial}{\partial t} - eA_0 \quad , \quad \hat{\boldsymbol{p}} \Rightarrow -\mathrm{i}\hbar \boldsymbol{\nabla} - \frac{e}{c}\boldsymbol{A} \quad ,$$

which can be compressed to the four-dimensional and covariant form as

$$\hat{p}^\mu \Rightarrow \hat{p}^\mu - \frac{e}{c} A^\mu \quad \text{or} \quad \hat{p}_\mu \Rightarrow \hat{p}_\mu - \frac{e}{c} A_\mu \quad . \tag{1.119}$$

With the same minimal coupling, the free Klein–Gordon equation (1.21) is transformed into the Klein–Gordon equation with an electromagnetic field (compare with the later passage on gauge invariance of the coupling),

$$\left(\hat{p}^\mu - \frac{e}{c} A^\mu\right) \left(\hat{p}_\mu - \frac{e}{c} A_\mu\right) \psi = m_0^2 c^2 \psi \tag{1.120}$$

or

$$\left[g^{\mu\nu} \left(\mathrm{i}\hbar \frac{\partial}{\partial x^\nu} - \frac{e}{c} A_\nu \right) \left(\mathrm{i}\hbar \frac{\partial}{\partial x^\mu} - \frac{e}{c} A_\mu \right) \right] \psi = m_0^2 c^2 \psi \quad , \tag{1.121}$$

and, explicitly,

$$\begin{aligned} \frac{1}{c^2} \left(\mathrm{i}\hbar \frac{\partial}{\partial t} - eA_0 \right)^2 \psi &= \left(\sum_{i=1}^3 \left(+\mathrm{i}\hbar \frac{\partial}{\partial x^i} + \frac{e}{c} A_i \right)^2 + m_0^2 c^2 \right) \psi \\ &= \left(\left(+\mathrm{i}\hbar \boldsymbol{\nabla} + \frac{e}{c} \boldsymbol{A} \right)^2 + m_0^2 c^2 \right) \psi \quad . \end{aligned} \tag{1.122}$$

In order to examine the charge and current densities, we start with (1.121), multiply by ψ^* from the lhs and subtract the complex conjugate. These operations can be symbolized by $\psi^*(xx) - \psi(xx)^*$, where (xx) denotes (1.121), and they result in

$$
\begin{aligned}
0 &= \psi^*\left[-g^{\mu\nu}\left(\frac{\partial}{\partial x^\nu}+\frac{\mathrm{i}e}{\hbar c}A_\nu\right)\left(\frac{\partial}{\partial x^\mu}+\frac{\mathrm{i}e}{\hbar c}A_\mu\right)\right]\psi \\
&\quad -\psi\left[-g^{\mu\nu}\left(\frac{\partial}{\partial x^\nu}-\frac{\mathrm{i}e}{\hbar c}A_\nu\right)\cdot\left(\frac{\partial}{\partial x^\mu}-\frac{\mathrm{i}e}{\hbar c}A_\mu\right)\right]\psi^* \\
&= g^{\mu\nu}\left[\psi\frac{\partial}{\partial x^\nu}\frac{\partial}{\partial x^\mu}\psi^*-\psi^*\frac{\partial}{\partial x^\nu}\frac{\partial}{\partial x^\mu}\psi-\psi^*\frac{\partial}{\partial x^\nu}\frac{\mathrm{i}e}{\hbar c}A_\mu\psi\right. \\
&\quad \left.-\psi\frac{\partial}{\partial x^\nu}\frac{\mathrm{i}e}{\hbar c}A_\mu\psi^*-\psi^*\frac{\mathrm{i}e}{\hbar c}A_\nu\frac{\partial}{\partial x^\mu}\psi-\psi\frac{\mathrm{i}e}{\hbar c}A_\nu\frac{\partial}{\partial x^\mu}\psi^*\right] \\
&= g^{\mu\nu}\left[\frac{\partial}{\partial x^\mu}\left(\psi\frac{\partial}{\partial x^\nu}\psi^*-\psi^*\frac{\partial}{\partial x^\nu}\psi\right)-2\frac{\partial}{\partial x^\mu}\left(\psi\frac{\mathrm{i}e}{\hbar c}A_\nu\psi^*\right)\right] \quad .
\end{aligned}
$$

One calls

$$
j_\nu=\frac{\mathrm{i}\hbar e}{2m_0}\left(\psi^*\frac{\partial}{\partial x^\nu}\psi-\psi\frac{\partial}{\partial x^\nu}\psi^*\right)-\frac{e^2}{m_0c}A_\nu\psi\psi^*=\{c\varrho',-\boldsymbol{j}'\} \tag{1.123}
$$

the four-current density in the electromagnetic field A_ν. Obviously, the following relation holds:

$$
g^{\mu\nu}\frac{\partial}{\partial x^\mu}j_\nu=\frac{\partial}{\partial x^\mu}j^\mu=0 \quad . \tag{1.124}
$$

This is a continuity equation for the four-current density, usually its meaning is that of a conservation law. Suitable normalization [like in (1.123)] yields charge conservation. Note that (1.123) is identical to the previous result, (1.34), if we exclude the term proportional to A_ν. To make this fact more evident we write explicitly,

$$
\varrho'=\frac{\mathrm{i}\hbar e}{2m_0c^2}\left(\psi^*\frac{\partial}{\partial t}\psi-\psi\frac{\partial}{\partial t}\psi^*\right)-\frac{e^2}{m_0c^2}A_0\psi\psi^* \tag{1.125}
$$

and

$$
\boldsymbol{j}'=-\frac{\mathrm{i}\hbar e}{2m_0}(\psi^*\boldsymbol{\nabla}\psi-\psi\boldsymbol{\nabla}\psi^*)-\frac{e^2}{m_0c}\boldsymbol{A}\psi\psi^* \tag{1.126}
$$

and compare these expressions with (1.35) and (1.36). It is remarkable that electromagnetic potentials appear in the charge and current densities (1.125) and (1.126). To obtain a deeper understanding of this fact, we consider a negatively charged particle in a central Coulomb potential, which is given by

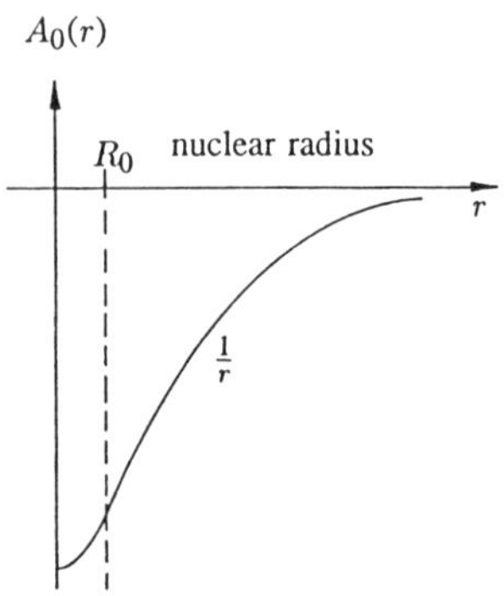

Fig. 1.3. The Coulomb potential of a negatively charged particle in the vicinity of the atomic nucleus

$$
eA_0(r)=Ze^2V(r) \quad , \quad \boldsymbol{A}=0 \quad , \tag{1.127}
$$

$V(r)\sim 1/r$ for large r and rounded off as for an oscillator within the range of the nucleus (see Fig. 1.3). A stationary state of the Klein–Gordon equation has the form

$$
\psi(\boldsymbol{r},t)=\psi(\boldsymbol{r})\exp(-\mathrm{i}\varepsilon t/\hbar) \quad , \tag{1.128}
$$

where $|\varepsilon|$ is the energy of the particle (see Exercises 1.10 and 1.11). The charge density (1.125) can be calculated as follows:

$$\varrho'(\boldsymbol{r}) = e\frac{[\varepsilon - eA_0(\boldsymbol{r})]}{m_0c^2}\psi\psi^*(\boldsymbol{r}) = e\frac{[\varepsilon + Ze^2V(r)]}{m_0c^2}\psi\psi^*(\boldsymbol{r}) \quad . \tag{1.129}$$

Therefore, the charge density becomes

$$\varrho' > 0 \quad \text{for} \quad \varepsilon > eA_0(\boldsymbol{x}) \quad , \quad \varrho' < 0 \quad \text{for} \quad \varepsilon < eA_0(\boldsymbol{x}) \quad . \tag{1.130}$$

In the first case the charge density has the same sign as the charge, e, of the particle; in the second it is the other way round. The charge density has the opposite sign to the particle charge e, wherever the potential energy has values so that $\varepsilon < eA_0(\boldsymbol{x})$. The physical meaning of this change of sign of ϱ' in strong fields can only be understood within the frame of the field theory where the number of particles becomes variable. One can imagine that in the areas of strong fields, particle–antiparticle pairs will be produced and that the potential term in the charge density (1.129) may be interpreted as simulating such many-body aspects.

Now let us carry over what we have said about normalization and the energy factor ε for the Klein–Gordon equation to our present case. The charge normalization is carried out according to $\int \varrho'(\boldsymbol{x})\,\mathrm{d}^3x = \pm e$ (e being the charge of the electron). For π^- mesons ($\varepsilon < 0$) we take the positive sign, for bound π^+ mesons we choose the negative sign. Hence, the charge density for π^- mesons becomes

$$\varrho'(\boldsymbol{x}) = e\frac{[\varepsilon + Ze^2V(\boldsymbol{x})]}{m_0c^2}\psi^*(\boldsymbol{x})\psi(\boldsymbol{x}) \quad . \tag{1.131}$$

Figures 1.3 and 1.4 illustrate strong Coulomb potentials and the corresponding charge density.[11]

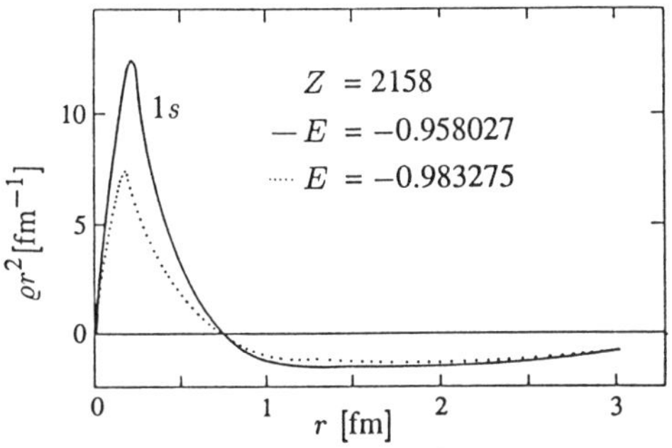

Fig. 1.4. The radial density ϱr^2 of a π meson in the $1s$ state of an exponential potential of the form $V(r) = Z\alpha\exp(-r/a)$ (see Exercise 1.15). The range parameter a of the potential is chosen to be $a = 0.2\lambda$, where $\lambda_\pi = \hbar/m_\pi c$ is the Compton wavelength of the pion. The coupling strength is fixed by $Z = 2158$. Such an extreme central charge is chosen to make various new features of the Klein–Gordon equation especially obvious. The energy eigenvalue of the $1s$ pion bound in this extremely short range and deep potential is $E = -0.958027$ measured in units of the rest energy $m_\pi c^2$ of the pion. It is evident that the charge density is not positive definite, and that antiparticles can also be bound in this potential. The radial density of antiparticle in the same potential is shown by the *dashed line* – the energy eigenvalue of a bound antiparticle is $E = -0.983275\,m_\pi c^2$

[11] See W. Fleischer, G. Soff: Z. Naturforsch. **39a**, 703 (1984).

An interpretation of the positive part of the charge density is that, in strong fields, some fraction of π^+ mesons is always mixed with the π^- meson. This indicates the great difficulties of the single particle interpretation in strong potentials.

EXERCISE

1.10 Separation of Angular and Radial Parts of the Wave Function for the Stationary Klein–Gordon Equation with a Coulomb Potential

Problem. Separate the angular and the radial part of the wave function for the stationary Klein–Gordon equation for spherically symmetric Coulomb potentials.

Solutions. We put $eA_0 = V(r)$ and $\boldsymbol{A} = 0$. Then the stationary Klein–Gordon equation reads

$$\left[(\varepsilon - V(r))^2 - m_0^2c^4 + \hbar^2c^2\nabla^2\right]\psi(\boldsymbol{r}) = 0 \tag{1}$$

or, explicitly,

$$\begin{aligned} -\hbar^2c^2\left[\frac{1}{r^2}\frac{\partial}{\partial r}\left(r^2\frac{\partial}{\partial r}\right) + \frac{1}{r^2\sin\theta}\frac{\partial}{\partial\theta}\left(\sin\theta\frac{\partial}{\partial\theta}\right) + \frac{1}{r^2\sin^2\theta}\frac{\partial^2}{\partial\phi^2}\right]\psi(\boldsymbol{r}) \\ = \left[(\varepsilon - V(r))^2 - m_0^2c^4\right]\psi(\boldsymbol{r}) \quad . \end{aligned} \tag{2}$$

For the wave function $\psi(\boldsymbol{r})$ we make the following separation ansatz

$$\psi(\boldsymbol{r}) = u(r)Y(\theta,\phi) \quad . \tag{3}$$

This leads to

$$\begin{aligned} &\hbar^2c^2\frac{1}{r^2}\frac{\partial}{\partial r}r^2\frac{\partial u}{\partial r} + \left\{(\varepsilon - V(r))^2 - m_0^2c^4 - \frac{\lambda}{r^2}\right\}u = 0 \quad , \\ &\frac{1}{\sin\theta}\frac{\partial}{\partial\theta}\sin\theta\frac{\partial Y}{\partial\theta} + \frac{1}{\sin^2\theta}\frac{\partial^2 Y}{\partial\phi^2} + \lambda Y = 0 \quad , \end{aligned} \tag{4}$$

where λ is the separation constant. The solutions of the last equation are the spherical harmonics Y_{lm} with $\lambda = l(l+1)$, $l = 0, 1, 2\dots$ and $m = 0\pm1, \pm2, \pm3\dots$[12] With this the radial differential equation follows as

$$\left[-\frac{1}{r^2}\frac{\partial}{\partial r}\left(r^2\frac{\partial}{\partial r}\right) + \frac{l(l+1)}{r^2}\right]u(r) = \frac{(\varepsilon - V)^2 - m_0^2c^4}{\hbar^2c^2}u(r) \quad . \tag{5}$$

Hence, using the common ansatz $u(r) = R(r)/r$, follows the transformation of (5) into

$$\left[\frac{\mathrm{d}^2}{\mathrm{d}r^2} - \frac{l(l+1)}{r^2} + k^2\right]R(r) = 0 \quad , \tag{6}$$

where

$$k^2 = \frac{(\varepsilon - V(r))^2 - m_0^2c^4}{\hbar^2c^2} \quad . \tag{7}$$

[12] See W. Greiner: *Quantum Mechanics – An Introduction*, 3rd ed. (Springer, Berlin, Heidelberg 1994).

EXERCISE

1.11 Pionic Atom with Point-Like Nucleus

Problem. Find the solution of the Klein–Gordon equation for the π^- meson in a Coulomb potential and discuss the energy eigenvalues. The pion has the mass $m_\pi c^2 = 139.577\,\mathrm{MeV}$ and spin 0. It obeys the Klein–Gordon equation.

Solution. We use the result of the previous Exercise 1.10. The attractive Coulomb potential $A_0(r) = -Ze/r$ is coupled as the 0 component of the four-potential in the Klein–Gordon equation ($\alpha = e^2/\hbar c = 1/137.03602$ is the fine-structure constant), giving

$$\left[\frac{\mathrm{d}^2}{\mathrm{d}r^2} - \frac{l(l+1)-(Z\alpha)^2}{r^2} + \frac{2\varepsilon Z\alpha}{\hbar cr} - \frac{m_0^2c^4-\varepsilon^2}{\hbar^2c^2}\right] R_l(r) = 0 \quad . \tag{1}$$

Now the energies and the wave functions for the bound states in the energy range $-m_0c^2 < \varepsilon < m_0c^2$ are to be calculated. Therefore, we first transform the above differential equation with the help of the substitutions

$$\begin{aligned} \beta &= 2\frac{\left[m_0^2c^4-\varepsilon^2\right]^{1/2}}{\hbar c} \quad , \\ \varrho &= \beta r \quad \text{with} \quad 0<\varrho<\infty \quad , \end{aligned} \tag{2}$$

$$\mu = \sqrt{\left(l+\frac{1}{2}\right)^2 - (Z\alpha)^2} \quad \text{and} \quad \lambda = \frac{2Z\alpha\varepsilon}{\hbar c\beta}$$

and obtain

$$\left[\frac{\mathrm{d}^2}{\mathrm{d}\varrho^2} - \frac{\mu^2-1/4}{\varrho^2} + \frac{\lambda}{\varrho} - \frac{1}{4}\right] R_l(\varrho) = 0 \quad . \tag{3}$$

In order to find an ansatz for the solution it is useful to first study the limits $\varrho \to \infty$ and $\varrho \to 0$. In the case $\varrho \to \infty$ we can neglect the terms proportional to ϱ^{-1} and ϱ^{-2} in (3) and, therefore,

$$\left(\frac{\mathrm{d}^2}{\mathrm{d}\varrho^2} - \frac{1}{4}\right) R_l(\varrho) = 0 \quad . \tag{4}$$

Its solution can easily be determined with the help of the exponential function

$$R_l(\varrho) \underset{\varrho\to\infty}{=} a\mathrm{e}^{-\varrho/2} + b\mathrm{e}^{+\varrho/2} \quad . \tag{5}$$

$b = 0$ follows from the requirement of normalization of the wave function. In the case $\varrho \to 0$ one can neglect the last two terms in the above radial equation and is led to

$$\left(\frac{\mathrm{d}^2}{\mathrm{d}\varrho^2} - \frac{\mu^2-1/4}{\varrho^2}\right) R_l(\varrho) = 0 \quad (\text{for } \varrho \to 0) \quad . \tag{6}$$

Exercise 1.11.

With

$$R(\varrho) = a\varrho^{\nu} \quad , \tag{7}$$

it follows that

$$a\nu(\nu-1)\varrho^{\nu-2} - a\left(\mu^2 - \tfrac{1}{4}\right)\varrho^{\nu-2} = 0 \quad , \tag{8}$$

which is a determining equation for the power ν, given by

$$\nu_{\pm} = \tfrac{1}{2} \pm \sqrt{\tfrac{1}{4} + \mu^2 - \tfrac{1}{4}} = \tfrac{1}{2} \pm \mu \quad . \tag{9}$$

Since μ can, in principle, take all positive values, and the wave function must not have any nonintegrable divergence at the origin, this unambiguously fixes ν,[13]

$$\nu = \tfrac{1}{2} + \mu \quad . \tag{10}$$

Furthermore, we see that the radial function $u(r) = R(r)/r$ has a singularity at the origin ($r = 0$) when $l = 0$, though this singularity can be integrated; therefore, the wave function can be normalized in spite of the singularity. This behaviour of the wave function for $r \to 0$ is quite novel compared to the Schrödinger wave function and characteristic of relativistic s wave functions in the Coulomb potential. Because of the definition of μ we further see that, for $l = 0$, only real wave functions can be found for $Z\alpha < 1/2$. For larger values of Z the parameter μ becomes imaginary.

Because of (5) and (7) we now choose

$$R_l(\varrho) = N\varrho^{1/2+\mu}\mathrm{e}^{-\varrho/2}f(\varrho) \tag{11}$$

as an ansatz for the full radial equation, where the still unknown function $f(\varrho)$ should be constant for $\varrho \to 0$ and should guarantee the normalization for $\varrho \to \infty$. Inserting (11) into the radial equation (3) we find the following differential equation for $f(\varrho)$:

$$\frac{\mathrm{d}^2 f}{\mathrm{d}\varrho^2} + \left(\frac{2\mu+1}{\varrho} - 1\right)\frac{\mathrm{d}f}{\mathrm{d}\varrho} - \frac{\mu + 1/2 - \lambda}{\varrho} f(\varrho) = 0 \quad . \tag{12}$$

For simplification let us introduce the new abbreviations

$$2\mu + 1 = c \quad , \quad \mu + \tfrac{1}{2} - \lambda = a \quad , \tag{13}$$

so that

$$\frac{\mathrm{d}^2 f}{\mathrm{d}\varrho^2} + \left(\frac{c}{\varrho} - 1\right)\frac{\mathrm{d}f(\varrho)}{\mathrm{d}\varrho} - \frac{a}{\varrho} f = 0$$

results.

Now we try to solve the differential equation for $f(\varrho)$ with the help of a power series expansion,

[13] This is not true in the special case $l = 0$. Here the solution with ν_- can also be normalized, and one must set another criterion in order to exclude this case. For example, one can demand that an expectation value of the kinetic energy should exist. See Exercises 9.8 and 9.9 in which the analogue problem for the Coulomb solutions of the Dirac equation is carefully discussed.

Exercise 1.11.

$$f(\varrho) = \sum_{n'=0}^{\infty} a_{n'} \varrho^{n'} \quad . \tag{14}$$

Inserting (14) into (13) yields

$$\sum_{n'=2}^{\infty} a_{n'} n'(n'-1)\varrho^{n'-2} + c \sum_{n'=1}^{\infty} n' a_{n'} \varrho^{n'-2} - \sum_{n'=1}^{\infty} n' a_{n'} \varrho^{n'-1} - a \sum_{n'=0}^{\infty} a_{n'} \varrho^{n'-1} = 0 \quad . \tag{15}$$

A comparison of coefficients of equal powers in ϱ gives

$$\begin{aligned} a_1 &= \frac{a a_0}{c} \\ a_2 &= \frac{a_1(a+1)}{2(c+1)} \quad , \\ &\vdots \end{aligned} \tag{16}$$

or, generally

$$a_m = \frac{a_{m-1}}{m} \frac{a+m-1}{c+m-1} \quad . \tag{17}$$

Therefore, $f(\varrho)$ can be written as

$$f(\varrho) = a_0 \left(1 + \frac{a}{c}\varrho + \frac{a}{c}\frac{a+1}{c+1}\frac{\varrho^2}{2} + \dots\right) = a_0 \sum_{n'=0}^{\infty} \frac{(a)_{n'}}{(c)_{n'}} \frac{\varrho^{n'}}{n'!} \quad . \tag{18}$$

The function defined by the series (18) is called a *confluent hypergeometric function* ${}_1F_1(a, c; \varrho)$. Note, there is always a generalized faculty function in the nominator and denominator of each term of the sum. However, the confluent hypergeometric function diverges for $\varrho \to \infty$ as

$${}_1F_1(a, c; \varrho \to \infty) \Rightarrow \frac{\Gamma(c)}{\Gamma(a)} \varrho^{a-c} \mathrm{e}^{\varrho} \quad . \tag{19}$$

Thus, the normalization condition can only be fulfilled if it is certain that the series breaks off at a fixed value of n'. If $a + n' = 0$ holds, i.e., if a is a negative integer, then all terms of higher order, $m > n'$, are also equal to zero.

The energy eigenvalue for bound spin-0 particles in a Coulomb potential can be calculated from (13). Starting from

$$\lambda = \mu + \frac{1}{2} + n' = \frac{Z\alpha\varepsilon}{(m_0^2c^4 - \varepsilon^2)^{1/2}} \quad ,$$

it follows that

$$\varepsilon_{n'l} = -m_0c^2 \left[1 + \frac{(Z\alpha)^2}{\left(n' + \frac{1}{2} + [(l+\frac{1}{2})^2 - (Z\alpha)^2]^{1/2}\right)^2}\right]^{-1/2} \quad . \tag{20}$$

Exercise 1.11.

Thereby, we have chosen the negative square root, because in the case of no fields ($Z\alpha \to \infty$) the free solution, i.e. the negative time evolution factors for negatively charged particles, must be obtained. The energy itself is given by

$$E_{nl} = \int T_{00}\,\mathrm{d}^3x = m_0c^2\left[1+\frac{(Z\alpha)^2}{\left(n'+\frac{1}{2}+[(l+\frac{1}{2})^2-(Z\alpha)^2]^{1/2}\right)^2}\right]^{-1/2} \tag{21}$$

(see Exercise 1.12). Defining the principal quantum number as

$$n = n' + l + 1 \quad , \tag{22}$$

we may rewrite (21) in the following form:

$$E_{nl} = m_0c^2\left[1+\frac{(Z\alpha)^2}{\left(n-l-\frac{1}{2}+[(l+\frac{1}{2})^2-(Z\alpha)^2]^{1/2}\right)^2}\right]^{-1/2}$$
$$n = 1,2,3\ldots \quad , \quad l = 0,1,2,\ldots(n-1) \quad . \tag{21a}$$

By setting $N' = Na_0$, the eigenfunction can be written as

$$R_l(\varrho) = N'\varrho^{\mu+1/2}\mathrm{e}^{-\varrho/2}\,{}_1F_1\left(\mu+\tfrac{1}{2}-\lambda, 2\mu+1;\varrho\right) = N'W_{\lambda,\mu}(\varrho) \quad , \tag{23}$$

where $W_{\lambda,\mu}(\varrho)$ are called *Whittaker functions*. Expanding, finally, the above eigenvalue in a series of powers of Z yields

$$E_{nl} = m_0c^2\left\{\underbrace{1}_{\text{rest energy}} - \underbrace{\frac{Z^2\alpha^2}{2n^2}}_{\text{Schrödinger eigenvalue}} - \underbrace{\frac{Z^4\alpha^4}{2n^4}\left(\frac{a}{l+1/2}-\frac{3}{4}\right)}_{\text{relative correction}} + \ldots\right\} \quad . \tag{24}$$

The binding energy of a bound Klein–Gordon particle is defined as $E_\mathrm{b} = E - m_0c^2$. Some numerical values for the binding energy of 1s pions ($n = 1$, $l = 0$) with mass $m_\pi c^2 = 139.577\,\mathrm{MeV}$ are listed in the following table.

Binding energies of 1s pions for various central charges Z

Z	E_b/MeV	Z	E_b/MeV
10	−0.374	50	−11.51
20	−1.528	60	−19.391
30	−3.568	65	−26.34
40	−6.725		

From the energy relation (21a), one learns that for $Z\alpha > 1/2$, there is no energy eigenvalue for the 1s state ($n = 1$, $l = 0$), because the energy becomes imaginary. The energy eigenvalue for 1s pions at $Z = \alpha^{-1}/2 \approx 68$ are $E_{nl} = m_\pi c^2(2)^{-1/2}$ (see next figure). The corresponding binding energy is about $E_\mathrm{b} \sim -40\,\mathrm{MeV}$.

To determine the energy eigenvalues for s states for which $Z\alpha > 1/2$, we must investigate the relativistic case of extended atomic nuclei. Therefore, in Exercise 1.13 we will solve the Klein–Gordon equation for the Coulomb potential rounded off by an oscillator potential. Furthermore, one has to take into account that pions

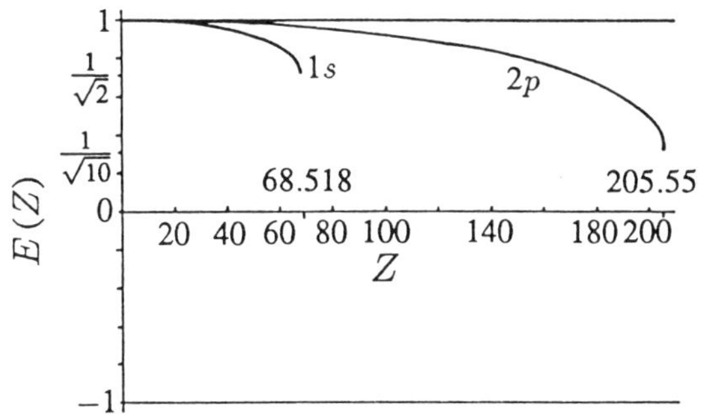

Solution of the Klein–Gordon equation for a central Coulomb potential. The energy is given in units of the rest energy $m_\pi c^2$. For the $1s$ level there are no solutions for $Z > 68$, and, similarly, for the $2p$ level when $Z > (3/2)(1/\alpha) \approx 205$. This changes if the finite size of the central nucleus is taken into account (Example 1.13)

and the nucleus interact strongly. Since the pion is 270 times more massive than an electron, its wave function already has a considerable overlap with the nucleus. Therefore, corrections to equation (21) are to be expected, based on the strong interaction, which can be quite massive.

1.10 Gauge Invariance of the Coupling

We know that electrodynamics is gauge invariant. By that we mean that the electrodyncamic laws (Maxwell equations) do not change under gauge transformations

$$A'_\mu(x) = A_\mu(x) + \frac{\partial \chi(x)}{\partial x^\mu} \quad . \tag{1.132}$$

The argument x represents all space-time coordinates, i.e.

$$x = \left\{x^0, x^1, x^2, x^3\right\} = \{ct, x, y, z\} \quad .$$

We now investigate the consequences of the gauge transformation (1.132) for the Klein–Gordon equation (1.121), the transformation producing the change,

$$g^{\mu\nu}\left(\mathrm{i}\hbar\frac{\partial}{\partial x^\nu} - \frac{e}{c}A'_\nu\right)\left(\mathrm{i}\hbar\frac{\partial}{\partial x^\mu} - \frac{e}{c}A'_\mu\right)\psi - m_0^2c^2\psi = 0 \quad ,$$

or

$$g^{\mu\nu}\left(\mathrm{i}\hbar\frac{\partial}{\partial x^\nu} - \frac{e}{c}A_\nu - \frac{e}{c}\frac{\partial\chi}{\partial x^\nu}\right)\left(\mathrm{i}\hbar\frac{\partial}{\partial x^\mu} - \frac{e}{c}A_\mu - \frac{e}{c}\frac{\partial\chi}{\partial x^\mu}\right)\psi - m_0^2c^2\psi = 0 \quad ,$$

or

$$g^{\mu\nu}\left(\mathrm{i}\hbar\frac{\partial}{\partial x^\nu} - \frac{e}{c}A_\nu\right)\left(\mathrm{i}\hbar\frac{\partial}{\partial x^\mu} - \frac{e}{c}A_\mu\right)\psi' - m_0^2c^2\psi' = 0 \quad , \tag{1.133}$$

with

$$\psi' = \exp\left(\frac{\mathrm{i}e}{\hbar c}\chi\right)\psi \quad . \tag{1.134}$$

In other words the gauge transformation (1.131) changes the phase of the wave function only. The phase factor $\exp[(\mathrm{i}e/\hbar c)\chi]$ is the same for all states ψ. Since all physical observables are represented by bilinear forms of the structure $\psi_m^* \dots \psi_n$, a common, identical phase factor does not play any role in the physics. Thus, we say that the Klein–Gordon equation is gauge invariant for the minimal coupling

$$\left[\left(\hat{p}_\nu - \frac{e}{c}A'_\nu\right)\psi\right]\exp\left(\frac{\mathrm{i}e}{\hbar c}\chi\right) = \left[\left(\hat{p}_\nu - \frac{e}{c}A_\nu - \frac{e}{c}\frac{\partial\chi}{\partial x^\nu}\right)\psi\right]\exp\left(\frac{\mathrm{i}e}{\hbar c}\chi\right)$$
$$= \left[\left(\mathrm{i}\hbar\frac{\partial}{\partial x^\nu} - \frac{e}{c}A_\nu\right)\psi\exp\left(\frac{\mathrm{i}e}{\hbar c}\chi\right)\right] \qquad (1.135)$$

and therefore also for arbitrary powers

$$\left[\left(\hat{p}_\nu - \frac{e}{c}A'_\nu\right)^n\psi\right]\exp\left(\frac{\mathrm{i}e}{\hbar c}\chi\right) = \left[\left(\hat{p}_\nu - \frac{e}{c}A_\nu\right)^n\psi\exp\left(\frac{\mathrm{i}e}{\hbar c}\chi\right)\right]$$
$$= \left[\left(\hat{p}_\nu - \frac{e}{c}A_\nu\right)^n\right]\psi' \quad , \qquad (1.136)$$

and, consequently, also for arbitrary operator functions $f(\hat{p}_\nu - (e/c)A_\nu)$, which are expandable in power series of $(\hat{p}_\nu - (e/c)A_\nu)$, giving

$$\left[f\left(\hat{p}_\nu - \frac{e}{c}A'_\nu\right)\psi\right]\exp\left(\frac{\mathrm{i}e}{\hbar c}\chi\right) = \left[f\left(\hat{p}_\nu - \frac{e}{c}A_\nu\right)\psi'\right] \quad . \qquad (1.137)$$

Thus, minimal coupling is gauge invariant in this very general sense. The parentheses [...] used in the above formulae shall help to illustrate the action of the $\hat{p}$ operators more clearly.

1.11 The Nonrelativistic Limit with Fields

We can study the nonrelativistic limit of the Klein–Gordon equation in the form of (1.122) with the help of the transformation

$$\psi(\boldsymbol{x},t) = \varphi(\boldsymbol{x},t)\exp\left(-\frac{\mathrm{i}}{\hbar}m_0c^2t\right) \quad . \qquad (1.138)$$

$\varphi(\boldsymbol{x},t)$ characterizes the nonrelativistic part of the wave function, for which the relations

$$\left|\mathrm{i}\hbar\frac{\partial\varphi}{\partial t}\right| \ll m_0c^2|\varphi| \quad , \quad |eA_0\varphi| \ll m_0c^2|\varphi| \qquad (1.139)$$

must be valid. The first expresses the smallness of the nonrelativistic energy ($\mathrm{i}\hbar(\partial/\partial t)$ is the energy operator) compared to the rest energy; the second condition means that the potentials involved have to be flat compared to the rest energy: potentials that are too deep would lead to increased binding energies and, finally, to spontaneous pair creation (see Chaps. 12 and 13). This would make the nonrelativistic limit impossible. Next we find that

$$\left(\mathrm{i}\hbar\frac{\partial}{\partial t} - eA_0\right)\psi = \left(\mathrm{i}\hbar\frac{\partial\varphi}{\partial t} - eA_0\varphi + m_0c^2\varphi\right)\exp\left(-\frac{\mathrm{i}}{\hbar}m_0c^2t\right)$$

and

$$\begin{aligned}\left(\mathrm{i}\hbar\frac{\partial}{\partial t}-eA_0\right)^2\psi &= \left(-\hbar^2\frac{\partial^2\varphi}{\partial t^2}-\mathrm{i}\hbar e\frac{\partial A_0}{\partial t}\varphi-\mathrm{i}\hbar eA_0\frac{\partial\varphi}{\partial t}+\mathrm{i}\hbar m_0c^2\frac{\partial\varphi}{\partial t}\right.\\ &\quad -\mathrm{i}\hbar e\frac{\partial\varphi}{\partial t}A_0+eA_0^2\varphi-eA_0m_0c^2\varphi\\ &\quad \left.+\mathrm{i}\hbar m_0c^2\frac{\partial\varphi}{\partial t}-eA_0\varphi m_0c^2+m_0^2c^4\varphi\right)\exp\left(-\frac{\mathrm{i}}{\hbar}m_0c^2t\right)\\ &\approx \exp\left(-\frac{\mathrm{i}}{\hbar}m_0c^2t\right)\left(m_0^2c^4-2m_0c^2eA_0+2m_0c^2\mathrm{i}\hbar\frac{\partial}{\partial t}\right.\\ &\quad \left.-\mathrm{i}\hbar e\frac{\partial A_0}{\partial t}\right)\varphi \quad .\end{aligned}$$

With regard to (1.139), the small quadratic terms have been omitted in the last step. Inserting this into (1.122) yields

$$\begin{aligned}&\exp\left(-\frac{\mathrm{i}}{\hbar}m_0c^2t\right)\left(m_0^2c^2-2m_0eA_0+2m_0\mathrm{i}\hbar\frac{\partial}{\partial t}-\frac{\mathrm{i}\hbar e}{c^2}\frac{\partial A_0}{\partial t}\right)\varphi\\ &\quad =\exp\left(-\frac{\mathrm{i}}{\hbar}m_0c^2t\right)\left[\left(+\mathrm{i}\hbar\boldsymbol{\nabla}+\frac{e}{c}\boldsymbol{A}\right)^2+m_0^2c^2\right]\varphi \quad ,\end{aligned}$$

or

$$\begin{aligned}\mathrm{i}\hbar\frac{\partial\varphi}{\partial t} &= \left[\frac{1}{2m_0}\left(+\mathrm{i}\hbar\boldsymbol{\nabla}+\frac{e}{c}\boldsymbol{A}\right)^2+\varepsilon A_0+\frac{\mathrm{i}\hbar e}{2m_0c^2}\frac{\partial A_0}{\partial t}\right]\varphi\\ &= \left[\frac{\hat{\boldsymbol{p}}^2}{2m_0}-\frac{e}{m_0c}\boldsymbol{A}\cdot\hat{\boldsymbol{p}}+eA_0+\frac{\mathrm{i}\hbar e}{2m_0c}(\boldsymbol{\nabla}\cdot\boldsymbol{A})+\frac{\mathrm{i}\hbar e}{2m_0c^2}\frac{\partial A_0}{\partial t}\right]\varphi . \qquad (1.140)\end{aligned}$$

This is the Schrödinger equation for electromagnetic potentials. But be careful: In studying radiation problems we generally choose the *Coulomb gauge*, defined by

$$\operatorname{div}\boldsymbol{A}=0 \quad .$$

In this case the term proportional to $\partial A_0/\partial t$ must be included. Only in the *Lorentz gauge*, for which

$$\frac{1}{c}\frac{\partial A_0}{\partial t}+\operatorname{div}\boldsymbol{A}=0$$

is valid, do both of the last terms in (1.140) vanish.

EXERCISE

1.12 Lagrange Density and Energy-Momentum Tensor for a lein–Gordon Particle in an Electromagnetic Field

Problem. Specify the Lagrange density for a Klein–Gordon particle in the electromagnetic field A_μ. Determine the canonical energy-momentum tensor.

Exercise 1.12.

Solution. The Lagrange density for the coupled system of the Maxwell field and the Klein–Gordon field is

$$\mathcal{L} = -\frac{1}{4}F_{\mu\nu}F^{\mu\nu} + \frac{1}{2m_0}\left[\left(\mathrm{i}\hbar\partial_\mu - \frac{e}{c}A_\mu\right)\psi^* \times\left(-\mathrm{i}\hbar\partial^\mu - \frac{e}{c}A^\mu\right)\psi - m_0^2c^2\psi^*\psi\right] \quad ,$$
$$F_{\mu\nu} = \partial_\mu A_\nu - \partial_\nu A_\mu \quad . \tag{1}$$

The variation of $I = \int \mathcal{L}\,\mathrm{d}^4x$ with respect to ψ^* yields the Klein–Gordon equation for a ψ field, minimally coupled to the electromagnetic field, i.e.

$$\frac{\delta I}{\delta\psi^*} = 0 \tag{2}$$

$$\Rightarrow \int\left\{\delta\left(\mathrm{i}\hbar\partial_\mu\psi^* - \frac{e}{c}A_\mu\psi^*\right)\left(-\mathrm{i}\hbar\partial^\mu - \frac{e}{c}A^\mu\right)\psi - m_0^2c^2\psi\delta\psi^*\right\}\mathrm{d}^4x = 0 \quad .$$

Using $\delta\partial_\mu\psi^* = \partial_\mu\delta\psi^*$ and with partial integration of the first term, we obtain under the assumption that $\delta\psi^*$ vanishes at the boundaries of integration,

$$\int\left\{\left(-\mathrm{i}\hbar\partial_\mu - \frac{e}{c}A_\mu\right)\left(-\mathrm{i}\hbar\partial^\mu - \frac{e}{c}A^\mu\right)\psi - m_0^2c^2\psi\right\}\delta\psi^*\,\mathrm{d}^4x = 0 \quad . \tag{3}$$

Due to the free choice of $\delta\psi^*$, this yields the Klein–Gordon equation

$$\left(\hat{p}^\mu - \frac{e}{c}A^\mu\right)\left(\hat{p}_\mu - \frac{e}{c}A_\mu\right)\psi = m_0^2c^2\psi \quad . \tag{4}$$

Variation of I with respect to A_μ yields the Maxwell equation in an analogous way

$$\partial^\mu F_{\mu\nu} = j_\nu = \frac{\mathrm{i}e\hbar}{2m_0}\left\{\psi^*\left(\partial_\nu + \frac{\mathrm{i}e}{\hbar c}A_\nu\right)\psi - \psi\left(\partial_\nu - \frac{\mathrm{i}e}{\hbar c}A_\nu\right)\psi^*\right\} \quad . \tag{5}$$

The canonical energy-momentum tensor is calculated according to (1) from

$$T^\mu{}_\nu = \frac{\partial\mathcal{L}}{\partial(\partial_\mu\psi)}\partial_\nu\psi + \frac{\partial\mathcal{L}}{\partial(\partial_\mu\psi^*)}\partial_\nu\psi^* + \frac{\partial\mathcal{L}}{\partial(\partial_\mu A_\varrho)}\partial_\nu A_\varrho - \delta^\mu_\nu\mathcal{L} \quad , \tag{6}$$

giving

$$T^\mu{}_\nu = \frac{1}{2m_0}(-\mathrm{i}\hbar\partial_\nu\psi)\left(\mathrm{i}\hbar\partial^\mu\psi^* - \frac{e}{c}A^\mu\psi^*\right) + \frac{1}{2m_0}(\mathrm{i}\hbar\partial_\nu\psi^*)\left(-\mathrm{i}\hbar\partial^\mu\psi - \frac{e}{c}A^\mu\psi\right) - \partial_\nu A_\varrho F^{\varrho\mu} - \delta^\mu_\nu\mathcal{L} \tag{7}$$

$$= \frac{1}{4}\delta^\mu_\nu F_{\sigma\varrho}F^{\sigma\varrho} - \partial_\nu A_\varrho F^{\varrho\mu} + \frac{1}{2m_0} \times\left[\left(\mathrm{i}\hbar\partial^\mu - \frac{e}{c}A^\mu\right)\psi^*(-\mathrm{i}\hbar\partial_\nu\psi) + (\mathrm{i}\hbar\partial_\nu\psi^*) \times\left(-\mathrm{i}\hbar\partial^\mu - \frac{e}{c}A^\mu\right)\psi - \delta^\mu_\nu\left(\mathrm{i}\hbar\partial_\varrho - \frac{e}{c}A_\varrho\right)\psi^* \times\left(-\mathrm{i}\hbar\partial^\varrho - \frac{e}{c}A^\varrho\right)\psi + \delta^\mu_\nu m_0^2c^2\psi^*\psi\right] \tag{8}$$

$$\equiv \frac{1}{4}\delta^\mu_\nu F_{\sigma\varrho}F^{\sigma\varrho} - \partial_\nu A_\varrho F^{\varrho\mu} + (T^\mu{}_\nu)_\psi \quad . \tag{9}$$

Here $(T^{\mu}{}_{\nu})_{\psi}$ represents the energy-momentum tensor for a Klein–Gordon particle with the minimal coupled electromagnetic field [last three terms of (1)], while the first two terms are standing for the free electromagnetic field.

Exercise 1.12.

EXAMPLE

1.13 Solution of the Klein–Gordon Equation for the Potential of an Homogeneously Charged Sphere

We solve the Klein–Gordon equation for a sphere with radius a since we consider this model to be an improved approximation for a realistic pionic atom. The results are subsequently discussed. Inside the sphere the corresponding potential is given by

$$\varepsilon A_0(r) = -\frac{Ze^2}{2a}\left(3 - \frac{r^2}{a^2}\right) \quad \text{for} \quad r \leq a \quad . \tag{1}$$

This is an oscillator potential which yields, after insertion into the radial Klein–Gordon equation [see Exercise 1.10, (6) and (7)],

$$\left[\frac{\mathrm{d}^2}{\mathrm{d}r^2} - \frac{l(l+1)}{r^2} + \left(\frac{E}{\hbar c} + \frac{3Z\alpha}{2a}\right)^2 - \frac{m_0^2c^2}{\hbar^2} \right. \\ \left. - 2\left(\frac{E}{\hbar c} + \frac{3Z\alpha}{2a}\right)\frac{Z\alpha}{2a}\frac{r^2}{a^2} + \left(\frac{Z\alpha}{2a^3}\right)^2 r^4\right]R(r) = 0 \quad . \tag{2}$$

This may be further simplified by introducing the following abbreviations:

$$A = \frac{E}{\hbar c} + \frac{3Z\alpha}{2a} \quad , \quad B = A^2 - \frac{m_0^2c^2}{\hbar^2} \quad , \quad C = \frac{Z\alpha}{2a^3} \quad . \tag{3}$$

For the total radial wave function, the following ansatz in the form of an infinite series is introduced:

$$R = r^{l+1}\sum_{n'=0}^{\infty} b_{n'}r^{2n'} \quad . \tag{4}$$

Insertion into the differential equation leads to

$$\sum_{n'} b_{n'}\left(2n'+l+1\right)\left(2n'+l\right)r^{2n'+l-1} \\ - \sum_{n'} l(l+1)b_{n'}r^{2n'+l-1} + B\sum_{n'} b_{n'}r^{2n'+l+1} \\ - 2AC\sum_{n'} r^{2n'+l+3} + C\sum_{n'} b_{n'}r^{2n'+l+5} = 0 \quad . \tag{5}$$

Example 1.13.

Ordering with respect to equal powers in r and comparing the coefficients, we find

$$b_1 = -\frac{Bb_0}{2(2l+3)} \quad ,$$
$$b_2 = -\frac{Bb_1 - 2ACb_0}{2(4l+10)} \quad ,$$
$$b_3 = -\frac{Bb_2 - 2ACb_1 + C^2 b_0}{2(6l+21)} \tag{6}$$

or, generally,

$$b_{n'} = \frac{Bb_{n'-1} - 2ACb_{n'-2} + C^2 b_{n'-3}}{4n'l + (2n'+1)2n'} \quad . \tag{7}$$

The energy is determined by specifying that the wave functions and their first derivatives should be continuous at $r = a$. In the region $r \geq a$ the wave functions are given by the Whittaker function $W_{\lambda\mu}(\varrho)$ with $\varrho = \beta r$ [see Exercise 1.11, (23)]. Thus, and by using (4), the energy can be fixed by

$$\frac{W'_{\lambda,\mu}(\beta a)}{W_{\lambda,\mu}(\beta a)} = \frac{\sum\limits_{n'} b_{n'}(2n'+l+1)a^{2n'+l}}{\sum\limits_{n'} b_{n'}a^{2n'+l+1}} \quad , \tag{8}$$

where

$$\beta = 2\left(m_0^2c^4 - E^2\right)^{1/2}/\hbar c \quad ,$$
$$\lambda = Z\alpha E\left(m_0^2c^4 - E^2\right)^{-1/2} \quad ,$$
$$\mu = \left[\left(l+\tfrac{1}{2}\right)^2 - (Z\alpha)^2\right]^{1/2} \quad . \tag{9}$$

For the point nucleus there exist bound solutions (for $l = 0$) only up to $Z = 1/2\alpha \approx 68$. In contrast to this, equation (8), containing the finite extension of the nucleus, also yields solutions for larger values of Z (see figure below). Choosing the nuclear radius to be $a \sim 10\,\text{fm} = 10^{-12}\,\text{cm}$, we get a nearly parabolic growth with Z for the binding energy.

At $Z \sim 1500$ the energy eigenvalue for a $1s$ pion approaches $E = 0$. For larger Z this value even becomes negative. At $Z \approx Z_{\text{critical}} \sim 3000$ we find $E = -m_\pi c^2$. An extrapolation of this behaviour for still larger Z would make the state dive into

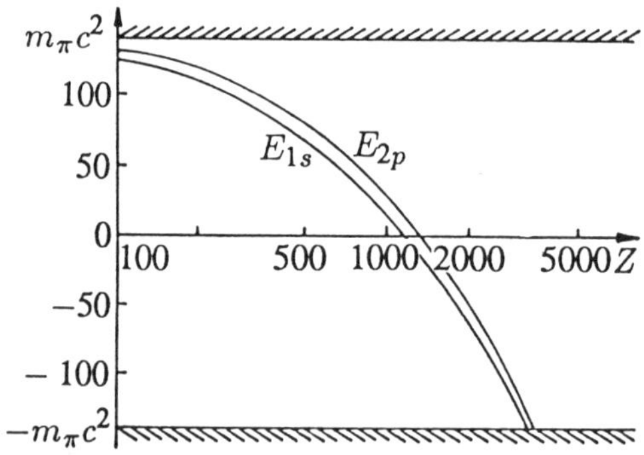

Energy eigenvalues for a π^- meson in the Coulomb potential of an extended nucleus. The charge distribution of the nucleus is assumed to be a Fermi distribution [see (10)]

Example 1.13.

the lower continuum of $E < -m_\pi c^2$. But at this point at least the single particle interpretation collapses totally because the new degree of freedom of pair creation has to be considered (see the later discussion, together with the Dirac equation, Chap. 12). One also has to take into account that, for pions (being bosons), the Pauli principle is not valid; thus, in principle, a state may be occupied by any number of pions. This is only true as far as the pion–pion interaction is neglected.

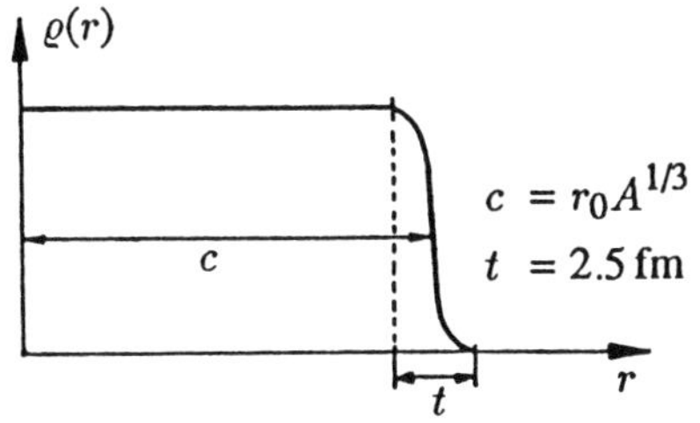

Fermi charge distribution of a nucleus

The interaction among the pions will, in fact, inhibit the infinite spontaneous $\pi^+\pi^-$ production. That this critical region, in which spontaneous pion production is possible, can be reached only for very large Z values may at once be estimated by considering the condition $|V| > |E_0| > 2m_\pi c^2$. Only for $Z \sim 1500$ does the depth of the potential reach the value $|V| \sim 2m_\pi c^2 \sim 280\,\mathrm{MeV}$. In the above figure the energies for $1s$ and $2p$ pions, calculated for the potential of a Fermi charge distribution, are shown. The Fermi charge distribution has the form

$$\varrho_{\text{nucleus}}(r) = \frac{N}{1 + \exp(4\ln 3[(r-c)/t])} \quad , \tag{10}$$

with the constant of normalization N given by

$$4\pi \int_0^\infty \varrho(r) r^2\,\mathrm{d}r = Z \quad , \tag{11}$$

and is shown in the second figure. Here c is the *half-density radius*. The *surface thickness* t characterizes the region in which the density drops from 90% to 10% of its value at the origin ($r = 0$). This also determines the constant, $4\ln 3$. The potential results as a solution of the Poisson equation and is given by

$$V(r) = -4\pi e^2 \left\{ \frac{1}{r} \int_0^r \varrho(r') r'^2\,\mathrm{d}r' + \int_r^\infty \varrho(r') r'\,\mathrm{d}r' \right\} \quad . \tag{12}$$

The value $E = -m_\pi c^2$ is reached at $Z = 3280$ for the $1s$ state and at $Z = 3425$ for the $2p$ state.

EXAMPLE

1.14 The Solution of the Klein–Gordon Equation for a Square-Well Potential

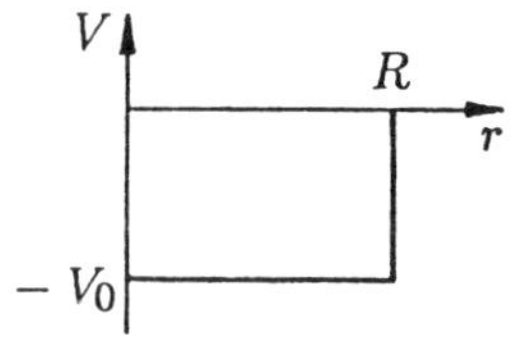

Squaro-well potential

This situation corresponds to minimal coupling for V, which is described by (see figure on the left)

$$V = \begin{cases} -V_0 & \text{for} \quad r \le R \\ 0 & \text{for} \quad r > R \end{cases} \quad .$$

We set

$$k_i = \frac{1}{\hbar c}\sqrt{(\varepsilon + V_0)^2 - m_0^2 c^4} \quad \text{for} \quad r \le R$$

and

$$k_o = \frac{1}{\hbar c}\sqrt{\varepsilon^2 - m_0^2 c^4} \quad \text{for} \quad r > R \quad . \tag{1}$$

With this the Klein–Gordon equation reads

$$\left[\frac{\mathrm{d}}{\mathrm{d}r}\left(r^2 \frac{\mathrm{d}}{\mathrm{d}r}\right) - l(l+1) + k^2 r^2\right] u(r) = 0 \quad . \tag{2}$$

The substitution $\varrho = kr$ leads to

$$\frac{\mathrm{d}}{\mathrm{d}r} = k\frac{\mathrm{d}}{\mathrm{d}\varrho} \quad \text{and} \quad \frac{\mathrm{d}^2}{\mathrm{d}r^2} = k^2 \frac{\mathrm{d}^2}{\mathrm{d}\varrho^2} \quad .$$

Hence, one finally arrives at *Bessel's equation*

$$\left(2\varrho\frac{\mathrm{d}}{\mathrm{d}\varrho} + \varrho^2 \frac{\mathrm{d}^2}{\mathrm{d}\varrho^2} - l(l+1) + \varrho^2\right) u(\varrho) = 0 \quad . \tag{3}$$

The solution for $r \le R$ is $u(\varrho) = N_{j_l}(k_i r)$. The *Neumann function* $n_l(k_i r)$ must be excluded as a solution because it is irregular at $\varrho = 0$. For $V = 0$, we always get

$$u(\varrho) = N_{j_l}(kr) \quad \text{with} \quad k = \frac{1}{\hbar c}\sqrt{\varepsilon^2 - m_0^2 c^4}$$

and where N is the normalization factor. Obviously, the solution is symmetric for positive and negative values of ε. The Bessel functions can easily be calculated, and one gets

$$j_0(\varrho) = \frac{\sin\varrho}{\varrho} \quad , \quad j_1(\varrho) = \frac{\sin\varrho}{\varrho^2} - \frac{\cos\varrho}{\varrho} \quad .$$

Furthermore, one can use the general recurrence relation

$$f_{n-1}(\varrho) + f_{n+1}(\varrho) = (2n+1)\varrho^{-1} f_n(\varrho) \quad , \tag{4}$$

where f_n stands for the Bessel function j_n, the Neumann function $n_n(\varrho)$ or the *Hankel functions*

Example 1.14.

$$h_n^{(1)}(\varrho) = j_n(\varrho) + \mathrm{i} n_n(\varrho) \quad \text{or}$$
$$h_n^{(2)}(\varrho) = j_n(\varrho) - \mathrm{i} n_n(\varrho) \quad .$$

In the outer region we set

$$k^2 = \frac{m_0^2 c^4 - \varepsilon^2}{\hbar^2 c^2} = -k_0^2 \quad . \tag{5}$$

With the substitution $\varrho = \mathrm{i}kr$ the differential equation (3) becomes

$$\frac{\mathrm{d}^2 u}{\mathrm{d}\varrho^2} + 2\varrho \frac{\mathrm{d}u}{\mathrm{d}\varrho} + \left[1 - \frac{l(l+1)}{\varrho^2}\right] u(\varrho) = 0 \tag{6}$$

and the general solution of this differential equation is

$$h_l^{(1)}(\varrho) = j_l(\varrho) + \mathrm{i} n_l(\varrho) \quad . \tag{7}$$

We see that $h_n^{(2)}$ cannot be a solution for $r > R$, since that function increases exponentially and so cannot be normalized. For an imaginary argument the Hankel function of the first kind is given by[14]

$$h_0^{(1)}(\mathrm{i}\beta r) = -\frac{1}{\beta r} \mathrm{e}^{-\beta r} \quad ,$$

$$h_1^{(1)}(\mathrm{i}\beta r) = \mathrm{i}\left(\frac{1}{\beta r} + \frac{1}{\beta^2 r^2}\right) \mathrm{e}^{-\beta r} \quad ,$$

$$h_2^{(1)}(\mathrm{i}\beta r) = \left(\frac{1}{\beta r} + \frac{3}{\beta^2 r^2} + \frac{3}{\beta^3 r^3}\right) \mathrm{e}^{-\beta r} \quad . \tag{8}$$

Higher orders can be calculated by use of the above recurrence relation (4), and the derivative can in general be calculated to yield

$$(2n+1)\frac{\mathrm{d}}{\mathrm{d}z} f_n(z) = n f_{n-1}(z) - (n+1) f_{n+1}(z) \quad . \tag{9}$$

The determination of the energy value ε requires the equality of the logarithmic derivatives of the solutions at $r = R$, so that

$$\frac{1}{u_\mathrm{i}} \frac{\mathrm{d}u_\mathrm{i}}{\mathrm{d}r} = \frac{1}{u_\mathrm{o}} \frac{\mathrm{d}u_\mathrm{o}}{\mathrm{d}r} \quad \text{for} \quad r = R \quad , \tag{10}$$

where the normalization constants cancel. The eigenvalues ε can be derived from

$$\frac{j_l'(k_\mathrm{i}R)}{j_l(k_\mathrm{i}R)} = \frac{h_l^{(1)\prime}(\mathrm{i}kR)}{h_l^{(1)}(\mathrm{i}kR)} \quad . \tag{11}$$

Now consider the special case of s states for the determination of ε. For this case one obtains after insertion of the definition of j_0 and $h_0^{(1)}$ and calculation of (11)

$$k_\mathrm{i} \cot(k_\mathrm{i}R) = -k \quad . \tag{12}$$

[14] See M. Abramowitz, I.A. Stegun: *Handbook of Mathematical Functions* (Dover, New York 1965), p. 438.

Example 1.14.

From this transcendent equation one can iteratively determine the eigenvalue ε. Numerical solutions of that problem are shown in the following three figures.

In the first figure the depth of the potential is set to $V_0 = -Ze^2/R$, where the nuclear radius is $R = r_0 A^{1/3}$ with $r_0 = 1.2\,\text{fm}$ and the number of nucleons $A = 2.5Z$. At $Z = 5150$, ε reaches the critical value $\varepsilon_{\text{crit}} = -m_\pi c^2$. There the bound pion level dives into the lower (negative energy) continuum.

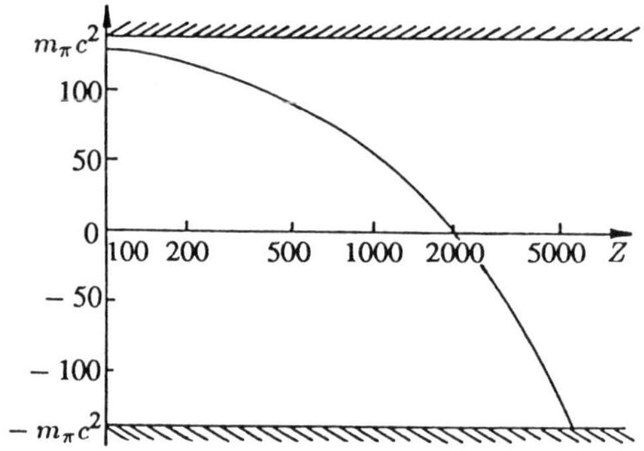

The energy of a π^- meson in the $1s$ state as a function of the nuclear charge Z. The radius of the nucleus is increased according to $R = r_0 A^{1/2}$ where $A = 2.5Z$ is the number of nucleons in the nucleus

In the second figure R was fixed (independent of Z). For smaller radii $\varepsilon_{\text{crit}}$ is shifted to smaller values of Z.

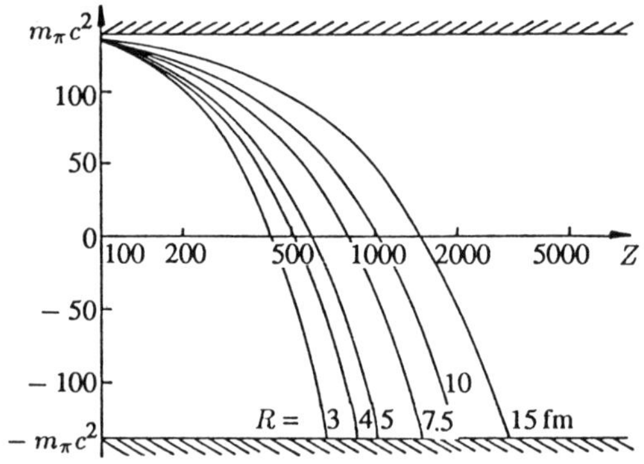

The energy of a π^- meson in the $1s$ state as a function of the nuclear charge Z. The radius was fixed at $R = 3$, 4, 5 fm, etc.

In the final figure the radius is compressed to $R = 0.25\,\text{fm}$. Furthermore, the charge symmetric solutions to the potential are plotted. These curves show reflection symmetry about the $\varepsilon = 0$ axis. It is essential in this case that the curves exhibit "critical points" for $R \ll 1\,\text{fm}$, characterized by the infinite slope of the binding energy as a function of Z. The first critical value is at $\varepsilon \approx -0.8 m_0 c^2$.

Here the solutions coming from the lower continuum into the region of bound states (π^+ states) meet with the solutions from the upper continuum (π^- states). This leads to the possibility of the spontaneous production of many $\pi^+ - \pi^-$ pairs (a true pion condensate). The deeper meaning of the phenomenon will be more clearly understood after the discussion of the phase transition of the electron-positron vacuum in supercritical fields. It is remarkable that a given short-range potential is able to simultaneously bind π^- as well as π^+ particles. This *Schiff–*

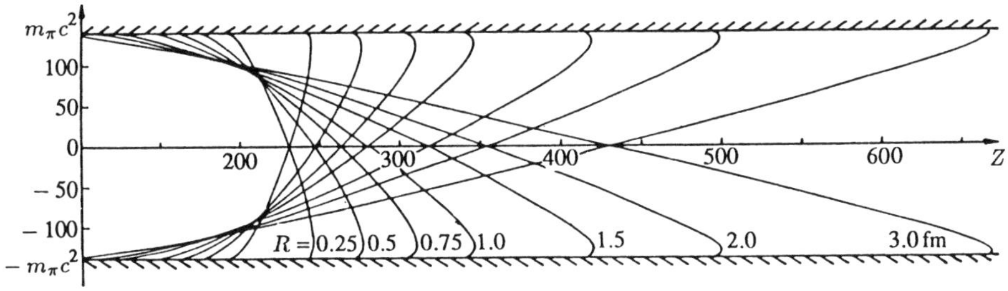

The energy of π mesons as a function of the nuclear charge Z, if the radius R of the nucleus is compressed

Snyder–Weinberg effect has its physical basis in the pionic charge density $\varrho'(\boldsymbol{r})$, the definition of which includes positive as well as negative frequencies. Therefore, the pion is not a particle with a charge distribution of only one sign [see our discussion in connection with (1.129)].

EXERCISE

1.15 Solution of the Klein–Gordon Equation for an Exponential Potential

Problem. Solve the Klein–Gordon equation for an exponential potential of the form

$$V(r) = -Z\alpha \mathrm{e}^{-r/a} \tag{1}$$

with $\alpha = m_0c^2e^2/\hbar c$. In natural units ($\hbar = c = m_0 = 1$) α is equivalent to the Sommerfeld fine structure constant, i.e. $\alpha \simeq 1/137$. α characterizes the range of the potential. Restrict yourself to s states ($l = 0$) only.

Solution. Under the restriction to s states, and, thus, the neglection of the centrifugal term, the Klein–Gordon equation assumes a very simple form. From (6) and (7) in Exercise 1.10 we have

$$\left(\frac{\mathrm{d}^2}{\mathrm{d}r^2} + k^2\right) R(r) = 0 \tag{2}$$

with

$$k^2 = \frac{[\varepsilon - V(r)]^2 - m_0^2c^4}{\hbar^2c^2} \quad . \tag{3}$$

With the aid of the separation ansatz

$$R(r) = \mathrm{e}^{r/2a} w(t) \tag{4}$$

and the substitution

$$t = 2\mathrm{i}Z\alpha\frac{a}{\hbar c}\mathrm{e}^{-r/a} \quad , \tag{5}$$

Exercise 1.15.

we can transform the differential equation (2) into

$$\frac{\mathrm{d}^2 w}{\mathrm{d}t^2}+\left\{-\frac{1}{4}-\frac{\mathrm{i}\varepsilon a}{\hbar ct}+\frac{1/4-p^2a^2}{t^2}\right\}w=0 \tag{6}$$

with

$$p^2=\frac{m_0^2c^4-\varepsilon^2}{\hbar c^2} \quad . \tag{7}$$

Equation (6) corresponds to the *Whittaker differential equation*[15] for which the regular solution for $r\to\infty$ (i.e. $t=0$) is, thus,

$$w(t)=NW_{\lambda,\mu}(t)=N\mathrm{e}^{-t/2}\,t^{1/2+\mu}\,{}_1F_1\left(\tfrac{1}{2}+\mu-\lambda,1+2\mu;t\right) \quad , \tag{8}$$

with

$$\lambda=-\frac{\mathrm{i}\varepsilon a}{\hbar c} \quad , \quad \text{and} \tag{9}$$

$$\mu=pa \quad . \tag{10}$$

N is the normalization constant. For the radial $1s$ wave function we can write, in summary,

$$R(r)=N\mathrm{e}^{r/2a}\,W_{\lambda,\mu}\left(2\mathrm{i}Z\alpha\frac{a}{\hbar c}\mathrm{e}^{-r/a}\right) \quad . \tag{11}$$

To obtain the energy eigenvalues we demand that $R(r)$ vanishes at the origin ($r=0$). This is the only way to guarantee the normalization of the radial wave functions. This condition leads to the eigenvalue equation

$${}_1F_1\left(\frac{1}{2}+\mu-\lambda,\ 1+2\mu;\ 2\mathrm{i}Z\alpha\frac{a}{\hbar c}\right)=0 \tag{12}$$

which is an implicit equation for the determination of the energy eigenvalues ε for s states. ε can be obtained from (12) only by using numerical methods. The calculated energy eigenvalue ε of an $1s$ pion in an exponential potential are shown in the next figure as a function of the coupling strength Z. The range constant α of the potential is chosen to be $a=1\cdot\lambda_\pi$, where $\lambda_\pi=\hbar/m_\pi c$ denotes the Compton wavelength of the pion. At $Z_{\mathrm{cr}}\simeq 778$ the $1s$ state reaches the boundary ($\varepsilon=-m_\pi c^2$) of the negative energy continuum. In a second numerical calculation the range constant is changed to $a=0.2\cdot\lambda_\pi$.

This causes a rise in the critical value to $Z\simeq 2158$, which can be seen in the last figure of this exercise. Additionally, we find here, for the domain $2150<Z<2158$, a bound state for the antiparticle. The corresponding energy eigenvalues can be taken from the dashed line in the insert to the figure. The appearance of these bound antiparticle states is correlated with the short range of the potential and the fact that the radial density ϱr^2 is not positive definite. The radial densities are depicted in Fig. 1.3 following (1.131).

[15] M. Abramowitz, I.A. Stegun: *Handbook of Mathematical Functions* (Dover, New York 1965).

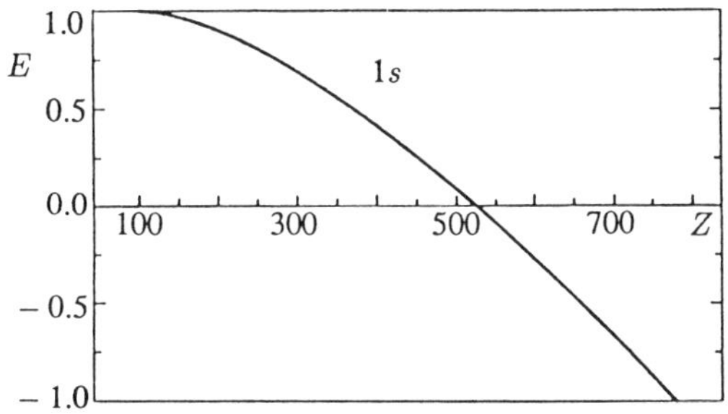

The energy eigenvalue of a $1s$ pion in an exponential potential as a function of the coupling strength parameter Z. The constant a characterizing the range of the potential is $a = 1\lambda_\pi$. The critical charge is $Z_{cr} \simeq 778$. E is given in units of the rest-energy $m_\pi c^2$

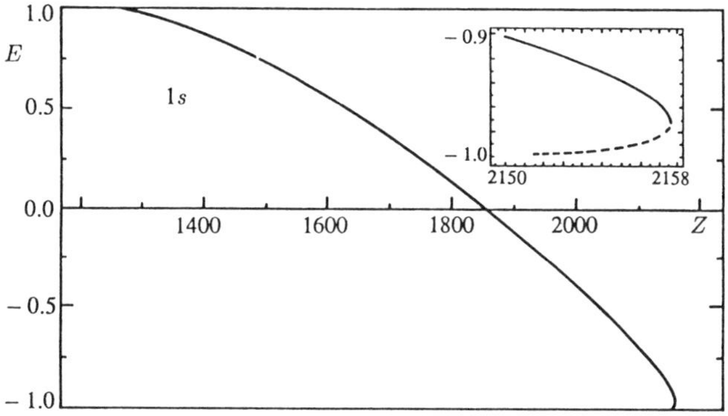

The same as in the foregoing figure for $a = 0.2\lambda_\pi$. The energy eigenvalues of bound antiparticles are also calculated. $Z_{cr} \simeq 2158$. The insert in the figure shows ε for the domain $2150 \leq Z \leq 2158$. The *dashed line* characterizes the energy of the antiparticle state

EXERCISE

1.16 Solution of the Klein–Gordon Equation for a Scalar $1/r$ Potential

Problem. Solve the Klein–Gordon equation for a scalar interaction of the form

$$W(r) = -Z\alpha/r \tag{1}$$

which is coupled to the square of the mass. For the present let α be an arbitrary constant, the coupling strength is determined by Z.

Solution. As an example of scalar coupling, we want to solve the Klein–Gordon equation for a long-range $1/r$ interaction. We introduced the Coulomb potential into the Klein–Gordon equation by minimal coupling ($\hat{p}_\mu \to \hat{p}_\mu - (e/c)A_\mu$). In contrast to this we now couple a scalar interaction $U(r)$ to the square of the mass in the equation of motion, i.e. we perform the substitution $m_0^2c^4 \to m_0^2c^4 + U^2(r)$. A direct coupling to the mass would yield mixing terms of the form $2m_0c^2U(r)$; however, we do not want to investigate this case in greater detail here. Thus, the radial Klein–Gordon equation with an arbitrary scalar interaction reads

$$\left(\frac{d^2}{dr^2} - \frac{l(l+1)}{r^2} + \frac{\varepsilon^2}{\hbar^2c^2} - \frac{m_0^2c^4}{\hbar^2c^2} - \frac{U^2(r)}{\hbar^2c^2}\right) R(r) = 0 \quad . \tag{2}$$

The scalar interaction is independent of the charge of the spin-0 particle considered, i.e. it has the same effect on particles and antiparticles respectively. However, since

Exercise 1.16.

there is no experimental evidence for such a long-range interaction, our calculations are only of academic interest. Nevertheless, it is instructive to pursue the formal solution of the Klein–Gordon equation for this unusual type of potential and to determine possible critical values of the coupling strength. So we divide (2) by $m_0^2c^4/\hbar^2c^2$ and substitute

$$r' = r\frac{m_0c^2}{\hbar c} \quad , \tag{3}$$

r' is a dimensionless quantity. This yields

$$\left(\frac{\mathrm{d}^2}{\mathrm{d}r'^2} - \frac{l(l+1)}{r'^2} + \frac{\varepsilon^2}{m_0^2c^4} - 1 - \frac{U^2(r')}{m_0^2c^4}\right)R(r') = 0 \quad . \tag{4}$$

Now we specify the interaction by

$$\frac{U^2(r')}{m_0^2c^4} = W(r') = -\frac{Z\alpha}{r'} \tag{5}$$

and, further, by

$$b^2 = 1 - \frac{\varepsilon^2}{m_0^2c^4} \quad . \tag{6}$$

Thus, we get

$$\left(\frac{\mathrm{d}^2}{\mathrm{d}r'^2} - \frac{l(l+1)}{r'^2} - b^2 + \frac{Z\alpha}{r'}\right)R(r') = 0 \quad . \tag{7}$$

Substituting again with

$$\varrho = 2br' \tag{8}$$

we now have

$$\left(\frac{\mathrm{d}^2}{\mathrm{d}\varrho^2} - \frac{l(l+1)}{\varrho^2} - \frac{1}{4} + \frac{\varrho}{c}\right)R(\varrho) = 0 \quad , \tag{9}$$

with

$$c = \frac{Z\alpha}{2b} \quad . \tag{10}$$

Now consider the asymptotics $\varrho \to \infty$ and $\varrho \to 0$. For $\varrho \to \infty$ it follows from (9) that

$$\left(\frac{\mathrm{d}^2}{\mathrm{d}\varrho^2} - \frac{1}{4}\right)R(\varrho) = 0 \tag{11}$$

which yields immediately

$$R(\varrho) \propto \mathrm{e}^{-\varrho/2} \quad . \tag{12}$$

Analogously for $\varrho \to 0$, *Exercise 1.16.*

$$\left(\frac{d^2}{d\varrho^2} - \frac{l(l+1)}{\varrho^2}\right) R(\varrho) = 0 \tag{13}$$

with the solution

$$R(\varrho) \propto \varrho^{l+1} \tag{14}$$

that may be normalized. Thus, for the solution we choose

$$R(\varrho) = N\varrho^{l+1} F(\varrho) e^{-\varrho/2} \quad , \tag{15}$$

where $F(\varrho)$ must still be determined. Putting (15) into the differential equation (9), we get

$$\varrho \frac{d^2 F(\varrho)}{d\varrho^2} + [(2l+2) - \varrho] \frac{dF(\varrho)}{d\varrho} [c - (l+1)] F(\varrho) = 0 \quad . \tag{16}$$

This is *Kummer's differential equation*,[16] with the solution

$$F(\varrho) = {}_1F_1(l+1-c, 2l+2, \varrho) \quad . \tag{17}$$

The confluent hypergeometric series allows normalization only if the first arguments equals a negative integer or zero. Hence, the condition for the determination of the energy eigenvalue is

$$l + 1 - c = -n_r \quad \text{with} \tag{18}$$

$$n_r = 0, 1, 2, \ldots \quad . \tag{19}$$

With the definition of the principal quantum number

$$n = l + 1 + n_r \quad , \tag{20}$$

we get from (18)

$$\varepsilon = \pm \left\{ 1 - \frac{(Z\alpha)^2}{4n^2} \right\}^{1/2} m_0 c^2 \quad . \tag{21}$$

In analogy to Schrödinger's equation for a Coulomb potential the energy eigenvalue ε does not depend on the orbital angular momentum quantum number l. Additionally, ε is symmetric for particle and antiparticle states, as expected. The critical value for the coupling strength Z is easily determined; it follows that

$$Z_{cr} = \frac{2n}{\alpha} \quad . \tag{22}$$

Inserting the value 1/137 for Sommerfeld's fine structure constant, we get $Z_{cr}(1s) \simeq 274.07$ and $Z_{cr}(2s) = Z_{cr}(2p) \simeq 548.14$, and for these values the derivative of the energy as a function of the coupling strength is

[16] M. Abramowitz, I.A. Stegun: *Handbook of Mathematical Functions* (Dover, New York 1965).

Exercise 1.16.

$$\left.\frac{d\varepsilon}{dZ}\right|_{Z=Z_{cr}} = \pm\infty \quad . \tag{23}$$

In summary, the complete radial wave function has the form

$$R(\varrho) = N\varrho^{l+1}\,{}_1F_1(-n+l+1, 2l+2, \varrho)\,e^{-\varrho/2} \tag{24}$$

where $R(\varrho)$ can be expressed by the generalized *Laguerre polynomials*

$$R(\varrho) = N\varrho^{l+1}\frac{(n-l-1)!}{(n+l)!}(2l+1)!L_{n-l-1}^{(2l+1)}(\varrho)\,e^{-\varrho/2} \quad . \tag{25}$$

The normalization factor N of the radial wave function is determined by

$$\int_0^\infty \varrho(r)r^2\,dr = \pm 1 \quad , \tag{26}$$

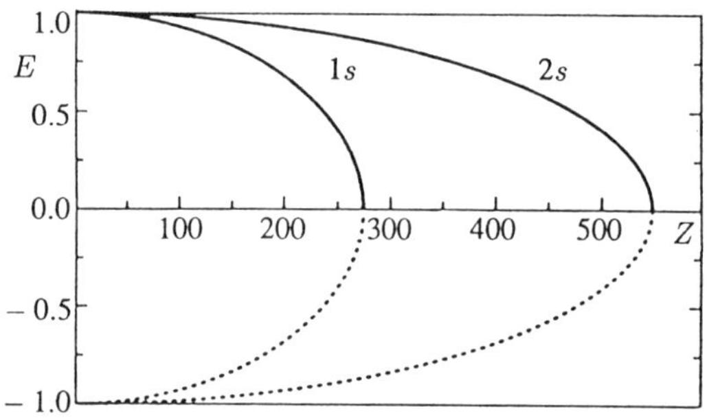

1s and 2s energies of the Klein–Gordon particles (*straight lines*) and antiparticles (*dashed lines*) in a scalar $1/r$ potential

with the radial density

$$\varrho(r)r^2 = \varepsilon R^2(r) \quad . \tag{27}$$

The plus sign in (26) is valid for those states which with decreasing coupling strength enter from the upper continuum ($\varepsilon > m_0c^2$) into the energy domain of the bound states, while the minus sign is valid for antiparticles which enter from the lower continuum ($\varepsilon < -m_0c^2$) into the domain of bound states. The evaluation of the normalization integral yields

$$N = \sqrt{\frac{(n+l)!}{2|\varepsilon|n(n-l-1)!}\,\frac{1}{(2l+1)!}(2p)^{3/2}} \quad . \tag{28}$$

The above figure shows the energy eigenvalue of the 1s state and the 2s state as a function of $Z\alpha \simeq 1/137$. The values for particle (straight line) and antiparticle (dashed line) states are given. $\varepsilon(Z)$ shows the square root behaviour due to (21).

EXAMPLE

1.17 Basics of Pionic Atoms

Pionic atoms consist of a nucleus and one (or more) π^- mesons which "circle" around it. The pions are generated in an inelastic proton scattering process (e.g.: $\mathrm{p}+\mathrm{p}\to\mathrm{p}+\mathrm{p}+\pi^-+\pi^+$), slowed down, filtered out of the beam and are incident on the elements under investigation as slow pions. Passing near an atom, the pion is captured by a simultaneous emission of electrons (Auger capture). The probability that two or more pions are simultaneously captured is extremely low. Thus, in general pionic atoms have only one pion, as occurs similarly with muonic atoms. Since pions have spin 0, they are described by the Klein–Gordon equation. The interaction of a pion with a nucleus consists of two essential parts: first the electromagnetic interaction, which is described by the Coulomb potential $A_0(x)$, and second the strong interaction. The latter consists of a real and an imaginary potential. A pion interacts with the nucleons in the nucleus according to the reactions:

$$\begin{aligned} &\pi^- + \mathrm{p} \to \mathrm{n} \quad , \\ &\pi^+ + \mathrm{n} \to \mathrm{p} \quad , \\ &\pi + 2\mathrm{N} \to \pi + 2\mathrm{N} \quad . \end{aligned} \tag{1}$$

For example, a π^- can completely disappear in a nucleus if it is caught in the conversion of a proton into a neutron. This capture process leads to an imaginary potential, by which it is globally simulated. Another possibility is a kind of "chain reaction" of the form

$$\pi^- + \mathrm{p} \to \mathrm{n} \to \pi^- + \mathrm{p} \to \mathrm{n} \to \pi^- + \mathrm{p} \to \dots \quad . \tag{2}$$

We abbreviate this – as above – by $\pi + 2\mathrm{N} \to \pi + 2\mathrm{N}$. This chain reaction leads to an effective, real, optical potential for the pion, which is known as the *Kisslinger potential*,

$$V_\pi^{\mathrm{Kissl.}}(\boldsymbol{x}) = A_0\varrho(\boldsymbol{x}) - A_1\boldsymbol{\nabla}\cdot(\varrho(\boldsymbol{x})\boldsymbol{\nabla}) \quad . \tag{3}$$

An improved form of this potential, the *Ericson–Ericson potential*, is given by

$$V_\pi^{\mathrm{Kissl.–Er–Er}}(\boldsymbol{x}) = A_0\varrho(\boldsymbol{x}) - A_1\boldsymbol{\nabla}\cdot\left(\frac{\varrho(\boldsymbol{x})}{1+\alpha\varrho(\boldsymbol{x})/3}\boldsymbol{\nabla}\right) \quad , \tag{4}$$

where $\varrho(\boldsymbol{x})$ denotes the nucleon density and A_0, A_1 and α are constants which are fitted to a large number of nuclei (π atoms). The imaginary potential is also a function of the nucleon density, namely,

$$V_\pi^{\mathrm{Imag.}}(\boldsymbol{x}) = \mathrm{i}\left(B_0\varrho^2(\boldsymbol{x}) - B_1\nabla\cdot\left(\varrho^2(\boldsymbol{x})\nabla\right)\right) \quad , \tag{5}$$

where B_0 and B_1 are again constants.[17]

[17] See, for example, Y.N. Kim: *Mesonic Atoms and Nuclear Structure* (North-Holland, Amsterdam 1971); J.M. Eisenberg, W. Greiner: *Nuclear Theory, Vol. 2: Reaction Mechanisms*, 3rd ed. (North-Holland, Amsterdam 1987).

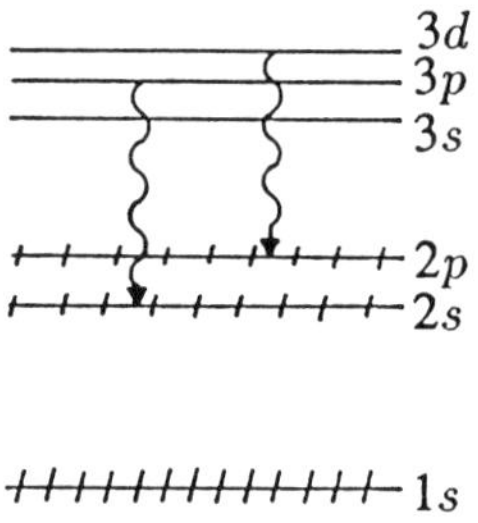

Schematic spectrum of a pionic atom. The hatching indicates the broadening of the inner levels due to pion capture by the nucleus. The $3p \to 2s-$ and $3d \to 2s-$ radiative transitions are also shown

Because of these nuclear potentials, which we derive and confirm more rigorously in Examples 1.22–24, there are deviations from the spectrum of a pionic atom, calculated on the basis of the Coulomb potential for a point nucleus in Exercises 1.11 and 1.13. Since the Kisslinger potential is short range [its range is reduced to the nucleus because of the dependence on $\varrho(\boldsymbol{x})$], then mainly the $1s$, $2s$ and $2p$ states are influenced, because only the wave functions of these inner states overlap considerably with the nucleus. Finally, the imaginary potential generates a broadening of these inner states, which expresses the pion capture by the nucleus. The figure on the left illustrates the situation qualitatively.

Behaviour of Solutions of the Klein–Gordon Equation Under Lorentz Transformation

Our acceptance of the special theory of relativity requires that relativistic wave equations have to be form invariant under Lorentz transformations, i.e.

$$x'^{\mu} = a^{\mu}{}_{\nu} x^{\nu} \quad , \tag{6}$$

with $a^{\mu}{}_{\nu} a^{\nu}{}_{\sigma} = \delta^{\mu}_{\sigma}$. For further investigations of the wave function ψ it is convenient to use the covariant notation (1.21) of the Klein–Gordon equation (see Exercise 1.5). Since the square of the four-vector $\hat{p}^{\mu}$, i.e. $\hat{p}^2 = \hat{p}^{\mu}\hat{p}_{\mu}$, does not change under Lorentz transformations, one can see that the wave function is just multiplied by a factor of absolute value 1 in these transformations. In other words, in the case of the coordinate transformations (6), which are simply abbreviated by

$$x \to x' = \hat{a}x \quad , \tag{7}$$

the transformation law of the wave function of the Klein–Gordon equation (1.21) reads as

$$\psi(x) \to \psi'(x') = \lambda\psi(x) \quad , \tag{8}$$

with $|\lambda| = 1$. If the Lorentz transformation is a continuous one (rotation through an arbitrary angle in the four-dimensional space), i.e. if the transformation matrix $a^{\mu}{}_{\nu}$ depends continuously on variable parameters $\alpha_1, \alpha_2, \ldots$, then the factor λ is equal to one, because $\lambda = 1$ holds for the identity transformation characterized by $\alpha_1 = \alpha_2 = \ldots = 0$.

Let us now consider space inversion, which is a discrete Lorentz transformation:

$$\begin{aligned} x'^{i} &= -x^{i} \quad , \\ x'^{0} &= x^{0} \quad . \end{aligned} \tag{9}$$

Two-fold performance of space inversion leads to the identical transformation. Since the wave function should be unique, one has

$$\lambda^2 = 1 \quad \text{or} \quad \lambda = \pm 1 \quad . \tag{10}$$

This means that in the case $\lambda = 1$ (8) yields

$$\psi'(\boldsymbol{x}', t') = \psi'(-\boldsymbol{x}, t) = \psi(\boldsymbol{x}, t) \quad . \tag{11}$$

Thus, the ψ function is then a *scalar*. For $\lambda = -1$ is

$$\psi'(\boldsymbol{x}', t') = \psi'(-\boldsymbol{x}, t) = -\psi(\boldsymbol{x}, t) \quad , \tag{12}$$

Example 1.17.

and one is dealing with a *pseudoscalar* wave function. We conclude: solutions $\psi(\boldsymbol{x})$ of the Klein–Gordon equation are either scalar or pseudoscalar functions, i.e. functions which are both invariant under spatial rotations and proper Lorentz transformations and also which are invariant (scalars) or change sign (pseudoscalars) under space inversions.

The transformation laws of the wave functions (8) are an essential characteristic for the properties of particles described by the Klein–Gordon equation (the same is true for particles which satisfy other wave equations). Wave functions which do not change under spatial rotations describe particles with spin 0, which is, so to say, a group theoretical argument[18] for the fact that the Klein–Gordon equation describes particles with spin 0. *The absence of level splitting due to spin interactions in pionic atoms* (no $2p_{1/2} - 2p_{3/2}, 3d_{3/2} - 3d_{5/2}, \ldots$ splitting, but only $2p-, 3d-, \ldots$ levels) *uniquely leads to the conclusion that pions have spin zero.* It has turned out that the K mesons ($m_{\mathrm{K}} \approx 960\, m_{\mathrm{e}}$) also have spin zero. The question whether pions are described by scalar or pseudoscalar wave functions has to be decided experimentally.

The Pion's Pseudoscalar Character

To complete our knowledge of the pion inner wave function, people have searched for the *creation of two neutrons in the capture of a slow π^- meson by a deuteron* and this reaction has, indeed, been observed. We now show that this proves the pseudoscalar character of the pionic wave function. The first phase of the process is the creation of a π-mesic deuterium atom in the $1s$ state. The spins of the deuteron and the π^- meson are 1 and 0, respectively. Therefore, the total angular momentum of this starting state, i.e. the pionic deuteron atom in its ground state, is 1. Its parity is equal to the *inner parity* of the pion, because the inner parities of both nucleons (which make up the deuteron) can be assumed to be equal, and the parity of the $1s$ relative wave function equals 1. In the final state two neutrons have been created and the pion has been absorbed. Due to Pauli's principle a system of two neutrons can only be in the following antisymmetrical states (pay attention to the spin!):

$$^1S_0\, ^3P_0\, ^3P_1\, ^3P_2\, ^1D_2 \ldots \quad . \tag{13}$$

During this reaction parity and total angular momentum are conserved. Since the total angular momentum in the starting state was 1, only those states shown in (13) which have spin 1 can appear in the final state. The only one satisfying this demand is $^3P_1$ with $L = S = J = 1$ and, because of $L = 1$, this state is of negative parity P. Consequently, the starting state of the reaction is of negative parity, too. This is only possible with a negative inner parity of the pion. Panofsky et al.[19] verified this with the observation of the reaction

$$\pi^- + \mathrm{d} \to 2\mathrm{n} \quad .$$

[18] See the detailed discussion of this point in W. Greiner, B. Müller: *Quantum Mechanics – Symmetries*, 2nd ed. (Springer, Berlin, Heidelberg 1994), Chap. 1.

[19] W.K.H. Panofsky, R.L. Adenot, J. Halley: Phys. Rev. **81**, 565 (1951).

1.12 Interpretation of One-Particle Operators in Relativistic Quantum Mechanics

This a rather general subject. However, in this section we want to study the possibilities of measuring eigenvalues of one-particle operators, especially for the case of spin-zero particles described by the Klein–Gordon equation. This point is rather important, because we will see that the naive one-particle interpretation encounters problems. We have already seen this fact in the charge distribution (1.131) of a pionic atom. Nevertheless, we shall try to maintain the one-particle interpretation as far as possible.

Let us return to the *Schrödinger form of the free Klein–Gordon equation*,

$$\mathrm{i}\hbar\frac{\partial\Psi}{\partial t} = \hat{H}_{\mathrm{f}}\Psi \quad , \tag{1.141}$$

with

$$\Psi(\boldsymbol{x},t) = \begin{pmatrix}\phi(\boldsymbol{x},t)\\ \chi(\boldsymbol{x},t)\end{pmatrix} \tag{1.142}$$

and the Hamiltonian

$$\hat{H}_{\mathrm{f}} = (\hat{\tau}_3 + \mathrm{i}\hat{\tau}_2)\,\frac{\hat{\boldsymbol{p}}^2}{2m_0} + m_0c^2\hat{\tau}_3 \quad . \tag{1.143}$$

With the help of (1.141), the column vector $\Psi(\boldsymbol{x},t)$ may be evaluated at any later time t, if the values $\Psi(\boldsymbol{x},0)$ are known at $t = 0$. This may be expressed by the transformation

$$\Psi(\boldsymbol{x},t) = \hat{S}(t)\Psi(\boldsymbol{x},0) \quad . \tag{1.144}$$

The transformation operator reads

$$\hat{S}(t) = \exp\left(-\frac{\mathrm{i}}{\hbar}\hat{H}_{\mathrm{f}}t\right) = 1 + \left(\frac{-\mathrm{i}\hat{H}_{\mathrm{f}}}{\hbar}\right)t + \left(\frac{-\mathrm{i}\hat{H}_{\mathrm{f}}}{\hbar}\right)^2\frac{t^2}{2!} + \dots \tag{1.145}$$

and is Φ *unitary* because of the property

$$\hat{S}^{\mathrm{H}}(t) = \hat{\tau}_3\hat{S}^{\dagger}(t)\hat{\tau}_3 = \exp\left(-\frac{\mathrm{i}}{\hbar}\hat{H}_{\mathrm{f}}t\right) = \hat{S}^{-1} \quad . \tag{1.146}$$

As in nonrelativistic quantum mechanics, time dependence in the relativistic case need not be expressed by the state vectors $\Psi(\boldsymbol{x},t)$ (*Schrödinger picture*) but can also be incorporated in the operators. This picture, where the operators and not the state vectors are time-dependent, is referred to as the ***Heisenberg** picture*. The change from the Schrödinger to Heisenberg picture is performed by the transformations

$$\Psi_{\mathrm{H}}(\boldsymbol{x}) = \hat{S}^{-1}(t)\Psi(\boldsymbol{x},t) \tag{1.147}$$

and

$$\hat{F}_{\mathrm{H}}(t) = \hat{S}^{-1}(t)\hat{F}(0)\hat{S}(t) \quad . \tag{1.148}$$

For the scalar products it yields

$$\begin{aligned}\big\langle\Psi(\boldsymbol{x},t)|\hat{F}(0)|\Psi'(\boldsymbol{x},t)\big\rangle &= \big\langle\hat{S}(t)\Psi_{\mathrm{H}}(\boldsymbol{x})|\hat{F}(0)|\hat{S}(t)\Psi'_{\mathrm{H}}(\boldsymbol{x})\big\rangle \\ &= \big\langle\Psi_{\mathrm{H}}(\boldsymbol{x})|\hat{S}^{\mathrm{H}}(t)\hat{F}(0)\hat{S}(t)|\psi'_{\mathrm{H}}(\boldsymbol{x})\big\rangle \\ &= \big\langle\Psi_{\mathrm{H}}(\boldsymbol{x})|\hat{F}_{\mathrm{H}}(t)|\Psi'_{\mathrm{H}}(\boldsymbol{x})\big\rangle \quad ,\end{aligned} \tag{1.149}$$

i.e. the change from the Schrödinger to the Heisenberg picture leaves the Φ scalar product invariant. In doing so we have made use of (1.146). From (1.148), it follows directly that for time independent $\hat{H}_{\mathrm{f}}$,

$$\begin{aligned}\mathrm{i}\hbar\frac{\mathrm{d}\hat{F}}{\mathrm{d}t} &= \mathrm{i}\hbar\frac{\mathrm{d}}{\mathrm{d}t}\left\{\mathrm{e}^{\mathrm{i}\hat{H}_{\mathrm{f}}t/\hbar}\hat{F}(0)\mathrm{e}^{-\mathrm{i}\hat{H}_{\mathrm{f}}t/\hbar}\right\} \\ &= -\hat{H}_{\mathrm{f}}\hat{F}+\hat{F}\hat{H}_{\mathrm{f}} = \left[\hat{F},\hat{H}_{\mathrm{f}}\right]_{-} \quad ,\end{aligned} \tag{1.150}$$

in analogy to nonrelativistic quantum mechanics.[20] From this relation it follows that the physical observables F, whose corresponding operator $\hat{F}$ commutes with $\hat{H}_{\mathrm{f}}$, are constants of motion. This means that the expectation values of these operators are constant in time. One of the basic postulates in nonrelativistic quantum mechanics states that the eigenvalues of an operator describe the measurable values of the corresponding classical observable (physical quantity) in a state of the system. To satisfy this postulate in the relativistic theory we must modify the definitions of some of the operators.

We illustrate this for the well-known example of energy. The eigenvalues and eigenstates of the operators $\hat{H}_{\mathrm{f}}$ (1.143) are determined – in the case of free motion with momentum $\boldsymbol{p}$ – by the equation

$$\hat{H}_{\mathrm{f}}\Psi = E\Psi \quad . \tag{1.151}$$

We know from our earlier discussions [see (1.74–79 and 114)] and from Exercise 1.9, that (1.151) has two solutions

$$\Psi_\lambda(\boldsymbol{x}) = \frac{1}{\sqrt{L^3}}\begin{pmatrix}\varphi_{0\lambda}\\ \chi_{0\lambda}\end{pmatrix}\mathrm{e}^{\mathrm{i}\boldsymbol{p}\cdot\boldsymbol{x}/\hbar} \quad , \quad \lambda=\pm 1 \quad , \tag{1.152}$$

for the corresponding "energies"

$$E_\lambda = \lambda E_p = \lambda c\sqrt{\boldsymbol{p}^2+m_0^2c^2} \quad . \tag{1.153}$$

E_{-1} is negative and, therefore, we cannot interpret it as a one-particle energy, which must always be positive. Here we have a remember the double meaning of the energy eigenvalues of the Hamiltonian in nonrelativistic quantum mechanics: first they represent the energy of stationary states and second they characterize the time dependence (time evolution) of the wave functions. We have already learned that the eigenvalues E_λ of $\hat{H}_{\mathrm{f}}$ also represent the time dependence of the wave functions in the relativistic theory [the time factors $\exp(\pm\mathrm{i}E_pt/\hbar)$ in (1.78) and (1.79)]:

[20] See W. Greiner: *Quantum Mechanics – An Introduction*, 3rd ed. (Springer, Berlin, Heidelberg 1994), Chap. 15.

$$\begin{aligned}\Psi_\lambda(\boldsymbol{x},t) &= \exp\left(-\mathrm{i}\hat{H}_\mathrm{f}t/\hbar\right)\Psi_\lambda(\boldsymbol{x}) = \exp\left(-\mathrm{i}\lambda E_p t/\hbar\right)\Psi_\lambda(\boldsymbol{x}) \\ &= \Psi_\lambda(\boldsymbol{x})\exp\left(-\mathrm{i}\lambda E_p t/\hbar\right) \quad . \end{aligned} \tag{1.154}$$

The energy of these states is always positive and, hence, λ independent. Previously, we derived this result by using the canonical formalism [see (1.61)]. We can also see this from the following statement: The energy ε of a system in a stationary state is identical with the mean value of the energy, i.e.

$$\varepsilon_\lambda = \int \Psi_\lambda^\dagger \hat{\tau}_3 \hat{H}_\mathrm{f} \Psi_\lambda \, \mathrm{d}^3x \quad . \tag{1.155}$$

We proved this in Exercise 1.7 by making use of the canonical formalism. Now with

$$\hat{H}_\mathrm{f}\Psi_\lambda = E_\lambda \Psi_\lambda = \lambda E_p \Psi_\lambda \quad \text{and} \quad \int \Psi_\lambda^\dagger \hat{\tau}_3 \Psi_\lambda \, \mathrm{d}^3x = \lambda \quad ,$$

we have

$$\varepsilon_\lambda = \lambda E_p \int \Psi_\lambda^\dagger \hat{\tau}_3 \Psi_\lambda \, \mathrm{d}^3x = \lambda^2 E_p = E_p \quad . \tag{1.156}$$

The energy is always positive and independent of λ. Thus, the problem of the energy is solved. To resume: The dual character of the eigenvalues of $\hat{H}_\mathrm{f}$, i.e. as a characteristic factor of the time evolution and as an energy, evolves quite naturally in the relativistic quantum theory. We can give the correct interpretation of the energies of the states by making use of the canonical formalism. Hence, the energy operator is not $\hat{H}_\mathrm{f}$ but $\hat{\tau}_3\hat{H}_\mathrm{f}$ [see (1.155)].

In nonrelativistic quantum mechanics there is always a correspondence between a relation of operators and that of classical objects (measurable values). For example, Newton's classical equation of motion corresponds to the operator equation

$$\frac{\mathrm{d}\hat{\boldsymbol{p}}}{\mathrm{d}t} = \frac{1}{\mathrm{i}\hbar}\left[\hat{\boldsymbol{p}}, \hat{H}\right] = -\boldsymbol{\nabla} U \quad ,$$

with $\hat{H} = \hat{\boldsymbol{p}}^2/2m + U(\boldsymbol{x})$ (Ehrenfest's theorem). Another example is given by the operator relation

$$\frac{\mathrm{d}\boldsymbol{x}}{\mathrm{d}t} = \frac{1}{\mathrm{i}\hbar}\left[\boldsymbol{x}, \hat{H}\right] = \frac{\hat{\boldsymbol{p}}}{m_0} \quad , \tag{1.157}$$

which corresponds to the classical relation between the velocity and linear momentum. Because these operator equations are of the same form as the classical equations, it is certain that the quantum-mechanical mean values satisfy the classical equations of motion.[21] In relativistic quantum theory the situation is different. For instance, in the last example the expression for the "velocity operator" of a relativistic spin-0 particle was

$$\frac{\mathrm{d}\boldsymbol{x}}{\mathrm{d}t} = \frac{1}{\mathrm{i}\hbar}\left[\boldsymbol{x}, \hat{H}_\mathrm{f}\right] = (\hat{\tau}_3 + \mathrm{i}\hat{\tau}_2)\frac{\hat{\boldsymbol{p}}}{m_0} \quad , \tag{1.158}$$

[21] See W. Greiner: *Quantum Mechanics – An Introduction*, 3rd ed. (Springer, Berlin, Heidelberg 1994).

while the classical relativistic velocity is given by

$$\frac{\mathrm{d}\boldsymbol{x}}{\mathrm{d}t} = \frac{\boldsymbol{p}}{M} = \frac{c^2\boldsymbol{p}}{Mc^2} = \frac{c^2\boldsymbol{p}}{E} \tag{1.159}$$

where $M = m_0/(1 - v^2/c^2)^{1/2}$ denotes the relativistic mass, i.e.

$$\begin{aligned} E = Mc^2 &= \frac{m_0c^2}{\sqrt{1-v^2/c^2}} = \frac{m_0c^2}{\sqrt{1-v^2/c^2}}\sqrt{\frac{v^2}{c^2} + \left(1 - \frac{v^2}{c^2}\right)} \\ &= c\sqrt{\frac{m_0^2v^2 + m_0^2c^2(1-v^2/c^2)}{1-v^2/c^2}} \\ &= c\sqrt{\left(\frac{m_0v}{(1-v^2/c^2)^{1/2}}\right)^2 + m_0^2c^2} = c\sqrt{\boldsymbol{p}^2 + m_0^2c^2} \end{aligned}$$

is the total energy of a free particle with rest mass m_0. Obviously, the rhs of (1.158) and (1.159) are different. Furthermore, we notice that the eigenvalues of the matrix

$$\hat{\tau}_3 + \mathrm{i}\hat{\tau}_2 \begin{pmatrix} 1 & 1 \\ -1 & -1 \end{pmatrix}$$

are zero! This also means that the eigenvalues of the velocity operator (1.158) are zero, too. Hence, we again notice that, in general within a relativistic theory, the eigenvalues (expectation values) of a reasonably constructed operator are not the same as the values of the corresponding classical quantity. Therefore, we conclude that not all operators of the nonrelativistic theory can be transferred to the relativistic one-particle theory. The reason for this is the *restriction to the one-particle concept*. In relativistic quantum mechanics the consistency of the one-particle description is limited. This may be specified more precisely: From a mathematical point of view the formulation of the relativistic theory within a one-particle concept implies the condition that the only valid operators are those which do not mix different charge states. Such operators are called *even operators* or *true one-particle operators*. More formally, an operator $[\hat{F}]$ is called *even*, if

$$[\hat{F}]\,\Psi^{(+)} = \Psi'^{(+)} \quad , \quad [\hat{F}]\,\Psi^{(-)} = \Psi'^{(-)} \tag{1.160}$$

is valid. $\Psi'^{(\pm)}$ are functions with positive and negative frequencies, respectively. Similarly, an operator $\{\hat{F}\}$ is called *odd* if it satisfies the conditions

$$\{\hat{F}\}\,\Psi^{(+)} = \Psi'^{(-)} \quad , \quad \{\hat{F}\}\,\Psi^{(-)} = \Psi'^{(+)} \quad . \tag{1.161}$$

Therefore, the Hamiltonian of the free Klein–Gordon equation in the Schrödinger representation $\hat{H}_\mathrm{f}$ and the momentum operator $\hat{\boldsymbol{p}} = -\mathrm{i}\hbar\boldsymbol{\nabla}$ are even operators. This means that

$$\hat{H}_\mathrm{f} = [\hat{H}_\mathrm{f}] \quad , \quad \hat{\boldsymbol{p}} = [\hat{\boldsymbol{p}}] \quad . \tag{1.162}$$

An operator can generally be split into an even and an odd part, e.g.

$$\hat{F} = [\hat{F}] + \{\hat{F}\} \quad . \tag{1.163}$$

Therefore, one can separate from any given operator $\hat{F}$ a true one-particle operator $[\hat{F}]$.

The investigation of even and odd operators can be simplified in the Feshbach–Villars representation, especially if one uses momentum eigenfunctions ($\hat{\boldsymbol{p}}$ representation). In the Φ representation the wave functions of the two charge states are given by (1.99) and (1.100), i.e.

$$\Phi^{(+)}(\boldsymbol{p}) = \frac{1}{\sqrt{L^3}}\begin{pmatrix}1\\0\end{pmatrix}\exp\left[\frac{\mathrm{i}}{\hbar}(\boldsymbol{p}\cdot\boldsymbol{x} - E_p t)\right] \quad ,$$
$$\Phi^{(-)}(\boldsymbol{p}) = \frac{1}{\sqrt{L^3}}\begin{pmatrix}0\\1\end{pmatrix}\exp\left[\frac{\mathrm{i}}{\hbar}(\boldsymbol{p}\cdot\boldsymbol{x} + E_p t)\right] \quad . \tag{1.164}$$

In this representation an even operator is diagonal because of the column vectors $\binom{1}{0}$ and $\binom{0}{1}$ for the two charge states. Therefore, we can deduce that $\hat{H}_\Phi$ is an even operator, because

$$\hat{H}_\Phi = \hat{\tau}_3 E_p = \begin{pmatrix}1 & 0\\ 0 & -1\end{pmatrix} E_p = \left[\hat{H}_\Phi\right] \quad . \tag{1.165}$$

Also, for $\Phi(\boldsymbol{p})$ (1.164), the momentum operator $\hat{\boldsymbol{p}}$ is diagonal, since with (1.96) and (1.112), we find that

$$\hat{\boldsymbol{p}}_\Phi = \hat{U}\hat{\boldsymbol{p}}\hat{U}^{-1} = \hat{\boldsymbol{p}}\begin{pmatrix}1 & 0\\ 0 & 1\end{pmatrix} = \hat{\boldsymbol{p}} = [\hat{\boldsymbol{p}}] \quad . \tag{1.166}$$

An even operator has to be a diagonal matrix in the Φ representation; therefore, it will be especially simple to separate the even part of any operator $\hat{F}$,

$$\hat{F}_\Phi = \begin{pmatrix}\hat{F}_{11} & \hat{F}_{12}\\ \hat{F}_{21} & \hat{F}_{22}\end{pmatrix} \quad , \tag{1.167}$$

according to

$$\hat{F}_\Phi = \begin{pmatrix}\hat{F}_{11} & 0\\ 0 & \hat{F}_{22}\end{pmatrix} + \begin{pmatrix}0 & \hat{F}_{12}\\ \hat{F}_{21} & 0\end{pmatrix} \equiv \left[\hat{F}_\Phi\right] + \left\{\hat{F}_\Phi\right\} \quad .$$

Having completed the introduction to odd and even operators, we now apply this method to the $\hat{\boldsymbol{x}}$ operator which caused problems in (1.158) and (1.159). In the $\boldsymbol{p}$ representation the $\hat{\boldsymbol{x}}$ operator is given by

$$\hat{\boldsymbol{x}} = \mathrm{i}\hbar\left\{\frac{\partial}{\partial p_x}, \frac{\partial}{\partial p_y}, \frac{\partial}{\partial p_z}\right\} = \mathrm{i}\hbar\nabla_p \quad , \tag{1.168}$$

which can be calculated to be

$$\hat{\boldsymbol{x}}_\Phi = \hat{U}\,\mathrm{i}\hbar\boldsymbol{\nabla}_p\hat{U}^{-1} = \mathrm{i}\hbar\boldsymbol{\nabla}_p\mathbb{1} - \frac{\mathrm{i}\hbar\boldsymbol{p}\hat{\tau}_1}{2(p^2 + m_0^2c^2)} \tag{1.169}$$

in the Φ representation (cf. Exercise 1.18). Since $\hat{\tau}_1$ is nondiagonal, the true one-particle position operator in Φ representation reads

$$[\hat{\boldsymbol{x}}_\Phi] = \mathrm{i}\hbar\boldsymbol{\nabla}_p\mathbb{1} \quad . \tag{1.170}$$

Clearly, this is the *canonical conjugate operator of the momentum operator* since, due to (1.166),

$$\left[[\hat{\boldsymbol{x}}_\Phi]_i\,,[\hat{\boldsymbol{p}}]_j\right]_- = \mathrm{i}\hbar\left[\frac{\partial}{\partial p_i}, p_j\right]_- = \mathrm{i}\hbar\delta_{ij} \quad , \tag{1.171}$$

as it should according to quantum mechanics.[22] With the help of the true position operator (1.170) we can calculate the *velocity operator*:

$$\begin{aligned}\frac{\mathrm{d}}{\mathrm{d}t}[\hat{\boldsymbol{x}}_\Phi] &= \frac{1}{\mathrm{i}\hbar}\left[\mathrm{i}\hbar\boldsymbol{\nabla}_p\mathbb{1}, \hat{H}_\Phi\right]_- = \left[\boldsymbol{\nabla}_p\mathbb{1}, \hat{\tau}_3 E_p\right]_- \\ &= \hat{\tau}_3\left[\boldsymbol{\nabla}_p, c\sqrt{\boldsymbol{p}^2+m_0^2c^2}\right]_- = \hat{\tau}_3\frac{c\boldsymbol{p}}{\sqrt{\boldsymbol{p}^2+m_0^2c^2}} = \hat{\tau}_3\frac{c^2\boldsymbol{p}}{E_p} \quad ,\end{aligned} \tag{1.172}$$

with eigenvalues

$$+\frac{c^2\boldsymbol{p}}{E_p} \quad \text{and} \quad -\frac{c^2\boldsymbol{p}}{E_p} \quad . \tag{1.173}$$

We have already stated the corresponding eigenfunctions in (1.164); therefore, we get

$$\frac{\mathrm{d}[\hat{\boldsymbol{x}}_\Phi]}{\mathrm{d}t}\Phi^{(\pm)}(\boldsymbol{p}) = \frac{c^2\boldsymbol{p}}{E_p}\hat{\tau}_3\Phi^{(\pm)}(\boldsymbol{p}) = \pm\frac{c^2\boldsymbol{p}}{E_p}\Phi^{(\pm)}(\boldsymbol{p}) \quad . \tag{1.174}$$

EXERCISE

1.18 Calculation of the Position Operator in the Φ Representation

Problem. Show that the x operator in Φ representation is given by

$$\hat{\boldsymbol{x}}_\Phi = \mathrm{i}\hbar\boldsymbol{\nabla}_p\mathbb{1} - \frac{\mathrm{i}\hbar\boldsymbol{p}\hat{\tau}_1}{2(\boldsymbol{p}^2+m_0^2c^2)} \quad .$$

Solution. We know that in the Schrödinger picture the position operator in momentum space is just the derivative with respect to $\boldsymbol{p}$, i.e.

$$\hat{\boldsymbol{x}} = \mathrm{i}\hbar\boldsymbol{\nabla}_p \quad .$$

By substituting the Feshbach–Villars representation, all operators change according to (1.112); hence,

$$\hat{\boldsymbol{x}}_\Phi = \hat{U}(\mathrm{i}\hbar\boldsymbol{\nabla}_p)\hat{U}^{-1} \quad , \quad \text{where}$$

$$\hat{U} = \frac{\left(m_0c^2+E_p\right)\mathbb{1} - \left(m_0c^2-E_p\right)\hat{\tau}_1}{\sqrt{4m_0c^2E_p}} \quad .$$

[22] See W. Greiner: *Quantum Mechanics – An Introduction*, 3rd ed. (Springer, Berlin, Heidelberg 1994).

Exercise 1.18.

Therefore, we get

$$\hat{\boldsymbol{x}}_\Phi = \left[\frac{\left(m_0c^2+E_p\right)\mathbb{1}-\left(m_0c^2-E_p\right)\hat{\tau}_1}{\sqrt{4m_0c^2E_p}}\right]\left(\mathrm{i}\hbar\boldsymbol{\nabla}_p\right) \times\left[\frac{\left(m_0c^2+E_p\right)\mathbb{1}+\left(m_0c^2-E_p\right)\hat{\tau}_1}{\sqrt{4m_0c^2E_p}}\right] .$$

Since

$$E_p=\sqrt{m_0^2c^4+\boldsymbol{p}^2c^2} \quad , \quad \text{and} \quad \boldsymbol{\nabla}_pE_p=\frac{c^2}{E_p}\left(\boldsymbol{p}\cdot\boldsymbol{\nabla}_p\right)\boldsymbol{p}=\frac{\boldsymbol{p}c^2}{E_p} \quad ,$$

we get

$$\begin{aligned}
\boldsymbol{\nabla}_p\hat{U}^{-1} &= \frac{1}{\sqrt{4m_0c^2E_p}}(\mathbb{1}-\hat{\tau}_1)\boldsymbol{\nabla}_pE_p-\hat{U}^{-1}\frac{1}{2(4m_0c^2E_p)}4m_0c^2\boldsymbol{\nabla}_pE_p\\
&= \frac{1}{\sqrt{4m_0c^2E_p}}(\mathbb{1}-\hat{\tau}_1)\frac{\boldsymbol{p}c^2}{E_p}\\
&\quad -\frac{1}{\sqrt{4m_0c^2E_p}}\left\{(\mathbb{1}+\hat{\tau}_1)m_0c^2+(\mathbb{1}-\hat{\tau}_1)E_p\right\}\frac{\boldsymbol{p}c^2}{2E_p^2}\\
&= -\frac{\boldsymbol{p}c^2}{2E_p^2}\left\{(\mathbb{1}+\hat{\tau}_1)m_0c^2-(\mathbb{1}-\hat{\tau}_1)E_p\right\}\frac{1}{\sqrt{4m_0c^2E_p}}\\
&= -\frac{\boldsymbol{p}c^2\hat{\tau}_1}{2E_p^2}\left\{\frac{\left(m_0c^2+E_p\right)\mathbb{1}+\left(m_0c^2-E_p\right)\hat{\tau}_1}{\sqrt{4m_0c^2E_p}}\right\}\\
&= -\hat{U}^{-1}\frac{\boldsymbol{p}c^2\hat{\tau}_1}{2E_p^2} \quad .
\end{aligned}$$

If $\boldsymbol{x}_\Phi$ acts on Φ the result is

$$\begin{aligned}
\hat{\boldsymbol{x}}_\Phi\Phi &= \hat{U}\left(\mathrm{i}\hbar\boldsymbol{\nabla}_p\right)\hat{U}^{-1}\Phi=\hat{U}\left[\mathrm{i}\hbar\boldsymbol{\nabla}_p\hat{U}^{-1}\right]\Phi+\hat{U}\hat{U}^{-1}\mathrm{i}\hbar\boldsymbol{\nabla}_p\Phi\\
&= -\hat{U}\hat{U}^{-1}\mathrm{i}\hbar\frac{\boldsymbol{p}c^2\hat{\tau}_1}{2E_p^2}+\hat{U}\hat{U}^{-1}\mathrm{i}\hbar\boldsymbol{\nabla}_p\Phi\\
&= \left\{\mathrm{i}\hbar\boldsymbol{\nabla}_p-\frac{\mathrm{i}\hbar\boldsymbol{p}\hat{\tau}_1}{2\left(m_0^2c^2+\boldsymbol{p}^2\right)}\right\}\Phi \quad ,
\end{aligned}$$

as required.

For states of positive charge (time evolution factor $E=+E_p$) the same relation between velocity and momentum holds as in classical relativistic mechanics, (1.159). For states of negative charge this is true only for the absolute values. Antiparticles with momentum $\boldsymbol{p}$ move in the direction opposite to $\boldsymbol{p}$. This is not unreasonable, especially if we think of the current density of the charge, $e(\mathrm{d}[\boldsymbol{x}_\Phi/\mathrm{d}t])=e(c^2\boldsymbol{p}/E_p)\hat{\tau}_3$, which must change its sign for antiparticles because of their negative charges. Therefore, the operator $[\hat{\boldsymbol{x}}_\Phi]$ fulfills a number of plausible conditions and can be accepted as a true one-particle operator. Since the position operator $[\hat{\boldsymbol{x}}_\Phi]$ and the charge operator $\hat{Q}_\Phi = e\hat{\tau}_3$ commute, there exists a set of simultaneous eigenfunctions $\Phi_{\boldsymbol{x}}^{(\pm)}(\boldsymbol{p})$ with

$$[\hat{\boldsymbol{x}}_\Phi]\,\Phi_{\boldsymbol{x}}^{(\pm)}(\boldsymbol{p}) = \mathrm{i}\hbar\boldsymbol{\nabla}_p\Phi_{\boldsymbol{x}}^{(\pm)}(\boldsymbol{p}) = \boldsymbol{x}\Phi_{\boldsymbol{x}}^{(\pm)}(\boldsymbol{p}) \quad , \quad \text{and} \tag{1.175}$$

$$\hat{Q}_\Phi\Phi_{\boldsymbol{x}}^{(\pm)}(\boldsymbol{p}) = \pm e\Phi_{\boldsymbol{x}}^{(\pm)}(\boldsymbol{p}) \quad .$$

The $\Phi_{\boldsymbol{x}}^{(\pm)}(\boldsymbol{p})$, which are normalized to δ functions, are given by

$$\Phi_{\boldsymbol{x}}^{(+)}(\boldsymbol{p}) = \frac{1}{\sqrt{(2\pi\hbar)^3}}\begin{pmatrix}1\\0\end{pmatrix}\exp(\mathrm{i}\boldsymbol{p}\cdot\boldsymbol{x}/\hbar) \quad ,$$

$$\Phi_{\boldsymbol{x}}^{(-)}(\boldsymbol{p}) = \frac{1}{\sqrt{(2\pi\hbar)^3}}\begin{pmatrix}0\\1\end{pmatrix}\exp(\mathrm{i}\boldsymbol{p}\cdot\boldsymbol{x}/\hbar) \quad . \tag{1.176}$$

This is the $\Phi - \boldsymbol{p}$ representation. In order to derive the $\Psi - \boldsymbol{p}$ representation (i.e. the usual Schrödinger momentum representattion) we must apply $\hat{U}^{-1}$ to $\Phi_{\boldsymbol{x}}^{(\pm)}(\boldsymbol{p})$ according to (1.95), so that

$$\begin{aligned}\Psi_{\boldsymbol{x}}^{(\pm)}(\boldsymbol{p}) &= \hat{U}^{-1}\Phi_{\boldsymbol{x}}^{(\pm)}(\boldsymbol{p})\\ &= \frac{\left(m_0c^2+E_p\right)\mathbb{1} + \left(m_0c^2-E_p\right)\hat{\tau}_1}{\sqrt{4m_0c^2E_p}}\Phi_{\boldsymbol{x}}^{(\pm)}(\boldsymbol{p})\\ &= \frac{\left(m_0c^2+E_p\right)\begin{pmatrix}1&0\\0&1\end{pmatrix} + \left(m_0c^2-E_p\right)\begin{pmatrix}0&1\\1&0\end{pmatrix}}{\sqrt{4m_0c^2E_p}}\Phi_{\boldsymbol{x}}^{(\pm)}(\boldsymbol{p}) \quad ,\end{aligned} \tag{1.177}$$

where $\hat{U}^{-1}$ is given by (1.97). This can immediately be calculated, yielding the explicit form

$$\Psi_{\boldsymbol{x}}^{(\pm)}(\boldsymbol{p}) = \frac{1}{\sqrt{2\pi\hbar^3}}\frac{1}{\sqrt{4m_0c^2E_p}}\begin{pmatrix}m_0c^2 \pm E_p\\ m_0c^2 \mp E_p\end{pmatrix}\exp(\mathrm{i}\boldsymbol{p}\cdot\boldsymbol{x}/\hbar) \quad . \tag{1.178}$$

As we have already noted these are the eigenfunctions (1.176) in Schrödinger momentum representation. We perform the transition from $\boldsymbol{p}$ representation to $\boldsymbol{x}$ representation by[23]

$$\begin{aligned}\Psi_{\boldsymbol{x}}^{(\pm)}(\boldsymbol{x}') &= \int\langle\boldsymbol{x}'|\boldsymbol{p}\rangle\Psi_{\boldsymbol{x}}^{(\pm)}(\boldsymbol{p})\,\mathrm{d}^3p\\ &= \int\frac{\exp(-\mathrm{i}\boldsymbol{p}\cdot\boldsymbol{x}'/\hbar)}{\sqrt{2\pi\hbar^3}}\Psi_{\boldsymbol{x}}^{(\pm)}(\boldsymbol{p})\,\mathrm{d}^3p\\ &= \frac{1}{(2\pi\hbar)^3}\frac{1}{\sqrt{4m_0c^2}}\int\begin{pmatrix}m_0c^2 \pm E_p\\ m_0c^2 \mp E_p\end{pmatrix}\frac{\exp(-\mathrm{i}\boldsymbol{p}\cdot(\boldsymbol{x}'-\boldsymbol{x})/\hbar)}{\sqrt{E_p}}\,\mathrm{d}^3p \quad .\end{aligned} \tag{1.179}$$

[23] See W. Greiner: *Quantum Mechanics – An Introduction*, 3rd ed. (Springer, Berlin, Heidelberg 1994), Chap. 15.

A lengthy calculation, which we shall perform in Exercise 1.20, yields

$$\Psi_{\boldsymbol{x}}^{(\pm)}(\boldsymbol{x}') = \begin{pmatrix} A \pm B \\ A \mp B \end{pmatrix} \underset{z \gg 1}{\sim} \begin{pmatrix} 1/z^{7/4} \pm 1/z^{9/4} \\ 1/z^{7/4} \mp 1/z^{9/4} \end{pmatrix} \mathrm{e}^{-z} \quad , \tag{1.180}$$

where

$$z = \frac{|\boldsymbol{x} - \boldsymbol{x}'|}{\hbar/m_0 c} \quad .$$

The meaning of this result is that the eigenfunctions of the true position operator $[\hat{\boldsymbol{x}}]$ in Schrödinger position representation are not $\delta(\boldsymbol{x} - \boldsymbol{x}')$, as we would have expected from nonrelativistic quantum mechanics, but a kind of smeared-out δ function. According to (1.180) the functions are smeared out over a region of the order

$$z \sim 1 \quad , \quad \text{implying} \quad |\boldsymbol{x} - \boldsymbol{x}'| \sim \frac{\hbar}{m_0 c} \quad , \tag{1.181}$$

i.e. the dimension of the Compton wavelength of the particle.

EXERCISE

1.19 Calculation of the Current Density in the Φ Representation for Particles and Antiparticles

Problem. Calculate the current density (1.36) in the Φ representation for particles and antiparticles.

Solution. By means of the Klein–Gordon field the current density reads

$$\boldsymbol{j} = -\frac{\mathrm{i}e\hbar}{2m_0}\left(\psi^* \boldsymbol{\nabla}\psi - \psi \boldsymbol{\nabla}\psi^*\right)$$

$$\Psi = \begin{pmatrix} \varphi \\ \chi \end{pmatrix} \quad , \quad \varphi + \chi = \psi \quad ,$$

$$\boldsymbol{j} = -\frac{\mathrm{i}e\hbar}{2m_0}\left\{\Psi^\dagger \hat{\tau}_3 \left(\hat{\tau}_3 + \mathrm{i}\hat{\tau}_2\right) \boldsymbol{\nabla}\Psi - (\boldsymbol{\nabla}\Psi^\dagger)\hat{\tau}_3 \left(\hat{\tau}_3 + \mathrm{i}\hat{\tau}_2\right)\Psi\right\} \quad ,$$

which can be verified by direct calculation,

$$\begin{aligned} &\Psi^\dagger \hat{\tau}_3 \left(\hat{\tau}_3 + \mathrm{i}\hat{\tau}_2\right) \boldsymbol{\nabla}\Psi - (\boldsymbol{\nabla}\Psi^\dagger)\hat{\tau}_3 \left(\hat{\tau}_3 + \mathrm{i}\hat{\tau}_2\right)\Psi \\ &\quad = (\varphi^*, \chi^*) \begin{pmatrix} 1 & 1 \\ 1 & 1 \end{pmatrix} \boldsymbol{\nabla} \begin{pmatrix} \varphi \\ \chi \end{pmatrix} - \left[\boldsymbol{\nabla}(\varphi^*, \chi^*)\right] \begin{pmatrix} 1 & 1 \\ 1 & 1 \end{pmatrix} \begin{pmatrix} \varphi \\ \chi \end{pmatrix} \\ &\quad = \left(\varphi^* + \chi^*\right) \boldsymbol{\nabla}(\varphi + \chi) - \boldsymbol{\nabla}\left(\varphi^* + \chi^*\right) \cdot (\varphi + \chi) \\ &\quad = \psi^* \boldsymbol{\nabla}\psi - \psi \boldsymbol{\nabla}\psi^* \quad . \end{aligned}$$

Using

$$\Psi = \hat{U}^{-1}\Phi \quad , \quad \Psi^\dagger = \Phi^\dagger(\hat{U}^\dagger)^{-1} \quad ,$$

where $\hat{U}$ is given by (1.96), and, because of *Exercise 1.19.*

$$(\hat{U}^{\dagger})^{-1} = \left(\hat{\tau}_3\hat{U}^{-1}\hat{\tau}_3\right)^{-1} = \hat{\tau}_3^{-1}\hat{U}\hat{\tau}_3^{-1} = \hat{\tau}_3\hat{U}\hat{\tau}_3 \quad ,$$

$$\Psi^{\dagger}(\mathbb{1}+\hat{\tau}_1)\nabla\Psi = \Phi^{\dagger}\hat{\tau}_3\hat{U}\hat{\tau}_3(\mathbb{1}+\hat{\tau}_1)\hat{U}^{-1}\nabla\Phi \quad .$$

Furthermore, $\hat{\tau}_3(\hat{\tau}_3+\mathrm{i}\hat{\tau}_2) = \mathbb{1}+\hat{\tau}_1$ is valid. We commute $\hat{U}$ with $\hat{\tau}_3$ to give

$$\begin{aligned}\hat{U}\hat{\tau}_3 &= \frac{\left(E_p+m_0c^2\right)\mathbb{1}+\hat{\tau}_1\left(E_p-m_0c^2\right)}{\sqrt{4m_0c^2E_p}}\hat{\tau}_3\\ &= \hat{\tau}_3\frac{\left(E_p+m_0c^2\right)\mathbb{1}-\left(E_p-m_0c^2\right)\hat{\tau}_1}{\sqrt{4m_0c^2E_p}}\\ &= \hat{\tau}_3\hat{U}^{-1} \quad ,\end{aligned}$$

and we get

$$\Psi^{\dagger}(\mathbb{1}+\hat{\tau}_1)\nabla\Psi = \Phi^{\dagger}(\mathbb{1}+\hat{\tau}_1)\left(\hat{U}^{-1}\right)^2\nabla\Phi \quad ,$$

because $\hat{U}^{-1}$ commutes with $1+\hat{\tau}_1$. Since

$$\begin{aligned}\left(\hat{U}^{-1}\right)^2 &= \frac{1}{4m_0c^2E_p}\left(\left(E_p+m_0c^2\right)\mathbb{1}-\left(E_p-m_0c^2\right)\hat{\tau}_1\right)^2\\ &= \frac{1}{4m_0c^2E_p}\left(\left(E_p+m_0c^2\right)^2\mathbb{1}+\left(E_p-m_0c^2\right)^2\mathbb{1}-2\left(E_p^2-m_0^2c^4\right)\hat{\tau}_1\right)\\ &= \frac{1}{4m_0c^2E_p}\left(2\left(E_p^2+m_0^2c^4\right)\mathbb{1}-2\left(E_p^2-m_0^2c^4\right)\hat{\tau}_1\right)\\ &= \frac{1}{2m_0c^2E_p}\left(m_0^2c^4(\mathbb{1}+\hat{\tau}_1)+E_p^2(\mathbb{1}-\hat{\tau}_1)\right) \quad ,\end{aligned}$$

and

$$\begin{aligned}(1+\hat{\tau}_1)\hat{U}^{-2} &= \frac{1}{2m_0c^2E_p}\left\{m_0^2c^4(\mathbb{1}+\hat{\tau}_1)^2+E_p^2(\mathbb{1}+\hat{\tau}_1)(\mathbb{1}-\hat{\tau}_1)\right\}\\ &= \frac{m_0c^2}{E_p}(\mathbb{1}+\hat{\tau}_1) \quad ,\end{aligned}$$

then the current density is

$$\begin{aligned}\boldsymbol{j} &= -\frac{\mathrm{i}e\hbar}{2m_0}\left\{\Phi^{\dagger}\frac{m_0c^2}{E_p}(\mathbb{1}+\hat{\tau}_1)\nabla\Phi-\nabla\Phi^{\dagger}\frac{m_0c^2}{E_p}(\mathbb{1}+\hat{\tau}_1)\Phi\right\}\\ &= -\frac{\mathrm{i}e\hbar c^2}{2E_p}\left\{\Phi^{\dagger}(\mathbb{1}+\hat{\tau}_1)\nabla\Phi-\nabla\Phi^{\dagger}(\mathbb{1}+\hat{\tau}_1)\Phi\right\} \quad .\end{aligned}$$

For a free particle solution $\Phi_p^{(+)} = 1/\sqrt{V}\binom{1}{0}\exp[\mathrm{i}(\boldsymbol{p}\cdot\boldsymbol{x}-E_pt)/\hbar]$ we, thus, get for the *current density*

$$\boldsymbol{j}^{(+)} = \frac{1}{V}\frac{e\boldsymbol{p}c^2}{E_p} \quad ,$$

and the *current*

Exercise 1.19.

$$\boldsymbol{J}^{(+)} = \int \boldsymbol{j}^{(+)}\,\mathrm{d}^3r = e\frac{c^2\boldsymbol{p}}{E_p} \quad .$$

For a free antiparticle solution with momentum $\boldsymbol{p}$, $\Phi_{\boldsymbol{p}}^{(-)} = 1/\sqrt{V}\binom{0}{1}\exp[\mathrm{i}(\boldsymbol{p}\cdot\boldsymbol{x}) + E_p t]$ the current is then given by

$$\boldsymbol{J}^{(-)} = \int \mathrm{d}^3r \left(\frac{1}{V}e\frac{c^2\boldsymbol{p}}{E_p}\right) = e\frac{c^2\boldsymbol{p}}{E_p} = \boldsymbol{J}^{(+)} \quad .$$

Thus, we see at once that

$$\begin{aligned}\boldsymbol{J}^{\pm} &= \int \mathrm{d}^3r\, \Phi_p^{(\pm)\dagger}\hat{\tau}_3 e\frac{\mathrm{d}[\hat{\boldsymbol{x}}_\Phi]}{\mathrm{d}t}\Phi_p^{(\pm)} \\ &= \int \mathrm{d}^3r\, \Phi_p^{(\pm)\dagger} e\frac{c^2\boldsymbol{p}}{E_p}\Phi_p^{(\pm)} \quad ,\end{aligned}$$

i.e. the current is equal to the expectation value of the operator

$$e\frac{\mathrm{d}[\hat{\boldsymbol{x}}_\Phi]}{\mathrm{d}t} \equiv e\hat{\boldsymbol{v}} \quad .$$

But why are the currents for particle and antiparticle with energy E_p and momentum $\boldsymbol{p}$ equal although the charges are different? The reason is that for antiparticles with momentum $\boldsymbol{p}$ the velocity operator is $-c^2\boldsymbol{p}/E_p$, i.e. velocity and momentum have opposite directions. The expectation value of the operator, however, has the same direction as $\boldsymbol{p}$ because the norm or "charge" of the wave function is included. To get the physical velocity we have to divide the expectation value of $\boldsymbol{v}$ by the norm of the state. This velocity again has opposite direction to the momentum. One says that the antiparticles move "backwards in time".

EXERCISE

1.20 Calculation of the Position Eigenfunctions in the Coordinate Representation

Problem. Carry out explicitly the calculation leading from the exact expression (1.179) for the eigenfunctions of position

$$\Psi_{\boldsymbol{x}(\pm)}(\boldsymbol{x}') = \frac{1}{(2\pi\hbar)^3}\frac{1}{\sqrt{4m_0c^2}}\int \begin{pmatrix} m_0c^2 \pm E_p \\ m_0c^2 \mp E_p \end{pmatrix} \frac{\exp[\mathrm{i}\boldsymbol{p}\cdot(\boldsymbol{x}-\boldsymbol{x}')/\hbar]}{\sqrt{E_p}}\,\mathrm{d}^3p$$

to the approximation (1.180).

Solution. In the analysis of (1.179) we face, in particular, two integrals:

$$I_1 = \int \sqrt{E_p}\exp[\mathrm{i}\boldsymbol{p}\cdot(\boldsymbol{x}-\boldsymbol{x}')/\hbar]\,\mathrm{d}^3p \quad ,$$

$$I_2 = \int \frac{1}{\sqrt{E_p}}\exp[\mathrm{i}\boldsymbol{p}\cdot(\boldsymbol{x}-\boldsymbol{x}')/\hbar]\,\mathrm{d}^3p \quad .$$

With *Exercise 1.20.*

$$\boldsymbol{p}\cdot(\boldsymbol{x}-\boldsymbol{x}') = |\boldsymbol{p}|\,|\boldsymbol{x}-\boldsymbol{x}'|\cos\vartheta \equiv pr\cos\vartheta \quad , \quad p = |\boldsymbol{p}| \quad ,$$

and ϑ being the angle between $\boldsymbol{p}$ and $(\boldsymbol{x}-\boldsymbol{x}')$, we can introduce polar coordinates. With the z axes in the direction of $(\boldsymbol{x}-\boldsymbol{x}')$, the volume element is given by $\mathrm{d}^3p = \mathrm{d}\varphi\,\mathrm{d}(\cos\vartheta)p^2\,\mathrm{d}p$, and we have

$$\begin{aligned}
I_1 &= \int_0^\infty \int_{-1}^1 \int_0^{2\pi} \sqrt[4]{m_0^2c^4+p^2c^2}\,\mathrm{e}^{(\mathrm{i}/\hbar)pr\cos\vartheta}\,\mathrm{d}\varphi\,\mathrm{d}(\cos\vartheta)p^2\,\mathrm{d}p \\
&= 2\pi \int_0^\infty \int_{-1}^1 \sqrt[4]{m_0^2c^4+p^2c^2}\,\mathrm{e}^{(\mathrm{i}/\hbar)pr\cos\vartheta}\,\mathrm{d}\cos\vartheta p^2\,\mathrm{d}p \\
&= 2\pi \int_0^\infty \sqrt[4]{m_0^2c^4+p^2c^2}\,\frac{\mathrm{e}^{(\mathrm{i}/\hbar)pr}-\mathrm{e}^{-(\mathrm{i}/\hbar)pr}}{\mathrm{i}pr/\hbar}p^2\,\mathrm{d}p \\
&= -\mathrm{i}\frac{2\pi\hbar}{r}\int_0^\infty \sqrt[4]{m_0^2c^4+p^2c^2}\,2\mathrm{i}\sin(pr/\hbar)p\,\mathrm{d}p \\
&= \frac{4\pi\hbar}{r}\int_0^\infty \sqrt[4]{m_0^2c^4+p^2c^2}\,\sin(pr/\hbar)p\,\mathrm{d}p \quad .
\end{aligned}$$

We write

$$\left(m_0^2c^4+p^2c^2\right)^{1/4} = \sqrt{m_0c^2}\left(1+\frac{p^2}{m_0^2c^2}\right)^{1/4} = \sqrt{m_0c^2}\,\left(1+q^2\right)^{1/4}$$

with

$$q = \frac{p}{m_0} \quad , \quad \frac{pr}{\hbar} = qz \quad \text{and} \quad z = \frac{m_0c}{\hbar}r \equiv k_0r \quad ,$$

and thus obtain

$$I_1 = \frac{4\pi(m_0)^3}{z}\sqrt{m_0c^2}\int_0^\infty \left(1+q^2\right)^{1/4} q\sin(qz)\,\mathrm{d}q \equiv 16\pi^3\hbar^3\sqrt{m_0c^2}B$$

with

$$B = \frac{k_0^3}{4\pi^2z}\int_0^\infty q\left(1+q^2\right)^{1/4}\sin(qz)\,\mathrm{d}q \quad .$$

In an analogous way we find

$$\begin{aligned}
I_2 &= \frac{4\pi\hbar}{r}\int_0^\infty \left(m_0^2c^4+p^2c^2\right)^{-1/4}\sin(pr/\hbar)p\,\mathrm{d}p \\
&= \frac{4\pi(m_0c)^3}{z}\frac{1}{\sqrt{m_0c^2}}\int_0^\infty \left(1+q^2\right)^{-1/4} q\sin(qz)\,\mathrm{d}q \\
&\equiv 16\pi^3\hbar^3\frac{1}{\sqrt{m_0c^2}}A \quad ,
\end{aligned}$$

with

$$A = \frac{k_0^3}{4\pi^2z}\int_0^\infty q\left(1+q^2\right)^{-1/4}\sin(qz)\,\mathrm{d}q \quad .$$

Exercise 1.20.

Thus,

$$\begin{aligned}\Psi_{\boldsymbol{x}(\pm)}(\boldsymbol{x}') &= \frac{1}{(2\pi\hbar)^3}\frac{1}{\sqrt{4m_0c^2}}\begin{pmatrix} m_0c^2I_2 \pm I_1 \\ m_0c^2I_2 \mp I_1\end{pmatrix} \\ &= \frac{1}{8\pi^3\hbar^3}\frac{1}{2}\begin{pmatrix}16\pi^3\hbar^3(A\pm B)\\ 16\pi^3\hbar^3(A\mp B)\end{pmatrix} \\ &= \begin{pmatrix} A\pm B \\ A\mp B\end{pmatrix} \quad .\end{aligned}$$

The integrals A und B can be expressed by modified spherical Bessel functions (see Example 1.21), giving

$$\begin{aligned} A &= \frac{k_0^3}{4\pi^2 z}\frac{\mathrm{d}}{\mathrm{d}z}\left(\frac{\mathrm{d}^2}{\mathrm{d}z^2}-1\right)\left\{\frac{z^{3/4}\sqrt{\pi}}{\Gamma(5/4)}K_{3/4}(z)\right\} \quad , \\ B &= \frac{k_0^3}{4\pi^2 z}\frac{\mathrm{d}}{\mathrm{d}z}\left(\frac{\mathrm{d}^2}{\mathrm{d}z^2}-1\right)\left\{\frac{z^{1/4}\sqrt{\pi}}{\Gamma(3/4)}K_{1/4}(z)\right\} \quad .\end{aligned}$$

Here we have used the *Basset formula*

$$\int_0^\infty \frac{\cos(qz)\,\mathrm{d}q}{(q^2+1)^{\nu+1/2}} = \frac{z^\nu\sqrt{\pi}}{\Gamma(\nu+1/2)}K_\nu(z) \quad .$$

These expressions for A and B can at once be verified by substitution. For large z the modified Bessel functions can be expanded as

$$K_\nu(z) = \sqrt{\frac{\pi}{2z}}\,\mathrm{e}^{-z}\left(1+\frac{4\nu^2-1}{8z}+\dots\right) \quad .$$

For $\nu = 3/4$ it follows immediately that

$$K_{3/4}(z) \underset{z\gg 1}{\longrightarrow} \sqrt{\frac{\pi}{2z}}\,\mathrm{e}^{-z}\left(1+\frac{5}{32z}+\dots\right) \quad ,$$

$$\begin{aligned}\frac{\mathrm{d}}{\mathrm{d}z}\left(z^{3/4}K_{3/4}(z)\right) &\underset{z\gg 1}{\longrightarrow} \sqrt{\frac{\pi}{2}}\frac{\mathrm{d}}{\mathrm{d}z}\left[\mathrm{e}^{-z}\left(z^{1/4}+\frac{5}{32}z^{-3/4}+\dots\right)\right] \\ &= \sqrt{\frac{\pi}{2}}\,\mathrm{e}^{-z}\left(-z^{1/4}+\frac{3}{32}z^{-3/4}\dots\right) \quad ,\end{aligned}$$

$$\begin{aligned}\frac{\mathrm{d}^2}{\mathrm{d}z^2}\left(z^{3/4}K_{3/4}(z)\right) &\underset{z\gg 1}{\longrightarrow} \sqrt{\frac{\pi}{2}}\,\mathrm{e}^{-z}\left(z^{1/4}-\frac{11}{32}z^{-3/4}\dots\right) \quad , \\ \frac{\mathrm{d}^3}{\mathrm{d}z^3}\left(z^{3/4}K_{3/4}(z)\right) &\underset{z\gg 1}{\longrightarrow} \sqrt{\frac{\pi}{2}}\,\mathrm{e}^{-z}\left(-z^{1/4}+\frac{19}{32}z^{-3/4}\dots\right) \quad .\end{aligned}$$

Thus we have

$$A \underset{z\gg 1}{\longrightarrow} \frac{k_0^3}{4\pi^2}\sqrt{\frac{\pi^2}{2}}\frac{1}{\Gamma(5/4)}\frac{1}{2}z^{-7/4}\mathrm{e}^{-z} \quad .$$

For $\nu = 1/4$ we get

Exercise 1.20.

$$K_{1/4}(z) \underset{z\gg 1}{\longrightarrow} \sqrt{\frac{\pi}{2z}}\,\mathrm{e}^{-z}\left(1 - \frac{3}{32}\frac{1}{z} + \dots\right) \quad ,$$

$$\frac{\mathrm{d}}{\mathrm{d}z}\left(z^{1/4}K_{1/4}(z)\right) \underset{z\gg 1}{\longrightarrow} \sqrt{\frac{\pi}{2}}\frac{\mathrm{d}}{\mathrm{d}z}\left[\mathrm{e}^{-z}\left(z^{-1/4} - \frac{3}{32}z^{-5/4} + \dots\right)\right]$$
$$= \sqrt{\frac{\pi}{2}}\,\mathrm{e}^{-z}\left(-z^{-1/4} - \frac{5}{32}z^{-5/4} + \dots\right) \quad ,$$

$$\frac{\mathrm{d}^2}{\mathrm{d}z^2}\left(z^{1/4}K_{1/4}(z)\right) \underset{z\gg 1}{\longrightarrow} \sqrt{\frac{\pi}{2}}\left[\mathrm{e}^{-z}\left(z^{-1/4} + \frac{13}{32}z^{-5/4} + \dots\right)\right] \quad ,$$

$$\frac{\mathrm{d}^3}{\mathrm{d}z^3}\left(z^{1/4}K_{1/4}(z)\right) \underset{z\gg 1}{\longrightarrow} \sqrt{\frac{\pi}{2}}\,\mathrm{e}^{-z}\left(-z^{-1/4} - \frac{21}{32}z^{-5/4} + \dots\right) \quad ,$$

and, therefore,

$$B \underset{z\gg 1}{\longrightarrow} \frac{k_0^3}{4\pi^2}\sqrt{\frac{\pi^2}{2}}\frac{1}{\Gamma(3/4)}\left(-\frac{1}{2}\right)\mathrm{e}^{-z}z^{-9/4}$$

with the result that

$$\psi_{\boldsymbol{x}(\pm)}(\boldsymbol{x}') \xrightarrow[|\boldsymbol{x}-\boldsymbol{x}'|\gg\hbar/m_0c]{} \frac{k_0^3}{8\sqrt{2\pi}}\,\mathrm{e}^{-z}\begin{pmatrix} \dfrac{1}{\Gamma(5/4)}z^{-7/4} \mp \dfrac{1}{\Gamma(3/4)}z^{-9/4} \\ \dfrac{1}{\Gamma(5/4)}z^{-7/4} \pm \dfrac{1}{\Gamma(3/4)}z^{-9/4} \end{pmatrix} \quad .$$

EXAMPLE

1.21 Mathematical Supplement: Modified Bessel Functions of the Second Type, $K_\nu(z)$

The *modified Bessel functions* are defined by the differential equation

$$z^2\frac{\mathrm{d}^2y}{\mathrm{d}z^2} + z\frac{\mathrm{d}y}{\mathrm{d}z} - \left(z^2 + \nu^2\right)y = 0 \quad , \tag{1}$$

while the *ordinary Bessel functions* obey

$$z^2\frac{\mathrm{d}^2y}{\mathrm{d}z^2} + z\frac{\mathrm{d}y}{\mathrm{d}z} + \left(z^2 - \nu^2\right)y = 0 \quad . \tag{2}$$

(1) can be transformed into (2) by the substitution $z \to \mathrm{i}z$, that is the modified Bessel functions are identical to the ordinary Bessel functions with imaginary arguments.

Let us denote the solutions of (2) by $J_\nu(z)$ and $J_{-\nu}(z)$, and the solutions of (1) by $I_\nu(z)$ and $I_{-\nu}(z)$. For $z \to 0$, I_ν, J_ν are the regular, and $J_{-\nu}$, $I_{-\nu}$ the irregular solutions. Then the following relation

$$I_\nu(z) = \mathrm{e}^{-\mathrm{i}\nu\pi/2}J_\nu(z) \tag{3}$$

Example 1.21.

holds. $J_\nu(z)$ and $I_\nu(z)$ are real functions of z, whereas $J_\nu(\mathrm{i}z)$ is not. This is the reason for the special choice of the phase factor in (3). The second type of functions K_ν are defined by

$$K_\nu(z) = \frac{\pi}{2}\frac{I_{-\nu}(z) - I_\nu(z)}{\sin \nu\pi} \quad . \tag{4}$$

The limit $\nu \to n$ ($n \in \mathbb{N}$) exists (though will not be proved here) and is

$$K_n(z) = \frac{\pi}{2}\lim_{\nu\to n}\frac{I_{-\nu}(z) - I_\nu(z)}{\sin \nu\pi} \quad (n \in \mathbb{N}) \quad . \tag{5}$$

A few examples are:

$$\begin{aligned}
J_{1/2}(z) &= \sqrt{\frac{2z}{\pi}}\frac{\sin z}{z} \quad , \\
J_{-1/2}(z) &= \sqrt{\frac{2z}{\pi}}\frac{\cos z}{z} \quad , \\
J_{3/2}(z) &= \sqrt{\frac{2z}{\pi}}\left(\frac{\sin z}{z^2} - \frac{\cos z}{z}\right) \quad , \\
J_{-3/2}(z) &= \sqrt{\frac{2z}{\pi}}\left(-\frac{\cos z}{z^2} - \frac{\sin z}{z}\right) \quad , \\
J_{5/2}(z) &= \sqrt{\frac{2z}{\pi}}\left[\left(\frac{3}{z^3} - \frac{1}{z}\right)\sin z - \frac{3}{z^2}\cos z\right] \quad , \\
J_{-5/2}(z) &= \sqrt{\frac{2z}{\pi}}\left[\left(\frac{3}{z^3} - \frac{1}{z}\right)\cos z + \frac{3}{z^2}\sin z\right] \quad .
\end{aligned} \tag{6}$$

One may express the Bessel functions $J_{n+1/2}(z)$ in terms of the *spherical Bessel functions* $j_n(z)$. The relation[24] is given by

$$j_n(z) = \sqrt{\frac{\pi}{2z}}\,J_{n+1/2}(z) \quad . \tag{7}$$

In analogy to (7) also the following relations hold for the $I_\nu(z)$ and $K_\nu(z)$:

$$\begin{aligned}
I_{1/2}(z) &= \sqrt{\frac{2z}{\pi}}\frac{\sinh z}{z} \quad , \\
I_{-1/2}(z) &= \sqrt{\frac{2z}{\pi}}\frac{\cosh z}{z} \quad , \\
I_{3/2}(z) &= \sqrt{\frac{2z}{\pi}}\left(-\frac{\sinh z}{z^2} + \frac{\cosh z}{z}\right) \quad , \\
I_{-3/2}(z) &= \sqrt{\frac{2z}{\pi}}\left(+\frac{\sinh z}{z} - \frac{\cosh z}{z^2}\right) \quad , \\
I_{5/2}(z) &= \sqrt{\frac{2z}{\pi}}\left[\left(\frac{3}{z^3} + \frac{1}{z}\right)\sinh z - \frac{3}{z^2}\cosh z\right] \quad ,
\end{aligned}$$

[24] See, e.g. G. Afken: *Mathematical Methods for Physicists*, 2nd ed. (Academic Press, New York 1970), p. 522. For a more extensive discussion of Bessel functions, see G.N. Watson: *Theory of Bessel Functions* (Cambridge University Press, Cambridge 1966).

$$I_{-5/2}(z) = \sqrt{\frac{2z}{\pi}}\left[-\frac{3}{z^3}\sinh z + \left(\frac{3}{z^3}+\frac{1}{z}\right)\cosh z\right] \quad ,$$

Example 1.21.

$$K_{1/2}(z) = \sqrt{\frac{\pi}{2z}}\,\mathrm{e}^{-z} \quad ,$$

$$K_{3/2}(z) = \sqrt{\frac{\pi}{2z}}\,\mathrm{e}^{-z}\left(1+\frac{1}{z}\right) \quad ,$$

$$K_{5/2}(z) = \sqrt{\frac{\pi}{2z}}\,\mathrm{e}^{-z}\left(1+\frac{3}{z}+\frac{3}{z^2}\right) \quad . \tag{8}$$

EXAMPLE

1.22 The Kisslinger Potential

The calculation of bound states of pions in atomic nuclei or the scattering of pions at atomic nuclei are complicated many-body problems which cannot be solved without simplifications. However, we will now derive a kind of effective potential to describe the strong interaction between the pion and the nucleus. The Lagrangian of the nonrelativistic pion-nucleon interaction (i.e. the interaction between pion and nucleon describing the processes $\pi^- + \mathrm{p} \to \mathrm{n}$ and $\pi^+ + \mathrm{n} \to \mathrm{p}$) is

$$\mathcal{L}_{\mathrm{int}} = g\overline{\psi}_{\mathrm{n}}\boldsymbol{\sigma}\cdot\boldsymbol{\nabla}\phi\psi_{\mathrm{p}} + \mathrm{h.c.} = g(\overline{\psi}_n\boldsymbol{\sigma}\psi_{\mathrm{p}})\cdot(\boldsymbol{\nabla}\phi) + \mathrm{h.c.} \quad , \tag{1}$$

where ψ_{n} and ψ_{p} are the neutron and the proton wave functions. The spin vector which is porportional to $\boldsymbol{\sigma}$ acts on the nucleon wave function, whereas the gradient acts on the pion wave function ϕ. This so-called *Chew–Low interaction* describes quite well the low-energy pion–nucleon experiments. The π^- meson (described by the wave function ϕ) creates the transition of a proton into a neutron; the Hermitian conjugate term of (1) describes the transition of a neutron into a proton by a π^+ meson (described by the wave function ϕ^*). g is the coupling constant for these reactions. The operator product $\boldsymbol{\sigma}\cdot\boldsymbol{\nabla}$ is rotationally invariant but not parity invariant. The parity invariance of the whole interaction is ensured by the pseudoscalar $\phi(\boldsymbol{r})$.

In the following we discuss the problem of how to describe the interaction between one pion and many bound nucleons, starting with this free pion–nucleon interaction (1). The simplest approximation is to describe the interaction between the pion and nucleon by the free interaction between them. This is called the *impulse approximation* and is reasonable for pions interacting with only *one* of the nucleons inside the nucleus, i.e. the other nucleons just appear in kinematic factors (e.g. altered energy–momentum balance of the nucleon). Since the π–N interaction is of short range ($\sim$ 1 fm), and the mean distance λ between the nucleons is larger than 1 fm, the impulse approximation should be applicable at least for low energies. To estimate the potential energy we assume that, on its way through the nucleus, the pion has several interactions in series, but that every interaction is only between the pion and one "free" nucleon. The model based on these assumptions is called

Example 1.22.

the *multiple scattering model.*[25] We shall now study the multiple scattering of pion and nucleus according to this model. The incoming state $|\mathrm{in}\rangle$ and the the outgoing state $|\mathrm{out}\rangle$ of the pion–nucleus system are related by the S matrix $\hat{S}$ via

$$|\mathrm{out}\rangle = \hat{S}\,|in\rangle \quad . \tag{2}$$

It is convenient to introduce a matrix $\hat{R} = \hat{S} - \mathbb{1}$, i.e. the case for no interaction ($|\mathrm{out}\rangle = |\mathrm{in}\rangle$) is subtracted. We consider the scattering of a particle with momentum $\boldsymbol{p} \to \boldsymbol{p}'$. Then, because of energy conservation, we have

$$\langle \boldsymbol{p}'|\hat{R}|\boldsymbol{p}\rangle = \delta\left(E_{p'} - E_p\right) t\left(\boldsymbol{p} \to \boldsymbol{p}'\right) \quad , \tag{3a}$$

where

$$t\left(\boldsymbol{p} \to \boldsymbol{p}'\right) = \lim_{\varepsilon \to 0} \langle \boldsymbol{p}'|\hat{T}(E_p + \mathrm{i}\varepsilon)|\boldsymbol{p}\rangle \quad . \tag{3b}$$

The T matrix (transition matrix) has a complex argument $z = E_p + \mathrm{i}\varepsilon$. To understand the intention behind this for the present, purely mathematical, manipulation, we consider the theory of Hermitian operators.

Let $\hat{H}_0 = \hat{\boldsymbol{p}}^2/2m_0$ and $\hat{H} = \hat{H}_0 + V$. The corresponding *Green's operators or resolvents* are defined by

$$\begin{aligned} \hat{G}_0(z) &= \left(z - \hat{H}_0\right)^{-1} \quad , \\ \hat{G}(z) &= \left(z - \hat{H}\right)^{-1} \quad , \end{aligned} \tag{4a}$$

provided the inverse exists. The name is easily understood since

$$\left(z - \hat{H}_0\right)\hat{G}_0(z) = \mathbb{1} \quad . \tag{4b}$$

Taking the matrix elements of $\hat{H}_0$ in coordinate space, we have, since $\hat{H}_0 = -\hbar^2\boldsymbol{\nabla}^2/2m_0$,

$$\langle \boldsymbol{x}|\hat{H}_0|\psi\rangle = -\frac{\hbar^2\boldsymbol{\nabla}^2}{2m_0}\langle \boldsymbol{x}|\psi\rangle \quad . \tag{5a}$$

Equation (4a) yields

$$\langle \boldsymbol{x}|\left(z - \hat{H}_0\right)\hat{G}_0(z)|\boldsymbol{x}'\rangle = \langle \boldsymbol{x}|\mathbb{1}|\boldsymbol{x}'\rangle \tag{5b}$$

and, after insertion of a complete set $|\boldsymbol{x}''\rangle$.

$$\int \mathrm{d}^3x'' \langle \boldsymbol{x}|\left(z - \hat{H}_0\right)|\boldsymbol{x}''\rangle \langle \boldsymbol{x}''|\hat{G}_0(z)|\boldsymbol{x}'\rangle = \delta^{(3)}(\boldsymbol{x} - \boldsymbol{x}')$$

or

$$\begin{aligned} &\int \mathrm{d}^3x'' \left(z - \hat{H}_0(\boldsymbol{x})\right) \langle \boldsymbol{x}|\boldsymbol{x}''\rangle \langle \boldsymbol{x}''|\hat{G}_0(z)|\boldsymbol{x}'\rangle \\ &\quad = \left(z - \hat{H}_0(\boldsymbol{x})\right)\langle \boldsymbol{x}|\hat{G}_0(z)|\boldsymbol{x}'\rangle = \delta^{(3)}(\boldsymbol{x} - \boldsymbol{x}') \quad . \end{aligned}$$

[25] K.M. Watson: Phys. Rev. **89**, 575 (1953).

Example 1.22.

Since z is a complex number, (5b) results, using (5a) (with $|\psi\rangle = |\boldsymbol{x}\rangle$), in

$$\left(z - \hat{H}_0\right)\left\langle \boldsymbol{x}|\hat{G}_0(z)|\boldsymbol{x}'\right\rangle = \delta^{(3)}(\boldsymbol{x}-\boldsymbol{x}')$$

and, thus,

$$\left(\frac{\hbar^2\boldsymbol{\nabla}^2}{2m_0} + z\right)\left\langle \boldsymbol{x}|\hat{G}_0(z)|\boldsymbol{x}'\right\rangle = \delta^{(3)}(\boldsymbol{x}-\boldsymbol{x}') \quad . \tag{6a}$$

In conventional notation (6a) reads

$$\left(\frac{\hbar^2\boldsymbol{\nabla}^2}{2m_0} + z\right) G_0(z\,;\boldsymbol{x},\boldsymbol{x}') = \delta^{(3)}(\boldsymbol{x}-\boldsymbol{x}') \quad . \tag{6b}$$

Thus $\hat{G}_0$ is indeed a Green's function, i.e. the solution of a differential equation where the inhomogeneity is a point-like source (δ function). From the definition (4a) we see that $\hat{G}_0$ and $\hat{G}$ do not exist for all values of z: If, say, E_n is an eigenvalue of $\hat{H}$, i.e. $(E_n - \hat{H})|n\rangle = 0$, then $\hat{G}$ does not exist for $z = E_n$! Obviously, the poles of $\hat{G}$ signal the eigenvalues of the system. Addition of an imaginary term to E, i.e. $z = E + \mathrm{i}\varepsilon$ defines the movement of the integration around the pole.[26] In particular it may be shown how, thereby, perturbations (waves) propagate into the future, thus ensuring causality. If $\hat{H}$ has a purely discrete spectrum, E_n are the eigenenergies and $\{|n\rangle\}$ the orthonormal basis of eigenvectors, then, because of $\mathbb{1} = \sum_n |n\rangle\langle n|$, we obtain

$$\hat{G}(z) = (z - \hat{H})^{-1} = \sum_n \frac{|n\rangle\langle n|}{z - E_n} = \sum_n \frac{|n\rangle\langle n|}{E - E_n + \mathrm{i}\varepsilon} \quad . \tag{7}$$

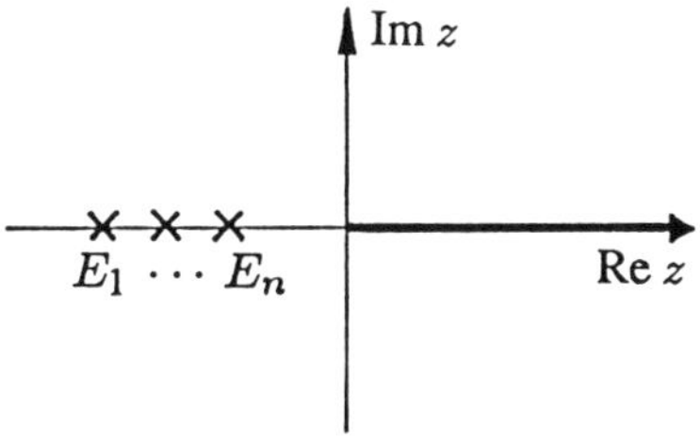

Illustration of a typical distribution of the eigenvalues of a Hamiltonian with discrete and continuous spectrum

If, in addition to the discrete spectrum, $\hat{H}$ also has a continuous spectrum, then a branch cut appears in the z plane from $E = 0$ to $E = \infty$, because every $E > 0$ is an eigenvalue. The branch cut runs along the whole z axis (see above figure). The operator identity

$$\hat{A}^{-1} = \hat{B}^{-1} + \hat{B}^{-1}(\hat{B} - \hat{A})\hat{A}^{-1} \quad , \tag{8}$$

where $\hat{A} = z - \hat{H}$ and $\hat{B} = z - \hat{H}_0$, yields

$$\hat{G}(z) = \hat{G}_0(z) + \hat{G}_0(z) V \hat{G}(z) \tag{9}$$

which relates $\hat{G}$ and $\hat{G}_0$.

[26] A discussion of this subject may be found in W. Greiner, J. Reinhardt: *Quantum Electrodynamics*, 2nd ed. (Springer, Berlin, Heidelberg 1994), Chap. 2.

Example 1.22.

If $\hat{A}$ and $\hat{B}$ are exchanged, the following relation can also be derived:

$$\hat{G}(z) = \hat{G}_0(z) + \hat{G}(z)V\hat{G}_0(z) \quad . \tag{10}$$

Equations (9) and (10) are called the *Lippmann–Schwinger equations.*

Next we define the operator

$$\hat{T}(z) = V + V\hat{G}(z)V \quad . \tag{11}$$

Multiplication with $\hat{G}_0$ from the lhs yields

$$\hat{G}_0\hat{T} = \left(\hat{G}_0 + \hat{G}_0V\hat{G}\right)V \tag{12}$$

or, because of (9),

$$\hat{G}_0(z)\hat{T}(z) = \hat{G}(z)V \quad . \tag{13}$$

Inserting this result into (11), one arrives at the Lippmann–Schwinger equation for $\hat{T}(z)$, namely,

$$\hat{T}(z) = V + V\hat{G}_0(z)\hat{T}(z) \quad . \tag{14a}$$

For sufficiently small values of V this equation can be solved iteratively, starting with the so-called *Born approximation* $\hat{T} \approx V$. Substituting this approximation into (14a) we obtain $\hat{T} \approx V + V\hat{G}_0V$. Continuation of this procedure yields the *infinite Born series* given by

$$\hat{T} = V + V\hat{G}_0V + V\hat{G}_0V\hat{G}_0V + \dots \quad . \tag{14b}$$

We should bear in mind that this series does not necessarily converge for all V ! However, further considerations of specific examples are based on the assumption of proper convergence of the Born series (14b), with which we have a method to handle the pion–nucleus problem: Let the Hamiltonian be of the form

$$\hat{H} = \hat{T}_\pi + \hat{H}_{\rm nuc} + \sum_{\alpha=1}^{A} v_\alpha = \hat{H}_0 + V \quad . \tag{15}$$

Here $\hat{T}_\pi$ is the free-pion Hamiltonian and $\hat{H}_{\rm nuc}$ describes the nucleus, which we assume to be known. The Lippmann–Schwinger equation for $\hat{T}$ is now given by (14b), using (4) for $G_0(E)$, so that

$$\hat{G}_0 = (E - \hat{H}_0 + \mathrm{i}\varepsilon)^{-1} \quad , \quad \varepsilon \to 0^+ \quad . \tag{16}$$

As we have introduced $\hat{T}$ as a scattering operator, we can write the multiple scattering solution (14b) as follows

$$\hat{T} = \sum_{\alpha}\hat{t}'_\alpha + \sum_{\alpha_1,\alpha_2}{}'\hat{t}'_{\alpha_1}\hat{G}_0\hat{t}'_{\alpha_2} + \sum_{\alpha_1,\alpha_2,\alpha_3}{}'\hat{t}'_{\alpha_1}\hat{G}_0\hat{t}'_{\alpha_2}\hat{G}_0 t'_{\alpha_3} + \dots \quad . \tag{17}$$

The first term corresponds to the scattering at a single nucleon (summed over all nucleons), the second term describes the scattering at two nucleons, etc. The notation $\sum'$ means that $\alpha_2 \neq \alpha_1$, $\alpha_3 \neq \alpha_2$, etc., so that we exclude multiple scattering at the same nucleon. These terms are already contained in the definition

Example 1.22.

of the $\hat{t}'$ matrix for the scattering of a pion at a *bound* nucleon, which, according to (14), reads

$$\hat{t}'_\alpha = v_\alpha + v_\alpha \hat{G}_0 \hat{t}'_\alpha \quad . \tag{18}$$

$\hat{t}'_\alpha$ contains the *total* nuclear Hamiltonian through $\hat{G}_0$, and the free pion–nucleon amplitude $\hat{t}_\alpha$ obeys an equivalent equation, in which $\hat{H}_{\text{nuc}}$ in $\hat{G}_0$ is replaced by a free one-nucleon Hamiltonian. Thus, $\hat{t}'_\alpha$ is a complicated many-body operator, whereas $\hat{t}_\alpha$ is not! The elastic scattering is given by

$$T_{\text{el}} = \langle 0|\hat{T}|0\rangle \quad . \tag{19}$$

Here $|0\rangle$ represents the ground state of the nucleus. In (19) the integration over the coordinates of the nucleus has to be performed. Therefore, T_{el} depends only on the coordinates of the pion. Analogously to (14), we define the *optical potential* of the pion as

$$\hat{T}_{\text{el}} = V_\pi + V_\pi \hat{G}_0 \hat{T}_{\text{el}} \tag{20}$$

for which, according to (17),

$$\hat{T}_{\text{el}} = \sum_\alpha \left(\hat{t}'_\alpha\right)_{\text{el}} + \sum_{\alpha_1,\alpha_2}{}' \left(\hat{t}'_{\alpha_1} G_0 \hat{t}'_{\alpha_2}\right)_{\text{el}} + \dots \tag{21}$$

follows. Consequently, the pion wave function obeys the one-particle equation

$$\left(\hat{T}_\pi + V_\pi\right)\phi = \hat{H}_\pi \phi = \hat{E}\phi \quad . \tag{22}$$

In principle this optical potential always exists, but in general it is energy dependent and nonlocal, as we will see in the following.

In order to proceed without calculation, we have to make a number of approximations, which we shall now discuss:

(1) The Impulse Approximation. According to our previous considerations, the scattering inside the nucleus is the same as free scattering and, therefore,

$$\hat{t}'_\alpha = \hat{t}_\alpha \quad , \tag{23}$$

i.e. the complicated many-particle operator for the bound nucleon will be replaced by the one-particle operator $\hat{t}_\alpha$.

(2) The Approximation of an Uncorrelated Nucleus. Nucleons are fermions. Thus, the wave function of the nucleus has to be antisymmetrical with respect to the coordinates of the nucleons. Instead, we use here products of the one particle wave functions which are not antisymmetrized (i.e. uncorrelated). For different nucleons ($\alpha_1 \neq \alpha_2$) this means that the nucleon, which is excited from the ground state $|0\rangle$ to the state $|n\rangle$ by $\hat{t}_{\alpha_2}$ and then propagates freely through the action of G_0, cannot be scattered back from the state $|n\rangle$ into the ground state by $\hat{t}_{\alpha_1}$, i.e.

$$\langle 0|\hat{t}_{\alpha_1}\rangle \, \hat{G}_0 \, \langle n|\hat{t}_{\alpha_2}|0\rangle = 0 \quad .$$

Example 1.22.

Only for $|n\rangle = |0\rangle$ (no scattering) is this equation not valid. At the same time this means that, for an elastic final amplitude, all the intermediate processes must be elastic too, so that the double scattering term in (21) becomes

$$\sum_{\alpha_1 \neq \alpha_2} \left(\hat{t}_{\alpha_1}\hat{G}_0\hat{t}_{\alpha_2}\right)_{\text{el}} = \sum_{\alpha_1 \neq \alpha_2} \hat{t}_{\alpha_1\,\text{el}}\hat{G}_0\hat{t}_{\alpha_2\,\text{el}} \quad . \tag{24}$$

(3) Assumption of Coherence. This assumption implies that the nucleus remains in the ground state *the whole time* during the scattering process, or, in other words, intermediate states of the excited nucleus do not exist; the assumption of an uncorrelated nucleus (24) implies for $\alpha_1 \neq \alpha_2 \neq \alpha_3 \ldots$

$$\left(\hat{t}_{\alpha_1}\hat{G}_0\hat{t}_{\alpha_2}\hat{G}_0\hat{t}_{\alpha_3\ldots}\right)_{\text{el}} = \hat{t}_{\alpha_1\,\text{el}}\hat{G}_0\hat{t}_{\alpha_2\,\text{el}}\hat{G}_0\hat{t}_{\alpha_2\,\text{el}}\hat{G}_0\hat{t}_{\alpha_3\,\text{el}}\ldots \quad . \tag{25}$$

In this sense coherence means that, though the pion scatters from nucleon 1 at the space point $\boldsymbol{x}_1$, the probability of colliding again with the same nucleon at $\boldsymbol{x}_1'$ is still given by the ground-state wave function. However, the slowly moving nucleon will stay close to $\boldsymbol{x}_1$. The probability of colliding with the nucleon twice without exciting the nucleus is very small, because the consequence would be a large momentum transfer to the nucleon in the intermediate state. Nevertheless, we also assume that for repeated scattering at the same nucleon ($\alpha_1, \alpha_2, \alpha_3 \ldots$, not all α_i are different) (25) is still valid, and

$$\left(\hat{t}_{\alpha_1}\hat{G}_0\hat{t}_{\alpha_2}\hat{G}_0\hat{t}_{\alpha_1}\right)_{\text{el}} \approx \hat{t}_{\alpha_1\,\text{el}}\hat{G}_0\hat{t}_{\alpha_2\,\text{el}}\hat{G}_0\hat{t}_{\alpha_1\,\text{el}} \quad . \tag{26}$$

According to previous statements, this assumption must be used with care, because one wouid actually expect that the backward scattering terms (connected with large momentum transfer) are negligible. With these three assumptions, (17) reads as

$$\hat{T}_{\text{el}} = \sum_{\alpha}\hat{t}_{\alpha\,\text{el}} + \sum_{\alpha_1,\alpha_2}\hat{t}_{\alpha_1\,\text{el}}\hat{G}_0\hat{t}_{\alpha_2\,\text{el}} + \sum_{\alpha_1,\alpha_2,\alpha_3}\hat{t}_{\alpha_1\,\text{el}}\hat{G}_0\hat{t}_{\alpha_2\,\text{el}}\hat{G}_0\hat{t}_{\alpha_3\,\text{el}} + \ldots \quad . \tag{27}$$

(4) Assumption of a Heavy Nucleus. If the number A of nucleons is very large, we abandon the restrictions in the sums (27) i.e. we allow $\alpha_i = \alpha_k$ (coherence). Furthermore, the number of the excited states is then very large, since the sums over α_i can separately be performed, with the result

$$\begin{aligned}\hat{T}_{\text{el}} = &\left(\sum \hat{t}_{\alpha\,\text{el}}\right) + \left(\sum \hat{t}_{\alpha\,\text{el}}\right)\hat{G}_0\left(\sum \hat{t}_{\alpha\,\text{el}}\right)\\ &+ \left(\sum \hat{t}_{\alpha\,\text{el}}\right)\hat{G}_0\left(\sum \hat{t}_{\alpha\,\text{el}}\right)\hat{G}_0\left(\sum \hat{t}_{\alpha\,\text{el}}\right) + \ldots\\ &\times\left(\sum \hat{t}_{\alpha\,\text{el}}\right) + \left(\sum \hat{t}_{\alpha\,\text{el}}\right)\hat{G}_0\hat{T}_{\text{el}} \quad .\end{aligned} \tag{28}$$

Comparing the result with the definition of V_π in (20), $\hat{T}_{\text{el}} = V_\pi + V_\pi\hat{G}_0\hat{T}_{\text{el}}$, it follows that

$$V_\pi = \sum \hat{t}_{\alpha\,\text{el}} \equiv \langle 0|\sum \hat{t}_\alpha|0\rangle \quad , \tag{29}$$

because the elastic amplitude of the ground-state expectation value is $\hat{T}_{\text{el}} = \langle 0|\hat{T}|0\rangle$. The expression (29) is called the *optical potential of the lowest order*, and is the first term of an infinite series for V_π.

Example 1.22.

(5) Absence of Recoil. The final approximation is to take the position of the nucleon after scattering a pion as unchanged (no recoil), i.e. if $\boldsymbol{r}$, $\boldsymbol{r}'$ are the pion coordinates and $\boldsymbol{R}_1$, $\boldsymbol{R}_1'$ the coordinates of the first nucleon, we assume that

$$
\begin{aligned}
&\left\langle \boldsymbol{r}', \boldsymbol{R}_1' \right| \hat{t}_1 \left| \boldsymbol{r}, \boldsymbol{R}_1 \right\rangle \\
&\quad \cong \langle \boldsymbol{r}' - \boldsymbol{R}_1 | \hat{t}_1 | \boldsymbol{r} - \boldsymbol{R}_1 \rangle \, \delta^{(3)} \left(\boldsymbol{R}_1' - \boldsymbol{R}_1 \right) \\
&\quad = \delta^{(3)} \left(\boldsymbol{R}_1' - \boldsymbol{R}_1 \right) \iint \frac{\mathrm{d}^3 k}{(2\pi)^3} \frac{\mathrm{d}^3 k'}{(2\pi)^3} \\
&\quad \times \mathrm{e}^{\mathrm{i}\boldsymbol{k}' \cdot (\boldsymbol{r} - \boldsymbol{R}_1)} \left\langle \boldsymbol{k}' | \hat{t}_1 | \boldsymbol{k} \right\rangle \mathrm{e}^{-\mathrm{i}\boldsymbol{k} \cdot (\boldsymbol{r} - \boldsymbol{R}_1)} \quad .
\end{aligned} \tag{30}
$$

Here $\langle \boldsymbol{k}' | \hat{t} | \boldsymbol{k} \rangle$ is the Fourier amplitude of the scattering of a pion at a free nucleon. From this we can now derive the optical potential of the lowest order, and, for that purpose, we integrate over R_1 and R_1' to obtain

$$
\begin{aligned}
&\langle \boldsymbol{r}' | \hat{t}_{1\,\mathrm{el}} | \boldsymbol{r} \rangle \\
&\quad = \iint \mathrm{d}^3 R_1 \, \mathrm{d}^3 R_1' \psi_1^* \left(R_1' \right) \left\langle \boldsymbol{r}', \boldsymbol{R}_1' | \hat{t}_1 | \boldsymbol{r}, \boldsymbol{R}_1 \right\rangle \psi_1 (\boldsymbol{R}_1) \\
&\quad = \iint \frac{\mathrm{d}^3 k}{(2\pi)^3} \frac{\mathrm{d}^3 k'}{(2\pi)^3} \mathrm{e}^{+\mathrm{i}\boldsymbol{k}' \cdot \boldsymbol{r}'} \left\langle \boldsymbol{k}' | \hat{t}_1 | \boldsymbol{k} \right\rangle \\
&\quad \times \mathrm{e}^{-\mathrm{i}\boldsymbol{k} \cdot \boldsymbol{r}} \int \mathrm{d}^3 R_1 |\psi_1(\boldsymbol{R}_1)|^2 \mathrm{e}^{+\mathrm{i}(\boldsymbol{k} - \boldsymbol{k}') \cdot \boldsymbol{R}_1} \quad .
\end{aligned} \tag{31}
$$

Summing over all nucleons for the calculation of

$$
\sum_\alpha \langle \boldsymbol{r}' | \hat{t}_{\alpha\,\mathrm{el}} | \boldsymbol{r} \rangle = \langle \boldsymbol{r}' | \sum_\alpha \hat{t}_{\alpha\,\mathrm{el}} | \boldsymbol{r} \rangle \quad ,
$$

$$
\begin{aligned}
&\sum_{\alpha=1}^{A} \int |\psi_\alpha(\boldsymbol{R}_\alpha)|^2 \mathrm{e}^{+\mathrm{i}(\boldsymbol{k} - \boldsymbol{k}') \cdot \boldsymbol{R}_\alpha} \mathrm{d}^3 R_\alpha \\
&\quad = A \int \varrho(\boldsymbol{R}) \, \mathrm{e}^{+\mathrm{i}(\boldsymbol{k} - \boldsymbol{k}') \cdot \boldsymbol{R}} \, \mathrm{d}^3 R = A \varrho(\boldsymbol{k} - \boldsymbol{k}') \quad .
\end{aligned} \tag{32}
$$

Here $\varrho(R)$ is the nuclear density distribution per nucleon (therefore, the factor A for the number of nucleons), and $\varrho(\boldsymbol{k} - \boldsymbol{k}')$ its Fourier transform (form factor). Thereby, results the generally nonlocal potential according to (29),

$$
\begin{aligned}
V_\pi(\boldsymbol{r}, \boldsymbol{r}') &= \langle \boldsymbol{r}' | V_\pi | \boldsymbol{r} \rangle = \langle \boldsymbol{r}' | \sum_\alpha \hat{t}_{\alpha\,\mathrm{el}} | \boldsymbol{r} \rangle \\
&= A \iint \frac{\mathrm{d}^3 k}{(2\pi)^3} \frac{\mathrm{d}^3 k'}{(2\pi)^3} \mathrm{e}^{+\mathrm{i}\boldsymbol{k}' \cdot \boldsymbol{r}'} \varrho(\boldsymbol{k} - \boldsymbol{k}') \langle \boldsymbol{k}' | \hat{t} | \boldsymbol{k} \rangle \mathrm{e}^{-\mathrm{i}\boldsymbol{k} \cdot \boldsymbol{r}} \\
&= \iint \frac{\mathrm{d}^3 k}{(2\pi)^3} \frac{\mathrm{d}^3 k'}{(2\pi)^3} \mathrm{e}^{+\mathrm{i}\boldsymbol{k}' \cdot \boldsymbol{r}'} \langle \boldsymbol{k}' | V_\pi | \boldsymbol{k} \rangle \mathrm{e}^{-\mathrm{i}\boldsymbol{k} \cdot \boldsymbol{r}} \quad .
\end{aligned} \tag{33}
$$

Conversely, in momentum representation it clearly follows that

$$
\langle \boldsymbol{k}' | V_\pi | \boldsymbol{k} \rangle = A \varrho(\boldsymbol{k} - \boldsymbol{k}') \langle \boldsymbol{k}' | \hat{t} | \boldsymbol{k} \rangle \quad . \tag{34}
$$

For a better understanding of this result, first consider the simplest case: Let the incoming pion wave function be a plane wave with the momentum $\boldsymbol{k}_0$. Then the total wave function within the nucleus is a wave packet, centred at $\boldsymbol{k}_0$, which

Example 1.22.

we call $\psi_{k_0}(\boldsymbol{k})$. The form factor of the *nucleus* $A\varrho(\boldsymbol{k}-\boldsymbol{k}')$ has an extension of $\approx 1/R_{\text{nucleus}}$ in momentum space, while the scattering amplitude for scattering at one nucleon is $\langle \boldsymbol{k}'|\hat{t}|\boldsymbol{k}\rangle \approx 1/R_{\text{nucleon}}$. By multiplication of $\psi_{k_0}(\boldsymbol{k})$ with $\varrho(\boldsymbol{k}-\boldsymbol{k}')$, whose width is given by $1/R_{\text{nucleus}} \ll 1/R_{\text{nucleon}}$, only the momenta $k \sim k' \sim k_0$ contribute to the scattering amplitude, because the pion wave function $\psi_{k_0}(\boldsymbol{k})$ is centred at $\boldsymbol{k} \approx \boldsymbol{k}_0$. Therefore, in (34) we can put

$$\langle \boldsymbol{k}'|V_\pi|\boldsymbol{k}\rangle \cong A\langle \boldsymbol{k}_0|\hat{t}|\boldsymbol{k}_0\rangle \varrho(\boldsymbol{k}-\boldsymbol{k}') \quad . \tag{35a}$$

According to (33), in coordinate space the Fourier-transformed expression follows as

$$\begin{aligned}
V_\pi &= A\langle \boldsymbol{k}_0|\hat{t}|\boldsymbol{k}_0\rangle \iint \frac{\mathrm{d}^3k}{(2\pi)^3}\frac{\mathrm{d}^3k'}{(2\pi)^3}\,\mathrm{e}^{+\mathrm{i}\boldsymbol{k}'\cdot\boldsymbol{r}'}\,\mathrm{e}^{-\mathrm{i}\boldsymbol{k}\cdot\boldsymbol{r}}\,\varrho(\boldsymbol{k}-\boldsymbol{k}')\\
&= A\langle \boldsymbol{k}_0|\hat{t}|\boldsymbol{k}_0\rangle \iint \frac{\mathrm{d}^3k}{(2\pi)^3}\frac{\mathrm{d}^3k'}{(2\pi)^3}\,\mathrm{e}^{+\mathrm{i}\boldsymbol{k}'\cdot(\boldsymbol{r}-\boldsymbol{r}')}\,\mathrm{e}^{-\mathrm{i}(\boldsymbol{k}-\boldsymbol{k}')\cdot\boldsymbol{r}}\,\varrho(\boldsymbol{k}-\boldsymbol{k}')\\
&= A\langle \boldsymbol{k}_0|\hat{t}|\boldsymbol{k}_0\rangle \int \frac{\mathrm{d}^3k'}{(2\pi)^3}\,\mathrm{e}^{+\mathrm{i}\boldsymbol{k}'\cdot(\boldsymbol{r}-\boldsymbol{r}')}\int \frac{\mathrm{d}^3K}{(2\pi)^3}\,\mathrm{e}^{-\mathrm{i}\boldsymbol{K}\cdot\boldsymbol{r}}\,\varrho(\boldsymbol{K})\\
&= A\langle \boldsymbol{k}_0|\hat{t}|\boldsymbol{k}_0\rangle \delta(\boldsymbol{r}-\boldsymbol{r}')\varrho(\boldsymbol{r}) \quad .
\end{aligned} \tag{35b}$$

The nonlocal potential acts on the pion wave function $\psi_\pi(\boldsymbol{r}')$ in coordinate space so that, e.g. the stationary Schrödinger equation for nonrelativistic pions reads

$$-\frac{\hbar^2}{2m_0}\boldsymbol{\nabla}^2\psi_\pi(\boldsymbol{r}) + \int \mathrm{d}^3r'\,V_\pi(\boldsymbol{r},\boldsymbol{r}')\psi_\pi(\boldsymbol{r}') = E\psi_\pi(\boldsymbol{r}) \quad , \tag{36a}$$

or the Klein–Gordon equation,

$$\left(\frac{\hbar^2}{c^2}\frac{\partial^2}{\partial t^2} - \hbar^2\boldsymbol{\nabla}^2 + m_0^2c^2\right)\psi(\boldsymbol{r},t) + m_0\int \mathrm{d}^3r'\,V_\pi(\boldsymbol{r},\boldsymbol{r}')\psi(\boldsymbol{r}',t) = 0 \quad . \tag{36b}$$

Clearly, the simple approximation (35a) has the effect that the generally nonlocal potential of the pion–nucleus interaction is approximated by a *local* one. The δ function in (35b) naturally ensures this. Furthermore, the $\delta(\boldsymbol{r}-\boldsymbol{r}')$ function disappears by integrating over $\boldsymbol{r}'$, which has to be done in (36a,b). The result is a "common" Schrödinger or Klein–Gordon equation, with a local potential that reads

$$V_\pi(\boldsymbol{r}) = A\langle \boldsymbol{k}_0|\hat{t}|\boldsymbol{k}_0\rangle \varrho(\boldsymbol{r}) \quad . \tag{37}$$

The resulting potential is local, and proportional to the so-called pion–nucleon *forward scattering amplitude*

$$a_0 = \langle \boldsymbol{k}_0|\hat{t}|\boldsymbol{k}_0\rangle \quad . \tag{38a}$$

However, the potential $V_\pi(r) = Aa_0\varrho(\boldsymbol{r})$ gives an insufficient description of the pion–nucleus scattering experiments and it is a small step to assume, analogously to (38a),[27] that

$$\langle \boldsymbol{k}'|\hat{t}|\boldsymbol{k}\rangle = a_0 + a_1(\boldsymbol{k}'\cdot\boldsymbol{k}) \quad . \tag{38b}$$

---

[27] L.S. Kisslinger: Phys. Rev. **98**, 761 (1955).

Example 1.22.

This is the next best generalization to (38a), keeping rotational invariance. Insertion into (34) results (after Fourier transformation into coordinate space, see Exercise 1.23) in

$$V_\pi^k \psi = A a_0 \varrho \psi - A a_1 \boldsymbol{\nabla} \cdot (\varrho \boldsymbol{\nabla} \psi) \quad , \tag{39}$$

which represents the so-called *Kisslinger potential.* It is obviously nonlocal because it contains momentum operators which are proportional to $\boldsymbol{\nabla}$. In view of a term proportional to $\boldsymbol{\nabla}\varrho$ one may deduce a strong surface sensitivity in pion–nucleon scattering, since the gradient of the nuclear density contributes mainly at the surface of the nucleus. By means of the Kisslinger potential most of the data of pion–nucleus scattering are quite well described, although further evaluation is still plagued with certain difficulties (divergencies in the wave functions).

EXERCISE

1.23 Evaluation of the Kisslinger Potential in Coordinate Space

Problem. Evaluate the Kisslinger potential in coordinate space, starting with

$$\langle \boldsymbol{k}' | V_\pi | \boldsymbol{k} \rangle = A \varrho(\boldsymbol{k} - \boldsymbol{k}') \langle \boldsymbol{k}' | \hat{t} | \boldsymbol{k} \rangle \quad ,$$

together with (see Example 1.22)

$$\langle \boldsymbol{k}' | \hat{t} | \boldsymbol{k} \rangle = a_0 + a_1 (\boldsymbol{k}' \cdot \boldsymbol{k}) \quad .$$

Solution. In coordinate space one has

$$V \psi_\pi = A \int (a_0 + a_1 \boldsymbol{k}' \cdot \boldsymbol{k}) \varrho(z_\alpha) \psi_\pi(z') \mathrm{e}^{\mathrm{i}\boldsymbol{k}' \cdot (\boldsymbol{z} - \boldsymbol{z}_\alpha)} \times \mathrm{e}^{-\mathrm{i}\boldsymbol{k} \cdot (z' - z_\alpha)} \frac{\mathrm{d}^3 k}{(2\pi)^3} \frac{\mathrm{d}^3 k'}{(2\pi)^3} \mathrm{d}^3 z' \, \mathrm{d}^3 z_\alpha \quad ,$$

where z_α denote the space vectors of a single nucleon [see (31) and (32) of Example 1.22]. Evaluating the first term a_0 leads to (37) of the previous example. Considering the second term we reexpress the momenta $\boldsymbol{k}$, $\boldsymbol{k}'$ by $\boldsymbol{k} = \mathrm{i}\boldsymbol{\nabla}_{z'}$, and $\boldsymbol{k}' = -\mathrm{i}\boldsymbol{\nabla}_z$, where the gradients act on $\boldsymbol{z}$ and $\boldsymbol{z}'$, respectively:

$$V' \psi_\pi = A a_1 \int \varrho(\boldsymbol{z}_\alpha) \psi_\pi(\boldsymbol{z}') \boldsymbol{\nabla}_z \mathrm{e}^{\mathrm{i}\boldsymbol{k}' \cdot (\boldsymbol{z} - \boldsymbol{z}_\alpha)} \cdot \boldsymbol{\nabla}_{z'} \mathrm{e}^{-\mathrm{i}\boldsymbol{k} \cdot (\boldsymbol{z}' - \boldsymbol{z}_\alpha)} \frac{\mathrm{d}^3 k}{(2\pi)^3} \frac{\mathrm{d}^3 k'}{(2\pi)^3} \mathrm{d}^3 z' \, \mathrm{d}^3 z_\alpha \quad .$$

Since there is no integration over $\boldsymbol{z}$, we can take $\boldsymbol{\nabla}_z$ out of the integrand and substitute $\boldsymbol{\nabla}_{z'} = -\boldsymbol{\nabla}_{z_\alpha}$,

$$V' \psi_\pi = A a_1 \boldsymbol{\nabla}_z \cdot \int \varrho(\boldsymbol{z}_\alpha) \psi_\pi(\boldsymbol{z}') \mathrm{e}^{\mathrm{i}\boldsymbol{k}' \cdot (\boldsymbol{z} - \boldsymbol{z}_\alpha)} \times \left(-\boldsymbol{\nabla}_{z_\alpha}\right) \mathrm{e}^{-\mathrm{i}\boldsymbol{k} \cdot (\boldsymbol{z}' - \boldsymbol{z}_\alpha)} \frac{\mathrm{d}^3 k}{(2\pi)^3} \frac{\mathrm{d}^3 k'}{(2\pi)^3} \mathrm{d}^3 z' \, \mathrm{d}^3 z_\alpha \quad .$$

Exercise 1.23.

Evaluating the integral over $\boldsymbol{k}'$,

$$V'\psi_\pi = Aa_1 \boldsymbol{\nabla}_z \cdot \int \varrho(z_\alpha)\psi_\pi(z')(-1)\delta(\boldsymbol{z}-\boldsymbol{z}_\alpha)\left(-\boldsymbol{\nabla}_{z_\alpha}\right) \times \mathrm{e}^{-\mathrm{i}\boldsymbol{k}\cdot(\boldsymbol{z}'-\boldsymbol{z}_\alpha)}\frac{\mathrm{d}^3k}{(2\pi)^3}\,\mathrm{d}^3z'\,\mathrm{d}^3z_\alpha$$

$$= Aa_1 \boldsymbol{\nabla}_z \cdot \int \varrho(\boldsymbol{z})\psi_\pi(\boldsymbol{z}')(+\boldsymbol{\nabla}_z)\,\mathrm{e}^{-\mathrm{i}\boldsymbol{k}\cdot(\boldsymbol{z}'-\boldsymbol{z})}\frac{\mathrm{d}^3k}{(2\pi)^3}\,\mathrm{d}^3z'$$

$$= Aa_1 \boldsymbol{\nabla}_z \cdot \left(\varrho(\boldsymbol{z})\int \psi_\pi(\boldsymbol{z}')(+\boldsymbol{\nabla}_z)\,\mathrm{e}^{-\mathrm{i}\boldsymbol{k}\cdot(\boldsymbol{z}'-\boldsymbol{z})}\frac{\mathrm{d}^3k}{(2\pi)^3}\,\mathrm{d}^3z'\right) \quad .$$

Next, performing the integration over $\boldsymbol{k}$, one obtains

$$V'\psi_\pi = Aa_1 \boldsymbol{\nabla}_z \cdot \left(\varrho(\boldsymbol{z})\int \psi_\pi(\boldsymbol{z}')(+\boldsymbol{\nabla}_z)\delta(\boldsymbol{z}-\boldsymbol{z}')\,\mathrm{d}^3z'\right) \quad ,$$

and, after partial integration,

$$V'\psi_\pi = a_1 A \boldsymbol{\nabla}_z \cdot \left(\varrho(\boldsymbol{z})\left\{+\int_{\text{surface}} \psi_\pi(\boldsymbol{z}')\delta(\boldsymbol{z}-\boldsymbol{z}')\,\mathrm{d}^3z' - \int \left(\boldsymbol{\nabla}\psi_\pi(\boldsymbol{z}')\right)\delta(\boldsymbol{z}-\boldsymbol{z}')\,\mathrm{d}^3z'\right\}\right) \quad .$$

The surface term vanishes if the surface tends to infinity, and we are left with

$$V'\psi_\pi = -a_1 A \boldsymbol{\nabla}\cdot(\varrho\boldsymbol{\nabla}\psi_\pi) \quad ,$$

i. e. altogether

$$V\psi_\pi = Aa_0\varrho(\boldsymbol{r})\psi_\pi(\boldsymbol{r}) - Aa_1\boldsymbol{\nabla}\cdot(\varrho(\boldsymbol{r})\boldsymbol{\nabla}\psi_\pi(\boldsymbol{r})) \quad .$$

EXAMPLE

1.24 Lorentz–Lorenz Effect in Electrodynamics and Its Analogy in Pion–Nucleon Scattering (the Ericson–Ericson Correction)

In order to understand our motivation and the physical nature here, let us first recall the electrodynamical Lorentz–Lorenz effect. For this reason we consider the propagation of light in a dielectric medium. The electric polarization $\boldsymbol{P}$ is related to the electric field $\boldsymbol{E}$ via $\boldsymbol{E} = \alpha\boldsymbol{P}$, α being the polarizibility. Accordingly, the $\boldsymbol{D}$ field is given by $\boldsymbol{D} = \boldsymbol{E} + 4\pi\boldsymbol{P} = (1 + 4\pi/\alpha)\boldsymbol{E}$. Considering the propagation along the z axis only, the fields $\boldsymbol{E}$, $\boldsymbol{D}$ and $\boldsymbol{H}$ are proportional to $\exp[\mathrm{i}(qz - \omega t)]$. The fields $\boldsymbol{E}$ and $\boldsymbol{D}$ have only the components E_x and D_x, i. e. the magnetic field consists only of the component H_y. Maxwell's equations

$$\frac{1}{c}\frac{\partial D}{\partial t} = \nabla \times \hat{H} \quad \text{and} \tag{1}$$

Example 1.24.

$$\nabla \times \boldsymbol{E} = -\frac{1}{c}\frac{\partial \boldsymbol{H}}{\partial t}$$

imply

$$\begin{aligned} -\mathrm{i}\frac{\omega}{c}D_x &= -\mathrm{i}qH_y \quad , \\ \mathrm{i}qE_x &= \mathrm{i}\frac{\omega}{c}H_x \quad , \end{aligned} \tag{2}$$

and elimination of the component H_y produces

$$D_x = c^2\frac{q^2}{\omega^2}E_x \quad . \tag{3}$$

Using the relation $D_x = (1 + 4\pi/\alpha)E_x$ one thus obtains the refractive index as

$$n^2 = \frac{c^2}{v^2} = \varepsilon = 1 + \frac{4\pi}{\alpha} \quad . \tag{4}$$

Also the phase velocity in a medium $v = \omega/q$ is obtained from (3). In order to determine α we must have a closer look at the details of the scattering process. We choose electrons as the centres of the scattering process, assuming that they are harmonically bound by means of a restoring force f. The classical equation of motion of an electron moving in the z direction thus reads

$$m_0\frac{\mathrm{d}^2\xi}{\mathrm{d}t^2} = -f\xi \quad , \tag{5}$$

where ξ denotes the elongation of the electron from its equilibrium position. The frequency of this harmonic oscillator is then given by $\omega_R^2 = f/m_0$, and the electric dipole moment per electron, $p_x = e\xi$, leads to a moment per unit volume of

$$P_x = e\varrho\xi \quad , \tag{6}$$

where ϱ stands for the electric charge density of the electron. Under the influence of the external oscillating field $\boldsymbol{E} = \alpha\boldsymbol{P}$, the electron performs forced oscillations, i. e. the total force acting on the electron is not given by the external field $\boldsymbol{E}$ alone, but by the resulting *local* field[28] $\boldsymbol{E} + (4\pi/3)\boldsymbol{P}$. This local field is the origin of the Lorentz–Lorenz correction!

To make this more evident let us assume that the oscillating electron be surrounded by a small sphere (see figure on the right). The surface cuts some polarization vectors, which causes a charge per surface element $\mathrm{d}\sigma$. The surface charge can be written as $\varrho_\sigma = \boldsymbol{P} \cdot \mathrm{d}\boldsymbol{\sigma}$, since $\nabla \cdot \boldsymbol{D} = 0$ is valid if real external charges are absent. One can conclude, in accord with Gauß's law, that the surface charges on the inner and outer side of the sphere are equal, i. e. $(\boldsymbol{E} \cdot \mathrm{d}\boldsymbol{S})_{\text{inside}} = [(\boldsymbol{E} + 4\pi\boldsymbol{P}) \cdot \mathrm{d}\boldsymbol{S}]_{\text{outside}}$. Averaging over all directions $\overline{P\cos^2\theta} = P/3$ leads to $(E_x)_{\text{inside}} = (E_x)_{\text{outside}} + (4\pi/3)P_x$ in the x direction. Accordingly, the equation of motion of the electron becomes

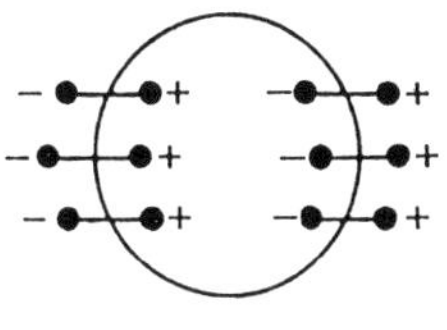

Illustration of the polarization in the surrounding of an electron under consideration

$$m_0\,\mathrm{d}^2\xi/\mathrm{d}t^2 = e(E_z + (4\pi/3)P_x)f\xi \quad ,$$

[28] See J.D. Jackson: *Classical Electrodynamics*, 2nd ed. (Wiley, New York 1975).

Example 1.24.

which can be written in a more convenient form using the relation $P_x = e\varrho\xi$ [see (6)]:

$$\frac{m_0}{\varrho e^2}\frac{d^2 P_x}{dt^2} = E_x + \frac{4\pi}{3}\pi P_x - \frac{f}{\varrho e^2}P_x \quad . \tag{7}$$

The requirement that the electrons should oscillate harmonically, i. e.

$$\ddot{P}_x + \omega^2 P_x = 0 \quad ,$$

leads to the condition

$$E_x = \left(\frac{f}{\varrho e^2} - \frac{m_0\omega^3}{\varrho e^3} - \frac{4\pi}{3}\right)P_x = \alpha P_x$$

which determines the polarizability α as

$$\alpha = \left(\frac{f}{\varrho e^2} - \frac{m_0\omega^2}{\varrho e^2} - \frac{4\pi}{3}\right) = \left(\frac{m_0\omega_R^2}{\varrho e^2} - \frac{m_0\omega^2}{\varrho e^2} - \frac{4\pi}{3}\right) \quad . \tag{8}$$

For the harmonic case the frequency is given by $\omega_R^2 = f/m_0$. Considering $n^2 = 1 + 4\pi/\alpha$ we, further, obtain

$$n^2 - 1 = \frac{4\pi\varrho e^2/m_0}{\left(\omega_R^2 - \omega^2\right) - (4\pi/3)\varrho e^2/m_0} \quad . \tag{9}$$

In the above derivation the radiative damping, which leads to a complex refractive index, has been neglected. If polarization corrections [2nd terms in the denominator of (9)] are ignored, the refractive index diverges when the frequency of the external field reaches the resonance frequency (resonance catastrophe), i. e. $n^2 \to \infty$ for $\omega \to \omega_R$. Thus the Lorentz–Lorenz correction prevents this catastrophe for $\omega = \omega_R$.

Relation (9) is of the form used in optics,

$$n^2 - 1 = -\frac{4\pi\varrho}{k_0^2}f(k,\omega) \quad , \tag{10}$$

where $f(k,\omega)$ represents the forward-scattering amplitude of particles with momentum k and frequency ω propagating in a medium which is characterized by a density of scattering centers ϱ. The momentum of the incoming particle (light) is denoted by k_0. This formulation suggests, in analogy to light scattering inside a medium with density ϱ, the consideration of the scattering of a pion (initial momentum k_0) at a nucleus with density ϱ by means of optical methods.

In analogy to the Lorentz–Lorenz correction $-4\pi\varrho e^2/3m_0^2$ in (9), let us now derive the higher-order correction in ϱ of the "optical" potential of a pion.[29] For this purpose we assume that a nucleon – as a scatterer – pushes off all other nucleons in its vicinity, with the effect that, in a sphere with radius R enclosing the particular nucleon, there is no other nuclear matter in it (see figure on the left). In what follows we have to show, as in the electrostatic case, that the result is independent of R. Therefore, we first rewrite the Klein–Gordon equation for the pion containing the Kisslinger potential ($\partial/\partial t \to \mathrm{i}\omega$),

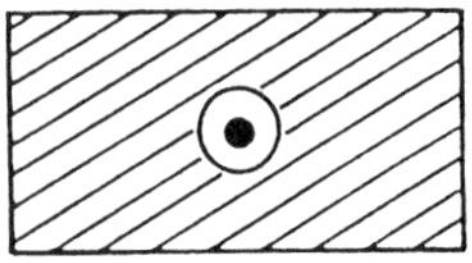

Nucleon surrounded by a sphere free of nuclear matter

[29] G.E. Brown, W. Weise: Phys. Rep. **22**, 279 (1975).

Example 1.24.

$$\left(\omega^2 - m_\pi^2 + \boldsymbol{\nabla}^2 - \overline{a}_1 \boldsymbol{\nabla} \cdot \varrho \boldsymbol{\nabla}\right) \phi_\pi = 0 \quad (\alpha > 0) \quad , \tag{11}$$

where $\overline{a}_1 = A a_1$ according to (39) of example (1.22). The first, less important term of the Kisslinger potential (39) has been neglected altogether. This can be expressed as

$$\boldsymbol{\nabla} \cdot (1 - \overline{a}_1 \varrho) \boldsymbol{\nabla} \phi_\pi = -\left(\omega^2 - m_\pi^2\right) \phi_\pi \quad . \tag{12}$$

The field ϕ_π must be continuous on each boundary. Thus, we can conclude that the rhs of (12) must be continuous at a boundary. In analogy to the continuity of $\boldsymbol{D}$ in the electromagnetic case, the normal component of

$$(1 - \overline{a}_1 \varrho) \boldsymbol{\nabla} \phi_\pi \tag{13}$$

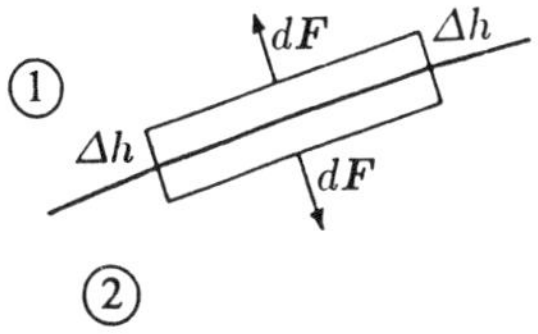

Illustration of the cylinder with faces $\mathrm{d}\boldsymbol{F}$ and height Δh surrounding the boundary. The different media at both sides are indicated by (1) and (2) respectively

must also be continuous. To make this explicit we enclose each boundary by a (cylindrical) volume, as in the figure on the right. In the limiting case of the vanishing height of the cylinder, i. e. $\Delta h \to 0$,

$$\int_r \boldsymbol{\nabla} \cdot (1 - \overline{a}_1 \varrho) \boldsymbol{\nabla} \phi_\pi \, \mathrm{d}\tau = (1 - \overline{a}_1 \varrho) \int \boldsymbol{\nabla} \phi_\pi \cdot \mathrm{d}\boldsymbol{F} \to 0 \quad . \tag{14}$$

Since the rhs of (12) is continuous at the boundary, the expression $(1 - \overline{a}_1 \varrho) \boldsymbol{\nabla} \phi_\pi \cdot \mathrm{d}\boldsymbol{F}$ has to be equal on both sides of the boundary. There exists a direct analogy between

$$(1 - \overline{a}_1 \varrho) \boldsymbol{\nabla} \phi_\pi \to \boldsymbol{D} \quad \text{and}$$

$$\boldsymbol{\nabla} \phi_\pi \to \boldsymbol{E} \quad . \tag{15}$$

In order to pursue this further, we discuss the Chew–Low model, which is based on the Lagrangian [see Example 1.22, (1)]

$$\mathcal{L}_{\text{int}} = g \overline{\psi} \boldsymbol{\sigma} \cdot \boldsymbol{\nabla} \psi \phi_\pi \quad . \tag{16}$$

For the pion it leads to the equation of motion

$$\left(-\frac{\partial^2}{\partial t^2} + \boldsymbol{\nabla}^2 - m_\pi^2\right) \phi_\pi(\boldsymbol{r}, t) = g \boldsymbol{\nabla} \cdot \boldsymbol{\sigma}(\boldsymbol{r}) \tag{17}$$

where

$$\boldsymbol{\sigma}(\boldsymbol{r}) = \overline{\psi} \boldsymbol{\sigma} \psi(\boldsymbol{r}) \quad , \quad (\hbar = c = 1) \quad .$$

Accordingly, the pion field couples to the spin density $\boldsymbol{\sigma}(\boldsymbol{r})$, and the ϕ_π field should be stationary, i. e.

$$\phi_\pi(\boldsymbol{r}, t) = \phi_\pi(\boldsymbol{r}) \mathrm{e}^{-\mathrm{i}\omega t} \quad . \tag{18}$$

We further assume that $\omega \sim m_\pi$, i. e. we study the case where nuclear effects may represent only a small correction to the binding. This is realized in pionic atoms; thus,

$$\frac{\partial^2}{\partial t^2} \phi_\pi(r, t) = (-\mathrm{i} m_\pi)^2 \phi_\pi(r, t) = -m_\pi^2 \phi_\pi(r, t) \quad .$$

Example 1.24.

Inserted into (17) this leads to the Laplace equation

$$\Delta\phi_\pi = g\boldsymbol{\nabla}\cdot\boldsymbol{\sigma}(\boldsymbol{r}) \quad . \tag{17a}$$

As known from electrostatics,[30] the Laplace equation (17a) can be solved by means of the Green function with the result

$$\phi_\pi(\boldsymbol{r}) = -\frac{g}{4\pi}\int\frac{1}{|\boldsymbol{r}-\boldsymbol{r}'|}\boldsymbol{\nabla}_{r'}\cdot\sigma(\boldsymbol{r}')\,\mathrm{d}^3r' \quad . \tag{19}$$

$\boldsymbol{\nabla}_{r'}$ is the gradient acting on $\boldsymbol{r}'$. This shows explicitly that $(g/4\pi)\boldsymbol{\nabla}\cdot\boldsymbol{\sigma}(\boldsymbol{r}')$ plays the role of a charge density as in the electrostatic case. Analogously to the electric field $\boldsymbol{E}$, (14) yields

$$(\boldsymbol{\nabla}\phi_\pi - \overline{a}_1\varrho\boldsymbol{\nabla}\phi_\pi)_{\mathrm{exterior}}\cdot\mathrm{d}\boldsymbol{F} = (\boldsymbol{\nabla}\phi_\pi)_{\mathrm{interior}}\cdot\mathrm{d}\boldsymbol{F} \quad , \tag{20}$$

which implies a surface charge

$$\varrho_{\mathrm{s}} = -\overline{a}_1\varrho(\boldsymbol{\nabla}\phi_{\pi\perp})_{\mathrm{exterior}} \quad . \tag{21}$$

$\boldsymbol{\nabla}\phi_{\pi\perp}$ represents the component of $\boldsymbol{\nabla}\phi_\pi$ perpendicular to the boundary and gives rise to a contribution to $\boldsymbol{\nabla}\phi_\pi(\boldsymbol{r}\in$ surface), which we denote by $(\boldsymbol{\nabla}\phi_\pi)_{\mathrm{local}}$,

$$\delta(\boldsymbol{\nabla}\phi_\pi)_{\mathrm{local}} \equiv \langle\boldsymbol{\nabla}\phi_{\pi_{\mathrm{interior}}} - \boldsymbol{\nabla}\phi_{\pi_{\mathrm{exterior}}}\rangle = -\frac{1}{3}\overline{a}_1\varrho(\boldsymbol{\nabla}\phi_\pi)_{\mathrm{exterior}} \quad , \tag{22}$$

where the averaging $\langle\ \rangle$ over all angles, i.e. over the small sphere around the nucleon, has already been carried out. This term corresponds to the $(4\pi/3)\boldsymbol{P}$ term in electrostatics. Equation (22) is correct to the lowest order in $\overline{a}_1\varrho$. However, note that if $(\boldsymbol{\nabla}\phi_\pi)_{\mathrm{local}}$ has to be corrected by the term (22), then the same has to be done for the Kisslinger term $-\overline{a}_1\varrho\boldsymbol{\nabla}\phi_\pi$, i.e. we must write

$$\delta(\boldsymbol{\nabla}\phi_\pi)_{\mathrm{local}} = -\frac{1}{3}\overline{a}_1\varrho(\boldsymbol{\nabla}\phi_\pi)_{\mathrm{local}} \quad . \tag{23}$$

The influence of the nucleon on the optical potential of the pion happens locally at the nucleon. From (23),

$$(\boldsymbol{\nabla}\phi_\pi)_{\mathrm{local}} = (\boldsymbol{\nabla}\phi_\pi) - \frac{1}{3}\overline{a}_1\varrho(\boldsymbol{\nabla}\phi_\pi)_{\mathrm{local}} \tag{24}$$

and thus

$$(\boldsymbol{\nabla}\phi_\pi)_{\mathrm{local}} = \frac{(\boldsymbol{\nabla}\phi_\pi)}{1+\overline{a}_1\varrho/3} \quad . \tag{25}$$

Here we have renamed $(\boldsymbol{\nabla}\phi_\pi)_{\mathrm{exterior}} = \boldsymbol{\nabla}\phi_\pi$, because this is just the pion field inside a medium. Furthermore the corrections of the correction, etc. have been summed up in terms of a geometrical series. The expression (25) represents the Ericson–Ericson correction[31] to the gradient of the pion field. Replacing $(\boldsymbol{\nabla}\phi_\pi)_{\mathrm{Kisslinger}}$ by $(\boldsymbol{\nabla}\phi_\pi)_{\mathrm{local}}$ in the Kisslinger potential leads to

[30] See J.D. Jackson: *Classical Electrodynamics*, 2nd ed. (Wiley, New York 1975) or W. Greiner: *Theoretische Physik III, Klassische Elektrodynamik* (Harri Deutsch, Frankfurt 1989), Chap. 1.

[31] M. Ericson, T.E.O. Ericson: Ann. of Physics **36**, 323 (1966).

$$V_{\pi}^{\text{K.-E.-E.}}\phi_{\pi}(r,t) = A\left(a_0\varrho - \nabla\cdot\left\{\frac{a_1\varrho}{1+\overline{a}_1\varrho/3}\nabla\right\}\right)\phi_{\pi}(r,t) \qquad (26)$$

Example 1.24.

as the new optical potential for pions inside a nucleus; and thereby, $\overline{a}_1 = Aa_1$ is proportional to a_1. A more detailed analysis shows that the potential (26) is valid only in the limiting case $k \to 0$, because in the above derivation we restricted ω of $\omega \sim m_{\pi}$. The dependence of ω and k on a_0 is quite complicated, and for this reason $\overline{a}_1$ is fitted to experimental data. Until now no unique fit has been found,[32] because the corresponding experimental effects represent very small corrections.

1.13 Biographical Notes

KLEIN, Oskar Benjamin, Swedish physicist, * 15.9.1894, † 1984. Professor for Theoretical Physics at the University Stockholm from 1931 to 1962. Important achievements apart from the formulation of the Klein–Gordon equation: Klein's paradox (see Chap. 13), Klein–Nishina formula, Kaluza–Klein theory. In 1960 he was awarded with the Max-Planck Medaille of the Deutsche Physikalische Gesellschaft (German Physical Society).

GORDON, Walter, * 3.8.1893 in Apolda (Thüringen, Germany), † 24.12.1939 in Stockholm, formulated in 1926 independently from O. Klein a relativistic wave equation for free particle without spin (Klein–Gordon equation) and derived in 1928 simultaneously with C. G. Darwin the fine structure of the hydrogen spectrum from Dirac's wave equation [BR].

LAGRANGE, Joseph Louis, * 25.01.1736 in Torino, † 10.04.1813 in Paris. L. originated from a French-Italian family, and in 1755 became professor in Torino. In the year 1766 he went to Berlin as the director of the mathematical-physics department. In 1786, after the death of Frederick II, he went to Paris, where he gave considerable support to the reform of the measuring system, and where he was professor at several universities. His very extensive work contains a new foundation of variational calculus (1760) and its application on dynamics, contributions to the three-body problem (1772), the application of the theory of chain fractions on the solution of equations (1767), number-theoretical problems, and an unsuccessful reduction of infinitesimal calculus on algebra. With his "Mécanique Analytique" (1788), L. became the initiator of analytic mechanics. Important for function theory is his "Théorie des Fonctions Analytiques, Contenant les Principes du Calcul Différentiel" (1789), and for algebra his "Traité de la Résolution des Equations Numériques de tous Degrés" (1798).

SCHRÖDINGER, Erwin, * 08.12.1887 in Vienna, † 01.04.1961 ibidem, professor in Zürich, Berlin, Oxford, Graz, Dublin and Vienna; worked in theoretical physics, especially on quantum theory. Based on L. de Broglie's idea of matter waves. S. developed the wave mechanics in 1926, established the wave equation named after him and proved the equivalence between the latter and the matrix mechanics of Heisenberg. With these principles S. created a homogenous base of quantum and atomic theory. In fact, S. found first the relativistic wave equation known today as Klein–Gordon equation, but he dismissed it. Later S. worked on problems of relativistic quantum theory, gravitation theory and on a new field theory. He also was concerned with philosophical questions. In 1933 he received the Nobel Prize in physics, together with P. A. M. Dirac.

[32] E. Friedman, A. Gal, V.B. Mandelzweig: Phys. Rev. Lett. **41**, 794 (1978).

FESHBACH, Herman, * 02.02.1917 in New York, from 1955 university professor at the Massachusetts Institute of Technology (MIT) in Boston, from 1973 head of the MIT physics department. He is one of the most distinguished theoretical nuclear scientists of his times. Among other things he is distinguished by his works in the field of nuclear reactions.

VILLARS, Felix Marc, * 06.01.1921 in Biel (Switzerland), student of W. Pauli, university professor at the Massachusetts Institute of Technology (MIT) in Boston. Among others he is distinguished by his work on the microscopic structure of collective motions in nuclei. The Pauli–Villars regularization procedure is discussed in W. Greiner, J. Reinhardt: *Quantum Electrodynamics* (Springer, Berlin, Heidelberg 1994).

SCHIFF, Leonard Isaac, * 29.03.1915 in Fall River (MA), † 19.01.1971 in Stanford. From 1941 at the University of Pennsylvania, becoming associate professor in 1944. In 1947 he moved to the Physics Department, Stanford (where he was Head of Physics from 1948 to 1966). Main fields: scattering theory, general relativity. A main part of his academic work was the education of young students. His excellent book on quantum mechanics made him world famous. In 1966 he received the Oerstedt-medal of the American Association of Physics Teachers.

WEINBERG, Steven, * 03.05.1933 in New York, today at the University of Texas, before this at Havard University. W. worked on cosmology and together with S. L. Glashow and A. Salam propounded the unified theory of weak and electromagnetic interaction ("standard model"), which, after this theory, only appear as different forms of one single force. For this they were awarded the Nobel Prize in physics in 1979.

HEISENBERG, Werner Karl, * 05.12.1901 in Würzburg, † 01.02.1976 in München, from 1927 to 1941 professor at Leipzig University, from 1941–45 director of the Kaiser-Wilhelm Institut für Physik in Berlin, 1946–57 director of the newly founded Max-Planck-Institut für Physik in Göttingen and 1958–70 in München. H. demanded in 1925, that only "principally measurable quantities" should be used to describe atomic phenomena (positivistic principle). With his "multiplication rules for quadratic schemes" he initiated the new "Göttinger Matrix Mechanics", which was further established in 1925 by M. Born, P. Jordan and H. In collaboration with Niels Bohr he succeeded in explaining the physical and philosophical background of the new formalism. Heisenberg's uncertainty relation (1927) became the basis of the "Kopenhagen interpretation" of quantum theory. After the discovery of the neutron by J. Chadwick in 1932 Heisenberg recognized that this particle is, in addition to the proton, an element of the nucleus, and he developed a theory of nuclear structure (concept of iso-spin). From 1953 Heisenberg worked on a unified theory of matter ("world formula"), which should be able to describe all elementary particles occuring in nature and all natural laws. In 1932 he received the Nobel Prize in physics and 1957 the peace class of the order "Pour le merite" [BR].

2. A Wave Equation for Spin-$\frac{1}{2}$ Particles: The Dirac Equation

We follow the historical approach of ***Dirac*** who, in 1928, searched for a relativistic covariant wave equation of the Schrödinger form

$$\mathrm{i}\hbar\frac{\partial\psi}{\partial t} = \hat{H}\psi \tag{2.1}$$

with positive definite probability density. At that time there were doubts concerning the Klein–Gordon equation, which did not yield such probability density [see (1.29)]. The charge density interpretation was not known at that time and would have made little physical sense, because π^+ and π^- mesons as charged spin-0 particles had not yet been discovered.

Since an equation in the form (2.1) is linear in the time derivative, it is natural to try to construct a Hamiltonian that is also linear in the spatial derivatives (equality of spatial and temporal coordinates). Hence, the desired equation (2.1) has to be of the form

$$\mathrm{i}\hbar\frac{\partial\psi}{\partial t} = \left[\frac{\hbar c}{\mathrm{i}}\left(\hat{\alpha}_1\frac{\partial}{\partial x^1} + \hat{\alpha}_2\frac{\partial}{\partial x^2} + \hat{\alpha}_3\frac{\partial}{\partial x^3}\right) + \hat{\beta}m_0c^2\right]\psi \equiv \hat{H}_\mathrm{f}\psi \quad . \tag{2.2}$$

The – yet unknown – coefficients $\hat{\alpha}_i$ cannot be simple numbers, otherwise (2.2) would not be form invariant with respect to simple spatial rotations. We suspect that the $\hat{\alpha}_i$ are matrices and indicate this by the operator sign $\wedge$. Then ψ cannot be a simple scalar, but has to be a column vector

$$\psi = \begin{pmatrix} \psi_1(\boldsymbol{x},t) \\ \psi_2(\boldsymbol{x},t) \\ \vdots \\ \psi_N(\boldsymbol{x},t) \end{pmatrix} \quad , \tag{2.3}$$

from which a positive definite density of the form

$$\varrho(x) = \psi^\dagger\psi(x) = (\psi_1^*,\psi_2^*,\dots,\psi_N^*)\begin{pmatrix} \psi_1 \\ \psi_2 \\ \vdots \\ \psi_N \end{pmatrix} = \sum_{i=1}^{N}\psi_i^*\psi_i(x) \tag{2.4}$$

can be constructed immediately. We still have to show that $\varrho(x)$ is the temporal component of a four-vector (current) for which a continuity equation must exist so that the spatial integral $\int\varrho\,\mathrm{d}^3x$ becomes constant in time. Only then is the probability interpretation of ϱx ensured. It is clear that the wave function ψ in (2.3) is a column vector analogous to the spin wave functions of the Pauli equation.[1] Hence, we shall call them spinors, specifying this name later. The dimension N of the spinor is not yet known, but we will be able to decide this soon. The coefficients $\hat{\alpha}_i$ and $\hat{\beta}$ must obviously be quadratic $N \times N$ (2.2). Thus the Schrödinger-like equation (2.1) and (2.3) represents a system of N coupled first-order differential equations of the spinor components ψ_i, $i = 1, 2, \ldots, N$. We also indicate this point in the notation and write (2.2) in the form

$$\begin{aligned} \mathrm{i}\hbar \frac{\partial \psi_\sigma}{\partial t} &= \frac{\hbar c}{\mathrm{i}} \sum_{\tau=1}^{N} \left(\hat{\alpha}_1 \frac{\partial}{\partial x^1} + \hat{\alpha}_2 \frac{\partial}{\partial x^2} + \hat{\alpha}_3 \frac{\partial}{\partial x^3} \right)_{\sigma\tau} \psi_\tau + m_0 c^2 \sum_{\tau=1}^{N} \hat{\beta}_{\sigma\tau} \psi_\tau \\ &\equiv \sum_{\tau=1}^{N} (\hat{H}_\mathrm{f})_{\sigma\tau} \psi_\tau \quad . \end{aligned} \tag{2.5}$$

Equation (2.2) is a short form of (2.5), in which the four $N \times N$ matrices $(\hat{\alpha}_i)_{\sigma\tau}$ $(i = 1, 2, 3)$ and $\hat{\beta}_{\sigma\tau}$ are expressed in the usual abbreviated form for matrices by $\hat{\alpha}_i$ $(i = 1, 2, 3)$ and $\hat{\beta}$ respectively. To continue, we demand the following natural properties:

(a) the correct energy-momentum relation for a relativistic free particle

$$E^2 = \boldsymbol{p}^2 c^2 + m_0^2 c^4 \quad , \tag{2.6}$$

(b) the continuity equation for the density (2.4), and
(c) the Lorentz covariance (i. e. Lorentz form-invariance) for (2.2) and (2.5), respectively.

To fulfill requirement (a), every single component ψ_σ of the spinor ψ has to satisfy the Klein–Gordon equation,[2] i. e.

$$-\hbar^2 \frac{\partial^2 \psi_\sigma}{\partial t^2} = \left(-\hbar^2 c^2 \nabla^2 + m_0^2 c^4 \right) \psi_\sigma \quad . \tag{2.7}$$

On the other hand, from (2.2) it follows by iteration that

---

[1] See W. Greiner: *Quantum Mechanics – An Introduction*, 3rd ed. (Springer, Berlin, Heidelberg 1994), Chaps. 12, 13.

[2] Notice that the analogy to classical electrodynamics, where the six electromagnetic fields E_x, E_y, E_z, H_x, H_y, H_z satisfy the first-order differential equations (Maxwell equations)

$$(\boldsymbol{\nabla} \times \boldsymbol{H}) = \frac{\partial \boldsymbol{E}}{c\partial t} \quad , \quad (\boldsymbol{\nabla} \times \boldsymbol{E}) = -\frac{\partial \boldsymbol{H}}{c\partial t} \quad , \quad \boldsymbol{\nabla} \cdot \boldsymbol{E} = 0 \quad , \quad \boldsymbol{\nabla} \cdot \boldsymbol{B} = 0$$

in a vacuum. Each single component E_i and H_i satifies simultaneously the differential equation of the second order (wave equation)

$$\left(\boldsymbol{\nabla}^2 - \frac{1}{c^2} \frac{\partial^2}{\partial t^2} \right) E_i = 0 \quad \text{and} \quad \left(\boldsymbol{\nabla}^2 - \frac{1}{c^2} \frac{\partial^2}{\partial t^2} \right) H_i = 0 \quad .$$

For further discussion of this analogy see Exercise 2.1.

$$
\begin{aligned}
-\hbar^2\frac{\partial^2\psi}{\partial t^2} = &-\hbar^2c^2\sum_{i,j=1}^{3}\frac{\hat{\alpha}_i\hat{\alpha}_j+\hat{\alpha}_j\hat{\alpha}_i}{2}\frac{\partial^2\psi}{\partial x^i\partial x^j}\\
&+\frac{\hbar m_0c^3}{\mathrm{i}}\sum_{i=1}^{3}(\hat{\alpha}_i\hat{\beta}+\hat{\beta}\hat{\alpha}_i)\frac{\partial\psi}{\partial x^i}+\hat{\beta}^2m_0^2c^4\psi \quad .
\end{aligned}
$$

Comparison with (2.7) shows the following requirements for the matrices $\hat{\alpha}_i$, $\hat{\beta}$:

$$
\begin{aligned}
&\hat{\alpha}_i\hat{\alpha}_j+\hat{\alpha}_j\hat{\alpha}_i = 2\delta_{ij}\,\mathbb{1} \quad ,\\
&\hat{\alpha}_i\hat{\beta}+\hat{\beta}\hat{\alpha}_i \quad = 0 \quad ,\\
&\hat{\alpha}_i^2 = \hat{\beta}^2 \qquad = \mathbb{1} \quad .
\end{aligned} \tag{2.8}
$$

These anticommutation relations define an algebra for the ψ matrices. In order to establish hermiticity of the Hamiltonian $\hat{H}_\mathrm{f}$ in (2.2), the matrices $\hat{\alpha}_i$, $\hat{\beta}$ also have to be Hermitian; thus,

$$
\hat{\alpha}_i^\dagger = \hat{\alpha}_i \quad , \quad \hat{\beta}^\dagger = \hat{\beta} \quad . \tag{2.9}
$$

Therefore, the eigenvalues of the matrices are real. Since, according to (2.8), one has $\hat{\alpha}_i^2 = 1$ and $\hat{\beta}^2 = 1$, it follows that the eigenvalues can only have the values ± 1. Because the eigenvalues are independent of the special representation,[3] this can best be shown in the diagonal representation of the single matrices. For example, $\hat{\alpha}_i$ in its eigenrepresentation has the form

$$
\hat{\alpha}_i = \begin{pmatrix} A_1 & 0 & 0 & \cdots & 0\\ 0 & A_2 & 0 & \cdots & 0\\ 0 & 0 & A_3 & \cdots & 0\\ \vdots & \vdots & \vdots & \ddots & \vdots\\ 0 & 0 & 0 & \cdots & A_N \end{pmatrix} \quad ,
$$

with the eigenvalues $A_1,\ldots,A_N$, and (2.8) now yields

$$
\hat{\alpha}_i^2 = \mathbb{1} = \begin{pmatrix} 1 & 0 & 0 & \cdots\\ 0 & 1 & 0 & \cdots\\ 0 & 0 & 1 & \cdots\\ .. & .. & .. & \ddots \end{pmatrix} = \begin{pmatrix} A_1^2 & 0 & \cdots\cdots\\ 0 & A_2^2 & \cdots\cdots\\ \vdots & & \ddots\\ .. & .. & ...A_N^2 \end{pmatrix} \quad ,
$$

from which

$$
A_k^2 = 1 \quad , \quad \text{i.e.} \quad A_k = \pm 1 \quad . \tag{2.10}
$$

---

[3] This follows, because $\hat{A}\psi_\alpha = \alpha\psi_\alpha$ implies that

$$
\hat{U}\hat{A}\hat{U}^{-1}\hat{U}\psi_\alpha = \alpha\hat{U}\psi_\alpha \quad ,
$$

and, therefore,

$$
\hat{A}'\left(\hat{U}\psi_\alpha\right) = \alpha\left(\hat{U}\psi_\alpha\right) \quad .
$$

The solutions of the rotated matrix $\hat{A}' = \hat{U}\hat{A}\hat{U}^{-1}$ are just the rotated vectors $\psi'_\alpha = \hat{U}\psi_\alpha$ with the same eigenvalues α.

Furthermore, from the anticommutation relations (2.8) it follows that the trace (i. e. the sum of the diagonal elements of the matrix) of each $\hat{\alpha}_i$ and of $\hat{\beta}$ has to be zero. Namely, according to (2.8) one has

$$\hat{\alpha}_i = -\hat{\beta}\hat{\alpha}_i\hat{\beta} \quad .$$

Because of the identity

$$\mathrm{tr}\,\hat{A}\hat{B} = -\,\mathrm{tr}\,\hat{B}\hat{A} \quad ,$$

one concludes that

$$\mathrm{tr}\,\hat{\alpha}_i = \mathrm{tr}\,\hat{\beta}^2\hat{\alpha}_i = \mathrm{tr}\,\hat{\beta}\hat{\alpha}_i\hat{\beta} = -\,\mathrm{tr}\,\hat{\alpha}_i \Rightarrow \mathrm{tr}\,\hat{\alpha}_i = 0 \quad . \tag{2.11}$$

The trace of a matrix is always equal to the sum of its eigenvalues, which can be seen if $\hat{U}$ transforms the matrix $\hat{\alpha}_i$ into its diagonal form,

$$\begin{pmatrix} A_1 & 0 & \cdots\cdots \\ 0 & A_2 & \cdots\cdots \\ \vdots & & \vdots \\ .. & .. & ...A_N \end{pmatrix} = \hat{U}\,\hat{\alpha}_i\,\hat{U}^{-1} \quad .$$

Then

$$\mathrm{tr}\begin{pmatrix} A_1 & 0 & \cdots\cdots \\ 0 & A_2 & \cdots\cdots \\ \vdots & & \ddots \\ .. & .. & ...A_N \end{pmatrix} = \sum_{k=1}^{N} A_k = \mathrm{tr}\,\hat{U}\,\hat{\alpha}_i\,\hat{U}^{-1} = \mathrm{tr}\,\hat{\alpha}_i\,\hat{U}\,\hat{U}^{-1} = \mathrm{tr}\,\hat{\alpha}_i \quad ,$$

which proves the above statement. Because the eigenvalues of $\hat{\alpha}_i$ and $\hat{\beta}$ are equal to ± 1, each matrix $\hat{\alpha}_i$ and $\hat{\beta}$ has to possess as many positive as negative eigenvalues, and therefore has to be of even dimension. The smallest even dimension, $N = 2$, cannot be right, because only three anticommuting matrices exist, namely the three Pauli matrices[4] $\hat{\sigma}_i$. Therefore, the smallest dimension for which the requirements (2.8) can be fulfilled is $N = 4$. We now study this case in more detail and indicate immediately one possible explicit representation of the Dirac matrices, i. e.

$$\hat{\alpha}_i = \begin{pmatrix} 0 & \hat{\sigma}_i \\ \hat{\sigma}_i & 0 \end{pmatrix} \quad , \quad \hat{\beta} = \begin{pmatrix} 1\!\!1 & 0 \\ 0 & -1\!\!1 \end{pmatrix} \quad , \tag{2.12}$$

where $\hat{\sigma}_i$ are Pauli's 2×2 matrices and $1\!\!1$ is the 2×2 unit matrix. With the explicit form of the Pauli matrices of (1.65), we have, in detail,

$$\hat{\alpha}_1 = \begin{pmatrix} 0 & 0 & 0 & 1 \\ 0 & 0 & 1 & 0 \\ 0 & 1 & 0 & 0 \\ 1 & 0 & 0 & 0 \end{pmatrix} \quad , \quad \hat{\alpha}_2 = \begin{pmatrix} 0 & 0 & 0 & -\mathrm{i} \\ 0 & 0 & i & 0 \\ 0 & -\mathrm{i} & 0 & 0 \\ i & 0 & 0 & 0 \end{pmatrix} \quad ,$$

$$\hat{\alpha}_3 = \begin{pmatrix} 0 & 0 & 1 & 0 \\ 0 & 0 & 0 & -1 \\ 1 & 0 & 0 & 0 \\ 0 & -1 & 0 & 0 \end{pmatrix} \quad , \quad \hat{\beta} = \begin{pmatrix} 1 & 0 & 0 & 0 \\ 0 & 1 & 0 & 0 \\ 0 & 0 & -1 & 0 \\ 0 & 0 & 0 & -1 \end{pmatrix} \quad . \tag{2.13}$$

[4] See W. Greiner: *Quantum Mechanics – An Introduction*, 3rd ed. (Springer, Berlin, Heidelberg 1994), Chaps. 12, 13 and especially Exercise 13.1.

Indeed, we can easily check the validity of the relations (2.8). For example,

$$\begin{aligned}
\hat{\alpha}_i\hat{\alpha}_j + \hat{\alpha}_j\hat{\alpha}_i &= \begin{pmatrix} 0 & \hat{\sigma}_i \\ \hat{\sigma}_i & 0 \end{pmatrix}\begin{pmatrix} 0 & \hat{\sigma}_j \\ \hat{\sigma}_j & 0 \end{pmatrix} + \begin{pmatrix} 0 & \hat{\sigma}_j \\ \hat{\sigma}_j & 0 \end{pmatrix}\begin{pmatrix} 0 & \hat{\sigma}_i \\ \hat{\sigma}_i & 0 \end{pmatrix} \\
&= \begin{pmatrix} \hat{\sigma}_i\hat{\sigma}_j & 0 \\ 0 & \hat{\sigma}_i\hat{\sigma}_j \end{pmatrix} + \begin{pmatrix} \hat{\sigma}_j\hat{\sigma}_i & 0 \\ 0 & \hat{\sigma}_j\hat{\sigma}_i \end{pmatrix} \\
&= \begin{pmatrix} \hat{\sigma}_i\hat{\sigma}_j + \hat{\sigma}_j\hat{\sigma}_i & 0 \\ 0 & \hat{\sigma}_i\hat{\sigma}_j + \hat{\sigma}_j\hat{\sigma}_i \end{pmatrix} = \begin{pmatrix} 2\delta_{ij}\,1\!\!1 & 0 \\ 0 & 2\delta_{ij}\,1\!\!1 \end{pmatrix} \\
&= 2\delta_{ij}\begin{pmatrix} 1\!\!1 & 0 \\ 0 & 1\!\!1 \end{pmatrix} \quad ,
\end{aligned}$$

holds. Here we have used the relation for the Pauli matrices[5]

$$\hat{\sigma}_i\hat{\sigma}_j + \hat{\sigma}_j\hat{\sigma}_i = 2\delta_{ij}\,1\!\!1 \quad . \tag{2.14}$$

We also notice that (2.12) describes just one possible choice of the Dirac matrices $\hat{\alpha}_i$, $\hat{\beta}$. Each set $\hat{\alpha}'_i = \hat{U}\hat{\alpha}_i\hat{U}^{-1}$, $\hat{\beta}' = \hat{U}\hat{\beta}\hat{U}^{-1}$, which is obtained from the original $\hat{\alpha}_i$, $\hat{\beta}$ of (2.13) by a unitary transformation $\hat{U}$, can be used equally as well as the one introduced here [see (2.21)]. In Example 3.1 it will be shown that all representations of the Dirac algebra are unitarily equivalent to each other. Therefore, physical results do not depend on the special choice of the Dirac matrices $\hat{\alpha}_i$ and $\hat{\beta}$, but the calculations can become particularly simple in a certain representation.

Next we want to construct the four-current density and the equation of continuity. For that we multiply (2.2) from the left by $\psi^\dagger = (\psi_1^*, \psi_2^*, \psi_3^*, \psi_4^*)$ and obtain

$$\mathrm{i}\hbar\psi^\dagger\frac{\partial}{\partial t}\psi = \frac{\hbar c}{\mathrm{i}}\sum_{k=1}^{3}\psi^\dagger\hat{\alpha}_k\frac{\partial}{\partial x^k}\psi + m_0c^2\psi^\dagger\hat{\beta}\psi \quad . \tag{2.15a}$$

Furthermore, we form the Hermitian conjugate of (2.12), i.e.

$$-\mathrm{i}\hbar\frac{\partial\psi^\dagger}{\partial t} = -\frac{\hbar c}{\mathrm{i}}\sum_{k=1}^{3}\frac{\partial\psi^\dagger}{\partial x^k}\hat{\alpha}_k^\dagger + m_0c^2\psi^\dagger\hat{\beta}^\dagger \quad ,$$

and multiply this equation from the right by ψ, taking into consideration the hermicity of the Dirac matrices ($\hat{\alpha}_i^\dagger = \hat{\alpha}_i$, $\hat{\beta}^\dagger = \hat{\beta}$), to give

$$-\mathrm{i}\hbar\frac{\partial\psi^\dagger}{\partial t}\psi = -\frac{\hbar c}{\mathrm{i}}\sum_{k=1}^{3}\frac{\partial\psi^\dagger}{\partial x^k}\hat{\alpha}_k\psi + m_0c^2\psi^\dagger\hat{\beta}\psi \quad . \tag{2.15b}$$

Then, subtraction of (2.15b) from (2.15a) yields

$$\mathrm{i}\hbar\frac{\partial}{\partial t}(\psi^\dagger\psi) = \frac{\hbar c}{\mathrm{i}}\sum_{k=1}^{3}\frac{\partial}{\partial x^k}(\psi^\dagger\hat{\alpha}_k\psi) \tag{2.16}$$

or

[5] This relation is covered in detail in W. Greiner: *Quantum Mechanics – An Introduction*, 3rd ed. (Springer, Berlin, Heidelberg 1994).

$$\frac{\partial \varrho}{\partial t} + \operatorname{div} \boldsymbol{j} = 0 \quad , \tag{2.17}$$

where

$$\varrho = \psi^\dagger \psi = \sum_{i=1}^{4} \psi_i^* \psi_i \tag{2.18a}$$

is the positive definite density (2.4) and

$$j^k = c\psi^\dagger \hat{\alpha}^k \psi \quad \text{or} \quad \boldsymbol{j} = c\psi^\dagger \hat{\boldsymbol{\alpha}} \psi \tag{2.18b}$$

is the *current density*. Here we have symbolically introduced the three-vector

$$\hat{\boldsymbol{\alpha}} = \{\hat{\alpha}^1, \hat{\alpha}^2, \hat{\alpha}^3\} = \{-\hat{\alpha}_1, -\hat{\alpha}_2, -\hat{\alpha}_3\} \tag{2.19}$$

and introduced the upper and lower indices according to our former convention [see (1.5) and (1.6)]. From (2.17) the conservation law follows immediately in the usual way

$$\frac{\partial}{\partial t} \int_V \mathrm{d}^3x\, \psi^\dagger \psi = -\int_V \operatorname{div} \boldsymbol{j}\, \mathrm{d}^3x = -\int_F \boldsymbol{j} \cdot \mathrm{d}\boldsymbol{f} = 0 \tag{2.20}$$

where V denotes a certain volume and F its surface. Since ϱ is positive definite and because of the conservation law (2.17) we can accept the interpretation of ϱ as a probability density [in contrast to the density ϱ obtained for the Klein–Gordon equation, see (1.29) which was not positive definite]. Accordingly, we call $\boldsymbol{j}$ the *probability current density*. Here we have presumed that $\boldsymbol{j}$ is a vector, i.e. that its components (2.18b) transform under spatial rotations as the components of a three-vector. This still has to be shown. Furthermore, $\{c\varrho, \boldsymbol{j}\}$ should form a four-vector. Hence, it should transform from one inertial system into another one by a Lorentz transformation. This point and, in addition, the form invariance of the Dirac equation (2.2) with respect to Lorentz transformations (we also call the form invariance *covariance*) have still to be shown, before we can regard the Dirac equation as an acceptable relativistic wave equation.

We also notice that we have achieved a special representation with (2.12). The choice of the matrices (2.12) is not unequivocal. One recognizes immediately that each unitary transformation $\hat{S}$ yields the matrices

$$\hat{\alpha}_i' = \hat{S}\hat{\alpha}_i\hat{S}^{-1} \quad , \quad \hat{\beta}' = \hat{S}\hat{\beta}\hat{S}^{-1} \tag{2.21}$$

which also satisfy the algebra (2.8). We check this for the first commutator (2.8), as an example:

$$\begin{aligned} &\hat{S}\hat{\alpha}_i\hat{S}^{-1}\hat{S}\hat{\alpha}_j\hat{S}^{-1} + \hat{S}\hat{\alpha}_j\hat{S}^{-1}\hat{S}\hat{\alpha}_i\hat{S}^{-1} = 2\delta_{ij}\hat{S}\,\mathbb{1}\hat{S}^{-1} \\ &\quad \Rightarrow \hat{\alpha}_i'\hat{\alpha}_j' + \hat{\alpha}_j'\hat{\alpha}_i' = 2\delta_{ij}\,\mathbb{1} \quad . \end{aligned} \tag{q.e.d.}$$

EXERCISE

2.1 Representation of the Maxwell Equations in the Form of the Dirac Equation

Problem. Write the Maxwell equations

$$\operatorname{curl} \boldsymbol{E} + \frac{1}{c}\frac{\partial \boldsymbol{H}}{\partial t} = 0 \quad , \quad \operatorname{curl} \boldsymbol{H} - \frac{1}{c}\frac{\partial \boldsymbol{E}}{\partial t} = \frac{4\pi}{c}\boldsymbol{j} \quad , \tag{a}$$

in the form analogous to the Dirac equation (spinor equation):

$$-\frac{1}{\mathrm{i}}\sum_{j=0}^{3} \hat{\alpha}^j \frac{\partial}{\partial x^j}\psi = -\frac{4\pi}{c}\Phi \quad . \tag{b}$$

Determine the matrices $\hat{\alpha}^j$ and their commutation relations and deduce the wave equation for ψ from (b).

Solution. We define the four-component column vectors

$$\psi = \begin{pmatrix} \psi_0 \\ \psi_1 \\ \psi_2 \\ \psi_3 \end{pmatrix} \quad \text{and} \quad \Phi = \begin{pmatrix} \phi_0 \\ \phi_1 \\ \phi_2 \\ \phi_3 \end{pmatrix} \quad , \tag{1}$$

where $\phi_0 = c\varrho$, $\phi_1 = j_1 = j_x$, $\phi_2 = j_2 = j_y$, $\phi_3 = j_3 = j_z$. Furthermore, we have $x^0 = x_0 = ct$, $x^1 = -x_1 = x$, $x^2 = -x_2 = y$, $x^3 = -x_3 = z$. Now we define the components of ψ as

$$\psi_0 \equiv 0 \quad , \quad \psi_1 = H_1 - \mathrm{i}E_1 \quad , \quad \psi_2 = H_2 - \mathrm{i}E_2 \quad , \quad \psi_3 = H_3 - \mathrm{i}E_3 \quad . \tag{2}$$

From this definition it follows that the matrices $\hat{\alpha}^i$ have partly real and partly pure imaginary matrix elements. From (b) we get the equation

$$-\frac{1}{i}\left(\hat{\alpha}^0\frac{1}{c}\frac{\partial}{\partial t}\psi + \hat{\alpha}^1\frac{\partial}{\partial x}\psi + \hat{\alpha}^2\frac{\partial}{\partial y}\psi + \hat{\alpha}^3\frac{\partial}{\partial z}\psi\right) = -\frac{4\pi}{c}\Phi \quad . \tag{3}$$

Denoting now the matrix elements of the matrices $\hat{\alpha}^j$ by α^j_{ik} we can write down the components of (3) explicitly. In the same manner we write the components of the Maxwell equations (a) and compare the coefficients of both *systems of equations*. In order to obtain the correct signs in the Maxwell equations using (2) we infer, as there appears a factor $-1/\mathrm{i}$, that the α^j_{ik} only take values ± 1 or $\pm \mathrm{i}$.

Important Remark. As $\psi_0 = 0$, this procedure determines the columns $1, 2, 3$, but not column 0. We can fix the remaining column by requiring that $\hat{\alpha}^j$ is Hermitian and that $(\hat{\alpha}^j)^2 = \mathbb{1}$. For the matrices $\hat{\alpha}^j$ we now find

Exercise 2.1.

$$\hat{\alpha}^0 = \mathbb{1} = \begin{pmatrix} 1 & 0 & 0 & 0 \\ 0 & 1 & 0 & 0 \\ 0 & 0 & 1 & 0 \\ 0 & 0 & 0 & 1 \end{pmatrix} \quad ,$$

$$\hat{\alpha}^1 = \begin{pmatrix} 0 & -1 & 0 & 0 \\ -1 & 0 & 0 & 0 \\ 0 & 0 & 0 & -\mathrm{i} \\ 0 & 0 & \mathrm{i} & 0 \end{pmatrix} \quad , \tag{4}$$

$$\hat{\alpha}^2 = \begin{pmatrix} 0 & 0 & -1 & 0 \\ 0 & 0 & 0 & \mathrm{i} \\ -1 & 0 & 0 & 0 \\ 0 & -\mathrm{i} & 0 & 0 \end{pmatrix} \quad ,$$

$$\hat{\alpha}^3 = \begin{pmatrix} 0 & 0 & 0 & -1 \\ 0 & 0 & -\mathrm{i} & 0 \\ 0 & \mathrm{i} & 0 & 0 \\ -1 & 0 & 0 & 0 \end{pmatrix} \quad ,$$

and the operators $\hat{\alpha}^j$ are Hermitian. We see immediately that trace $\hat{\alpha}^j = 0$ holds for $j = 1, 2, 3$ and, thus, we obtain the commutation relations

$$\begin{aligned} &\hat{\alpha}^1\hat{\alpha}^2 + \hat{\alpha}^2\hat{\alpha}^1 = 0 \quad , \quad \hat{\alpha}^1\hat{\alpha}^2 = \mathrm{i}\hat{\alpha}^3 \quad , \\ &\hat{\alpha}^2\hat{\alpha}^3 + \hat{\alpha}^3\hat{\alpha}^2 = 0 \quad , \quad \hat{\alpha}^2\hat{\alpha}^3 = \mathrm{i}\hat{\alpha}^1 \quad , \\ &\hat{\alpha}^3\hat{\alpha}^1 + \hat{\alpha}^1\hat{\alpha}^3 = 0 \quad , \quad \hat{\alpha}^3\hat{\alpha}^1 = \mathrm{i}\hat{\alpha}^2 \quad \text{and} \end{aligned} \tag{5}$$

$$(\hat{\alpha}^1)^2 = (\hat{\alpha}^2)^2 = (\hat{\alpha}^3)^2 = \mathbb{1} \quad . \tag{6}$$

Using these, we infer from (b) that

$$\begin{aligned} &\left(-\frac{1}{\mathrm{i}}\sum_{j=0}^{3}\hat{\alpha}^j\frac{\partial}{\partial x^j}\right)\left(-\frac{1}{\mathrm{i}}\sum_{k=0}^{3}\hat{\alpha}^k\frac{\partial}{\partial x^k}\right)\psi \\ &= -\left\{(\hat{\alpha}^0)^2\frac{\partial^2}{\partial(ct)^2} - \sum_{i=1}^{3}(\hat{\alpha}^i)^2\frac{\partial^2}{\partial x^{i2}}\right\}\psi \\ &= \left\{\nabla^2 - \frac{\partial^2}{c^2\partial t^2}\right\}\psi = \Box\psi \quad , \end{aligned} \tag{7}$$

as the mixing terms are proportional to the anticommutators (5) and hence vanish. This means, from (3) and (7),

$$\Box\psi = \frac{4\pi}{\mathrm{i}}\sum_{j=0}^{3}\hat{\alpha}^j\frac{\partial}{\partial x^j}\Phi \quad . \tag{8}$$

Hence, if there are no source terms ($\Phi = 0$) the components of ψ, that is the components of the electromagnetic field, obey a wave equation. If there are sources present, then for the upper component of (8) with $\psi_0 = 0$

$$\sum_{j=0}^{3}\frac{\partial\phi j}{\partial x^j} = \frac{\partial\varrho}{\partial t} + \operatorname{div}\boldsymbol{j} = 0 \tag{9}$$

follows as a necessary condition of a solution of (b). Obviously, (9) is just the continuity equation. We also can recognize the analogue to the Dirac equation in the Schrödinger form:

$$\left(-\frac{1}{\mathrm{i}}\frac{\partial}{\partial t}-\hat{H}_0\right)\psi=-4\pi\Phi \quad , \tag{10}$$

where $\hat{H}_0=(1/\mathrm{i})\sum_{k=1}^{3}\hat{\alpha}^k\partial/\partial x^k$ has the same form as the Dirac Hamiltonian for vanishing mass m_0.

2.1 Free Motion of a Dirac Particle

We examine the solution of the free Dirac equation (2.2) (that is, the Dirac equation without potentials) and again write it in the form

$$\mathrm{i}\hbar\frac{\partial\psi}{\partial t}=\hat{H}_\mathrm{f}\psi=(c\hat{\boldsymbol{\alpha}}\cdot\hat{\boldsymbol{p}}+m_0c^2\hat{\beta})\psi \quad . \tag{2.22}$$

Its stationary states are found with the ansatz

$$\psi(\boldsymbol{x},t)=\psi(\boldsymbol{x})\exp[-(\mathrm{i}/\hbar)\varepsilon t] \quad , \tag{2.23}$$

which transforms (2.2) into

$$\varepsilon\psi(\boldsymbol{x})=\hat{H}_\mathrm{f}\psi(\boldsymbol{x}) \quad . \tag{2.24}$$

Again the quantity ε describes the time evolution of the stationary state $\psi(\boldsymbol{x})$. For many applications it is useful to split up the four-component spinor into two two-component spinors ϕ and χ, i. e.

$$\psi=\begin{pmatrix}\psi_1\\\psi_2\\\psi_3\\\psi_4\end{pmatrix}=\begin{pmatrix}\varphi\\\chi\end{pmatrix} \quad \text{with} \tag{2.25a}$$

$$\varphi=\begin{pmatrix}\psi_1\\\psi_2\end{pmatrix} \quad \text{and} \quad \chi=\begin{pmatrix}\psi_3\\\psi_4\end{pmatrix} \quad . \tag{2.25b}$$

Using the explicit form (2.12) for the $\hat{\alpha}$ and $\hat{\beta}$ matrices (2.24) can be written as

$$\varepsilon\begin{pmatrix}\varphi\\\chi\end{pmatrix}=c\begin{pmatrix}0&\hat{\boldsymbol{\sigma}}\\\hat{\boldsymbol{\sigma}}&0\end{pmatrix}\cdot\hat{\boldsymbol{p}}\begin{pmatrix}\varphi\\\chi\end{pmatrix}+m_0^2c^2\begin{pmatrix}\mathbb{1}&0\\0&-\mathbb{1}\end{pmatrix}\begin{pmatrix}\varphi\\\chi\end{pmatrix}$$

or

$$\begin{aligned}\varepsilon\varphi&=c\hat{\boldsymbol{\sigma}}\cdot\hat{\boldsymbol{p}}\chi+m_0c^2\varphi \quad ,\\ \varepsilon\chi&=c\hat{\boldsymbol{\sigma}}\cdot\hat{\boldsymbol{p}}\varphi-m_0c^2\chi \quad .\end{aligned} \tag{2.26}$$

States with definite momentum $\boldsymbol{p}$ are

$$\begin{pmatrix} \varphi \\ \chi \end{pmatrix} = \begin{pmatrix} \varphi_0 \\ \chi_0 \end{pmatrix} \exp[(\mathrm{i}/\hbar)\boldsymbol{p}\cdot\boldsymbol{x}] \quad . \tag{2.27}$$

The equations (2.26) are transformed into the same equations for ϕ_0 and χ_0, but replacing the operators $\hat{\boldsymbol{p}}$ by the eigenvalues $\boldsymbol{p}$. Ordering with respect to ϕ_0 and χ_0 results in the system of equations

$$\begin{aligned} &\left(\varepsilon - m_0c^2\right) \mathbb{1}\varphi_0 - c\hat{\boldsymbol{\sigma}}\cdot\boldsymbol{p}\chi_0 = 0 \quad , \\ &-c\hat{\boldsymbol{\sigma}}\cdot\boldsymbol{p}\varphi_0 + \left(\varepsilon + m_0c^2\right) \mathbb{1}\chi_0 = 0 \quad . \end{aligned} \tag{2.28}$$

This linear homogenous system of equations for ϕ_0 and χ_0 has nontrivial solutions only in the case of a vanishing determinant of the coefficients, that is

$$\begin{vmatrix} (\varepsilon - m_0c^2)\mathbb{1} & -c\hat{\boldsymbol{\sigma}}\cdot\boldsymbol{p} \\ -c\hat{\boldsymbol{\sigma}}\cdot\boldsymbol{p} & (\varepsilon + m_0c^2)\mathbb{1} \end{vmatrix} = 0 \quad . \tag{2.29}$$

Using the relation[6]

$$(\hat{\boldsymbol{\sigma}}\cdot\boldsymbol{A})(\hat{\boldsymbol{\sigma}}\cdot\boldsymbol{B}) = \boldsymbol{A}\cdot\boldsymbol{B}\mathbb{1} + \mathrm{i}\hat{\boldsymbol{\sigma}}\cdot(\boldsymbol{A}\times\boldsymbol{B}) \quad , \tag{2.30}$$

equation (2.29) transforms into

$$\begin{aligned} &\left(\varepsilon^2 - m_0^2c^4\right) \mathbb{1} - c^2(\hat{\boldsymbol{\sigma}}\cdot\boldsymbol{p})(\hat{\boldsymbol{\sigma}}\cdot\boldsymbol{p}) = 0 \quad , \\ &\varepsilon^2 = m_0^2c^4 + c^2\boldsymbol{p}^2 \quad , \end{aligned}$$

from which follows

$$\varepsilon = \pm E_p \quad , \quad E_p = +c\sqrt{\boldsymbol{p}^2 + m_0^2c^2} \quad . \tag{2.31}$$

The two signs of the time-evolution factor ε correspond to two types of solutions of the Dirac equation. We call them *positive* and *negative* solutions, respectively. From (2.28), for fixed ε,

$$\chi_0 = \frac{c(\hat{\boldsymbol{\sigma}}\cdot\boldsymbol{p})}{m_0c^2 + \varepsilon}\varphi_0 \quad . \tag{2.32}$$

Let us denote the two-spinor φ_0 in the form

$$\varphi_0 = U = \begin{pmatrix} U_1 \\ U_2 \end{pmatrix} \quad , \tag{2.33}$$

with the normalization

$$U^\dagger U = U_1^*U_1 + U_2^*U_2 = 1 \quad ,$$

where U_1, U_2 are complex. Using (2.27) and (2.23) we obtain the complete set of *positive and negative free solutions of the Dirac equation* as

[6] Encountered previously in W. Greiner: *Quantum Mechanics – An Introduction*, 3rd ed. (Springer, Berlin, Heidelberg 1994), Exercise 13.2.

$$\Psi_{p\lambda}(\boldsymbol{x},t) = N \begin{pmatrix} U \\ \dfrac{c(\hat{\boldsymbol{\sigma}}\cdot\boldsymbol{p})}{m_0c^2+\lambda E_p}U \end{pmatrix} \frac{\exp[\mathrm{i}(\boldsymbol{p}\cdot\boldsymbol{x} - \overbrace{\lambda E_p}^{\varepsilon}\,t)/\hbar]}{\sqrt{2\pi\hbar}^3} \quad . \tag{2.34}$$

Here $\lambda = \pm 1$ characterizes the positive and negative solutions with the time evolution factor $\varepsilon = \lambda E_p$. The normalization factor N is determined from the condition

$$\int \Psi^\dagger_{p\lambda}(\boldsymbol{x},t)\Psi_{p'\lambda'}(\boldsymbol{x},t)\,\mathrm{d}^3x = \delta_{\lambda\lambda'}\delta(\boldsymbol{p}-\boldsymbol{p}') \quad . \tag{2.35}$$

Hence,

$$N^2\left(U^\dagger U + U^\dagger \frac{c^2(\hat{\boldsymbol{\sigma}}\cdot\boldsymbol{p})(\hat{\boldsymbol{\sigma}}\cdot\boldsymbol{p})}{(m_0c^2+\lambda E_p)^2}U\right) = 1$$

or, using (2.30)

$$N^2\left(1 + \frac{\hat{\boldsymbol{\sigma}}^2\boldsymbol{p}^2}{(m_0c^2+\lambda E_p)^2}\right) = 1$$

$$\begin{aligned}\Rightarrow N &= \sqrt{\frac{(m_0c^2+\lambda E_p)^2}{(m_0c^2+\lambda E_p)^2+c^2\boldsymbol{p}^2}} \\ &= \sqrt{\frac{(m_0c^2+\lambda E_p)^2}{(m_0^2c^4+c^2\boldsymbol{p}^2)+2m_0c^2\lambda E_p+E_p^2}} \\ &= \sqrt{\frac{(m_0c^2+\lambda E_p)^2}{2(m_0c^2+\lambda E_p)\lambda E_p}} \\ &= \sqrt{\frac{(m_0c^2+\lambda E_p)}{2\lambda E_p}} \quad .\end{aligned} \tag{2.36}$$

The spectrum of $\varepsilon_{p\lambda} = \lambda E_p$, corresponding to the spinors $\Psi_{p\lambda}(\boldsymbol{x},t)$, is shown in Fig. 2.1. There appears – as in the case of the Klein–Gordon equation – a domain of positive and negative frequencies ("energy eigenvalues"). We will discuss the interpretation of the states with $\lambda = -1$ in detail later on. We now recognize that all states (2.34) are eigenfunctions of momentum

$$\hat{\boldsymbol{p}}\Psi_{p\lambda} = \boldsymbol{p}\Psi_{p\lambda}(\boldsymbol{x},t) \quad . \tag{2.37}$$

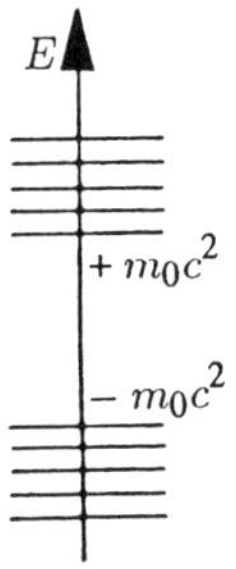

Fig. 2.1. Spectrum of the eigenvalues of the free Dirac equation

For every momentum $\boldsymbol{p}$ there are two different kinds of solutions, those with $\lambda = +1(\varepsilon = +E_p)$ and those with $\lambda = -1(\varepsilon = -E_p)$. We will now show that another quantum number, the *helicity*, can be used to classify the free one-particle states (2.34). First we note that the operator

$$\hat{\boldsymbol{\Sigma}}\cdot\hat{\boldsymbol{p}} = \begin{pmatrix} \hat{\boldsymbol{\sigma}} & 0 \\ 0 & \hat{\boldsymbol{\sigma}} \end{pmatrix}\cdot\hat{\boldsymbol{p}} \tag{2.38}$$

commutes with the free Dirac–Hamiltonian operator $\hat{H}_\mathrm{f}$ [cf. (2.2)]. Here

$$\hat{\boldsymbol{S}} = \frac{\hbar}{2}\hat{\boldsymbol{\Sigma}} = \frac{\hbar}{2}\begin{pmatrix} \hat{\boldsymbol{\sigma}} & 0 \\ 0 & \hat{\boldsymbol{\sigma}} \end{pmatrix} \tag{2.39}$$

is the four-dimensional generalization of the *spin vector operator*.[7] We calculate

$$\left[\hat{H}_{\mathrm{f}}, \hat{\boldsymbol{\Sigma}}\cdot\hat{\boldsymbol{p}}\right]_{-} = \left[c\hat{\boldsymbol{\alpha}}\cdot\hat{\boldsymbol{p}} + \beta m_0 c^2, \hat{\boldsymbol{\Sigma}}\cdot\hat{\boldsymbol{p}}\right]_{-} = c\,[\hat{\boldsymbol{\alpha}}\cdot\hat{\boldsymbol{p}}, \hat{\boldsymbol{\Sigma}}\cdot\hat{\boldsymbol{p}}]_{-} \quad ,$$

as $\hat{\beta}$ is a diagonal matrix and hence $[\hat{\beta}, \hat{\boldsymbol{\Sigma}}]_{-} = 0$. Furthermore, we obtain

$$\begin{aligned}
&(\hat{\boldsymbol{\alpha}}\cdot\hat{\boldsymbol{p}})(\hat{\boldsymbol{\Sigma}}\cdot\hat{\boldsymbol{p}}) - (\hat{\boldsymbol{\Sigma}}\cdot\hat{\boldsymbol{p}})(\hat{\boldsymbol{\alpha}}\cdot\hat{\boldsymbol{p}}) \\
&\quad = \begin{pmatrix} 0 & \hat{\boldsymbol{\sigma}}\cdot\hat{\boldsymbol{p}} \\ \hat{\boldsymbol{\sigma}}\cdot\hat{\boldsymbol{p}} & 0 \end{pmatrix}\begin{pmatrix} \hat{\boldsymbol{\sigma}}\cdot\hat{\boldsymbol{p}} & 0 \\ 0 & \hat{\boldsymbol{\sigma}}\cdot\hat{\boldsymbol{p}} \end{pmatrix} \\
&\qquad - \begin{pmatrix} \hat{\boldsymbol{\sigma}}\cdot\hat{\boldsymbol{p}} & 0 \\ 0 & \hat{\boldsymbol{\sigma}}\cdot\hat{\boldsymbol{p}} \end{pmatrix}\begin{pmatrix} 0 & \hat{\boldsymbol{\sigma}}\cdot\hat{\boldsymbol{p}} \\ \hat{\boldsymbol{\sigma}}\cdot\hat{\boldsymbol{p}} & 0 \end{pmatrix} \\
&\quad = \begin{pmatrix} 0 & (\hat{\boldsymbol{\sigma}}\cdot\hat{\boldsymbol{p}})^2 \\ (\hat{\boldsymbol{\sigma}}\cdot\hat{\boldsymbol{p}})^2 & 0 \end{pmatrix} - \begin{pmatrix} 0 & (\hat{\boldsymbol{\sigma}}\cdot\hat{\boldsymbol{p}})^2 \\ (\hat{\boldsymbol{\sigma}}\cdot\hat{\boldsymbol{p}})^2 & 0 \end{pmatrix} = 0 \quad .
\end{aligned}$$

Hence,

$$\left[\hat{H}_{\mathrm{f}}, \hat{\boldsymbol{\Sigma}}\cdot\hat{\boldsymbol{p}}\right]_{-} = 0 \tag{2.40a}$$

and naturally

$$[\hat{\boldsymbol{p}}, \hat{\boldsymbol{\Sigma}}\cdot\hat{\boldsymbol{p}}]_{-} = 0 \quad . \tag{2.40b}$$

This means that $\hat{\boldsymbol{\Sigma}}\cdot\hat{\boldsymbol{p}}$, $\hat{H}_{\mathrm{f}}$ and $\hat{\boldsymbol{p}}$ can be diagonalized together. The same holds for the *helicity operator*

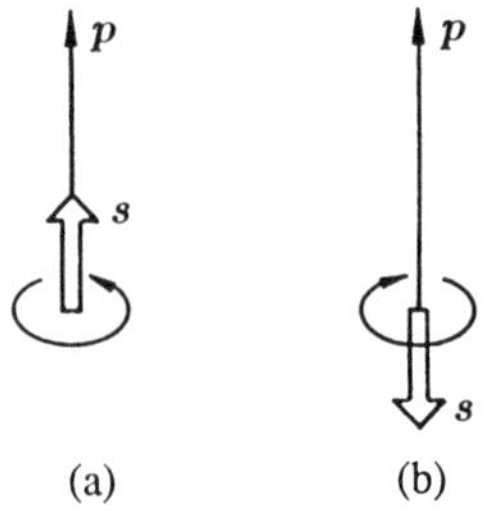

Fig. 2.2. Electrons with positive (**a**) and negative (**b**) helicity. The *double arrow* denotes spin. ↻ and ↺ symbolize the two possible rotations of the electron

$$\hat{\Lambda}_S = \frac{\hbar}{2}\hat{\boldsymbol{\Sigma}}\cdot\frac{\hat{\boldsymbol{p}}}{|\boldsymbol{p}|} = \hat{\boldsymbol{S}}\cdot\frac{\hat{\boldsymbol{p}}}{|\boldsymbol{p}|} \tag{2.41}$$

as we can immediately see by repeating the calculations that led us to (2.40a, b). Helicity has an obvious interpretation: it is the projection of the spin onto the direction of momentum, as illustrated in Fig. 2.2.

If the electron wave propagates into the direction of the z axis, we have

$$\boldsymbol{p} = \{0, 0, p\}$$

and, because of (2.41),

$$\hat{\Lambda}_S = \hat{S}_z = \frac{\hbar}{2}\hat{\Sigma}_z = \frac{\hbar}{2}\begin{pmatrix} 1 & 0 & 0 & 0 \\ 0 & -1 & 0 & 0 \\ 0 & 0 & 1 & 0 \\ 0 & 0 & 0 & -1 \end{pmatrix} \tag{2.42}$$

with the eigenvalues $\pm\hbar/2$. Clearly, the eigenvectors of $\hat{\Lambda}_S$ are

$$\begin{pmatrix} u_1 \\ 0 \end{pmatrix} \;,\; \begin{pmatrix} u_{-1} \\ 0 \end{pmatrix} \;,\; \begin{pmatrix} 0 \\ u_1 \end{pmatrix} \;,\; \begin{pmatrix} 0 \\ u_{-1} \end{pmatrix} \quad \text{with} \tag{2.43}$$

$$u_1 = \begin{pmatrix} 1 \\ 0 \end{pmatrix} \quad \text{and} \quad u_{-1} = \begin{pmatrix} 0 \\ 1 \end{pmatrix} \quad .$$

[7] See W. Greiner: *Quantum Mechanics – An Introduciion*, 3rd ed. (Springer, Berlin, Heidelberg 1994), Chaps. 12, 13.

Now we can classify completely the free Dirac waves propagating in the z direction; we denote them by $\Psi_{p_z,\lambda,s_z}(\boldsymbol{x},t)$ and write explicitly

$$\Psi_{p,\lambda,+1/2} = N \begin{pmatrix} \begin{pmatrix} 1 \\ 0 \end{pmatrix} \\ \dfrac{c\hat{\sigma}_z p}{m_0c^2 + \lambda E_p} \begin{pmatrix} 1 \\ 0 \end{pmatrix} \end{pmatrix} \exp[\mathrm{i}(pz - \lambda E_p t)/\hbar] \quad , \tag{2.44a}$$

$$\Psi_{p,\lambda,-1/2} = N \begin{pmatrix} \begin{pmatrix} 0 \\ 1 \end{pmatrix} \\ \dfrac{c\hat{\sigma}_z p}{m_0c^2 + \lambda E_p} \begin{pmatrix} 0 \\ 1 \end{pmatrix} \end{pmatrix} \exp[\mathrm{i}(pz - \lambda E_p t)/\hbar] \quad . \tag{2.44b}$$

From (2.35) one immediately recognizes the validity of the orthonormality relations

$$\int \Psi^{\dagger}_{p_z\lambda s_z} \Psi_{p'_z\lambda' s'_z}\, \mathrm{d}^3x = \delta_{\lambda\lambda'}\delta_{s_z s'_z}\delta(p_z - p'_z) \quad . \tag{2.45}$$

2.2 Single-Particle Interpretation of the Plane (Free) Dirac Waves

In this section we shall examine the single-particle interpretation of the free Dirac equation and its solutions in greater detail. Many of our considerations will be similar to those carried out when discussing the Klein–Gordon equation and, indeed, we can already remark at this stage that we will find that a single-particle interpretation can be performed to a large extent after a suitable modification of the operators. However, such an interpretation will not stand a rigorous treatment.

First we shall discuss the interpretation of the energy. From (2.31) we know that the eigenvalues of the free Dirac Hamiltonian $\hat{H}_\mathrm{f}$ (2.2) are $\varepsilon = \pm E_p$. As in the previous discussion of the Klein–Gordon equation we have to find out whether the time evolution factors can be interpreted as energies. In order to answer this question we use the canonical (Lagrange) formalism. In the following Exercise 2.2 it is shown that the Lagrange density, leading to the free Dirac equation (2.2) is given by

$$\mathcal{L} = \mathrm{i}\hbar\Psi^{\dagger}\frac{\partial}{\partial t}\Psi + \mathrm{i}\hbar c\Psi^{\dagger}\nabla\cdot\hat{\boldsymbol{\alpha}}\Psi - m_0c^2\Psi^{\dagger}\hat{\beta}\Psi \quad , \tag{2.46}$$

and, using this result, we will also calculate, as previously in (1.59–61), the energy related to the free solutions (2.44) of the Dirac equation. The result is that

$$E = \varepsilon = \pm E_p \tag{2.47}$$

is identical with the energy of the states. This is different from the case of the Klein–Gordon equation: The free Dirac equation possesses positive and negative energy solutions, whose energies are given by (2.47). What is their interpretation since there is none in the framework of a one-particle theory? With the proposal of the hole theory, Dirac showed the following way out of this difficulty: let us assume that real electrons are described only by positive energy states (2.47), these are the

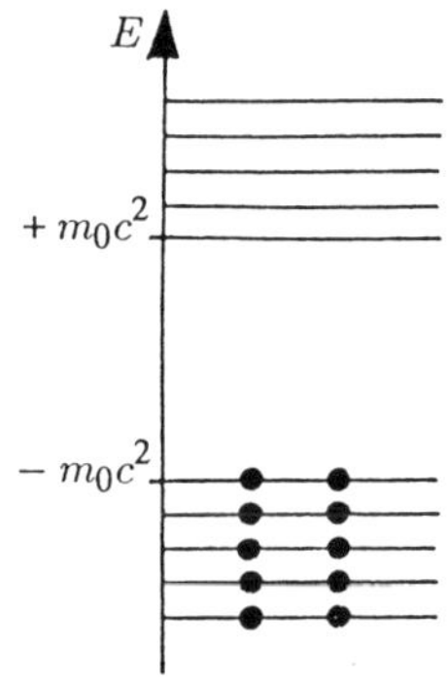

Fig. 2.3. The negative energy states with $E < m_0c^2$ are occupied by electrons and form the "Dirac sea". As we shall see later on in greater detail, they represent the vacuum state and are unobservable; whereas real, i.e. observable, electrons in general exist only in states of possitive energy

states with $E = +E_p$. All states of negative energy are occupied by electrons, one electron in each state of negative energy and given spin projection [see (2.44)]. This is illustrated in Fig. 2.3.

In this way a real electron of positive energy is also prevented from falling into energetically lower and lower states by radiation emission. A radiation catastrophe of this kind is averted by the effective Pauli principle which simply does not allow these transitions. On the other hand the question arises as to the meaning of a hole in this occupied "sea of negative states". Later on in Chap. 12, we will see that this leads to a meaningful description of the positron (the antiparticle of the electron). One thing should already have become clear here: The interpretation of the negative energy states of the Dirac equation takes us out of the one-particle picture and into the many-particle picture (more precisely, infinitely many particles are necessary for the formation of the Dirac sea). Therefore a consistent single-particle interpretation of the Dirac theory is not possible, a fact which is contrary to Dirac's initial intentions. Figure 2.3 gives an especially clear view of this aspect.

However, we pursue the single-particle aspect further and devote ourselves to the investigation of modified one-particle operators and to Ehrenfest's theorems. This way, perhaps, a consistent single-particle description can be obtained. For this it is helpful to introduce the *sign operator* $\hat{\Lambda}$,

$$\hat{\Lambda} = \frac{\hat{H}_\mathrm{f}}{\sqrt{\hat{H}_\mathrm{f}^2}} = \frac{c\hat{\boldsymbol{\alpha}}\cdot\hat{\boldsymbol{p}} + \hat{\beta}m_0c^2}{c\sqrt{\hat{\boldsymbol{p}}^2 + m_0^2c^2}} \quad . \tag{2.48}$$

Of course it commutes with the free Dirac Hamiltonian $\hat{H}_\mathrm{f}$. Furthermore, $\hat{\Lambda}$ is Hermitian and *unitary*, i. e.

$$\hat{\Lambda} = \hat{\Lambda}^\dagger = \hat{\Lambda}^{-1} \quad . \tag{2.49}$$

In momentum representation $\hat{\Lambda}$ has an especially simple form, that is

$$\hat{\Lambda} = \frac{c(\hat{\boldsymbol{\alpha}}\cdot\hat{\boldsymbol{p}}) + \hat{\beta}m_0c^2}{E_p} \quad . \tag{2.50}$$

The name "*sign operator*" comes from the fact that

$$\hat{\Lambda}\Psi_{p,\lambda,s_z} = \frac{\varepsilon}{E_p}\Psi_{p,\lambda,s_z} = \frac{\lambda E_p}{E_p}\Psi_{p,\lambda,s_z} = \lambda\Psi_{p,\lambda,s_z} \quad . \tag{2.51}$$

$\hat{\Lambda}$ has as eigenvalue the sign $\lambda(=\pm 1)$ of the time-evolution factor. $\lambda = +1$ means positive-energy states, $\lambda = -1$ negative-energy states. An arbitrary state with fixed λ can be written in the form

$$\Psi_\lambda = \sum_{s_z}\int A_{s_z}(\boldsymbol{p})\Psi_{p,\lambda,s_z}\,\mathrm{d}^3p \quad . \tag{2.52}$$

Then

$$\hat{\Lambda}\Psi_\lambda = \sum_{s_z}\int A_{s_z}(\boldsymbol{p})\frac{\hat{H}_\mathrm{f}}{E_p}\Psi_{p,\lambda,s_z}\,\mathrm{d}^3\boldsymbol{p} = \lambda\Psi_\lambda \quad . \tag{2.53}$$

We can use $\hat{\Lambda}$ in order to introduce the *projection operators* $\hat{\Lambda}_\pm$ by

$$\hat{\Lambda}_\pm = \tfrac{1}{2}(1 \pm \hat{\Lambda}) \tag{2.54}$$

with the useful properties

$$\begin{aligned}
&\hat{\Lambda}_+ \Psi_{\lambda=+1} = \Psi_{\lambda=+1} \quad , \\
&\hat{\Lambda}_+ \Psi_{\lambda=-1} = 0 \quad , \\
&\hat{\Lambda}_- \Psi_{\lambda=+1} = 0 \quad , \\
&\hat{\Lambda}_- \Psi_{\lambda=-1} = \Psi_{\lambda=-1} \quad .
\end{aligned} \tag{2.55}$$

The operators $\hat{\Lambda}_\pm$ split off the positive (or negative) parts of the state to which they are applied. As earlier, in the context of the Klein–Gordon equation [cf. (1.160, 161)], we call operators "*even*" or "*odd*" if they transform positive (negative) functions into positive (negative) or negative (positive) functions, respectively. The product of two even or two odd operators is always an even operator, and the product of an even and an odd operator is always an odd operator. Since all positive functions ($\lambda = \pm 1$) are orthogonal with respect to all negative ($\lambda = -1$) functions, the expectation value of an odd operator with states of fixed λ is always zero; hence,

$$\langle \Psi_\lambda | \hat{A}_{\text{odd}} | \Psi_\lambda \rangle = 0 \quad .$$

A consistent one-particle theory can only use states with a specified sign (either $\lambda = +1$ or $\lambda = -1$), because the energy can only be defined meaningfully in that way (cf. our preceding discussion). However, from that it follows that *in a consistent one-particle theory all physical quantities must necessarily be defined by even operators*. In the following we shall see that for Dirac's theory, Ehrenfest's theorems follow under this condition too, i.e. the quantum-mechanical operator equations and the corresponding classical equations become identical. Once again this means that the mean values comply with the classical equations, a fact which is quite significant. We formalize these considerations by splitting up every operator $\hat{A}$ into an even $[\hat{A}]$ and an odd $\{\hat{A}\}$ part

$$\hat{A} = [\hat{A}] + \{\hat{A}\} \quad . \tag{2.56}$$

If we simply write in short form $\Psi_\pm$ for $\Psi_{\lambda=\pm 1}$ we obtain

$$\hat{A}\Psi_+ = [\hat{A}]\Psi_+ + \{\hat{A}\}\Psi_+ \quad , \tag{2.57a}$$

$$\hat{A}\Psi_- = [\hat{A}]\Psi_- + \{\hat{A}\}\Psi_- \quad , \tag{2.57b}$$

$$\hat{\Lambda}\hat{A}\hat{\Lambda}\Psi_+ = \hat{\Lambda}\hat{A}\Psi_+ = [\hat{A}]\Psi_+ - \{\hat{A}\}\Psi_+ \quad , \tag{2.57c}$$

$$\hat{\Lambda}\hat{A}\hat{\Lambda}\Psi_- = -\hat{\Lambda}\hat{A}\Psi_- = [\hat{A}]\Psi_- - \{\hat{A}\}\Psi_- \quad . \tag{2.57d}$$

Hence, it follows by, say, addition and subtraction of (2.57a) and (2.57c) or, also, (2.57b) and (2.57d) that

$$[\hat{A}] = \tfrac{1}{2}(\hat{A} + \hat{\Lambda}\hat{A}\hat{\Lambda}) \quad , \quad \{\hat{A}\} = \tfrac{1}{2}(\hat{A} - \hat{\Lambda}\hat{A}\hat{\Lambda}) \quad . \tag{2.58}$$

We immediately recognize that the free Dirac Hamiltonian $\hat{H}_\text{f}$ is an even operator since

$$\hat{\Lambda}\hat{H}_\text{f}\hat{\Lambda} = \hat{H}_\text{f} \quad \text{and therefore} \quad [\hat{H}_\text{f}] = \tfrac{1}{2}(\hat{H}_\text{f} + \hat{\Lambda}\hat{H}_\text{f}\hat{\Lambda}) = \hat{H}_\text{f} \quad .$$

EXERCISE

2.2 Lagrange Density and Energy-Momentum Tensor of the Free Dirac Equation

Problem. Determine the Lagrangian density of the free Dirac field. Calculate the energy-momentum tensor and interpret the individual results.

Solution. We claim that the free Dirac Lagrange density has the form

$$\mathcal{L} = \overline{\psi}(ci\hbar\gamma^\mu \partial_\mu - m_0c^2)\psi \quad . \tag{1}$$

$\overline{\psi} := \psi^\dagger\gamma^0$ is called the spinor adjoint to ψ and the abbreviation γ^μ stands for $\gamma^0 = \hat{\beta}$, $\gamma^i = \hat{\beta}\hat{\alpha}^i$ [cf. Chap. 3 (3.8)]. Also $\partial_\mu := \partial/\partial x^\mu$ is a shorthand notation. Straight away we introduce these γ^μ matrices instead of the $\hat{\alpha}^i$ and $\hat{\beta}$, but will revert to the $\hat{\alpha}^i$, $\hat{\beta}$ representation in all important sections. The γ^μ are appropriate for the covariant formulation of the Dirac equation. The Lagrangian density (1) can therefore be rewritten as

$$\mathcal{L} = \psi^\dagger i\hbar\partial_t\psi + c\psi^\dagger i\hbar\gamma^0\gamma^i\partial_i\psi - m_0c^2\psi^\dagger\gamma_0\psi \quad , \tag{2}$$

since $\gamma_0 = \gamma^0$. Because of $\partial_i = \partial/\partial x^i = (\nabla)_i$, this yields

$$\begin{aligned} \mathcal{L} &= \psi^\dagger\left(i\hbar\partial_t + i\hbar c\hat{\boldsymbol{\alpha}}\cdot\boldsymbol{\nabla} - m_0c^2\gamma_0\right)\psi \\ &= \psi^\dagger\left(i\hbar\partial_t - c\hat{\boldsymbol{\alpha}}\cdot\hat{\mathbf{p}} - \hat{\beta}m_0c^2\right)\psi \quad , \end{aligned} \tag{3}$$

which uses the $\hat{\alpha}^i$, $\hat{\beta}$ matrices instead of the γ^μ. We realize that this is the correct Lagrangian density if we determine the equations of motion by variation. Variation with respect to $\overline{\psi}$ yields

$$\begin{aligned} \frac{\delta\int\mathcal{L}\mathrm{d}^4x}{\delta\overline{\psi}} = 0 &\Rightarrow \frac{\partial\mathcal{L}}{\partial\overline{\psi}} - \partial_\mu\frac{\partial\mathcal{L}}{\partial(\partial_\mu\overline{\psi})} = 0 \\ &\Rightarrow \left(ci\hbar\gamma^\mu\partial_\mu - m_0c^2\right)\psi = 0 \quad . \end{aligned} \tag{4}$$

This is Dirac's equation

$$\begin{aligned} i\hbar\partial_t\psi &= (\hat{\boldsymbol{\alpha}}\cdot\hat{\mathbf{p}} + \hat{\beta}m_0c^2)\psi \\ &\equiv \hat{H}_\mathrm{f}\psi \end{aligned} \tag{5}$$

with the free Hamiltonian

$$\hat{H}_\mathrm{f}(c\hat{\boldsymbol{\alpha}}\cdot\hat{\mathbf{p}} + \hat{\beta}m_0c^2) \quad . \tag{6}$$

We recognize that for the solutions of the equations of motion (5)

$$\delta\mathcal{L}(\psi) \equiv 0 \quad . \tag{7}$$

Variation with respect to ψ yields

Exercise 2.2.

$$\frac{\delta \int \mathcal{L}\,\mathrm{d}^4x}{\delta\psi} = 0 \Rightarrow \frac{\partial\mathcal{L}}{\partial\psi} - \partial_\mu \frac{\partial\mathcal{L}}{\partial(\partial_\mu\psi)} = 0$$
$$\Rightarrow \overline{\psi}\left(\mathrm{i}\hbar\gamma_\mu \overleftarrow{\partial}_\mu c + m_0c^2\right) = 0 \quad , \tag{8}$$

where $\overleftarrow{\partial}_\mu$ acts to the left on $\overline{\psi}$. One can easily calculate the canonical energy-momentum tensor from the Lagrangian density $\mathcal{L}$:

$$T^\mu{}_\nu = \frac{\partial\mathcal{L}}{\partial(\partial_\mu\psi)}\partial_\nu\psi + \frac{\partial\mathcal{L}}{\partial(\partial_\mu\overline{\psi})}\partial_\nu\overline{\psi} - \delta^\mu_\nu\mathcal{L} \quad , \tag{9}$$

which follows explicitly with (1) as

$$T^\mu{}_\nu = \overline{\psi}\mathrm{i}\hbar c\gamma^\mu\gamma_\nu\psi - \delta^\mu_\nu\overline{\psi}\mathrm{i}\hbar c\gamma^\sigma\partial_\sigma\psi + \delta^\mu_\nu m_0c^2\overline{\psi}\psi \quad . \tag{10}$$

From that one obtains the energy density $T^0{}_0$,

$$T^0{}_0 = -\psi^\dagger\mathrm{i}\hbar\hat{\boldsymbol{\alpha}}\cdot\boldsymbol{\nabla}c\psi + m_0c^2\overline{\psi}\psi$$
$$= \psi^\dagger\left(\hat{\boldsymbol{\alpha}}\cdot\hat{\boldsymbol{p}}c + \hat{\beta}m_0c^2\right)\psi = \psi^\dagger\hat{H}_\mathrm{f}\psi \quad . \tag{11}$$

Consequently

$$\int T^0{}_0\,\mathrm{d}^3x = \langle\psi|\hat{H}_\mathrm{f}|\psi\rangle \tag{12}$$

(that is, the energy), is just the expectation value of $\hat{H}_\mathrm{f}$ in the state ψ. By analogy the momentum density $T^0{}_i$ is given by

$$T^0{}_i = \overline{\psi}\mathrm{i}\hbar c\gamma^0\partial_i\psi = \psi^\dagger(\hat{\boldsymbol{p}})_i c\psi \quad , \tag{13}$$

in other words, $p_i \equiv (1/c)\int T^0{}_i\,\mathrm{d}^3x = \langle\psi|(\hat{\boldsymbol{p}})_i|\psi\rangle$ is the expectation value of the momentum operator in the state ψ. The components

$$T^i{}_j = \overline{\psi}\mathrm{i}\hbar c\left(\gamma^i\partial_j - \delta^i_j\gamma^\mu\partial_\mu\right)\psi + \delta^i_j m_0c^2\overline{\psi}\psi \quad , \tag{14}$$

which can simply be written for each solution of the equation of motion as

$$T^i{}_j = \overline{\psi}\mathrm{i}\hbar c\gamma^i\partial_j\psi = -\overline{\psi}\gamma^i\hat{p}_jc\psi = -\psi^\dagger\hat{\alpha}^i\hat{p}_jc\psi \quad , \tag{15}$$

are called the components of the *stress-strain tensor*. The trace of $T^\mu{}_\nu$ is given by

$$T \equiv T^\mu{}_\mu = \overline{\psi}\mathrm{i}\hbar c\left(\gamma^\mu\partial_\mu - 4\gamma^\sigma\partial_\sigma\right)\psi + 4m_0c^2\overline{\psi}\psi$$
$$= -3\overline{\psi}\left(\mathrm{i}\hbar c\gamma^\mu\partial_\mu - m_0c^2\right)\psi + m_0c^2\overline{\psi}\psi \tag{16}$$

which, for every solution of the equation of motion, just becomes

$$T = m_0c^2\overline{\psi}\psi \quad . \tag{17}$$

One should notice that this is not proportional to the charge density

$$\varrho = e\overline{\psi}\gamma_0\psi = e\psi^\dagger\psi \quad . \tag{18}$$

EXERCISE

2.3 Calculation of the Energies of the Solutions to the Free Dirac Equation in the Canonical Formalism

Problem. Calculate in the framework of the canonical formalism the energies of the solutions (2.44) of the free Dirac equation.

Solution. We can determine the energy as the integral

$$E = \int_V T^0{}_0 \, \mathrm{d}^3x \quad . \tag{1}$$

If we insert (11) of Exercise 2.2 as well as the solutions (2.44a) for ψ into this, we obtain

$$\begin{aligned} E = N^2 \int_V & \left(1, 0, \frac{pc}{m_0c^2 + \lambda E_p}, 0\right) \\ & \times \left(\hat{\boldsymbol{\alpha}} \cdot \hat{\boldsymbol{p}}c + \hat{\beta} m_0 c^2\right) \begin{pmatrix} 1 \\ 0 \\ \dfrac{pc}{m_0c^2 + \lambda E_p} \\ 0 \end{pmatrix} \mathrm{d}^3x \quad . \end{aligned} \tag{2}$$

Since for (2.44) $\boldsymbol{p} = (0, 0, p)$ holds, we get by calculation

$$\begin{aligned} E &= N^2 V \left(1, 0, \frac{pc}{m_0c^2 + \lambda E_p}, 0\right) \\ &\quad \times \left(pc \begin{pmatrix} 0 & \sigma_z \\ \sigma_z & 0 \end{pmatrix} + m_0c^2 \begin{pmatrix} 1\!\!1 & 0 \\ 0 & -1\!\!1 \end{pmatrix}\right) \begin{pmatrix} 1 \\ 0 \\ \dfrac{pc}{m_0c^2 + \lambda E_p} \\ 0 \end{pmatrix} \\ &= N^2 V \left(1, 0, \frac{pc}{m_0c^2 + \lambda E_p}, 0\right) \\ &\quad \times \left(pc \begin{pmatrix} \dfrac{pc}{m_0c^2 + \lambda E_p} \\ 0 \\ 1 \\ 0 \end{pmatrix} + m_0c^2 \begin{pmatrix} 1 \\ 0 \\ \dfrac{-pc}{m_0c^2 + \lambda E_p} \\ 0 \end{pmatrix} \right) \\ &= N^2 V \left(\frac{2p^2c^2}{m_0c^2 + \lambda E_p} + m_0c^2 - \frac{m_0c^2p^2c^2}{(m_0c^2 + \lambda E_p)^2} \right) \\ &= N^2 V \left(m_0c^2 + \lambda E_p\right)^{-2} \left(2p^2c^2m_0c^2 + 2p^2c^2\lambda E_p \right. \\ &\quad \left. + m_0^3c^6 + 2m_0^2c^4\lambda E_p + m_0^3c^6 + m_0c^2p^2c^2 - m_0c^2p^2c^2\right) \\ &= N^2 V \left(m_0c^2 + \lambda E_p\right)^{-2} \left(2(\lambda E_p)^3 + 2m_0^3c^6 + 2m_0c^2p^2c^2\right) \\ &= N^2 V \left(m_0c^2 + \lambda E_p\right)^{-1} 2\lambda^2 E_p^2 \quad . \end{aligned} \tag{3}$$

Exercise 2.3.

Here we have used wave functions which are normalized with respect to a finite spherical volume V. If we furthermore insert the normalization factor N from (2.36)

$$N^2 = V^{-1}\left(\frac{2\lambda E_p}{m_0c^2 + \lambda E_p}\right)^{-1} \quad , \tag{4}$$

we finally obtain

$$E = \lambda E_p \quad ,$$

that means

$$E = +\sqrt{p^2c^2 + m_0^2c^4}$$

for the upper energy continuum ($\lambda = +1$) and

$$E = -\sqrt{p^2c^2 + m_0^2c^4}$$

for the lower energy continuum ($\lambda = -1$). Hence, the free Dirac equation leads to states with positive and negative energy.

As for the Dirac Hamiltonian $\hat{H}_{\mathrm{f}}$, for the momentum operator $\hat{\boldsymbol{p}}$ also

$$\hat{\Lambda}\hat{\boldsymbol{p}}\hat{\Lambda} = \hat{\boldsymbol{p}} \quad , \quad \text{so that} \quad [\hat{\boldsymbol{p}}] = \hat{\boldsymbol{p}} \quad .$$

Let us also determine the even part of the $\hat{\boldsymbol{\alpha}}$ operator. We have

$$\hat{\Lambda}\hat{\alpha}_i\hat{\Lambda} = \frac{c\hat{\boldsymbol{\alpha}}\cdot\hat{\boldsymbol{p}} + \hat{\beta}m_0c^2}{c\sqrt{\hat{\boldsymbol{p}}^2 + m_0^2c^2}}\hat{\alpha}_i\frac{c(\hat{\boldsymbol{\alpha}}\cdot\hat{\boldsymbol{p}}) + \hat{\beta}m_0c^2}{c\sqrt{\hat{\boldsymbol{p}}^2 + m_0^2c^2}} = -\hat{\alpha}_i + 2c\hat{p}_i\frac{\hat{\Lambda}}{c\sqrt{\hat{\boldsymbol{p}}^2 + m_0^2c^2}} \quad ,$$

and therefore

$$[\hat{\alpha}_i] = \frac{1}{2}(\hat{\alpha}_i + \hat{\Lambda}\hat{\alpha}_i\hat{\Lambda}) = c\hat{p}_i\frac{\hat{\Lambda}}{c\sqrt{\hat{\boldsymbol{p}}^2 + m_0^2c^2}} \quad . \tag{2.59}$$

In a similar way one calculates

$$[\hat{\beta}] = \frac{m_0c^2}{E_p}\hat{\Lambda} = m_0c^2\frac{\hat{\Lambda}}{c\sqrt{\hat{\boldsymbol{p}}^2 + m_0^2c^2}} \quad . \tag{2.60}$$

Now we determine the *velocity operator* in the Dirac theory. Since we are concerned with an equation of Schrödinger type [cf. (2.2)], the theorems for the time derivatives of operators which we formulated in (1.157) are also valid here and we obtain

$$\begin{aligned}\frac{\mathrm{d}\hat{\boldsymbol{x}}}{\mathrm{d}t} &= \frac{1}{\mathrm{i}\hbar}\left[\hat{\boldsymbol{x}}, \hat{H}_{\mathrm{f}}\right]_- = \frac{1}{\mathrm{i}\hbar}\left[\hat{\boldsymbol{x}}, c(\hat{\boldsymbol{\alpha}}\cdot\hat{\boldsymbol{p}}) + \hat{\beta}m_0c^2\right]_- \\ &= \frac{c}{\mathrm{i}\hbar}[\hat{\boldsymbol{x}}, \hat{\boldsymbol{\alpha}}\cdot\hat{\boldsymbol{p}}]_- = c\hat{\boldsymbol{\alpha}} \equiv \hat{\boldsymbol{v}} \quad .\end{aligned} \tag{2.61}$$

Since the eigenvalues of $\hat{\boldsymbol{\alpha}}$ have the values ± 1 we obtain here the *paradoxical result* that the absolute value of the velocity of a relativistic spin-$\frac{1}{2}$ particle always equals the velocity of light.[8] Moreover, since the $\hat{\alpha}_i$ do not commute with each other the components of the velocity $\mathrm{d}x_i/\mathrm{d}t$ would not be simultaneously measurable. Of course, this is nonsense and certainly would not yield the classical relations for the mean values (Ehrenfest's theorems). According to previous statements, however, we find, using (2.59), the *true velocity operator*, that is the even part of $\mathrm{d}\boldsymbol{x}/\mathrm{d}t$, to be

$$\left[\frac{\mathrm{d}\hat{\boldsymbol{x}}}{\mathrm{d}t}\right] = c[\hat{\boldsymbol{\alpha}}] = \frac{c^2\hat{\boldsymbol{p}}\hat{\Lambda}}{c\sqrt{\hat{\boldsymbol{p}}^2 + m_0^2c^2}} \quad . \tag{2.62}$$

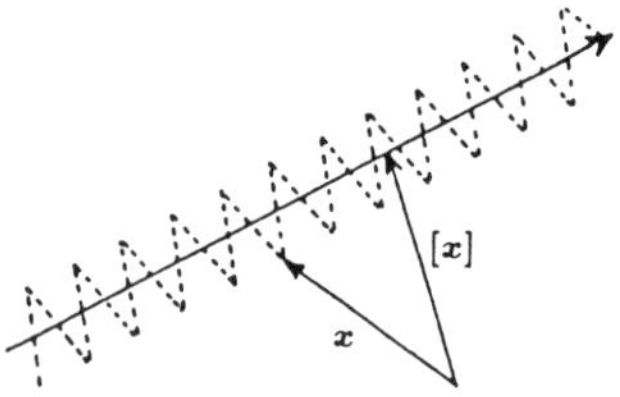

Fig. 2.4. Illustration of the *Zitterbewegung* in the forced one-particle picture. The location $\boldsymbol{x}$ performs a Zitterbewegung around the mean (classical) trajectory $[\boldsymbol{x}]$

Accordingly, the true velocity equals $c^2\hat{\boldsymbol{p}}/E_p$ for positive and $-c^2\hat{\boldsymbol{p}}/E_p$ for negative free solutions of the Dirac equation. This exactly corresponds, for positive solutions, to the classical picture. For waves with negative energy the result is paradoxical in the first instance since their velocity is directed against their momentum. Therefore particles with negative energy behave formally as if they would have a negative mass. Later on we shall comprehend these facts better in the context of the hole theory (Chap. 12). Accordingly, the motion of a single particle can be visualized as follows (cf. Fig. 2.4). The particle has to perform – if we adhere to the one particle interpretation – a kind of *Zitterbewegung* (trembling motion) around the classical location $[\boldsymbol{x}]$ (classical trajectory).

Now we investigate the result (2.62) from yet another point of view. If we consider wave packets built up from free Dirac waves, we show that only the mean of the centre of the wave packet generally follows a classical trajectory.

In order to understand this we integrate the *equations of motion in the Heisenberg representation*:

$$\frac{\mathrm{d}\hat{\boldsymbol{x}}}{\mathrm{d}t} = \frac{1}{\mathrm{i}\hbar}\left[\hat{\boldsymbol{x}},\hat{H}_\mathrm{f}\right]_- = c\hat{\boldsymbol{\alpha}} \quad , \tag{2.63}$$

$$\frac{\mathrm{d}\hat{\boldsymbol{\alpha}}}{\mathrm{d}t} = \frac{1}{\mathrm{i}\hbar}\left[\hat{\boldsymbol{\alpha}},\hat{H}_\mathrm{f}\right]_- = \frac{\mathrm{i}}{\hbar}\left[\hat{H}_\mathrm{f}\hat{\boldsymbol{\alpha}} + \hat{\boldsymbol{\alpha}}\hat{H}_\mathrm{f}\right] - \frac{2\mathrm{i}}{\hbar}\hat{\boldsymbol{\alpha}}\hat{H}_\mathrm{f} = \frac{2\mathrm{i}c\hat{\boldsymbol{p}}}{\hbar} - \frac{2\mathrm{i}}{\hbar}\hat{\boldsymbol{\alpha}}\hat{H}_\mathrm{f} \quad . \tag{2.64}$$

Since $\hat{\boldsymbol{p}}$ and $\hat{H}_\mathrm{f}$ are constant in time (because $[\hat{\boldsymbol{p}},\hat{H}_\mathrm{f}]_- = 0 = [\hat{H}_\mathrm{f},\hat{H}_\mathrm{f}]_-$), one can easily integrate the former equation and obtain

$$\hat{\boldsymbol{\alpha}}(t) = \left(\hat{\boldsymbol{\alpha}}(0) - \frac{c\hat{\boldsymbol{p}}}{\hat{H}_\mathrm{f}}\right)\exp\left(-2\mathrm{i}\hat{H}_\mathrm{f}t/\hbar\right) + \frac{c\hat{\boldsymbol{p}}}{\hat{H}_\mathrm{f}} \quad . \tag{2.65}$$

If this result is inserted into (2.63), then $\hat{\boldsymbol{x}}(t)$ can be explicitly calculated with the result

$$\hat{\boldsymbol{x}}(t) = \hat{\boldsymbol{x}}(0) + \frac{c^2\hat{\boldsymbol{p}}}{\hat{H}_\mathrm{f}}t + \left(\hat{\boldsymbol{\alpha}}(0) - \frac{c\hat{\boldsymbol{p}}}{\hat{H}_\mathrm{f}}\right)\frac{\mathrm{i}\hbar c}{2\hat{H}_\mathrm{f}}\exp\left(-2\mathrm{i}\hat{H}_\mathrm{f}t/\hbar\right) \quad . \tag{2.66}$$

This result determines the space operator $\hat{\boldsymbol{x}}(t)$ in the Heisenberg representation at time t. Of course, the operator $\hat{\boldsymbol{x}}(t)$ also depends on $\hat{\boldsymbol{\alpha}}(0)$ and $\hat{\boldsymbol{x}}(0)$. Making use of (2.66), the equation of motion of the centre $\langle\hat{\boldsymbol{x}}(t)\rangle$ of each wave packet can be

[8] This was first recognized by Gregory Breit in 1928.

determined. It is interesting to compare this with the classical equation of motion, which reads

$$\boldsymbol{x}_{\rm cl}(t) = \boldsymbol{x}_{\rm cl}(0) + \left(\frac{c^2\boldsymbol{p}}{E_p}\right)_{\rm cl} t \quad . \tag{2.67}$$

The comparison of (2.67) with (2.66) shows that the free wave packet indeed follows the uniform classical motion [first two terms in (2.66)], but that moreover a *rapidly oscillating motion* is superposed, namely

$$\left\langle \left| \mathrm{i}\left(\hat{\boldsymbol{\alpha}}(0) - \frac{c\hat{\boldsymbol{p}}}{\hat{H}_{\rm f}}\right)\frac{\hbar c}{2\hat{H}_{\rm f}} \exp\left(-2\mathrm{i}\hat{H}_{\rm f}t/\hbar\right)\right| \right\rangle \quad . \tag{2.68}$$

The amplitude and frequency of these additional oscillations are of the order $\hbar/2m_0c$ and $2m_0c^2/\hbar$, respectively. This oscillating motion is the previously mentioned *Zitterbewegung*. It vanishes if wave packets with exclusively positive or negative energy are considered. This becomes clear if we calculate, with (2.54),

$$\hat{\Lambda}_\pm\left(\hat{\boldsymbol{\alpha}} - \frac{c\hat{\boldsymbol{p}}}{\hat{H}_{\rm f}}\right)\frac{\exp(-2\mathrm{i}\hat{H}_{\rm f}t/\hbar)}{2\hat{H}_{\rm f}}\hat{\Lambda}_\pm = 0 \quad , \tag{2.69}$$

where $\hat{\Lambda}_\pm$ are the projection operators for the states of positive and negative energy, respectively. Namely,

$$\left[\hat{H}_{\rm f}, \hat{\boldsymbol{\alpha}}\right]_- = 2c\hat{\boldsymbol{p}} - 2\hat{\boldsymbol{\alpha}}\hat{H}_{\rm f} \quad , \tag{2.70a}$$

$$\hat{H}_{\rm f}\hat{\boldsymbol{\alpha}} + \hat{\boldsymbol{\alpha}}\hat{H}_{\rm f} = 2c\hat{\boldsymbol{p}} \quad ; \tag{2.70b}$$

$$\begin{aligned}\left[\hat{\Lambda}_\pm, \hat{\boldsymbol{\alpha}}\right]_- &= \frac{1}{2}\left[(1 \pm \hat{\Lambda}), \hat{\boldsymbol{\alpha}}\right]_- = \pm\frac{1}{2}[\hat{\Lambda}, \hat{\boldsymbol{\alpha}}] \\ &= \pm\frac{1}{2}\left[\frac{\hat{H}_{\rm f}}{E_p}, \hat{\boldsymbol{\alpha}}\right]_- = \pm\frac{1}{2}\frac{1}{E_p}\left[\hat{H}_{\rm f}, \hat{\boldsymbol{\alpha}}\right]_- = \pm\frac{c\hat{\boldsymbol{p}}}{E_p} \mp \hat{\boldsymbol{\alpha}}\frac{\hat{H}_{\rm f}}{E_p} \quad ;\end{aligned} \tag{2.70c}$$

$$\left[\hat{H}_{\rm f}, \hat{\Lambda}_\pm\right]_- = 0 \quad , \tag{2.70d}$$

$$\hat{H}_{\rm f}\hat{\Lambda}_\pm = \pm E_p\hat{\Lambda}_\pm \quad . \tag{2.70e}$$

With this we can write

$$\hat{\Lambda}_\pm\left(\hat{\boldsymbol{\alpha}} - \frac{c\hat{\boldsymbol{p}}}{\hat{H}_{\rm f}}\right)\hat{\Lambda}_\pm\frac{\exp(-2\mathrm{i}\hat{H}_{\rm f}t/\hbar)}{2\hat{H}_{\rm f}} \equiv \hat{A}\frac{\exp(-2\mathrm{i}\hat{H}_{\rm f}t/\hbar)}{2\hat{H}_{\rm f}} \quad ,$$

and, because of the preceding calculation (2.70c), one gets

$$\hat{\Lambda}_\pm\hat{\boldsymbol{\alpha}} = \hat{\boldsymbol{\alpha}}\hat{\Lambda}_\pm \pm \frac{c\hat{\boldsymbol{p}}}{E_p} \mp \hat{\boldsymbol{\alpha}}\frac{\hat{H}_{\rm f}}{E_p} \quad .$$

Thus one obtains

$$\hat{A} = \left[\hat{\boldsymbol{\alpha}}\hat{\Lambda}_\pm \pm \frac{c\hat{\boldsymbol{p}}}{E_p} \mp \hat{\boldsymbol{\alpha}}\frac{\hat{H}_{\rm f}}{E_p} - \frac{c\hat{\boldsymbol{p}}}{\hat{H}_{\rm f}}\right]\hat{\Lambda}_\pm \quad ,$$

and the identity $\hat{\Lambda}_\pm\hat{\Lambda}_\pm = \hat{\Lambda}_\pm$ for the projection operators is recovered in a trivial way, finally resulting in

$$\hat{A} = \left[\hat{\alpha} \pm \frac{c\hat{p}}{E_p} \mp \hat{\alpha}\frac{(\pm E_p)}{E_p} - \frac{c\hat{p}}{\hat{H}_\mathrm{f}}\right]\hat{\Lambda}_\pm = 0 \quad .$$

Hence, the *Zitterbewegung* is caused by the interference between the positive and negative energy compounds of a wave packet. It demonstrates that in a real sense a single-particle theory is not possible; it can only be approximately obtained if the associated wave packets can be restricted to one energy range.

2.3 Nonrelativistic Limit of the Dirac Equation

Before we proceed further with the extension of the Dirac theory, it is important to check whether the Dirac equation yields physically reasonable results in the nonrelativistic limiting case. First we study the case of an *electron at rest*; in this case we obtain the Dirac equation by setting $\hat{\boldsymbol{p}}\psi = 0$ in (2.2),

$$\mathrm{i}\hbar\frac{\partial\psi}{\partial t} = \hat{\beta}m_0c^2\psi \quad . \tag{2.71}$$

In the particular representation (2.12) with

$$\hat{\beta} = \begin{pmatrix} \mathbb{1} & 0 \\ 0 & -\mathbb{1} \end{pmatrix}$$

we are able to write down the four solutions

$$\psi^{(1)} = \begin{pmatrix} 1 \\ 0 \\ 0 \\ 0 \end{pmatrix} \exp[-\mathrm{i}(m_0c^2/\hbar)t] \quad , \quad \psi^{(2)} = \begin{pmatrix} 0 \\ 1 \\ 0 \\ 0 \end{pmatrix} \exp[-\mathrm{i}(m_0c^2/\hbar)t] \quad ,$$
$$\psi^{(3)} = \begin{pmatrix} 0 \\ 0 \\ 1 \\ 0 \end{pmatrix} \exp[+\mathrm{i}(m_0c^2/\hbar)t] \quad , \quad \psi^{(4)} = \begin{pmatrix} 0 \\ 0 \\ 0 \\ 1 \end{pmatrix} \exp[+\mathrm{i}(m_0c^2/\hbar)t] \quad . \tag{2.72}$$

The first two wave functions correspond to positive, the last two to negative energy values. The interpretation of the solutions with negative energy still causes problems and is postponed [cf. (2.47)]; however, the correct interpretation leads to a considerable triumph of relativistic quantum theory by forecasting and describing antiparticles. At first therefore we restrict ourselves to solutions of positive energy. In order to show that the Dirac equation reproduces the two-component Pauli equation[9] in the nonrelativistic limit, we introduce the electromagnetic four-potential

$$A^\mu = \{A_0(\boldsymbol{x}), \boldsymbol{A}(\boldsymbol{x})\} \tag{2.73}$$

into the Dirac equation (2.2). We know that the *minimal coupling*

$$\hat{p}^\mu \to \hat{p}^\mu - \frac{e}{c}A^\mu \equiv \hat{\Pi}^\mu$$

[9] See W. Greiner: *Quantum Mechanics – An Introduction*, 3rd ed. (Springer, Berlin, Heidelberg 1994), Chap. 12.

ensures gauge invariance of the theory, where $\hat{\Pi}^\mu$ is the kinetic momentum and $\hat{p}^\mu$ the canonical momentum. So we are inevitably guided to the *Dirac equation with electromagnetic potentials*

$$c\left(\mathrm{i}\hbar\frac{\partial}{\partial ct}-\frac{e}{c}A_0\right)\psi=\left(c\hat{\boldsymbol{\alpha}}\cdot\left(\hat{\boldsymbol{p}}-\frac{e}{c}\boldsymbol{A}\right)+\hat{\beta}m_0c^2\right)\psi$$

or

$$\mathrm{i}\hbar\frac{\partial}{\partial t}\psi=\left(c\hat{\boldsymbol{\alpha}}\cdot\left(\hat{\boldsymbol{p}}-\frac{e}{c}\boldsymbol{A}\right)+eA_0+\hat{\beta}m_0c^2\right)\psi \quad . \tag{2.74}$$

This contains the interaction with the electromagnetic field

$$\hat{H}'=-\frac{e}{c}c\hat{\boldsymbol{\alpha}}\cdot\boldsymbol{A}+eA_0=-\frac{e}{c}\hat{\boldsymbol{v}}\cdot\boldsymbol{A}+eA_0 \quad , \tag{2.75}$$

where

$$\hat{\boldsymbol{v}}=\frac{\mathrm{d}\hat{\boldsymbol{x}}}{\mathrm{d}t}=c\hat{\boldsymbol{\alpha}}$$

is the relativistic velocity operator. The expression (2.75) corresponds to the classical interaction of a moving charged point-like particle with the electromagnetic field. The velocity operator, however, is the formal operator $\hat{\boldsymbol{v}}$ from (2.61) which contains the *Zitterbewegung*.

EXERCISE

2.4 Time Derivative of the Position and Momentum Operators for Dirac Particles in an Electromagnetic Field

Problem. Calculate the time derivative of the position operator $\hat{\boldsymbol{x}}$ and of the kinetic momentum operator $\hat{\boldsymbol{\Pi}}=\hat{\boldsymbol{p}}-(e/c)\boldsymbol{A}$ for Dirac particles in an electromagnetic field. Compare the result with the corresponding classical expressions and discuss Ehrenfest's theorems.

Solution. The Hamiltonian of a Dirac particle in the electromagnetic field is [setting $A_0(\boldsymbol{x})$ equal to $\phi(\boldsymbol{x})$]

$$\hat{H}=c\hat{\boldsymbol{\alpha}}\cdot\left(\hat{\boldsymbol{p}}-\frac{e}{c}\boldsymbol{A}\right)+\hat{\beta}m_0c^2+e\phi \quad . \tag{1}$$

Furthermore, the equation of motion of an arbitrary operator $\hat{F}$ is given by

$$\frac{\mathrm{d}\hat{F}}{\mathrm{d}t}=\frac{\partial\hat{F}}{\partial t}+\frac{\mathrm{i}}{\hbar}[\hat{H},\hat{F}]_- \quad , \tag{2}$$

and for the position operator we obtain

$$\frac{\mathrm{d}\hat{\boldsymbol{x}}}{\mathrm{d}t}=\frac{\mathrm{i}}{\hbar}[\hat{H},\hat{\boldsymbol{x}}] \tag{3}$$

Exercise 2.4.

since $\partial\hat{\boldsymbol{x}}/\partial t = 0$. With (1), the commutator reads

$$[\hat{H},\hat{\boldsymbol{x}}]_- = c[\hat{\boldsymbol{\alpha}}\cdot\hat{\boldsymbol{p}},\hat{\boldsymbol{x}}]_- - e[\hat{\boldsymbol{\alpha}}\cdot\boldsymbol{A},\hat{\boldsymbol{x}}]_- + m_0c^2[\hat{\beta},\hat{\boldsymbol{x}}]_- + e[\phi,\hat{\boldsymbol{x}}]_- \quad , \tag{4}$$

ϕ is the Coulomb potential, i.e. $[\phi,\hat{\boldsymbol{x}}]_- = 0$. The position operator $\hat{\boldsymbol{x}}$ is nothing but a simple multiplication operator $\hat{\boldsymbol{x}}\psi = \boldsymbol{x}\psi$ (i.e. it is diagonal with respect to the spinor indices) and contains no differentiation. Hence $[\hat{\beta},\hat{\boldsymbol{x}}]_- = 0 = [\hat{\boldsymbol{\alpha}},\hat{\boldsymbol{x}}]_-$. Furthermore we will use the identity $[\hat{A}\hat{B},\hat{C}]_- = [\hat{A},\hat{C}]_-\hat{B} + \hat{A}[\hat{B},\hat{C}]_-$ to show that

$$\begin{aligned}[\hat{\boldsymbol{\alpha}}\cdot\hat{\boldsymbol{p}},\hat{\boldsymbol{x}}]_- &= \sum_{j,k}\{[\hat{\alpha}_j,\hat{x}_k]_-\hat{p}_j\boldsymbol{e}_k + \hat{\alpha}_j[\hat{p}_j,\hat{x}_k]_-\boldsymbol{e}_k\}\\ &= \sum_j \frac{\hbar}{\mathrm{i}}\hat{\alpha}_j\boldsymbol{e}_j = \frac{\hbar}{\mathrm{i}}\hat{\boldsymbol{\alpha}} \quad , \quad \text{since} \quad [\hat{p}_j,\hat{x}_k]_- = \frac{\hbar}{\mathrm{i}}\delta_{jk} \quad .\end{aligned} \tag{5}$$

$[\hat{\boldsymbol{\alpha}}\cdot\hat{\boldsymbol{A}},\hat{\boldsymbol{x}}]_- = 0$ since $\hat{\boldsymbol{x}}$ commutes with $\hat{\boldsymbol{\alpha}}$ as well as with $\boldsymbol{A}$: we obtain $[\hat{H},\hat{\boldsymbol{x}}]_- = (\hbar/\mathrm{i})c\hat{\boldsymbol{\alpha}}$, and hence

$$\frac{\mathrm{d}\hat{\boldsymbol{x}}}{\mathrm{d}t} = c\hat{\boldsymbol{\alpha}} \equiv \hat{\boldsymbol{v}} \quad ; \tag{6}$$

hence the velocity operator of a Dirac particle is given by

$$\hat{\boldsymbol{v}} = c\hat{\boldsymbol{\alpha}} \quad . \tag{7}$$

Let us now see how this operator acts on a Dirac spinor. Considering the single components, this reads as

$$\hat{v}^j\psi = c\hat{\alpha}_j\psi = \pm c\psi$$

since the operator $\hat{\boldsymbol{\alpha}}$ has the eigenvalues $\tilde{\alpha}_j = \pm 1$. This result means that a Dirac particle always moves with the speed of light and it is clear that this finding has no classical analogy [cf. the discussion following (2.61)]. The equation of motion for the kinetic momentum $\hat{\boldsymbol{\Pi}} = \hat{\boldsymbol{p}} - (e/c)\boldsymbol{A}$ is

$$\frac{\mathrm{d}\hat{\boldsymbol{\Pi}}}{\mathrm{d}t} = \frac{\partial\hat{\boldsymbol{\Pi}}}{\partial t} + \frac{\mathrm{i}}{\hbar}[\hat{H},\hat{\boldsymbol{\Pi}}]_- = \frac{\mathrm{i}}{\hbar}[\hat{H},\hat{\boldsymbol{\Pi}}]_- - \frac{e}{c}\frac{\partial\boldsymbol{A}}{\partial t} \quad , \tag{8}$$

because the potential $\boldsymbol{A}(\boldsymbol{r},t)$ can depend on time explicitly. The commutator is given by

$$[\hat{H},\hat{\boldsymbol{\Pi}}]_- = [\hat{H},\hat{\boldsymbol{p}}]_- - \frac{e}{c}[\hat{H},\boldsymbol{A}]_- \quad . \tag{9}$$

First we calculate the single commutators

$$[\hat{H},\hat{\boldsymbol{p}}]_- = c[\hat{\boldsymbol{\alpha}}\cdot\hat{\boldsymbol{p}},\hat{\boldsymbol{p}}]_- - e[\hat{\boldsymbol{\alpha}}\cdot\boldsymbol{A},\hat{\boldsymbol{p}}]_- + m_0c^2[\hat{\beta},\hat{\boldsymbol{p}}]_- + e[\phi,\hat{\boldsymbol{p}}] \quad , \tag{10}$$

and $\hat{\boldsymbol{p}} = -\mathrm{i}\hbar\boldsymbol{\nabla}$, it follows that i.e. $[\hat{\beta},\hat{\boldsymbol{p}}] = 0$ since $\hat{\beta}$ does not depend on space coordinates. Furthermore it holds that

$$e[\phi,\hat{\boldsymbol{p}}]_- = \mathrm{i}e\hbar[\boldsymbol{\nabla},\phi]_- = \mathrm{i}e\hbar(\boldsymbol{\nabla}\phi - \phi\boldsymbol{\nabla}) \quad .$$

Exercise 2.4.

Making use of this relation we obtain

$$\begin{aligned}e[\phi,\hat{\boldsymbol{p}}]_-\psi &= \mathrm{i}e\hbar(\boldsymbol{\nabla}\phi-\phi\boldsymbol{\nabla})\psi = \mathrm{i}e\hbar(\boldsymbol{\nabla}\phi\psi-\phi\boldsymbol{\nabla}\psi)\\ &= \mathrm{i}e\hbar\{\psi\boldsymbol{\nabla}\phi+\phi\boldsymbol{\nabla}\psi-\phi\boldsymbol{\nabla}\psi\} = \mathrm{i}e\hbar(\boldsymbol{\nabla}\phi)\psi\end{aligned}$$

for any wave function ψ, i.e. $e[\phi,\hat{\boldsymbol{p}}]_- = \mathrm{i}e\hbar(\boldsymbol{\nabla}\phi)$. Further we find

$$c[\hat{\boldsymbol{\alpha}}\cdot\hat{\boldsymbol{p}},\hat{\boldsymbol{p}}]_- = c\sum_{i,j}\{[\hat{\alpha}_i,\hat{p}_j]_-\hat{p}_i\boldsymbol{e}_j+\hat{\alpha}_i[\hat{p}_i,\hat{p}_j]_-\boldsymbol{e}_j\} = 0 \quad ,$$

because $[\hat{p}_i,\hat{p}_j]_- = 0$ and $[\hat{\alpha}_i,\hat{p}_j]_- = 0$ (since $\hat{\alpha}_i$ does not depend on the space coordinates). This leaves

$$\begin{aligned}-e\,[\hat{\boldsymbol{\alpha}}\cdot\boldsymbol{A},\boldsymbol{p}]_- &= -e\sum_{i,j}\left\{\left[\hat{\alpha}_i,\hat{p}_j\right]_-A_i\boldsymbol{e}_j+\hat{\alpha}_i\left[A_i,\hat{p}_j\right]_-\boldsymbol{e}_j\right\}\\ &= -e\sum_{i,j}\hat{\alpha}_i[A_i,\hat{p}_j]_-\boldsymbol{e}_j \quad .\end{aligned} \tag{11}$$

The second commutator in (9) yields

$$\begin{aligned}-\frac{e}{c}\left[\hat{H},\boldsymbol{A}\right]_- = -\frac{e}{c}\Big\{&c\,[\hat{\boldsymbol{\alpha}}\cdot\hat{\boldsymbol{p}},\boldsymbol{A}]_- - e\,[\hat{\boldsymbol{\alpha}}\cdot\boldsymbol{A},\boldsymbol{A}]_-\\ &+m_0c^2[\hat{\beta},\boldsymbol{A}]_- + e[\phi,\boldsymbol{A}]_-\Big\} \quad .\end{aligned}$$

Since $[\hat{\beta},\boldsymbol{A}]_- = [\phi,\boldsymbol{A}]_i = 0$ and $[\hat{\boldsymbol{\alpha}}\cdot\boldsymbol{A},\boldsymbol{A}]_- = 0$ ($\boldsymbol{A}$ commutes with $\hat{\boldsymbol{\alpha}}$ and also with itself), we get

$$\begin{aligned}-e[\hat{\boldsymbol{\alpha}}\cdot\hat{\boldsymbol{p}},\boldsymbol{A}]_- &= -e\left[\sum_{i,j}\left\{\left[\hat{\alpha}_i,A_j\right]_-\hat{p}_i\boldsymbol{e}_j+\hat{\alpha}_i\left[\hat{p}_i,A_j\right]_-\boldsymbol{e}_j\right\}\right]\\ &= -e\sum_{i,j}\hat{\alpha}_i\left[\hat{p}_i,A_j\right]_-\boldsymbol{e}_j \quad .\end{aligned} \tag{12}$$

In total we have

$$\begin{aligned}\frac{\mathrm{d}\hat{\boldsymbol{\Pi}}}{\mathrm{d}t} &= \frac{\mathrm{i}}{\hbar}\left[\mathrm{i}e\hbar(\boldsymbol{\nabla}\phi)-e\sum_{i,j}\hat{\alpha}_i\left\{[A_i,\hat{p}_j]_-+\left[\hat{p}_i,A_j\right]_-\right\}\boldsymbol{e}_j\right]-\frac{e}{c}\frac{\partial\boldsymbol{A}}{\partial t}\\ &= +e\underbrace{\left(-\frac{1}{c}\frac{\partial\boldsymbol{A}}{\partial t}-\boldsymbol{\nabla}\phi\right)}_{\boldsymbol{E}}+\frac{\mathrm{i}e}{\hbar}\sum_{i,j}\hat{\alpha}_i\left\{-\left[A_i,\hat{p}_j\right]_- - \left[\hat{p}_i,A_j\right]_-\right\}\boldsymbol{e}_j \,.\end{aligned} \tag{13}$$

Additionally,

$$\begin{aligned}\left[\hat{p}_i,A_j\right]_-\psi &= -\mathrm{i}\hbar\left(\boldsymbol{\nabla}_iA_j-A_j\boldsymbol{\nabla}_i\right)\psi\\ &= -\mathrm{i}\hbar\left(\boldsymbol{\nabla}_i(A_j\psi)-A_j\boldsymbol{\nabla}_i\psi\right)\\ &= -\mathrm{i}\hbar\left\{\psi\boldsymbol{\nabla}_iA_j+A_j\boldsymbol{\nabla}_i\psi-A_j\boldsymbol{\nabla}_i\psi\right\}\\ &= -\mathrm{i}\hbar\left(\boldsymbol{\nabla}_iA_j\right)\psi \quad ,\end{aligned}$$

where the gradients act on $\boldsymbol{A}$ only! Hence $[\hat{p}_j,A_i]_- - [\hat{p}_i,A_j]_- = \mathrm{i}\hbar(\boldsymbol{\nabla}_jA_i-\boldsymbol{\nabla}_iA_j)$, and thus we find

Exercise 2.4.

$$\frac{\mathrm{d}\hat{\boldsymbol{\Pi}}}{\mathrm{d}t} = e\boldsymbol{E} + \frac{e}{c}\sum_{i,j} c\hat{\alpha}_i \left(\boldsymbol{\nabla}_j A_i - \boldsymbol{\nabla}_i A_j\right) e_j = e\left(\boldsymbol{E} + \frac{1}{c}c\hat{\alpha} \times \operatorname{curl}\boldsymbol{A}\right) \quad . \quad (14)$$

The final step is verified by performing some simple algebra, leading to

$$\frac{\mathrm{d}\hat{\boldsymbol{\Pi}}}{\mathrm{d}t} = e\left(\boldsymbol{E} + \frac{1}{c}\hat{\boldsymbol{v}} \times \boldsymbol{B}\right) \quad . \quad (15)$$

In the classical case this is just the *Lorentz force*. Thus, we realize that the operator $\hat{\boldsymbol{x}}$ does not satisfy classical equations of motion whereas a "classical" equation of motion can be set up for the operator $\hat{\Pi}$, in which the formal velocity operator (7) appears. Hence (15) seems to coincide, at least formally, with the corresponding classical equation, although we always have to bear in mind that any expectation values of (15) are not very useful because $\hat{\boldsymbol{v}}$ contains the *Zitterbewegung*. Only the projection of the even contributions of (15) produces results which are relevant within the scope of a (classical) single-particle description. Obviously at this point we reach the limits of the single-particle interpretation.

The nonrelativistic limiting case of (2.74) can be most efficiently studied in the representation

$$\psi = \begin{pmatrix} \tilde{\varphi} \\ \tilde{\chi} \end{pmatrix} \quad , \quad (2.76)$$

where the four-component spinor ψ is decomposed into two two-component spinors $\tilde{\varphi}$ and $\tilde{\chi}$. Then the Dirac equation (2.74) becomes

$$\mathrm{i}\hbar\frac{\partial}{\partial t}\begin{pmatrix} \tilde{\varphi} \\ \tilde{\chi} \end{pmatrix} = \begin{pmatrix} c\hat{\boldsymbol{\sigma}} \cdot \hat{\boldsymbol{\Pi}} & \tilde{\chi} \\ c\hat{\boldsymbol{\sigma}} \cdot \hat{\boldsymbol{\Pi}} & \tilde{\varphi} \end{pmatrix} + eA_0 \begin{pmatrix} \tilde{\varphi} \\ \tilde{\chi} \end{pmatrix} + m_0c^2 \begin{pmatrix} \tilde{\varphi} \\ -\tilde{\chi} \end{pmatrix} \quad , \quad (2.77)$$

the $\hat{\alpha}_i$ as well as the $\hat{\beta}$ matrices having been inserted according to (2.12). If the rest energy m_0c^2, as the largest occuring energy, is additionally separated by

$$\begin{pmatrix} \tilde{\varphi} \\ \tilde{\chi} \end{pmatrix} = \begin{pmatrix} \varphi \\ \chi \end{pmatrix} \exp[-\mathrm{i}(m_0c^2/\hbar)t] \quad , \quad (2.78)$$

then (2.77) takes the form

$$\mathrm{i}\hbar\frac{\partial}{\partial t}\begin{pmatrix} \varphi \\ \chi \end{pmatrix} = \begin{pmatrix} c\hat{\boldsymbol{\sigma}} \cdot \hat{\boldsymbol{\Pi}} & \chi \\ c\hat{\boldsymbol{\sigma}} \cdot \hat{\boldsymbol{\Pi}} & \varphi \end{pmatrix} + eA_0 \begin{pmatrix} \varphi \\ \chi \end{pmatrix} - 2m_0c^2 \begin{pmatrix} 0 \\ \chi \end{pmatrix} \quad . \quad (2.79)$$

Let us consider first the lower (second) of the above equations. For the conditions $|\mathrm{i}\hbar\partial\chi/\partial t| \ll |m_0c^2\chi|$ and $|eA_0\chi| \ll |m_0c^2\chi|$ (i. e. if the kinetic energy as well as the potential energy are small compared to the rest energy) we obtain from the lower component of (2.79)

$$\chi = \frac{\hat{\boldsymbol{\sigma}} \cdot \hat{\boldsymbol{\Pi}}}{2m_0c}\varphi \quad . \quad (2.80)$$

This means that χ represents the small components of the wave function ψ, a result we already know from (2.32), while φ represents the large components $\chi \sim$

$(v/2c)\varphi$. Insertion of (2.80) into the first equation (2.79) results in a nonrelativistic wave function for φ

$$\mathrm{i}\hbar\frac{\partial\varphi}{\partial t}=\frac{(\hat{\boldsymbol{\sigma}}\cdot\hat{\boldsymbol{\Pi}})(\hat{\boldsymbol{\sigma}}\cdot\hat{\boldsymbol{\Pi}})}{2m_0}\varphi+eA_0\varphi \quad . \tag{2.81}$$

With the help of (2.30) we continue the calculation,

$$\begin{aligned}(\hat{\boldsymbol{\sigma}}\cdot\hat{\boldsymbol{\Pi}})(\hat{\boldsymbol{\sigma}}\cdot\hat{\boldsymbol{\Pi}})&=\hat{\boldsymbol{\Pi}}^2+\mathrm{i}\hat{\boldsymbol{\sigma}}\cdot(\hat{\boldsymbol{\Pi}}\times\hat{\boldsymbol{\Pi}})\\&=\left(\hat{\boldsymbol{p}}-\frac{e}{c}\boldsymbol{A}\right)^2+\mathrm{i}\hat{\boldsymbol{\sigma}}\cdot\left[\left(-\mathrm{i}\hbar\boldsymbol{\nabla}-\frac{e}{c}\boldsymbol{A}\right)\times\left(-\mathrm{i}\hbar\boldsymbol{\nabla}-\frac{e}{c}\boldsymbol{A}\right)\right]\\&=\left(\hat{\boldsymbol{p}}-\frac{e}{c}\boldsymbol{A}\right)^2-\frac{e}{c}\hbar\hat{\boldsymbol{\sigma}}\cdot(\boldsymbol{\nabla}\times\boldsymbol{A})\\&=\left(\hat{\boldsymbol{p}}-\frac{e}{c}\boldsymbol{A}\right)^2-\frac{e\hbar}{c}\hat{\boldsymbol{\sigma}}\cdot\boldsymbol{B}\end{aligned}$$

and finally obtain (2.81) in the form

$$\mathrm{i}\hbar\frac{\partial\varphi}{\partial t}=\left[\left(\hat{\boldsymbol{p}}-\frac{e}{c}\boldsymbol{A}\right)^2\Big/2m_0-\frac{e\hbar}{2m_0c}\hat{\boldsymbol{\sigma}}\cdot\boldsymbol{B}+eA_0\right]\varphi \quad . \tag{2.82}$$

This is, as it should be, the *Pauli equation.*[10] The two components of φ, therefore, describe the spin degrees of freedom, which we have already dealt with in the section dealing with free Dirac waves [cf. (2.38–44)]. From the former discussion on the Pauli equation we know that this equation, and hence also (2.82), yields the correct gyromagnetic factor of $g=2$ for a free electron. This can be demonstrated once again by turning on a weak, homogeneous magnetic field

$$\boldsymbol{B}=\operatorname{curl}\boldsymbol{A} \quad , \quad \boldsymbol{A}=\tfrac{1}{2}\boldsymbol{B}\times\boldsymbol{x} \quad ,$$

where the quadratic terms of $\boldsymbol{A}$ in (2.82) have been neglected. With

$$\begin{aligned}\left(\hat{\boldsymbol{p}}-\frac{e}{c}\boldsymbol{A}\right)^2=\left(\hat{\boldsymbol{p}}-\frac{e}{2c}\boldsymbol{B}\times\boldsymbol{x}\right)^2&\approx\hat{\boldsymbol{p}}^2-\frac{e}{c}(\boldsymbol{B}\times\boldsymbol{x})\cdot\hat{\boldsymbol{p}}\\&=\hat{\boldsymbol{p}}^2-\frac{e}{c}\boldsymbol{B}\cdot\boldsymbol{L} \quad ,\end{aligned}$$

where $\hat{\boldsymbol{L}}=\boldsymbol{x}\times\hat{\boldsymbol{p}}$ is the operator of orbital angular momentum, and

$$\hat{\boldsymbol{S}}=\tfrac{1}{2}\hbar\hat{\boldsymbol{\sigma}}$$

is the spin operator, it follows for the Pauli equation (2.82) that

$$\mathrm{i}\hbar\frac{\partial}{\partial t}\varphi=\left[\frac{\hat{\boldsymbol{p}}^2}{2m_0}-\frac{e}{2m_0c}(\hat{\boldsymbol{L}}+2\hat{\boldsymbol{S}})\cdot\boldsymbol{B}+eA_0\right]\varphi \quad . \tag{2.83}$$

This form shows explicitly the g factor 2. However, the most important result is that, in the nonrelativistic limit, the Dirac equation transforms into the Pauli equation, i. e. to the proper nonrelativistic wave equation for spin-$\frac{1}{2}$ particles. Since spin

[10] This is well known from W. Greiner: *Quantum Mechanics – An Introduction*, 3rd ed. (Springer, Berlin, Heidelberg 1994) Chap. 12.

exists both at low as well as at high velocities, this implies that *the Dirac equation describes particles with spin* $\frac{1}{2}$. In contrast to the Klein–Gordon equation [cf. (1.140), Chap. 1], valid for spin-0 particles, we have now found a relativistic wave equation for spin-$\frac{1}{2}$ particles. Clearly spin comes into the theory by linearization of the second-order differential equation (Klein–Gordon equation).[11]

2.4 Biographical Notes

DIRAC, Paul Adrien Maurice, British physicist, * 8.8.1902 in Bristol, † 2.10.1984 also in Bristol. With his fundamental investigations he contributed essentially to the formulation of quantum mechanics and quantum electrodynamics. In 1933 Dirac was awarded the Nobel prize in physics, together with E. Schrödinger. With many original contributions, he initiated modern developments in physics (e. g. magnetic monopoles, path integrals). He was one of the really great physicists!

[11] cf. in this context the linearization of the Schrödinger equation in W. Greiner: *Quantum Mechanics – An Introduction*, 3rd ed. (Springer, Berlin, Heidelberg 1994) Chap. 13.

3. Lorentz Covariance of the Dirac Equation

A proper relativistic theory has to be Lorentz covariant, i.e. its form has to be invariant under a transition from one inertial system to another one. To establish this we will first restate the essentials of **Lorentz** transformations and also refer to Chap. 14 for supporting group theoretical arguments.

Two observers, A and B, in different inertial systems describe the same physical event with their particular, different space-time coordinates. Let the coordinates of the event be x^μ for observer A and x'^μ for observer B. Both coordinates are connected by means of the *Lorentz transformation*

$$(x')^\nu = \sum_{\mu=0}^{3} a^\nu{}_\mu x^\mu \equiv a^\nu{}_\mu x^\mu \equiv (\hat{a}\,\vec{\vec{x}}\,)^\nu \quad . \tag{3.1}$$

$\hat{a}$ denotes the abbreviated version of the transformation matrix and $\vec{\vec{x}}$ the four-dimensional world vector. Equation (3.1) is a linear, homogeneous transformation and the coefficients a^ν_μ depend only on the relative velocities and spatial orientations of the reference frames. The distance between two space-time points is invariant under the Lorentz transformations (3.1), which can be expressed differentially by means of the invariance of the line element (see Fig. 3.1)

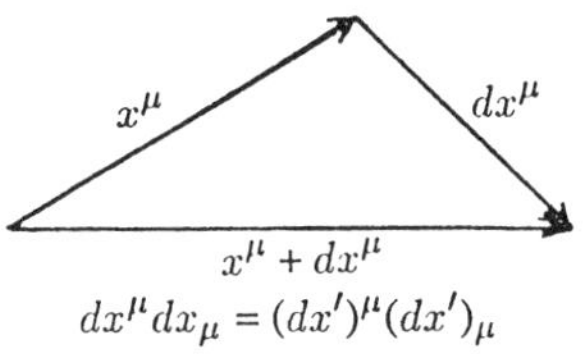

Figure 3.1.

$$ds^2 = dx^\mu\, dx_\mu = g_{\mu\nu}\, dx^\mu\, dx^\nu \quad . \tag{3.2}$$

Hence

$$dx^\mu\, dx_\mu = (dx')^\mu (dx')_\mu \quad . \tag{3.3}$$

This may also be deduced from the empirical fact that the velocity of light is the same in every inertial system.[1] Now, from (3.3) and (3.1) it follows that

$$(dx')^\mu (dx')_\mu = \alpha^\mu{}_\nu a_\mu{}^\sigma\, dx^\nu\, dx_\sigma \stackrel{!}{=} dx^\nu\, dx_\nu = \delta^\mu_\nu\, dx^\nu\, dx_\sigma \quad ;$$

hence

$$a^\mu{}_\nu a_\mu{}^\sigma = \delta^\sigma_\nu \quad . \tag{3.4}$$

These are the *orthogonality relations for Lorentz transformations*. We now distinguish between *proper* and *improper* Lorentz transformations. Namely, from (3.4) follows

[1] See J.D. Jackson: *Classical Electrodynamics*, 2nd ed. (Wiley, New York 1975).

$$\left[\det\left(a^{\mu}_{\nu}\right)\right]^2 = 1 \quad , \quad \text{i.e.} \quad \det\left(a^{\mu}_{\nu}\right) = \pm 1 \quad . \tag{3.5}$$

For the *proper Lorentz transformations* the determinant of the transformation coefficients is

$$\det\left(a^{\mu}_{\nu}\right) = +1 \quad . \tag{3.6}$$

These proper Lorentz transformations can be obtained from the identity (which has determinant +1 too) by an infinite number of successive, infinitesimal Lorentz transformations. They consist of the group of all transformations of coordinates from one coordinate system into another one which moves with *constant velocity in an arbitrary direction.* Normal three-dimensional rotations and translations belong to the proper Lorentz transformation too. The improper Lorentz transformations contain a (discrete) reflection either in space or in time. Such discrete transformations can not be obtained from the identity by successive infinitesimal transformations. The determinant of the transformation coefficients of the *improper* Lorentz transformation is

$$\det\left(a^{\mu}_{\nu}\right) = -1 \quad . \tag{3.7}$$

It will now be our task to find a relation between the measurements of observer A and those of observer B which have been performed by both of them in their respective inertial systems. More precisely, we have to find a relation between (Fig. 3.2)

$$\left(\mathrm{i}\hbar\gamma^{\mu}\frac{\partial}{\partial x^{\mu}} - m_0c\right)\psi(\vec{x}) = 0 \quad \text{and} \quad \left(\mathrm{i}\hbar\gamma'^{\mu}\frac{\partial}{\partial x'^{\mu}} - m_0c\right)\psi'(\vec{x}') = 0 \quad .$$

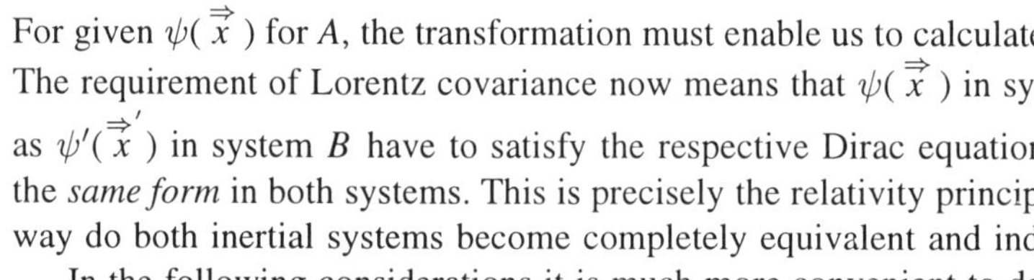

Figure 3.2.

For given $\psi(\vec{x})$ for A, the transformation must enable us to calculate $\psi'(\vec{x}')$ for B. The requirement of Lorentz covariance now means that $\psi(\vec{x})$ in system A as well as $\psi'(\vec{x}')$ in system B have to satisfy the respective Dirac equations, which have the *same form* in both systems. This is precisely the relativity principle: only in this way do both inertial systems become completely equivalent and indistinguishable.

In the following considerations it is much more convenient to denote the Dirac equation in four-dimensional notation to show the symmetry between the time-coordinate ct and the space coordinates x^i. Therefore we start with [from (2.2)]

$$\frac{\hat{\beta}}{c}\Bigg| \left(\mathrm{i}\hbar\frac{\partial}{\partial t} + \mathrm{i}\hbar c\sum_{k=1}^{3}\hat{\alpha}_k\frac{\partial}{\partial x^k} - \hat{\beta}m_0c^2\right)\psi = 0 \quad ,$$

which we multiply by $\hat{\beta}/c$ from the lhs to obtain

$$\left(\hat{\beta}\mathrm{i}\hbar\frac{\partial}{\partial ct}+\sum_{k=1}^{3}\hat{\beta}\hat{\alpha}_k\mathrm{i}\hbar\frac{\partial}{\partial x^k}-m_0c\right)\psi=0 \quad .$$

With the definitions

$$\gamma^0=\hat{\beta} \quad , \quad \gamma^i=\hat{\beta}\hat{\alpha}_i \quad , \quad i=1,2,3 \tag{3.8}$$

this can finally be written in the form

$$\mathrm{i}\hbar\left(\gamma^0\frac{\partial}{\partial x^0}+\gamma^1\frac{\partial}{\partial x^1}+\gamma^2\frac{\partial}{\partial x^2}+\gamma^3\frac{\partial}{\partial x^3}\right)\psi-m_0c\psi=0 \quad . \tag{3.9}$$

From now on we will write the matrices γ^μ without the operator sign $\hat{}$; likewise we will denote the four-dimensional radius vector (world vector) by x; hence

$$\overset{\Rightarrow}{x}\equiv x \quad , \tag{3.10}$$

which will simplify the notation in the following considerably and will scarcely lead to confusion. A more elegant formulation of the anticommutation relations (2.8) is possible using the γ matrices. Because of $\hat{\beta}^2=1\!\!1$ it is obvious that they now read

$$\gamma^\mu\gamma^\nu+\gamma^\nu\gamma^\mu=2g^{\mu\nu}1\!\!1 \quad , \tag{3.11}$$

"$1\!\!1$" being the 4×4 unit matrix. The γ^i $(i=1,2,3)$ *are unitary* $\left[\left(\gamma^i\right)^{-1}=\gamma^{i\dagger}\right]$ *and anti-Hermitian* $\left[\left(\gamma^i\right)^{\dagger}=-\gamma^i\right]$; indeed,

$$\left(\gamma^i\right)^2=-1\!\!1=-\gamma^i\gamma^{i\dagger} \quad , \quad i=1,2,3 \quad \Rightarrow \quad \left(\gamma^i\right)^{-1}=\gamma^{i\dagger} \quad .$$

The fact that they are anti-Hermitian follows directly from (2.8) and (2.9)

$$\gamma^{i\dagger}=(\hat{\beta}\hat{\alpha}_i)^\dagger=\hat{\alpha}_i^\dagger\hat{\beta}^\dagger=\hat{\alpha}_i\hat{\beta}=-\hat{\beta}\hat{\alpha}_i=-\gamma^i \quad . \tag{3.12a}$$

On the other hand γ^0 is *unitary and Hermitian*,

$$\left(\gamma^0\right)^2=+1\!\!1=+\gamma^0\gamma^{0\dagger} \quad \Rightarrow \quad \gamma^{0\dagger}=\gamma^0 \quad . \tag{3.12b}$$

In the so-far-used standard representation the γ^μ can be written down explicitly using (2.13)

$$\gamma^i=\begin{pmatrix}0 & \hat{\sigma}^i\\ -\hat{\sigma}^i & 0\end{pmatrix} \quad , \quad \gamma^0=\begin{pmatrix}1\!\!1 & 0\\ 0 & -1\!\!1\end{pmatrix} \quad . \tag{3.13}$$

A further short hand notation is often convenient: It is the so-called ***Feynman-dagger*** notation, i.e. for example

$$A\!\!\!/\equiv\gamma^\mu A_\mu=g_{\mu\nu}\gamma^\mu A^\nu=\gamma^0A^0-\sum_{i=1}^{3}\gamma^iA^i=\gamma^0A^0-\boldsymbol{\gamma}\cdot\boldsymbol{A} \quad . \tag{3.14}$$

Another example is the *nabla dagger*:

$$\nabla\!\!\!/ \equiv \gamma^\mu \frac{\partial}{\partial x^\mu} = \gamma^0 \frac{\partial}{\partial ct} + \sum_{i=1}^{3} \gamma^i \frac{\partial}{\partial x^i} = \frac{\gamma^0}{c}\frac{\partial}{\partial t} + \boldsymbol{\gamma}\cdot\boldsymbol{\nabla} \quad . \tag{3.15}$$

With this the Dirac equation (3.9) can be written in the very concise form

$$(\mathrm{i}\hbar \nabla\!\!\!/ - m_0 c)\psi = 0 \quad , \tag{3.16}$$

or, with $\hat{p}_\mu = \mathrm{i}\hbar\partial/\partial x^\mu$,

$$(\hat{p}\!\!\!/ - m_0 c)\psi = 0 \quad . \tag{3.17}$$

Introducing the electromagnetic potentials using minimal (gauge-invariant) coupling yields

$$\left(\hat{p}\!\!\!/ - \frac{e}{c}A\!\!\!/ - m_0 c\right)\psi = 0 \quad . \tag{3.18}$$

Both $\hat{p}^\mu$ and A^μ are four-vectors; hence the difference $\hat{p}^\mu - (e/c)A^\mu$ is a four-vector too. While discussing the covariance in the following part we can thus confine ourselves to the free equations (3.16) and (3.17).

3.1 Formulation of Covariance (Form Invariance)

Covariance of the Dirac equation means two different things:

(1) There must be an explicit rule to enable observer B to calculate his $\psi'(x')$ if $\psi(x)$ of observer A is given. Hence $\psi'(x')$ of B describes the same physical state as $\psi(x)$ of A.

(2) According to the principle of relativity, which states that physics (i.e. the basic equations of physics) is the same in every inertial system, $\psi'(x')$ must be a solution of a Dirac equation which has the form (3.16)

$$\left(\mathrm{i}\hbar\gamma'^\mu \frac{\partial}{\partial x'^\mu} - m_0 c\right)\psi'(x') = 0 \tag{3.19}$$

in the primed system, too. Additionally the γ'^μ have also to satisfy the anti-commutation relations (3.11). This is a requirement of the principle of relativity, as otherwise A and B could distinguish their inertial systems. Therefore,

$$\gamma'^\mu\gamma'^\nu + \gamma'^\nu\gamma'^\mu = 2g^{\mu\nu}\,\mathbb{1} \tag{3.20}$$

and

$$\gamma'^{0\dagger} = \gamma'^0 \tag{3.21a}$$

$$\gamma'^{i\dagger} = -\gamma'^i \quad , \quad i = 1,2,3 \quad . \tag{3.21b}$$

Within the sense of the principle of relativity these conditions of Hermiticity and anti-hermiticity must hold in all inertial systems, something we can also understand in the following: Let us first write (3.19) in the form

$$\mathrm{i}\hbar\gamma'^0 \frac{\partial\psi'(x')}{\partial ct'} = \left(-\gamma'^k \mathrm{i}\hbar \frac{\partial}{\partial x'^k} + m_0 c\right)\psi' \quad \text{or}$$

$$\mathrm{i}\hbar\gamma'^0\frac{\partial\psi'(x')}{\partial t'} = \left(-c\gamma'^k + \mathrm{i}\hbar\frac{\partial}{\partial x'^k} + m_0c^2\right)\psi' \equiv \hat{\hat{H}}'\psi' \quad \text{or} \tag{3.22a}$$

$$\mathrm{i}\hbar\frac{\partial\psi'(x')}{\partial t'} = \left[-c\gamma'^0\gamma'^k\left(\mathrm{i}\hbar\frac{\partial}{\partial x'^k}\right) + \gamma'^0 m_0c^2\right]\psi'(x') \equiv \hat{H}'\psi' \quad . \tag{3.22b}$$

The last form (3.22b) is Schrödinger type; the former (3.22a) is *not* Schrödinger type, because on the lhs the factor γ'^0 appears with the time derivative. Here we have used $(\gamma'^0)^2 = 1$ from (3.20). The Hamiltonian $\hat{H}'$ from (3.22b), given by

$$\hat{H}' = -c\gamma'^0\gamma'^k\left(\mathrm{i}\hbar\frac{\partial}{\partial x'^k}\right) + \gamma'^0 m_0c^2 \quad , \tag{3.23}$$

must be Hermitian in order to have real eigenvalues

$$\left(\hat{H}'\right)^\dagger = \hat{H}' \quad .$$

This is perhaps a more evident requirement of the principle of relativity because it ensures that both observers see real energy eigenvalues. Now the momentum operators

$$\hat{p}'_\mu = \mathrm{i}\hbar\frac{\partial}{\partial x'^\mu}$$

are Hermitian and commute with the γ' matrices. Hence, both γ'^0 as well as the products $\gamma'^0\gamma'^k$ from (3.23) must be Hermitian, which means that

$$\left(\gamma'^0\right)^\dagger = \gamma'^0 \quad ,$$

$$\left(\gamma'^0\gamma'^k\right)^\dagger = \left(\gamma'^k\right)^\dagger\left(\gamma'^0\right)^\dagger = \left(\gamma'^k\right)^\dagger\gamma'^0 \stackrel{!}{=} \gamma'^0\gamma'^k$$
$$\Rightarrow \left(\gamma'^k\right)^\dagger = \gamma'^0\gamma'^k\gamma'^0 = -\gamma'^k \quad .$$

Again we have used the relations $\gamma'^0\gamma'^k = -\gamma'^k\gamma'^0$ and $(\gamma'^0)^2 = 1$ which are included in (3.20) and we stress the point that only $\hat{H}'$ from (3.23) has to be Hermitian, but not

$$\hat{\hat{H}}' \equiv -c\gamma'^\kappa\left(\mathrm{i}\hbar\frac{\partial}{\partial x'^k}\right) + m_0c^2$$

from (3.22a), the first part of which is indeed anti-Hermitian. Therefore, we have

$$\left(\hat{\hat{H}}'\right)^\dagger \neq \hat{\hat{H}}' \quad .$$

It can be shown by means of a rather long algebraic proof (see Example 3.1 for better understanding) that all 4×4 matrices γ'^μ which satisfy (3.20) and (3.21) are identical up to a unitary transformation $\hat{U}$, i.e.

$$\gamma'^\mu = \hat{U}^\dagger\gamma^\mu\hat{U} \quad , \quad \hat{U}^\dagger = \hat{U}^{-1} \quad . \tag{3.24}$$

Hence it follows, because unitary transformations do not change the physics, that without loss of generality we can use *the same* γ matrices in the Lorentz system of observer B as in the Lorentz system of physicist A, i.e. the matrices (3.13).[2] *Therefore, we shall no longer differentiate between γ'^{μ} and γ^{μ}, instead we rewrite* (3.19) as

$$\left(\hat{p}\!\!/' - m_0 c\right)\psi'(x') = 0 \quad , \tag{3.25}$$

where now

$$\hat{p}\!\!/' = \mathrm{i}\hbar\gamma^{\nu}\frac{\partial}{\partial x'^{\nu}} \tag{3.26}$$

holds.

EXERCISE

3.1 Pauli's Fundamental Theorem: The 16 Complete 4 × 4 Matrices $\hat{\Gamma}_A$

Problem. Show that all representations of the Dirac algebra $\gamma^{\mu}\gamma^{\nu}+\gamma^{\nu}\gamma^{\mu} = 2g^{\mu\nu}\mathbb{1}$ for 4×4 matrices which satisfy $\gamma^{0\dagger} = \gamma^0$, $\gamma^{i\dagger} = -\gamma^i$ are unitary equivalent!

Solution. The proof[3] is divided into two parts:

(i) Proof of the fundamental theorem for Dirac matrices. For two four-dimensional representations $\gamma_{\mu}, \gamma'_{\mu}$ of the Dirac algebra there exists a nonsingular 4×4 matrix $\hat{S}$ with $\gamma'_{\mu} = \hat{S}\gamma_{\mu}\hat{S}^{-1}$.

(ii) If additionally $\gamma_0 = \gamma_0^{\dagger}$, $\gamma^i = -\gamma^{i\dagger}$, $\gamma'_0 = \gamma_0'^{\dagger}$, $\gamma'_i = -\gamma_i'^{\dagger}$, then $\hat{S}$ can be chosen to be unitary.

The proof uses the 16 following 4×4 matrices $\hat{\Gamma}_A$ $(A = 1, \dots, 16)$ (see also Chap. 5):

$$\begin{aligned}\hat{\Gamma}_A = \mathbb{1} \quad &, \\ &\gamma_0, \mathrm{i}\gamma_1, \mathrm{i}\gamma_2, \mathrm{i}\gamma_3 \quad , \\ &\mathrm{i}\gamma_2\gamma_3, \mathrm{i}\gamma_3\gamma_1, \mathrm{i}\gamma_1\gamma_2, \gamma_1\gamma_0, \gamma_2\gamma_0, \gamma_3\gamma_0 \quad , \\ &\gamma_1\gamma_2\gamma_3, \mathrm{i}\gamma_1\gamma_2\gamma_0, \mathrm{i}\gamma_3\gamma_1\gamma_0, \mathrm{i}\gamma_2\gamma_3\gamma_0 \quad , \\ &\mathrm{i}\gamma_1\gamma_2\gamma_3\gamma_0 \quad ,\end{aligned} \tag{1}$$

for which

$$\hat{\Gamma}_A^2 = 1 \quad (A = 1, \dots, 16) \quad . \tag{2}$$

Let us denote $\mathrm{i}\gamma_1\gamma_2\gamma_3\gamma_0 = \gamma_5$. It can easily be proven that γ_5 anticommutes with γ_{μ} $(\mu = 0, \dots, 3)$,

$$\gamma_5\gamma_{\mu} + \gamma_{\mu}\gamma_5 = 0 \quad (\mu = 0, \dots, 3) \quad . \tag{3}$$

[2] This fact can also be considered to be a consequence of the principle of relativity, since otherwise the different structure of the γ matrices in the inertial systems of A and B would indicate to A and B in which of the systems they are.

[3] This argument stems from R.H. Good: Rev. Mod. Phys. **27**, 187 (1955).

Now let us prove the fundamental theorem for Dirac matrices: *Exercise 3.1.*

(1) *For all* $\hat{\Gamma}_A$ *but the* $1\!\!1$ *there exists a* $\hat{\Gamma}_B$ *with*

$$\hat{\Gamma}_B \hat{\Gamma}_A \hat{\Gamma}_B = -\hat{\Gamma}_A \quad . \tag{4}$$

We will specify for every $\hat{\Gamma}_A$ a corresponding $\hat{\Gamma}_B$:

A	2	3	4	5	6	7	8	9	10	11	12	13	14	15	16
B	9	4	3	3	4	5	3	2	2	2	2	6	6	7	2

(5)

One verifies easily by explicit calculation that (4) holds.

(2) *The traces of all* $\hat{\Gamma}_A$ $(A = 2, \ldots, 16)$ *are zero.* From (2) and (4)

$$-\operatorname{tr}\left(\hat{\Gamma}_A\right) = \operatorname{tr}\left(\hat{\Gamma}_B \hat{\Gamma}_A \hat{\Gamma}_B\right) = \operatorname{tr}\left(\hat{\Gamma}_B^2 \hat{\Gamma}_A\right) = +\operatorname{tr}\left(\hat{\Gamma}_A\right) \quad .$$

(3) *The* $\hat{\Gamma}_A$ *are linearly independent.* From

$$\sum_{A=1}^{16} a_A \hat{\Gamma}_A = 0 \quad \text{follows}$$

$$a_A = 0 \quad (A = 1, \ldots, 16) \quad , \tag{6}$$

because if we multiply the sum by $\hat{\Gamma}_B$,

$$a_B\, 1\!\!1 + \sum_{A \neq B} a_A \hat{\Gamma}_A \hat{\Gamma}_B = 0 \tag{7}$$

and

$$4 a_B + \sum_{A \neq B} a_A \operatorname{tr}\left(\hat{\Gamma}_A \hat{\Gamma}_B\right) = 0 \quad . \tag{8}$$

Equation (1) implies that $\hat{\Gamma}_A \hat{\Gamma}_B = \text{const}\, \hat{\Gamma}_C$, i.e. by means of the algebra a product of two matrices of (1) can be expressed in terms of another such matrix. In the case where $A \neq B$ then $\hat{\Gamma}_C \neq 1\!\!1$; therefore the sum in (8) vanishes and consequently $a_B = 0$. This conclusion is valid for all $B = 1, \ldots, 16$. The linear independence of the matrices $\hat{\Gamma}_A$ is most important and will frequently be used in the following

(4) *Each* 4×4 *matrix can be expanded in terms of the* $\hat{\Gamma}_a$,

$$\hat{\chi} = \sum_{A=1}^{16} x_A \hat{\Gamma}_A \quad . \tag{9}$$

This is evident since the 16 linearly independent $\hat{\Gamma}_A$ generate a 16-dimensional space, that is the space of the 4×4 matrices. Then

$$x_B = \tfrac{1}{4} \operatorname{tr}\left(\hat{\Gamma}_B \hat{\chi}\right) \quad . \tag{10}$$

(5) *Each* 4×4 *matrix which commutes with all* $\hat{\Gamma}_A$ *is a multiple of* $1\!\!1$ *(Schur's Lemma).* Consider the matrix $\hat{\chi}$,

Exercise 3.1.

$$\hat{\chi} = x_B \hat{\Gamma}_B + \sum_{A \neq B} x_A \hat{\Gamma}_A \quad , \tag{11}$$

where we have singled out a particular matrix $\hat{\Gamma}_B$ on which we shall focus our attention. We will show that $x_B = 0$. With respect to (4) we first choose $\hat{\Gamma}_C$ such that if fulfills

$$\hat{\Gamma}_C \hat{\Gamma}_B \hat{\Gamma}_C = -\hat{\Gamma}_B \quad , \tag{12}$$

and, since $\hat{\chi}$ commutes with all $\hat{\Gamma}_A$ and therefore also with $\hat{\Gamma}_C$,

$$\hat{\chi} = \hat{\Gamma}_C \hat{\chi} \hat{\Gamma}_C$$

$$\Rightarrow \; x_B \hat{\Gamma}_B + \sum_{A \neq B} x_A \hat{\Gamma}_A = x_B \hat{\Gamma}_C \hat{\Gamma}_B \hat{\Gamma}_C + \sum_{A \neq C} x_A \hat{\Gamma}_C \hat{\Gamma}_A \hat{\Gamma}_C \quad , \tag{13}$$

$$x_B \hat{\Gamma}_B + \sum_{A \neq B} x_A \hat{\Gamma}_A = -x_B \hat{\Gamma}_B + \sum_{A \neq B} (\pm) x_A \hat{\Gamma}_A \quad , \tag{14}$$

due to the fact that $\hat{\Gamma}_C$ and $\hat{\Gamma}_A$ commute or anticommute. Next we multiply with $\hat{\Gamma}_B$, take the trace and obtain

$$x_B = -x_B \quad . \tag{15}$$

Thus, $x_B = 0$ as claimed.

(6) *If γ_μ and γ'_μ are two representations of the Dirac algebra and $\hat{\Gamma}_A$, $\hat{\Gamma}'_A$ are the corresponding 16-dimensional bases, then*

$$\hat{\Gamma}'_A \hat{S} = \hat{S} \hat{\Gamma}_A \quad , \tag{16}$$

where

$$\hat{S} = \sum_{B=1}^{16} \hat{\Gamma}'_B \hat{F} \hat{\Gamma}_B \tag{17}$$

with an arbitrary 4×4 matrix $\hat{F}$. To understand this, consider the matrix

$$\hat{\Gamma}'_A \hat{S} \hat{\Gamma}_A = \sum_{B=1}^{16} \hat{\Gamma}'_A \hat{\Gamma}'_B \hat{F} \hat{\Gamma}_B \hat{\Gamma}_A \quad . \tag{18}$$

According to (1) we have $\hat{\Gamma}_B \hat{\Gamma}_A = \varepsilon_C \hat{\Gamma}_C$ with $\varepsilon_C \in \{\pm 1, \pm \mathrm{i}\}$. For fixed A, if B runs from 1 to 16 then C takes all values from 1 to 16. This is so, since otherwise we would have

$$\hat{\Gamma}_B \hat{\Gamma}_A = \varepsilon_C \hat{\Gamma}_C$$

and

$$\hat{\Gamma}_D \hat{\Gamma}_A = \delta_C \hat{\Gamma}_C \quad (\hat{\Gamma}_D \neq \hat{\Gamma}_B) \quad , \tag{19}$$

leading to $\hat{\Gamma}_B = \varepsilon_C \hat{\Gamma}_C \hat{\Gamma}_A = (\varepsilon_C / \delta_C) \hat{\Gamma}_D$, which contradicts the linear independence of the $\hat{\Gamma}_A$. Since the $\hat{\Gamma}'_A$ are constructed from the γ'_μ in the same way as the $\hat{\Gamma}_A$ from the γ_μ, we also have the relation

$$\hat{\Gamma}'_B \hat{\Gamma}'_A = \varepsilon_C \hat{\Gamma}'_C \tag{20}$$

Exercise 3.1.

with the same ε_C. Inverting (20) yields

$$\hat{\Gamma}'_A \hat{\Gamma}'_B = \frac{1}{\varepsilon_C} \hat{\Gamma}'_C \quad , \tag{21}$$

and substituting (21) into (18) then gives

$$\hat{\Gamma}'_A S \hat{\Gamma}_A = \sum_{C=1}^{16} \frac{1}{\varepsilon_C} \hat{\Gamma}'_C \hat{F} \varepsilon_C \hat{\Gamma}_C = \sum_{C=1}^{16} \hat{\Gamma}'_C \hat{F} \hat{\Gamma}_C = \hat{S} \quad , \tag{22}$$

thus proving the relation (16).

(7) *The matrix $\hat{F}$ can be chosen such that $\hat{S}$ does not vanish.* If $\hat{S} = 0$ did hold for all $\hat{F}$, then by a special choice of $\hat{F}$ such that a single element has the value 1, all the remaining elements being set to zero (i.e. $F_{\nu\sigma} = 1$, all other $F_{\alpha\beta} = 0$), we could infer from (17) that

$$\sum_{B=1}^{16} \left(\hat{\Gamma}'_B\right)_{\mu\nu} \left(\hat{\Gamma}_B\right)_{\sigma\varrho} = 0 \tag{23}$$

which must hold for all ν, σ, due to the arbitrary choice of $\hat{F}$. From the relations (23) for the various combinations ν, σ we infer the matrix equation

$$\sum_{B=1}^{16} \left(\hat{\Gamma}'_B\right)_{\mu\nu} \hat{\Gamma}_B = 0 \quad . \tag{24}$$

Since $\hat{\Gamma}'^2_B = 1$, the $(\hat{\Gamma}'_B)_{\mu\nu}$ cannot vanish simultaneously, so that (24) is in contradiction to the linear independence of the $\hat{\Gamma}_B$.

(8) *The matrix $\hat{F}$ can be chosen such that $\hat{S}$ is not singular.* To prove this lemma, we construct

$$\hat{T} = \sum_{B=1}^{16} \hat{\Gamma}_B \hat{G} \hat{\Gamma}'_B \tag{25}$$

which arbitrary $\hat{G}$, specified below. In analogy to point 6 we can show that

$$\hat{\Gamma}_A \hat{T} = \hat{T} \hat{\Gamma}'_A \quad , \tag{26}$$

which, together with (16) leads to

$$\hat{\Gamma}_A \hat{T} \hat{S} = \hat{T} \hat{\Gamma}'_A \hat{S} = \hat{T} \hat{S} \hat{\Gamma}_A \quad , \tag{27}$$

i.e. $(\hat{T}\hat{S})$ commutes with all $\hat{\Gamma}_A$ and is therefore, according to point 5, a multiple of the unit matrix

$$\hat{T}\hat{S} = K \hat{1} \quad . \tag{28}$$

Here we choose $\hat{G}$ such that $\hat{T} \neq 0$ (cf. point 7). This enables us to choose $\hat{F}$ in (17) such that $K \neq 0$. This is so because the assumption $K = 0$ for all $\hat{F}$ would, with respect to (28) and (17), result in

Exercise 3.1.

$$\sum_{B=1}^{16} \hat{T}\hat{\Gamma}'_B\hat{F}\hat{\Gamma}_B = 0 \quad . \tag{29}$$

With a choice of $\hat{F}$ in line with point 7, we have, from (29),

$$\sum_{B=1}^{16} \left(\hat{T}\hat{\Gamma}'_B\right)_{\mu\nu} \left(\hat{\Gamma}_B\right)_{\varrho\sigma} = 0 \quad \text{or} \tag{30}$$

$$\sum_{B=1}^{16} \left(\hat{T}\hat{\Gamma}'_B\right)_{\mu\nu} \hat{\Gamma}_B = 0 \quad . \tag{31}$$

The $(\hat{T}\hat{\Gamma}'_B)_{\mu\nu}$ do not vanish simultaneously since $\hat{T} \neq 0$ and $\{\hat{\Gamma}_B\}$ contain the unit matrix $\mathbb{1}$. Consequently this yields a contradiction to the linear independence of the $\hat{\Gamma}_B$. Thus $\hat{S}$ is not singular, and

$$\gamma'_\mu = \hat{S}\gamma_\mu\hat{S}^{-1} \quad , \tag{32}$$

completing part (i) of the proof.

(ii) Now we will show that in case of

$$\gamma^\dagger_\mu = g_{\mu\mu}\gamma_\mu \quad , \quad \gamma'^\dagger_\mu = g_{\mu\mu}\gamma'_\mu \quad . \tag{33}$$

$\hat{S}$ can be chosen as a unitary operator. To see this let

$$\gamma'_\mu = \hat{V}\gamma_\mu\hat{V}^{-1} \quad , \quad \det\hat{V} = 1 \quad , \tag{34}$$

i.e.

$$\hat{V} = \left(\det\hat{S}\right)^{-1}\hat{S} \quad . \tag{35}$$

Except for an arbitrary factor ± 1, $\pm\mathrm{i}$, due to $\det\hat{V} = \det(\pm\hat{V}) = \det(\pm\mathrm{i}\hat{V}) = 1$, the matrix $\hat{V}$ is *completely determined.* To assume in contrast to this that

$$\gamma'_\mu = \hat{V}_1\gamma_\mu\hat{V}_1^{-1} = \hat{V}_2\gamma_\mu\hat{V}_2^{-1} \quad , \tag{36}$$

would lead to

$$\hat{V}_2^{-1}\hat{V}_1\gamma_\mu = \gamma_\mu\hat{V}_2^{-1}\hat{V}_1 \tag{37}$$

and, with respect to point above, $\hat{V}_2^{-1}\hat{V}_1 = k \cdot \mathbb{1}$; hence

$$\hat{V}_1 = k\hat{V}_2 \quad , \quad k = \pm 1, \ \pm\mathrm{i} \quad . \tag{38}$$

Therefore, taking the Hermitian conjugate of (34), we obtain

$$\gamma'^\dagger_\mu = \left(\hat{V}^{-1}\right)^\dagger \gamma^\dagger_\mu\hat{V}^\dagger \tag{39}$$

and, by means of (33),

$$\gamma'_\mu = \left(\hat{V}^\dagger\right)^{-1}\gamma_\mu\hat{V}^\dagger \quad , \tag{40}$$

i.e. $(\hat{V}^\dagger)^{-1}$ likewise fulfills (35) as does $\hat{V}$. From (38), it follows that

$$\left(\hat{V}^{\dagger}\right)^{-1} = \mathrm{e}^{\mathrm{i}m\pi/2}\hat{V} \quad , \quad \hat{V}^{\dagger} = \mathrm{e}^{-\mathrm{i}m\pi/2}\hat{V}^{-1} \quad , \tag{41}$$

$$\hat{V}^{\dagger}\hat{V} = \mathrm{e}^{-\mathrm{i}m\pi/2} \quad . \tag{42}$$

Since the diagonal elements of the product of a matrix and its Hermitian conjugated counterpart have the form

$$\left(\hat{V}^{\dagger}\hat{V}\right)_{ii} = \sum_j \left(\hat{V}^{\dagger}\right)_{ij}\left(\hat{V}\right)_{ji} = \sum_j V_{ji}^{*}V_{ji} \quad , \tag{43}$$

and $V_{ji}^{*}V_{ji}$ is positive definite and real, the factor $\mathrm{e}^{-\mathrm{i}m\pi/2}$ in (42) must be positive definite and real too, implying $m = 0$. Hence,

$$\hat{V}^{\dagger}\hat{V} = 1\!\!1 \quad , \tag{44}$$

which was to be proven.

We will now explicitly construct the transformation between $\psi(x)$ and $\psi'(x')$. *This transformation is required to be linear, since both the Dirac equation as well as the Lorentz transformation* (3.1) *are linear in the space-time coordinates*. Hence it must have the form

$$\psi'(x') = \psi'(\hat{a}x) = \hat{S}(\hat{a})\psi(x) = \hat{S}(\hat{a})\psi\left(\hat{a}^{-1}x'\right) \quad , \tag{3.27}$$

where $\hat{a}$ denotes the matrix of the Lorentz transformation a_{μ}^{ν} of (3.1) and $\hat{S}(\hat{a})$ is a 4×4 matrix which is a function of the parameters of the Lorentz transformation $\hat{a}$ and operates upon the four components of the bispinor $\psi(x)$. Through $\hat{a}$ it depends on the relative velocities and spatial orientations of the observers A and B. The principle of relativity, stating the invariance of physical laws for all inertial systems, implies the existence of the inverse operator $\hat{S}^{-1}(\hat{a})$ that enables the observer A to construct his wave function $\psi(x)$ from the $\psi'(x')$ of observer B. Therefore, it must hold that

$$\psi(x) = \hat{S}^{-1}(\hat{a})\psi'(x') = \hat{S}^{-1}(\hat{a})\psi'(\hat{a}x) \quad . \tag{3.28a}$$

Because of (3.27) we can also write

$$\psi(x) = \hat{S}\left(\hat{a}^{-1}\right)\psi'(x') = \hat{S}\left(\hat{a}^{-1}\right)\psi'(\hat{a}x) \tag{3.28b}$$

and, comparing with (3.28a), we find

$$\hat{S}^{-1}(\hat{a}) = \hat{S}\left(\hat{a}^{-1}\right) \quad . \tag{3.29}$$

Our aim is to construct $\hat{S}$ fulfilling (3.27–29). Starting from the Dirac equation (3.16) of the observer A, i.e.

$$\left(\mathrm{i}\hbar\gamma^{\mu}\frac{\partial}{\partial x^{\mu}} - m_0 c\right)\psi(x) = 0 \quad ,$$

and expressing $\psi(x)$ by means of (3.28a) yields

$$\left(\mathrm{i}\hbar\gamma^{\mu}\hat{S}^{-1}(\hat{a})\frac{\partial}{\partial x^{\mu}} - m_0 c\hat{S}^{-1}(\hat{a})\right)\psi'(x') = 0 \quad .$$

Multiplication with $\hat{S}(\hat{a})$ from the left and using $\hat{S}(\hat{a})\hat{S}^{-1}(\hat{a}) = \mathbb{1}$ then gives

$$\left(\mathrm{i}\hbar\hat{S}(\hat{a})\gamma^{\mu}\hat{S}^{-1}(\hat{a})\frac{\partial}{\partial x^{\mu}} - m_0 c\right)\psi'(x') = 0 \quad . \tag{3.30}$$

With regard to (3.1) we transform $\partial/\partial x^{\mu}$ to the coordinates of the system B,

$$\frac{\partial}{\partial x^{\mu}} = \frac{\partial x'^{\nu}}{\partial x^{\mu}}\frac{\partial}{\partial x'^{\nu}} = a^{\nu}{}_{\mu}\frac{\partial}{\partial x'^{\nu}} \quad , \tag{3.31}$$

so that (3.30) becomes

$$\left[\mathrm{i}\hbar\left(\hat{S}(\hat{a})\gamma^{\mu}\hat{S}^{-1}(\hat{a})a^{\nu}{}_{\mu}\right)\frac{\partial}{\partial x'^{\nu}} - m_0 c\right]\psi'(x') = 0 \quad . \tag{3.32}$$

This has to be identical with the Dirac equation (3.25), since form invariance of the equations of motion is required, i.e. $\hat{S}(\hat{a})$ must have the property

$$\hat{S}(\hat{a})\gamma^{\mu}\hat{S}^{-1}(\hat{a})a^{\nu}{}_{\mu} = \gamma^{\nu} \tag{3.33}$$

or equivalently

$$\hat{S}(\hat{a})\gamma^{\nu}\hat{S}^{-1}(\hat{a}) = a_{\mu}{}^{\nu}\gamma^{\mu} \quad . \tag{3.34}$$

This is the fundamental relation determining the operator $\hat{S}$: To find $\hat{S}$ means solving (3.34) which holds for discrete as well as continuous Lorentz transformations since the above deduction does not depend on $\det a_{\mu}{}^{\nu} = \pm 1$. Once we have shown that there exists a solution $\hat{S}(\hat{a})$ of (3.34) and have found it, we will have proven the covariance of the Dirac equation. We may already now specify more precisely the definition of a spinor which, we have previously introduced somewhat inaccurately as a four-component column vector: *In general, a wave function is termed a four-component Lorentz spinor if it transforms according to* (3.27) *by means of the fundamental relation* (3.34). Such a four-component spinor is also frequently called a *bispinor*, since it consists of two two-component spinors, known to us from the Pauli equation.[4]

In determining $\hat{S}(\hat{a})$ we expect to deal with novel features which are not present in ordinary tensor calculus, since bilinear forms of ψ, e.g. the current four-vector of (2.18)

$$\{c\varrho, j^k\} \quad \text{with} \quad \varrho = \psi^{\dagger}\psi \quad \text{and} \quad j^k = \psi^{\dagger}\alpha^k\psi \quad ; \quad k = 1,2,3 \quad ,$$

have to transform like four-vectors; we will now discuss these properties. As we have already learned,[5] in general it is simplest to generate a continuous group

[4] See W. Greiner: *Quantum Mechanics – An Introduction*, 3rd ed. (Springer, Berlin, Heidelberg 1994).

[5] See W. Greiner, B. Müller: *Quantum Mechanics – Symmetries*, 2nd ed. (Springer, Berlin, Heidelberg 1994).

transformation by constructing the group operators for infinitesimal transformations and then composing operators for finite rotations, translations, etc., by connecting the infinitesimal operators in series. Following the same pattern in our case of Lorentz transformations, we first construct the operator $\hat{S}(\hat{a})$. for *infinitesimal proper Lorentz transformations* given by

$$a^{\nu}{}_{\mu} = \delta^{\nu}_{\mu} + \Delta\omega^{\nu}{}_{\mu} \qquad (3.35)$$

with

$$\Delta\omega^{\nu\mu} = -\Delta\omega^{\mu\nu} \quad . \qquad (3.36a)$$

The antisymmetric form (3.36) follows from (3.4) by neglecting quadratic terms [of the order $(\Delta\omega)^2$]:

$$\begin{aligned} a^{\mu}{}_{\nu}a_{\mu}{}^{\sigma} &= \delta^{\sigma}_{\nu} = \left(\delta^{\mu}_{\nu} + \Delta\omega^{\mu}{}_{\nu}\right)\left(\delta^{\sigma}_{\mu} + \Delta\omega_{\mu}{}^{\sigma}\right) \\ &\approx \delta^{\mu}_{\nu}\delta^{\sigma}_{\mu} + \delta^{\mu}_{\nu}\Delta\omega_{\mu}{}^{\sigma} + \delta_{\mu}{}^{\sigma}\Delta\omega^{\mu}{}_{\nu} = \delta^{\sigma}_{\nu} + \Delta\omega_{\nu}{}^{\sigma} + \Delta\omega^{\sigma}{}_{\nu} \quad . \end{aligned} \qquad (3.36b)$$

Hence,

$$\Delta\omega_{\nu}{}^{\sigma} + \Delta\omega^{\sigma}{}_{\nu} = 0 \quad , \quad \text{or} \quad g^{\mu\nu}\left(\Delta\omega_{\nu}{}^{\sigma} + \Delta\omega^{\sigma}{}_{\nu}\right) = 0 = \Delta\omega^{\mu\sigma} + \Delta\omega^{\sigma\mu} \quad .$$

Consequently, there are six independent non-vanishing parameters $\Delta\omega^{\mu\nu}$. Each of these *group-parameters* (rotation angle in the four-dimensional Minkowski space) generates an infinitesimal Lorentz transformation. We will now give two examples pointing out the physical significance of the $\Delta\omega^{\mu\nu}$:

(a) $\Delta\omega^{10} = \Delta\omega^{01} \equiv -\Delta\beta \neq 0 \quad , \quad$ all other $\quad \Delta\omega^{\mu\nu} = 0 \quad .$

This implies that

$$\Delta\omega_1{}^0 = g_{1\sigma}\Delta\omega^{\sigma 0} = g_{11}\Delta\omega^{10} = -\Delta\omega^{10} = +\Delta\beta = +\Delta\omega^{01} = \Delta\omega_0{}^1 = -\Delta\omega^0{}_1$$

and

$$\Delta\omega_i{}^1 = g_{i\sigma}\Delta\omega^{\sigma 1} = 0 \quad \text{for all} \quad i = 1,2,3 \quad .$$

According to (3.1) we then find

$$\begin{aligned} (x')^{\nu} &= \left(\delta^{\nu}_{\mu} + \Delta\omega^1{}_0\delta^{\nu}_1\delta^0_{\mu} + \Delta\omega^0{}_1\delta^{\nu}_0\delta^1_{\mu}\right)x^{\mu} \\ &= \left(\delta^{\nu}_{\mu} - \Delta\beta\delta^{\nu}_1\delta^0_{\mu} - \Delta\beta\delta^{\nu}_0\delta^1_{\mu}\right)x^{\mu} \quad , \end{aligned} \qquad (3.37a)$$

or, explicitly,

$$\begin{aligned} (x')^0 &= x^0 - \Delta\beta x^1 = x^0 - \frac{\Delta v}{c}x^1 \quad , \\ (x')^1 &= -\Delta\beta x^0 + x^1 = -\frac{\Delta v}{c}x^0 + x^1 \quad , \\ (x')^2 &= x^2 \quad , \\ (x')^3 &= x^3 \quad . \end{aligned} \qquad (3.37b)$$

Therefore the inertial system (′) of the observer B moves along the positive x^1 axis relative to the system of observer A according to

$$(x')^1 = 0 \to \frac{x^1}{x^0} = \frac{\Delta v}{c} = \Delta\beta \quad .$$

This means that case (a) describes an (infinitesimal) Lorentz transformation for a motion parallel to the x' axis with a velocity $\Delta v = \Delta\beta c$, (see Fig. 3.3).

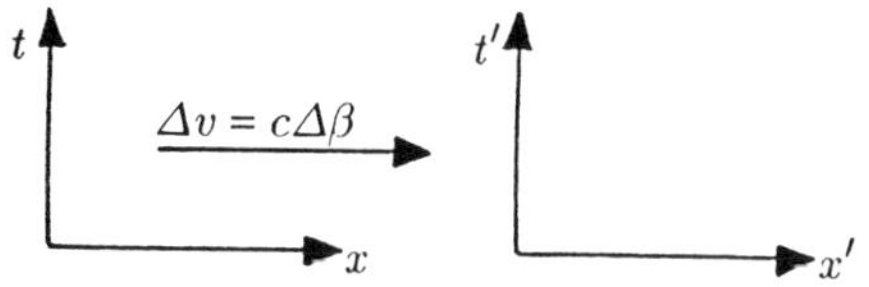

Fig. 3.3. The x' system moves relative to the x system with velocity $\Delta v = c\Delta\beta$

$$\text{(b)} \quad \Delta\omega^1_2 = -\Delta\omega^{12} = \Delta\omega^{21} \equiv \Delta\varphi \quad , \quad \text{all other} \quad \Delta\omega^{\mu\nu} = 0 \quad .$$

According to (3.1)

$$(x')^\nu = \left[\delta^\nu_\mu + \delta^\nu_1 \delta^2_\mu \Delta\varphi + \delta^\nu_2 \delta^1_\mu (-\Delta\varphi)\right] x^\mu \tag{3.38a}$$

or explicitly

$$\begin{aligned}(x')^0 &= x^0 \quad , \\ (x')^1 &= x^1 + \Delta\varphi x^2 \quad , \\ (x')^2 &= -\Delta\varphi x^1 + x^2 \quad , \\ (x')^3 &= x^3 \quad .\end{aligned} \tag{3.38b}$$

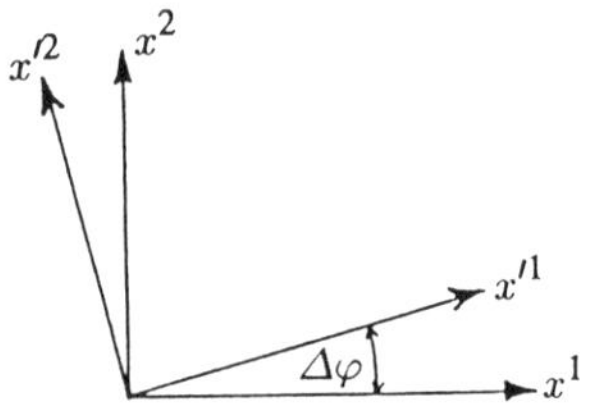

Fig. 3.4. An infinitesimal rotation of the coordinate system around the z axis

Clearly this transformation generates an infinitesimal rotation about the z axis by an angle $\Delta\varphi$ (Fig. 3.4). For finite rotations it would read

$$\begin{aligned}(x')^1 &= x^1 \cos\varphi + x^2 \sin\varphi \\ (x')^2 &= -x^1 \sin\varphi + x^2 \cos\varphi \quad .\end{aligned} \tag{3.38c}$$

3.2 Construction of the $\hat{S}$ Operator for Infinitesimal Lorentz Transformations

We now return to our original aim of determining the operator $\hat{S}(\hat{a}) = \hat{S}(\Delta\omega^{\mu\nu})$, by expanding $\hat{S}$ in powers of $\Delta\omega^{\mu\nu}$ and keeping only the linear terms of the infinitesimal generators $\Delta\omega^{\mu\nu}$; hence we write

$$\begin{aligned}\hat{S}\,(\Delta\omega^{\mu\nu}) &= 1\!\!1 - \frac{\mathrm{i}}{4}\hat{\sigma}_{\mu\nu}\Delta\omega^{\mu\nu} \quad , \\ \hat{S}^{-1}\,(\Delta\omega^{\mu\nu}) &= 1\!\!1 + \frac{\mathrm{i}}{4}\hat{\sigma}_{\mu\nu}\Delta\omega^{\mu\nu} \quad , \quad (\hat{\sigma}_{\mu\nu} = -\hat{\sigma}_{\nu\mu}) \quad .\end{aligned} \tag{3.39}$$

The factors $\mathrm{i}/4$ in the second term of the rhs are chosen such that the six coefficients $\hat{\sigma}_{\mu\nu}$ can be expressed in a convenient form, as will become clear later. *Each of the six coefficients $\hat{\sigma}_{\mu\nu}$ is a 4×4 matrix, which is indicated by the operator sign* ^! Of course, the same holds for the operator $\hat{S}$ and the unit matrix $1\!\!1$. By finding the $\hat{\sigma}_{\mu\nu}$, we can determine the operator $\hat{S}$. The second relation (3.39) results from (3.29), which means that the inverse operator is obtained by substituting $\Delta\omega^\nu_\mu \to -\Delta\omega^\nu_\mu$. Inserting (3.39) and (3.35) into (3.34) which determines $\hat{S}$, we find

$$\left(\delta^\nu{}_\mu + \Delta\omega_\mu{}^\nu\right)\gamma^\mu = \left(\mathbb{1} - \frac{\mathrm{i}}{4}\hat{\sigma}_{\alpha\beta}\Delta\omega^{\alpha\beta}\right)\gamma^\mu\left(\mathbb{1} + \frac{\mathrm{i}}{4}\hat{\sigma}_{\alpha\beta}\Delta\omega^{\alpha\beta}\right)$$

or, omitting the quadratic terms in $\Delta\omega^{\mu\nu}$,

$$\Delta\omega_\mu{}^\nu\gamma^\mu = -\frac{\mathrm{i}}{4}\Delta\omega^{\alpha\beta}\left(\hat{\sigma}_{\alpha\beta}\gamma^\nu - \gamma^\nu\hat{\sigma}_{\alpha\beta}\right) \quad . \tag{3.40}$$

Taking into account the antisymmetry (3.36) $\Delta\omega^{\mu\nu} = -\Delta\omega^{\nu\mu}$, the lhs of (3.40) becomes

$$\begin{aligned}
-\frac{\mathrm{i}}{4}\Delta\omega^{\alpha\beta}\left(\hat{\sigma}_{\alpha\beta}\gamma^\nu - \gamma^\nu\hat{\sigma}_{\alpha\beta}\right) &= g^\nu{}_\sigma\Delta\omega_\mu{}^\sigma\gamma^\mu = \Delta\omega_\mu{}^\sigma g^\nu{}_\sigma\gamma^\mu \\
&= \Delta\omega_\beta{}^\alpha\left(g^\nu{}_\alpha\gamma^\beta\right) = \Delta\omega^{\beta\alpha}\left(g^\nu{}_\alpha\gamma_\beta\right) \\
&= \tfrac{1}{2}\Delta\omega^{\beta\alpha}\left(g^\nu{}_\alpha\gamma_\beta - g^\nu{}_\beta\gamma_\alpha\right) \\
&= -\tfrac{1}{2}\Delta\omega^{\alpha\beta}\left(g^\nu{}_\alpha\gamma_\beta - g^\nu{}_\beta\gamma_\alpha\right) \quad .
\end{aligned}$$

Hence, we end up with the relation

$$-2\mathrm{i}\left(g^\nu{}_\alpha\gamma_\beta - g^\nu{}_\beta\gamma_\alpha\right) = \left[\hat{\sigma}_{\alpha\beta}, \gamma^\nu\right]_- \quad . \tag{3.41}$$

The problem of constructing $\hat{S}$ according to the fundamental relation (3.34) is now reduced to that of determining the six matrices $\hat{\sigma}_{\alpha\beta}$. Since $\hat{\sigma}_{\alpha\beta}$ has to be antisymmetric in both indices, α and β, it is natural to try an antisymmetric product of two matrices:

$$\hat{\sigma}_{\alpha\beta} = \frac{\mathrm{i}}{2}\left[\gamma_\alpha, \gamma_\beta\right]_- \quad . \tag{3.42}$$

This form fulfills the requirement (3.41), which will be proven in the following exercise by taking into consideration the commutation relations (3.11).

EXERCISE

3.2 Proof of the Coefficients $\hat{\sigma}_{\alpha\beta}$ for the Infinitesimal Lorentz Transformation

Problem. Prove that the $\hat{\sigma}_{\alpha\beta} = \mathrm{i}[\gamma_\alpha, \gamma_\beta]_-/2$ fulfill (3.41),

$$\left[\gamma^\nu, \hat{\sigma}_{\alpha\beta}\right]_- = +2\mathrm{i}\left(g^\nu{}_\alpha\gamma_\beta - g^\nu{}_\beta\gamma_\alpha\right) \quad .$$

Solution. Inserting the expression for $\hat{\sigma}_{\alpha\beta}$ in (3.41), we get

$$\left[\gamma^\nu, \hat{\sigma}_{\alpha\beta}\right]_- = \frac{\mathrm{i}}{2}\left[\gamma^\mu, [\gamma_\alpha, \gamma_\beta]_-\right]_- = \frac{\mathrm{i}}{2}\left\{\left[\gamma^\nu, \gamma_\alpha\gamma_\beta\right]_- - \left[\gamma^\nu, \gamma_\beta\gamma_\alpha\right]_-\right\} \quad .$$

By means of the algebra (3.11) it then follows that

$$\left[\gamma^\nu, \hat{\sigma}_{\alpha\beta}\right]_- = \frac{\mathrm{i}}{2}\left\{2\left[\gamma^\nu, \gamma_\alpha\gamma_\beta\right]_- - 2\left[\gamma^\nu, g_{\alpha\beta}\right]_-\right\} = \mathrm{i}\left[\gamma^\nu, \gamma_{\alpha\beta}\right]_- \quad .$$

Exercise 3.2.

Furthermore we have

$$\begin{aligned}\mathrm{i}\left[\gamma^\nu,\gamma_\alpha\gamma_\beta\right]_- &= \mathrm{i}\left\{\gamma^\nu\gamma_\alpha\gamma_\beta - 2g^\nu{}_\beta\gamma_\alpha + \gamma_\alpha\gamma^\nu\gamma_\beta\right\}\\ &= \mathrm{i}\left\{\gamma^\nu\gamma_\alpha\gamma_\beta - 2g^\nu{}_\beta\gamma_\alpha + 2g^\nu{}_\alpha\gamma_\beta - \gamma^\nu\gamma_\alpha\gamma_\beta\right\}\\ &= 2\mathrm{i}\left(g^\nu{}_\alpha\gamma_\beta - g^\nu{}_\beta\gamma_\alpha\right) \quad ,\end{aligned}$$

which is the required form.

According to (3.39), the operator $\hat{S}(\Delta\omega^{\mu\nu})$ for infinitesimal proper Lorentz transformations is now

$$\hat{S}\left(\Delta\omega^{\mu\nu}\right) = 1\!\!1 - \frac{\mathrm{i}}{4}\hat{\sigma}_{\mu\nu}\Delta\omega^{\mu\nu} = 1\!\!1 + \frac{1}{8}\left[\gamma_\mu,\gamma_\nu\right]_-\Delta\omega^{\mu\nu} \quad . \tag{3.43}$$

The next step is to construct $\hat{S}(\hat{a})$ for *finite proper Lorentz transformations* by successive application of the infinitesimal operators (3.43). To construct the finite Lorentz transformation of (3.1) from the infinitesimal one (3.35), we write

$$\Delta\omega^\nu{}_\mu = \Delta\omega\left(\hat{I}_n\right)^\nu{}_\mu \quad . \tag{3.44}$$

Here $\Delta\omega$ is an infinitesimal parameter of the Lorentz group [infinitesimal (generalized) rotation angle] around an axis in the $\boldsymbol{n}$ direction. $(\hat{I}_{\boldsymbol{n}})^\nu{}_\mu$ is the 4×4 matrix (in space and time) for a *unit Lorentz rotation around the* $\boldsymbol{n}$ axis. For the Lorentz rotation (3.37), which – as we already know – corresponds to a Lorentz transformation along the x axis with velocity $\Delta v = c\Delta\beta$, we find

$$a^\nu{}_\mu = \delta^\nu_\mu + \Delta\omega^\nu{}_\mu = \delta^\nu_\mu + \Delta\omega\left(\hat{I}_x\right)^\nu{}_\mu = \delta^\nu_\mu - \Delta\beta\left(\delta^\nu_1\delta^0_\mu + \delta^\nu_0\delta^1_\mu\right) \quad . \tag{3.45}$$

With $\Delta\omega = \Delta\beta$,

$$\left(\hat{I}_x\right)^\nu{}_\mu = -\left(\delta^\nu_1\delta^0_\mu + \delta^\nu_0\delta^1_\mu\right) = \begin{pmatrix} 0 & -1 & 0 & 0\\ -1 & 0 & 0 & 0\\ 0 & 0 & 0 & 0\\ 0 & 0 & 0 & 0\end{pmatrix} \quad , \tag{3.46}$$

that only the matrix elements

$$\left(\hat{I}_x\right)^0{}_1 = \left(\hat{I}_x\right)^1{}_0 = -\left(\hat{I}_x\right)^{01} = +\left(\hat{I}_x\right)^{10} = -1$$

are different from zero. Besides, we can easily calculate the relations

$$\left(\hat{I}_x\right)^2 = \begin{pmatrix} 0 & -1 & 0 & 0\\ -1 & 0 & 0 & 0\\ 0 & 0 & 0 & 0\\ 0 & 0 & 0 & 0\end{pmatrix}\begin{pmatrix} 0 & -1 & 0 & 0\\ -1 & 0 & 0 & 0\\ 0 & 0 & 0 & 0\\ 0 & 0 & 0 & 0\end{pmatrix} = \begin{pmatrix} 1 & 0 & 0 & 0\\ 0 & 1 & 0 & 0\\ 0 & 0 & 0 & 0\\ 0 & 0 & 0 & 0\end{pmatrix}$$

and

$$\begin{aligned}\left(\hat{I}_x\right)^3 &= \begin{pmatrix} 1 & 0 & 0 & 0\\ 0 & 1 & 0 & 0\\ 0 & 0 & 0 & 0\\ 0 & 0 & 0 & 0\end{pmatrix}\begin{pmatrix} 0 & -1 & 0 & 0\\ -1 & 0 & 0 & 0\\ 0 & 0 & 0 & 0\\ 0 & 0 & 0 & 0\end{pmatrix}\\ &= \begin{pmatrix} 0 & -1 & 0 & 0\\ -1 & 0 & 0 & 0\\ 0 & 0 & 0 & 0\\ 0 & 0 & 0 & 0\end{pmatrix} = \hat{I}_x \quad .\end{aligned} \tag{3.47}$$

3.3 Finite Proper Lorentz Transformations

The algebraic properties (3.47) for $\hat{I}_x$ are of value, because we can use them, following (3.1) and (3.45), to construct the finite proper Lorentz transformations. Indeed we get by successive application of the infinitesimal Lorentz transformations

$$
\begin{aligned}
(x')^\nu &= \lim_{N\to\infty}\left(\mathbb{1}+\frac{\omega}{N}\hat{I}_x\right)^\nu_{\ \nu_1}\left(\mathbb{1}+\frac{\omega}{N}\hat{I}_x\right)^{\nu_1}_{\ \nu_2}\dots x^{\nu_N} \\
&= \lim_{N\to\infty}\left[\left(\mathbb{1}+\frac{\omega}{N}\hat{I}_x\right)^N\right]^\nu_{\ \mu} x^\mu
\end{aligned}
$$

(The last summation index ν_N of the matrix multiplication has been renamed μ in the last step.)

$$
\begin{aligned}
&= \left(\mathrm{e}^{\omega\hat{I}_x}\right)^\nu_{\ \mu} x^\mu \\
&= \left(\cosh(\omega\hat{I}_x)+\sinh(\omega\hat{I}_x)\right)^\nu_{\ \mu} x^\mu \\
&= \left[\left(\mathbb{1}+\frac{(\omega\hat{I}_x)^2}{2!}+\frac{(\omega\hat{I}_x)^4}{4!}+\dots\right)+\left(\frac{\omega\hat{I}_x}{1!}+\frac{(\omega\hat{I}_x)^3}{3!}+\dots\right)\right]^\nu_{\ \mu} x^\mu \\
&= \left[\left(\mathbb{1}+\frac{\omega^2}{2!}(\hat{I}_x)^2+\frac{\omega^4}{4!}(\hat{I}_x)^2+\dots\right)+\left(\frac{\omega}{1!}+\frac{\omega^3}{3!}+\dots\right)\hat{I}_x\right]^\nu_{\ \mu} x^\mu \\
&= \left[\mathbb{1}-(\hat{I}_x)^2+(\cosh\omega)(\hat{I}_x)^2+(\sinh\omega)\hat{I}_x\right]^\nu_{\ \mu} x^\mu \quad ,
\end{aligned}
\tag{3.48}
$$

or, explicitly,

$$
\begin{pmatrix} x'^0 \\ x'^1 \\ x'^2 \\ x'^3 \end{pmatrix} = \begin{pmatrix} \cosh\omega & -\sinh\omega & 0 & 0 \\ -\sinh\omega & \cosh\omega & 0 & 0 \\ 0 & 0 & 1 & 0 \\ 0 & 0 & 0 & 1 \end{pmatrix} \begin{pmatrix} x^0 \\ x^1 \\ x^2 \\ x^3 \end{pmatrix}
\tag{3.49a}
$$

or

$$
\begin{aligned}
x'^0 &= x^0\cosh\omega - x^1\sinh\omega = \cosh\omega(x^0 - x^1\tanh\omega) \quad , \\
x'^1 &= -x^0\sinh\omega + x^1\cosh\omega = \cosh\omega(x^1 - x^0\tanh\omega) \quad , \\
x'^2 &= x^2 \quad , \\
x'^3 &= x^3 \quad .
\end{aligned}
\tag{3.49b}
$$

Now the finite Lorentz rotation angle ω can easily be related to the relative velocity $v_x = c\beta_x$ of both inertial systems. Let us observe the origin of the x' system and its motion from the x system,

$$
x'^1 = 0 = \cosh\omega(x^1 - x^0\tanh\omega) \quad .
$$

From this we can deduce the velocity of the x' system along the x axis of the x system.

$$
\frac{x^1}{x^0} = \frac{x^1}{ct} = \frac{v_x}{c} = \tanh\omega = \beta
\tag{3.50a}
$$

and so by use of the relation $\cosh^2\omega - \sinh^2\omega = 1$ we get

$$\cosh\omega = \frac{\cosh\omega}{\sqrt{\cosh^2\omega - \sinh^2\omega}} = \frac{1}{\sqrt{1-\tanh^2\omega}} = \frac{1}{\sqrt{1-\beta^2}} \quad , \tag{3.50b}$$

so that (3.49b) results in the known Lorentz transformations

$$x'^0 = \frac{x^0 - \beta x^1}{\sqrt{1-\beta^2}} \quad , \quad x'^1 = \frac{x^1 - \beta x^0}{\sqrt{1-\beta^2}} \quad , \quad x'^2 = x^2 \quad , \quad x'^3 = x^3 \, . \tag{3.51}$$

This procedure to construct Lorentz transformations can be generalized to motions in an arbitrary direction or to arbitrary spatial rotations. There exist six different matrices (generators) $(\hat{I}_n)^\nu{}_\mu$ corresponding to the six independent Lorentz transformations. They represent the four-dimensional genralization of the generators of spatial rotations known from nonrelativistic quantum mechanics.[6] We refer to Chap. 16 for a deeper discussion of the group theoretical structure of the Lorentz transformations.

3.4 The $\hat{S}$ Operator for Proper Lorentz Transformations

Now we can construct the spinor transformation operator $\hat{S}(\hat{\alpha})$ for a finite "rotation angle". We start from the infinitesimal transformation (3.39) with the rotation angle (3.44) around the $\boldsymbol{n}$ axis and apply N of these infinitesimal transformations.

$$\begin{aligned} \psi'(x') = \hat{S}(\hat{a})\psi(x) &= \lim_{N\to\infty}\left(\mathbb{1} - \frac{\mathrm{i}}{4}\frac{\omega}{N}\hat{\sigma}_{\mu\nu}(\hat{I}_n)^{\mu\nu}\right)^N \psi(x) \\ &= \mathrm{e}^{-(\mathrm{i}/4)\omega\hat{\sigma}_{\mu\nu}(\hat{I}_n)^{\mu\nu}}\psi(x) \quad . \end{aligned} \tag{3.52}$$

In particular for the rotation (3.46) which corresponds to a Lorentz transformation along the x axis we get

$$\begin{aligned} \psi'(x') &= \exp\left\{-(\mathrm{i}/4)\omega\left[\hat{\sigma}_{01}(\hat{I}_x)^{01} + \hat{\sigma}_{10}(\hat{I}_x)^{10}\right]\right\}\psi(x) \\ &= \exp\left\{-(\mathrm{i}/4)\omega\left[\hat{\sigma}_{01}(+1) + \hat{\sigma}_{10}(-1)\right]\right\}\psi(x) \\ &= \exp\left[-(\mathrm{i}/2)\omega\hat{\sigma}_{01}\right]\psi(x) \quad , \end{aligned} \tag{3.53}$$

because following (3.42) we have $\hat{\sigma}_{01} = -\hat{\sigma}_{10}$ and following (3.46), $(\hat{I}_x)^{01} = 1$ and $(\hat{I}_x)^{01} = -1$. Analogously we can write for a *rotation around the z axis of angle* φ (to avoid confusion we use the notation $\hat{I}_x, \hat{I}_y, \hat{I}_z$ for the Lorentz transformation into a moving system and $\hat{I}_1, \hat{I}_2, \hat{I}_3$ for spatial rotations, according to (3.28) and (3.44)),

$$\Delta\omega^\nu_\mu = \Delta\varphi(\hat{I}_3)^\nu{}_\mu \quad , \tag{3.54}$$

where

$$(\hat{I}_3)^\nu{}_\mu = \begin{pmatrix} 0 & 0 & 0 & 0 \\ 0 & 0 & 1 & 0 \\ 0 & -1 & 0 & 0 \\ 0 & 0 & 0 & 0 \end{pmatrix} \quad . \tag{3.55}$$

[6] See the discussion of angular momentum operators in W. Greiner, B. Müller: *Quantum Mechanics – Symmetries*, 2nd ed. (Springer, Berlin, Heidelberg 1994).

Thus only the elements $(\hat{I}_3)^{12} = -(\hat{I}_3)^{21}$ are non-zero, in which case we get

$$\begin{aligned}\psi'(x') &= \exp\left[-(\mathrm{i}/4)\varphi\hat{\sigma}_{\mu\nu}(\hat{I}_3)^{\mu\nu}\right]\psi(x)\\ &= \exp\left[-(\mathrm{i}/4)\varphi\left(\hat{\sigma}_{12}(\hat{I}_3)^{12} + \hat{\sigma}_{21}(\hat{I}_3)^{21}\right)\right]\psi(x)\\ &= \exp\left[(\mathrm{i}/4)\varphi\left(\hat{\sigma}_{12}(-1) + \hat{\sigma}_{21}(+1)\right)\right]\psi(x)\\ &= \exp\left[(\mathrm{i}/2)\varphi\hat{\sigma}_{12}\right]\psi(x) = \exp\left[(\mathrm{i}/2)\varphi\hat{\sigma}^{12}\right]\psi(x) \quad ,\end{aligned} \tag{3.56}$$

corresponding to (3.52). Using the explicit representations (3.13) we obtain, according to (3.42),

$$\begin{aligned}\hat{\sigma}^{12} &= \frac{\mathrm{i}}{2}\left[\gamma^1,\gamma^2\right]_-\\ &= \frac{\mathrm{i}}{2}\left[\begin{pmatrix}0 & \hat{\sigma}^1\\ -\hat{\sigma}^1 & 0\end{pmatrix}\begin{pmatrix}0 & \hat{\sigma}^2\\ -\hat{\sigma}^2 & 0\end{pmatrix} - \begin{pmatrix}0 & \hat{\sigma}^2\\ -\hat{\sigma}^2 & 0\end{pmatrix}\begin{pmatrix}0 & \hat{\sigma}^1\\ -\hat{\sigma}^1 & 0\end{pmatrix}\right]\\ &= \frac{\mathrm{i}}{2}\left[\begin{pmatrix}-\hat{\sigma}^1\hat{\sigma}^2 & 0\\ 0 & -\hat{\sigma}^1\hat{\sigma}^2\end{pmatrix} - \begin{pmatrix}-\hat{\sigma}^2\hat{\sigma}^1 & 0\\ 0 & -\hat{\sigma}^2\hat{\sigma}^1\end{pmatrix}\right]\\ &= -\frac{\mathrm{i}}{2}\begin{pmatrix}\hat{\sigma}^1\hat{\sigma}^2 - \hat{\sigma}^2\hat{\sigma}^1 & 0\\ 0 & \hat{\sigma}^1\hat{\sigma}^2 - \hat{\sigma}^2\hat{\sigma}^1\end{pmatrix}\\ &= -\frac{\mathrm{i}}{2}\begin{pmatrix}2\mathrm{i}\hat{\sigma}^3 & 0\\ 0 & 2\mathrm{i}\hat{\sigma}^3\end{pmatrix} = \begin{pmatrix}\hat{\sigma}^3 & 0\\ 0 & \hat{\sigma}^3\end{pmatrix} = \hat{\Sigma}_3 \quad .\end{aligned} \tag{3.57}$$

Here $\hat{\sigma}^i$ are the well-known 2×2 Pauli matrices, in particular

$$\hat{\sigma}_3 = \begin{pmatrix}1 & 0\\ 0 & -1\end{pmatrix} \quad .$$

We recognize the similarly between the spinor transformation (3.56) and the rotation of the two-component Pauli spinor,

$$\varphi'(x') = \mathrm{e}^{(\mathrm{i}/2)\boldsymbol{\omega}\cdot\hat{\boldsymbol{\sigma}}}\varphi(x) \quad , \tag{3.58}$$

with which we are already acquainted.[7] We can call the $\omega^{\mu\nu}$ the *covariant angular variables*, because they arise in a similar way as did the components of the axial rotation vector $\boldsymbol{\omega}$ for the spatial three-dimensional rotations. The existence of half-angles in the transformation law (3.56) is a result of the *pecularity of the spinor-rotation laws*. A spinor is first transformed into itself by a rotation of 4π, not 2π, as one might expect. Therefore, physically observable quantities have to be bilinear in the spinors $\psi(x)$ in spinor theory (Dirac theory), and hence they have to be of even order in the fields $\psi(x)$. Only in this case do observables become identical under a rotation of 2π, a property of observables we know from experience.

Following (3.52) and (3.56) the operator $\hat{S}_R$ for *spatial rotations* of spinors is given by

$$\hat{S}_R(\omega_{ij}) = \exp\left[-(\mathrm{i}/4)\hat{\sigma}_{ij}\omega^{ij}\right] \quad , \quad i,j = 1,2,3 \quad . \tag{3.59}$$

Note that in the previous case[7] we subsequently used active rotations, whereas for the Lorentz transformations we have used passive rotations. It is unitary, because the $\hat{\sigma}_{ij}$ $(i,j = 1,2,3)$ are Hermitian and

[7] See W. Greiner, B. Müller: *Quantum Mechanics – Symmetries*, 2nd ed. (Springer, Berlin, Heidelberg 1994).

$$\hat{S}_R^\dagger = \exp\left[(\mathrm{i}/4)\hat{\sigma}_{ij}^\dagger\omega^{ij}\right] = \exp\left[(\mathrm{i}/4)\hat{\sigma}_{ij}\omega^{ij}\right] = \hat{S}_R^{-1} \quad . \tag{3.60}$$

For proper Lorentz transformations (e.g. transformation into a moving frame) this does not hold. For example, for a Lorentz transformation into a moving inertial system along the x axis (3.53) we find that

$$\hat{S}_\mathrm{L} = \mathrm{e}^{-(\mathrm{i}/2)\omega\hat{\sigma}_{01}} = \mathrm{e}^{+(\omega/2)\hat{\alpha}_1} = \hat{S}_\mathrm{L}^\dagger \neq \hat{S}^{-1}{}_\mathrm{L} \quad , \tag{3.61}$$

where we have followed from (3.42)

$$\begin{aligned}\hat{\sigma}_{01} &= \frac{\mathrm{i}}{2}(\gamma_0\gamma_1 - \gamma_1\gamma_0) = \frac{\mathrm{i}}{2}(\hat{\beta}\hat{\beta}\hat{\alpha}_1 - \hat{\beta}\hat{\alpha}_1\hat{\beta}) \\ &= \frac{\mathrm{i}}{2}(\hat{\alpha}_1 + \hat{\alpha}_1) = \mathrm{i}\hat{\alpha}_1 \quad .\end{aligned}$$

In Exercise 3.3 we shall prove that in this case

$$\hat{S}^{-1}{}_\mathrm{L} = \gamma_0\hat{S}_\mathrm{L}^\dagger\gamma_0 \tag{3.62}$$

holds and further that, because of

$$\left[\gamma_0, \hat{\sigma}_{ij}\right]_- = 0 \quad ,$$

we can combine (3.60) and (3.62). Hence, for any Lorentz transformation (i.e. for proper Lorentz transformations and for spatial rotations) we can write

$$\hat{S}^{-1} = \gamma_0\hat{S}^\dagger\gamma_0 \quad . \tag{3.63}$$

EXERCISE

3.3 Calculation of the Inverse Spinor Transformation Operators

Problem. Show that for $\hat{S} = \exp\left(-\frac{\mathrm{i}}{4}\omega\hat{\sigma}_{\mu\nu}(\hat{I}_n)^{\mu\nu}\right)$ the inverse operator is given by:

$$\hat{S}^{-1} = \gamma_0\hat{S}^\dagger\gamma_0 \quad . \tag{1}$$

Solution. We prove (1) separately for transformations on a moving coordinate system and for spatial rotations.

(i) Spatial rotations. For spatial rotations the time coordinate remains unchanged. So the components $\hat{I}^{0\mu}$, $\hat{I}^{0\nu}$ of the corresponding generators vanish identically. Thus we can write

$$\begin{aligned}\hat{S} &= \exp\left(-\frac{\mathrm{i}}{4}\omega^{ij}\hat{\sigma}_{ij}\right) \quad , \\ \hat{S}^\dagger &= \exp\left(\frac{\mathrm{i}}{4}\omega^{ij}\hat{\sigma}_{ij}^\dagger\right) = \exp\left(\frac{\mathrm{i}}{4}\omega^{ij}\hat{\sigma}_{ij}\right) \quad ,\end{aligned} \tag{2}$$

because the $\hat{\sigma}_{ij}$ are Hermitian and

$$\hat{\sigma}_{ij}^{\dagger} = -\frac{\mathrm{i}}{2}\left\{(\gamma_i\gamma_j)^{\dagger} - (\gamma_j\gamma_i)^{\dagger}\right\} = -\frac{\mathrm{i}}{2}\left\{\gamma_j\gamma_i - \gamma_i\gamma_j\right\} = \hat{\sigma}_{ij} \quad . \tag{3}$$

Exercise 3.3.

According to (3.41) γ_0 commutes with $\hat{\sigma}_{ij}$ and thus γ_0 commutes with $S^{\dagger}$ to 0. Hence we get

$$\gamma_0\hat{S}^{\dagger}\gamma_0 = \hat{S}^{\dagger} = \hat{S}^{-1} \quad . \tag{4}$$

(ii) Transformation onto a moving coordinate system. First we rotate the coordinate system such that the boost coincides with the x direction. For this transformation we have

$$\hat{S} = \exp\left(-\frac{\mathrm{i}}{2}\omega\hat{\sigma}_{01}\right)$$

[see (3.61)] and

$$\hat{S}^{\dagger} = \exp\left(\frac{\mathrm{i}}{2}\omega\hat{\sigma}_{01}^{\dagger}\right) = \exp\left(-\frac{\mathrm{i}}{2}\omega\hat{\sigma}_{01}\right) = \hat{S} \quad , \tag{5}$$

because

$$\hat{\sigma}_{01}^{\dagger} = -\frac{\mathrm{i}}{2}\left\{(\gamma_0\gamma_1)^{\dagger} - (\gamma_1\gamma_0)^{\dagger}\right\} = \frac{\mathrm{i}}{2}(\gamma_1\gamma_0 - \gamma_0\gamma_1) = -\hat{\sigma}_{01} \quad . \tag{6}$$

From

$$\gamma_0\hat{\sigma}_{01} = \frac{\mathrm{i}}{2}(\gamma_0\gamma_0\gamma_1 - \gamma_0\gamma_1\gamma_0) = \frac{\mathrm{i}}{2}(\gamma_1\gamma_0\gamma_0 - \gamma_0\gamma_1\gamma_0) = \hat{\sigma}_{10}\gamma_0 = -\hat{\sigma}_{01}\gamma_0 \tag{7}$$

we get

$$\begin{aligned}
\gamma_0\hat{S}^{\dagger}\gamma_0 &= \gamma_0\left[\sum_{n=0}^{\infty}\left(-\frac{\mathrm{i}}{2}\omega\hat{\sigma}_{01}\right)^n\right]\gamma_0 = \sum_{n=0}^{\infty}\gamma_0\left(-\frac{\mathrm{i}}{2}\omega\hat{\sigma}_{01}\right)^n\gamma_0 \\
&= \sum_{n=0}^{\infty}\underbrace{\gamma_0\left(-\frac{\mathrm{i}}{2}\omega\hat{\sigma}_{01}\right)\gamma_0\gamma_0\left(-\frac{\mathrm{i}}{2}\omega\hat{\sigma}_{01}\right)\gamma_0\ldots\gamma_0\left(-\frac{\mathrm{i}}{2}\omega\hat{\sigma}_{01}\right)\gamma_0}_{n\text{-times}} \\
&= \sum_{n=0}^{\infty}\left(+\frac{\mathrm{i}}{2}\omega\hat{\sigma}_{01}\right)^n = \exp\left(\frac{\mathrm{i}}{2}\omega\hat{\sigma}_{01}\right) = \hat{S}^{-1} \quad .
\end{aligned} \tag{8}$$

Thus (1) holds for all Lorentz transformations $\hat{S}$.

3.5 The Four-Current Density

Next we prove the covariance of the continuity equation (2.16) and (2.17). According to (2.18) the probability current density reads

$$\{j^{\mu}\} = \{j^0, \boldsymbol{j}\} = \{c\psi^{\dagger}\psi, c\psi^{\dagger}\hat{\boldsymbol{\alpha}}\psi\} = \{c\psi^{\dagger}\gamma^0\gamma^{\mu}\psi\} \quad ,$$

i.e.

$$j_{\mu}(x) = c\psi^{\dagger}(x)\gamma^0\gamma^{\mu}\psi(x) \quad . \tag{3.64}$$

This *current density* transforms under the Lorentz transformation (3.1) as

$$\begin{aligned} j'^{\mu}(x') &= c\psi'^{\dagger}(x')\gamma^0\gamma^{\mu}\psi'(x') \\ &= c\psi^{\dagger}(x)\hat{S}^{\dagger}\gamma^0\gamma^{\mu}\hat{S}\,\psi(x) \\ &= c\psi^{\dagger}(x)\gamma^0\hat{S}^{-1}\gamma^{\mu}\hat{S}\,\psi(x) \quad \text{[because of (3.63)]} \\ &= ca^{\mu}{}_{\nu}\psi^{\dagger}(x)\gamma^0\gamma^{\nu}\psi(x) \quad \text{[because of (3.33)]} \\ &= a^{\mu}{}_{\nu}j^{\nu}(x) \end{aligned} \tag{3.65}$$

and is such recognized as a *four-vector*. This is the reason for calling $j^{\mu}(x)$ the *four-current density*. The continuity equation (2.17) can now be written in *Lorentz-invariant form*

$$\frac{\partial j^{\mu}(x)}{\partial x^{\mu}} = 0 \quad . \tag{3.66}$$

By this we have explicitly proven that the *probability density*

$$j^0(x) = c\varrho(x) = c\psi^{\dagger}\psi(x)$$

transforms like the time component of a (conserved) four-vector. Thus an invariant probability is guaranteed, because it holds for the Lorentz system of observer A

$$\frac{\partial}{\partial t}\int j^0(x)\,\mathrm{d}^3x = 0 \quad \Rightarrow \quad \int j^0(x)\,\mathrm{d}^3x = 1$$

and for observer B

$$\frac{\partial}{\partial t'}\int j'^0(x')\,\mathrm{d}^3x' = 0 \quad \Rightarrow \quad \int j'^0(x')\,\mathrm{d}^3x' = 1 \quad .$$

For further considerations it is useful to introduce the short-hand notation

$$\overline{\psi} \equiv \psi^{\dagger}\gamma^0 \tag{3.67}$$

for the combination $\psi^{\dagger}\gamma^0$, which occurs very often. $\overline{\psi}$ is called the *adjoint spinor* and is converted by Lorentz transformations as [using (3.63)]

$$\begin{aligned} \overline{\psi}'(x') &= \psi'^{\dagger}(x')\gamma^0 = \left(\hat{S}\Psi(x)\right)^{\dagger}\gamma^0 = \psi^{\dagger}(x)\hat{S}^{\dagger}\gamma^0 \\ &= \psi^{\dagger}(x)\gamma^0\hat{S}^{-1} = \overline{\psi}(x)\hat{S}^{-1} \quad . \end{aligned} \tag{3.68}$$

3.6 Biographical Notes

LORENTZ, Hendrik Antoon, Dutch physicist, * 18.7.1853 in Arnheim, † 4.2.1928 in Haarlem. Professor at Leiden, founded in 1895 the theory of electrons, with which he explained the Zeeman effect as well as the rotation of the plane of polarization of light in a magnetic field. He gave, furthermore, a first explanation of the results of the Michelson–Morley experiment (L. contraction) and established the Lorentz transformation. Together with P. Zeeman he was awarded the Nobel prize in physics in 1902.

FEYNMAN, Richard Phillips, * 11.5.1918 in New York, † 15.2.1988 in Pasadena, professor at the California Institute of Technology in Pasadena. F. developed the Feynman graphs for the mathematical treatment of quantum field theory. In 1965 he was awarded the Nobel prize in physics (together with J. Schwinger and S. Tomonaga) for the development of the theory of quantum electrodynamics.

4. Spinors Under Spatial Reflection

Next we explore the *improper Lorentz transformations of spatial reflections*, which are given by

$$\boldsymbol{x}' = -\boldsymbol{x} \quad ,$$
$$t' = t \quad , \tag{4.1}$$

with a corresponding transformation matrix

$$a^{\nu}{}_{\mu} = \begin{pmatrix} 1 & 0 & 0 & 0 \\ 0 & -1 & 0 & 0 \\ 0 & 0 & -1 & 0 \\ 0 & 0 & 0 & -1 \end{pmatrix} = g^{\mu\nu} \quad . \tag{4.2}$$

In this case the Dirac equation should be covariant too, because (4.1) is just a special case of the general Lorentz transformation (3.1). All of our considerations from Chap. 3 can therefore be used again here, except for those which are based on infinitesimal transformations. This is consequence of the fact that a spatial reflection cannot be generated by means of infinitesimal rotations acting on the identity element. Let us call the operator of the spinor transformation $\hat{P}$ (for parity). For $\hat{S} = \hat{P}$ holds also the defining equation (3.34) which now reads with (4.2) as

$$a^{\nu}{}_{\mu}\gamma^{\mu} = \hat{P}\gamma^{\nu}\hat{P}^{-1} \quad \text{or}$$
$$a^{\sigma}{}_{\nu}a^{\nu}{}_{\mu}\gamma^{\mu} = \hat{P}a^{\sigma}{}_{\nu}\gamma^{\nu}\hat{P}^{-1}$$
$$\Leftrightarrow \delta^{\sigma}_{\mu}\gamma^{\mu} = \hat{P}\sum_{\nu=0}^{3} g^{\sigma\nu}\gamma^{\nu}\hat{P}^{-1}$$
$$\Leftrightarrow \hat{P}^{-1}\gamma^{\sigma}\hat{P} = g^{\sigma\sigma}\gamma^{\sigma} \quad . \tag{4.3}$$

Notice that on the rhs there is *no summation over* σ, because conventionally summation is only defined over identical indices occuring simultaneously as subscript and superscript in an expression. Equation (4.3) has the simple solution

$$\hat{P} = \mathrm{e}^{\mathrm{i}\varphi}\gamma^{0} \quad , \quad \hat{P}^{-1} = \mathrm{e}^{-\mathrm{i}\varphi}\gamma^{0} \quad , \tag{4.4}$$

where for the time being φ is an unobservable arbitrary phase. It can e.g. be chosen in the following way. In analogy to the proper Lorentz transformations for which a rotation of 4π reproduces the original spinor, we postulate that *four space inversions will reproduce the spinor*, i.e.

$$\hat{P}^{4}\psi = \psi = \mathrm{e}^{\mathrm{i}4\varphi}(\gamma^{0})^{4}\psi = \mathrm{e}^{\mathrm{i}4\varphi}\psi \quad , \tag{4.5}$$

This implies that

$$(\mathrm{e}^{\mathrm{i}\varphi})^4 = 1$$

and thus

$$\mathrm{e}^{\mathrm{i}\varphi} = \pm 1, \pm \mathrm{i} \quad . \tag{4.6}$$

The operator $\hat{P}$ given by (4.4) is unitary,

$$\hat{P}^{-1} = \mathrm{e}^{-\mathrm{i}\varphi}\gamma^0 = \hat{P}^\dagger \quad , \tag{4.7}$$

and, as will be seen shortly, also fulfills

$$\hat{P}^{-1} = \gamma^0 \hat{P}^\dagger \gamma^0 \tag{4.8}$$

which is analogous to (3.63). We are now able to write down explicitly the transformation of the spinor under space inversion, [cf. (3.27)]

$$\begin{aligned} \psi'(x') = \psi'(\boldsymbol{x}', t') = \psi'(-\boldsymbol{x}, t) = \hat{P}\psi(x) = \mathrm{e}^{\mathrm{i}\varphi}\gamma^0\psi(\boldsymbol{x}, t) \\ = \hat{P}\psi(-\boldsymbol{x}', t') \quad . \end{aligned} \tag{4.9}$$

In the nonrelativistic limit

$$\psi(x) = \begin{pmatrix} \phi(x) \\ 0 \end{pmatrix} \quad ,$$

and thus ϕ is an eigenvector of $\hat{P}$. For an electron at rest [see (2.71)],

$$\begin{aligned} \hat{P}\psi^{(1)} &= \mathrm{e}^{\mathrm{i}\varphi}\psi^{(1)} \quad , \\ \hat{P}\psi^{(2)} &= \mathrm{e}^{\mathrm{i}\varphi}\psi^{(2)} \quad , \\ \hat{P}\psi^{(3)} &= -\mathrm{e}^{\mathrm{i}\varphi}\psi^{(3)} \quad , \\ \hat{P}\psi^{(4)} &= -\mathrm{e}^{\mathrm{i}\varphi}\psi^{(4)} \quad . \end{aligned} \tag{4.10}$$

Therefore the eigenfunctions of positive energy $\psi^{(1)}$ and $\psi^{(2)}$ have a $\hat{P}$ eigenvalue ("inertial parity") opposite to that of the states with negative energy $\psi^{(3)}$ and $\psi^{(4)}$.

Another improper transformation is *time reversal* but since this is more complicated we shall deal with it later, in Sect. 12.5. The parity of spinors is also discussed in Example 9.3 and is further illustrated in Sect. 12.5.

5. Bilinear Covariants of the Dirac Spinors

There must exist 16 linearly independent 4×4 matrices, which we denote by $(\hat{\Gamma}^n)_{\alpha\beta}$. It turns out that one can construct 16 (a complete set) of these $\hat{\Gamma}_n$ ($n = 1, \dots, 16$) from the Dirac matrices and their products. We write

$$\underset{(1)}{\hat{\Gamma}^{\mathrm{S}} = \mathbb{1}} \quad , \quad \underset{(4)}{\hat{\Gamma}^{\mathrm{V}}_\mu = \gamma_\mu} \quad , \quad \underset{(6)}{\hat{\Gamma}^{\mathrm{T}}_{\mu\nu} = \hat{\sigma}_{\mu\nu}} = -\hat{\sigma}_{\nu\mu} \quad ,$$
$$\underset{(1)}{\hat{\Gamma}^{\mathrm{P}} = \mathrm{i}\gamma^0\gamma^1\gamma^2\gamma^3} = \gamma_5 \equiv \gamma^5 \quad , \quad \underset{(4)}{\hat{\Gamma}^{\mathrm{A}}_\mu = \gamma_5\gamma_\mu} \tag{5.1}$$

and verify step by step the postulated properties of the $\hat{\Gamma}^n$ as well as some extra ones (also cf. Example 3.1). First we shall prove that in (5.1) there are indeed 16 matrices. This is easily done by adding the values written in brackets below the symbols. The upper indices of the matrices ("S", "V", "T", "P", and "A") have the meaning *"scalar"*, *"vector"*, *"tensor"*, *"pseudovector"*, and *"axial vector"*, and these specifications will become clear in the following. Furthermore it holds that:

(a) For each $\hat{\Gamma}^n$ holds $(\hat{\Gamma}^n)^2 = \pm\mathbb{1}$. We shall prove this for some of the $\hat{\Gamma}^n$:

$$\left(\hat{\Gamma}^{\mathrm{S}}\right)^2 = \mathbb{1} \quad , \quad \left(\hat{\Gamma}^{\mathrm{V}}_\mu\right)^2 = \left(\gamma_\mu\right)^2 = \mathbb{1}_{g_{\mu\mu}} \quad , \tag{5.2}$$

due to the commutation relations (3.11),

$$\begin{aligned}
\left(\hat{\Gamma}^{\mathrm{T}}_{\mu\nu}\right)^2 &= \hat{\sigma}^2_{\mu\nu} = -\tfrac{1}{4}(\gamma_\mu\gamma_\nu - \gamma_\nu\gamma_\mu)^2 \\
&= -\tfrac{1}{4}(\gamma_\mu\gamma_\nu\gamma_\mu\gamma_\nu - \gamma_\mu\gamma_\nu\gamma_\nu\gamma_\mu - \gamma_\nu\gamma_\mu\gamma_\mu\gamma_\nu + \gamma_\nu\gamma_\mu\gamma_\nu\gamma_\mu) \\
&= -\tfrac{1}{4}\big[2g_{\mu\nu}\gamma_\mu\gamma_\nu - (\gamma_\mu)^2(\gamma_\nu)^2 - 2(\gamma_\mu)^2(\gamma_\nu)^2 \\
&\quad + 2g_{\mu\nu}\gamma_\nu\gamma_\mu - (\gamma_\nu)^2(\gamma_\mu)^2\big] \\
&= -\tfrac{1}{4}\left[2g_{\mu\nu}(\gamma_\mu\gamma_\nu + \gamma_\nu\gamma_\mu) - 4(\gamma_\nu)^2(\gamma_\mu)^2\right] \\
&= -\tfrac{1}{4}\left[4(g_{\mu\nu})^2\mathbb{1} - 4g_{\nu\nu}g_{\mu\mu}\mathbb{1}\right] \\
&= \begin{cases} g_{\nu\nu}g_{\mu\mu}\mathbb{1} & \text{for } \mu \neq \nu \\ 0 & \text{for } \mu = \nu\text{. In this case } \hat{\sigma}_{\mu\nu} = 0 \end{cases} \quad .
\end{aligned}$$

Similarly the conjecture is proven for the other matrices (5.1).

(b) To each $\hat{\Gamma}^n$ *except* $\hat{\Gamma}^{\mathrm{S}}$ there exists at least one corresponding $\hat{\Gamma}^m$ with

$$\hat{\Gamma}^n\hat{\Gamma}^m = -\hat{\Gamma}^m\hat{\Gamma}^n \quad . \tag{5.3}$$

This is easily proved explicitly, e.g.

$$\hat{\Gamma}^{\mathrm{V}}_\mu : \hat{\Gamma}^{\mathrm{V}}_\mu\hat{\Gamma}^{\mathrm{V}}_\nu = -\hat{\Gamma}^{\mathrm{V}}_\nu\hat{\Gamma}^{\mathrm{V}}_\mu \quad \text{if} \quad \nu \neq \mu$$

since this corresponds to the commutation relation (3.11)

$$\hat{\Gamma}^{\mathrm{T}}_{\mu\nu} : \hat{\Gamma}^{\mathrm{T}}_{\mu\nu}\hat{\Gamma}^{\mathrm{V}}_{\mu} = -\hat{\Gamma}^{\mathrm{V}}_{\mu}\hat{\Gamma}^{\mathrm{T}}_{\mu\nu} \quad ,$$

because for each μ, ν we have

$$\begin{aligned}\sigma_{\mu\nu}\gamma_\mu &= \frac{\mathrm{i}}{2}(\gamma_\mu\gamma_\nu - \gamma_\nu\gamma_\mu)\gamma_\mu = \frac{\mathrm{i}}{2}(-\gamma_\mu\gamma_\mu\gamma_\nu + 2\gamma_\mu g_{\mu\nu} + \gamma_\mu\gamma_\nu\gamma_\mu - 2\gamma_\mu g_{\mu\nu}) \\ &= -\frac{\mathrm{i}}{2}\gamma_\mu(\gamma_\mu\gamma_\nu - \gamma_\nu\gamma_\mu) = -\gamma_\mu\sigma_{\mu\nu} \quad .\end{aligned}$$

This shows that the $\hat{\Gamma}^{\mathrm{V}}_{\mu}$'s correspond to the $\hat{\Gamma}^{\mathrm{T}}_{\mu\nu}$ in the sense of relation (5.3), and vice versa, etc. For the other Γ^n matrices one proves the relation (5.3) in a similar way. From (5.3) and (5.2) in particular,

$$\pm\hat{\Gamma}^n = -\hat{\Gamma}^m\hat{\Gamma}^n\hat{\Gamma}^m = +\hat{\Gamma}^n(\hat{\Gamma}^m)^2$$

and therefore calculating the trace yields [making use of $\mathrm{tr}(\hat{A}\hat{B}) = \mathrm{tr}(\hat{B}\hat{A})$]

$$\pm\,\mathrm{tr}(\hat{\Gamma}^n) = -\,\mathrm{tr}\left(\hat{\Gamma}^n(\hat{\Gamma}^m)^2\right) = +\,\mathrm{tr}\left(\hat{\Gamma}^n(\hat{\Gamma}^m)^2\right) = 0 \quad . \tag{5.4}$$

Here we have a remarkable result: *All $\hat{\Gamma}^n$ matrices except $\hat{\Gamma}^{\mathrm{S}}$ have a vanishing trace*;

(c) For given $\hat{\Gamma}^a$ and $\hat{\Gamma}^b$ $(a \neq b)$ there exists a $\hat{\Gamma}^n \neq \hat{\Gamma}^{\mathrm{S}}$ with

$$\hat{\Gamma}^a\hat{\Gamma}^b = f^n_{ab}\hat{\Gamma}^n \quad , \tag{5.5}$$

so defining f^n_{ab} as a complex number. We check this property with some examples: For $\mu \neq \nu$ we have

$$\hat{\Gamma}^{\mathrm{V}}_{\mu}\hat{\Gamma}^{\mathrm{V}}_{\nu} = \gamma_\mu\gamma_\nu = \frac{\gamma_\mu\gamma_\nu - \gamma_\nu\gamma_\mu}{2} = -\mathrm{i}\hat{\sigma}_{\mu\nu} = -\mathrm{i}\hat{\Gamma}^{\mathrm{T}}_{\mu\nu} \quad .$$

In this case the f factor is $-\mathrm{i}$. As a second example we take a look at

$$\hat{\Gamma}^{\mathrm{V}}_{\mu}\hat{\Gamma}^{\mathrm{T}}_{\sigma\tau} = \frac{\mathrm{i}}{2}\gamma_\mu(\gamma_\sigma\gamma_\tau - \gamma_\tau\gamma_\sigma) = \begin{cases} -\mathrm{i}\gamma_\sigma g_{\mu\mu} = -\mathrm{i}g_{\mu\mu}\hat{\Gamma}^{\mathrm{V}}_{\sigma} & \text{for } \mu = \tau \quad , \\ +\mathrm{i}\gamma_\sigma g_{\mu\mu} = \mathrm{i}g_{\mu\mu}\hat{\Gamma}^{\mathrm{V}}_{\sigma} & \text{for } \mu = \sigma \quad , \\ \pm\mathrm{i}\hat{\Gamma}^{\mathrm{A}}_{\kappa}, \kappa \neq \mu, \sigma, \tau & \text{for } \mu \neq \tau \neq \sigma \quad , \end{cases}$$

etc. It can easily be shown that (5.5) holds in each case.

Now we show that the $\hat{\Gamma}^n$'s of (5.1) are linearly independent. For this purpose we suppose the existence of the relation

$$\sum_n a_n\hat{\Gamma}^n = 0 \quad , \tag{5.6}$$

and by multiplication with $\hat{\Gamma}^m \neq \hat{\Gamma}^{\mathrm{S}}$ we get

$$\begin{aligned}0 = \Sigma_n a_n\,\mathrm{tr}\left(\hat{\Gamma}^n\hat{\Gamma}^m\right) &= a_m\,\mathrm{tr}\left((\hat{\Gamma}^m)^2\right) + \Sigma_{n\neq m}a_n\,\mathrm{tr}\left(f^\nu_{nm}\hat{\Gamma}^\nu\right) \quad \text{[cf. (5.5)]} \\ &= a_m\,\mathrm{tr}\,\underbrace{\left((\hat{\Gamma}^m)^2\right)}_{=\pm\mathbb{1}\ \text{[cf. (5.2)]}} + \Sigma_{n\neq m}a_n f^\nu_{nm}\,\underbrace{\mathrm{tr}\left(\hat{\Gamma}^\nu\right)}_{=0\ \text{[cf. (5.4)]}} \\ &= \pm 4a_m \quad .\end{aligned} \tag{5.7a}$$

Thus $a_m = 0$ for all $m \neq S$, and in the case of $\hat{\Gamma}^m = \hat{\Gamma}^S$

$$0 = \mathrm{tr}\left(\sum_n a_n \hat{\Gamma}^S \hat{\Gamma}^n\right) = a_S \underbrace{\mathrm{tr}\left(\hat{\Gamma}^S\right)^2}_{=4\ [\mathrm{cf.}\ (5.1)]} + \sum_{n \neq S} a_n \underbrace{\mathrm{tr}\left(\hat{\Gamma}^n\right)}_{=0\ [\mathrm{cf.}\ (5.4)]} = 0 \quad , \tag{5.7b}$$

i.e. $a_S = 0$ also. Therefore all coefficients a_n of (5.6) must vanish and the linear independence of the $\hat{\Gamma}^n$'s is proved. Hence every 4×4 matrix can be expressed by the $\hat{\Gamma}^n$'s.

We now turn to the behaviour of the *bilinear expressions*

$$\overline{\psi}(x)\hat{\Gamma}\psi(x) \tag{5.8}$$

under Lorentz transformations. For this purpose we need

$$\gamma^\mu \gamma_5 + \gamma_5 \gamma^\mu = 0 \quad , \tag{5.9}$$

which may be easily verified and leads immediately to

$$\left[\gamma_5, \hat{\sigma}_{\mu\nu}\right]_- = \frac{\mathrm{i}}{2}\left(\gamma_5(\gamma_\mu\gamma_\nu - \gamma_\nu\gamma_\mu) - (\gamma_\mu\gamma_\nu - \gamma_\nu\gamma_\mu)\gamma_5\right) = 0 \quad . \tag{5.10}$$

According to the proper Lorentz transformations (3.52) the spinors are transformed by the operator

$$\hat{S}(\hat{a}) = \exp\left(-\frac{\mathrm{i}}{4}\hat{\sigma}_{\mu\nu}\omega_n^{\mu\nu}\right) = \exp\left(-\frac{\mathrm{i}}{4}\omega\hat{\sigma}_{\mu\nu}(\hat{I}_n)^{\mu\nu}\right) \quad . \tag{5.11}$$

Together with (5.10) we get directly

$$\left[\hat{S}(\hat{a}), \gamma_5\right]_- = 0 \quad . \tag{5.12}$$

According to (4.9) the space inversion (improper Lorentz transformation) of the spinors is accomplished by the operator

$$\hat{P} = \mathrm{e}^{\mathrm{i}\phi}\gamma^0 \quad . \tag{5.13}$$

For this operator, by virtue of (5.9),

$$\hat{P}\gamma_5 = -\gamma_5\hat{P} \quad \text{or} \tag{5.14a}$$

$$\left[\hat{P}, \gamma_5\right]_+ = 0 \quad . \tag{5.14b}$$

Now we consider bilinear quantities in $\overline{\psi}$ and ψ and can easily calculate that, e.g.

$$\begin{aligned} \overline{\psi}'(x')\psi'(x') &= \psi'^\dagger(x')\gamma^0\psi'(x') = \psi^\dagger(x)\hat{S}^\dagger\gamma^0\hat{S}\psi(x) \\ &= \psi^\dagger(x)\gamma^0\hat{S}^{-1}\hat{S}\psi(x) = \psi^\dagger(x)\gamma^0\psi(x) = \overline{\psi}(x)\psi(x) \quad . \end{aligned} \tag{5.15}$$

Thus the bilinear expression $\overline{\psi}(x)\psi(x)$ has the same value in every Lorentz frame. We therefore call

$$\overline{\psi}(x)\psi(x) \equiv \overline{\psi}(x)\hat{\Gamma}^S\psi(x) \tag{5.16}$$

a *scalar* under Lorentz transformations (or simply *Lorentz scalar*) and $\overline{\psi}(x)\psi(x)$ the *scalar density*. Similarly, in Exercise 5.1 we prove that

$$\overline{\psi}'(x')\gamma_5\psi'(x') = \overline{\psi}(x)\hat{S}^{-1}\gamma_5\hat{S}\,\psi(x) = \det(a)\overline{\psi}(x)\gamma_5\psi(x) \tag{5.17a}$$

is a *pseudoscalar*,

$$\overline{\psi}'(x')\gamma^\nu\psi'(x') = a^\nu{}_\mu\overline{\psi}(x)\gamma^\mu\psi(x) \tag{5.17b}$$

is a *vector*,

$$\overline{\psi}'(x')\gamma_5\gamma^\nu\psi'(x') = \det(a)a^\nu{}_\mu\overline{\psi}(x)\gamma_5\gamma^\mu\psi(x) \tag{5.17c}$$

is a *pseudovector*, and

$$\overline{\psi}'(x')\hat{\sigma}^{\mu\nu}\psi'(x') = a^\mu{}_\alpha a^\nu{}_\beta\overline{\psi}(x)\hat{\sigma}^{\alpha\beta}\psi(x) \tag{5.17d}$$

is a *tensor of second rank*. The prefix "pseudo" in pseudoscalar indicates that this quantity transforms as a Lorentz scalar but reverses its sign under improper Lorentz transformations. The same holds for the pseudovector (5.17c).

EXERCISE

5.1 Transformation Properties of Some Bilinear Expressions of Dirac Spinors

Problem. Investigate the transformation properties of the following bilinear expressions consisting of Dirac spinors:

$$(1)\ \overline{\psi}\psi \quad,\quad (2)\ \overline{\psi}\gamma_5\psi \quad,\quad (3)\ \overline{\psi}\gamma_\mu\psi \quad,\quad (4)\ \overline{\psi}\gamma_5\gamma_\mu\psi \quad,\quad (5)\ \overline{\psi}\hat{\sigma}^{\mu\nu}\psi \quad.$$

Solution. ψ changes to $\hat{S}\psi$ under Lorentz transformations and $\overline{\psi}$ to $\overline{\psi}\hat{S}^{-1}$ (since $\hat{S}^\dagger\gamma_0 = \gamma_0\hat{S}^{-1}$). It holds that $\hat{S}^{-1}\gamma^\mu\hat{S} = a^\mu{}_\nu\gamma^\nu$ and $\hat{S}\gamma_5 = \gamma_5\hat{S}\det|a|$. The latter relation is a useful summary of (5.12) and (5.14). Thus,

$$\begin{aligned}
(1)\quad & \overline{\psi}\psi \to \overline{\psi}\hat{S}^{-1}\hat{S}\psi \\
& = \overline{\psi}\psi \ \text{(scalar)} \quad, \\
(2)\quad & \overline{\psi}\gamma_5\psi \to \overline{\psi}\hat{S}^{-1}\gamma_5\hat{S}\psi \\
& = \overline{\psi}\det|a|\hat{S}^{-1}\hat{S}\gamma_5\psi \\
& = \det|a|\overline{\psi}\gamma_5\psi \ \text{(pseudoscalar)} \quad, \\
(3)\quad & \overline{\psi}\gamma^\mu\psi \to \overline{\psi}\hat{S}^{-1}\gamma^\mu\hat{S}\psi \\
& = a^\mu{}_\nu\overline{\psi}\gamma^\nu\psi \ \text{(vector)} \quad, \\
(4)\quad & \overline{\psi}\gamma_5\gamma^\mu\psi \to \overline{\psi}\hat{S}^{-1}\gamma_5\gamma^\mu\hat{S}\psi \\
& = \det|a|\overline{\psi}\gamma_5\hat{S}^{-1}\gamma^\mu\hat{S}\psi \\
& = \det|a|a^\mu{}_\nu\overline{\psi}\gamma_5\gamma^\nu\psi \ \text{(pseudovector)} \quad, \\
(5)\quad & \overline{\psi}\hat{\sigma}^{\mu\nu}\psi \to \overline{\psi}\hat{S}^{-1}\hat{\sigma}^{\mu\nu}\hat{S}\psi \\
& = \frac{\mathrm{i}}{2}\overline{\psi}\hat{S}^{-1}\left(\gamma^\mu\hat{S}\hat{S}^{-1}\gamma^\nu - \gamma^\nu\hat{S}\hat{S}^{-1}\gamma^\mu\right)\hat{S}\psi \\
& = \frac{\mathrm{i}}{2}\overline{\psi}\left(a^\mu{}_\varrho\gamma^\varrho a^\nu{}_\tau\gamma^\tau - a^\nu{}_\tau\gamma^\tau a^\mu{}_\varrho\gamma^\varrho\right)\psi \\
& = a^\mu{}_\varrho a^\nu{}_\tau\overline{\psi}\hat{\sigma}^{\varrho\tau}\psi \ \text{(second-rank tensor)} \quad.
\end{aligned}$$

EXERCISE

5.2 Majorana Representation of the Dirac Equation

Problem. Show that there exist four 4×4 matrices $\hat{\Gamma}_\mu$ such that

$$\begin{aligned}
&\mathrm{Re}\left(\hat{\Gamma}_\mu\right)_{\alpha\beta} = 0 \quad , \\
&\hat{\Gamma}_\mu \hat{\Gamma}_\nu + \hat{\Gamma}_\nu \hat{\Gamma}_\mu = 2g_{\mu\nu} \quad , \\
&\left(\mathrm{i}\hbar \hat{\Gamma}_\mu \partial^\mu - m_0 c\right)\psi = 0 \quad ,
\end{aligned}$$

i.e., the Dirac equation is real in this representation. Here $\partial^\mu = \partial/\partial x_\mu$.

Solution. From the common γ_μ matrices we know that they fulfill

$$\gamma_0 = \begin{pmatrix} 1\!\!1 & 0 \\ 0 & -1\!\!1 \end{pmatrix} \quad , \quad \gamma_i = \begin{pmatrix} 0 & \hat{\sigma}_i \\ -\hat{\sigma}_i & 0 \end{pmatrix} \quad ,$$

with

$$\hat{\sigma}_1 = \begin{pmatrix} 0 & 1 \\ 1 & 0 \end{pmatrix} \quad , \hat{\sigma}_2 = \begin{pmatrix} 0 & -\mathrm{i} \\ \mathrm{i} & 0 \end{pmatrix} \quad , \quad \hat{\sigma}_3 = \begin{pmatrix} 1 & 0 \\ 0 & -1 \end{pmatrix}$$

and

$$\gamma_\mu \gamma_\nu + \gamma_\nu \gamma_\mu = 2g_{\mu\nu} \quad , \quad \gamma_0^\dagger = \gamma_0 \quad , \quad \gamma_i^\dagger = -\gamma_i \quad .$$

We define

$$\begin{aligned}
\hat{\Gamma}_0 &= \gamma_0 \gamma_2 = \begin{pmatrix} 0 & \hat{\sigma}_2 \\ \hat{\sigma}_2 & 0 \end{pmatrix} \quad , \\
\hat{\Gamma}_1 &= \mathrm{i}\gamma_0 \gamma_1 = \mathrm{i} \begin{pmatrix} 0 & \hat{\sigma}_1 \\ \hat{\sigma}_1 & 0 \end{pmatrix} \quad , \\
\hat{\Gamma}_2 &= \mathrm{i}\gamma_0 = \mathrm{i} \begin{pmatrix} 1\!\!1 & 0 \\ 0 & -1\!\!1 \end{pmatrix} \quad , \\
\hat{\Gamma}_3 &= \mathrm{i}\gamma_0 \gamma_3 = \mathrm{i} \begin{pmatrix} 0 & \hat{\sigma}_3 \\ \hat{\sigma}_3 & 0 \end{pmatrix}
\end{aligned}$$

and recognize immediately that these $\hat{\Gamma}_\mu$ are purely imaginary matrices. On complex conjugation

$$\begin{aligned}
\hat{\Gamma}_0^* &= \gamma_0^* \gamma_2^* = -\gamma_0 \gamma_2 = -\hat{\Gamma}_0 \quad , \\
\hat{\Gamma}_1^* &= -\mathrm{i}\gamma_0^* \gamma_1^* = -\mathrm{i}\gamma_0 \gamma_1 = -\hat{\Gamma}_1 \quad , \\
\hat{\Gamma}_2^* &= -\mathrm{i}\gamma_0 = -\hat{\Gamma}_2 \quad , \\
\hat{\Gamma}_2^* &= -\mathrm{i}\gamma_0 \gamma_3 = -\hat{\Gamma}_3
\end{aligned}$$

since only γ_2 is imaginary. Now we check the anticommutation relations:

Exercise 5.2.

$$\begin{aligned}
g_{00} &= 1 = \left(\hat{\Gamma}_0\right)^2 = \gamma_0\gamma_2\gamma_0\gamma_2 \\
&= -(\gamma_0)^2(\gamma_2)^2 = 1 \quad ,
\end{aligned}$$

$$\begin{aligned}
2g_{01} &= 2g_{10} = 0 = -\mathrm{i}(\gamma_2\gamma_1 + \gamma_1\gamma_2) \\
&= -\mathrm{i}\left[(\gamma_0)^2\gamma_2\gamma_1 + (\gamma_0)^2\gamma_1\gamma_2\right] \\
&= -\mathrm{i}(\gamma_0\gamma_2\gamma_0\gamma_1 + \gamma_0\gamma_1\gamma_0\gamma_2) \\
&= \hat{\Gamma}_0\hat{\Gamma}_1 + \hat{\Gamma}_1\hat{\Gamma}_0 \quad ,
\end{aligned}$$

$$\begin{aligned}
2g_{02} &= 2g_{20} = 0 = \mathrm{i}(\gamma_2 - \gamma_2) = \mathrm{i}(\gamma_0\gamma_0\gamma_2 + \gamma_0\gamma_2\gamma_0) \\
&= \hat{\Gamma}_2\hat{\Gamma}_0 + \hat{\Gamma}_0\hat{\Gamma}_2 \quad ,
\end{aligned}$$

$$2g_{03} = 2g_{30} = 0 = \hat{\Gamma}_0\hat{\Gamma}_3 + \hat{\Gamma}_3\hat{\Gamma}_0 \quad \text{as for} \quad g_{01} \quad ,$$

$$\begin{aligned}
g_{11} &= -1 = (\gamma_0)^2(\gamma_1)^2 = -\gamma_0\gamma_1\gamma_0\gamma_1 \\
&= (\mathrm{i}\gamma_0\gamma_1)^2 = \left(\hat{\Gamma}_1\right)^2 \quad ,
\end{aligned}$$

$$\begin{aligned}
2g_{12} &= 2g_{21} = 0 = \hat{\Gamma}_1\hat{\Gamma}_2 + \hat{\Gamma}_2\hat{\Gamma}_1 \quad , \quad \text{as for} \quad g_{20} \quad , \\
2g_{13} &= 2g_{31} = 0 = \hat{\Gamma}_1\hat{\Gamma}_3 + \hat{\Gamma}_3\hat{\Gamma}_1 \quad , \quad \text{as for} \quad g_{10} \quad ,
\end{aligned}$$

$$g_{22} = -1 = (\mathrm{i}\gamma_0)^2 = \left(\hat{\Gamma}_2\right)^2 \quad ,$$

$$2g_{23} = 2g_{32} = 0 = \hat{\Gamma}_2\hat{\Gamma}_3 + \hat{\Gamma}_3\hat{\Gamma}_2 \quad , \quad \text{as for} \quad g_{20} \quad ,$$

$$g_{33} = -1 = \left(\hat{\Gamma}_3\right)^2 \quad , \quad \text{as for} \quad g_{11} \quad .$$

Thus we have proved all of the necessary properties of the $\hat{\Gamma}_\mu$ matrices. This $\hat{\Gamma}_\mu$ representation of the Dirac algebra is called the *Majorana representation.*

5.1 Biographical Notes

MAJORANA, Ettore, *05.08.1906 in Catania, dropped out of sight in 1938, went to the classical secondary school of Catania until the final examination in 1923. Afterwards he studied engineering sciences in Rome until the beginning of the last year of studies. 1928 transfer to the physics faculty (own desire) and in 1929 Ph.D. in theoretical physics at Fermi's. Title of the thesis: "Quantum Theory of Radioactive Atomic Nuclei". In the subsequent years free-lance collaborator at the Institute of Physics in Rome. In 1933 he went to Germany (Leipzig) for some years and worked with Heisenberg. This resulted in a publication on nuclear theory [Z. Phys. **82**, 137 (1933)]. In 1937 he published "The Symmetric Theory of Electron and Positron" and four years after his disappearence the "Significance of Statistical Laws for Physics and Social Sciences" was published.

6. Another Way of Constructing Solutions of the Free Dirac Equation: Construction by Lorentz Transformations

Our considerations in the last chapters showed that the free Dirac equation exhibits all of the properties of covariance. Moreover positive-energy solutions possess the correct behaviour in the nonrelativistic limit. We consider now the solutions of the free Dirac equation with a new approach better suited to the covariant formulation and, later on, for the field-theoretical apllications. Let us return to the solutions (2.71) of the free electron in its rest frame. We denote them in a more compact form:

$$\psi^r = \omega^r(0)\mathrm{e}^{-\mathrm{i}\varepsilon_r(m_0c^2/\hbar)t} \quad , \quad r = 1,2,3,4 \quad , \tag{6.1}$$

where

$$\varepsilon_r = \begin{cases} +1 & \text{for} \quad r = 1,\, 2 \\ -1 & \text{for} \quad 3,\, 4 \quad . \end{cases} \tag{6.2}$$

The x dependence of the spinors ψ^r in (6.1) reduces to a simple time dependence. There is no space dependence because the wavefunction is smeared out homogeneously in the whole of space. We also have

$$\omega^1(0) = \begin{pmatrix} 1 \\ 0 \\ 0 \\ 0 \end{pmatrix} \;,\; \omega^2(0) = \begin{pmatrix} 0 \\ 1 \\ 0 \\ 0 \end{pmatrix} \;,\; \omega^3(0) = \begin{pmatrix} 0 \\ 0 \\ 1 \\ 0 \end{pmatrix} \;,\; \omega^4(0) = \begin{pmatrix} 0 \\ 0 \\ 0 \\ 1 \end{pmatrix} \;. \tag{6.3}$$

The first and the second solutions $\psi^1(x)$ and $\psi^2(x)$ have positive energy and correspond to the spin degrees of freedom of the Schrödinger–Pauli electron. For the solutions $\psi^3(x)$ and $\psi^4(x)$ with negative energy we must still find a reasonable interpretation. All of the $\psi^r(x)$ of (6.1) are also eigenfunctions of

$$\hat{\Sigma}_3 = \hat{\sigma}_{12} = \begin{pmatrix} \hat{\sigma}_3 & 0 \\ 0 & \hat{\sigma}_3 \end{pmatrix} \tag{6.4}$$

[cf. (3.57)] with the eigenvalues ± 1:

$$\hat{\Sigma}_3\psi^r(x) = (\pm 1)\psi^r(x) \quad , \tag{6.5}$$

where the eigenvalues

$$+1 \quad \text{are valid for} \quad r = 1,\, 3 \quad \text{and} \quad -1 \quad \text{are valid for} \quad r = 2,\, 4 \quad .$$

Earlier [cf. (2.34)] we directly obtained the free solutions for finite momentum (finite velocity) by solving the free Dirac equation. Now we follow a different path: *By transforming to a coordinate system which moves with the velocity* $-\boldsymbol{v}$ *relative to the rest system, the free wave functions of the electron with velocity* $+\boldsymbol{v}$ *are constructed from the wave functions (6.1) of the electron at rest* (see Fig. 6.1). For this purpose we write first the exponent of the rest solution (6.1) in invariant form:

$$\mathrm{e}^{-\mathrm{i}\varepsilon_r (m_0 c^2/\hbar)t} = \mathrm{e}^{-\mathrm{i}\varepsilon_r (p_\mu x^\mu \hbar)} = \mathrm{e}^{-\mathrm{i}\varepsilon_r (p'_\mu x'^\mu \hbar)} \quad , \tag{6.6}$$

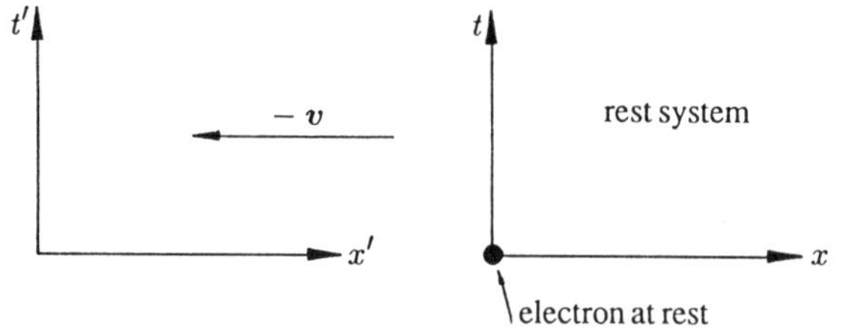

Fig. 6.1. In the $x't'$ system the electron moves with the velocity $+\boldsymbol{v}$. In the xt system the four-momentum is p^μ; in the $x't'$ system it is p'^μ

where

$$x'^\mu = a^\mu{}_\nu x^\nu \quad , \quad p'^\mu = a^\mu{}_\nu p^\nu = a^\mu{}_0 p^0 = a^\mu{}_0 m_0 c \quad ,$$
$$p^\nu = \{m_0 c, 0\} = \left\{\frac{m_0 c^2}{c}, 0\right\} \quad . \tag{6.7}$$

The primed quantities are valid in the moving reference system, the unprimed in the rest system of the electron. The zero-component of the four-momentum is always given by

$$p^0 = \frac{E}{c} = +\sqrt{\boldsymbol{p}^2 + m_0^2 c^2} > 0 \quad . \tag{6.8}$$

Next we remark that the solutions for positive and negative energy transform separately under proper Lorentz transformations and also under space inversion. This means that the solutions for positive and negative energy are not mixed under these transformations. One recognizes this from (6.6): The four-momentum of the free particle $p^\mu p_\mu = m_0^2 c^2 > 0$ is always timelike. Therefore, for free particles p^μ lies within the light cone in momentum space (Fig. 6.2). Under proper Lorentz transformation and also under space inversion (but not under time inversion), the future and past cones (i.e. the vectors with $p^0 > 0$ and $p^0 < 0$, respectively), and with them the solutions for positive and negative energy, remain strictly separated from each other.

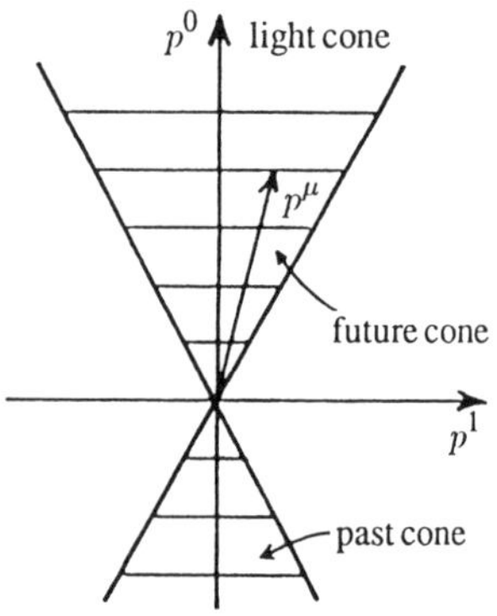

Fig. 6.2. The four-momentum of a free particle always lies within the light cone in momentum space

We transform the spinors ω^r according to (3.61) with the operator

$$\hat{S} = \mathrm{e}^{-\mathrm{i}\omega\hat{\sigma}_{01}/2} \tag{6.9}$$

if, for convenience, we choose the velocity parallel to the x axis. According to (3.50a),

$$\frac{v_x}{c} = \tanh\omega \quad .$$

Since in our case v_x should be negative (the moving inertial system has the velocity $\boldsymbol{v} = -v_x \boldsymbol{e}_1$ in the rest system), this relation is now

$$-\frac{v_x}{c} = \tanh\omega \quad , \tag{6.10a}$$

$$\omega = \tanh^{-1}\left(-\frac{v_x}{c}\right) = -\tanh^{-1}\left(\frac{v_x}{c}\right) \quad . \tag{6.10b}$$

Therefore (3.49b) and (3.51) now read as

$$\begin{aligned} x^{0'} &= \cosh\omega\left(x^0 - \tanh\omega x^1\right) = \frac{1}{\sqrt{1-\beta^2}}\left(x^0 + \frac{v_x}{c}x^1\right) \quad , \\ x^{1'} &= \cosh\omega\left(x^1 - \tanh\omega x^0\right) = \frac{1}{\sqrt{1-\beta^2}}\left(x^1 + \frac{v_x}{c}x^0\right) \end{aligned} \tag{6.11}$$

and the operator (6.9), with $\hat{\sigma}_{01} = \frac{\mathrm{i}}{2}(\gamma_0\gamma_1 - \gamma_1\gamma_0) = \mathrm{i}\gamma_0\gamma_1 = -\mathrm{i}\gamma^0\gamma^1 = -\mathrm{i}\gamma^0\gamma^0\hat{\alpha}_1 = -\mathrm{i}\hat{\alpha}_1$, becomes

$$\begin{aligned} \hat{S} &= \mathrm{e}^{-\mathrm{i}(\omega/2)\hat{\sigma}_{01}} = \mathrm{e}^{-(\omega/2)\hat{\alpha}_1} \\ &= 1\!\!1\left(1 + \frac{(-\omega/2)\hat{\alpha}_1}{1!} + \frac{(-\omega/2)^2(\hat{\alpha}_1)^2}{2!} + \frac{(-\omega/2)^3(\hat{\alpha}_1)^3}{3!} + \dots\right) \\ &= \left(1 + \frac{(\omega/2)^2}{2!} + \frac{(\omega/2)^4}{4!} + \dots\right)1\!\!1 - \left(\frac{\omega/2}{1!} + \frac{(\omega/2)^3}{3!} + \dots\right)\hat{\alpha}_1 \\ &= 1\!\!1\cosh\frac{\omega}{2} - \hat{\alpha}_1\sinh\frac{\omega}{2} \quad . \end{aligned} \tag{6.12}$$

Here we used $\hat{\alpha}_i^2 = 1$, according to (2.8). Now the spinor transformation (3.53) can be applied directly to the $\omega^r(0)$ from (6.1) to obtain

$$\begin{aligned} \omega^r(p_x) &= \mathrm{e}^{-\mathrm{i}(\omega/2)\hat{\sigma}_{01}}\omega^r(0) = \left(\cosh\frac{\omega}{2} - \hat{\alpha}_1\sinh\frac{\omega}{2}\right)\omega^r(0) \\ &= \cosh\frac{\omega}{2}\begin{pmatrix} 1 & 0 & 0 & -\tanh\frac{\omega}{2} \\ 0 & 1 & -\tanh\frac{\omega}{2} & 0 \\ 0 & -\tanh\frac{\omega}{2} & 1 & 0 \\ -\tanh\frac{\omega}{2} & 0 & 0 & 1 \end{pmatrix}\omega^r(0) \quad . \end{aligned} \tag{6.13}$$

Because the $\omega^r(0)$ have the simple form (6.3), the rth column of this transformation matrix (6.13) is identical to the spinor $\omega^r(\boldsymbol{p})$. More precisely,

$$\begin{aligned} \omega^1(p_x) &= \cosh\frac{\omega}{2}\begin{pmatrix} 1 \\ 0 \\ 0 \\ -\tanh\frac{\omega}{2} \end{pmatrix} , \quad \omega^2(p_x) = \cosh\frac{\omega}{2}\begin{pmatrix} 0 \\ 1 \\ -\tanh\frac{\omega}{2} \\ 0 \end{pmatrix} , \\ \omega^3(p_x) &= \cosh\frac{\omega}{2}\begin{pmatrix} 0 \\ -\tanh\frac{\omega}{2} \\ 1 \\ 0 \end{pmatrix} , \quad \omega^4(p_x) = \cosh\frac{\omega}{2}\begin{pmatrix} -\tanh\frac{\omega}{2} \\ 0 \\ 0 \\ 1 \end{pmatrix} . \end{aligned} \tag{6.14}$$

To return to physical quantities, we convert the rotation angle ω with the aid of (6.10),

$$\begin{aligned} \cosh x\sinh 2x &= \tfrac{1}{2}\sinh 2x \quad , \\ \cosh x\cosh x &= \tfrac{1}{2}(\cosh 2x + 1) \quad , \\ \sinh x\sinh x &= \tfrac{1}{2}(\cosh 2x - 1) \quad , \end{aligned} \tag{6.15}$$

and therefore

$$\tanh x = \frac{\sinh 2x}{\cosh 2x + 1} = \frac{\tanh 2x}{1 + 1/\cosh 2x} \quad . \tag{6.16}$$

Since

$$\frac{\sinh x}{\cosh x} = \tanh x = \frac{\sqrt{\cosh^2 x - 1}}{\cosh x} \quad , \quad \text{then} \quad \cosh x = \frac{1}{\sqrt{1 - \tanh^2 x}}$$

and hence (6.16) becomes

$$\tanh x = \frac{\tanh 2x}{1 + \sqrt{1 - \tanh^2 2x}} \quad . \tag{6.17}$$

With (6.10) we may now calculate

$$\begin{aligned} -\tanh\frac{\omega}{2} &= \frac{-\tanh\omega}{1 + \sqrt{1 - \tanh^2\omega}} = \frac{v_x/c}{1 + \sqrt{1 - (v_x/c)^2}} \\ &= \frac{[m_0c^2/\sqrt{1 - (v_x/c)^2}](v_x/c)}{[m_0c^2/\sqrt{1 - (v_x/c)^2}]\,[1 + \sqrt{1 - (v_x/c)^2}]} = \frac{p_x c}{E + m_0 c^2} \end{aligned} \tag{6.18}$$

and

$$\begin{aligned} \cosh\frac{\omega}{2} &= \frac{1}{\sqrt{1 - [\tanh^2\omega/(1 + \sqrt{1 - \tanh^2\omega})^2]}} \\ &= \frac{1}{\sqrt{1 - \left[(v_x/c)^2/\left(1 + \sqrt{1 - (v_x/c)^2}\right)^2\right]}} \\ &= \frac{1 + \sqrt{1 - (v_x/c)^2}}{\sqrt{1 + 2\sqrt{1 - (v_x/c)^2} + \left(1 - (v_x/c)^2\right) - (v_x/c)^2}} \\ &= \frac{1 + \sqrt{1 - (v_x/c)^2}}{\sqrt{2}\sqrt{1 - (v_x/c)^2 + \sqrt{1 - (v_x/c)^2}}} \\ &= \frac{\left[\left(1/\sqrt{1 - (v_x/c)^2}\right) + 1\right] m_0c^2}{\sqrt{2}\sqrt{1 + \left[1/\sqrt{1 - (v_x/c)^2}\right]} m_0c^2} \\ &= \frac{E + m_0c^2}{\sqrt{m_0c^2 + E}\sqrt{2m_0c^2}} = \sqrt{\frac{E + m_0c^2}{2m_0c^2}} \quad . \end{aligned} \tag{6.19}$$

Hence by (6.18) and (6.19), the transformation (6.13) can be expressed completely by physical observables (E, p_x, m_0c^2) and reads

$$\omega^r(p_x) = \sqrt{\frac{E + m_0c^2}{2m_0c^2}} \times \begin{pmatrix} 1 & 0 & 0 & \frac{p_x c}{E + m_0c^2} \\ 0 & 1 & \frac{p_x c}{E + m_0c^2} & 0 \\ 0 & \frac{p_x c}{E + m_0c^2} & 1 & 0 \\ \frac{p_x c}{E + m_0c^2} & 0 & 0 & 1 \end{pmatrix} \omega^r(0) \quad . \tag{6.20}$$

6.1 Plane Waves in Arbitrary Directions

The result (6.20) can be generalized to plane waves with velocity $\boldsymbol{v}$ in arbitrary directions. We write

$$\boldsymbol{v} = v\boldsymbol{n} = v\{\cos\alpha, \cos\beta, \cos\gamma\} \quad ; \tag{6.21}$$

thus $\boldsymbol{n}$ is the unit vector which specifies the direction of motion of the moving inertial system with respect to the rest system of the electron. Then according to (3.44) the "rotation angles" are

$$\Delta\omega^{\nu}{}_{\mu} = \Delta\omega(\hat{I}_n)^{\nu}{}_{\mu} = \Delta\omega \begin{pmatrix} 0 & -\cos\alpha & -\cos\beta & -\cos\gamma \\ -\cos\alpha & 0 & 0 & 0 \\ -\cos\beta & 0 & 0 & 0 \\ -\cos\gamma & 0 & 0 & 0 \end{pmatrix} \tag{6.22}$$

and with that, according to (3.45), the infinitesimal Lorentz transformation reads

$$x'^{\nu} = a^{\nu}{}_{\mu}x^{\mu} = x^{\nu} + \Delta\omega^{\nu}{}_{\mu}x^{\mu} \quad , \tag{6.23}$$

or, in expanded form,

$$\begin{aligned} x'^0 &= x^0 - \Delta\omega\cos\alpha x^1 - \Delta\omega\cos\beta x^2 - \Delta\omega\cos\gamma x^3 \quad , \\ x'^1 &= -\Delta\omega\cos\alpha x^0 + x^1 \quad , \\ x'^2 &= -\Delta\omega\cos\beta x^0 + x^2 \quad , \\ x'^3 &= -\Delta\omega\cos\gamma x^0 + x^3 \quad . \end{aligned} \tag{6.24}$$

Since $\boldsymbol{x}' = 0 = \{x'^1, x'^2, x'^3\}$ characterizes the origin of the moving inertial system, it follows immediately that it moves with

$$\Delta\boldsymbol{v} = \{\Delta v_1, \Delta v_2, \Delta v_3\} = c\left\{\frac{x^1}{x^0}, \frac{x^2}{x^0}, \frac{x^3}{x^0}\right\} = c\Delta\omega\{\cos\alpha, \cos\beta, \cos\gamma\} \tag{6.25}$$

relative to the rest coordinate system. From (6.24) easily follows the former special case of motion along the x axis if we set $\cos\alpha = 1$, $\cos\beta = 0$, $\cos\gamma = 0$. Just as simply, one obtains the movements along the y and z axes for $\cos\alpha = 0$, $\cos\beta = 1$, $\cos\gamma = 0$ and $\cos\alpha = 0$, $\cos\beta = 0$, $\cos\gamma = 1$, respectively, both of which are also contained in (6.24). Thus from (6.22) follows the coefficient matrix belonging to the direction of motion $\boldsymbol{n}$.

$$(\hat{I}_n)^{\nu}{}_{\mu} = \begin{pmatrix} 0 & -\cos\alpha & -\cos\beta & -\cos\gamma \\ -\cos\alpha & 0 & 0 & 0 \\ -\cos\beta & 0 & 0 & 0 \\ -\cos\gamma & 0 & 0 & 0 \end{pmatrix} \quad . \tag{6.26}$$

Setting $\Delta\omega = \omega/N$ we get, in complete analogy to (3.52),

$$\begin{aligned} \psi'(x') = \hat{S}(\hat{\alpha})\psi(x) &= \lim_{N\to\infty}\left(\mathbb{1} - \frac{\mathrm{i}}{4}\frac{\omega}{N}\hat{\sigma}_{\mu\nu}(\hat{I}_n)^{\mu\nu}\right)^N \psi(x) \\ &= \exp\left(-\mathrm{i}\omega\hat{\sigma}_{\mu\nu}(\hat{I}_n)^{\mu\nu}/4\right)\psi(x) \quad ; \end{aligned} \tag{6.27a}$$

thus

$$\hat{S}(\hat{\alpha}) = \exp\left(-\mathrm{i}\omega\hat{\sigma}_{\mu\nu}(\hat{I}_n)^{\mu\nu}/4\right) \quad . \tag{6.27b}$$

The expression in the exponent reads

$$\hat{\sigma}_{\mu\nu}(\hat{I}_n)^{\mu\nu} = 2\left(\hat{\sigma}_{01}(\hat{I}_n)^{01} + \hat{\sigma}_{02}(\hat{I}_n)^{02} + \hat{\sigma}_{03}(\hat{I}_n)^{03}\right) \quad ,$$

which can be converted by use of (6.26) and $g^{11} = g^{22} = g^{33} = -1$ into

$$\hat{\sigma}_{\mu\nu}(\hat{I}_n)^{\mu\nu} = 2\left(\hat{\sigma}_{01}\cos\alpha + \hat{\sigma}_{02}\cos\beta + \hat{\sigma}_{03}\cos\gamma\right) = -2\mathrm{i}\hat{\boldsymbol{\alpha}}\cdot\frac{\boldsymbol{v}}{v} \quad . \tag{6.28}$$

Here we have made use of

$$\frac{\boldsymbol{v}}{v} = \{\cos\alpha, \cos\beta, \cos\gamma\}$$

along with $v = |\boldsymbol{v}|$ and

$$\begin{aligned}\hat{\sigma}_{0i} &= \frac{\mathrm{i}}{2}[\gamma_0, \gamma_i]_- = \frac{\mathrm{i}}{2}(\gamma_0\gamma_i - \gamma_i\gamma_0) = \mathrm{i}\gamma_0\gamma_i \\ &= \mathrm{i}\gamma^0\gamma^i g_{ii} = -\mathrm{i}\gamma^0\gamma^i = -\mathrm{i}\gamma^0\gamma^0\hat{\alpha}_i = -\mathrm{i}\hat{\alpha}_i \quad .\end{aligned} \tag{6.29}$$

With this the spinor transformation (6.27b) for Lorentz transformations to inertial systems with direction of velocity $\boldsymbol{v}/v$ now becomes

$$\begin{aligned}\hat{S}(-\boldsymbol{v}) &= \hat{S}\left(-\frac{\boldsymbol{p}}{E}\right) = \mathrm{e}^{-(\omega/2)\hat{\boldsymbol{\alpha}}\cdot\boldsymbol{v}/v} \\ &= \sqrt{\frac{E+m_0c^2}{2m_0c^2}}\begin{bmatrix} 1 & 0 & \frac{p_zc}{E+m_0c^2} & \frac{p_-c}{E+m_0c^2} \\ 0 & 1 & \frac{p_+c}{E+m_0c^2} & \frac{-p_zc}{E+m_0c^2} \\ \frac{p_zc}{E+m_0c^2} & \frac{p_-c}{E+m_0c^2} & 1 & 0 \\ \frac{p_+c}{E+m_0c^2} & \frac{-p_zc}{E+m_0c^2} & 0 & 1 \end{bmatrix} \\ &= \left[\omega^1(\boldsymbol{p}), \omega^2(\boldsymbol{p}), \omega^3(\boldsymbol{p}), \omega^4(\boldsymbol{p})\right] \quad .\end{aligned} \tag{6.30}$$

Here we have set $p_\pm = p_x \pm \mathrm{i}p_y$, and the last line indicates that the individual column vectors of the $S(-\boldsymbol{v})$ matrix are identical with $\omega^r(\boldsymbol{p})$. We will calculate the final step of (6.30) in Exercise 6.1.

EXERCISE

6.1 Calculation of the Spinor Transformation Operator in Matrix Form

Problem. Calculate explicitly the spinor transformation operator

$$\hat{S}(-\boldsymbol{v}) = \mathrm{e}^{-(\omega/2)\hat{\boldsymbol{\alpha}}\cdot\boldsymbol{v}/v}$$

in matrix form.

Solution. We expand $\hat{S}$ in a series

$$\hat{S}(-\boldsymbol{v}) = 1 - \frac{\omega}{2}\frac{\hat{\boldsymbol{\alpha}}\cdot\boldsymbol{v}}{v} + \frac{1}{2}\frac{\omega^2}{4v^2}(\hat{\boldsymbol{\alpha}}\cdot\boldsymbol{v})^2 - \frac{1}{6}\frac{\omega^3}{8v^3}(\hat{\boldsymbol{\alpha}}\cdot\boldsymbol{v})^3 + \dots \tag{1}$$

and use

Exercise 6.1.

$$
\begin{aligned}
(\hat{\boldsymbol{\alpha}}\cdot\boldsymbol{v})^2 &= \hat{\alpha}^i\hat{\alpha}^j v_i v_j = \gamma^0\gamma^i\gamma^0\gamma^j v_i v_j = -\gamma^i\gamma^j v_i v_j \\
&= -\tfrac{1}{2}\left(\gamma^i\gamma^j v_i v_j + \gamma^j\gamma^i v_j v_i\right) = -\tfrac{1}{2}\left\{2g^{ij}\,\mathbb{1} v_i v_j\right) \\
&= +v^2\mathbb{1} \quad . \qquad (2)
\end{aligned}
$$

This yields

$$
\begin{aligned}
\hat{S}(-\boldsymbol{v}) &= \left\{1+\frac{1}{2}\frac{\omega^2}{4}\left(\frac{\boldsymbol{v}}{v}\right)^2+\frac{1}{24}\frac{\omega^4}{16}\left(\frac{\boldsymbol{v}}{v}\right)^4+\dots\right\}\cdot\mathbb{1} \\
&\quad -\frac{\hat{\boldsymbol{\alpha}}\cdot\boldsymbol{v}}{v}\left\{\frac{\omega}{2}+\frac{1}{6}\frac{\omega^3}{8}\left(\frac{\boldsymbol{v}}{v}\right)^2+\dots\right\} \\
&= \mathbb{1}\cosh\frac{\omega}{2}-\frac{\hat{\boldsymbol{\alpha}}\cdot\boldsymbol{v}}{v}\sinh\frac{\omega}{2} \quad . \qquad (3)
\end{aligned}
$$

The matrix $\hat{\boldsymbol{\alpha}}\cdot\boldsymbol{v}/v$ has the form

$$
\begin{aligned}
\frac{\hat{\boldsymbol{\alpha}}\cdot\boldsymbol{v}}{v} &= \hat{\alpha}_x\cdot\frac{v_x}{v}+\hat{\alpha}_y\cdot\frac{v_y}{v}+\hat{\alpha}_z\cdot\frac{v_z}{v} \\
&= \frac{p_x}{p}\begin{bmatrix} \cdot & \cdot & \cdot & 1 \\ \cdot & \cdot & 1 & \cdot \\ \cdot & 1 & \cdot & \cdot \\ 1 & \cdot & \cdot & \cdot \end{bmatrix} + \frac{\mathrm{i}p_y}{p}\begin{bmatrix} \cdot & \cdot & \cdot & -1 \\ \cdot & \cdot & 1 & \cdot \\ \cdot & -1 & \cdot & \cdot \\ 1 & \cdot & \cdot & \cdot \end{bmatrix} \\
&\quad + \frac{p_z}{p}\begin{bmatrix} \cdot & \cdot & 1 & \cdot \\ \cdot & \cdot & \cdot & -1 \\ 1 & \cdot & \cdot & \cdot \\ \cdot & -1 & \cdot & \cdot \end{bmatrix} \\
&= \frac{1}{p}\begin{bmatrix} \cdot & \cdot & p_z & p_- \\ \cdot & \cdot & p_+ & -p_z \\ p_z & p_- & \cdot & \cdot \\ p_+ & -p_z & \cdot & \cdot \end{bmatrix} \quad , \qquad (4)
\end{aligned}
$$

where $p_\pm = p_x \pm \mathrm{i}p_y$.

With the help of (6.19) for $\cosh(\omega/2)$, we obtain from (3) that

$$
\begin{aligned}
\hat{S}(-\boldsymbol{v}) &= \sqrt{\frac{E+m_0c^2}{2m_0c^2}}\begin{bmatrix} 1 & \cdot & \cdot & \cdot \\ \cdot & 1 & \cdot & \cdot \\ \cdot & \cdot & 1 & \cdot \\ \cdot & \cdot & \cdot & 1 \end{bmatrix} \\
&\quad - \sqrt{\frac{E+m_0c^2}{2m_0c^2}}\frac{\tanh\omega/2}{p}\begin{bmatrix} \cdot & \cdot & p_z & p_- \\ \cdot & \cdot & p_+ & -p_z \\ p_z & p_- & \cdot & \cdot \\ p_+ & -p_z & \cdot & \cdot \end{bmatrix} \\
&= \sqrt{\frac{E+m_0c^2}{2m_0c^2}}\begin{pmatrix} 1 & 0 & \dfrac{p_zc}{E+m_0c^2} & \dfrac{p_-c}{E+m_0c^2} \\ 0 & 1 & \dfrac{p_+c}{E+m_0c^2} & \dfrac{-p_zc}{E+m_0c^2} \\ \dfrac{p_zc}{E+m_0c^2} & \dfrac{p_-c}{E+m_0c^2} & 1 & 0 \\ \dfrac{p_+c}{E+m_0c^2} & \dfrac{-p_zc}{E+m_0c^2} & 0 & 1 \end{pmatrix} \qquad (5)
\end{aligned}
$$

Exercise 6.1.

According to (6.18),

$$-\tanh\frac{\omega}{2}=\frac{pc}{E+m_0c^2}\quad,\tag{6}$$

taking into account that in (6.18) p was substituted by p_x, because there we only considered motion in the x direction. Hence the matrices in (5) can be summarized by the expression (6.30).

EXERCISE

6.2 Calculation of the Spinor $u(p+q)$ by Means of a Lorentz Transformation from the Spinor $u(p)$ of a Free Particle

Problem. Given a free-particle spinor $u(p)$, express $u(p+q)$ by $u(p)$ for $q_\mu \to 0$, $pq \to 0$ by means of a Lorentz transformation.

Solution. The solution of the Dirac equation for a free particle with momentum p reads as

$$\psi(x)=u(p)\,\mathrm{e}^{-\mathrm{i}p\cdot x}\quad.\tag{1}$$

If we transform into a system x', which moves with the velocity $\boldsymbol{u}=-\boldsymbol{q}/m$ in the system chosen in (1), the particle (in system x') has the momentum $p+q$, and

$$\psi'(x')=u(p+q)\,\mathrm{e}^{-\mathrm{i}(p+q)\cdot x'}\quad.\tag{2}$$

Because of the invariance of the scalar product $p\cdot x$,

$$\mathrm{e}^{-\mathrm{i}p\cdot x}=\mathrm{e}^{-i(p+q)\cdot x'}\quad.\tag{3}$$

From (3.52) we obtain the transformation of $\psi(x)$ into the moving system

$$\psi'(x')=\exp\left(-\frac{\mathrm{i}}{4}\omega\hat{\sigma}_{\mu\nu}\hat{I}_n^{\mu\nu}\right)\psi(x)\quad.\tag{4}$$

By inserting (1) and (2) into (4) we get the general result

$$u(p+q)=\exp\left(-\frac{\mathrm{i}}{4}\omega\hat{\sigma}_{\mu\nu}\hat{I}_n^{\mu\nu}\right)u(p)\quad.\tag{5}$$

For $q_\mu \to 0$ the exponential function can be expanded;

$$\exp\left(-\frac{\mathrm{i}}{4}\omega\hat{\sigma}_{\mu\nu}\hat{I}_n^{\mu\nu}\right)\cong 1\!\!1-\frac{\mathrm{i}}{4}\omega\hat{\sigma}_{\mu\nu}\hat{I}_n^{\mu\nu}\quad.\tag{6}$$

If we choose the coordinate system in (1) with $\boldsymbol{q}$ pointing in the x direction and with $pq\cong 0$, (6) can be simplified as

$$\exp\left(-\frac{\mathrm{i}}{2}\omega\sigma_{01}\right)\cong 1\!\!1-\frac{\mathrm{i}\omega}{2}\sigma_{01}\quad,\tag{7}$$

where the connection between ω and $\boldsymbol{q} = (0, q_x, 0, 0)$ is

$$\tanh\omega = \frac{v}{c} = -\frac{q_x}{mc} \simeq \omega \quad . \tag{8}$$

Therefore one obtains

$$\begin{aligned}
u(p+q) &= \left(\mathbb{1} + \frac{\mathrm{i}q_x}{2mc}\sigma_{01}\right) u(p) \\
&= \left(\mathbb{1} + \frac{q_x}{4mc}(\gamma_1\gamma_0 - \gamma_0\gamma_1)\right) u(p) \\
&= \left\{\mathbb{1} + \frac{q_x}{2mc}\begin{bmatrix} \cdot & \cdot & \cdot & -1 \\ \cdot & \cdot & -1 & \cdot \\ \cdot & -1 & \cdot & \cdot \\ -1 & \cdot & \cdot & \cdot \end{bmatrix}\right\} u(p) \\
&= \begin{pmatrix} u_1(p) - \dfrac{q_x}{2mc}u_4(p) \\ u_2(p) - \dfrac{q_x}{2mc}u_3(p) \\ u_3(p) - \dfrac{q_x}{2mc}u_2(p) \\ u_4(p) - \dfrac{q_x}{2mc}u_1(p) \end{pmatrix} \quad .
\end{aligned} \tag{9}$$

6.2 The General Form of the Free Solutions and Their Properties

From our previous considerations (6.1), (6.6) and (6.30), the general free solution must have the form

$$\psi^r(x) = \omega^r(\boldsymbol{p})\,\mathrm{e}^{-\mathrm{i}\varepsilon_r p_\mu x^\mu/\hbar} \quad . \tag{6.31}$$

In the above, due to (6.3) and (6.30), the spinor $\omega^r(\boldsymbol{p})$ is identical to the row r of the matrix (6.30), because

$$\omega^r(\boldsymbol{p}) = \hat{S}\left(\frac{-\boldsymbol{p}}{E}\right)\omega^r(0) \quad . \tag{6.32}$$

However, we must take note of the fact that this result holds for the special representation of the γ matrices (3.13) only. The spinors $\omega^r(\boldsymbol{p})$ satisfy

$$(\not{p}' - \varepsilon_r m_0 c)\omega^r(\boldsymbol{p}) = 0 \quad , \tag{6.33a}$$

$$\overline{\omega}^r(\boldsymbol{p})(\not{p} - \varepsilon_r m_0 c) = 0 \quad , \tag{6.33b}$$

which are very useful later on. The first one is obvious and directly follows with (6.1) from (3.9) and the covariance of the Dirac equation. It represents the *Dirac equation of a free particle in momentum space.* We have $\varepsilon_r = +1$ for $r = 1, 2$ and therefore (6.33a) reads $(\not{p} - m_0 c)\omega^r(\boldsymbol{p}) = 0$. These are the solutions with positive energy. As already mentioned $\omega^1(\boldsymbol{p})$ and $\omega^2(\boldsymbol{p})$ are given by the two first rows of (6.30). Furthermore, we see directly that in the nonrelativistic limit of the representation of (6.30) the lower components of $\omega^1(\boldsymbol{p})$ and $\omega^2(\boldsymbol{p})$ become small; in the limit of the electron being at rest they finally change to the first two solutions (6.3). Similar considerations apply for the solutions of negative energy ($\varepsilon_r = -1$, $r = 3, 4$). According to (6.30) and (6.3) we have to exchange the large components with the small ones in this case. Equation (6.33b) follows from (6.33a) by Hermitian conjugation,

$$\begin{aligned}\left[(\not{p} - \varepsilon_r m_0 c)\omega^r(\boldsymbol{p})\right]^\dagger = 0 &= \omega^{r\dagger}(\boldsymbol{p})\left(\not{p} - \varepsilon_r m_0 c\right)^\dagger \\ &= \omega^{r\dagger}(\boldsymbol{p})\left(\not{p}^\dagger - \varepsilon_r m_0 c\right) \\ &= \omega^{r\dagger}(\boldsymbol{p})\left(p_\mu \gamma^{\mu\dagger} - \varepsilon_r m_0 c\right) \\ &= \omega^{r\dagger}(\boldsymbol{p})\left(p_0 \gamma^0 - p_k \gamma^k - \varepsilon_r m_0 c\right) \quad .\end{aligned}$$

Note that the p_μ are real numbers ($p_\mu^\dagger = p_\mu$), and in the last line (3.12a) was used, i.e. $\gamma^{0\dagger} = \gamma^0$ and $\gamma^{k\dagger} + -\gamma^k$. Multiplication with γ^0 from the right yields

$$\begin{aligned}\omega^{r\dagger}(\boldsymbol{p})\left(p_\mu \gamma^{\mu\dagger} - \varepsilon_r m_0 c\right)\gamma_0 = 0 &= \omega^{r\dagger}(\boldsymbol{p})\gamma^0\left(p_0\gamma^0 + p_k\gamma^k - \varepsilon_r m_0 c\right) \\ &= \overline{\omega}^r(\boldsymbol{p})\left(p_\mu\gamma^\mu - \varepsilon_r m_0 c\right) \\ &= \overline{\omega}^r(\boldsymbol{p})\left(\not{p} - \varepsilon_r m_0 c\right) \quad .\end{aligned}$$

We name (6.33b) the *adjoint wave equation* because it is valid for the adjoint spinor $\overline{\omega}^r(\boldsymbol{p}) = \omega^{r\dagger}(\boldsymbol{p})\gamma_0$. With the help of the explicit representation of the $\omega^r(\boldsymbol{p})$ in (6.30) we can now compute the normalization condition (see Exercise 6.3),

$$\overline{\omega}^r(\boldsymbol{p})\omega^{r'}(\boldsymbol{p}) = \delta_{rr'}\varepsilon_r \quad . \tag{6.34}$$

Then, in Exercise 6.4, we will prove the validity of

$$\omega^{r\dagger}(\varepsilon_r\boldsymbol{p})\omega^{r'}(\varepsilon_{r'}\boldsymbol{p}) = \frac{E}{m_0c^2}\delta_{rr'} \quad . \tag{6.35}$$

EXERCISE

6.3 Normalization of the Spinor $\omega^r(p)$

Problem. Prove explicitly the normalization condition

$$\overline{\omega}^r(\boldsymbol{p})\omega^{r'}(\boldsymbol{p}) = \delta_{rr'}\varepsilon_r$$

for the spinors given in (6.30).

Solution. From these equations we get

$$\omega^1(\boldsymbol{p}) = \begin{pmatrix} 1 \\ 0 \\ \dfrac{p_z c}{E + m_0c^2} \\ \dfrac{p_+ c}{E + m_0c^2} \end{pmatrix} \sqrt{\frac{E + m_0c^2}{2m_0c^2}} \quad ,$$

Exercise 6.3.

$$\omega^2(\boldsymbol{p}) = \begin{pmatrix} 0 \\ 1 \\ \dfrac{p_- c}{E + m_0c^2} \\ \dfrac{-p_z c}{E + m_0c^2} \end{pmatrix} \sqrt{\frac{E + m_0c^2}{2m_0c^2}} \quad ,$$

$$\omega^3(\boldsymbol{p}) = \begin{pmatrix} \dfrac{p_z c}{E + m_0c^2} \\ \dfrac{p_+ c}{E + m_0c^2} \\ 1 \\ 0 \end{pmatrix} \sqrt{\frac{E + m_0c^2}{2m_0c^2}} \quad ,$$

$$\omega^4(\boldsymbol{p}) = \begin{pmatrix} \dfrac{p_- c}{E + m_0c^2} \\ \dfrac{-p_z c}{E + m_0c^2} \\ 0 \\ 1 \end{pmatrix} \sqrt{\frac{E + m_0c^2}{2m_0c^2}} \quad .$$

Now we take, say, the product $\overline{\omega}^{1\dagger}(\boldsymbol{p})\omega^1(\boldsymbol{p})$. Note, that $p_+^\dagger = p_-$!

$$\begin{aligned}
\overline{\omega}^1(\boldsymbol{p})\omega^1(\boldsymbol{p}) &= \left(1, 0, -\frac{p_z c}{E + m_0c^2}, -\frac{p_- c}{E + m_0c^2}\right) \frac{E + m_0c^2}{2m_0c^2} \begin{pmatrix} 1 \\ 0 \\ \dfrac{p_z c}{E + m_0c^2} \\ \dfrac{p_+ c}{E + m_0c^2} \end{pmatrix} \\
&= \frac{E + m_0c^2}{2m_0c^2} \left\{1 - \frac{p_z^2 c^2}{(E + m_0c^2)^2} - \frac{p_+ p_- c^2}{(E + m_0c^2)^2}\right\} \\
&= \frac{1}{2m_0c^2} \left\{E + m_0c^2 - \frac{\left(p_x^2 + p_y^2 + p_z^2\right) c^2}{(E + m_0c^2)}\right\} \\
&= \frac{1}{2m_0c^2} \frac{1}{E + m_0c^2} \left\{E^2 + 2Em_0c^2 + m_0^2c^4 - \boldsymbol{p}^2c^2\right\} \\
&= \frac{1}{2m_0c^2(E + m_0c^2)} \left\{2Em_0c^2 + 2m_0^2c^4\right\} \\
&= \frac{2(E + m_0c^2)m_0c^2}{2m_0c^2(E + m_0c^2)} = 1 = \delta_{11}\varepsilon_1 \quad .
\end{aligned}$$

Next we calculate

$$\begin{aligned}
\overline{\omega}^2(\boldsymbol{p})\omega^3(\boldsymbol{p}) &= \frac{E + m_0c^2}{2m_0c^2} \left(0, 1, -\frac{p_+ c}{E + m_0c^2}, +\frac{p_z c}{E + m_0c^2}\right) \begin{pmatrix} \dfrac{p_z}{E + m_0c^2} \\ \dfrac{p_+ c}{E + m_0c^2} \\ 1 \\ 0 \end{pmatrix} \\
&= \frac{E + m_0c^2}{2m_0c^2} \left\{\frac{p_+ c}{E + m_0c^2} - \frac{p_+ c}{E + m_0c^2}\right\} \\
&= 0 = \delta_{23}\varepsilon_2 \quad .
\end{aligned}$$

Exercise 6.3.

The third example we give is

$$\begin{aligned}\bar{\omega}^4(\boldsymbol{p})\omega^4(\boldsymbol{p}) &= \frac{E+m_0c^2}{2m_0c^2}\left(\frac{p_+c}{E+m_0c^2}, \frac{-p_zc}{E+m_0c^2}, 0, -1\right)\begin{pmatrix}\dfrac{p_-c}{E+m_0c^2}\\ \dfrac{-p_zc}{E+m_0c^2}\\ 0\\ 1\end{pmatrix}\\ &= \frac{1}{2m_0c^2(E+m_0c^2)}\left\{\left(p_x^2+p_y^2+p_z^2\right)c^2-(E+m_0c^2)^2\right\}\\ &= \frac{1}{2m_0c^2(E+m_0c^2)}\left\{\boldsymbol{p}^2c^2-E^2-2Em_0c^2-m_0^2c^2\right\}\\ &= -1 = \delta_{44}\varepsilon_4 \quad .\end{aligned}$$

All other combinations can be calculated in the same way. However, there is a more elegant procedure available: $\bar{\omega}^r(\boldsymbol{p})\omega^{r'}(\boldsymbol{p})$ is a Lorentz scalar (cf. Exercise 5.1), and hence

$$\bar{\omega}^r(\boldsymbol{p})\omega^{r'}(\boldsymbol{p}) = \bar{\omega}^r(0)\omega^{r'}(0) = \omega^{r\dagger}(0)\gamma^0\omega^{r'}(0) = \delta_{rr'}\varepsilon_r \quad ,$$

which we see at once because of (6.3). But note: *The probability density $\omega^{r\dagger}(\boldsymbol{p})\omega^r(\boldsymbol{p})$ is not Lorentz invariant.* This quantity is only the fourth component of the four-vector.

EXERCISE

6.4 Proof of the Relation $\omega^{r\dagger}(\varepsilon_r\boldsymbol{p})\omega^{r'}(\varepsilon_{r'}\boldsymbol{p}) = \delta_{rr'}(E/m_0c^2)$

Problem. Show with the help of (6.30) that the following relation holds:

$$\omega^{r\dagger}(\varepsilon_r\boldsymbol{p})\omega^{r'}(\varepsilon_{r'}\boldsymbol{p}) = \delta_{rr'}\frac{E}{m_0c^2} \quad .$$

Solution. Again we calculate some examples

(a) $r=1,\ r'=1$:

$$\begin{aligned}&\left(1, 0, \frac{p_zc}{E+m_0c^2}, \frac{p_-c}{E+m_0c^2}\right)\frac{E+m_0c^2}{2m_0c^2}\begin{pmatrix}1\\ 0\\ \dfrac{p_zc}{E+m_0c^2}\\ \dfrac{p_+c}{E+m_0c^2}\end{pmatrix}\\ &= \frac{E+m_0c^2}{2m_0c^2}\left\{1+\frac{\left(p_x^2+p_y^2+p_z^2\right)c^2}{(E+m_0c^2)^2}\right\}\\ &= \frac{E+m_0c^2}{2m_0c^2}\frac{\left((E+m_0c^2)^2+\boldsymbol{p}^2c^2\right)}{(E+m_0c^2)^2}\\ &= \frac{2E^2+2m_0c^2E}{2m_0c^2(E+m_0c^2)} = \frac{E}{m_0c^2}\delta_{11} \quad .\end{aligned}$$

Exercise 6.4.

(b) $r = 2$, $r' = 3$:

$$\left(0, 1, +\frac{p_+c}{E+m_0c^2}, -\frac{p_zc}{E+m_0c^2}\right)\frac{E+m_0c^2}{2m_0c^2}\begin{pmatrix} -\dfrac{p_zc}{E+m_0c^2} \\ -\dfrac{p_+c}{E+m_0c^2} \\ 1 \\ 0 \end{pmatrix}$$

$$= \frac{E+m_0c^2}{2m_0c^2}\left\{-\frac{p_+c}{E+m_0c^2} + \frac{p_+c}{E+m_0c^2}\right\} = 0 \quad .$$

(c) $r = 4$, $r' = 4$:

$$\left(-\frac{p+c}{E+m_0c^2}, +\frac{p_zc}{E+m_0c^2}, 0, 1\right)\frac{E+m_0c^2}{2m_0c^2}\begin{pmatrix} -\dfrac{p_-c}{E+m_0c^2} \\ +\dfrac{p_zc}{E+m_0c^2} \\ 0 \\ 1 \end{pmatrix}$$

$$= \frac{E+m_0c^2}{2m_0c^2}\left\{\frac{\left(p_x^2+p_y^2+p_z^2\right)c^2}{(E+m_0c^2)^2} + 1\right\}$$

$$= \frac{1}{2m_0c^2(E+m_0c^2)}\left\{\boldsymbol{p}^2c^2 + E^2 + 2Em_0c^2 + m_0^2c^4\right\}$$

$$= \frac{E}{m_0c^2}\delta_{44} \quad .$$

The remaining combinations can be calculated similarly.

The factor

$$\frac{E}{m_0c^2} = \frac{m_0c^2}{\sqrt{1-\beta^2}}\frac{1}{m_0c^2} = \frac{1}{\sqrt{1-\beta^2}}$$

appearing in (6.35) just cancels the Lorentz contraction of the volume element in the direction of motion

$$\Delta V' = \Delta x'\Delta y'\Delta z' = \Delta x\sqrt{1-\beta^2}\Delta y\Delta z = \Delta V\sqrt{1-\beta^2} \quad .$$

Consequently the probability in the volume $\Delta V'$ becomes invariant, i.e.

$$\begin{aligned} \omega^{r'\dagger}(\boldsymbol{p})\omega^{r'}(\boldsymbol{p})\Delta V' &= \omega^{r\dagger}(0)\omega^{r}(0)\Delta V \\ &= \frac{1}{\sqrt{1-\beta^2}}\sqrt{1-\beta^2}\Delta V = 1\cdot\Delta V \quad . \end{aligned} \tag{6.36}$$

We expect this property to hold for a proper normalization. Let us clearly point out the difference between the orthogonalization relations (6.34) and (6.35): In (6.34) the spinor $\omega^{r'}(\boldsymbol{p})$ is orthogonal to the adjoint spinor $\overline{\omega}^{r}(\boldsymbol{p})$ with the *same momentum argument.* On the other hand, for instance in (6.35), the spinor $\omega^{r'}(\varepsilon_{r'}\boldsymbol{p})$ with positive energy ($r' = 1, 2$) is orthogonal to the Hermitian conjugate spinor $\omega^{r\dagger}(\varepsilon_r\boldsymbol{p})$

($r = 3, 4$) belonging to negative energy and the *reverse momentum argument.* When considering the latter property we have to bear in mind that the sign of the momentum dependent term in the plane wave factor is also reversed for the negative energy solutions. Therefore two plane Dirac waves with the same spatial momentum $\boldsymbol{p}$ but opposite energy are orthogonal. This follows simply from the definition of the plane Dirac waves given in (6.6), (6.13) and (6.33) [compare also with (6.31)],

$$\psi^r_{\boldsymbol{p}}(x) = \omega^r(\boldsymbol{p})\,\mathrm{e}^{-\mathrm{i}\varepsilon_r p_\mu x^\mu/\hbar} = \omega^r(\boldsymbol{p})\,\mathrm{e}^{-\mathrm{i}\varepsilon_r p_0 x^0/\hbar}\mathrm{e}^{+\mathrm{i}\varepsilon_r \boldsymbol{p}\cdot\boldsymbol{x}/\hbar} \quad ,$$
$$p_0 = +\sqrt{\boldsymbol{p}^2 + m_0^2c^2} > 0 \quad .$$

Accordingly, we have, for a wave with $r = 1, 2$, the energy $E = +\sqrt{\boldsymbol{p}^2 + m_0^2c^2} = p_0c$ and momentum $+\boldsymbol{p}$,

$$\psi^{1,2}_{\boldsymbol{p}}(x) = \omega^{1,2}(\boldsymbol{p})\,\mathrm{e}^{-\mathrm{i}p_0x^0/\hbar}\mathrm{e}^{\mathrm{i}\boldsymbol{p}\cdot\boldsymbol{x}/\hbar} = \omega^{1,2}(\boldsymbol{p})\,\mathrm{e}^{-\mathrm{i}p_\mu x^\mu/\hbar} \quad , \tag{6.37a}$$

and for one with $r = 3, 4$, the energy $E = -\sqrt{\boldsymbol{p}^2 + m_0^2c^2} = -p_0c$ and momentum $+\boldsymbol{p}$

$$\psi^{3,4}_{-\boldsymbol{p}}(x) = \omega^{3,4}(-\boldsymbol{p})\,\mathrm{e}^{+\mathrm{i}p_0x^0/\hbar}\mathrm{e}^{\mathrm{i}\boldsymbol{p}\cdot\boldsymbol{x}/\hbar} = \omega^{3,4}(-\boldsymbol{p})\,\mathrm{e}^{+\mathrm{i}p_0x^0/\hbar}\mathrm{e}^{-\mathrm{i}(-\boldsymbol{p}\cdot\boldsymbol{x}/\hbar)} \quad . \tag{6.37b}$$

Clearly we learn that both waves have the same momentum $\boldsymbol{p}$ (the same factor $\mathrm{e}^{\mathrm{i}\boldsymbol{p}\cdot\boldsymbol{x}/\hbar}$) but the opposite energy ($\mathrm{e}^{-\mathrm{i}p_0x^0/\hbar}$ and $\mathrm{e}^{+\mathrm{i}p_0x^0/\hbar}$, respectively). Hence the spinors $\omega^{1,2}(\boldsymbol{p})$ belong to positive energy and momentum $+\boldsymbol{p}$, whereas $\omega^{3,4}(-\boldsymbol{p})$ correspond to negative energy and the same momentum $+\boldsymbol{p}$. These spinors are orthogonal according to (6.35), and thus (6.37) holds. In fact, the relation (6.35) is required to ensure that any two plane waves $\psi^{r'}_{\boldsymbol{p}'}(x)$ and $\psi^r_{\boldsymbol{p}}(x)$ are orthogonal in the sense of the following scalar product:

$$\langle\psi^{r'}_{\boldsymbol{p}'}|\psi^r_{\boldsymbol{p}}\rangle = \int \mathrm{d}^3x\, \psi^{r'\dagger}_{\boldsymbol{p}'}(x)\psi^r_{\boldsymbol{p}}(x) \quad . \tag{6.38}$$

Inserting the plane wave spinors of (6.31) we obtain

$$\begin{aligned}\langle\psi^{r'}_{\boldsymbol{p}'}|\psi^r_{\boldsymbol{p}}\rangle &= \int \mathrm{d}^3x\, \omega^{r'\dagger}(\boldsymbol{p}')\omega^r(\boldsymbol{p})\exp\left[-\mathrm{i}\left(\varepsilon_r p_\mu x^\mu/\hbar - \varepsilon_{r'}p'_\mu x^\mu/\hbar\right)\right] \\ &= \omega^{r'\dagger}(\boldsymbol{p}')\omega^r(\boldsymbol{p})\exp\left[-\mathrm{i}\left(\varepsilon_{r'}p'_0x^0/\hbar - \varepsilon_r p_0 x^0 x^0/\hbar\right)\right] \\ &\quad\times (2\pi)^3\delta^3\left(\varepsilon_{r'}\boldsymbol{p}'/\hbar - \varepsilon_r\boldsymbol{p}/\hbar\right) \quad .\end{aligned} \tag{6.39}$$

The integration over the spatial coordinates has led to a delta function containing the momentum vectors $\boldsymbol{p}$ and $\boldsymbol{p}'$. Thus the scalar product vanishes unless $\boldsymbol{p}' = +\boldsymbol{p}$ (if the energies have equal sign) or unless $\boldsymbol{p}' = -\boldsymbol{p}$ (for opposite signs of the energy, $\varepsilon_{r'} \neq -\varepsilon_r$). This is just the condition for which the orthogonality relation of the unit spinors ω^r (6.35) applies! Thus we obtain for any two plane waves

$$\langle\psi^{r'}_{\boldsymbol{p}'}|\psi^r_{\boldsymbol{p}}\rangle = \frac{E}{m_0c^2}(2\pi)^3\delta^3\left((\boldsymbol{p}' - \boldsymbol{p})/\hbar\right)\delta_{rr'} \quad . \tag{6.40}$$

Up to a normalization factor this is just the orthogonality property one would expect for plane wave states.

The *closure relation* is important too:

$$\sum_{r=1}^{4} \varepsilon_r \omega^r_\alpha(\boldsymbol{p}) \overline{\omega}^r_\beta(\boldsymbol{p}) = \delta_{\alpha\beta} \quad . \tag{6.41}$$

Here the sum extends over all four spinors, taking from the first one only the component α and from the second adjoint one solely the component β. Therefore, the closure relation (6.39) expresses *a kind of row orthonormality of the matrix* (6.30). Due to (6.13), in the rest frame ($\boldsymbol{p} = 0$) of the electron clearly

$$\sum_{r=1}^{4} \varepsilon_r \omega^r_\alpha(0) \overline{\omega}^r_\beta(0) = \delta_{\alpha\beta} \tag{6.42}$$

is valid. For this reason (6.41) can be traced back to (6.42). Indeed with the help of (6.32) we can calculate it directly:

$$\begin{aligned}
\sum_{r=1}^{4} \varepsilon_r \omega^r_\alpha(\boldsymbol{p}) \overline{\omega}^r_\beta(\boldsymbol{p}) &= \sum_{r=1,\gamma,\lambda}^{4} \varepsilon_r \hat{S}_{\alpha\gamma}\left(-\frac{\boldsymbol{p}}{E}\right) \omega^r_\gamma(0) \overline{\omega}^r_\lambda(0) \hat{S}^{-1}_{\lambda\beta}\left(-\frac{\boldsymbol{p}}{E}\right) \\
&= \sum_{\gamma,\lambda} \hat{S}_{\alpha\gamma}\left(-\frac{\boldsymbol{p}}{E}\right) \hat{S}^{-1}_{\lambda\beta}\left(-\frac{\boldsymbol{p}}{E}\right) \sum_{r=1}^{4} \varepsilon_r \omega^r_\gamma(0) \overline{\omega}^r_\lambda(0) \\
&= \sum_{\gamma,\lambda} \hat{S}_{\alpha\gamma}\left(-\frac{\boldsymbol{p}}{E}\right) \hat{S}^{-1}_{\lambda\beta}\left(-\frac{\boldsymbol{p}}{E}\right) \delta_{\gamma\lambda} \\
&= \sum_{\gamma} \hat{S}_{\alpha\gamma}\left(-\frac{\boldsymbol{p}}{E}\right) \hat{S}^{-1}_{\gamma\beta}\left(-\frac{\boldsymbol{p}}{E}\right) = \delta_{\alpha\beta} \quad ,
\end{aligned} \tag{6.43}$$

where in the first line we have used the relation

$$\hat{S}^\dagger = \gamma^0 \hat{S}^{-1} \gamma^0 \tag{6.44}$$

from (3.63). This is again the reason why the adjoint spinor $\overline{\omega}^r = \omega^{r\dagger}\gamma_0$ appears in the closure relation (6.41), but not the Hermitian conjugate $\omega^{r\dagger}$. It is in accordance with the fact that the Lorentz transformation of the spinors is not unitary as expressed in (6.44).

EXERCISE

6.5 Independence of the Closure Relation from the Representation of the Dirac Spinors

Problem. Show that the relation (6.41)

$$\sum_{r=1}^{4} \varepsilon_r \omega^r_\alpha(\boldsymbol{p}) \overline{\omega}^r_\beta(\boldsymbol{p}) = \delta_{\alpha\beta}$$

is independent of the special representation from the Dirac spinors.

Exercise 6.5.

Solution. Let another representation $\tilde{\gamma}_\mu = \hat{U}\gamma_\mu\hat{U}^{-1}$ of the Dirac algebra ($\hat{U}^\dagger = \hat{U}^{-1}$) be given, and let $v^r(0)$ denote the spinors of the free particles at rest, constructed accordingly. We will prove that from (6.41) it follows that

$$\sum_r \varepsilon_r v^r_\alpha(\boldsymbol{p})\overline{v}^r_\beta(\boldsymbol{p}) = \delta_{\alpha\beta} \quad . \tag{1}$$

As seen previously [see (6.43)], one can show that (1) is equivalent to

$$\sum_r \varepsilon_r v^r_\alpha(0)\overline{v}^r_\beta(0) = \delta_{\alpha\beta} \quad , \tag{2}$$

and therefore we prove (2).

With the change of the representation $\tilde{\gamma}_\mu = \hat{U}\gamma_\mu\hat{U}^{-1}$, the spinors become $v^r(0) = \hat{U}\omega^r(0)$. Hence we have

$$\begin{aligned}
\sum_r \varepsilon_r v^r_\alpha\overline{v}^r_\beta &= \sum_r \varepsilon_r(\hat{U}\omega^r)_\alpha(\overline{\hat{U}\omega^r})\beta = \sum_{r,\gamma}\varepsilon_r\hat{U}_{\alpha\gamma}\omega^r_\gamma\left((\hat{U}\omega^r)^\dagger\tilde{\gamma}_0\right)_\beta \\
&= \sum_{r,\gamma}\varepsilon_r\hat{U}_{\alpha\gamma}\omega^r_\gamma\left(\omega^{r\dagger}\hat{U}^{-1}\tilde{\gamma}_0\right)_\beta = \sum_{r,\gamma,\delta}\varepsilon_r\hat{U}_{\alpha\gamma}\omega^r_\gamma\left(\omega^{r\dagger}\gamma_0\right)_\delta(\hat{U}^{-1})^{\delta\beta} \\
&= \sum_{r,\gamma,\delta}\varepsilon_r\omega^r_\gamma\overline{\omega}^r_\delta\hat{U}_{\alpha\gamma}(\hat{U}^{-1})_{\delta\beta} = \sum_{\gamma,\delta}\delta_{\gamma\delta}\hat{U}_{\alpha\gamma}(\hat{U}^{-1})_{\delta\beta} \\
&= \sum_\delta \hat{U}_{\alpha\beta}(\hat{U}^{-1})_{\delta\beta} = \delta_{\alpha\beta} \quad ,
\end{aligned}$$

which is the form required.

EXERCISE

6.6 Proof of Another Closure Relation for Dirac Spinors

Problem. Prove directly the closure relation

$$\sum_{r=1}^{4}\omega^r_\alpha(\varepsilon_r\boldsymbol{p})\omega^{r\dagger}_\beta(\varepsilon_r\boldsymbol{p}) = \frac{E}{m_0c^2}\delta_{\alpha\beta} \quad . \tag{1}$$

Solution. We use the equation

$$\omega^r(\varepsilon_r\boldsymbol{p}) = \hat{S}\left(-\frac{\varepsilon_r\boldsymbol{p}}{E}\right)\omega^r(0) \quad , \tag{2}$$

insert the expression for $\hat{S}$ given by

$$\begin{aligned}
\hat{S}\left(-\frac{\varepsilon_r\boldsymbol{p}}{E}\right) &= 1\!\!1\cosh\frac{\omega}{2} - \frac{\varepsilon_r\hat{\boldsymbol{\alpha}}\cdot\boldsymbol{p}}{|\boldsymbol{p}|}\sinh\frac{\omega}{2} \\
&= 1\!\!1\cosh\frac{\omega}{2} - \sum_i \varepsilon_r\frac{\hat{\alpha}_i p_i}{p}\sinh\frac{\omega}{2}
\end{aligned} \tag{3}$$

and use this to transform (1) into

Exercise 6.6.

$$A \equiv \sum_r \omega^r_\alpha(\varepsilon_r \boldsymbol{p})\omega^{r\dagger}_\beta(\varepsilon_r \boldsymbol{p})$$

$$= \sum_{r,\gamma,\delta}\left[\left(\cosh\left(\frac{\omega}{2}\right)\delta_{\alpha\gamma} - \sum_i \frac{\varepsilon_r(\hat{\alpha}_i)_{\alpha\gamma}p_i}{p}\sinh\frac{\omega}{2}\right)\omega^r_\gamma(0)\right]$$

$$\times\left[\omega^{r\dagger}_\delta(0)\left(\cosh\left(\frac{\omega}{2}\right)\delta_{\delta\beta} - \sum_i \frac{\varepsilon_r(\hat{\alpha}^\dagger_i)_{\delta\beta}p_i}{p}\sinh\frac{\omega}{2}\right)\right] \tag{4}$$

The matrices $\hat{\alpha}_i$ are Hermitian $[\hat{\alpha}^\dagger_i = \hat{\alpha}_i]$ and by explicit multiplication we get

$$A = \cosh^2\frac{\omega}{2}\sum_r \omega^r_\alpha(0)\omega^{r\dagger}_\beta(0)$$

$$- \cosh\frac{\omega}{2}\sinh\frac{\omega}{2}\sum_{r,i,\gamma,\delta}\left[\frac{(\hat{\alpha}_i)_{\alpha\gamma}p_i}{p}\varepsilon_r\omega^r_\gamma(0)\omega^{r\dagger}_\beta(0) + \varepsilon_r\omega^r_\alpha(0)\omega^{r\dagger}_\delta(0)\frac{(\hat{\alpha}_i)_{\delta\beta}p_i}{p}\right]$$

$$+ \sinh^2\frac{\omega}{2}\sum_{r,i,j,\gamma,\delta}\omega^r_\gamma(0)\omega^{r\dagger}_\delta(0)\frac{p_ip_j}{p^2}(\hat{\alpha}_i)_{\alpha\gamma}(\hat{\alpha}_j)_{\delta\beta} \quad . \tag{5}$$

In the last term we have used

$$\left[-\varepsilon_r\frac{\hat{\boldsymbol{\alpha}}\cdot\boldsymbol{p}}{p}\right]_{\alpha\gamma}\left[-\varepsilon_r\frac{\hat{\boldsymbol{\alpha}}\cdot\boldsymbol{p}}{r}\right]_{\delta\beta} = \sum_{ij}\frac{p_ip_j}{p^2}(\hat{\alpha}_i)_{\alpha\gamma}(\hat{\alpha}_j)_{\delta\beta} \quad . \tag{6}$$

For the $\omega^r(0)$

$$\gamma_0\omega^r(0) = \varepsilon_r\omega^r(0) \tag{7a}$$

$$\Leftrightarrow \sum_r(\gamma_0)_{\sigma\tau}\omega^r_\tau(0) = \varepsilon_r\omega^r_\sigma(0) \tag{7b}$$

is valid; hence,

$$\omega^{r\dagger}(0) = \overline{\omega}^r(0)\gamma_0 = \varepsilon_r\overline{\omega}^r(0) \quad . \tag{8}$$

Therefore (5) turns out to be

$$\cosh^2\frac{\omega}{2}\sum_r \varepsilon_r\omega^r_\alpha(0)\overline{\omega}^r_\beta(0) - \sinh\frac{\omega}{2}\cosh\frac{\omega}{2}$$

$$\times \underbrace{\sum_{r,i,\gamma,\delta}\left[\frac{(\hat{\alpha}_i)_{\alpha\gamma}p_i}{p}\omega^r_\gamma(0)\overline{\omega}^r_\beta(0) + \omega^r_\alpha(0)\overline{\omega}^r_\delta(0)\frac{(\hat{\alpha}_i)_{\delta\beta}p_i}{p}\right]}_{=B}$$

$$+ \sinh^2\frac{\omega}{2}\sum_{r,i,j,\gamma,\delta}\varepsilon_r\omega^r_\gamma(0)\overline{\omega}^r_\delta(0)\frac{p_ip_j}{p^2}(\hat{\alpha}_i)_{\alpha\gamma}(\hat{\alpha}_j)_{\delta\beta} \quad . \tag{9}$$

Because of the closure relation (6.41) one can sum over r in the first and the last sum to obtain

$$A = \cosh^2\left(\frac{\omega}{2}\right)\delta_{\alpha\beta} - \sinh\left(\frac{\omega}{2}\right)\cosh\left(\frac{\omega}{2}\right)B$$

$$+ \sinh^2\left(\frac{\omega}{2}\right)\sum_{i,j,\gamma}\frac{p_ip_j}{p^2}(\hat{\alpha}_i)_{\alpha\gamma}(\hat{\alpha}_j)_{\gamma\beta} \quad . \tag{10}$$

Exercise 6.6.

According to Exercise 6.1, (2) one has

$$\sum_{i,j,\gamma} p_i p_j (\hat{\alpha}_i)_{\alpha\gamma} (\hat{\alpha}_j)_{\gamma\beta} = \sum_{i,j} (p_i p_j \hat{\alpha}_i \hat{\alpha}_j)_{\alpha\beta} = p^2 \delta_{\alpha\beta} \quad ,$$

and consequently (10) changes to

$$\begin{aligned} A &= \cosh^2\left(\frac{\omega}{2}\right)\delta_{\alpha\beta} + \sinh^2\left(\frac{\omega}{2}\right)\delta_{\alpha\beta} - B\sinh\left(\frac{\omega}{2}\right)\cosh\left(\frac{\omega}{2}\right) \\ &= \delta_{\alpha\beta}\left(2\cosh^2\left(\frac{\omega}{2}\right) - 1\right) - B\sinh\left(\frac{\omega}{2}\right)\cosh\left(\frac{\omega}{2}\right) \\ &= \delta_{\alpha\beta}\left(2\frac{E + m_0c^2}{2m_0c^2} - 1\right) - B\sinh\left(\frac{\omega}{2}\right)\cosh\left(\frac{\omega}{2}\right) \\ &= \delta_{\alpha\beta}\frac{E}{m_0c^2} - B\sinh\left(\frac{\omega}{2}\right)\cosh\left(\frac{\omega}{2}\right) \quad . \end{aligned} \tag{11}$$

To finish the proof, we need to show that B vanishes. With the help of (7a) and (9) this term reads

$$\begin{aligned} B &= \sum_{r,i,\gamma,\delta}\left[\frac{(\hat{\alpha}_i)_{\alpha\gamma}p_i}{p}\omega^r_\gamma(0)\overline{\omega}^r_\beta(0) + \omega^r_\alpha(0)\overline{\omega}^r_\delta(0)\frac{(\hat{\alpha}_i)_{\delta\beta}p_i}{p}\right] \\ &= \sum_{r,i,\gamma,\delta}\left[\frac{(\hat{\alpha}_i)_{\alpha\gamma}(\gamma_0)_{\gamma\sigma}p_i}{p}\varepsilon_r\omega^r_\sigma(0)\overline{\omega}^r_\beta(0) + \varepsilon_r\omega^r_\alpha(0)\overline{\omega}^r_\tau(0)\frac{(\gamma_0)_{\tau\delta}(\hat{\alpha}_i)_{\delta\beta}p_i}{p}\right] \\ &= \sum_i \frac{p_i}{p}\left[(\hat{\alpha}_i\gamma_0)_{\alpha\beta} + (\gamma_0\hat{\alpha}_i)_{\alpha\beta}\right] \quad , \end{aligned} \tag{12}$$

where again we have used (6.41). But now it holds that $\hat{\alpha}_i\gamma_0 = -\gamma_0\hat{\alpha}_i$ and consequently (12) vanishes. Hence we have in fact:

$$\sum_r \omega^r_\alpha(\varepsilon_r \boldsymbol{p})\omega^{r\dagger}_\beta(\varepsilon_r \boldsymbol{p}) = \frac{E}{m_0c^2}\delta_{\alpha\beta} \quad .$$

6.3 Polarized Electrons in Relativistic Theory

Electrons at rest are described by the spinors (6.3) and (6.13). These electrons are polarized in the z direction. For example, the spinor

$$\omega^2(0) = \begin{pmatrix} 0 \\ 1 \\ 0 \\ 0 \end{pmatrix}$$

describes an electron at rest with spin projection $-\hbar/2$ on the z direction. If the *rotation operator* [(3.56), (3.57)]

$$\hat{S}_R = \mathrm{e}^{\mathrm{i}\varphi\hat{\boldsymbol{\Sigma}}\cdot\boldsymbol{s}/2} \quad , \tag{6.45}$$

where $\boldsymbol{s}$ is the *unit vector* along the rotation axis, acts on the states (6.3), one will obtain states which are – according to $\boldsymbol{s}$ –polarized in any arbitrary direction. A state (spinor) ω, which is polarized in the direction $\boldsymbol{s}$ is an eigenstate of the spin operator $\hat{\boldsymbol{S}} \cdot \boldsymbol{s}$ in this direction:

$$\hat{\boldsymbol{S}} \cdot \boldsymbol{s}\omega = \frac{\hbar}{2}\omega$$

or, with $\hat{\boldsymbol{S}} = \hbar\hat{\boldsymbol{\Sigma}}/2$,

$$\hat{\boldsymbol{\Sigma}} \cdot \boldsymbol{s}\omega = \omega \quad . \tag{6.46}$$

Our next task is to turn this equation into a covariant form. This means we are searching for the generalization of a (three-)spin vector into a (four-)spin vector. First we note:

$$\hat{\boldsymbol{\Sigma}} = \begin{pmatrix} \hat{\boldsymbol{\sigma}} & 0 \\ 0 & \hat{\boldsymbol{\sigma}} \end{pmatrix}$$

contains the Pauli matrices $\hat{\boldsymbol{\sigma}}$ in the main diagonal; thus the form of the 4-spinor equation (6.46) will be similar to the 2-component Pauli theory. In the relativistic theory one commonly uses a *different notation.* The spinor of a (free) solution of the Dirac equation with *positive energy, momentum* p^μ and *spin vector* s^μ is denoted by

$$u(p,s) \quad .$$

Hence $u(p,s)$ satisfies the equation

$$(\not{p} - m_0c)_{\alpha\beta}u_\beta(p,s) = 0 \quad . \tag{6.47}$$

Now we have to clarify the meaning of a *four-spin vector* s^μ. To do this we start with the polarization vector $\boldsymbol{s}$ in the rest system, which is a unit vector ($\boldsymbol{s} \cdot \boldsymbol{s} = 1$), and write

$$(s^\nu)_{\mathrm{R.S.}} \equiv (s^\nu)_{\text{rest sytem}} = (0, \boldsymbol{s}) \quad . \tag{6.48}$$

In an arbitrary inertial system we can get the four-spinor s^μ by a Lorentz transformation of $(s^\mu)_{\mathrm{R.S.}}$ from the rest sytem:

$$s^\mu = \alpha^\mu{}_\nu (s^\nu)_{\mathrm{R.S.}} \quad . \tag{6.49}$$

The $a^\mu{}_\nu$ are the coefficients of the Lorentz transformation from the moving system to the rest sytem. Hence it also holds, e.g. for the momentum, that

$$p^\mu = a^\mu{}_\nu (p^\nu)_{\mathrm{R.S.}} \quad , \quad \text{where} \tag{6.50}$$

$$(p^\nu)_{\mathrm{R.S.}} = (m_0c, 0, 0, 0) \quad . \tag{6.51}$$

Because of the Lorentz invariance of the four-dimensional scalar product it now follows immediately that

$$s_\mu s^\mu = (s_\mu)_{\mathrm{R.S.}}(s^\mu)_{\mathrm{R.S.}} = -\boldsymbol{s} \cdot \boldsymbol{s} = -1 \tag{6.52}$$

and

$$p^\mu s_\mu = (p^\mu)_{\text{R.S.}}(s_\mu)_{\text{R.S.}} = (m_0c, 0, 0, 0)\cdot\begin{pmatrix}0\\-s_x\\-s_y\\-s_z\end{pmatrix} = 0 \quad . \tag{6.53}$$

In the rest system u satisfies (6.46), and hence

$$\hat{\boldsymbol{\Sigma}}\cdot \boldsymbol{s}u\big((p)_{\text{R.S.}},(s)_{\text{R.S.}}\big) = u\big((p)_{\text{R.S.}},(s)_{\text{R.S.}}\big) \quad . \tag{6.54}$$

In the moving system we obtain $u(p,s)$ from $u\,((p)_{\text{R.S.}},(s)_{\text{R.S.}})$ by the Lorentz transformation $\hat{S}(\hat{a})$ for spinors. Let us consider a solution $v(p,s)$ of the free Dirac equation for negative energy. Because of (6.33a) it satisfies

$$(\not{p} + m_0c)v(p,s) = 0 \quad . \tag{6.55}$$

We *require* that the solution $v(p,s)$ has the polarization $-\boldsymbol{s}$ in the rest sytem, i.e. in the rest system

$$\hat{\boldsymbol{\Sigma}}\cdot \boldsymbol{s}v\big((p)_{\text{R.S.}},(s)_{\text{R.S.}}\big) = -v\big((p)_{\text{R.S.}},(s)_{\text{R.S.}}\big) \tag{6.56}$$

is valid. The minus sign on the rhs, which on first sight seems to be paradoxical, is remarkable. Later we will interpret this within the hole theory. The spinors $\omega^r(\boldsymbol{p})$, which arise from the rest spinors (6.13) obviously fulfill the demands (6.54) and (6.56), if we define $\boldsymbol{s} = \boldsymbol{e}_z$ and use the following relations:

$$\begin{aligned} u(p,u_z) &= \omega^1(\boldsymbol{p}) \quad , & v(p,u_z) &= \omega^4(\boldsymbol{p}) \quad , \\ u(p,-u_z) &= \omega^2(\boldsymbol{p}) \quad , & v(p,-u_z) &= \omega^3(\boldsymbol{p}) \quad . \end{aligned} \tag{6.57}$$

There u_z^ν is that four-vector in the moving system which arises from the Lorentz transformation of the unit vector $\boldsymbol{u}_z$ in the z direction, which is defined in the rest system by

$$\{u_z^\nu\}_{\text{R.S.}} = (0,\boldsymbol{u}_z) = (0,0,0,1) \quad . \tag{6.58}$$

One may wonder why the spin projection (6.56) for electrons with negative energy is defined with the opposite sign. The reason for this becomes obvious later when we interpret an electron with negative energy, momentum $-\boldsymbol{p}$ and spin direction minus $= \downarrow$ as a positron with positive energy, momentum $+\boldsymbol{p}$ and spin direction plus $= \uparrow$. Then the relations given in (6.56) and (6.57) can be understood in the positron language! The spinor $v\,((p)_{\text{R.S.}},(s)_{\text{R.S.}})$ describes a positron with the opposite spin to that which might first seem to be the case (in the "electron language"). We note that, according to this, an arbitrary spinor is characterized by its momentum p_μ, the sign of the energy and the polarization $(s^\mu)_{\text{R.S.}}$ in the rest system. [See the last part of Chap. 7 and also Chap. 12 (Hole Theory).]

7. Projection Operators for Energy and Spin

In practical calculations of quantum electrodynamic (QED) processes we will become acquainted with a technique of calculation which allows the simple treatment of complicated expressions; especially the calculation of traces of products of many γ matrices. It is based on a *projection procedure*, i.e. a method to project a spinor with a given sign of energy and fixed polarization out of a general wave function or a wave packet. The appropriate operators which achieve this are called *projection operators*. In the nonrelativistic case, say,

$$\hat{P}_\pm = \frac{1 \pm \hat{\sigma}_z}{2} \tag{7.1}$$

is a *projection operator for spin* up $(+)$ or down $(-)$. Acting on an arbitrary state $\hat{P}_\pm$ just takes out the corresponding parts. Now we want to generalize this concept to the relativistic case and to *search for four operators* which *project those four independent parts belonging to positive or negative energy with spin up or spin down, and with the same momentum* $\boldsymbol{p}$, out of an arbitrary free solution of the Dirac equation (i.e. out of a plane wave with momentum $\boldsymbol{p}$). The four projection operators are denoted by $\hat{P}_r(\boldsymbol{p})$. They should, of course, be in covariant form so that they can be given in any Lorentz system (by transformation) in an easy way. We denote those four projectors explicitly as

$$\hat{P}_r(\boldsymbol{p}) \equiv \hat{P}(p_\mu, u_z, \varepsilon) \quad .$$

They should obey the conditions

$$\hat{P}_r(\boldsymbol{p})\omega^{r'}(\boldsymbol{p}) = \delta_{rr'}\omega^{r'}(\boldsymbol{p}) \quad \text{and} \tag{7.2}$$

$$\hat{P}_r(\boldsymbol{p})\hat{P}_{r'}(\boldsymbol{p}) = \delta_{rr'}\hat{P}_r(\boldsymbol{p}) \quad . \tag{7.3}$$

Both equations show the projection properties of $\hat{P}_r(\boldsymbol{p})$ very clearly. If we now recall (6.33a), i.e.

$$(\not{p} - \varepsilon_r m_0 c)\omega^r(\boldsymbol{p}) = 0 = (\varepsilon_r \not{p} - m_0 c)\omega^r(p) \quad , \tag{7.4}$$

we at once find the *projection operator for eigenstates with positive or negative energy*:

$$\hat{\Lambda}(p) = \frac{\varepsilon_r \not{p} + m_0 c}{2m_0 c} \quad . \tag{7.5}$$

Indeed, with the help of (7.4) we find

$$\hat{\Lambda}_r(p)\omega^r(p) = \frac{\varepsilon_r \not{p} + m_0 c}{2m_0 c}\omega^r(p) = \frac{2m_0 c}{2m_0 c}\omega^r(p) = \omega^r(p) \quad , \tag{7.6}$$

and furthermore because of

$$\begin{aligned} \not{p}\not{p} = \gamma_\mu \gamma_\nu p^\mu p^\nu &= \frac{1}{2}(\gamma_\mu\gamma_\nu + \gamma_\nu\gamma_\mu)p^\mu p^\nu = g_{\mu\nu}p^\mu p^\nu = p^2 = \frac{E^2}{c^2} - \boldsymbol{p}^2 \\ &= \frac{m_0^2 c^4 + c^2 \boldsymbol{p}^2}{c^2} - \boldsymbol{p}^2 = m_0^2 c^2 \quad , \end{aligned}$$

then

$$\begin{aligned} \hat{\Lambda}_r(p)\hat{\Lambda}_{r'}(p) &= \frac{(\varepsilon_r \not{p} + m_0 c)(\varepsilon_{r'}\not{p} + m_0 c)}{4m_0^2 c^2} \\ &= \frac{\varepsilon_r \varepsilon_{r'} \not{p}\not{p} + m_0^2 c^2 + (\varepsilon_r + \varepsilon_{r'}) m_0 c \not{p}}{4m_0^2 c^2} \\ &= \frac{m_0^2 c^2 (1 + \varepsilon_r \varepsilon_{r'}) + m_0 c \not{p}(\varepsilon_r + \varepsilon_{r'})}{4m_0^2 c^2} \\ &= \frac{m_0^2 c^2 (1 + \varepsilon_r \varepsilon_{r'}) + m_0 c \not{p}\varepsilon_r (1 + \varepsilon_r \varepsilon_{r'})}{4m_0^2 c^2} \\ &= \frac{(1 + \varepsilon_r \varepsilon_{r'})}{2} m_0 c \frac{(m_0 c + \varepsilon_r \not{p})}{2m_0^2 c^2} = \frac{(1 + \varepsilon_r \varepsilon_{r'})}{2}\hat{\Lambda}_r(p) \quad . \end{aligned} \tag{7.7}$$

Obviously all the operators (7.5) decompose into only two types, namely those with $\varepsilon_r = +1$ (which we call $\hat{\Lambda}_+$) and the ones with $\varepsilon_r = -1$ (which we call $\hat{\Lambda}_-$):

$$\hat{\Lambda}_\pm(p) = \frac{\pm\not{p} + m_0 c}{2m_0 c} \quad . \tag{7.8}$$

Notice that $\not{p}$ is written here, and not $\tilde{\not{p}}$. Hence the momentum (with real numbers as components) and not the momentum operator occurs in the projector (7.8). We may present the relations (7.7) more transparently by writing explicitly

$$\begin{aligned} &\text{(a)}\quad (\hat{\Lambda}_+)^2 = \hat{\Lambda}_+ \quad , \\ &\text{(b)}\quad (\hat{\Lambda}_-)^2 = \hat{\Lambda}_- \quad , \\ &\text{(c)}\quad \hat{\Lambda}_+\hat{\Lambda}_- = 0 \quad , \quad \text{and also} \\ &\text{(d)}\quad \hat{\Lambda}_+ + \hat{\Lambda}_- = 1 \quad . \end{aligned} \tag{7.9}$$

These are the typical properties an *energy projection operator* must have. Besides, the expression (7.5) is also covariant, so that all required conditions are fulfilled.

Now we consider the *spin–projection operator*. Again we consider the rest frame, where the spin can be easily described. We already know from (7.1) that in this case the projection operator for "spin up" or "spin down" is given in the nonrelativistic limit by

$$\hat{P}_\pm = \frac{1 \pm \hat{\sigma}_3}{2} \quad .$$

If we define the spin-projection operator with respect to an arbitrary axis given by the unit vector $\boldsymbol{u}$ ($\boldsymbol{u} \cdot \boldsymbol{u} = 1$) in a nonrelativistic theory, then (7.1) is generalized to

$$\hat{P}(\boldsymbol{u}) = \frac{1 + \hat{\boldsymbol{\sigma}} \cdot \boldsymbol{u}}{2} \quad . \tag{7.10}$$

It is obvious that with $u_z = (0, 0, +1)$, the special case (7.1) is contained within the general expression (7.10). But this is still nonrelativistic. In the Dirac theory we need the relativistic covariant generalization of the operators (7.1) or (7.10). For that purpose we make use of the four-component vector

$$u_z^\nu \tag{7.11}$$

which is given in the rest system of the electron by

$$\left(u_z^\nu\right)_{\text{R.S.}} = (0, 0, 0, 1) = (0, \boldsymbol{u}_z) \tag{7.12}$$

[cf. (6.58)]. Thus, by Lorentz transformation into arbitrary inertial systems it follows that

$$u_z^\nu = a^\nu{}_\mu \left(u_z^\mu\right)_{\text{R.S.}} \quad . \tag{7.13}$$

With (7.12) we can write the spin-projection operator in the rest system, which is now extended to the fourth dimension and labelled by $\hat{\Sigma}(u_z)$. This generalization is achieved by first denoting $\hat{\Sigma}(u_z^3)$ where u_z^3 is the third component of $\boldsymbol{u}_z$ of (7.12):

$$\hat{\Sigma}\left(u_z^3\right) = \frac{1 + \hat{\Sigma}_3}{2} = \frac{1 + \gamma_5 \gamma_3 \left(u_z^3\right)_{\text{R.S.}} \gamma_0}{2} = \frac{1 + \gamma_5 \left(\not{u}_z\right)_{\text{R.S.}} \gamma_0}{2} \quad . \tag{7.14}$$

This is because $(\not{u}_z)_{\text{R.S.}} = (u_z^\nu \gamma_\nu)_{\text{R.S.}} = (u_z^3)_{\text{R.S.}} \gamma_3$ and therefore

$$\begin{aligned}
\gamma_5 \gamma_3 \left(u_z^3\right)_{\text{R.S.}} \gamma_0 &= \gamma_5 \gamma_3 \gamma_0 = \mathrm{i} \gamma^0 \gamma^1 \gamma^2 \gamma^3 \gamma_3 \gamma_0 \\
&= -\mathrm{i} \gamma^0 \gamma^1 \gamma^2 \underbrace{\gamma^3 \gamma^3}_{-\mathbb{1}} \gamma^0 = +\mathrm{i} \gamma^0 \gamma^1 \gamma^2 \gamma^0 \\
&= +\mathrm{i} \gamma^1 \gamma^2 = \mathrm{i} \begin{pmatrix} 0 & \hat{\sigma}_1 \\ -\hat{\sigma}_1 & 0 \end{pmatrix} \begin{pmatrix} 0 & \hat{\sigma}_2 \\ -\hat{\sigma}^2 & 0 \end{pmatrix} \\
&= \mathrm{i} \begin{pmatrix} -\hat{\sigma}_1 \hat{\sigma}_2 & 0 \\ 0 & -\hat{\sigma}_1 \hat{\sigma}_2 \end{pmatrix} = \begin{pmatrix} \hat{\sigma}_3 & 0 \\ 0 & \hat{\sigma}_3 \end{pmatrix} \equiv \hat{\Sigma}_3 \quad ,
\end{aligned} \tag{7.15}$$

and[1] $\hat{\sigma}_i \hat{\sigma}_j = \mathrm{i} \varepsilon_{ijk} \hat{\sigma}_k + \delta_{ij}$. Now the question arises how to generalize (7.14) to a covariant form. The factor γ_0 is disturbing, while $\gamma_5 \not{u}_z$ is a covariant expression. In the rest frame the effect of γ_0 on the rest spinors $\omega^r(0)$ is given solely by the factor ± 1. For the spinors $\omega^{1,2}(0)$ we can therefore easily omit γ_0 in (7.14). If we do the same for the spionors $\omega^{3,4}(0)$, the effet in the rest frame is

$$\begin{aligned}
\hat{\Sigma}\left(u_z^3\right) \omega^{3,4}(0) &= \frac{1 + \gamma_5 \left(\not{u}_z^3\right)_{\text{R.S.}}}{2} \omega^{3,4}(0) = \frac{1 + \gamma_5 \left(\not{u}_z^3\right)_{\text{R.S.}} \gamma^0 \gamma^0}{2} \omega^{3,4}(0) \\
&= \frac{1 - \gamma_5 \left(\not{u}_z^3\right)_{\text{R.S.}} \gamma_0}{2} \omega^{3,4}(0) = \frac{1 - \hat{\Sigma}_3}{2} \omega^{3,4}(0) \\
&= \begin{cases} 0 \cdot \omega^3(0) \\ 1 \cdot \omega^4(0) \end{cases} \quad .
\end{aligned} \tag{7.16}$$

[1] See W. Greiner: *Quantum Mechanics – An Introduction*, 3rd ed. (Springer, Berlin, Heidelberg 1994), Exercise 13.2.

This would be exactly opposite to what we may naively expect. Analogously, one easily verifies that

$$\Sigma\left(-u_z^3\right)\omega^{3,4}(0) = \begin{cases} 1\cdot\omega^3(0) \\ 0\cdot\omega^4(0) \end{cases} . \tag{7.17}$$

Now we see that the *covariant spin-projection operator*

$$\hat{\Sigma}(u_z) = \frac{1+\gamma_5 \not{u}_z}{2} \tag{7.18}$$

satisfies the following relations in the rest system [denoting $\hat{\Sigma}((u_z)_{\mathrm{R.S.}}) \equiv \hat{\Sigma}(u_z)(0)$]:

$$\begin{aligned}
&\hat{\Sigma}(u_z)(0)\omega^1(0) = \omega^1(0) \quad , \quad \hat{\Sigma}(-u_z)(0)\omega^1(0) = 0 \quad , \\
&\hat{\Sigma}(u_z)(0)\omega^2(0) = 0 \quad , \quad \hat{\Sigma}(-u_z)(0)\omega^2(0) = \omega^2(0) \quad , \\
&\hat{\Sigma}(u_z)(0)\omega^3(0) = 0 \quad , \quad \hat{\Sigma}(-u_z)(0)\omega^3(0) = \omega^3(0) \quad , \\
&\hat{\Sigma}(u_z)(0)\omega^4(0) = \omega^4(0) \quad , \quad \hat{\Sigma}(-u_z)(0)\omega^4(0) = 0 \quad ,
\end{aligned} \tag{7.19a}$$

which we can rewrite with the help of the definitions (6.57):

$$\begin{aligned}
&\hat{\Sigma}(u_z)(0)u(0,u_z) = u(0,u_z) \quad , \quad \hat{\Sigma}(-u_z)(0)u(0,u_z) = 0 \quad , \\
&\hat{\Sigma}(u_z)(0)u(0,-u_z) = 0 \quad , \quad \hat{\Sigma}(-u_z)(0)u(0,-u_z) = u(0,-u_z) \quad , \\
&\hat{\Sigma}(u_z)(0)v(0,-u_z) = 0 \quad , \quad \hat{\Sigma}(-u_z)(0)v(0,-u_z) = v(0,-u_z) \quad , \\
&\hat{\Sigma}(u_z)(0)v(0,u_z) = v(0,u_z) \quad , \quad \hat{\Sigma}(-u_z)(0)v(0,u_z) = 0 \quad .
\end{aligned} \tag{7.19b}$$

Notice, the operator $\hat{\Sigma}(u_z)$ should not be confused with the operator $\hat{\Sigma}$! Because (7.18) is covariant, these relations are also valid in a moving system in which the spinors are given by $u(p,u_z)\ldots$:

$$\begin{aligned}
&\hat{\Sigma}(u_z)u(p,u_z) \quad = u(p,u_z) \quad , \\
&\hat{\Sigma}(u_z)v(p,u_z) \quad = v(p,u_z) \quad , \\
&\hat{\Sigma}(-u_z)u(p,u_z) = \hat{\Sigma}(-u_z)v(p,u_z) = 0 \quad .
\end{aligned} \tag{7.20}$$

The covariant generalization (7.18) of the nonrelativistic spin-projection operator (7.14) corresponds naturally to the convention, given by (6.56) and (6.57). Therein the spin projections of negative-energy states are contragredient (i.e. opposite) to those of normal "states" of positive energy [see (6.54) and (6.56)]. The positron interpretation of the hole theory, which we have already preliminary considered in (6.57), comes up here in a natural way through the covariant spin-projection operator. Furthermore, we generalize the spin-projection operator (7.18) for an arbitrary spin vector s^μ with $s^\mu p_\mu = 0$ [see (6.52) and (5.53)]:

$$\hat{\Sigma}(s) = \frac{\mathbb{1}+\gamma_5 \not{s}}{2} . \tag{7.21}$$

Because of the covariance, it follows that as a generalization of (7.20) the $u(p,s)$ and $v(p,s)$ obey the relations [see (6.48)–(6.57)]

$$\begin{aligned}
&\hat{\Sigma}(s)u(p,s) \quad = u(p,s) \quad , \\
&\hat{\Sigma}(s)v(p,s) \quad = v(p,s) \quad , \\
&\hat{\Sigma}(-s)u(p,s) = \hat{\Sigma}(-s)v(p,s) = 0 \quad .
\end{aligned} \tag{7.22}$$

7.1 Simultaneous Projections of Energy and Spin

With the projection operators $\hat{\Lambda}_{\pm}(p)$ for the energy and $\hat{\Sigma}(s)$ for the spin we can easily construct projectors for energy and spin. As we already know, the motion of a free particle is completely determined by

the four momentum p_μ ,
the sign of the energy ε ,
and the polarization s^μ ,

with $s^\mu p_\mu = 0$. Therefore the *energy–spin projectors*

$$\begin{aligned} \hat{P}_1(\boldsymbol{p}) &= \hat{\Lambda}_+(p)\hat{\Sigma}(u_z) \quad , \qquad & \hat{P}_2(\boldsymbol{p}) &= \hat{\Lambda}_+(p)\hat{\Sigma}(-u_z) \quad , \\ \hat{P}_3(\boldsymbol{p}) &= \hat{\Lambda}_-(p)\hat{\Sigma}(-u_z) \quad , \qquad & \hat{P}_4(\boldsymbol{p}) &= \hat{\Lambda}_-(p)\hat{\Sigma}(u_z) \quad , \end{aligned} \tag{7.23}$$

which are composed of (7.8) and (7.18), determine the free Dirac waves, these being projected out of superposed wave functions (wave packets). The order of the energy and spin projection in (7.23) does not matter because

$$\left[\hat{\Sigma}(s), \hat{\Lambda}_{\pm}(p)\right]_- = 0 \quad , \tag{7.24}$$

as $s^\mu p_\mu = 0$ is valid. This is easy to verify, because the commutator (7.24) is equivalent to the commutator

$$\left[\not{p}, \gamma_5 \not{s}\right]_- = 0 \quad , \tag{7.25}$$

which we immediately prove:

$$\begin{aligned} \not{p}\gamma_5\not{s} &= \gamma^\mu p_\mu \mathrm{i}\gamma^0\gamma^1\gamma^2\gamma^3\gamma^\nu s_\nu = -\mathrm{i}\gamma^0\gamma^1\gamma^2\gamma^3\gamma^\mu\gamma^\nu s_\nu p_\mu \\ &= -\gamma_5(2g^{\mu\nu} - \gamma^\nu\gamma^\mu)s_\nu p_\mu = \gamma_5\not{s}\not{p} \quad . \end{aligned}$$

Here $g^{\mu\nu}s_\nu p_\mu = s^\mu p_\mu = s \cdot p = 0$ according to (6.53). Because of (7.2), (7.6) and (7.20) or (7.19a), we easily confirm that the defining equations (7.2), i.e.

$$\hat{P}_r(\boldsymbol{p})\omega^{r'}(p) = \delta_{rr'}\omega^{r'}(p) \tag{7.26}$$

are fulfilled!

In the following[2] we will appreciate the usefulness of these projection operators, which at first seem to be a little artificial. With their help we will often perform practical calculations, without explicit use of γ matrices and free spinors. Indeed, we will see that an explicit calculation, component by component, can be replaced by a rather elegant one through the use of the projectors.

According to (6.31) the general free solution was given by

$$\psi^r(x) = \omega^r(\boldsymbol{p})\,\mathrm{e}^{-\mathrm{i}\varepsilon_r p_\mu x^\mu/\hbar} \quad .$$

For $r = 1, 2$ these are free waves of positive energy ($\varepsilon_r p_0 = p_0$) and momentum $\hat{p}_i\psi^{1,2}(x) = p_i\psi^{1,2}(x)$. For $r = 3, 4$ these waves have nagative energy ($\varepsilon_r p_0 = -p_0$) and the eigenvalues of the momentum operators are

[2] See in particular W. Greiner, J. Reinhardt: *Quantum Electrodynamics*, 2nd ed. (Springer, Berlin, Heidelberg 1994).

$$\hat{p}_i \psi^{3,4}(x) = \varepsilon_{3,4} p_i \psi^{3,4}(x) = -p_i \psi^{3,4}(x) \quad ,$$

that is, they are negative ($-p_i$). We call $\psi^{3,4}(x)$ "spinors with negative energy". They describe antiparticles of positive energy and positive momentum p_i. This striking redefinition of plane waves with negative energy has already occured for the spin [see discussion following (6.57)]. To this end the eigenvalues of the momentum and spin operators agree with the momentum or spin direction of the waves of positive energy. For waves with negative energy we define it the other way round. In other words: a particle with negative energy and momentum eigenvalue $-p_i$ has momentum $+p_i$, and a particle with spin eigenvalue $-u_z$ has the spin $+u_z$. This seemingly strange definition has a deeper reason: in the framework of the hole theory, which will follow soon, we *interpret an electron with negative energy, momentum $-p_i$ and spin $-u_z$ as a positron (i.e. the antiparticle of the electron) with positive energy, momentum $+p_i$ and spin u_z*. This occurs as a natural result. Thus we hold on to our definition. Accordingly $\psi^{3,4}(x)$ is an electron wave function with negative energy, negative momentum and negative spin projection [see (7.19a)] or – and this is the physically correct description – a positron with positive energy, positive momentum and positive spin projection. For the later property the redefinition (6.57) is necessary, which means the change of $r = 3,4 \Leftrightarrow r = 4,3$.

8. Wave Packets of Plane Dirac Waves

Already in Sect. 2.3 we dealt with the single particle aspects of Dirac waves. In order to gain a deeper understanding and a possible interpretation of the free solutions, we study wave packets. These are superpositions of plane waves which yield localized wave functions in space time. Since the Dirac equation is a linear wave equation, the wave packets are also solutions of the free Dirac equation, which is just the superposition principle. A wave packet of plane waves with *positive energy* has the form

$$\psi^{(+)}(\boldsymbol{x},t) = \int \frac{\mathrm{d}^3p}{\sqrt{2\pi\hbar}^3}\sqrt{\frac{m_0c^2}{E}}\sum_{\pm s} b(p,s)u(p,s)\,\mathrm{e}^{-\mathrm{i}(p_0-p_0')x^0/\hbar} \quad . \tag{8.1}$$

The amplitudes $b(p,s)$ determine the admixture of the plane waves $u(p,s)\cdot \mathrm{e}^{-\mathrm{i}p_\mu x^\mu/\hbar}$ to the wave packet. The "(+)" indicates that a superposition of only positive-energy plane waves is taken. Normalizing to unity implies

$$\begin{aligned}
\int \psi^{(+)\dagger}(\boldsymbol{x},t)\psi^{(+)}(\boldsymbol{x},t)\,\mathrm{d}^3x &\overset{!}{=} 1 \overset{!}{=} \int \mathrm{d}^3p \int \mathrm{d}^3p \sum_{\pm s}\sum_{\pm s'} \\
&\times \sqrt{\frac{m_0c^2}{E}}\sqrt{\frac{m_0c^2}{E'}}\,b^\dagger(p,s)b(p',s')u^\dagger(p,s)u(p',s')\,\mathrm{e}^{-\mathrm{i}(p_0-p_0')x^0/\hbar} \\
&\times \underbrace{\int \frac{\mathrm{e}^{\mathrm{i}(\boldsymbol{p}-\boldsymbol{p}')\cdot\boldsymbol{x}/\hbar}}{\sqrt{2\pi\hbar^3}\sqrt{2\pi\hbar^3}}\,\mathrm{d}^3x}_{\delta(\boldsymbol{p}-\boldsymbol{p}')} \\
&= \int \mathrm{d}^3p \sum_{\pm s}\sum_{\pm s'}\frac{m_0c^2}{E}b^\dagger(p,s)b(p,s') \quad \underbrace{u^\dagger(p,s)u(p,s')}_{\delta_{ss'}(E/m_0c^2)\ \text{because of (6.35)}} \\
&= \int \mathrm{d}^3p \sum_{\pm s} b^\dagger(p,s)b(p,s) \\
&= \int \mathrm{d}^3p \sum_{\pm s} |b(p,s)|^2 \overset{!}{=} 1 \quad .
\end{aligned} \tag{8.2}$$

The factor

$$\exp\left(-\frac{\mathrm{i}}{\hbar}(p_0-p_0')x^0\right) = \exp\left(-\frac{\mathrm{i}}{\hbar}(E-E')\frac{x^0}{c}\right)$$

is equal to 1 because of

$$E = c\sqrt{p^2 + m_0c^2} = c\sqrt{p'^2 + m_0c^2} = E'$$

utilizing the property of $\delta(\boldsymbol{p} - \boldsymbol{p}')$. The next step is to find the current of such a packet, given the velocity operator $c\hat{\boldsymbol{\alpha}}$; thus we obtain

$$\boldsymbol{J}^{(+)} = \int \psi^{(+)\dagger} c\hat{\boldsymbol{\alpha}} \psi^{(+)} \, \mathrm{d}^3x \quad . \tag{8.3}$$

In the calculation of this expression (shown in Examples 8.1 and 8.2), we will learn a new and elegant mathematical technique, i.e. the Gordon decomposition. The result is

$$J_i^{(+)} = \int \mathrm{d}^3p \frac{p_i c^2}{E} \sum_{\pm s} |b(p,s)|^2 = \left\langle \frac{c^2 p_i}{E} \right\rangle_+ = \langle (v_{\mathrm{gr}})_i \rangle_+ \quad . \tag{8.4}$$

Because of the normalization (8.2), in the second step we have written $J_i^{(+)}$ as an expectation value with respect to the wave packet of positive energy and labelled this $\langle\ \rangle_+$. In this way the mean current of an arbitrary wave packet of plane waves of positive energy is equal to the expectation value of the *classical group velocity* $v_{\mathrm{gr}} = c^2\boldsymbol{p}/E$. This corresponds to the Ehrenfest theorem of the Schrödinger theory and agrees with our earlier considerations (2.62) and (2.67), where we restricted ourselves to the even part of the velocity operator to get the same result. Obviously the restriction to the even part of an operator is equivalent to the restriction to those wave packets which are constructed from one sort of solution (of positive or negative energy) only.

EXERCISE

8.1 The Gordon Decomposition

ψ_1 and ψ_2 are considered to be two arbitrary solutions of the free Dirac equation, that is

$$\begin{aligned} (\hat{\not{p}} - m_0c)\psi_1 &= 0 \quad , & \qquad \bar{\psi}_1(-\overleftarrow{\hat{\not{p}}} - m_0c) &= 0 \quad , \\ & & \text{and} \qquad & \\ (\hat{\not{p}} - m_0c)\psi_2 &= 0 \quad , & \qquad \bar{\psi}(-\overleftarrow{\hat{\not{p}}} - m_0c) &= 0 \quad . \end{aligned} \tag{1}$$

The second set of equations holds for the adjoint spinors, as will be proven below. Then

$$c\bar{\psi}_2\gamma^\mu\psi_1 = \frac{1}{2m_0}\left[\bar{\psi}_2\hat{p}^\mu\psi_1 - \left(\hat{p}^\mu\bar{\psi}_2\right)\psi_1\right] - \frac{\mathrm{i}}{2m_0}\hat{p}_\nu\left(\bar{\psi}_2\hat{\sigma}^{\mu\nu}\psi_1\right) \tag{2}$$

is valid. We prove this with the help of the relation

$$\begin{aligned} \not{a}\not{b} &= \gamma^\mu a_\mu \gamma^\nu b_\nu = a_\mu b_\nu \gamma^\mu\gamma^\nu \\ &= a_\mu b_\nu \left\{ \frac{1}{2}\underbrace{(\gamma^\mu\gamma^\nu + \gamma^\nu\gamma^\mu)}_{2g^{\mu\nu}} + \frac{1}{2}\underbrace{(\gamma^\mu\gamma^\nu - \gamma^\nu\gamma^\mu)}_{-\mathrm{i}\hat{\sigma}^{\mu\nu}} \right\} \\ &= a_\mu b^\mu - \mathrm{i}a^\mu b^\nu \hat{\sigma}_{\mu\nu} \quad , \end{aligned} \tag{3}$$

Exercise 8.1.

which is valid for two arbitrary four-vectors a^μ and b^μ. Furthermore, we introduce the definition

$$\overline{\psi}\overleftarrow{\hat{p}\!\!\!/} \underset{\text{def}}{\equiv} (\hat{p}_\mu\overline{\psi})\gamma^\mu \quad , \tag{4}$$

which means that the momentum operator acts to the left onto the function $\overline{\psi}(x)$. Because of (1) we obtain

$$0 = \overline{\psi}_2\big(-\overleftarrow{\hat{p}\!\!\!/} - m_0c\big)a\!\!\!/\psi_1 + \overline{\psi}_2 a\!\!\!/\big(\overrightarrow{\hat{p}\!\!\!/} - m_0c\big)\psi_1 \quad . \tag{5}$$

This happens because the terms on the rhs are zero [according to (1)]. In particular the first term on the rhs can be rewritten as

$$\big(-\hat{p}_\mu\overline{\psi}_2\gamma^\mu - m_0c\overline{\psi}_2\big)\, a\!\!\!/\psi_1 \quad .$$

That it vanishes is obvious from the fourth equation (1). Let us deduce this from the second equation (1). It follows by multiplication with γ_0 and Hermitian conjugation, that

$$\begin{aligned}\big(\gamma^0(\hat{p}\!\!\!/ - m_0c)\psi_2\big)^\dagger &= 0 = \psi_2^\dagger\big(\gamma^{\mu\dagger}\overleftarrow{\hat{p}}{}^\dagger_\mu - m_0c\big)\gamma^{0\dagger}\\ &= +\hat{p}^\dagger_\mu\psi_2^\dagger\gamma^{\mu\dagger}\gamma^{0\dagger} - m_0c\psi_2^\dagger\gamma^{0\dagger}\\ &= -\hat{p}_\mu\psi_2^\dagger\gamma^{\mu\dagger}\gamma^{0\dagger} - m_0c\psi_2^\dagger\gamma^{0\dagger} \quad .\end{aligned} \tag{6}$$

Because of $\gamma^{i\dagger} = -\gamma^i, \gamma^{0\dagger} = \gamma^0$, we can write

$$\gamma^0\gamma^{\nu\dagger}\gamma^0 = \gamma^\nu \quad \text{or} \quad \gamma^0\gamma^{\nu\dagger} = \gamma^\nu\gamma^0 \quad \text{or} \quad \gamma^{\nu\dagger}\gamma^0 = \gamma^0\gamma^\nu \tag{7}$$

and for (6) we achieve

$$\begin{aligned}&-\hat{p}_\mu\psi_2^\dagger\gamma^0\gamma^\mu - m_0c\psi_2^\dagger\gamma^0\\ &\quad = 0 = -\hat{p}_\mu\overline{\psi}_2\gamma^\mu - m_0c\overline{\psi}_2 = \overline{\psi}_2\big(-\overleftarrow{\hat{p}\!\!\!/} - m_0^2c\big) \quad .\end{aligned} \tag{8}$$

Thus we have proven (5), from which we can furthermore conclude that

$$\begin{aligned}0 &= -2m_0c\overline{\psi}_2 a\!\!\!/\psi_1 + \overline{\psi}_2\big[a\!\!\!/\overrightarrow{\hat{p}\!\!\!/} - \overleftarrow{\hat{p}\!\!\!/}a\!\!\!/\big]\psi_1\\ &= -2m_0c\overline{\psi}_2 a\!\!\!/\psi_1 + \overline{\psi}_2\big[a^\mu\overrightarrow{\hat{p}}{}^\mu - \mathrm{i}a^\mu\overrightarrow{\hat{p}}{}^\nu\hat{\sigma}_{\mu\nu} - \overleftarrow{\hat{p}\!\!\!/}{}^\mu a_\mu + \mathrm{i}\overleftarrow{\hat{p}\!\!\!/}{}^\mu a^\nu\hat{\sigma}_{\mu\nu}\big]\psi_1\\ &= a_\mu\Big\{-2m_0c\overline{\psi}_2\gamma^\mu\psi_1 + \overline{\psi}_2\big[\overrightarrow{\hat{p}\!\!\!/}{}^\mu - \overleftarrow{\hat{p}\!\!\!/}{}^\mu - \mathrm{i}\overrightarrow{\hat{p}\!\!\!/}{}^\nu\hat{\sigma}^\mu{}_\nu - \mathrm{i}\overleftarrow{\hat{p}\!\!\!/}{}^\nu\hat{\sigma}^\mu{}_\nu\big]\psi_1\Big\} \quad .\end{aligned} \tag{9}$$

Because a^μ was arbitrary the coefficient of a^μ must vanish, that is

$$c\overline{\psi}_2\gamma^\mu\psi_1 = \frac{1}{2m_0}\left[\overline{\psi}_2\hat{p}^\mu\psi_1 - \left(\hat{p}^\mu\overline{\psi}_2\right)\psi_1\right] - \frac{\mathrm{i}}{2m_0}\hat{p}^\nu\left(\overline{\psi}_2\hat{\sigma}^\mu{}_\nu\psi_1\right) \quad .$$

This is just the statement of (2). This splitting of the matrix $\overline{\psi}_2\gamma^\mu\psi_1$ is called *Gordon decomposition.*[1] Its physical meaning is that the Dirac current density $c\overline{\psi}_2\gamma^\mu\psi_1$ can be split up into a *convection current density*

[1] Surprisingly, this procedure was derived by Walter Gordon in Z. Phys. **50**, 630 (1928).

Exercise 8.1.

$$\frac{1}{2m_0}\left[\overline{\psi}_2\hat{p}^\mu\psi_1 - \left(\hat{p}^\mu\overline{\psi}_2\right)\psi_1\right] \tag{10}$$

and *spin-current density*

$$\frac{-\mathrm{i}}{2m_0}\hat{p}^\nu\left(\overline{\psi}_2\hat{\sigma}_{\mu\nu}\psi_1\right) \quad . \tag{11}$$

EXERCISE

8.2 Calculation of the Expectation Value of a Velocity Operator

Problem. Calculate the expectation value of the velocity operator of a wave packet consisting of plane waves with positive energy, using the Gordon decomposition.

Solution. A wave packet of Dirac plane waves of positive energy is given by

$$\psi^{(+)}(\boldsymbol{x},t) = \int\frac{\mathrm{d}^3p}{(2\pi\hbar)^{3/2}}\sqrt{\frac{m_0c^2}{E}}\sum_{\pm s}b(p,s)u(p,s)\,\mathrm{e}^{\mathrm{i}p_\mu x^\mu/\hbar} \quad . \tag{1}$$

This gives the expectation value of the velocity operator (do not confuse the given symbol for wave packets of positive energy (+) with that for Hermitian conjugation †):

$$\begin{aligned} J^{i(+)} &= \int\psi^{(+)\dagger}c\hat{\alpha}_i\psi^{(+)}\,\mathrm{d}^3x \\ &= \int\psi^{(+)\dagger}\gamma^0c\gamma^i\psi^{(+)}\,\mathrm{d}^3x = c\int\overline{\psi}^{(+)}\gamma^i\psi^{(+)}\,\mathrm{d}^3x \\ &= \int\mathrm{d}^3x\iint\frac{\mathrm{d}^3p\,\mathrm{d}^3p'}{(2\pi\hbar)^{3/2}(2\pi\hbar)^{3/2}}\sqrt{\frac{m_0c^2}{E}}\sqrt{\frac{m_0c^2}{E'}} \\ &\quad\times\sum_{\pm s,\pm s'}b^*(p',s')b(p,s)\,\mathrm{e}^{(\mathrm{i}/\hbar)(p'^\mu-p^\mu)x_\mu}\overline{u}(p',s')c\gamma^iu(p,s) \quad . \end{aligned} \tag{2}$$

Via the Gordon decomposition [Example 8.1, (2)] of the spinor matrix element this becomes

$$\begin{aligned} J^{i(+)} &= \iint\mathrm{d}^3p\,\mathrm{d}^3p'\frac{m_0c^2}{\sqrt{EE'}}\sum_{\pm s,\pm s'}b^*(p',s')b(p,s) \\ &\quad\times\int\frac{\mathrm{e}^{(\mathrm{i}/\hbar)(p'^\mu-p^\mu)x_\mu}}{(2\pi\hbar)^3}\,\mathrm{d}^3x\cdot\frac{1}{2m_0}\overline{u}(p',s') \\ &\quad\times\left[\left(p'^i+p^i\right)-\mathrm{i}\left(-p'^\nu+p^\nu\right)\hat{\sigma}^i{}_\nu\right]u(p,s) \quad . \end{aligned}$$

Exercise 8.2.

Because of the identity

$$\int \mathrm{e}^{(\mathrm{i}/\hbar)(p_0'-p_0)x^0} \frac{\mathrm{e}^{-(\mathrm{i}/\hbar)(p'^i-p^i)x_i}}{\sqrt{(2\pi\hbar)}^3\sqrt{(2\pi\hbar)}^3}\,\mathrm{d}^3x$$
$$= \delta\left(p'^i - p^i\right) \underbrace{\exp\left[\frac{\mathrm{i}}{\hbar}\left(\frac{E'}{c}-\frac{E}{c}\right)x^0\right]}_{=1} \quad , \tag{3}$$

this evaluates to be

$$J^{i(+)} = \int \mathrm{d}^3p \frac{m_0c^2}{E}\frac{2p^i}{2m_0} \sum_{\pm s,\pm s'} b^*(p,s')b(p,s)\,\underbrace{\overline{u}(p,s')u(p,s)}_{=\delta_{ss'}\ \text{[from (6.34)]}}$$
$$= \int \mathrm{d}^3p \frac{p_ic^2}{E}\sum_{\pm s}|b(p,s)|^2 \quad .$$

As we can see, the spin-dependent part of the current vanishes.

Let us resume our earlier discussion [see (2.61) to (2.70)] on the velocity operator for *one* particle in relativistic theory: In Schrödinger's theory the velocity operator $\hat{v} = \hat{p}/m$ was proportional to momentum, but this is no longer the case in Dirac's relativistic theory. The velocity operator for free particles $c\hat{\boldsymbol{\alpha}}$ is no longer a temporal constant, because of

$$\frac{\mathrm{d}c\hat{\boldsymbol{\alpha}}}{\mathrm{d}t} = \frac{c}{\mathrm{i}\hbar}\left[\hat{\boldsymbol{\alpha}}, \hat{H}_f\right]_- \neq 0 \quad . \tag{8.5}$$

From (8.4) we can even conclude that one needs the solutions of negative energy to construct eigenfunctions of $c\hat{\boldsymbol{\alpha}}$. Further, wave packets consisting of plane waves with only positive energy have the expectation value of the velocity $|\langle c\hat{\alpha}^i\rangle| \sim |\langle c^2p_i/E\rangle| < c$, whereas the eigenvalues of $c\hat{\alpha}^i$ are exactly $\pm c$. This motivates us to consider *wave packets made up of the complete set of plane Dirac waves*, i.e. plane waves with both positive and negative energy. Instead of (8.1) we now write

$$\psi(\boldsymbol{x},t) = \int \frac{\mathrm{d}^3p}{\sqrt{(2\pi\hbar)}^3}\sqrt{\frac{m_0c^2}{E}}$$
$$\times \sum_{\pm s}\left[b(p,s)u(p,s)\,\mathrm{e}^{-\mathrm{i}p_\mu x^\mu/\hbar} + d^*(p,s)v(p,s)\,\mathrm{e}^{+\mathrm{i}p_\mu x^\mu/\hbar}\right] \quad . \tag{8.6}$$

The coefficient $b(p,s)$ are the probability amplitudes for waves with positive energy, whereas $d^*(p,s)$ are those for negative energy. The probability of finding a particle anywhere must be one (see Exercise 8.3):

$$\int \mathrm{d}^3x\,\psi^\dagger(\boldsymbol{x},t)\psi(\boldsymbol{x},t) = \int \mathrm{d}^3p\sum_{\pm s}\left[|b(p,s)|^2 + |d(p,s)|^2\right] = 1 \quad . \tag{8.7}$$

From this we calculate the current of the wave packet (see Exercise 8.4):

$$\begin{aligned} J^k &= \int d^3x\, \psi^\dagger(\boldsymbol{x},t) c\hat{\alpha}_k \psi(\boldsymbol{x},t) \\ &= \int d^3p \Bigg\{ \sum_{\pm s} \left(|b(p,s)|^2 + |d(p,s)|^2 \right) \frac{p^k c^2}{E} \\ &\quad + \mathrm{i}c \sum_{\pm s, \pm s'} b^*(-p,s') d^*(p,s)\, \mathrm{e}^{2\mathrm{i}x_0 p_0/\hbar} \overline{u}(-p,s') \hat{\sigma}^{k0} v(p,s) \\ &\quad - \mathrm{i}c \sum_{\pm s, \pm s'} d(-p,s') b(p,s)\, \mathrm{e}^{-2\mathrm{i}x_0 p_0/\hbar} \overline{v}(-p,s') \hat{\sigma}^{k0} u(p,s) \Bigg\} \quad . \end{aligned} \tag{8.8}$$

Here $u(-p,s)$ means $u\left(\sqrt{\boldsymbol{p}^2 + m_0c^2}, -\boldsymbol{p}, s\right)$. This abbreviation is not particularly obvious, but it is absolutely clear what is meant. In the following we use this notation for the coefficients b, d^* and for $v(-p,s)$ too.

If we compare (8.8) with (8.4) we see that additional terms appear in (8.8). The first term in (8.8) represents the time-independent group velocity that also appears in (8.4). The second and third terms are interferences of solutions with positive and negative energy, which oscillate time-dependently because of the factors $\exp(\pm 2\mathrm{i}p_0x_0/\hbar)$. The frequency of this *Zitterbewegung*[2] is

$$\frac{2p_0c}{\hbar c} > \frac{2m_0c^2}{\hbar c} \approx 2 \times 10^{21}\,\mathrm{s}^{-1} \quad , \tag{8.9}$$

and its strength is proportional to the amplitudes $d(p,s)$ of the waves with negative energy in the wave packet. See Exercise 8.5!

EXERCISE

8.3 Calculation of the Norm of a Wave Packet

Problem. Calculate the norm of the general wave packet built out of plane Dirac waves of positive and negative energies

$$\begin{aligned} \psi(\boldsymbol{x},t) = \int \frac{d^3p}{(2\pi\hbar)^{3/2}} \sqrt{\frac{m_0c^2}{E}} \\ \times \sum_{\pm s} \left[b(p,s) u(p,s)\, \mathrm{e}^{-\mathrm{i}p_\mu x^\mu/\hbar} + d^*(p,s) v(p,s)\, \mathrm{e}^{+\mathrm{i}p_\mu x^\mu/\hbar} \right] \quad . \end{aligned}$$

Solution. The current is defined as

$$J^\mu = c \int d^3x\, \overline{\psi}(\boldsymbol{x},t) \gamma^\mu \psi(\boldsymbol{x},t) \quad , \tag{1}$$

and the wave packet reads (see above)

[2] This name stems from E. Schrödinger: Sitzungsber. Preuß. Akad. Wiss., Phys.-Math. **24**, 418 (1930) and in German means literally “trembling motion”.

Exercise 8.3.

$$\psi(\boldsymbol{x},t) = \int \frac{d^3p}{(2\pi\hbar)^{3/2}}\sqrt{\frac{m_0c^2}{E}} \times \sum_{\pm s}\left[b(p,s)u(p,s)\,e^{-ip\cdot x/\hbar} + d^*(p,s)v(p,s)\,e^{+ip\cdot x/\hbar}\right] \quad . \tag{2}$$

The zero component of the current (probability) is then evaluated as

$$\begin{aligned} W &\equiv \frac{1}{c}J^0 \\ &= \int d^3x\, \psi^\dagger(\boldsymbol{x},t)\psi(\boldsymbol{x},t) = \iiint \frac{d^3x\, d^3p\, d^3p'}{(2\pi\hbar)^3}\frac{m_0c^2}{\sqrt{EE'}} \\ &\times \sum_{\pm s,s'}\Big[b^*(p',s')b(p,s)u^\dagger(p',s')u(p,s)\,e^{-i(p-p')\cdot x/\hbar} \\ &+ b^*(p',s')d^*(p,s)u^\dagger(p',s')v(p,s)\,e^{i(p+p')\cdot x/\hbar} \\ &+ d(p',s')b(p,s)v^\dagger(p',s')u(p,s)\,e^{-i(p+p')\cdot x/\hbar} \\ &+ d(p',s')d^*(p,s)v^\dagger(p',s')v(p,s)\,e^{i(p-p')\cdot x/\hbar}\Big] \quad . \end{aligned} \tag{3}$$

Integrating over $d^3x/(2\pi\hbar)^3$ leads to δ functions $\delta^{(3)}(\boldsymbol{p}' \pm \boldsymbol{p})$. Because $p_0 = E = \sqrt{m_0^2c^4 + \boldsymbol{p}^2c^2}$, the energies $p_0 = p_0'(E = E')$ are identical if the momenta $|\boldsymbol{p}|$ and $|\boldsymbol{p}'|$ are the same. Then we can also integrate over d^3p' and obtain

$$\begin{aligned} \varrho = \int d^3p \frac{m_0c^2}{E}\sum_{\pm s,s'}\Big\{&b^*(p,s')b(p,s)u^\dagger(p,s')u(p,s) \\ &+ d(p,s')d^*(p,s)v^\dagger(p,s')v(p,s) \\ &+ b^*(-p,s')d^*(p,s)u^\dagger(-p,s')v(p,s)\,e^{2ip_0x_0/\hbar} \\ &+ d(-p,s')b(p,s)v^\dagger(-p,s')u(p,s)\,e^{-2ip_0x_0/\hbar}\Big\} \quad . \end{aligned} \tag{4}$$

From (6.35) we have

$$u^\dagger(p,s')u(p,s) = v^\dagger(p,s')v(p,s) = \frac{E}{m_0c^2}\delta_{ss'} \quad , \tag{5}$$

$$u^\dagger(-p,s')v(p,s) = v^\dagger(-p,s')u(p,s) = 0 \quad , \tag{6}$$

and therefore

$$W = \int d^3p \sum_{\pm s}\left\{|b(p,s)|^2 + |d(p,s)|^2\right\} = 1 \quad . \tag{7}$$

In the very last step we were careful to normalize the total probability W to 1.

EXERCISE

8.4 Calculation of the Current for a Wave Packet

Problem. Calculate the current

$$J^k = c\int \psi^\dagger(\boldsymbol{x},t)\hat{\alpha}_k\psi(\boldsymbol{x},t)\,\mathrm{d}^3x = c\int \overline{\psi}(\boldsymbol{x},t)\gamma^k\psi(\boldsymbol{x},t)\,\mathrm{d}^3x$$

for the general wave packet of Exercise 8.3.

Solution. The space components are calculated in exactly the same manner as in the last Exercise (8.3):

$$\begin{aligned} J^k &= c\int \mathrm{d}^3x\,\overline{\psi}(\boldsymbol{x},t)\gamma^k\psi(\boldsymbol{x},t) \\ &= \frac{1}{2m_0}\int \mathrm{d}^3x\,\left\{\overline{\psi}\hat{p}^k\psi - (\hat{p}^k\overline{\psi})\psi - \mathrm{i}\hat{p}_\nu(\overline{\psi}\hat{\sigma}^{k\nu}\psi)\right\} \quad , \end{aligned} \tag{1}$$

where we used the Gordon decomposition (Example 8.1). Inserting (2) of Exercise 8.3 yields

$$\begin{aligned} J^k = \frac{1}{2m_0}\iiint &\frac{\mathrm{d}^3x\,\mathrm{d}^3p\,\mathrm{d}^3p'}{(2\pi\hbar)^3}\frac{m_0c^2}{\sqrt{EE'}} \\ \times \sum_{\pm s,s'}\Big\{ &b^*(p',s')b(p,s)\overline{u}(p',s')u(p,s)p^k\,\mathrm{e}^{-\mathrm{i}(p-p')\cdot\boldsymbol{x}/\hbar} \\ &- b^*(p',s')d^*(p,s)\overline{u}(p',s')v(p,s)p^k\mathrm{e}^{\mathrm{i}(p+p')\cdot\boldsymbol{x}/\hbar} \\ &+ d(p',s')b(p,s)\overline{v}(p',s')u(p,s)p^k\mathrm{e}^{-\mathrm{i}(p+p')\cdot\boldsymbol{x}/\hbar} \\ &- d(p',s')d^*(p,s)\overline{v}(p',s')v(p,s)p^k\mathrm{e}^{-\mathrm{i}(p-p')\cdot\boldsymbol{x}/\hbar} \\ &+ b^*(p',s')b(p,s)\overline{u}(p',s')u(p,s)p'^k\mathrm{e}^{-\mathrm{i}(p-p')\cdot\boldsymbol{x}/\hbar} \\ &+ b^*(p',s')d^*(p,s)\overline{u}(p',s')v(p,s)p'^k\mathrm{e}^{+\mathrm{i}(p+p')\cdot\boldsymbol{x}/\hbar} \\ &- d(p',s')b(p,s)\overline{v}(p',s')u(p,s)p'^k\mathrm{e}^{-\mathrm{i}(p+p')\cdot\boldsymbol{x}/\hbar} \\ &- d(p',s')d^*(p,s)\overline{v}(p',s')v(p,s)p'^k\mathrm{e}^{\mathrm{i}(p-p')\cdot\boldsymbol{x}/\hbar} \\ &- \mathrm{i}b^*(p',s')b(p,s)\overline{u}(p',s')\hat{\sigma}^{k\nu}u(p,s)(p_\nu - p'_\nu)\,\mathrm{e}^{-\mathrm{i}(p-p')\cdot\boldsymbol{x}/\hbar} \\ &+ \mathrm{i}b^*(p',s')d^*(p,s)\overline{u}(p',s')\hat{\sigma}^{k\nu}v(p,s)(p_\nu + p'_\nu)\,\mathrm{e}^{\mathrm{i}(p+p')\cdot\boldsymbol{x}/\hbar} \\ &- \mathrm{i}d(p',s')b(p,s)\overline{v}(p',s')\hat{\sigma}^{k\nu}u(p,s)(p_\nu + p'_\nu)\,\mathrm{e}^{-\mathrm{i}(p+p')\cdot\boldsymbol{x}/\hbar} \\ &+ \mathrm{i}d(p',s')d^*(p,s)\overline{v}(p',s')\hat{\sigma}^{k\nu}v(p,s)(p_\nu - p'_\nu)\,\mathrm{e}^{\mathrm{i}(p-p')\cdot\boldsymbol{x}/\hbar}\Big\} \quad . \end{aligned} \tag{2}$$

Integrating again over $\mathrm{d}^3x/(2\pi\hbar)^3$ we get δ functions $\delta^{(3)}(\boldsymbol{p}'\pm\boldsymbol{p})$. The integration over d^3p' then yields

Exercise 8.4.

$$
\begin{aligned}
j^k = \frac{1}{2m_0}\int \mathrm{d}^3p\,\frac{m_0c^2}{E}
&\times \sum_{\pm s,s'}\Big\{2p^k\big[b^*(p,s')b(p,s)\overline{u}(p,s')u(p,s)\\
&- d(p,s')d^*(p,s)\overline{v}(p,s')v(p,s)\big]\\
&+ 2p^k\big[d(-p,s')b(p,s)\overline{v}(-p,s')u(p,s)\,\mathrm{e}^{-2\mathrm{i}p_0x_0/\hbar}\\
&- b^*(-p,s')d^*(p,s)\overline{u}(-p,s')v(p,s)\,\mathrm{e}^{2\mathrm{i}p_0x_0/\hbar}\big]\\
&+ \frac{2\mathrm{i}E}{c}\big[b^*(-p,s')d^*(p,s)\overline{u}(-p,s')\hat{\sigma}^{k0}v(p,s)\,\mathrm{e}^{2\mathrm{i}p_0x_0/\hbar}\\
&- d(-p,s')b(p,s)\overline{v}(-p,s')\hat{\sigma}^{k0}u(p,s)\,\mathrm{e}^{-2\mathrm{i}p_0x_0/\hbar}\big]\Big\} \quad . \qquad (3)
\end{aligned}
$$

From the orthogonality relation (6.34) we get

$$
\begin{aligned}
&\overline{u}(p,s')u(p,s) = \delta_{ss'} = -\overline{v}(p,s')v(p,s) \quad ,\\
&\overline{u}(-p,s')v(p,s) = 0 = \overline{v}(-p,s')u(p,s) \quad . \qquad (4)
\end{aligned}
$$

Therefore, from (3) only

$$
\begin{aligned}
J^k = \int \mathrm{d}^3p\Big\{&\sum_{\pm s}\big[|b(p,s)|^2 + |d(p,s)|^2\big]\frac{p^kc^2}{E}\\
&+ \mathrm{i}c\sum_{\pm s,\pm s'}\big[b^*(-p,s')d^*(p,s)\overline{u}(-p,s')\hat{\sigma}^{k0}v(p,s)\,\mathrm{e}^{2\mathrm{i}p_0x_0/\hbar}\\
&- d(-p,s')b(p,s)\overline{v}(-p,s')\hat{\sigma}^{k0}u(p,s)\,\mathrm{e}^{-2\mathrm{i}p_0x_0/\hbar}\big]\Big\} \qquad (5)
\end{aligned}
$$

remains. This is the earlier used result (8.8).

EXERCISE

8.5 Temporal Development of a Wave Packet with Gaussian Density Distribution

Problem. At time $t = 0$ the following wave packet with Gaussian density distribution is defined as

$$
\psi'(\boldsymbol{x},0,s) = \frac{1}{(\pi d^2)^{3/4}}\,\mathrm{e}^{-|x|^2/2d^2}\,\omega^1(0) \quad . \qquad (1)
$$

Determine the wave packet at time t developing from (1). Consider the intensity of the negative energy solutions in the wave packet. What does one learn in general about the applicability of the one-particle interpretation of the Dirac equation?

Exercise 8.5.

Solution. Equation (1) is a boundary condition for the general wave packet

$$\psi(\boldsymbol{x},t) = \int \frac{d^3p}{(2\pi\hbar)^{3/2}} \sqrt{\frac{m_0/c^2}{E}} \sum_{s'=\pm s} \left[b(p,s')u(p,s')\,e^{-i(Et-\boldsymbol{p}\cdot\boldsymbol{x})/\hbar} \right.$$
$$\left. + d^*(p,s')v(p,s')\,e^{i(Et-\boldsymbol{p}\cdot\boldsymbol{x})/\hbar} \right] \quad . \tag{2}$$

The requirement

$$\psi(\boldsymbol{x},0) = \psi'(\boldsymbol{x},0,s) \tag{3}$$

leads to equations that determine the coefficients $b(p,s')$ and $d^*(p,s')$. In particular, it follows from

$$\int \frac{d^3p}{(2\pi\hbar)^{3/2}} \sqrt{\frac{m_0c^2}{E}} \sum_{s'=\pm s} d^*(p,s')v(p,s')\,e^{-i\boldsymbol{p}\cdot\boldsymbol{x}/\hbar}$$
$$= \int \frac{d^3p}{(2\pi\hbar)^{3/2}} \sqrt{\frac{m_0c^2}{E}} \sum_{s'=\pm s} d^*(p',s')v(p',s')\,e^{i\boldsymbol{p}\cdot\boldsymbol{x}/\hbar} \tag{4}$$

(where we have inserted $\boldsymbol{p} = -\boldsymbol{p}'$ and $p_0 = -p_0'$) that

$$\int \sqrt{\frac{m_0c^2}{E}} \sum_{s'=\pm s} \left[b(p,s')u(p,s') + d^*(p',s')v(p',s') \right]$$
$$\times\, e^{i\boldsymbol{p}\cdot\boldsymbol{x}/\hbar} \frac{d^3p}{(2\pi\hbar)^{3/2}} = \frac{1}{(\pi d^2)^{3/4}}\, e^{-|x|^2/2d^2}\,\omega^1(0) \quad . \tag{5}$$

The rhs of (5) is just the Fourier transform of the braced expression on the left. Therefore the inverse transformation reads:

$$\sqrt{\frac{m_0c^2}{E}} \sum_{s'=\pm s} \left[b(p,s')u(p,s') + d^*(p',s')v(p',s') \right]$$
$$= \int \frac{d^3x}{(2\pi\hbar)^{3/2}}\, e^{-i\boldsymbol{p}\cdot\boldsymbol{x}/\hbar} \left[\frac{1}{(\pi d^2)^{3/4}}\, e^{-|x|^2/2d^2}\,\omega^1(0) \right] \quad . \tag{6}$$

On the rhs of (6) we have the Fourier transform of a Gaussian distribution. With the identity

$$\int \frac{d^3x}{(2\pi\hbar)^{3/2}}\, e^{-i\boldsymbol{p}\cdot\boldsymbol{x}/\hbar} \frac{1}{(\pi d^2)^{3/4}}\, e^{-|x|^2/2d^2}$$
$$= \frac{1}{\pi^{3/4}} \left(\frac{d}{\hbar}\right)^{3/2} e^{-|p|^2d^2/2\hbar^2} \stackrel{\text{def}}{=} g(|\boldsymbol{p}|) \quad , \tag{7}$$

we rewrite (6) as

$$\sqrt{\frac{m_0c^2}{E}} \sum_{s'=\pm s} \left[b(p,s')u(p,s') + d^*(p',s')v(p',s') \right] = g(|\boldsymbol{p}|)\omega^1(0) \quad . \tag{8}$$

Expanding the sum and using the definitions

Exercise 8.5.

$$
\begin{aligned}
u(p,s'=+s) &= \omega^1(\boldsymbol{p}) \quad , \qquad u(p,-s)=\omega^2(\boldsymbol{p}) \quad , \\
v(p',+s) &= \omega^4(\boldsymbol{p}') \quad , \qquad v(p',-s)=\omega^3(\boldsymbol{p}') \quad ,
\end{aligned} \tag{9}
$$

we get

$$
\begin{aligned}
&\sqrt{\frac{m_0c^2}{E}}\left[b(p,s)\omega^1(\boldsymbol{p})+b(p,-s)\omega^2(\boldsymbol{p})+d^*(p',s)\omega^4(\boldsymbol{p}')+d^*(p',-s)\omega^3(\boldsymbol{p}')\right] \\
&\quad = g(|\boldsymbol{p}|)\omega^1(0) \quad .
\end{aligned} \tag{10}
$$

Now we multiply (10), in turn, by $[\omega^1(\boldsymbol{p})]^\dagger$, $[\omega^2(\boldsymbol{p})]^\dagger$, $[\omega^3(-\boldsymbol{p})]^\dagger$ and $[\omega^4(-\boldsymbol{p})]^\dagger$. With [see (6.30)]

$$
\begin{aligned}
\left[\omega^1(\boldsymbol{p})\right]^\dagger &= \sqrt{\frac{E+m_0c^2}{2m_0c^2}}\left(1,0,\frac{p_zc}{E+m_0c^2},\frac{p_-c}{E+m_0c^2}\right) \quad , \\
\left[\omega^2(\boldsymbol{p})\right]^\dagger &= \sqrt{\frac{E+m_0c^2}{2m_0c^2}}\left(1,0,\frac{p_+c}{E+m_0c^2},\frac{-p_zc}{E+m_0c^2}\right) \quad , \\
\left[\omega^3(-\boldsymbol{p})\right]^\dagger &= \sqrt{\frac{E+m_0c^2}{2m_0c^2}}\left(\frac{-p_zc}{E+m_0c^2},\frac{-p_-c}{E+m_0c^2},1,0\right) \quad , \\
\left[\omega^4(-\boldsymbol{p})\right]^\dagger &= \sqrt{\frac{E+m_0c^2}{2m_0c^2}}\left(\frac{-p_+c}{E+m_0c^2},\frac{p_zc}{E+m_0c^2},0,1\right) \quad ,
\end{aligned} \tag{11}
$$

the following relations are valid:

$$
\begin{aligned}
\left[\omega^i(\boldsymbol{p})\right]^\dagger \omega^j(-\boldsymbol{p}) &= \left[w^j(-\boldsymbol{p})\right]^\dagger \omega^i(\boldsymbol{p}) = 0 && \text{for } i=1,2 \text{ and } j=3,4 \quad ; \\
\left[\omega^i(\boldsymbol{p})\right]^\dagger \omega^j(\boldsymbol{p}) &= \frac{E}{m_0c^2}\delta_{ij} && \text{for } i,j=1,2 \quad ; \\
\left[\omega^i(-\boldsymbol{p})\right]^\dagger \omega^j(-\boldsymbol{p}) &= \frac{E}{m_0c^2}\delta_{ij} && \text{for } i,j=3,4 \quad .
\end{aligned}
$$

Hence, we get the four equations:

$$
\begin{aligned}
&\sqrt{\frac{m_0c^2}{E}}b(p,s)\frac{E}{m_0c^2} = g(|\boldsymbol{p}|)\left[\omega^1(\boldsymbol{p})\right]^\dagger\omega^1(0) \\
&\qquad \Rightarrow\; b(p,s) = \sqrt{\frac{E+m_0c^2}{2E}}g(|\boldsymbol{p}|)\cdot 1 \quad , \\
&\sqrt{\frac{m_0c^2}{E}}b(p,-s)\frac{E}{m_0c^2} = g(|\boldsymbol{p}|)\left[\omega^2(\boldsymbol{p})\right]^\dagger\omega^1(0) \\
&\qquad \Rightarrow\; b(p,-s) = 0 \quad , \\
&\sqrt{\frac{m_0c^2}{E}}d^*(p',-s)\frac{E}{m_0c^2} = g(|\boldsymbol{p}|)\left[\omega^3(-\boldsymbol{p})\right]^\dagger\omega^1(0) \\
&\qquad \Rightarrow\; d^*(p',-s) = \sqrt{\frac{E+m_0c^2}{2E}}g(|\boldsymbol{p}|)\frac{(-p_z)c}{E+m_0c^2} \quad , \\
&\sqrt{\frac{m_0c^2}{E}}d^*(p',s)\frac{E}{m_0c^2} = g(|\boldsymbol{p}|)\left[\omega^4(-\boldsymbol{p})\right]^\dagger\omega^1(0) \\
&\qquad \Rightarrow\; d^*(p',s) = \sqrt{\frac{E+m_0c^2}{2E}}g(|\boldsymbol{p}|)\frac{(-p_+)c}{E+m_0c^2} \quad .
\end{aligned} \tag{12}
$$

Exercise 8.5.

We now insert these equations, as well as (9) and (4), into (2), which yields:

$$\psi(\boldsymbol{x},t) = \int \frac{\mathrm{d}^3p}{(2\pi\hbar)^{3/2}}\,\mathrm{e}^{\mathrm{i}\boldsymbol{p}\cdot\boldsymbol{x}/\hbar}\sqrt{\frac{m_0c^2}{E}}\sqrt{\frac{E+m_0c^2}{2E}}\,g(|\boldsymbol{p}|)\sqrt{\frac{E+m_0c^2}{2m_0c^2}}$$

$$\times\left[\begin{pmatrix} 1 \\ 0 \\ \dfrac{p_zc}{E+m_0c^2} \\ \dfrac{p_+c}{E+m_0c^2} \end{pmatrix}\mathrm{e}^{-\mathrm{i}Et/\hbar} + \begin{pmatrix} \dfrac{p_z^2c^2}{(E+m_0c^2)2} \\ \dfrac{p_zp_+c^2}{(E+m_0c^2)^2} \\ \dfrac{-p_zc}{E+m_0c^2} \\ 0 \end{pmatrix}\right.$$

$$\left.\times\,\mathrm{e}^{\mathrm{i}Et/\hbar} + \begin{pmatrix} \dfrac{(p_x^2+p_y^2)c^2}{(E+m_0c^2)^2} \\ \dfrac{-p_zp_+c^2}{(E+m_0c^2)^2} \\ 0 \\ \dfrac{-p_+c}{E+m_0c^2} \end{pmatrix}\mathrm{e}^{\mathrm{i}Et/\hbar}\right]$$

$$= \int \frac{\mathrm{d}^3p}{(2\pi\hbar)^{3/2}}\,\mathrm{e}^{\mathrm{i}\boldsymbol{p}\cdot\boldsymbol{x}/\hbar}g(|\boldsymbol{p}|)$$

$$\times\begin{pmatrix} \dfrac{E+m_0c^2}{2E}\,\mathrm{e}^{-\mathrm{i}Et/\hbar} + \dfrac{\boldsymbol{p}^2c^2}{2E(E+m_0c^2)}\,\mathrm{e}^{\mathrm{i}Et/\hbar} \\ 0 \\ \dfrac{p_zc}{2E}\left(\mathrm{e}^{-\mathrm{i}Et/\hbar}-\mathrm{e}^{\mathrm{i}Et/\hbar}\right) \\ \dfrac{p+c}{2E}\left(\mathrm{e}^{-\mathrm{i}Et/\hbar}-\mathrm{e}^{\mathrm{i}Et/\hbar}\right) \end{pmatrix}$$

$$= \int \frac{\mathrm{d}^3p}{(2\pi\hbar)^{3/2}}\,\mathrm{e}^{\mathrm{i}\boldsymbol{p}\cdot\boldsymbol{x}/\hbar}g(|\boldsymbol{p}|)$$

$$\times\begin{pmatrix} \dfrac{E+m_0c^2}{2E}\,\mathrm{e}^{-\mathrm{i}Et/\hbar} + \dfrac{E-m_0c^2}{2E}\,\mathrm{e}^{\mathrm{i}Et/\hbar} \\ 0 \\ -\dfrac{p_zc}{2E}2\mathrm{i}\sin\frac{Et}{\hbar} \\ -\dfrac{p+c}{E}\mathrm{i}\sin\frac{Et}{\hbar} \end{pmatrix}$$

$$= \int \frac{\mathrm{d}^3p}{(2\pi\hbar)^{3/2}}\,\mathrm{e}^{\mathrm{i}\boldsymbol{p}\cdot\boldsymbol{x}/\hbar}g(|\boldsymbol{p}|)\left[\left(\cos\frac{Et}{\hbar}-\mathrm{i}\frac{m_0c^2}{E}\sin\frac{Et}{\hbar}\right)\omega^1(0)\right.$$

$$\left. -\,\mathrm{i}\frac{p_zc}{E}\sin\left(\frac{Et}{\hbar}\right)\omega^3(0) - \mathrm{i}\frac{p_+c}{E}\sin\left(\frac{Et}{\hbar}\right)\omega^4(0)\right] \quad . \tag{13}$$

At an arbitrary time $t \neq 0$, $\psi(\boldsymbol{x},t)$ is thus composed of the following three parts:

$$\psi(\boldsymbol{x},t) = c_1(\boldsymbol{x},t)\omega^1(0) + c_3(\boldsymbol{x},t)\omega^3(0) + c_4(\boldsymbol{x},t)\omega^4(0) \quad .$$

The functions $c_1(\boldsymbol{x})$, $c_3(\boldsymbol{x})$, $c_4(\boldsymbol{x})$ can be calculated only by use of numerical methods:

Exercise 8.5.

$$
\begin{aligned}
c_1(\boldsymbol{x},t) &:= \int \frac{\mathrm{d}^3 p}{(2\pi\hbar)^{3/2}}\,\mathrm{e}^{\mathrm{i}\boldsymbol{p}\cdot\boldsymbol{x}/\hbar}\frac{1}{\pi^{3/4}}\left(\frac{d}{\hbar}\right)^{3/2}\mathrm{e}^{-|p|^2 d^2/2\hbar^2} \\
&\quad\times\left(\cos\frac{\sqrt{m_0^2c^4+\boldsymbol{p}^2c^2}}{\hbar}t-\mathrm{i}\frac{m_0c^2}{\sqrt{m_0^2c^4+\boldsymbol{p}^2c^2}}\sin\frac{\sqrt{m_0^2c^4+\boldsymbol{p}^2c^2}}{\hbar}t\right) \quad , \\
c_3(\boldsymbol{x},t) &:= -\mathrm{i}\int \frac{\mathrm{d}^3 p}{(2\pi\hbar)^{3/2}}\,\mathrm{e}^{\mathrm{i}\boldsymbol{p}\cdot\boldsymbol{x}/\hbar}\frac{1}{\pi^{3/4}}\left(\frac{d}{\hbar}\right)^{3/2}\mathrm{e}^{-|p|^2 d^2/2\hbar^2} \\
&\quad\times\frac{p_z c}{\sqrt{m_0^2c^4+\boldsymbol{p}^2c^2}}\sin\frac{\sqrt{m_0^2c^4+\boldsymbol{p}^2c^2}}{\hbar}t \quad , \\
c_4(\boldsymbol{x},t) &:= \int \frac{\mathrm{d}^3 p}{(2\pi\hbar)^{3/2}}\,\mathrm{e}^{\mathrm{i}\boldsymbol{p}\cdot\boldsymbol{x}/\hbar}\frac{1}{\pi^{3/4}}\left(\frac{d}{\hbar}\right)^{3/2}\mathrm{e}^{-|p|^2 d^2/2\hbar^2} \\
&\quad\times\frac{p_y-\mathrm{i}p_x}{\sqrt{m_0^2c^4+\boldsymbol{p}^2c^2}}\sin\frac{\sqrt{m_0^2c^4+\boldsymbol{p}^2c^2}}{\hbar}t \quad .
\end{aligned} \tag{14}
$$

This result can be understood in the following way: To obtain the wave-packet of (1) a superposition of plane waves of positive as well as of negative energy is necessary. Here the the coefficients occur in such a way that for $t=0$ the $\omega^3(0)$ and $\omega^4(0)$ parts of the partial waves of positive energies and those of negative energies cancel each other. This can be clearly seen for $t=0$ in (13): The third and fourth components of the sum of the three spinors vanish. Since the partial waves of positive and negative energy behave differently in time [$\exp(-\mathrm{i}Et/\hbar)$ and $\exp(\mathrm{i}Et/\hbar)$, respectively] this is no longer valid for times $t\neq 0$. This implies that (1) obviously cannot be regarded as a localized electron of spin $+s$, as it would have been in the framework of a single-particle interpretation of the Dirac equation. One may wonder why a spin-up electron in time gets admixtures of components of negative energy with spin down. The reason for that comes from the fact, that

$$
\left[\hat{H}_f,\boldsymbol{\sigma}\right]_- = 2\hat{\boldsymbol{\alpha}}\times\hat{\boldsymbol{p}}\neq 0 \quad\text{while}\quad \left[\hat{H}_f,\hat{\boldsymbol{p}}\cdot\hat{\boldsymbol{\sigma}}\right]_- = 0 \quad .
$$

Hence spin is not conserved; only the projection of spin along the momentum $\hat{\boldsymbol{p}}$ is a conserved quantity.

Let us briefly return to (12), from which we get the relative intensities of the partial waves of positive and negative energy:

$$
\begin{aligned}
R_b &= \frac{|b(p,s)|^2}{|d^*(p',s)|^2+|d^*(p',-s)|^2+|b(p,s)|^2} \\
&= \frac{1}{p^2c^2/(E+m_0c^2)^2+1} = \frac{1}{(E-m_0c^2)/(E+m_0c^2)+1} \\
&= \frac{n+1}{n-1+n+1} = \frac{n+1}{2n} \quad\text{with}\quad n=\frac{E}{m_0c^2} \quad ,
\end{aligned}
$$

$$
R_d = \frac{|d^*(p',s)|^2+|d^*(p',-s)|^2}{|d^*(p',s)|^2+|d^*(p',-s)|^2+|b(p,s)|^2} = 1-\frac{n+1}{2n} = \frac{n-1}{2n} \quad . \tag{15}
$$

Exercise 8.5.

Because of the Gaussian distribution $g(|\boldsymbol{p}|)$ [see (7)] only partial waves with $|\boldsymbol{p}| \leq \hbar/d$ contributee noticeably, so that

$$n = \frac{E_0}{m_0c^2} = \frac{\sqrt{(m_0c^2)^2 + c^2p^2}}{m_0c^2} = \sqrt{1 + \frac{p^2}{m_0^2c^2}} = \sqrt{1 + \left(\frac{d_c}{d}\right)^2} \quad , \tag{16}$$

where we have set $p = \hbar/d$ and $d_c = \hbar/m_0c$. In the figure below we recognize that only for $d \leq d_c = \hbar/m_0c$, i.e. only if the width of the wave packet is compressed to a size of about a Compton wavelength of the electron, do the partial waves of negative energies have an appreciable effect.

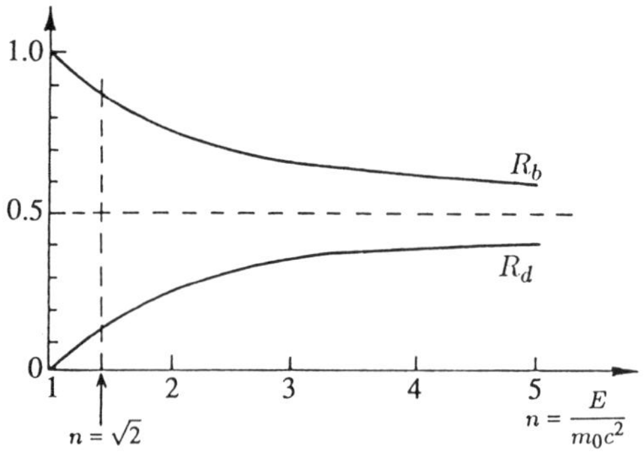

Relative intensities of partial waves of positive (R_b) and negative (R_d) energy is a Gaussian wave packet of width d

9. Dirac Particles in External Fields: Examples and Problems

EXAMPLE

9.1 Eigenvalue Spectrum of a Dirac Particle in a One-Dimensional Square-Well Potential

We calculate the spectrum of eigenvalues for Dirac particles in a square-well potential of depth $V_0 \leq 0$ and width a.

For that purpose we decompose the real axis into three domains I, II, III:

$$\text{I: } z \leq -a/2 \quad ; \quad \text{II: } -a/2 \leq z \leq a/2 \quad ; \quad \text{III: } z \geq a/2 \quad .$$

The Dirac equation within these domains reads (see Fig. 9.1)

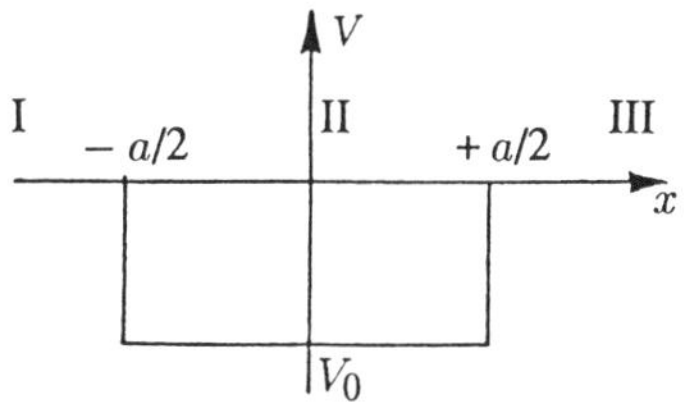

I, III:

$$(\hat{\alpha} \cdot \hat{p}c + \hat{\beta} m_0 c^2)\psi = E\psi \quad ,$$

II:

$$(\hat{\alpha} \cdot \hat{p}c + \hat{\beta} m_0 c^2)\psi = (E - V_0)\psi$$
$$(V_0 < 0) \quad . \tag{1}$$

As the relevant coordinate we choose z. Therefore the spinor ψ is only a function of z, e.g. $\psi = \psi(z)$.

While the wave functions in I and III are just the free solutions of the Dirac equation, the solutions in II are obtained from the free solution by the substitution $E \to E - V_0$. Since no spin-flip occurs at the border of the well, we can restrict our discussion to solutions with spin up. The energies of the solutions can take all allowed values from $-\infty$ up to $+\infty$, so that we can describe particles as well as antiparticles.

Example 9.1.

The solution of the Dirac equation (1) are thus given by:

I.

$$\psi_{\mathrm{I}}(z) = A\,\mathrm{e}^{\mathrm{i}p_1 z/\hbar} \begin{pmatrix} 1 \\ 0 \\ \dfrac{p_1 c}{E + m_0 c^2} \\ 0 \end{pmatrix} + A'\,\mathrm{e}^{-\mathrm{i}p_1 z/\hbar} \begin{pmatrix} 1 \\ 0 \\ \dfrac{-p_1 c}{E + m_0 c^2} \\ 0 \end{pmatrix}$$

$$p_1^2 + \frac{E^2}{c^2} - m_0^2 c^2 \quad , \qquad z \leq -\frac{a}{2} \quad , \tag{2}$$

II.

$$\psi_{\mathrm{II}}(z) = B\,\mathrm{e}^{\mathrm{i}p_2 z/\hbar} \begin{pmatrix} 1 \\ 0 \\ \dfrac{p_2 c}{E - V_0 + m_0 c^2} \\ 0 \end{pmatrix} + B'\,\mathrm{e}^{-\mathrm{i}p_2 z/\hbar} \begin{pmatrix} 1 \\ 0 \\ \dfrac{-p_2 c}{E - V_0 + m_0 c^2} \\ 0 \end{pmatrix} \tag{3}$$

$$p_2^2 = \frac{(E - V_0)^2}{c^2} - m_0^2 c^2 \quad , \qquad -\frac{a}{2} \leq z \leq \frac{a}{2} \quad ,$$

III:

$$\psi_{\mathrm{III}}(z) = C\,\mathrm{e}^{\mathrm{i}p_1 z/\hbar} \begin{pmatrix} 1 \\ 0 \\ \dfrac{p_1 c}{E + m_0 c^2} \\ 0 \end{pmatrix} + C'\,\mathrm{e}^{-\mathrm{i}p_1 z/\hbar} \begin{pmatrix} 1 \\ 0 \\ \dfrac{-p_1 c}{E + m_0 c^2} \\ 0 \end{pmatrix} \quad , \tag{4}$$

$$p_1^2 = \frac{E^2}{c^2} - m_0^2 c^2 \quad , \qquad z \geq \frac{a}{2} \quad .$$

At the borders of the well the wave function must be continuous (because of the current conservation $\partial_\mu j^\mu = 0$). Therefore we get the condition at the boundaries

$z = -\frac{a}{2}$:

$$\psi_{\mathrm{I}}\left(-\frac{a}{2}\right) = \psi_{\mathrm{II}}\left(-\frac{a}{2}\right)$$

$z = +\frac{a}{2}$:

$$\psi_{\mathrm{II}}\left(\frac{a}{2}\right) = \psi_{\mathrm{III}}\left(\frac{a}{2}\right) \quad .$$

This means in particular:

$z = -\frac{a}{2}$:

$$A\,\mathrm{e}^{-\mathrm{i}p_1 a/2\hbar} + A'\,\mathrm{e}^{\mathrm{i}p_1 a/2\hbar} = B\,\mathrm{e}^{-\mathrm{i}p_2 a/2\hbar} + B'\,\mathrm{e}^{\mathrm{i}p_2 a/2\hbar} \quad , \tag{5}$$

$$\left(A\,\mathrm{e}^{-\mathrm{i}p_1 a/2\hbar} - A'\,\mathrm{e}^{\mathrm{i}p_1 a/2\hbar}\right)\frac{p_1 c}{E + m_0 c^2} = \left(B\,\mathrm{e}^{-\mathrm{i}p_2 a/2\hbar} - B'\,\mathrm{e}^{\mathrm{i}p_2 a/2\hbar}\right)\frac{p_2 c}{E - V_0 m_0 c^2} \quad . \tag{6}$$

$z = +\frac{a}{2}$:

$$B\,\mathrm{e}^{\mathrm{i}p_2 a/2\hbar} + B'\,\mathrm{e}^{-\mathrm{i}p_2 a/2\hbar} = C\,\mathrm{e}^{\mathrm{i}p_1 a/2\hbar} + C'\,\mathrm{e}^{-\mathrm{i}p_1 a/2\hbar} \quad , \tag{7}$$

$$\left(B\,\mathrm{e}^{\mathrm{i}p_2 a/2\hbar} - B'\,\mathrm{e}^{-\mathrm{i}p_2 a/2\hbar}\right)\frac{p_2 c}{E - V_0 + m_0 c^2} = \left(C\,\mathrm{e}^{\mathrm{i}p_1 a/2\hbar} - C'\,\mathrm{e}^{-\mathrm{i}p_1 a/2\hbar}\right)\frac{p_1 c}{E + m_0 c^2} \quad . \tag{8}$$

Example 9.1.

If we define γ to be

$$\gamma = \frac{p_1 c}{E + m_0c^2}\frac{E - V_0 + m_0c^2}{p_2 c} = \sqrt{\frac{(E - m_0c^2)\,(E - V_0 + m_0c^2)}{(E + m_0c^2)\,(E - V_0 - m_0c^2)}} \quad , \tag{9}$$

we can write (5–8) in the following matrix form:

$$\begin{pmatrix} A \\ A' \end{pmatrix} = \frac{1}{2}\begin{pmatrix} \frac{\gamma+1}{\gamma}\,\mathrm{e}^{\mathrm{i}(p_1-p_2)a/2\hbar} & \frac{\gamma-1}{\gamma}\,\mathrm{e}^{\mathrm{i}(p_1+p)a/2\hbar} \\ \frac{\gamma-1}{\gamma}\,\mathrm{e}^{-\mathrm{i}(p_1+p_2)a/2\hbar} & \frac{\gamma+1}{\gamma}\,\mathrm{e}^{\mathrm{i}(p_2-p_1)a/2\hbar} \end{pmatrix}\begin{pmatrix} B \\ B' \end{pmatrix} \quad , \tag{10}$$

$$\begin{pmatrix} B \\ B' \end{pmatrix} = \frac{1}{2}\begin{pmatrix} (1+\gamma)\,\mathrm{e}^{\mathrm{i}(p_1-p_2)a/2\hbar} & (1-\gamma)\,\mathrm{e}^{-\mathrm{i}(p_1+p_2)a/2\hbar} \\ (1-\gamma)\,\mathrm{e}^{\mathrm{i}(p_1+p_2)a/2\hbar} & (1+\gamma)\,\mathrm{e}^{\mathrm{i}(p_2-p_1)a/2\hbar} \end{pmatrix}\begin{pmatrix} C \\ C' \end{pmatrix} \quad , \tag{11}$$

and by inserting (11) into (10) we get in summary

$$\begin{aligned} \begin{pmatrix} A \\ A' \end{pmatrix} &= \frac{1}{4\gamma} \\ &\times \begin{pmatrix} (1+\gamma)^2\,\mathrm{e}^{\mathrm{i}(p_1-p_2)a/\hbar} - (1-\gamma)^2\,\mathrm{e}^{\mathrm{i}(p_1+p_2)a/\hbar} & (1-\gamma^2)(\mathrm{e}^{-\mathrm{i}p_2a/\hbar} - \mathrm{e}^{\mathrm{i}p_2a\hbar}) \\ -(1-\gamma)^2(\mathrm{e}^{\mathrm{i}p_2a/\hbar} - \mathrm{e}^{\mathrm{i}p_2a/\hbar}) & (1+\gamma)^2\,\mathrm{e}^{-\mathrm{i}(p_1-p_2)a/\hbar} - (1-\gamma)^2\,\mathrm{e}^{-\mathrm{i}(p_1+p_2)a/\hbar} \end{pmatrix} \\ &\times \begin{pmatrix} C \\ C' \end{pmatrix} \quad . \end{aligned} \tag{12}$$

Thus we have two equations with four unknown coefficients A, A', C, C'. The normalization condition is a third equation:

$$\int \psi^\dagger \psi \,\mathrm{d}z = 1 \quad , \tag{13}$$

Therefore one of the four coefficients A, A', C, C' can, in general, be arbitrarily chosen.

Now we can discuss the solutions to different energies. Let us look at Figs. 9.1, 2 where typical behaviour of the large component of the wave function is plotted for different energies. In Fig. 9.1 we have assumed $|V_0|$ to be smaller than $2m_0c^2$; therefore we have to consider four energy ranges.

(1) $E > m_0c^2$: Free electrons are moving from left to right and are scattered by the (attractive) potential. If in domain II the width of the potential is an integer multiple of the wavelength, there exists a potential resonance, i.e. one gets an especially large probability for finding the electron in the well.

(2) $m_0c^2 + V_0 < E < m_0c^2$: Here we find bound states with an exponentially decreasing probability for finding an electron in domains I and III.

(3) $-m_0c^2 + V_0 < E < -m_0^2c^2$: Incoming positrons (continuum states of negative energy) "feel" a repulsive potential (because of their opposite charge – these wave functions actually describe positrons, as well shall see in Chap. 12 on hole theory) and will be scattered at this potential. Since the probability of finding an electron in domain II decreases exponentially, a large proportion of the positrons will be reflected at the repulsive potential. (The transmission decreases with a and increases with $|E|$.)

(4) $E < m_0c^2 + V_0$: The positrons are scattered at the repulsive potential; again, there can exist potential resonances if a is an integer multiple of the wavelength in the domain II.

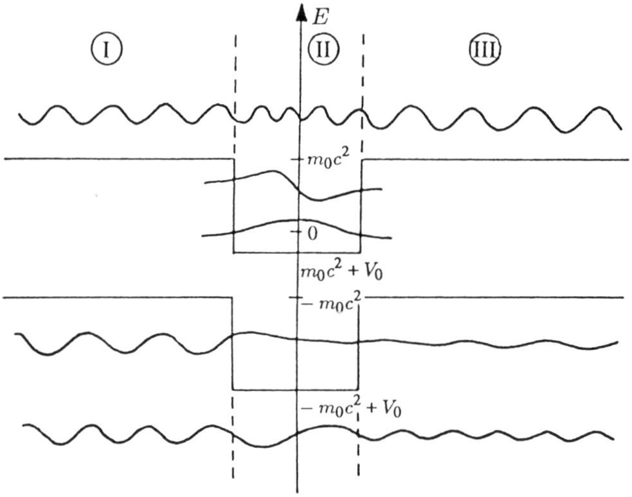

Fig. 9.1. The large component of some wave functions for different energies. In this case the depth is $|V_0| < 2m_0c^2$

On the other hand in Fig. 9.2 the potential is assumed to be $|V_0| > 2m_0c^2$. In this case an additional energy domain appears, showing a new behaviour of the wave function: $m_0c^2 + V_0 < E < -m_0c^2$. In this domain, bound electronic states are possible; however their wave functions do not decrease exponentially, in the domains I, III, but they join a continuum wave of the same energy $E < -m_0c^2$. Therefore the probability of finding the particle far away from the well is not equal to zero. We interpret this observation using hole theory (see Chap. 12). It means that a hole in that state will travel away (as a positron) from the well to infinity, and if a state is unoccupied, it will be slowly occupied by an electron of the filled Dirac sea. Thus a spontaneous creation of an electron–positron pair is possible. An empty bound state will spontaneously be occupied by an electron from the negative continuum, whereas the positron will move away to infinity. (Because of the spin degeneracy of the state there will even be two positrons: one with spin up and one with spin down.) One usually says that in this case *the potential well is overcritical* with respect to spontaneous $e^- - e^+$ pair creation. While bound states in the energy domain $-m_0c^2 < E < m_0c^2$ can remain empty without causing any instability to the system, it is impossible to keep bound states that "dive" into the negative continuum empty for a long time. They will be spontaneously filled up, i.e. the hole in this state has a finite decay width. We will see that, in fact, no sharp energy levels of bound states exist in the domain $m_0c^2 + V_0 < E < -m_0c^2$ (as was the case for $-m_0c^2 < E < m_0c^2$), but these wave functions will have a reasonating structure that peaks around the expected binding energy of the bound state.

Following these qualitative considerations, we shall now show how these statements result from the solutions (10)–(12) and we distinguish the two cases:

$$\text{(a)} \quad |E| > m_0c^2 \quad , \quad \text{i.e. } p_1 \text{ is real} \quad .$$
$$\text{(b)} \quad |E| < m_0c^2 \quad , \quad \text{i.e. } p_1 \text{ is imaginary} \quad .$$

Solutions for (a) are in general called *scattering states*,, solutions for (b) are *bound states.*

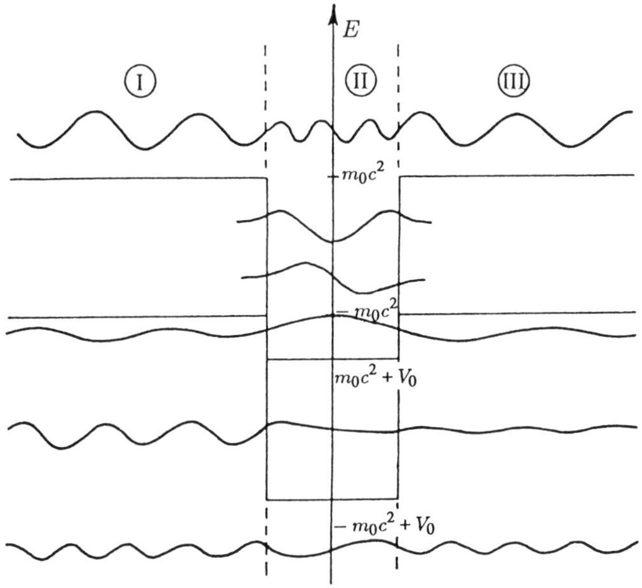

Fig. 9.2. The large component of some wave functions for different energies. In this case the depth is $|V_0| > 2m_0c^2$

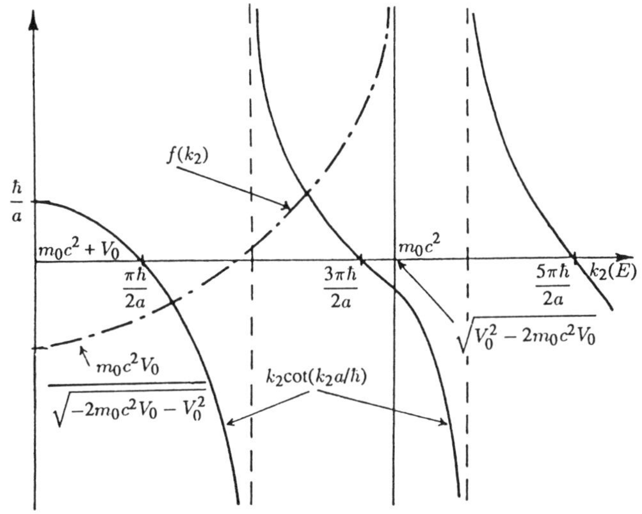

Fig. 9.3. The graphical solution of (19)

First we consider case (b). Here (12) are significantly simplified since A and C' have to vanish, so that ψ_{I} and ψ_{III} do not increase exponentially and are therefore normalizable; thus the first of (12) has the form:

$$0 = \frac{1}{4\gamma}\,\mathrm{e}^{\mathrm{i}p_1a/\hbar}C\left((1+\gamma)^2\,\mathrm{e}^{-\mathrm{i}p_2a/\hbar} - (1-\gamma)^2\,\mathrm{e}^{\mathrm{i}p_2a/\hbar}\right) \quad . \tag{14}$$

Since $C \neq 0$ (otherwise the whole wave function will be zero), we obtain

$$\frac{1+\gamma}{1-\gamma}\,\mathrm{e}^{\mathrm{i}p_2a/\hbar} = \frac{1-\gamma}{1+\gamma}\,\mathrm{e}^{\mathrm{i}p_2a/\hbar} \quad . \tag{15}$$

Example 9.1.

As long as p_2 is real (in Fig. 9.1 for $m_0c^2 + V_0 < E < m_0c^2$, in Fig. 9.2 in the whole domain $-m_0c^2 < E < m_0c^2$), γ is imaginary (since p_1 is imaginary). Thus (15) means that

$$\left[\frac{1+\gamma}{1-\gamma}\,\mathrm{e}^{-\mathrm{i}p_2a/\hbar}\right]^* = \frac{1+\gamma}{1-\gamma}\,\mathrm{e}^{-\mathrm{i}p_2a/\hbar} \tag{16}$$

whereby "*" denotes complex conjugation, or

$$\mathrm{Im}\left(\frac{1+\gamma}{1-\gamma}\,\mathrm{e}^{-\mathrm{i}p_2a/\hbar}\right) = 0 \quad . \tag{17}$$

With $\gamma = \mathrm{i}\Gamma$ ($\Gamma \in \mathbb{R}$) follows

$$\frac{2\Gamma}{1-\Gamma^2} = \tan\left(p_2\frac{a}{\hbar}\right) \quad . \tag{18}$$

Inserting Γ form (9), we finally have

$$cp_2\cot\left(p_2\frac{a}{\hbar}\right) = f(p_2) \quad \text{with}$$

$$f(p_2) = -\frac{EV_0}{c\kappa} - \kappa c \quad , \tag{19}$$

whereby $p_1 = \mathrm{i}\kappa$. This is the equation to determine the energy eigenvalues of the bound states, which will be discussed immediately. Before this, one should briefly mention that (15) has no solution, if in Fig. 9.1

$$-m_0c^2 < E < m_0c^2 + V_0 \quad ,$$

i.e. if p_2 is imaginary:

$$\frac{(1+\gamma)^2}{(1-\gamma)^2}\,\mathrm{e}^{2\kappa_2a/\hbar} \neq 1 \quad (p_2 = \mathrm{i}\kappa_2) \quad , \tag{20}$$

because $(1+\gamma)^2 > 1$, $(1-\gamma)^2 < 1$, $\mathrm{e}^{2\kappa_2a/\hbar} > 1$, which is the reason why the wave functions in this domain vanish identically. The condition (19) can be approximately solved graphically (take a look at Fig. 9.3); this is sufficient to get a qualitative overview about the behaviour of bound states. Let us do this first; afterwards we shall obtain the exact energy spectrum for bound states by numerical solution of (19).

First we note that

$$y := cp_2\cot\left(p_2\frac{a}{\hbar}\right) = \frac{\hbar c}{a}\left(p_2\frac{a}{\hbar}\right)\cot\left(p_2\frac{a}{\hbar}\right) \xrightarrow[p_2\to 0]{} \frac{\hbar c}{a} \quad ,$$

$$f(p_2) \xrightarrow[p_2\to 0]{} \frac{m_0c^2V_0}{\sqrt{-2m_0c^2V_0 - V_0^2}} \quad ,$$

$$f(p_2) \xrightarrow[cp_2\to\sqrt{V_0^2-2m_0c^2V_0}]{} +\infty \quad (\text{corresponding to } E \to +m_0c^2) \quad . \tag{21}$$

In the case of $V_0 < 2m_0c^2$, $f(p_2 \to 0)$ becomes imaginary. This looks, if graphically presented, somewhat like like Fig. 9.3. We note immediately, that a bound state always exists, independent of the depth and width of the potential well. This is

Example 9.1.

in accordance with the corresponding nonrelativistic problem, but is opposite to the corresponding three-dimensional problem. Because of the angular momentum barrier occuring in the latter case, not every three-dimensional potential well has a bound state, but only those below a certain depth V_0.

One can solve (19) numerically and thus determine the energy spectrum of the bound solutions for several sets of parameters V_0, a. This is depicted in Fig. 9.4 for the case of $a = 10\lambda\!\!\!^{-}_{e}$. We note that, with increasing V_0, more and more states appear. For $V_0 \simeq -2.04 m_0c^2$ the well becomes supercritical, the lowest bound states enter the lower continuum and can there be realized as a resonance in the transmission coefficient (s. b.).

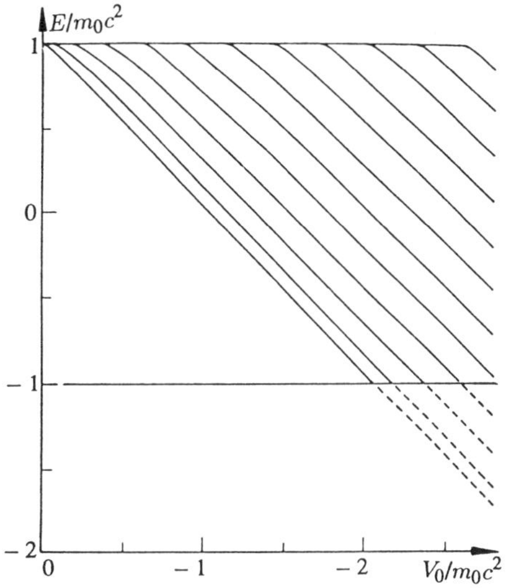

Fig. 9.4. Eigenvalue spectrum of bound electrons in a one-dimensional potential well of width $a = 10\lambda\!\!\!^{-}_{e}$. The energies of the dived states corresponding to resonances are depicted by *dashed lines*, as they can be extracted from the maxima of the transmission coefficients (see Fig. 9.5)

Now we want to consider the scattering states. Again there are several domains:

1. p_2 and γ are real. This is the case for $E > m_0c^2$ and for $E < -m_0c^2 + V_0$. In the overcritical case there is an additional domain $m_0c^2 + V_0 < E < -m_0c^2$.
2. p_2 and γ are imaginary. Obviously this is the case for $V_0 - m_0c^2 < E < V_0 + m_0c^2$.

We will discuss both cases successively. First we make use of the possibility that we can choose one of the coefficients A, A', C, C' freely. We assume that from the rhs no wave enters the potential; thus $C' = 0$ and C is interpretable as that part resulting from a wave with amplitude A which arrives from the left, travelling through the potential pocket or well. The term proportional to A' stems from the wave reflected at the potential. Now we can define a transmission coefficient T and a phase shift δ by

$$\frac{C}{A} = \sqrt{T}\,\mathrm{e}^{-\mathrm{i}\delta} \quad , \tag{22}$$

i.e. the amplitude of the outcoming wave is reduced by a factor $\sqrt{T}$ and shifted by the phase δ compared to the wave impacting from the left. From (12) we obtain for real p_2:

Example 9.1.

$$T = \left|\frac{C}{A}\right|^2 = \left[\cos^2\left(p_2\frac{a}{\hbar}\right) + \left(\frac{1+\gamma^2}{2\gamma}\right)^2 \sin^2\left(p_2\frac{a}{\hbar}\right)\right]^{-1}$$
$$= \left[1 + \left(\frac{1-\gamma^2}{2\gamma}\right)^2 \sin^2\left(p_2\frac{a}{\hbar}\right)\right]^{-1} \le 1 \quad . \tag{23}$$

The phase follows from [see (12)]

$$\frac{1}{\sqrt{T}} = \mathrm{e}^{\mathrm{i}\delta}\,\mathrm{e}^{-(\mathrm{i}/\hbar)p_1 a} = \cos\left(p_2\frac{a}{\hbar}\right) - \mathrm{i}\frac{1+\gamma^2}{2\gamma}\sin\left(p_2\frac{a}{\hbar}\right) \tag{24}$$

and

$$\frac{1}{\sqrt{T}}\left[\mathrm{i}\sin\left(\delta - \frac{p_1 a}{\hbar}\right) + \cos\left(\delta - \frac{p_1 a}{\hbar}\right)\right]$$
$$= \cos\left(\frac{p_2 a}{\hbar}\right) - \mathrm{i}\frac{1+\gamma^2}{2\gamma}\sin\left(\frac{p_2 a}{\hbar}\right) \quad . \tag{25}$$

The separation of real and imaginary parts and the elimination of $\sqrt{T}$ yields

$$-\tan\left(\frac{p_1 a}{\hbar} - \delta\right) = -\frac{1+\gamma^2}{2\gamma}\tan\left(\frac{p_2 a}{\hbar}\right) \quad , \tag{26}$$

$$\delta = \left(\frac{p_1 a}{\hbar}\right) - \arctan\left(\frac{1+\gamma^2}{2\gamma}\tan\left(\frac{p_2 a}{\hbar}\right)\right) \quad . \tag{27}$$

If, however, p_2 is imaginary, then instead of (23) we get from (12) the transmission coefficient

$$T = \left[1 + \left(\frac{1+\Gamma^2}{2\Gamma}\right)^2 \sinh^2\left(\frac{\kappa_2 a}{\hbar}\right)\right]^{-1} \le 1 \tag{28}$$

(whereby $\gamma = \mathrm{i}\Gamma$ and $p_2 = \mathrm{i}\kappa_2$ as before), and instead of (27) the phase shift reads

$$\delta = \left(\frac{p_1 a}{\hbar}\right) - \arctan\left(\frac{1-\gamma^2}{2\Gamma}\tanh\left(\frac{\kappa_2 a}{\hbar}\right)\right) \quad . \tag{29}$$

In Fig. 9.5 the transmission coefficient for a potential of the depth $V_0 = -3m_0c^2$ and the width $a = 300\hbar/m_0c$ is depicted. Now we choose an overcritical potential, since the undercritical case differs only by the omission of the domain $m_0c^2 + V_0 < E < -m_0c^2$. We have significant structures of resonance in the electron continuum for $E > m_0c^2$ and in the positron continuum for positron energies above the potential barrier $|E| > |V_0 - m_0c^2|$; $(E < -m_0c^2 + V_0)$. Positrons with lower kinetic energy $(-m_0c^2 + V_0 < E < m_0c^2 + V_0)$ only penetrate the barrier with a probability which decreases exponentially with the width a of the barrier. Hence T is about zero in our case. In the domain $m_0c^2 + V_0 < E < -m_0c^2$, however, there is the possibility that the incoming wave meets an overcritical, quasi-bound state, and thus penetrates the potential domain more or less unhindered. At the point where by extrapolation of the spectrum of the bound states in Fig. 9.4 one would expect the quasi-bound state, T is equal to 1. The dived bound state in this way becomes perceptible as a resonance in the scattering spectrum below $E = -m_0c^2$.

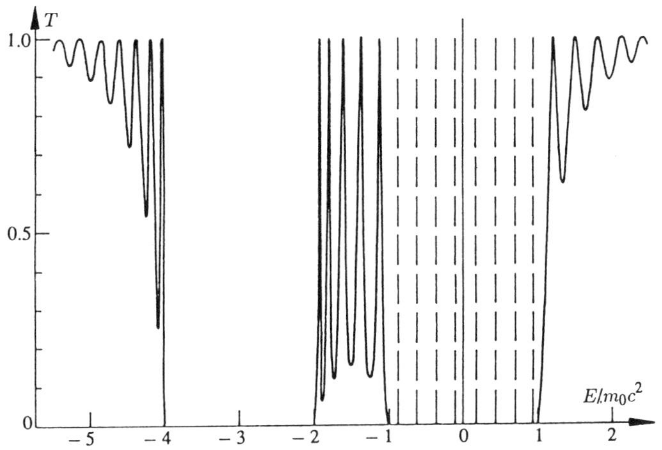

Fig. 9.5. The transmission coefficient for scattered states of a one-dimensional square-well potential of depth $V_0 = -3m_0c^2$ and width $a = 10\lambda\!\!\!^{-}_e$. The energies of the bound states are depicted by *dashed lines*

These resonances do not exist for subcritical potentials; their interpretation as a signature for spontaneous pair creation has already been discussed in the qualitative discussion above.[1]

Finally we look at the scattering phase in Fig. 9.6. As can already be seen from (23), (27), (28), (29), one gets $T = 1$ for $\delta - p_1(a/\hbar) = 0 (\mathrm{mod}\,\pi)$, and T becomes minimal for $\delta - p_1(a/\hbar) = \frac{\pi}{2} (\mathrm{mod}\,\pi)$. If T is minimal the reflection coefficient becomes maximal. Hence this statement is in agreement with the well-known statement from scattering theory that the scattering cross-section becomes maximal if the scattering phase passes through $\frac{\pi}{2}$. ("Scattering" in the one-dimensional case is identified with reflection at the potential well.)

Thus in the domain $m_0c^2 + V_0 < E < -m_0c^2$ we can determine the energies of the dived states, where $(\delta - p_1a/\hbar) = 0 (\mathrm{mod}\,\pi)$. These energies lie exactly where one would expect them by extrapolation of the bound spectrum (see Fig. 9.4).

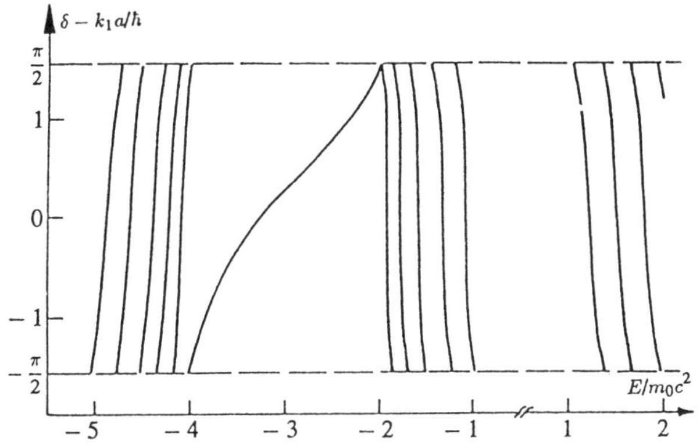

Fig. 9.6. The phase shifts of scattering states of a one-dimensional well of depth $V_0 = -3m_0c^2$ and with width $a = 10\lambda\!\!\!^{-}_e$ as a function of energy E. The energy E is given in units of m_0c^2

[1] For greater detail see W. Greiner, B. Müller, J. Rafelski: *Quantum Electrodynamics of Strong Fields* (Springer, Berlin, Heidelberg, New York 1985).

EXERCISE

9.2 Eigenvalues of the Dirac Equation in a One-Dimensional Square Potential Well with Scalar Coupling

Problem. Investigate the eigenvalues of the Dirac equation in a one-dimensional square-well potential, with depth V_0 and width a, if this potential is not coupled as a time-like component of a four-vector, but like a scalar.

Solution. As in Example 9.1 we define three domains I($z < a/2$), II($-a/2 \leq z \leq a/2$) and III($z > a/2$). In these cases the Dirac equation takes the following respective forms:

$$\begin{aligned}
\text{I, III}: &\quad \left(\hat{\alpha}\cdot\hat{p}c + \hat{\beta}m_0c^2\right)\psi = E\psi \quad , \\
\text{II}: &\quad \left(\hat{\alpha}\cdot\hat{p}c + \hat{\beta}(m_0c^2 + V_0)\right)\psi = E\psi \quad (V_0 < 0) \quad .
\end{aligned} \tag{1}$$

In contrast to the vector coupling, which in domain II leads to the replacement $E \to E - V_0$, we now must replace m_0c^2 by $m_0c^2 + V_0$ in II. While vector coupling acts differently on electron and positron states, respectively (if electrons in the square well are attracted, positrons are repelled, and vice versa) and thus the eigenvalue spectrum is not symmetric (bound states exist for only one of the two kinds of particles), scalar coupling acts equally on particles and antiparticles. Alternatively, one can say: For vector coupling the potential couples to the charge (which is different for particles and antiparticles); for scalar coupling the potential couples to the mass (which is equal for both particles and antiparticles). In the latter case we thus expect a symmetrical energy eigenvalue spectrum, i.e. for both electrons and positrons there will exist bound states. Hence we expect a supercritical behaviour even for $V_0 \leq -m_0c^2$. In this case, in principle electron and positrons states can cross. What happens then will be discussed in detail a little later. Now, though, let us proceed analogously to the case of vector coupling. Again the momentum in the regions I, II is given by

$$p_1^2 = \frac{E^2}{c^2} - m_0c^2 \quad . \tag{2}$$

In domain II, however, we get

$$p_2^2 = \frac{E^2}{c^2} - \left(m_0c + \frac{V_0}{c}\right)^2 \quad . \tag{3}$$

Again we can write down the conditions for continuity of the wave functions at $z = -a/2$ and $z = a/2$. Analogously to the case of vector coupling, we define

$$\gamma \equiv \frac{p_1c}{E + m_0c^2}\,\frac{E + m_0c^2 + V_0}{p_2c} = \sqrt{\frac{(E - m_0c^2)(E + m_0c^2 + V_0)}{(E + m_0c^2)(E - m_0c^2 - V_0)}} \quad . \tag{4}$$

One should note that $\gamma \to 1/\gamma$ if $E \to -E$. Using the same notation as in the case of vector coupling, we can work with (12) from Exercise 9.1, if we observe that γ is now given by (4):

Exercise 9.2.

$$\left[\begin{pmatrix} A \\ A' \end{pmatrix} = \frac{1}{4\gamma} \times \begin{pmatrix} (1+\gamma)^2 \mathrm{e}^{\mathrm{i}(p_1-p_2)a/\hbar} - (1-\gamma)^2 \mathrm{e}^{\mathrm{i}(p_1+p_2)a/\hbar} & (1-\gamma^2)(\mathrm{e}^{-\mathrm{i}p_2a/\hbar} - \mathrm{e}^{\mathrm{i}p_2a/\hbar}) \\ -(1-ga^2)(\mathrm{e}^{-\mathrm{i}p_2a/\hbar} - \mathrm{e}^{\mathrm{i}p_2a/\hbar}) & (1+\gamma)^2 \mathrm{e}^{-\mathrm{i}(p_1-p_2)a/\hbar} - (1-\gamma)^2 \mathrm{e}^{-\mathrm{i}(p_1+p_2)a/\hbar} \end{pmatrix} \times \begin{pmatrix} C \\ C' \end{pmatrix}\right] . \tag{5}$$

We divide the further discussion into several steps:

(1) $|E| < m_0c^2$; bound states
(2) $|E| > m_0c^2$; scattering states in the bound electron or positron continuum.

Let us first consider the bound states. For these the following eigenvalue equation holds:

$$\frac{1+\gamma}{1-\gamma}\,\mathrm{e}^{-\mathrm{i}p_2a/\hbar} = \frac{1-\gamma}{1+\gamma}\,\mathrm{e}^{\mathrm{i}p_2a/\hbar} . \tag{6}$$

This equation only has solutions if p_2 is real, i.e. $|E| > |+m_0c^2+V_0|$ [$E > m_0c^2+V_0$ for electrons, $E < -(m_0c^2+V_0)$ for positrons]. Thus, it again follows from (6) that

$$\tan\left(p_2\frac{a}{\hbar}\right) = \frac{2\Gamma}{1-\Gamma^2} , \quad \text{with} \tag{7}$$

$$\gamma = \mathrm{i}\Gamma \quad (\Gamma \in \mathbb{R}) \quad \text{and}$$

$$cp_2 \cot\left(p_2\frac{a}{\hbar}\right) = -\left(\frac{m_0c^2V_0}{\kappa_1 c} + \kappa_1 c\right) =: f(p_2) , \tag{8}$$

whereby $p_1 = \mathrm{i}\kappa_1$. Of course this equation is symmetric in $\pm E$, too. Furthermore the following relations hold:

$$f(p_2) = \frac{m_0c^2V_0 + V_0^2 + p_2^2c^2}{\sqrt{-2m_0c^2V_0 - V_0^2 - p_2^2c^2}} \quad \left(0 \le p_2 < \sqrt{-2m_0c^2V_0 - V_0^2}\right) . \tag{9}$$

$$f(p_2) \xrightarrow[p_2\to 0]{} \frac{m_0c^2V_0 + V_0^2}{\sqrt{-2m_0c^2V_0 - V_0^2}} \begin{cases} < 0 & \text{for } |V_0| < m_0c^2 \\ > 0 & \text{for } |V_0| > m_0c^2 \\ \to \infty & \text{for } V_0 \to -2m_0c^2 \end{cases} , \tag{10}$$

$$f(p_2) \xrightarrow[|E|\to m_0c^2]{} +\infty \quad \left(\text{corresponding to } p_2 \to \sqrt{-2m_0V_0 - V_0^2/c^2}\right) . \tag{11}$$

Since $V_0 < m_0c^2$, the graphical solution of (9) looks similar to the case of vector coupling (see Fig. 9.7), except that one has to imagine Fig. 9.3 in Example 9.1 to be continued to negative energies in a symmetric way.

If now $|V_0|$ increase, the intersection point of $f(p_2)$ with the ordinate in Fig. 9.7 is shifted upwards. Correspondingly the lowest eigenvalue decreases to smaller values of p_2. In case of

$$\frac{(m_0c^2V_0 + V_0^2)}{\sqrt{-2m_0c^2V_0 - V_0^2}} > \frac{\hbar c}{a} , .$$

Fig. 9.7. Graphical solution of (8)

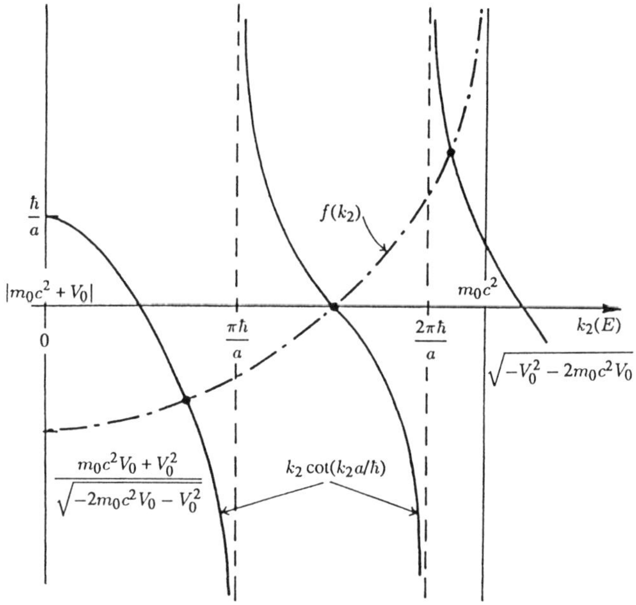

the lowest eigenvalue vanishes, namely at $p_2 = 0$ or $|E| = |m_0c^2 + V_0| > 0$. Consequently the deepest bound state never reaches the energy $E = 0$ and the corresponding electron and positron states never overlap. This does however not mean that the eigenstate does not exist any longer for larger values of $|V_0|$. Rather, now there exists a solution of (6) with imaginary momentum $p_2 = \mathrm{i}\kappa_2$, $\kappa_2 = \sqrt{(m_0c + V_0/c)^2 + E^2/c}$. The transcendental eigenvalue equation analogous to (8) reads in this case

$$c\kappa_2 \coth\left(\frac{\kappa_2 a}{\hbar}\right) = -\left(\frac{m_0c^2V_0}{\kappa_1 c} + \kappa_1 c\right) =: \tilde{f}(\kappa_2) \quad .$$

A graphical construction similar to the one in Fig. 9.1 reveals that there is always exactly one solution for

$$\frac{m_0c^2V_0 + V_0^2}{\sqrt{-2m_0c^2V_0 - V_0^2}} > \frac{\hbar c}{a} \quad .$$

For $V_0 \to \infty$ it monotonically decreasingly approaches the value $\kappa_2 \to -m_0c - V_0/c$ corresponding to $E \to \pm 0$.

If we further increase $|V_0|$, the ordinate cut of $f(p_2)$ still moves upwards and the higher states approach the eigenvalues $p_2 = n\pi\hbar/a$ with the eigenenergies $E^2 = (n\pi/a)^2(\hbar c)^2 + (m_0c^2 + V_0)^2$. For $|V_0| \to 2m_0c^2$, the quantity $f(p_2 = 0)$ diverges and the eigenenergies come close to the value $E^2 = (n\pi/a)^2(\hbar c)^2 + m_0^2c^4 > m_0^2c^4$. That means no more bound states exist for $|V_0| \to 2m_0c^2$; with increasing potential depth all bound states vanish, one after another. This behaviour is illustrated in Fig. 9.8, which was obtained by solving (8) numerically. The energy diagram shown is characteristic for the square-well potential. [One can also consider the same problem in three dimensions with a Coulomb-like potential of the form $-\alpha'/r$

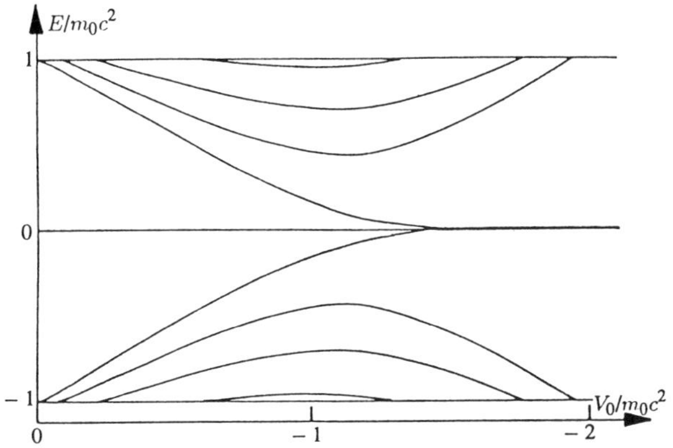

Fig. 9.8. Spectrum of eigenvalues of the Dirac equation with a one-dimensional square-well potential of width $a = 10\lambda\!\!\!^{-}_e$ as a function of the potential strength V_0 (scalar coupling)

(see Example 9.8); then, all states in the diagram are conserved for arbitrary high coupling strength α', too. However, their binding energies approach the value $|E| = 0$ only asymptotically (cf. Fig. 9.9) and also in this case electron and positron states do not cross.]

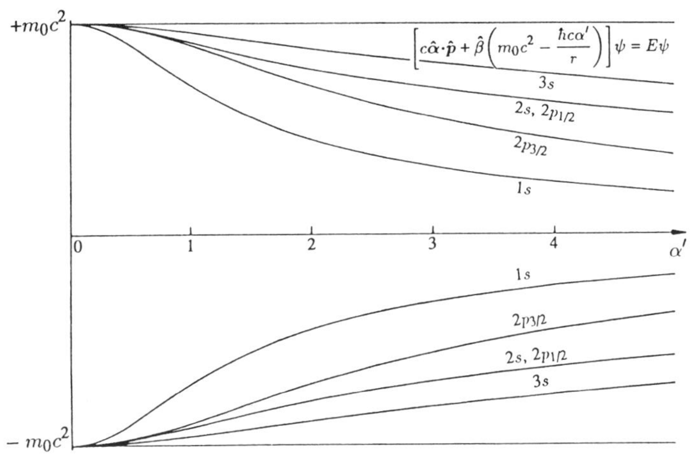

Fig. 9.9. Eigenvalue spectrum of the three-dimensional Dirac equation with α'/r potential (scalar coupling)

Therefore we find, in contrast to vector coupling, that in the case of scalar coupling spontaneous $e^+ - e^-$ pair creation never occurs, no matter how strong the potential chosen. This qualitatively different behaviour of the bound states in case of the α'/r potential is easily understood in the following way: Due to the scalar potential the electrons obtain an effictive mass $m_{\text{eff}}c^2 = m_0c^2 + V_0$. Figure 9.10 schematically shows this effective mass as a function of r for the α'/r potential, together with some bound states.

One sees that a region with $m_{\text{eff}} < m_0$ always exists, so that bound states are always possible (i.e. for all values of the coupling strength α'). With increasing parameter α' the wave functions are shifted to larger values of r. Simultaneously the potential bag is broadened and hence the energy eigenvalues $|E|$ decrease and become zero in the limit $\alpha' \to \infty$. Since $m_{\text{eff}} \geq 0$ always holds, then $|E| > 0$ is always valid, too, and electron and positron states can never cross. The energy

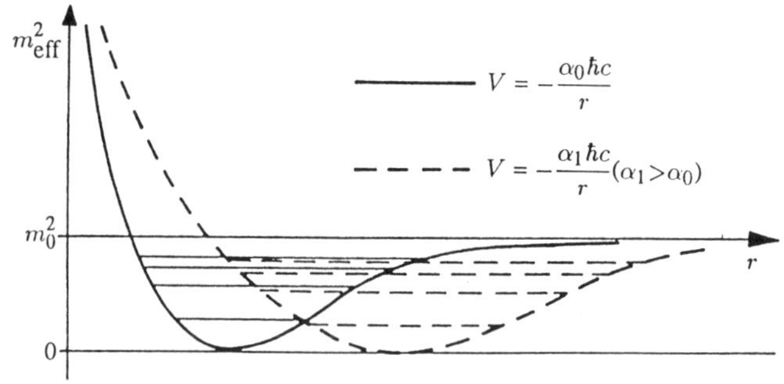

Fig. 9.10. Square of the effective mass $(m_{\rm eff}c^2)^2 = [m_0c^2 + V(r)]^2$ as a function of the radius r

gap between electrons and positrons becomes smaller and smaller, but never zero; therefore, $e^+ - e^-$ pair creation without energy (i.e. spontaneous pair creation) can never occur (see Figs. 9.8 and 9.9).

Finally we give a brief qualitative discussion of the scattering states $|E| > m_0c^2$. Again we consider the case $C' = 0$ and study the transmission coefficient $T = |C/A|^2$. Of course, the relation $|E| > m_0c^2 + V_0$ also holds in the energy region $|E| > m_0c^2$. For $V_0 < -2m_0c^2$ the electrons feel a potential barrier if they lie in the energy interval $m_0c^2 < E < |m_0c^2 + V_0|$ (similar statements hold for positrons). While for $|V_0| < 2m_0c^2$ resonances occur in both continua (in complete analogy to Fig. 9.5 in Example 9.1), in the region $E > m_0c^2$ one obtains for $|V_0| < 2m_0c^2$ an additional energy region, where electrons and positrons feel a potential barrier, and the transmission coefficient becomes very small (corresponding to the behaviour in the energy interval $-m_0c^2 + V_0 < E < m_0c^2 + V_0$ in Fig. 9.5).

EXAMPLE

9.3 Separation of the Variables for the Dirac Equation with Central Potential (Minimally Coupled)

The Dirac-Hamiltonian in this case reads

$$\hat{H}_{\rm D} = c\hat{\boldsymbol{\alpha}} \cdot \hat{\boldsymbol{p}} + \hat{\beta} m_0 c^2 + V(r) \quad , \tag{1}$$

where $V(r) = eA_0(r)$. Because of the spherical symmetry of the field,[2] the angular-momentum operator $\hat{\boldsymbol{J}}$ and the parity operator $\hat{P} = \mathrm{e}^{\mathrm{i}\varphi}\hat{\beta}(\boldsymbol{x} \to -\boldsymbol{x}) = \mathrm{e}^{\mathrm{i}\varphi}\gamma^0 \hat{P}_0$ [see (4.9)] with respect to the origin of the coordinate system commute with the Hamiltionian. Hence states with definite energy, angular momentum and parity occur. The corresponding wave functions are denoted by

$$\psi_{jm} = \begin{pmatrix} \varphi_{jlm}(\boldsymbol{x}, t) \\ \chi_{jl'm}(\boldsymbol{x}, t) \end{pmatrix} \quad . \tag{2}$$

Here φ_{jlm} and $\chi_{jl'm}$ are two-spinors which are to be determined. Snce ψ_{jm} must have good parity and the parity operator reads $\hat{P} = \mathrm{e}^{\mathrm{i}\varphi}\hat{\beta}\hat{P}_0$ ($\hat{P}_0$ changes $\boldsymbol{x}$ into $-\boldsymbol{x}$), we have

[2] See W. Greiner, B. Müller: *Quantum Mechanics – Symmetries*, 2nd ed. (Springer, Berlin, Heidelber 1994).

Example 9.3.

$$\psi'_{jm}(\boldsymbol{x}',t') \stackrel{!}{=} \lambda\psi_{jm}(\boldsymbol{x}',t') \quad \text{or} \quad \hat{P}\psi_{jm}(x') \stackrel{!}{=} \lambda\psi_{jm}(x') \quad , \tag{3a}$$

$$\begin{aligned} \hat{P}\psi_{jm} &= \mathrm{e}^{\mathrm{i}\varphi}\begin{pmatrix} 1\!\!1 & 0 \\ 0 & -1\!\!1 \end{pmatrix}\hat{P}_0\begin{pmatrix} \psi_{jlm}(\boldsymbol{x},t) \\ \chi_{jl'm}(\boldsymbol{x},t) \end{pmatrix} = \mathrm{e}^{\mathrm{i}\varphi}\begin{pmatrix} \hat{P}_0\varphi_{jlm}(\boldsymbol{x},t) \\ -\hat{P}_0\chi_{jl'm}(\boldsymbol{x},t) \end{pmatrix} \\ &\stackrel{!}{=} \lambda\psi_{jm} = \lambda\begin{pmatrix} \varphi_{jlm}(\boldsymbol{x},t) \\ \chi_{jl'm}(\boldsymbol{x},t) \end{pmatrix} \quad , \end{aligned} \tag{3b}$$

where $|\lambda| - 1$ [in (3b) we have replaced $\boldsymbol{x}'$ by $\boldsymbol{x}$]. One should clearly recognize the content of (3a): the demand for "good parity" of the wave function means that the wave function is an eigenfunction of the parity operator $\hat{P}$ [see (4.9)]. Equation (3b) shows that the parity of the two-spinor φ_{jlm} must be equal to the negative parity of $\chi_{jl'm}$. We can also understand this statement in the following way: Starting with the starionary Dirac equation $\hat{H}_{\mathrm{D}}\psi = E\psi$, we get with

$$\hat{\boldsymbol{\alpha}} = \begin{pmatrix} 0 & \hat{\boldsymbol{\sigma}} \\ \hat{\boldsymbol{\sigma}} & 0 \end{pmatrix}$$

that

$$\begin{aligned} c\,(\hat{\boldsymbol{\sigma}}\cdot\hat{\boldsymbol{p}})\,\chi + m_0c^2\varphi + V\varphi &= E\varphi \quad , \\ c\,(\hat{\boldsymbol{\sigma}}\cdot\hat{\boldsymbol{p}})\,\varphi - m_0c^2\chi + V\chi &= E\chi \quad , \end{aligned} \tag{4}$$

or

$$\begin{aligned} \left(E - m_0c^2 - V\right)\varphi &= c\,(\hat{\boldsymbol{\sigma}}\cdot\hat{\boldsymbol{p}})\,\chi \quad , \\ \left(E + m_0c^2 - V\right)\chi &= c\,(\hat{\boldsymbol{\sigma}}\cdot\hat{\boldsymbol{p}})\,\varphi \quad . \end{aligned} \tag{5}$$

Since the operator $\hat{\boldsymbol{\sigma}}\cdot\hat{\boldsymbol{p}}$ changes parity, these equations show that the two spinors φ and χ must have opposite parity.

Eigenfunctions of the angular momentum and the parity operator are the well-known spherical spinors. To avoid confusion with the complete wave function Ψ, we shall denote the spherical spinors here by Ω_{jlm}. They are defined by

$$\Omega_{jlm} = \sum_{m',m_s}\left(l\tfrac{1}{2}j\,|m'm_sm\right)Y_{lm'}\chi_{\frac{1}{2}m_s} \quad . \tag{6}$$

Here the two-spinors $\chi_{\frac{1}{2}m_s}$ are eigenfunctions of the spin operators $\hat{\boldsymbol{S}}^2 = \hbar^2\hat{\sigma}^2/4$ and $\hat{S}_3 = \hbar\hat{\sigma}_3/2$; they read explicitly:

$$\chi_{\frac{1}{2}\frac{1}{2}} = \begin{pmatrix} 1 \\ 0 \end{pmatrix} \quad , \quad \chi_{\frac{1}{2}-\frac{1}{2}} = \begin{pmatrix} 0 \\ 1 \end{pmatrix} \quad .$$

The parity of Ω_{jlm} is given by Y_{lm}:

$$\hat{P}_0\Omega_{jlm} = (-1)^l\,\Omega_{jlm} \quad . \tag{7}$$

We make the following ansatz

$$\begin{aligned} \varphi_{jlm} &= \mathrm{i}g(r)\Omega_{jlm}\left(\frac{\boldsymbol{r}}{r}\right) \quad , \\ \chi_{jl'm} &= -f(r)\Omega_{jl'm}\left(\frac{\boldsymbol{r}}{r}\right) \end{aligned} \tag{8}$$

with

Example 9.3.

$$l' = 2j - l =: \begin{cases} 2\left(l+\frac{1}{2}\right) - l = l+1 & \text{for } j = l+\frac{1}{2} \\ 2\left(l-\frac{1}{2}\right) - l = l-1 & \text{for } j = l-\frac{1}{2} \end{cases} \quad . \tag{9}$$

Let us repeat this once more: Because of (6) either $j = l + \frac{1}{2}$ or $j = l - \frac{1}{2}$. If $j = l + \frac{1}{2}$, the orbital angular momentum l' of $\chi_{jl'm}$ is $l' = l + 1$. This is the only way to realize the opposite parity of χ compared to φ. The value $l' = l - 1$ must be excluded, because no total angular momentum $j = l + \frac{1}{2}$ can be constructed by $l' = l - 1$ and $S = \frac{1}{2}$. The arguments follow a similar pattern for the second case of (9). One has

$$\begin{aligned} \hat{\boldsymbol{\sigma}} \cdot \hat{\boldsymbol{p}} \varphi_{jlm} &= \hat{\boldsymbol{\sigma}} \cdot \hat{\boldsymbol{p}} \big(\mathrm{i}g(r)\big) \Omega_{jlm} \left(\frac{\boldsymbol{r}}{r}\right) = \big(\hat{\boldsymbol{\sigma}} \cdot \hat{\boldsymbol{p}} \mathrm{i}g(r)\big) \Omega_{jlm} + \mathrm{i}g(r) \hat{\boldsymbol{\sigma}} \cdot \hat{\boldsymbol{p}} \Omega_{jlm} \\ &= \hbar \frac{dg(r)}{dr} \left(\hat{\boldsymbol{\sigma}} \cdot \frac{\boldsymbol{r}}{r}\right) \Omega_{jlm} + \mathrm{i}g(r) \hat{\boldsymbol{\sigma}} \cdot \hat{\boldsymbol{p}} \Omega_{jlm} \quad . \end{aligned} \tag{10}$$

With respect to (6) the spherical spinors are eigenfunctions of the operators $\hat{\boldsymbol{L}}^2$, $\hat{\boldsymbol{J}}^2$ and $\hat{\boldsymbol{S}}^2 = \left(\frac{1}{2}\hat{\boldsymbol{\sigma}}\right)^2$ with eigenvalues $l'(l'+1)\hbar^2$, $j(j+1)\hbar^2$ and $\frac{3}{4}\hbar^2$ respectively. To be complete we once more give the explicit form of the Ω_{jlm} for the useful cases $j = l + \frac{1}{2}$ and $j = l - \frac{1}{2}$ $\left(j \geq \frac{1}{2}\right)$:

$$\Omega_{\underbrace{l+\frac{1}{2}}_{j},l,m} = \begin{pmatrix} \sqrt{\dfrac{j+m}{2j}} Y_{l,m-\frac{1}{2}} \\ \sqrt{\dfrac{j-m}{2j}} Y_{l,m+\frac{1}{2}} \end{pmatrix} \quad , \tag{11a}$$

$$\Omega_{\underbrace{l-\frac{1}{2}}_{j},l,m} = \begin{pmatrix} -\sqrt{\dfrac{j-m+1}{2j+2}} Y_{l,m-\frac{1}{2}} \\ \sqrt{\dfrac{j+m+1}{2j+2}} Y_{l,m+\frac{1}{2}} \end{pmatrix} \quad . \tag{11b}$$

The root factors (11) are the Clebsch–Gordon coefficients in explicit form. Now we make use of the following relation between the spherical spinors

$$\left(\hat{\boldsymbol{\sigma}} \cdot \frac{\boldsymbol{r}}{r}\right) \Omega_{jlm} = -\Omega_{jl'm} \quad , \tag{12}$$

which is easily proven, because $(\hat{\boldsymbol{\sigma}} \cdot \boldsymbol{r}/r)$ is a scalar operator of negative parity [see (10.54) and the following]. With (12) we get

$$-(\hat{\boldsymbol{\sigma}} \cdot \hat{\boldsymbol{p}}) \Omega_{jlm} = (\hat{\boldsymbol{\sigma}} \cdot \hat{\boldsymbol{p}}) \left(\hat{\boldsymbol{\sigma}} \cdot \frac{\boldsymbol{r}}{r}\right) \Omega_{jl'm} \quad . \tag{13}$$

Now we take the already familiar relation

$$(\hat{\boldsymbol{\sigma}} \cdot \boldsymbol{A})(\hat{\boldsymbol{\sigma}} \cdot \boldsymbol{B}) = \boldsymbol{A} \cdot \boldsymbol{B} + \mathrm{i}\hat{\boldsymbol{\sigma}} \cdot (\boldsymbol{A} \times \boldsymbol{B}) \tag{14}$$

to change (13) into

$$-(\hat{\boldsymbol{\sigma}} - \hat{\boldsymbol{p}}) \Omega_{jlm} = \left(\hat{\boldsymbol{p}} \cdot \frac{\boldsymbol{r}}{r} + \mathrm{i}\hat{\boldsymbol{\sigma}} \cdot \left(\hat{\boldsymbol{p}} \times \frac{\boldsymbol{r}}{r}\right)\right) \Omega_{jl'm} \quad . \tag{15}$$

With $\hat{\boldsymbol{p}} = -\mathrm{i}\hbar\boldsymbol{\nabla}$ and $\hat{\boldsymbol{L}} = \boldsymbol{r} \times \hat{\boldsymbol{p}}$, (15) can further be transformed into

Example 9.3.

$$\begin{aligned}
&\left(\hat{\boldsymbol{p}}\cdot\boldsymbol{r}+\mathrm{i}\hat{\boldsymbol{\sigma}}\cdot(\hat{\boldsymbol{p}}\times\boldsymbol{r})\right)\frac{1}{r}\Omega_{jl'm}\\
&\quad=\left(-\mathrm{i}\hbar(\boldsymbol{\nabla}\cdot\boldsymbol{r})-\mathrm{i}\hbar\boldsymbol{r}\cdot\boldsymbol{\nabla}-\mathrm{i}\hat{\boldsymbol{\sigma}}\cdot(\boldsymbol{r}\times\hat{\boldsymbol{p}})\right)\frac{1}{r}\Omega_{jl'm}\\
&\quad=\left(-\mathrm{i}\hbar\frac{3}{r}-\mathrm{i}\hbar r\left(-\frac{1}{r^2}\right)-\mathrm{i}\frac{\hat{\boldsymbol{\sigma}}\cdot\hat{\boldsymbol{L}}}{r}\right)\Omega_{jl'm}\\
&\quad=-\mathrm{i}\left(\frac{2\hbar}{r}+\frac{1}{r}\hat{\boldsymbol{L}}\cdot\hat{\boldsymbol{\sigma}}\right)\Omega_{jl'm}\quad .
\end{aligned}\tag{16}$$

From

$$\hat{\boldsymbol{J}}^2=\left(\hat{\boldsymbol{L}}+\frac{\hbar}{2}\hat{\boldsymbol{\sigma}}\right)^2=\hat{\boldsymbol{L}}^2+\left(\frac{\hbar}{2}\hat{\boldsymbol{\sigma}}\right)^2+\hbar\hat{\boldsymbol{\sigma}}\cdot\hat{\boldsymbol{L}}\tag{17}$$

follows

$$\begin{aligned}
\hbar\hat{\boldsymbol{L}}\cdot\hat{\boldsymbol{\sigma}}\Omega_{jl'm}&=\left(\hat{\boldsymbol{J}}^2-\hat{\boldsymbol{L}}^2-\left(\frac{\hbar}{2}\hat{\boldsymbol{\sigma}}\right)^2\right)\Omega_{jl'm}\\
&=\left\{j(j+1)-l'(l'+1)-\tfrac{3}{4}\right\}\hbar\Omega_{jl'm}\quad .
\end{aligned}\tag{18}$$

Now and in the following it is convenient to define a quantum number κ by

$$\kappa=\mp\left(j+\tfrac{1}{2}\right)=\begin{cases}-(l+1) & \text{for}\quad j=l+\frac{1}{2}\\ l & \text{for}\quad j=l-\frac{1}{2}\end{cases}\quad .\tag{19a}$$

Obviously there is always

$$|\kappa|=j+\tfrac{1}{2}\quad\text{or}\quad j=|\kappa|-\tfrac{1}{2}\quad .\tag{19b}$$

With this and taking $l'=2j-1$ into account one can rewrite the expectation value on the rhs of (18). For $j=l+\frac{1}{2}$ one gets

$$\begin{aligned}
&\left(l+\tfrac{1}{2}\right)\left(l+\tfrac{1}{2}+1\right)-\left[2\left(l+\tfrac{1}{2}\right)-l\right]\left[2\left(l+\tfrac{1}{2}\right)-l+1\right]-\tfrac{3}{4}\\
&\quad=l^2+\tfrac{1}{2}l+l+\tfrac{1}{2}l+\tfrac{1}{4}+\tfrac{1}{2}-(l+1)(l+2)-\tfrac{3}{4}\\
&\quad=l^2+2l-l^2-2l-l-2=-(l+2)=\kappa-1\quad .
\end{aligned}\tag{20a}$$

Similarly for $j=l-\frac{1}{2}$:

$$\begin{aligned}
&\left(l-\tfrac{1}{2}\right)\left(l+\tfrac{1}{2}\right)-\left[2\left(1-\tfrac{1}{2}\right)-l\right]\left[2\left(1-\tfrac{1}{2}\right)-l+1\right]-\tfrac{3}{4}\\
&\quad=\quad l^2-\tfrac{1}{4}-(l-1)l-\tfrac{3}{4}=l-1=\kappa-1\quad .
\end{aligned}\tag{20b}$$

Now (16) can be written as

$$(2\hbar+\hat{\boldsymbol{L}}\cdot\hat{\boldsymbol{\sigma}})\Omega_{jl'm}=(1+\kappa)\hbar\Omega_{jl'm}\quad .\tag{21}$$

If in (18) we had started with Ω_{jlm} we would have obtained $(2\hbar+\hat{\boldsymbol{L}}\cdot\hat{\boldsymbol{\sigma}})\Omega_{jlm}=(1-\kappa)\hbar\Omega_{jlm}$ by the same procedure. In the literature[3] the following notation is often used for (21): One writes

$$\chi_{\kappa,m}\equiv\Omega_{jlm}\quad ,\quad \chi_{-\kappa,m}\equiv\Omega_{jl'm}\tag{21a}$$

---

[3] See, for example, M.E. Rose: *Relativistic Electron Theory* (Wiley, New York, London).

Example 9.3.

and defines the operator

$$\hat{\kappa} = \hbar + \hat{\boldsymbol{L}} \cdot \hat{\boldsymbol{\sigma}} \quad , \tag{21b}$$

so that due to (21) the eigenvalue equation

$$\hat{\kappa}\chi_{\kappa,m} = -\hbar\kappa\chi_{\kappa,m} \quad , \quad \hat{\kappa}\chi_{-\kappa,m} = \hbar\kappa\chi_{-\kappa,m} \tag{21c}$$

holds, where

$$\kappa = \begin{cases} -(l+1) = -\left(j+\frac{1}{2}\right) & \text{for } j = l + \frac{1}{2} \\ l = +\left(j+\frac{1}{2}\right) & \text{for } j = l - \frac{1}{2} \end{cases} \quad \text{and } |\kappa| = j + \tfrac{1}{2} \quad .$$

Consequently, the spherical spinors of (8) can also be denoted by $\chi_{\kappa,m} = \Omega_{jlm}$ and $\chi_{-\kappa,m} = \Omega_{jl'm}$ and we can therefore write for the four-spinor in a central field

$$\begin{aligned} \psi_{jm} &= \begin{pmatrix} \varphi_{jlm}(\boldsymbol{x},t) \\ \chi_{jl'm}(\boldsymbol{x},t) \end{pmatrix} = \begin{pmatrix} \mathrm{i}g(r)\Omega_{jlm}\left(\frac{\boldsymbol{r}}{r}\right) \\ -f(r)\Omega_{jl'm}\left(\frac{\boldsymbol{r}}{r}\right) \end{pmatrix} \\ &= \begin{pmatrix} \mathrm{i}g(r)\chi_{\kappa,m} \\ -f(r)\chi_{-\kappa,m} \end{pmatrix} = +\mathrm{i}\begin{pmatrix} g(r)\chi_{\kappa,m} \\ \mathrm{i}f(r)\chi_{-\kappa,m} \end{pmatrix} \quad . \end{aligned}$$

Sometimes we shall use this alternative notation. With this and (15) and (12), equation (10) finally takes the form

$$\hat{\boldsymbol{\sigma}} \cdot \hat{\boldsymbol{p}}\varphi_{jlm} = -\Omega_{jl'm}\left(\hbar\frac{\mathrm{d}g}{\mathrm{d}r} + \frac{\kappa+1}{r}\hbar g(r)\right) \quad . \tag{22}$$

Analogously one derives an expression for $(\hat{\boldsymbol{\sigma}} \cdot \hat{\boldsymbol{p}})\chi$, namely

$$(\hat{\boldsymbol{\sigma}} \cdot \hat{\boldsymbol{p}})\chi_{jl'm} = -\mathrm{i}\Omega_{jlm}\left(\hbar\frac{\mathrm{d}f(r)}{\mathrm{d}r} - \frac{\kappa-1}{r}\hbar f(r)\right) \quad . \tag{23}$$

Now the expression (22) and (23) are inserted into (5). The angular functions from both sides of the equation can be eliminated. So we obtain the differential equations for the radial functions f and g:

$$\begin{aligned} &\hbar c\frac{\mathrm{d}g(r)}{\mathrm{d}r} + (1+\kappa)\hbar c\frac{g(r)}{r} - \left[E + m_0c^2 - V(r)\right]f(r) = 0 \quad , \\ &\hbar c\frac{\mathrm{d}f(r)}{\mathrm{d}r} + (1-\kappa)\hbar c\frac{f(r)}{r} + \left[E - m_0c^2 - V(r)\right]g(r) = 0 \quad . \end{aligned} \tag{24}$$

With the substitution

$$\begin{aligned} &G = rg \quad \text{and} \quad F = rf \quad \text{with} \\ &\frac{\mathrm{d}G}{\mathrm{d}r} = g + r\frac{\mathrm{d}g(r)}{\mathrm{d}r} \quad \text{and} \quad \frac{\mathrm{d}F}{\mathrm{d}r} + f + r\frac{\mathrm{d}f(r)}{\mathrm{d}r} \quad , \end{aligned} \tag{25}$$

one finally gets

$$\begin{aligned} &\hbar c\frac{\mathrm{d}G(r)}{\mathrm{d}r} + \hbar c\frac{\kappa}{r}G(r) - \left[E + m_0c^2 - V(r)\right]F(r) = 0 \quad , \\ &\hbar c\frac{\mathrm{d}F(r)}{\mathrm{d}r} - \hbar c\frac{\kappa}{r}F(r) + \left[E - m_0c^2 - V(r)\right]G(r) = 0 \quad . \end{aligned} \tag{26}$$

These are the frequently used coupled differential equations for the radial wave functions F and G of the Dirac equation in the case of a spherically symmetric potential $V(r)$. In the literature one occasionally sets $G = u_1$ and $F = u_2$.

EXERCISE

9.4 Commutation of the Total Angular Momentum Operator with the Hamiltonian in a Spherically Symmetric Potential

Problem. Show that the operator of the total angular momentum $\hat{\boldsymbol{J}}$ commutes with the Hamiltonian of a Dirac particle in a spherically symmetric potential.

Solution. In analogy to the Pauli theory for electrons with spin we define the total angular momentum as the sum of the orbital angular momentum $\hat{\boldsymbol{L}}$ and the spin angular momentum $\frac{1}{2}\hbar\hat{\boldsymbol{\Sigma}}$ with

$$\hat{\boldsymbol{\Sigma}} = \begin{pmatrix} \hat{\boldsymbol{\sigma}} & 0 \\ 0 & \hat{\boldsymbol{\sigma}} \end{pmatrix} \quad , \tag{1}$$

$$\hat{\boldsymbol{J}} = \hat{\boldsymbol{L}} + \tfrac{1}{2}\hbar\hat{\boldsymbol{\Sigma}} \equiv \hat{\boldsymbol{L}} + \hat{\boldsymbol{S}} \quad . \tag{2}$$

The one-particle Dirac Hamiltonian for a spherically symmetric potential $A_0(\boldsymbol{r})$ which is minimally coupled as the fourth component of a four-potential reads

$$\hat{H}_\mathrm{D} = c\hat{\boldsymbol{\alpha}}\cdot\hat{\boldsymbol{p}} + \hat{\beta}m_0c^2 + \frac{e}{c}A_0(\boldsymbol{r}) \quad . \tag{3}$$

First we investigate the commutator $[\hat{\boldsymbol{L}}, \hat{H}_\mathrm{D}]_-$. Since $\hat{\boldsymbol{L}}$ commutes with $\hat{\beta}$ and the spherical symmetric potential $A_0(r)$, it only remains to calculate $[\hat{\boldsymbol{L}}, \hat{\boldsymbol{\alpha}}\cdot\hat{\boldsymbol{p}}]_-$. We restrict ourselves to the calculation of the commutator for the $\hat{L}_x$ component and get:

$$\left[\hat{L}_x, \hat{\boldsymbol{\alpha}}\cdot\hat{\boldsymbol{p}}\right]_- = \hat{L}_x\left(\hat{\alpha}_x, \hat{p}_x + \hat{\alpha}_y\hat{p}_y + \hat{\alpha}_z\hat{p}_z\right) - \left(\hat{\alpha}_x\hat{p}_x + \hat{\alpha}_y\hat{p}_y + \hat{\alpha}_z\hat{p}_z\right)\hat{L}_x \quad , \tag{4}$$

$$\hat{L}_x = y\hat{p}_z - z\hat{p}_y \quad . \tag{5}$$

Since $\hat{L}_x$ commutes with $\hat{p}_x$, it follows from (4) that

$$\begin{aligned}
\left[\hat{L}_x, \hat{\boldsymbol{\alpha}}\cdot\hat{\boldsymbol{p}}\right]_- &= \hat{\alpha}_y\left[\hat{L}_x, \hat{p}_y\right]_- + \hat{\alpha}_z\left[\hat{L}_x, \hat{p}_z\right]_- \\
&= \hat{\alpha}_y\left(y\hat{p}_z\hat{p}_y - z\hat{p}_y\hat{p}_y - \hat{p}_y y\hat{p}_z + \hat{p}_y z\hat{p}_y\right) \\
&\quad + \hat{\alpha}_z\left(y\hat{p}_z\hat{p}_z - z\hat{p}_y\hat{p}_z - \hat{p}_z y\hat{p}_z + \hat{p}_z z\hat{p}_y\right) \\
&= \hat{\alpha}_y\left(y\hat{p}_z\hat{p}_y - \hat{p}_y y\hat{p}_z\right) + \hat{\alpha}_z\left(\hat{p}_z z\hat{p}_y - z\hat{p}_y\hat{p}_z\right) \\
&= \hat{\alpha}_y\left(y\hat{p}_z\hat{p}_y - (\hat{p}_y y)p_z - y\hat{p}_y\hat{p}_z\right) \\
&\quad + \hat{\alpha}_z\left((\hat{p}_z z)\hat{p}_y + z\hat{p}_z\hat{p}_y - z\hat{p}_y\hat{p}_z\right) \\
&= \hat{\alpha}_y(-\hat{p}_y y)\hat{p}_z + \hat{\alpha}_z(\hat{p}_z z)\hat{p}_y \quad .
\end{aligned} \tag{6}$$

With $\hat{p}_i - \mathrm{i}\hbar\partial/\partial x_i$ we then have

$$\left[\hat{L}_x, \hat{\boldsymbol{\alpha}}\cdot\hat{\boldsymbol{p}}\right]_- = \mathrm{i}\hbar\left(\hat{\alpha}_y\hat{p}_z - \hat{\alpha}_z\hat{p}_y\right) \quad , \tag{7}$$

and generally

$$\left[\hat{L}_i, \hat{\boldsymbol{\alpha}}\cdot\hat{\boldsymbol{p}}\right]_- = \mathrm{i}\hbar\left(\hat{\alpha}_j\hat{p}_k - \hat{\alpha}_k\hat{p}_j\right) \quad .$$

Exercise 9.4.

By cyclical permutations of the indices the final result is

$$[\hat{\boldsymbol{L}}, \hat{\boldsymbol{\alpha}} \cdot \hat{\boldsymbol{p}}]_- = \mathrm{i}\hbar(\hat{\boldsymbol{\alpha}} \times \hat{\boldsymbol{p}}) \neq 0 \quad . \tag{8}$$

Equation (8) states that the orbital angular momentum $\hat{\boldsymbol{L}}$ does not commute with the Hamiltonian and hence it is not a constant of motion. Now we investigate the commutator of spin angular momentum with the Hamiltonian: $[\hbar\hat{\boldsymbol{\Sigma}}/2, \hat{H}_{\mathrm{D}}]_-$. In this case we restrict ourselves to the component $[\hbar\hat{\Sigma}_x/2]_-$, too. It is convenient to introduce the 4×4 matrix γ_5':

$$\gamma_5' \begin{pmatrix} 0 & -1\!\!1 \\ -1\!\!1 & 0 \end{pmatrix} \quad \text{with} \quad \gamma_5'^2 = 1\!\!1 \quad . \tag{9}$$

The following relations hold:

$$-\gamma_5'\hat{\boldsymbol{\alpha}} = \begin{pmatrix} 0 & 1\!\!1 \\ 1\!\!1 & 0 \end{pmatrix}\begin{pmatrix} 0 & \hat{\boldsymbol{\sigma}} \\ \hat{\boldsymbol{\sigma}} & 0 \end{pmatrix} = \begin{pmatrix} \hat{\boldsymbol{\sigma}} & 0 \\ 0 & \hat{\boldsymbol{\sigma}} \end{pmatrix} = \hat{\boldsymbol{\Sigma}} \quad , \tag{10}$$

$$-\hat{\boldsymbol{\alpha}}\gamma_5' = \begin{pmatrix} 0 & -\hat{\boldsymbol{\sigma}} \\ -\hat{\boldsymbol{\sigma}} & 0 \end{pmatrix}\begin{pmatrix} 0 & -1\!\!1 \\ -1\!\!1 & 0 \end{pmatrix} = \begin{pmatrix} \hat{\boldsymbol{\sigma}} & 0 \\ 0 & \hat{\boldsymbol{\sigma}} \end{pmatrix} = \hat{\boldsymbol{\Sigma}} \quad , \tag{11}$$

$$-\gamma_5'\hat{\boldsymbol{\Sigma}} = \begin{pmatrix} 0 & 1\!\!1 \\ 1\!\!1 & 0 \end{pmatrix}\begin{pmatrix} \hat{\boldsymbol{\sigma}} & 0 \\ 0 & \hat{\boldsymbol{\sigma}} \end{pmatrix} = \begin{pmatrix} 0 & \hat{\boldsymbol{\sigma}} \\ \hat{\boldsymbol{\sigma}} & 0 \end{pmatrix} = \hat{\boldsymbol{\alpha}} \quad , \tag{12}$$

$$-\hat{\boldsymbol{\Sigma}}\gamma_5' = \begin{pmatrix} -\hat{\boldsymbol{\sigma}} & 0 \\ 0 & -\hat{\boldsymbol{\sigma}} \end{pmatrix}\begin{pmatrix} 0 & -1\!\!1 \\ -1\!\!1 & 0 \end{pmatrix} = \begin{pmatrix} 0 & \hat{\boldsymbol{\sigma}} \\ \hat{\boldsymbol{\sigma}} & 0 \end{pmatrix} = \hat{\boldsymbol{\alpha}} \quad , \tag{13}$$

$$\hat{\beta}\hat{\boldsymbol{\Sigma}} = \begin{pmatrix} 1\!\!1 & 0 \\ 0 & -1\!\!1 \end{pmatrix}\begin{pmatrix} \hat{\boldsymbol{\sigma}} & 0 \\ 0 & \hat{\boldsymbol{\sigma}} \end{pmatrix} = \begin{pmatrix} \hat{\boldsymbol{\sigma}} & 0 \\ 0 & -\hat{\boldsymbol{\sigma}} \end{pmatrix} \quad , \tag{14}$$

$$\hat{\boldsymbol{\Sigma}}\hat{\beta} = \begin{pmatrix} \hat{\boldsymbol{\sigma}} & 0 \\ 0 & \hat{\boldsymbol{\sigma}} \end{pmatrix}\begin{pmatrix} 1\!\!1 & 0 \\ 0 & -1\!\!1 \end{pmatrix} = \begin{pmatrix} \hat{\boldsymbol{\sigma}} & 0 \\ 0 & -\hat{\boldsymbol{\sigma}} \end{pmatrix} \quad , \tag{15}$$

and therefore

$$[\hat{\beta}, \hat{\boldsymbol{\Sigma}}]_- = 0 \quad , \tag{16a}$$

$$\left[\gamma_5', \hat{\boldsymbol{\alpha}}\right]_- = 0 \quad \text{and} \tag{16b}$$

$$\left[\gamma_5', \hat{\boldsymbol{\Sigma}}\right]_- = 0 \quad . \tag{16c}$$

Thus there remains only the calculation of the following commutator:

$$\begin{aligned}
\tfrac{1}{2}\hbar\left\{\hat{\Sigma}_x, \hat{\boldsymbol{\alpha}} \cdot \hat{\boldsymbol{p}}\right]_- &\underbrace{=}_{\text{due to (16b)}} -\tfrac{1}{2}\hbar\gamma_5' \left[\hat{\alpha}_x, \hat{\boldsymbol{\alpha}} \cdot \hat{\boldsymbol{p}}\right] \\
&= -\tfrac{1}{2}\hbar\gamma_5'\left(\hat{\alpha}_x\left\{\hat{\alpha}_x\hat{p}_x + \hat{\alpha}_y\hat{p}_y + \hat{\alpha}_z\hat{p}_z\right\} - \left\{\hat{\alpha}_x\hat{p}_x + \hat{\alpha}_y\hat{p}_y + \hat{\alpha}_z\hat{p}_z\right\}\hat{\alpha}_x\right) \\
&= -\tfrac{1}{2}\hbar\gamma_5'\left\{\hat{\alpha}_x\hat{\alpha}_y\hat{p}_y + \hat{\alpha}_x\hat{\alpha}_z\hat{p}_z - \hat{\alpha}_y\hat{p}_y\hat{\alpha}_x - \hat{\alpha}_z\hat{p}_z\hat{\alpha}_x\right\} \\
&= -\tfrac{1}{2}\hbar\gamma_5'\left\{\left(\hat{\alpha}_x\hat{\alpha}_y - \hat{\alpha}_y\hat{\alpha}_x\right)\hat{p}_y + \left(\hat{\alpha}_x\hat{\alpha}_z - \hat{\alpha}_z\hat{\alpha}_x\right)\hat{p}_z\right\} \quad .
\end{aligned} \tag{17}$$

Because of the commutation relations of the α matrices,

$$\hat{\alpha}_x\hat{\alpha}_y = -\hat{\alpha}_y\hat{\alpha}_x \quad , \quad \hat{\alpha}_x\hat{\alpha}_z = -\hat{\alpha}_z\hat{\alpha}_x \quad , \tag{18}$$

and it follows that

Exercise 9.4.

$$\begin{aligned}\tfrac{1}{2}\hbar\left[\hat{\Sigma}_x,\hat{\boldsymbol{\alpha}}\cdot\hat{\boldsymbol{p}}\right]_- &= \hbar\gamma_5'\left\{\hat{\alpha}_y\hat{\alpha}_x\hat{p}_y+\hat{\alpha}_z\hat{\alpha}_x\hat{p}_z\right\}\\ &= \gamma_5'\hbar\left\{-\gamma_5'\hat{\Sigma}_y\left(-\gamma_5'\hat{\Sigma}_x\right)\hat{p}_y-\gamma_5'\hat{\Sigma}_z\left(-\gamma_5'\hat{\Sigma}_x\right)\hat{p}_z\right\}\\ &\underbrace{=}_{\text{due to (16c)}} \gamma_5'\hbar\left\{\hat{\Sigma}_y\hat{\Sigma}_x\hat{p}_y+\hat{\Sigma}_z\hat{\Sigma}_x\hat{p}_z\right\}\quad .\end{aligned}\tag{19}$$

Now the commutation relations of the Pauli matrices are

$$\begin{aligned}\hat{\sigma}_x\hat{\sigma}_y &= -\hat{\sigma}_y\hat{\sigma}_x = \mathrm{i}\hat{\sigma}_z\quad ,\\ \hat{\sigma}_y\hat{\sigma}_z &= -\hat{\sigma}_z\hat{\sigma}_y = \mathrm{i}\hat{\sigma}_x\quad ,\\ \hat{\sigma}_z\hat{\sigma}_x &= -\hat{\sigma}_x\hat{\sigma}_z = \mathrm{i}\hat{\sigma}_y\quad .\end{aligned}\tag{20}$$

Hence one gets

$$\tfrac{1}{2}\hbar\left[\hat{\Sigma}_x,\hat{\boldsymbol{\alpha}}\cdot\hat{\boldsymbol{p}}\right]_- = \hbar\gamma_5'\left(-\mathrm{i}\hat{\Sigma}_z\hat{p}_y+\mathrm{i}\hat{\Sigma}_y\hat{p}_z\right) = \mathrm{i}\hbar\left(\hat{\alpha}_z\hat{p}_y-\hat{\alpha}_y\hat{p}_z\right)\quad .\tag{21}$$

Comparing (21) with (7) we recognize that the component $\hat{J}_x$ of total angular momentum indeed commutes with the Hamiltonian. Cyclical permutation of the indices results in

$$\left[\hat{\boldsymbol{J}},\hat{H}_{\mathrm{D}}\right]_- = [\hat{\boldsymbol{J}},\hat{\boldsymbol{\alpha}}\cdot\hat{\boldsymbol{p}}]_- = 0\quad .\tag{22}$$

We know[4] that $\hat{\boldsymbol{J}}^2$ and any component of $\hat{\boldsymbol{J}}$, for example $\hat{J}_z$, can be simultaneously diagonalised with $\hat{H}$. The fact that $\hat{\boldsymbol{J}}$, but not $\hat{\boldsymbol{L}}$ and $\hat{\boldsymbol{S}}$, provide good quantum numbers in the Dirac theory (i.e. that $\hat{\boldsymbol{J}}$ commutes with $\hat{H}_{\mathrm{D}}$) is due to the spin-orbit coupling $\sim \hat{\boldsymbol{L}}\cdot\boldsymbol{S}$ contained in $\hat{H}_{\mathrm{D}}$ (see Chap. 11, the Foldy–Wouthuysen tranformation). It causes a coupling between spin and orbital angular momentum, and hence only the total angular momentum is a constant of motion.

EXERCISE

9.5 A Dirac Particle in a Spherical Potential Box

Problem. Establish the solutions of the Dirac equation

$$\left(\hat{\boldsymbol{\alpha}}\cdot\hat{\boldsymbol{p}}+\hat{\beta}m_0c^2\right)\psi(\boldsymbol{r}) = [E-V(r)]\,\psi(\boldsymbol{r})\tag{1}$$

in a spherical square-well potential

$$V(r) = \begin{cases}-V_0<0 & \text{for}\quad r\le R\\ 0 & \text{for}\quad r>R\end{cases}\quad .\tag{2}$$

[4] This is covered in detail in W. Greiner: *Quantum Mechanics – An Introduction*, 3rd ed. and in W. Greiner: *Quantum Mechanics – Symmetries*, 2nd ed. (Springer, Berlin, Heidelberg 1994).

Exercise 9.5.

Solution. First we write the operator $\hat{\boldsymbol{\sigma}} \cdot \hat{\boldsymbol{p}}$ of kinetic energy in spherical polar coordinates. This is achieved by using the identity

$$\boldsymbol{\nabla} = \boldsymbol{e}_r \cdot (\boldsymbol{e}_r \cdot \boldsymbol{\nabla}) - \boldsymbol{e}_r \times (\boldsymbol{e}_r \times \boldsymbol{\nabla}) = \boldsymbol{e}_r \cdot \frac{\partial}{\partial_r} - \frac{\mathrm{i}}{\hbar} \frac{\boldsymbol{e}_r}{r} \times \hat{\boldsymbol{L}} \tag{3}$$

with $\hat{\boldsymbol{L}} = -\mathrm{i}\hbar(\boldsymbol{r} \times \boldsymbol{\nabla})$. It follows that

$$\hat{\boldsymbol{\alpha}} \cdot \hat{\boldsymbol{p}} = -\mathrm{i}\hbar\hat{\alpha}_r \frac{\partial}{\partial r} - \frac{1}{r} \hat{\boldsymbol{\alpha}} \cdot (\boldsymbol{e}_r \times \hat{\boldsymbol{L}}) \quad . \tag{4}$$

Making use of

$$(\hat{\boldsymbol{\alpha}} \cdot \boldsymbol{A})(\hat{\boldsymbol{\alpha}} \cdot \boldsymbol{B}) = \boldsymbol{A} \cdot \boldsymbol{B} + \mathrm{i}\hat{\boldsymbol{\Sigma}} \cdot (\boldsymbol{A} \times \boldsymbol{B}) \quad , \tag{5}$$

we obtain (with $\boldsymbol{A} = \boldsymbol{e}_r$, $\boldsymbol{B} = \boldsymbol{L}$)

$$\hat{\alpha}_r \cdot (\hat{\boldsymbol{\alpha}} \cdot \hat{\boldsymbol{L}}) = \boldsymbol{e}_r \cdot \hat{\boldsymbol{L}} + \mathrm{i}\hat{\boldsymbol{\Sigma}} \cdot (\boldsymbol{e}_r \times \hat{\boldsymbol{L}}) = \mathrm{i}\hat{\boldsymbol{\Sigma}} \cdot (\boldsymbol{e}_r \times \hat{\boldsymbol{L}}) \quad . \tag{6}$$

Multiplying by

$$\gamma_5 = \begin{pmatrix} 0 & \mathbb{1} \\ \mathbb{1} & 0 \end{pmatrix}$$

from the right results in

$$\hat{\alpha}_r \cdot (\hat{\boldsymbol{\Sigma}} \cdot \hat{\boldsymbol{L}}) = \mathrm{i}\hat{\boldsymbol{\alpha}} \cdot (\boldsymbol{e}_r \times \hat{\boldsymbol{L}}) \quad . \tag{7}$$

Therefore (4) reads

$$\hat{\boldsymbol{\alpha}} \cdot \hat{\boldsymbol{p}} = -\mathrm{i}\hbar\hat{\alpha}_r \frac{\partial}{\partial r} + \mathrm{i}\frac{\hat{\alpha}_r}{r}(\hat{\boldsymbol{\Sigma}} \cdot \hat{\boldsymbol{L}}) = -\mathrm{i}\hat{\alpha}_r \left(\frac{\partial}{\partial r} + \frac{\hbar}{r} - \frac{\hat{\beta}}{r}\hat{K} \right) \quad , \tag{8}$$

where we have introduced

$$\hat{K} = \hat{\beta}(\hat{\boldsymbol{\Sigma}} \cdot \hat{\boldsymbol{L}} + \hbar) \quad . \tag{9}$$

We obtain the eigensolutions of (1) by a separation ansatz, i.e.

$$\psi(\boldsymbol{r}) = \begin{pmatrix} g(r)\chi_{\kappa,\mu}(\vartheta,\varphi) \\ \mathrm{i}f(r)\chi_{-\kappa,\mu}(\vartheta,\varphi) \end{pmatrix} \quad , \tag{10}$$

where the $\chi_{\kappa,\mu}$ are the eigenfunctions of the angular dependent part:

$$\begin{aligned} (\hat{\boldsymbol{\sigma}} \cdot \hat{\boldsymbol{L}} + \hbar)\,\chi_{\kappa,\mu} &= -\hbar\kappa\chi_{\kappa,\mu} \quad , \\ (\hat{\boldsymbol{\sigma}} \cdot \hat{\boldsymbol{L}} + \hbar)\,\chi_{-\kappa,\mu} &= \hbar\kappa\chi_{-\kappa,\mu} \quad , \\ \hat{J}_z\chi_{\kappa,\mu} &= \hbar\mu\chi_{\kappa,\mu} \quad , \quad \kappa = (-1,1,-2,2,-3,3,\dots) \quad , \end{aligned} \tag{11}$$

[see Example 9.3, (21) and also Chap. 10, (10.30–37)].

With the help of the results of Example 9.3, two coupled differential equations for $g(r)$ and $f(r)$ follow from (1), namely

$$\begin{aligned} \left[E - V(r) - m_0c^2\right] g(r) &= \hbar c \left[-\left(\frac{\mathrm{d}}{\mathrm{d}r} + \frac{1}{r} \right) + \frac{\kappa}{r} \right] f(r) \quad , \\ \left[E - V(r) + m_0c^2\right] f(r) &= \hbar c \left[\frac{\mathrm{d}}{\mathrm{d}r} + \frac{1}{r} + \frac{\kappa}{r} \right] g(r) \quad . \end{aligned} \tag{12}$$

Often it is more convenient to use

$$u_1(r) = rg(r) \quad , \quad u_2(r) = rf(r) \tag{13}$$

for which the differential equations read:

$$\frac{\mathrm{d}}{\mathrm{d}r}\begin{pmatrix} u_1(r) \\ u_2(r) \end{pmatrix} = \begin{bmatrix} -\frac{\kappa}{r} & \frac{1}{\hbar c}\left(E + m_0c^2 - V(r)\right) \\ -\frac{1}{\hbar c}\left(E - m_0c^2 - V(r)\right) & \frac{\kappa}{r} \end{bmatrix} \times \begin{pmatrix} u_1(r) \\ u_2(r) \end{pmatrix} \quad . \tag{14}$$

For constant values of V_0, (14) has the following solutions:

(1) If

$$\begin{aligned} &\hbar^2k^2c^2 \equiv (E + V_0)^2 - m_0^2c^4 > 0 \\ &u_1(r) = r\left(a_1 j_{l_\kappa}(kr) + a_2 y_{l_\kappa}(kr)\right) \quad , \\ &u_2(r) = \frac{\kappa}{|\kappa|}\frac{\hbar ckr}{E + V_0 + m_0c^2}\left(a_1 j_{l_{-\kappa}}(kr) + a_2 y_{l_{-\kappa}}(kr)\right) \end{aligned} \tag{15}$$

with

$$l_\kappa = \begin{cases} \kappa & \text{for } \kappa > 0 \\ -\kappa - 1 & \text{for } \kappa < 0 \end{cases} \quad ,$$

$$l_{-\kappa} = \begin{cases} -\kappa & \text{for } -\kappa > 0 \\ \kappa - 1 & \text{for } -\kappa < 0 \end{cases} \quad . \tag{16}$$

(ii) If

$$\begin{aligned} &\hbar^2K^2c^2 \equiv m_0^2c^4 - (E + V_0)^2 > 0 \\ &u_1(r) = \sqrt{\frac{2Kr}{\pi}}\left(b_1 K_{l_\kappa+1/2}(Kr) + b_2 I_{l_\kappa+1/2}(Kr)\right) \quad , \\ &u_2(r) = \frac{\hbar cK}{E + V_0 + m_0c^2}\sqrt{\frac{2Kr}{\pi}}\left(-b_1 K_{l_{-\kappa}+1/2}(Kr) + b_2 I_{l_{-\kappa}+1/2}(Kr)\right) \quad . \end{aligned} \tag{17}$$

The j_l and y_l are the spherical Bessel functions of the first and second kind[5] and the $K_{l+/12}$ are the modified spherical Bessel functions. Their asymptotic behaviour is:

$$\begin{aligned} &j_n(z) \xrightarrow[z\to 0]{} \frac{1}{(2n+1)!!}z^n \quad , \\ &y_n(z) \xrightarrow[z\to 0]{} -(2n-1)!!z^{-n-1} \quad , \end{aligned} \tag{18}$$

$$\begin{aligned} &\sqrt{\frac{\pi}{2z}}I_{n+1/2}(z) \xrightarrow[z\to 0]{} \frac{1}{(2n+1)!!}z^n \quad , \\ &\sqrt{\frac{\pi}{2z}}K_{n+1/2}(z) \xrightarrow[z\to 0]{} \sim (2n-1)!!z^{-n-1} \quad , \end{aligned} \tag{19}$$

[5] The irregular solutions y_l are also noted in the literature as spherical Neumann functions n_l.

Exercise 9.5.

$$\sqrt{\frac{\pi}{2z}} I_{n+1/2}(z) \xrightarrow[z\to\infty]{} \sim \frac{\mathrm{e}^{z}}{z} \quad ,$$
$$\sqrt{\frac{\pi}{2z}} K_{n+1/2}(z) \xrightarrow[z\to\infty]{} \sim \frac{\mathrm{e}^{-z}}{z} \quad . \tag{20}$$

Before continuing, we want to give the representation of the $\chi_{\kappa,\mu}(\vartheta,\varphi)$ in (11) [see Example 9.3 and Chap. 10, (10.32)]:

$$\chi_{\kappa,\mu}(\vartheta,\varphi) = \sum_{m=-1/2,1/2} \left(l_\kappa \tfrac{1}{2} j \,\middle|\, \mu - m, m \right) Y_{l_\kappa,\mu-m}(\vartheta,\varphi)\chi_{\frac{1}{2},m} \tag{21}$$

with

$$\chi_{\frac{1}{2},\frac{1}{2}} = \begin{pmatrix} 1 \\ 0 \end{pmatrix} \quad , \quad \chi_{\frac{1}{2},-\frac{1}{2}} = \begin{pmatrix} 0 \\ 1 \end{pmatrix} \quad . \tag{22}$$

Let us now find the bound states.[6] For these, $E > V_0 + m_0c^2$ and $-m_0c^2 < E < m_0c^2$. In the inner region of the potential field we must therefore take the solutions (15) and set $a_2 = 0$, in order that the wave functions remain normalizable at the origin. On the other hand in the outer region we must set $b_2 = 0$ in (17), so that the wave functions are normalizable at infinity. Both solutions must be joined at $r = R_0$. One can eliminate the normalizing constants a_1, b_1 by adjustment of the ratio u_1/u_2 at $r = R_0$. This gives

$$\frac{\kappa}{|\kappa|} \frac{R_0 j_{l_\kappa}(kR_0)}{kR_0 j_{l_{-\kappa}}(kR_0)} \left(E + V_0 + m_0c^2\right) = -\frac{R_0 K_{l_\kappa+\frac{1}{2}}(KR_0)}{KR_0 K_{l_{-\kappa}+\frac{1}{2}}(KR_0)} \left(E + m_0c^2\right) \tag{23}$$

and

$$\frac{j_{l_\kappa}(kR_0)}{j_{l_{-\kappa}}(kR_0)} = -\frac{\kappa}{|\kappa|} \frac{k}{K} \frac{K_{l_\kappa+\frac{1}{2}}(KR_0)}{K_{l_{-\kappa}+\frac{1}{2}}(KR_0)} \frac{E + m_0c^2}{E + V_0 + m_0c^2} \tag{24}$$

with

$$\hbar ck = \sqrt{(E + V_0)^2 - m_0^2c^4} \quad ,$$
$$\hbar cK = \sqrt{m_0^2c^4 - E^2} \quad . \tag{25}$$

For $|\kappa| = 1$ one can further simplify the equations analytically, and for s states ($\kappa = -1$, $l_\kappa = 0$, $l_{-\kappa} = 1$) this results in

$$\frac{kR_0 \sin kR_0}{\sin kR_0 - kR_0 \cos kR_0} = +\frac{k}{K} \frac{\mathrm{e}^{-KR_0}}{\mathrm{e}^{-KR_0}(1 + 1/KR_0)} \frac{E + m_0c^2}{E + V_0 + m_0c^2} \quad . \tag{26}$$

After some transformations one gets

$$\tan\left[\frac{R_0}{\hbar c}\sqrt{(E + V_0)^2 - m_0^2c^4}\right] \sqrt{\frac{E + V_0 + m_0c^2}{E + V_0 - m_0c^2}}$$
$$\times \left\{ \frac{\hbar c}{R_0} \left[\frac{1}{E + V_0 + m_0c^2} - \frac{1}{E + m_0c^2}\right] - \sqrt{\frac{m_0c^2 - E}{m_0c^2 + E}} \right\} = 1 \quad . \tag{27}$$

[6] W. Pieper, W. Greiner: Z. Phys. **218**, 327 (1969).

Exercise 9.5.

Analogously one obtains for $p_{1/2}$ states ($\kappa = 1$, $l_\kappa = 1$, $l_{-\kappa} = 0$):

$$\frac{kR_0 \sin kR_0}{\sin kR_0 - kR_0 \cos kR_0} = -\frac{K}{k}\frac{1}{1+1/KR_0}\frac{E+V_0+m_0c^2}{E+m_0c^2} \tag{28}$$

and

$$\tan\left[\frac{R_0}{\hbar c}\sqrt{(E+V_0)^2 - m_0^2c^4}\right]\sqrt{\frac{E+V_0-m_0c^2}{E+V_0+m_0c^2}} \times \left\{\frac{\hbar c}{R_0}\left[\frac{1}{E+V_0-m_0c^2}+\frac{1}{m_0c^2-E}\right]+\sqrt{\frac{m_0c^2+E}{m_0c^2-E}}\right\} = 1 \; . \tag{29}$$

Another form for (26) is (defining $\alpha = kR_0$)

$$\alpha\cot\alpha - 1 = \frac{(m_0c^2/\hbar c)R_0\sqrt{1-(E^2/m_0^2c^4)}+1}{1+E/m_0c^2} \times \left(1+\sqrt{1+\left(\frac{a\hbar c}{m_0c^2R_0}\right)^2}\right) \; , \tag{30}$$

and for (28) one can write:

$$1 - \alpha\cot\alpha = \frac{(m_0c^2/\hbar c)R_0\sqrt{1-(E^2/m_0^2c^4)}+1}{1-E/m_0c^2} \times \left(-1+\sqrt{1+\left(\frac{a\hbar c}{m_0c^2R_0}\right)^2}\right) \; . \tag{31}$$

From (27) and (29) we can now calculate the energy eigenvalues of $s_{\frac{1}{2}}$ and $p_{\frac{1}{2}}$ states. If we assume R_0 to be small ($m_0cR_0/\hbar \ll 1$), we may solve (27) and (29) approximately by expanding in terms of $m_0cR_0/\hbar$. A short calculation (which is left as an exercise for the reader) leads to the following Table 9.1, where $n = 1, 2, 3, \ldots$ labels the states. Similarly, one can find approximate solutions for (30) and (31)

Table 9.1. Energy eigenvalues for $s_{1/2}$ and $p_{1/2}$ states in a small potential box

E	$V_0(\kappa=-1)$	$V_0(\kappa=+1)$
m_0c^2	$\frac{\hbar}{m_0c}\frac{n\pi}{R_0} - 3m_0c^2$	$\frac{\hbar}{m_0c}\frac{n\pi}{R_0} - m_0c^2$
0	$\frac{\hbar}{m_0c}\frac{n\pi}{R_0} - m_0c^2$	$\frac{\hbar}{m_0c}\frac{n\pi}{R_0} + m_0c^2$
$-m_0c^2$	$\frac{\hbar}{m_0c}\frac{n\pi}{R_0} + m_0c^2$	$\frac{\hbar}{m_0c}\frac{n\pi}{R_0} + 3m_0c^2$

$n = 1, 2, 3, \ldots$ $(m_0cR_0/\hbar) \ll 1$

for the opposite limiting case of a very large potential box ($m_0cR_0/\hbar \gg 1$), and this is shown in Table 9.2. One sees that the $p_{\frac{1}{2}}$ states are energetically higher than

Exercise 9.5.

Table 9.2. Energy eigenvalues for $s_{1/2}$ and $p_{1/2}$ states in a large potential box

E	$V_0(\kappa = -1)$	$V_0(\kappa = +1)$
m_0c^2	$\frac{(n-1/2)^2\pi^2\hbar^2}{2m_0R_0^2}$	$\frac{n^2\pi^2\hbar^2}{2m_0R_0^2}$
0	$m_0c^2\left(1+\frac{n^2\pi^2\hbar^2}{2m_0^2c^2R_0^2}\right)$	$m_0c^2\left(1+\frac{(n+1/2)^2\pi^2\hbar^2}{2m_0^2c^2R_0^2}\right)$
$-m_0c^2$	$2m_0c^2\left(1+\frac{n^2\pi^2\hbar^2}{4m_0^2c^2R_0^2}\right)$	$2m_0c^2\left(1+\frac{(n+1/2)^2\pi^2\hbar^2}{4m_0^2c^2R_0^2}\right)$

$n = 1, 2, 3, \dots \ (m_0cR_0/\hbar) \gg 1$

the s states, which can be understood intuitively because of the orbital angular momentum $l = 1$ for the p states. But even for the s states (with $l = 0$), for a given R_0 a minimal potential depth V_0 is required in order to get at least one bound state, in contrast to the one-dimensional problem, where at least one bound state always exists. This is due to the fact that for the s state of a Dirac particle in a three-dimensional potential well there is an angular momentum barrier due to the spin. Indeed, this can be easily seen by decoupling (14), differentiating again with constant V_0 and reinserting:

$$
\begin{aligned}
g'' - \left\{\frac{1}{(\hbar c)^2}\left[(E+V_0)^2 - m_0^2c^4\right] - \frac{\kappa(\kappa+1)}{r^2}\right\} g &= 0 \quad , \\
f'' + \left\{\frac{1}{(\hbar c)^2}\left[(E+V_0)^2 - m_0^2c^4\right] - \frac{\kappa(\kappa-1)}{r^2}\right\} f &= 0 \quad .
\end{aligned} \tag{32}
$$

On the one hand, for s states ($\kappa = -1$) the angular momentum barrier is zero for the large components. On the other, the equation for f contains an angular momentum term, which increases the energy in the three-dimensional case even for s states.

In Fig. 9.11 the eigenvalues[7] [found numerically from (27)] for the $1s$ state in potential wells with different values of R_0 have been plotted. One sees that for $(m_0cR_0/\hbar) \ll 1$ as well as for $(m_0cR_0/\hbar) \gg 1$ the energy eigenvalue $E(V_0)$ grows almost linerarly with V_0. As in the one-dimensional case, we can determine the scattering phase shifts of the continuum. For the s waves this can be done with little effort, whereas for the waves with higher angular momentum the matching condition at $r = R_0$ cannot be evaluated easily. Let us therefore look at the scattering phase shifts of the s waves. First we have to match solutions of the interior region,

$$
\begin{aligned}
u_1^{\mathrm{i}}(r) &= a_1 \sin k_{\mathrm{i}} r \quad , \\
u_2^{\mathrm{i}}(r) &= -a_1 \sqrt{\frac{E+V_0-m_0c^2}{E+V_0+m_0c^2}} \left(\frac{\sin k_{\mathrm{i}} r}{k_{\mathrm{i}} r} - \cos k_{\mathrm{i}} r\right) \quad ,
\end{aligned} \tag{33}
$$

at $r = R_0$ to the solution $\left((\hbar c k_{\mathrm{o}})^2 = E^2 - m_0^2c^4\right)$ of the outside region:

[7] From J. Rafelski, L. Fulcher, A. Klein: Phys. Rep. **38**, 227 (1978).

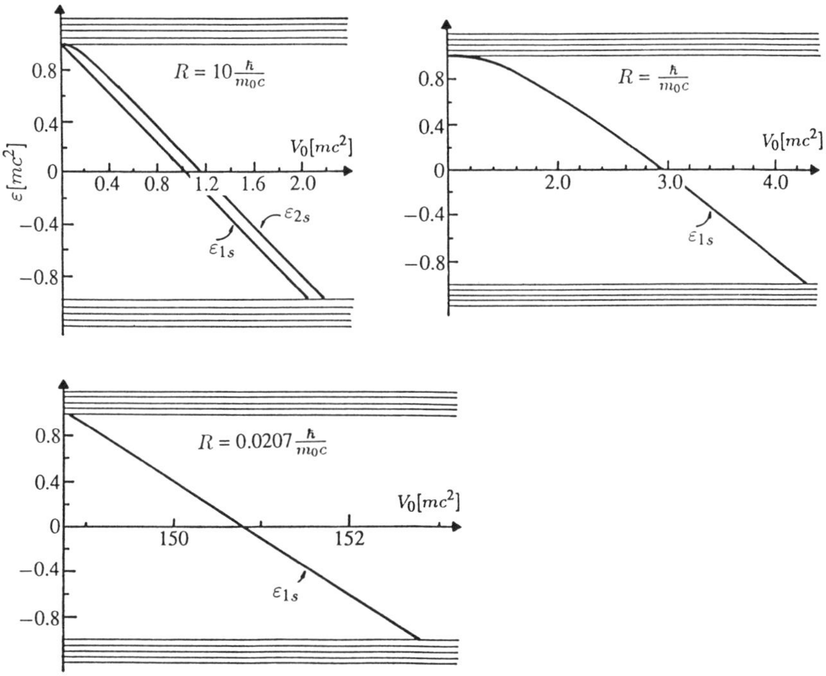

Fig. 9.11. $1s$ eigenvalues as a function of the potential strength for a spherical potential well. One example of the reults for the $2s$ level are also shown

$$u_1^{\mathrm{o}}(r) = b_1 \sin k_{\mathrm{o}} r - b_2 \cos k_{\mathrm{o}} r \quad ,$$

$$u_2^{\mathrm{o}}(r) = -\sqrt{\frac{E - m_0c^2}{E + m_0c^2}} \times \left[b_1 \left(\frac{\sin k_{\mathrm{o}} r}{k_{\mathrm{o}} r} - \cos k_{\mathrm{o}} r \right) - b_2 \left(\frac{\cos k_{\mathrm{o}} r}{k_{\mathrm{o}} r} + \sin k_{\mathrm{o}} r \right) \right] \quad . \tag{34}$$

Introducing

$$\gamma \equiv \sqrt{\frac{(E + V_0 - m_0c^2)(E + m_0c^2)}{(E + V_0 + m_0c^2)(E - m_0c^2)}} \quad , \tag{35}$$

and making use of the fact that the wave function is continuous at $r = R_0$, we conclude with

$$\Delta_{\mathrm{i}} \equiv k_{\mathrm{i}} R_0 \quad , \quad \Delta_{\mathrm{o}} \equiv k_{\mathrm{o}} R_0$$

that

$$b_1 = a_1 \left[-\gamma \frac{\sin \Delta_{\mathrm{i}}}{\Delta_{\mathrm{i}}} \cos \Delta_{\mathrm{o}} + \gamma \cos \Delta_{\mathrm{o}} \cos \Delta_{\mathrm{i}} + \sin \Delta_{\mathrm{i}} \frac{\cos \Delta_{\mathrm{o}}}{\Delta_{\mathrm{o}}} + \sin \Delta_{\mathrm{i}} \sin \Delta_{\mathrm{o}} \right] \quad ,$$

$$b_2 = a_1 \left[-\gamma \frac{\sin \Delta_{\mathrm{i}}}{\Delta_{\mathrm{i}}} \sin \Delta_{\mathrm{o}} + \gamma \sin \Delta_{\mathrm{o}} \cos \Delta_{\mathrm{i}} + \sin \Delta_{\mathrm{i}} \frac{\sin \Delta_{\mathrm{o}}}{\Delta_{\mathrm{o}}} - \sin \Delta_{\mathrm{i}} \cos \Delta_{\mathrm{o}} \right] \quad . \tag{36}$$

Exercise 9.5.

From the asymptotic behaviour of the outside solution one can derive a phase shift δ:

$$u_1^{\mathrm{o}}(r) \xrightarrow[r\to\infty]{} Aa_1 \sin(k_{\mathrm{o}} r + \delta) \quad ,$$

$$u_2^{\mathrm{o}}(r) \xrightarrow[r\to\infty]{} A\sqrt{\frac{E - m_0c^2}{E + m_0c^2}}\, a_1 \cos(k_{\mathrm{o}} + \delta) \quad . \tag{37}$$

Inserting (36) into (34) and comparing this with (37), one derives the equations

$$\cos\delta = \frac{b_1}{Aa_1} \quad , \quad \sin\delta = -\frac{b_2}{Aa_1} \tag{38}$$

and after some algebraic rearrangements the result

$$\begin{aligned}\delta &= -\Delta_{\mathrm{o}} + \operatorname{arccot}\left(\frac{1}{\Delta_{\mathrm{o}}} - \frac{\gamma}{\Delta_{\mathrm{i}}} + \gamma \cot \Delta_{\mathrm{i}}\right)\\ &= -k_{\mathrm{o}} R_0 + \operatorname{arccot}\left[\gamma \cot(k_{\mathrm{i}} R_0) + \frac{\hbar c}{R_0}\sqrt{\frac{E + m_0c^2}{E - m_0c^2}}\right.\\ &\quad \left. \times \left(\frac{1}{E + m_0c^2} - \frac{1}{E + V_0 + m_0c^2}\right)\right] \quad .\end{aligned} \tag{39}$$

As in the one-dimensional case the term $k_{\mathrm{o}} R_0$ is the same for all states and therefore this term is usually absorbed into the definition of the phase shift:

$$\begin{aligned}\delta'_{\kappa=-1} &= \delta + k_0 R_0\\ &= \operatorname{arccot}\left(\gamma \cot(k_{\mathrm{i}} R_0) + \frac{\hbar c}{R_0}\sqrt{\frac{E + m_0c^2}{E - m_0c^2}}\right.\\ &\quad \left. \times \left(\frac{1}{E + m_0c^2} - \frac{1}{E + V_0 + m_0c^2}\right)\right) \quad .\end{aligned} \tag{40}$$

In Fig. 9.12 the phase shift of the s states for a potential of depth $V_0 = 3m_0c^2$ and radius $R_0 = 10\lambda_e$ has been depicted. Once again the zeros of $\delta'_{\kappa=-1}$ and $\sin^2\delta'$ correspond to resonances; also resonances in the range $m_0c^2 - V_0 < E < -m_0c^2$

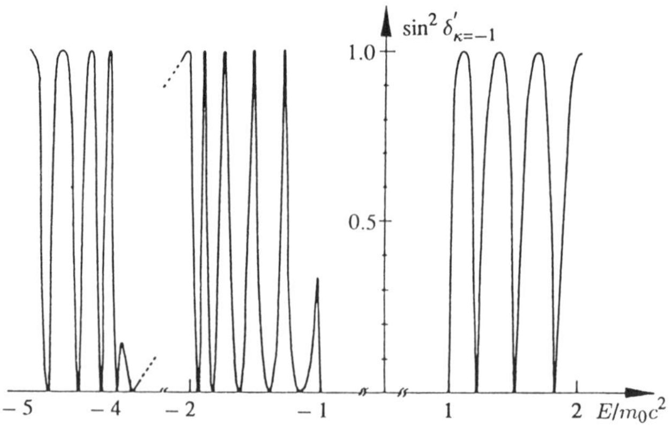

Figure 9.12.
Phase shift $\sin^2\delta'_{\kappa=-1}$ (for s waves) of a spherical potential box as a function of energy

of "dived" s states in a supercritical potential appear. A comparison with Fig. 9.11 shows that the positions of the resonances are just where one would expect them to be by extrapolating the curve $E(V_0)$ of the binding energies.

Exercise 9.5.

EXERCISE

9.6 Solution of the Radial Equations for a Dirac Particle in a Coulomb Potential

Problem. Solve the coupled radial equations for a Dirac particle in a Coulomb potential and determine the energy eigenvalues for the bound states.

Solution. The Coulomb interaction energy of a point nucleus and a particle of charge $-e$ is $V = -Ze^2/r$, so that the radial equations (cf. Example 9.3) for a Dirac particle read

$$\begin{aligned}\frac{\mathrm{d}G}{\mathrm{d}r} &= -\frac{\kappa}{r}G + \left[\frac{E + m_0c^2}{\hbar c} + \frac{Z\alpha}{r}\right]F(r) \quad ,\\ \frac{\mathrm{d}F}{\mathrm{d}r} &= \frac{\kappa}{r}F - \left[\frac{E - m_0c^2}{\hbar c} + \frac{Z\alpha}{r}\right]G(r) \quad ,\end{aligned} \tag{1}$$

where $\alpha = e^2/\hbar c \approx 1/137$ is the *fine-structure constant.* First we shall examine the solution of the equations (1) for small r, i.e. near the origin ($r \sim 0$). In this case the terms with $E \pm m_0c^2$ can be omitted and one gets

$$\begin{aligned}\frac{\mathrm{d}G}{\mathrm{d}r} + \frac{\kappa}{r}G - \frac{Z\alpha}{r}F(r) = 0 \quad ,\\ \frac{\mathrm{d}F}{\mathrm{d}r} + \frac{\kappa}{r}F + \frac{Z\alpha}{r}G(r) = 0 \quad .\end{aligned} \tag{2}$$

Using a power series expansion as an ansatz for the solution of (2), the first term of this series dominates in the region near the origin. This motivates the ansatz $G = ar^\gamma$ and $F = br^\gamma$. Using this it follows that

$$\begin{aligned}a\gamma r^{\gamma-1} + \kappa a r^{\gamma-1} - Z\alpha b r^{\gamma-1} = 0 \quad ,\\ b\gamma r^{\gamma-1} + \kappa b r^{\gamma-1} + Z\alpha a r^{\gamma-1} = 0\end{aligned} \tag{3}$$

or

$$a(\gamma + \kappa) - bZ\alpha = 0 \quad , \quad aZ\alpha + b(\gamma - \kappa) = 0 \quad . \tag{4}$$

The determinant of the coefficients must vanish, yielding

$$\begin{aligned}\gamma^2 &= \kappa^2 - (Z\alpha)^2 \quad ,\\ \gamma &= \pm\sqrt{\kappa^2 - (Z\alpha)^2} = \pm\sqrt{\left(j + \tfrac{1}{2}\right)^2 - Z^2\alpha^2} \quad .\end{aligned} \tag{5}$$

Since the wave function has to be normalizable we must choose the positive sign for γ. For the negative solution $\gamma = -|\gamma|$ it follows that $F^2 + G^2 \sim r^{-2|\gamma|}$ near $r = 0$, which would yield a divergent integral for the norm if $|\gamma| \leq \frac{1}{2}$. However,

Exercise 9.6.

we should mention that for $\kappa^2 = 1$ and $Z\alpha\sqrt{3}/2$ or $Z \gtrsim 118$, regular solutions with negative γ seem possible. Indeed, to have $F^2 + G^2 \sim r^{-2\sqrt{1-(Z\alpha)^2}}$ still integrable for $r \to 0$, the inequality $2\sqrt{1-(Z\alpha)^2} < 1$ should be fulfilled, which leads to $(Z\alpha)^2 > \frac{3}{4}$. In Exercise 9.12 we shall show that these "solutions" do, in fact, not exist. Here we only give a plausibility argument: *these solutions must be discarded* because of the following postulates: Not only the normalization integral $\int \psi^\dagger \hat{H}_\mathrm{D}\psi\, \mathrm{d}^3r$ must exist, but also the expectation value of each partial operator within $\hat{H}_\mathrm{D}$, especially the expectation value of the Coulomb energy:

$$\int \psi^\dagger \left(-\frac{Ze^2}{r}\right)\psi\, \mathrm{d}^3x = \int \left(F^2 + G^2\right)\left(-\frac{Ze^2}{r}\right)\mathrm{d}r \quad .$$

Note that the usual factor r^2 from the volume element, according to (25) of Example 9.3, is already contained in $(F^2 + G^2)$. The integrand behaves like $(r^{+2\gamma})/r\, \mathrm{d}r = r^{+2\gamma-1}\,\mathrm{d}r$ in the limit $r \to 0$, yielding a finite contribution only if $2\gamma - 1 > -1$, i.e. $\gamma > 0$. This, on the other hand, means that only the positive root in (5) is physically meaningful. Furthermore one can conclude from relation (5) that for states with $j + \frac{1}{2} = 1$ only solutions up to $Z = \alpha^{-1} \sim 137$ can be constructed. For larger values of Z the root becomes imaginary and the wave functions are no longer normalizable. For $(Z\alpha)^2 > \kappa^2$, in general the real part of the wave function shows an oscillatory behaviour of the form $\mathrm{Re}(F,G) \sim \cos(|\gamma|\ln r)$ for small r. We shall deal with this strange situation later when we treat electrons in the fields of extended nuclei and discuss the supercritical vacuum. In order to solve (1) we make the following substitutions:

$$\varrho = 2\lambda r \quad \text{with} \quad \lambda = \frac{\left(m_0c^2c^4 - E^2\right)^{1/2}}{\hbar c} \quad . \tag{6}$$

With $\mathrm{d}\varrho/\mathrm{d}r = 2\lambda$ and $\mathrm{d}/\mathrm{d}r = 2\lambda\,\mathrm{d}/\mathrm{d}\varrho$ and dividing by 2γ it follows that

$$\begin{aligned} \frac{\mathrm{d}G(\varrho)}{\mathrm{d}\varrho} &= -\frac{\kappa G(\varrho)}{\varrho} + \left[\frac{E + m_0c^2}{2\lambda\hbar c} + \frac{Z\alpha}{\varrho}\right]F(\varrho) \quad , \\ \frac{\mathrm{d}F(\varrho)}{\mathrm{d}\varrho} &= -\left[\frac{E - m_0c^2}{2\lambda\hbar c} + \frac{Z\alpha}{\varrho}\right]G(\varrho) + \frac{\kappa}{\gamma}F(\varrho) \quad . \end{aligned} \tag{7}$$

Using this form of the equations one can get the behaviour of $F(\varrho)$ and $G(\varrho)$ for $\varrho \to \infty$, since neglecting the terms proportional to $1/\varrho$ the differential equations (7) read

$$\begin{aligned} \frac{\mathrm{d}G(\varrho)}{\mathrm{d}\varrho} &= \frac{E + m_0c^2}{2\hbar c\lambda}F(\varrho) \quad , \\ \frac{\mathrm{d}F(\varrho)}{\mathrm{d}\varrho} &= -\frac{E - m_0c^2}{2\hbar c\lambda}G(\varrho) \quad . \end{aligned}$$

Combined with (6) it follows immediately that

$$\frac{\mathrm{d}^2G(\varrho)}{\mathrm{d}\varrho^2} = -\frac{\left(E^2 - m_0^2c^4\right)}{(2\hbar c\lambda)^2}G(\varrho) = \frac{1}{4}G(\varrho) \quad .$$

Exercise 9.6.

One gets two possible solutions with $G(\varrho) \sim e^{\pm/\varrho/2}$, but only the exponentially decreasing one can be used since only this one is normalizable. A similar result holds for $F(\varrho)$. This motivates the ansatz

$$\begin{aligned} G(\varrho) &= \left(m_0c^2 + E\right)^{1/2} e^{-\varrho/2}\left(\phi_1(\varrho) + \phi_2(\varrho)\right) \quad , \\ F(\varrho) &= \left(m_0c^2 - E\right)^{1/2} e^{-\varrho/2}\left(\phi_1(\varrho) - \phi_2(\varrho)\right) \quad , \end{aligned} \tag{8}$$

which we insert into (7) to give:

$$\begin{aligned} &\left(m_0c^2 + E\right)^{1/2} e^{-\varrho/2}\left[-\frac{1}{2}(\phi_1 + \phi_2) + \frac{d\phi_1}{d\varrho} + \frac{d\phi_2}{d\varrho}\right] \\ &\quad = \frac{-\kappa}{\varrho}\left(m_0c^2 + E\right)^{1/2} e^{-\varrho/2}(\phi_1 + \phi_2) \\ &\qquad + \left[\frac{E + m_0c^2}{2\hbar c\lambda} + \frac{Z\alpha}{\varrho}\right]\left(m_0c^2 - E\right)^{1/2} e^{-\varrho/2}(\phi_1 - \phi_2) \quad , \\ &\left(m_0c^2 - E\right)^{1/2} e^{-\varrho/2}\left[-\frac{1}{2}(\phi_1 - \phi_2) + \frac{d\phi_1}{d\varrho} - \frac{d\phi_2}{d\varrho}\right] \\ &\quad = -\left[\frac{E - m_0c^2}{2\hbar c\lambda} + \frac{Z\alpha}{\varrho}\right]\left(m_0c^2 + E\right)^{1/2} e^{-\varrho/2}(\phi_1 + \phi_2) \\ &\qquad + \frac{\kappa}{\varrho}\left(m_0c^2 - E\right)^{1/2} e^{-\varrho/2}(\phi_1 - \phi_2) \quad . \end{aligned} \tag{9}$$

Dividing by $e^{-\varrho/2}$ and furthermore the first equation by $(m_0c^2 + E)^{1/2}$ and the second one by $(m_0c^2 - E)^{1/2}$ yields the result

$$\begin{aligned} &-\frac{1}{2}(\phi_1 + \phi_2) + \frac{d\phi_1}{d\varrho} + \frac{d\phi_2}{d\varrho} \\ &\quad = -\frac{\kappa}{\varrho}(\phi_1 + \phi_2) + \left[\frac{E + m_0c^2}{2\hbar c\lambda} + \frac{Z\alpha}{\varrho}\right]\frac{\left(m_0c^2 - E\right)^{1/2}}{\left(m_0c^2 + E\right)^{1/2}}(\phi_1 - \phi_2) \quad , \\ &-\frac{1}{2}(\phi_1 - \phi_2) + \frac{d\phi_1}{d\varrho} - \frac{d\phi_2}{d\varrho} \\ &\quad = -\left[\frac{E - m_0c^2}{2\hbar c\lambda} + \frac{Z\alpha}{\varrho}\right]\frac{\left(m_0c^2 + E\right)^{1/2}}{\left(m_0c^2 - E\right)^{1/2}}(\phi_1 + \phi_2) + \frac{\kappa}{\varrho}(\phi_1 - \phi_2) \quad . \end{aligned} \tag{10}$$

On the other hand we have

$$\frac{\left(m_0c^2 - E\right)^{1/2}}{\left(m_0c^2 + E\right)^{1/2}} = \frac{m_0c^2 - E}{\hbar c\lambda}$$

and

$$\frac{\left(m_0c^2 + E\right)^{1/2}}{\left(m_0c^2 - E\right)^{1/2}} = \frac{m_0c^2 + E}{\hbar c\lambda} \tag{11}$$

and therefore

Exercise 9.6.

$$
\begin{aligned}
&-\frac{1}{2}(\phi_1+\phi_2)+\frac{\mathrm{d}\phi_1}{\mathrm{d}\varrho}+\frac{\mathrm{d}\phi_2}{\mathrm{d}\varrho}\\
&\quad=-\frac{\kappa}{\varrho}(\phi_1+\phi_2)+\left[\frac{E+m_0c^2}{2\hbar c\lambda}+\frac{Z\alpha}{\varrho}\right]\frac{\left(m_0c^2-E\right)^{1/2}}{\hbar c\lambda}(\phi_1-\phi_2)\quad,\\
&-\frac{1}{2}(\phi_1-\phi_2)+\frac{\mathrm{d}\phi_1}{\mathrm{d}\varrho}-\frac{\mathrm{d}\phi_2}{\mathrm{d}\varrho}\\
&\quad=-\left[\frac{E-m_0c^2}{2\hbar c\lambda}+\frac{Z\alpha}{\varrho}\right]\frac{m_0c^2+E}{\hbar c\lambda}(\phi_1+\phi_2)+\frac{\kappa}{\varrho}(\phi_1-\phi_2)\quad.
\end{aligned}\tag{12}
$$

Adding both equations of (12) yields

$$
\begin{aligned}
&-\phi_1+2\frac{\mathrm{d}\phi_1}{\mathrm{d}\varrho}\\
&\quad=-\frac{2\kappa}{\varrho}\phi_2+\phi_1+\frac{Z\alpha}{\varrho}\frac{\left(m_0c^2-E\right)}{\hbar c\lambda}(\phi_1-\phi_2)\\
&\qquad-\frac{Z\alpha}{\varrho}\frac{\left(m_0c^2+E\right)}{\hbar c\lambda}(\phi_1+\phi_2)\quad,
\end{aligned}\tag{13}
$$

whereas by subtracting them we get

$$
\begin{aligned}
&-\phi_2+2\frac{\mathrm{d}\phi_2}{\mathrm{d}\varrho}\\
&\quad=-\frac{2\kappa}{\varrho}\phi_1-\phi_2+\frac{Z\alpha}{\varrho}\frac{\left(m_0c^2-E\right)}{\hbar c\lambda}(\phi_1-\phi_2)\\
&\qquad+\frac{Z\alpha}{\varrho}\frac{\left(m_0c^2+E\right)}{\hbar c\lambda}(\phi_1+\phi_2)\quad.
\end{aligned}\tag{14}
$$

Summarizing all this yields

$$
\begin{aligned}
\frac{\mathrm{d}\phi_1}{\mathrm{d}\varrho}&=\left(1-\frac{Z\alpha E}{\hbar c\lambda\rho}\right)\phi_1-\left(\frac{\kappa}{\varrho}+\frac{Z\alpha m_0c^2}{\hbar c\lambda\rho}\right)\phi_2\quad,\\
\frac{\mathrm{d}\phi_2}{\mathrm{d}\varrho}&=\left(-\frac{\kappa}{\varrho}+\frac{Z\alpha m_0c^2}{\hbar c\lambda\rho}\right)\phi_1+\frac{Z\alpha E}{\hbar c\lambda\rho}\phi_2\quad.
\end{aligned}\tag{15}
$$

In order to find the solutions for ϕ_1 and ϕ_2 we make the ansatz of a power series expansion. Separating out a factor ϱ^γ, which describe the behaviour of the solution for $\varrho\to 0$, we write

$$
\phi_1=\varrho^\gamma\sum_{m=0}^{\infty}\alpha_m\varrho^m\quad,\quad\phi_2=\varrho^\gamma\sum_{m=0}^{\infty}\beta_m\varrho^m\quad.\tag{16}
$$

Inserting this into (15) yields

$$
\begin{aligned}
&\sum(m+\gamma)\alpha_m\varrho^{m+\gamma-1}\\
&\quad=\sum\alpha_m\varrho^{m+\gamma}-\frac{Z\alpha E}{\hbar c\lambda}\sum\alpha_m\varrho^{m+\gamma-1}\\
&\qquad-\left(\kappa+\frac{Z\alpha m_0c^2}{\hbar c\lambda}\right)\sum\beta_m\varrho^{m+\gamma-1}\quad,
\end{aligned}\tag{17}
$$

Exercise 9.6.

$$\sum \beta_m (m+\gamma)\varrho^{m+\gamma-1}$$
$$\left(-\kappa + \frac{Z\alpha m_0 c^2}{\hbar c\lambda}\right)\sum \alpha_m \varrho^{m+\gamma-1} + \frac{Z\alpha E}{\hbar c\lambda}\sum \beta_m \varrho^{m+\gamma-1} \quad .$$

Comparing the coefficients we conclude that

$$\alpha_m(m+\gamma) = \alpha_{m-1} - \frac{Z\alpha E}{\hbar c\lambda}\alpha_m - \left(\kappa + \frac{Z\alpha m_0 c^2}{\hbar c\lambda}\right)\beta_m \quad ,$$
$$\beta_m(m+\gamma) = \left(-\kappa + \frac{Z\alpha m_0 c^2}{\hbar c\lambda}\right)\alpha_m + \frac{Z\alpha E}{\hbar c\lambda}\beta_m \quad . \tag{18}$$

From the second equation of (18) it follows that

$$\frac{\beta_m}{\alpha_m} = \frac{-\kappa + Z\alpha m_0 c^2/\hbar c\lambda}{m+\gamma - Z\alpha E/\hbar c\lambda} = \frac{\kappa - Z\alpha m_0 c^2/\hbar c\lambda}{n'-m} \quad , \tag{19}$$

with

$$n' = \frac{Z\alpha E}{\hbar c\lambda} - \gamma \quad . \tag{20}$$

For $m = 0$ one gets

$$\frac{\beta_0}{\alpha_0} = \frac{\kappa - Z\alpha m_0 c^2/\hbar c\lambda}{n'} = \frac{\kappa - (n'+\gamma)m_0 c^2/E}{n'} \quad . \tag{21}$$

Inserting the result (19) into the first equation of (18) yields

$$\alpha_m\left[m+\gamma+\frac{Z\alpha E}{\hbar c\lambda} + \left(\kappa + \frac{Z\alpha m_0 c^2}{\hbar c\lambda}\right)\left(\frac{\kappa - Z\alpha m_0 c^2/\hbar c\lambda}{n'-m}\right)\right] = \alpha_{m-1} \tag{22}$$

or

$$\alpha_m\left[\left(m+\gamma+\frac{Z\alpha E}{\hbar c\lambda}\right)(n'-m) + \kappa^2 - \frac{Z^2\alpha^2 m_0^2 c^4}{\hbar^2 c^2\lambda^2}\right] = \alpha_{m-1}(n'-m) \quad . \tag{23}$$

We calculate both brackets on the lhs of (23):

$$\left(m+\gamma+\frac{Z\alpha E}{\hbar c\lambda}\right)\left(\frac{Z\alpha E}{\hbar c\lambda} - \gamma - m\right) = -2m\gamma - m^2 - \gamma^2 + \left(\frac{Z\alpha E}{\hbar c\lambda}\right)^2 , \tag{24}$$

with $\gamma^2 = \kappa^2 - (Z\alpha)^2$ and it follows that

$$\alpha_m\left[-m(2\gamma+m) + (Z\alpha)^2 + \left(\frac{Z\alpha E}{\hbar c\lambda}\right)^2 - \left(\frac{Z\alpha m_0 c^2}{\hbar c\lambda}\right)^2\right]$$
$$= \alpha_{m-1}(n'-m) \quad , \tag{25}$$

which can be further summarized as

$$\alpha_m = -\frac{(n'-m)}{m(2\gamma+m)}\alpha_{m-1} = \frac{(-1)^m(n'-1)\dots(n'-m)}{m!(2\gamma+1)\dots(2\gamma+m)}\alpha_0$$
$$= \frac{(1-n')(2-n')\dots(m-n')}{m!(2\gamma+1)\dots(2\gamma+m)}\alpha_0 \quad . \tag{26}$$

Exercise 9.6.

According to (19) β_m is found to be

$$\beta_m = \frac{\left(\kappa - Z\alpha m_0 c^2/\hbar c\lambda\right)}{n'-m} \frac{(-1)^m (n'-1)\dots(n'-m)}{m!(2\gamma+1)\dots(2\gamma+m)}\alpha_0 \quad . \tag{27}$$

Using (21) this yields

$$\beta_m = (-1)^m \frac{n'(n'-1)\dots(n'-m+1)}{m!(2\gamma+1)\dots(2\gamma+m)}\beta_0 \quad . \tag{28}$$

This power series turns out to be the *confluent hypergeometric function*

$$F(a,c;x) = 1 + \frac{a}{c}x + \frac{a(a+1)}{c(c+1)}\frac{x^2}{2!} + \dots \quad . \tag{29}$$

We thus find that

$$\begin{aligned} \phi_1 &= \alpha_0 \varrho^\gamma F(1-n', 2\gamma+1; \varrho) \quad , \\ \phi_2 &= \beta_0 \varrho^\gamma F(-n', 2\gamma+1; \varrho) \\ &= \left(\frac{\kappa - Z\alpha m_0 c^2/\hbar c\lambda}{n'}\right) \alpha_0 \varrho^\gamma F(-n', 2\gamma+1; \varrho) \quad . \end{aligned} \tag{30}$$

In order that the wave functions remain normalizable we must require that the series for ϕ_1 and ϕ_2 terminate; thus the hypergeometric functions have to be simple polynomials. This can only be achieved if n' is a non-negative integer, i.e. $n' = 0, 1, 2, \dots$.

We define *a principal quantum number*

$$n = n' + |\kappa| = n' + j + \tfrac{1}{2} \quad , \quad n = 1, 2, 3, \dots \quad . \tag{31}$$

With this we can calculate the energy eigenvalue from (20) of this example and obtain

$$\frac{Z\alpha E}{\left(m_0^2 c^4 - E^2\right)^{1/2}} = n' + \gamma = n - j - \frac{1}{2} + \gamma \tag{32}$$

and consequently

$$\left[(Z\alpha)^2 + \left(n - j - \tfrac{1}{2} + \gamma\right)^2\right] E^2 = m_0^2 c^4 \left(n - j - \tfrac{1}{2} + \gamma\right)^2 \quad , \tag{33}$$

$$\begin{aligned} E &= \underset{(-)}{+} m_0 c^2 \left[1 + \frac{(Z\alpha)^2}{\left[n - j - \frac{1}{2} + \left[\left(j+\frac{1}{2}\right)^2 - (Z\alpha)^2\right]^{1/2}\right]^2}\right]^{-1/2} \quad , \\ n &= 1, 2, 3 \dots \quad , \\ \kappa &= \pm\left(j + \tfrac{1}{2}\right) = \pm 1, \pm 2, \pm 3, \dots \quad . \end{aligned} \tag{34}$$

The negative sign in (34) must be excluded because, for positively charged nuclei ($Z\alpha > 0$), negative energies E do not fulfill the original equation (32), since its rhs is positive. We therefore write the negative sign in (34) in parenthesis, and we thus obtain the *Sommerfeld fine-structure formula* for the energy eigenvalues of electrons in atoms with a Coulomb potential and point nuclei.

Exercise 9.6.

Finally we want to quote the complete expression for the radial wave functions. Here the wave functions are normalized according to the prescription $\int \psi^\dagger \psi \, \mathrm{d}V = 1$, which explicitly implies for $f(r)$ and $g(r)$ that

$$\int_0^\infty \left(f^2 + g^2\right) r^2 \, \mathrm{d}r = 1 \quad . \tag{35}$$

This leads to the final expressions for the *normalized radial wave functions*:[8]

$$\begin{aligned}
\left.\begin{matrix} g(r) \\ f(r) \end{matrix}\right\} &= \frac{\pm(2\lambda)^{3/2}}{\Gamma(2\gamma+1)} \\
&\times \sqrt{\frac{\left(m_0c^2 \pm E\right)\Gamma(2\gamma+n'+1)}{4m_0c^2 \dfrac{(n'+\gamma)m_0c^2}{E}\left(\dfrac{(n'+\gamma)m_0c^2}{E} - \kappa\right) n'!}} \\
&\times (2\lambda r)^{\gamma-1}\, \mathrm{e}^{-\lambda r} \left\{ \left(\frac{(n'+\gamma)m_0c^2}{E} - \kappa\right) F(-n', 2\gamma+1; 2\lambda r) \right. \\
&\left. \mp n' F(1-n', 2\gamma+1; 2\lambda r) \right\} \quad .
\end{aligned} \tag{36}$$

EXERCISE

9.7 Discuss the Sommerfeld Fine-Structure Formula and the Classification of the Electron Levels in the Dirac Theory

Solution. In the preceding problem, for the electron eigenvalues in a Coulomb potential we derived

$$\begin{aligned}
E &= +m_0c^2 \left[1 + \frac{(Z\alpha)^2}{\left[n - j - \frac{1}{2} + \sqrt{\left(j+\frac{1}{2}\right)^2 - (Z\alpha)^2}\right]^2} \right]^{-1/2} \\
n' &= n - j - \tfrac{1}{2} = n - |\kappa| \quad , \\
n &= 1, 2, 3, \dots \\
j &= \tfrac{1}{2}, \tfrac{3}{2}, \tfrac{5}{2} \dots \quad .
\end{aligned} \tag{1}$$

The energy eigenvalues thus only depend on the principal quantum number n, on $|\kappa|$ and on Z. For a vanishing potential ($Z = 0$) the energy eigenvalue is $+m_0c^2$. The bound electron states thus adjoin the continuum of positive energy beginning at $+m_0c^2$. This is plausible because, due to a "switching on of the potential", that is, due to a continuous increase of the coupling strength $Z\alpha$ from $Z\alpha = 0$, electron states from the positive energy continuum can be "pulled" into the energy gap between m_0c^2 and $-m_0c^2$, thus becoming bound states. The limit of ionization of

[8] The integral of normalization can be derived by a lengthy but clearly stated calculation. See W. Greiner: *Quantum Mechanics – An Introduction*, 3rd ed. (Springer, Berlin, Heidelberg 1994) Chap. 7 (Exercise 7.1).

Exercise 9.7.

an electronic atom is obviously m_0c^2, and the ionization energy of an electron in the state n_j therefore reads

$$E_{\text{ioniz}} = m_0c^2\left\{1 - \left[1 + \frac{(Z\alpha)^2}{\left[n - |\kappa| + \sqrt{\kappa^2 - (Z\alpha)^2}\,\right]^2}\right]^{-1/2}\right\}$$
$$\approx m_0c^2(Z\alpha)^2\left\{\frac{1}{2n^2} + \frac{(Z\alpha)^2}{2n^3}\left(\frac{1}{|\kappa|} - \frac{3}{4n}\right)\right\} \quad , \tag{2}$$

which is equal to the negative binding energy, i.e. $E_{\text{bind}} = -E_{\text{ioniz}}$. We draw attention to the existence of bound states even for negative nuclear charge $Z\alpha < 0$ with energies corresponding to the negative of the solutions for $Z\alpha > 0$. This seemingly paradoxical result will be explained later in connection with charge conjugation in Chap. 12.

For states with $j = 1/2$, energy values can be calculated up to $Z\alpha \leq 1$, i.e. up to $Z \sim 137$; for $j = 1/2$, $n = 1$ we have

$$E = m_0c^2\left[1 + \frac{(Z\alpha)^2}{1 - Z^2\alpha^2}\right]^{-1/2} = m_0c^2\sqrt{1 - Z^2\alpha^2} \quad .$$

$Z\alpha = 1$ yields $E = 0$ or $E_{\text{bind}} = E - m_0c^2 = -m_0c^2$. With increasing Z the absolute value of the binding energy also increases, as it should do. The "slope" $\mathrm{d}E/\mathrm{d}Z$ approaches infinity for this $1s_{1/2}$ state ($n = 1, l = 0, j = 1/2$) at $Z\alpha = 1$, while for $Z\alpha > j + 1/2$ the energy becomes imaginary. Thus it seems that there exist no bound $ns_{1/2}$ or $np_{1/2}$ states for point nuclei with charge greater than $Z = 1/\alpha = 137$. This strange result can be understood in connection with the supercritical phenomena as a *collapse of the vacuum*; a new, fundamental process.

To calculate actual numbers, we replace m_0c^2 by the electron's rest mass $m_0c^2 = 0.5110041$ MeV. For $n' = 0$ the confluent series terminates only for $\kappa < 0$. For $\kappa > 0$ they diverge even for $n' = 0$. It thus follows that

$$n' = \begin{cases} 0, 1, 2, \ldots & \text{for} \quad \kappa < 0 \\ 1, 2, 3, \ldots & \text{for} \quad \kappa > 0 \end{cases} \quad .$$

For $Z\alpha \ll 1$ the energy formula can be expanded:

$$\frac{E - m_0c^2}{m_0c^2} = -(Z\alpha)^2\left\{\frac{1}{2n^2} + \frac{(Z\alpha)^2}{2n^3}\left(\frac{1}{j + \frac{1}{2}} - \frac{3}{4n}\right)\right\} \tag{3}$$

The degeneracy of levels with equal $|\kappa|$ but different l, already stated previously, remains unaffected. The first term in (3) represents the Bohr formula for the energy levels of the atom calculated according to the Schrödinger equation. Accordingly, relativistic corrections for the energy levels in a Coulomb field are of the order $(Z\alpha)^2$. These corrections are only significant for small principal quantum numbers and in heavy nuclei. Furthermore we note that the relativistic wave funtions $f(r)$ and $g(r)$ show for $|\kappa| = 1$ a weak (but quadratically integrable) divergence at the origin $r = 0$, in contrast to the nonrelativistic case. The states are classified in complete accordance to the levels of the hydrogen atom ($Z = 1$), as summarized in detail in the following Table 9.3. Table 9.4 shows the binding energy for the $1s_{1/2}$ electron as a function of Z. The situation is illustrated in Fig. 9.13.

Exercise 9.7.

Table 9.3. The classification of bound states of the electron according to the Dirac equation for $Z = 1$ (hyrogen atom)

Notation	n	l	j	n'	κ	$E_{\text{bind}}/\text{eV}$
$1s_{1/2}$	1	0	1/2	0	−1	−13.606
$2s_{1/2}$	2	0	1/2	1	−1	−3.402
$2p_{1/2}$	2	1	1/2	1	1	−3.402
$2p_{3/2}$	2	1	3/2	0	−2	−3.401
$3s_{1/2}$	3	0	1/2	2	−1	−1.512
$3p_{1/2}$	3	1	1/2	2	1	−1.512
$3p_{3/2}$	3	1	3/2	1	−2	−1.512
$3d_{3/2}$	3	2	3/2	1	2	−1.512
$3d_{5/2}$	3	2	5/2	0	−3	−1.512
$4s_{1/2}$	4	0	1/2	3	−1	−0.850
$4p_{1/2}$	4	1	1/2	3	1	−0.850
$4p_{3/2}$	4	1	3/2	2	−2	−0.850
$4d_{3/2}$	4	2	3/2	2	2	−0.850
$4d_{5/2}$	4	2	5/2	1	−3	−0.850
$4f_{5/2}$	4	3	5/2	1	3	−0.850
$4f_{7/2}$	4	3	7/2	0	−4	−0.850

Table 9.4. Table of binding energies for $1s_{1/2}$ electrons as a function of Z according to the Sommerfeld fine-structure formula

Z	E_{bind}/eV	Z	E_{bind}/eV
10	−1 362	80	−96 117
20	−5 472	90	−125 657
30	−12 396	100	−161 615
40	−22 254	110	−206 256
50	−35 229	120	−264 246
60	−51 585	130	−349 368
70	−71 699	137	−499 288

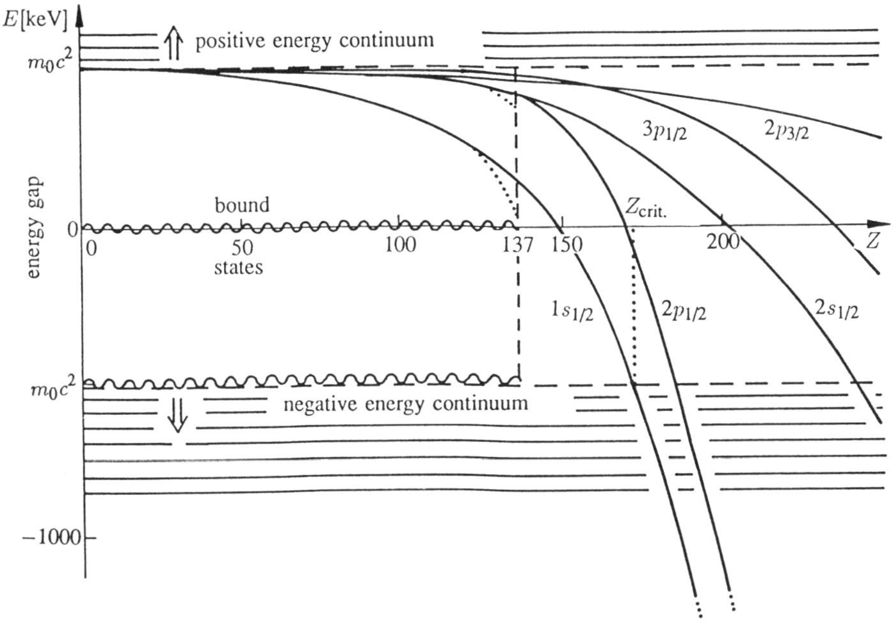

Fig. 9.13. The solutions of the Dirac equation for an electron in a Coulomb central potential: ... for point nuclei (for $s_{1/2}$ and $p_{1/2}$ existing only up to $Z = 137$), – for extended nuclei (see Example 9.9). The supercritical case $Z > Z_{\text{crit}} \approx 172$ will be discussed later in detail

EXAMPLE

9.8 Solution of the Dirac Equation for a Coulomb and a Scalar Potential

We shall solve the Dirac equation for a mixed potential consisting of a scalar potential and Coulomb potential. Both cases differ in the manner of coupling into the Dirac equation. In the case of the Coulomb potential minimal coupling is used as usual, whereas the scalar potential is added to the mass term of the Dirac equation. Therefore it can be interpreted as an effective, position-dependent mass. In the same way that the Coulomb potential is derived from the exchange of massless photons between the nucleus and the leptons orbiting around it, the scalar potential of the form $V_2 = -a'/r$ is created by the exchange of massless scalar mesons.[9] The σ-meson frequently quoted in the literature has a very high mass and therefore the corresponding potential has a very short range. Our investigations concerning the potential V_2 and its influence on the energy eigenvalues can therefore be regarded as a model study. The scalar potential can be interpreted as a Newtonian potential of the form $V_2 = V_\mathrm{N} = GM_0m_0/r \equiv -\hbar c\alpha'/r$.

For a mixed scalar and electrostatic potential, the Dirac equation thus reads

$$\left[c\hat{\alpha}\cdot\hat{p} + \hat{\beta}\left(m_0c^2 + V_2\right) - (E - V_1)\right]\psi = 0 \quad . \tag{1}$$

With the assumption of spherical symmetric potentials, we can again derive the coupled radial differential equations in the usual way:

$$\begin{aligned}
\frac{\mathrm{d}G}{\mathrm{d}r} &= -\frac{\kappa}{r}G + \frac{1}{\hbar c}\left[E + m_0c^2 + V_2 - V_1\right]F(r) \quad ,\\
\frac{\mathrm{d}F}{\mathrm{d}r} &= +\frac{\kappa}{r}F - \frac{1}{\hbar c}\left[E - m_0c^2 - V_2 - V_1\right]G(r) \quad .
\end{aligned} \tag{2}$$

Defining

$$V_1 = -\hbar c\frac{\alpha}{r} \quad \text{and} \quad V_2 = -\hbar c\frac{\alpha'}{r} \quad , \tag{3}$$

with α and α' being the electrostatic and the scalar coupling constants, respectively, we get

$$\begin{aligned}
\frac{\mathrm{d}G}{\mathrm{d}r} &= -\frac{\kappa}{r}G + \left[\frac{E + m_0c^2}{\hbar c} + \frac{1}{r}(\alpha - \alpha')\right]F(r) \quad ,\\
\frac{\mathrm{d}F}{\mathrm{d}r} &= \frac{\kappa}{r}F - \left[\frac{E - m_0c^2}{\hbar c} + \frac{1}{r}(\alpha + \alpha')\right]G(r) \quad .
\end{aligned} \tag{4}$$

In the subsequent calculations we follow a very similar procedure to that used for finding the solution of the Coulomb potential, so that it is not necessary to repeat every single step in detail (see Exercise 9.6 and 9.7). We first consider the region $r \sim 0$, where the constant terms proportional to mass and energy can be neglected:

[9] For a more detailed discussion, see W. Greiner, J. Reinhardt: *Quantum Electrodynamics*, 2nd ed. (Springer, Berlin, Heidelberg, 1994).

Example 9.8.

$$\frac{\mathrm{d}G}{\mathrm{d}r}=-\frac{\kappa}{r}G+\frac{(\alpha-\alpha')}{r}F(r) \quad ,$$
$$\frac{\mathrm{d}F}{\mathrm{d}r}=\frac{\kappa}{r}F-\frac{(\alpha+\alpha')}{r}G(r) \quad . \tag{5}$$

The ansatz

$$G=ar^{\gamma} \quad , \quad F=br^{\gamma} \quad , \tag{6}$$

yields

$$a\gamma r^{\gamma-1}+\kappa ar^{\gamma-1}-(\alpha-\alpha')br^{\gamma-1}=0 \quad ,$$
$$b\gamma r^{\gamma-1}-\kappa br^{\gamma-1}+(\alpha+\alpha')ar^{\gamma-1}=0 \tag{7}$$

or

$$a(\gamma+\kappa)-b(\alpha-\alpha')=0 \quad ,$$
$$a(\alpha+\alpha')+b(\gamma-\kappa)=0 \quad . \tag{8}$$

These are two linear homogeneous equations for a and b, and the coefficient determinant yields

$$\gamma^2=\kappa^2-\alpha^2+a'^2$$

or

$$\gamma=\pm\sqrt{\kappa^2-\alpha^2+\alpha'^2} \quad . \tag{9}$$

To allow for the normalization of the wave functions, we again choose the positive sign for γ and introduce the substitution

$$\varrho=2\lambda r \quad \text{with} \quad \lambda=\frac{\left(m_0c^2c^4-E^2\right)^{1/2}}{\hbar c} \quad . \tag{10}$$

The differential equations thus have the form

$$\frac{\mathrm{d}G}{\mathrm{d}\varrho}=-\frac{\kappa}{\varrho}G+\left[\frac{E+m_0c^2}{\hbar c2\lambda}+\frac{(\alpha-\alpha')}{\varrho}\right]F \quad ,$$
$$\frac{\mathrm{d}F}{\mathrm{d}\varrho}=-\left[\frac{E-m_0c^2}{\hbar c2\lambda}+\frac{(\alpha+\alpha')}{\varrho}\right]G+\frac{\kappa}{\varrho}F \quad , \tag{11}$$

which with the choice of

$$G=\left(m_0c^2+E\right)^{1/2}\mathrm{e}^{-\varrho/2}(\phi_1+\phi_2) \quad ,$$
$$F=\left(m_0c^2-E\right)^{1/2}\mathrm{e}^{-\varrho/2}(\phi_1-\phi_2) \quad , \tag{12}$$

leads to:

$$-\frac{1}{2}(\phi_1+\phi_2)+\frac{\mathrm{d}\phi_1}{\mathrm{d}\varrho}+\frac{\mathrm{d}\phi_2}{\mathrm{d}\varrho}=-\frac{\kappa}{\varrho}(\phi_1+\phi_2)$$
$$+\left[\frac{E+m_0c^2}{2\hbar c\lambda}+\frac{(\alpha-\alpha')}{\varrho}\right]\frac{m_0c^2-E}{\hbar c\lambda}(\phi_1-\phi_2) \quad ,$$

Example 9.8.

$$-\frac{1}{2}(\phi_1-\phi_2)+\frac{\mathrm{d}\phi_1}{\mathrm{d}\varrho}-\frac{\mathrm{d}\phi_2}{\mathrm{d}\varrho}$$
$$=-\left[\frac{E-m_0c^2}{2\hbar c\lambda}+\frac{(\alpha+\alpha')}{\varrho}\right]\frac{m_0c^2+E}{\hbar c\lambda}(\phi_1+\phi_2)+\frac{\kappa}{\varrho}(\phi_1-\phi_2) \quad .$$

Adding and subtracting the equations yields

$$-\phi_1+\frac{2\mathrm{d}\phi_1}{\mathrm{d}\varrho}=-\frac{2\kappa}{\varrho}\phi_2+\phi_1+\frac{(\alpha-\alpha')}{\varrho}\frac{m_0c^2-E}{\hbar c\lambda}(\phi_1-\phi_2)$$
$$-\frac{(\alpha+\alpha')}{\varrho}\frac{m_0c^2+E}{\hbar c\lambda}(\phi_1+\phi_2)$$

and

$$-\phi_2+\frac{2\mathrm{d}\phi_2}{\mathrm{d}\varrho}=-\frac{2\kappa}{\varrho}\phi_1-\phi_2+\frac{(\alpha-\alpha')}{\varrho}\frac{m_0c^2-E}{\hbar c\lambda}(\phi_1-\phi_2)$$
$$+\frac{(\alpha+\alpha')}{\varrho}\frac{m_0c^2+E}{\hbar c\lambda}(\phi_1+\phi_2) \quad ,$$

respectively. We now collect terms and obtain:

$$\frac{\mathrm{d}\phi_1}{\mathrm{d}\varrho}=\left(1-\frac{\alpha E}{\hbar c\lambda\rho}-\frac{\alpha' m_0c^2}{\hbar c\lambda\rho}\right)\phi_1-\left(\frac{\kappa}{\varrho}+\frac{\alpha m_0c^2}{\hbar c\lambda\rho}+\frac{\alpha' E}{\hbar c\lambda\rho}\right)\phi_2 \quad ,$$
$$\frac{\mathrm{d}\phi_2}{\mathrm{d}\varrho}=\left(-\frac{\kappa}{\varrho}+\frac{\alpha m_0c^2}{\hbar c\lambda\rho}+\frac{\alpha' E}{\hbar c\lambda\rho}\right)\phi_1+\left(\frac{\alpha E}{\hbar c\lambda\rho}+\frac{\alpha' m_0c^2}{\hbar c\lambda\rho}\right)\phi_2 \quad . \qquad (13)$$

For ϕ_1 and ϕ_2 we make the power series ansatz:

$$\phi_1=\varrho^\gamma\sum_{m=0}^{\infty}\alpha_m\varrho^m \quad , \quad \phi_2=\varrho^\gamma\sum_{m=0}^{\infty}\beta_m\varrho^m \qquad (14)$$

and insert them into the differential equations, giving

$$\sum(m+\gamma)\alpha_m\varrho^{m+\gamma-1}$$
$$=\sum\alpha_m\varrho^{m+\gamma}-\frac{\alpha E}{\hbar c\lambda}\sum\alpha_m\varrho^{m+\gamma-1}-\frac{\alpha' m_0c^2}{\hbar c\lambda}\sum\alpha_m\varrho^{m+\gamma-1}$$
$$-\left(\kappa+\frac{\alpha m_0c^2}{\hbar c\lambda}+\frac{\alpha' E}{\hbar c\lambda}\right)\sum\beta_m\varrho^{m+\gamma-1} \quad ;$$
$$\sum\beta_m(m+\gamma)\varrho^{m+\gamma-1}$$
$$=\left(-\kappa+\frac{\alpha m_0c^2}{\hbar c\lambda}+\frac{\alpha' E}{\hbar c\lambda}\right)\sum\alpha_m\varrho^{m+\gamma-1}$$
$$+\left(\frac{\alpha E}{\hbar c\lambda}+\frac{\alpha' m_0c^2}{\hbar c\lambda}\right)\sum\beta_m\varrho^{m+\gamma-1} \quad . \qquad (15)$$

A comparison of coefficients yields

$$\alpha_m(m+\gamma)=\alpha_{m-1}-\left(\frac{\alpha E}{\hbar c\lambda}+\frac{\alpha' m_0c^2}{\hbar c\lambda}\right)\alpha_m-\left(\kappa+\frac{\alpha m_0c^2}{\hbar c\lambda}+\frac{\alpha' E}{\hbar c\lambda}\right)\beta_m \quad ,$$
$$\beta_m(m+\gamma)=\left(-\kappa+\frac{\alpha m_0c^2}{\hbar c\lambda}+\frac{\alpha' E}{\hbar c\lambda}\right)\alpha_m+\left(\frac{\alpha E}{\hbar c\lambda}+\frac{\alpha' m_0c^2}{\hbar c\lambda}\right)\beta_m \quad , \qquad (16)$$

Example 9.8.

and from the second equation we get

$$\frac{\beta_m}{\alpha_m} = \frac{-\kappa + \alpha m_0c^2/\hbar c\lambda + \alpha' E/\hbar c\lambda}{m + \gamma - \alpha E/\hbar c\lambda - \alpha' m_0 c^2/\hbar c\lambda}$$
$$= \frac{\kappa - \alpha m_0c^2/\hbar c\lambda - \alpha' E/\hbar c\lambda}{n' - m} \quad , \quad \text{with} \tag{17}$$

$$n' = \frac{\alpha E}{\hbar c\lambda} + \frac{\alpha' m_0 c^2}{\hbar c\lambda} - \gamma \quad .$$

As in the case of a pure Coulomb potential, this is the equation for determining the energy eigenvalues. Only for $n' = 0, 1, 2, \dots$ do the resulting confluent hypergeometric functions for the wave functions degenerate to polynomials, and we get standing waves.

Defining

$$n = n' + j + \tfrac{1}{2} = 1, 2, 3, \dots \quad ,$$

we get

$$\frac{\alpha E + \alpha' m_0 c^2}{\left(m_0^2c^4 - E^2\right)^{1/2}} = n - j - \frac{1}{2} + \gamma \quad , \tag{18}$$

which leads to

$$\left[\alpha^2 + \left(n - j - \tfrac{1}{2} + \gamma\right)^2\right] E^2 + 2\alpha\alpha' m_0 c^2 E$$
$$= m_0^2 c^4 \left(n - j - \tfrac{1}{2} + \gamma\right)^2 - a'^2 m_0^2 c^4$$

or

$$E^2 + \frac{2\alpha\alpha' m_0 c^2}{\alpha^2 + \left(n - j - \frac{1}{2} + \gamma\right)^2} E$$
$$+ m_0^2 c^4 \frac{\alpha'^2 - \left(n - j - \frac{1}{2} + \gamma\right)^2}{\alpha'^2 + \left(n - j - \frac{1}{2} + \gamma\right)^2} = 0 \quad . \tag{19}$$

Finally we obtain the energy eigenvalue

$$E = m_0 c^2 \left\{ \frac{-\alpha\alpha'}{\alpha^2 + \left(n - j - \frac{1}{2} + \gamma\right)^2} \pm \left[\left(\frac{\alpha\alpha'}{\alpha^2 + \left(n - j - \frac{1}{2} + \gamma\right)^2} \right)^2 \right.\right.$$
$$\left.\left. - \frac{\alpha'^2 - \left(n - j - \frac{1}{2} + \gamma\right)^2}{\alpha^2 + \left(n - j - \frac{1}{2} + \gamma\right)^2} \right]^{1/2} \right\} \quad . \tag{20}$$

Let us now consider several special cases:

(1)

$$\alpha = 0 \quad , \quad \gamma = \sqrt{\kappa^2 + \alpha'^2} \quad ,$$
$$E = \pm m_0 c^2 \sqrt{1 - \frac{\alpha'^2}{\left(n - j - \frac{1}{2} + \gamma\right)^2}} \quad .$$

Example 9.8.

Obviously there exist two branches of solutions in the bound region: the solutions for positive and negative energies exhibit identical behaviour, which reflects the fact that the scalar interaction does not distinguish between positive and negative charges. The order of the levels is striking: $1s_{1/2}$, $2p_{3/2}$, $2s_{1/2}$ and $2p_{1/2}$, $3s_{1/2}$, and so on. For $\alpha' \to \infty$, E approaches the value $E = 0$ (see Fig. 9.9 in Exercise 9.2).

States with negative energies correspond to antiparticles. The particle and antiparticle states approach each other with increasing coupling constant, without touching. Critical behaviour, as in the case of the Coulomb potential, does not occur; the Dirac vacuum remains stable in the case of scalar potentials (see Exercise 9.2).

(2)

$$\alpha = 0 \quad , \quad \gamma = \sqrt{\kappa^2 - \alpha^2} \quad ,$$

$$E = m_0c^2 \left[1 + \frac{\alpha^2}{\left(n - j - \frac{1}{2} + \gamma\right)^2} \right]^{1/2} \quad .$$

We have here to choose the positive square root, because in the case of very weak fields the electron states dive from the upper continuum into the region of bound states. Furthermore the negative square root yields a contradiction to (18). The obtained result is, of course, identical with Sommerfeld's fine-structure formula (see Exercise 9.6 and 9.7).

(3) $\alpha = \alpha'$.
This yields $\gamma = |\kappa|$ and $n - j - \frac{1}{2} + \gamma = n$.

$$\begin{aligned} E &= m_0c^2 \left\{ \frac{-\alpha^2}{\alpha^2 + n^2} \pm \sqrt{\frac{\alpha^4}{(\alpha^2 + n^2)^2} - \frac{\alpha^2 - n^2}{\alpha^2 + n^2}} \right\} \\ &= m_0c^2 \left\{ \frac{-\alpha^2}{\alpha^2 + n^2} \pm \sqrt{\frac{\alpha^4 - \alpha^4 + n^2}{(\alpha^2 + n^2)^2}} \right\} \\ &= m_0c^2 \left\{ \frac{-\alpha^2}{\alpha^2 + n^2} \pm \frac{n^2}{\alpha^2 + n^2} \right\} \quad . \end{aligned}$$

The negative sign would seem to yield a solution $E_2 = -m_0c^2$, and the large and the small component in the potential term of the Hamiltonian decouple. However, this is an invalid solution, because $E = -m_0c^2$ contradicts (18). For the positive sign, it follows that

$$E = m_0c^2 \left\{ 1 - \frac{2\alpha^2}{\alpha^2 + n^2} \right\} \quad .$$

For $\alpha \to \infty$, E approaches the value $-m_0c^2$ asymptotically, but the state never dives into the negative energy continuum. For $n = 1$ and $\alpha = 1$ the energy becomes zero.

EXAMPLE

9.9 Stationary Continuum States of a Dirac Particle in a Coulomb Field

Again we start with the coupled radial differential equations for F and G:

$$\begin{aligned}\frac{\mathrm{d}G}{\mathrm{d}r} &= -\frac{\kappa}{r}G + \left[\frac{E+m_0c^2}{\hbar c} + \frac{Z\alpha}{r}\right]F \quad ,\\ \frac{\mathrm{d}F}{\mathrm{d}r} &= \frac{\kappa}{r}F + \left[\frac{E-m_0c^2}{\hbar c} + \frac{Z\alpha}{r}\right]G \quad ,\end{aligned} \tag{1}$$

and perform the substitution $x = 2\mathrm{i}pr$, where

$$p = \frac{\left[E^2 - (m_0c^2)^2\right]^{1/2}}{\hbar c} = \mathrm{i}\lambda \quad . \tag{2}$$

With $\mathrm{d}x/\mathrm{d}r = 2\mathrm{i}p$ followed by the division by $2\mathrm{i}p$, one gets

$$\begin{aligned}\frac{\mathrm{d}G}{\mathrm{d}x} &= -\frac{\kappa}{x}G + \left[\frac{E+m_0c^2}{2\mathrm{i}p\hbar c} + \frac{Z\alpha}{x}\right]F \quad ,\\ \frac{\mathrm{d}F}{\mathrm{d}x} &= +\frac{\kappa}{x}F - \left[\frac{E-m_0c^2}{2\mathrm{i}p\hbar c} + \frac{Z\alpha}{x}\right]G \quad .\end{aligned} \tag{3}$$

For F and G we make the following ansatz:

(a) for positive energies $E > m_0c^2$:

$$\begin{aligned}G &= \sqrt{E+m_0c^2}(\phi_1+\phi_2) \quad ,\\ F &= \mathrm{i}\sqrt{E-m_0c^2}(\phi_1+\phi_2) \quad ,\end{aligned} \tag{4}$$

and (b) for negative energies $E < -m_0c^2$:

$$\begin{aligned}G &= \sqrt{-E-m_0c^2}(\phi_1+\phi_2) \quad ,\\ F &= -\mathrm{i}\sqrt{-E+m_0c^2}(\phi_1-\phi_2) \quad .\end{aligned} \tag{5}$$

Inserting (4) into (3) and dividing by $\sqrt{E+m_0c^2}$ and $\mathrm{i}\sqrt{E-m_0c^2}$, respectively, yields

$$\begin{aligned}&\frac{\mathrm{d}\phi_1}{\mathrm{d}x} + \frac{\mathrm{d}\phi_2}{\mathrm{d}x}\\ &\quad = -\frac{\kappa}{x}(\phi_1+\phi_2) + \left[\frac{E+m_0c^2}{2\mathrm{i}p\hbar c} + \frac{Z\alpha}{x}\right]\frac{\mathrm{i}\sqrt{E-m_0c^2}}{\sqrt{E+m_0c^2}}(\phi_1-\phi_2) \quad ,\\ &\frac{\mathrm{d}\phi_1}{\mathrm{d}x} - \frac{\mathrm{d}\phi_2}{\mathrm{d}x}\\ &\quad = \frac{\kappa}{x}(\phi_1-\phi_2) - \left[\frac{E-m_0c^2}{2\mathrm{i}p\hbar c} + \frac{Z\alpha}{x}\right]\frac{\sqrt{E+m_0c^2}}{\mathrm{i}\sqrt{E-m_0c^2}}(\phi_1+\phi_2) \quad .\end{aligned} \tag{6}$$

Analogously, inserting equation (5) into (3) and dividing by $\sqrt{-E-m_0c^2}$ and $-\mathrm{i}\sqrt{-E+m_0c^2}$, respectively, yields

Example 9.9.

$$\frac{d\phi_1}{dx}+\frac{d\phi_2}{dx} = -\frac{\kappa}{x}(\phi_1+\phi_2)+\left[\frac{E+m_0c^2}{2\mathrm{i}p\hbar c}+\frac{Z\alpha}{x}\right]\frac{(-\mathrm{i})\sqrt{-E+m_0c^2}}{\sqrt{-E-m_0c^2}}(\phi_1-\phi_2) \quad ,$$

$$\frac{d\phi_1}{dx}-\frac{d\phi_2}{dx} = \frac{\kappa}{x}(\phi_1-\phi_2)-\left[\frac{E-m_0c^2}{2\mathrm{i}p\hbar c}+\frac{Z\alpha}{x}\right]\frac{\mathrm{i}\sqrt{-E-m_0c^2}}{\sqrt{-E+m_0c^2}}(\phi_1+\phi_2) \quad . \tag{7}$$

Let us now consider the factors on the rhs of (6):

$$\left[\frac{E+m_0c^2}{2\mathrm{i}p\hbar c}+\frac{Z\alpha}{x}\right]\frac{\mathrm{i}\left(E-m_0c^2\right)}{\hbar cp} = \frac{\mathrm{i}(\hbar c)^2p^2}{2\mathrm{i}p^2(\hbar c)^2}+\frac{Z\alpha}{x}\frac{\mathrm{i}E}{\hbar cp}-\frac{\mathrm{i}Z\alpha m_0c^2}{x\hbar cp} \quad ,$$

$$-\left[\frac{E-m_0c^2}{2\mathrm{i}p\hbar c}+\frac{Z\alpha}{x}\right]\frac{\left(E+m_0c^2\right)}{\mathrm{i}\hbar cp} = \frac{-(\hbar c)^2p^2}{2\mathrm{i}^2p^2(\hbar c)^2}-\frac{Z\alpha}{x}\frac{E}{\mathrm{i}\hbar cp}-\frac{Z\alpha m_0c^2}{\mathrm{i}\hbar cp} \,. \tag{8}$$

Adding the equations in (6) and division by 2 yields

$$\frac{d\phi_1}{dx} = -\frac{\kappa}{x}\phi_2+\frac{1}{2}\phi_1+\frac{\mathrm{i}Z\alpha E}{\hbar cpx}\phi_1+\frac{\mathrm{i}Z\alpha m_0c^2}{x\hbar cp}\phi_2 \quad . \tag{9a}$$

Similarly, subtracting the equations (6) from each other and dividing by 2 we get

$$\frac{d\phi_2}{dx} = -\frac{\kappa}{x}\phi_1-\frac{1}{2}\phi_2-\frac{\mathrm{i}Z\alpha m_0c^2}{\hbar cpx}\phi_1-\frac{\mathrm{i}Z\alpha E}{\hbar cpx}\phi_2 \quad , \tag{9b}$$

or together:

$$\frac{d\phi_1}{dx} = \left(\frac{1}{2}+\frac{\mathrm{i}Z\alpha E}{\hbar cpx}\right)\phi_1+\left(-\frac{\kappa}{x}+\frac{\mathrm{i}Z\alpha m_0c^2}{x\hbar cp}\right)\phi_2 \quad ,$$

$$\frac{d\phi_2}{dx} = \left(-\frac{\kappa}{x}-\frac{\mathrm{i}Z\alpha m_0c^2}{\hbar cpx}\right)\phi_1+\left(-\frac{1}{2}-\frac{\mathrm{i}Z\alpha E}{\hbar cpx}\right)\phi_2 \quad . \tag{10}$$

The factors on the rhs of (7) yield

$$\left[\frac{E+m_0c^2}{2\mathrm{i}p\hbar c}+\frac{Z\alpha}{x}\right]\frac{(-\mathrm{i})\sqrt{E^2-m_0^2c^4}}{-E-m_0c^2} = \frac{1}{2}+\frac{Z\alpha(-\mathrm{i})\hbar cp}{-\left(E+m_0c^2\right)x} \quad ,$$

$$\left[\frac{E-m_0c^2}{2\mathrm{i}p\hbar c}+\frac{Z\alpha}{x}\right]\frac{\sqrt{E^2-m_0^2c^4}}{\mathrm{i}\left(-E+m_0c^2\right)} = \frac{1}{2}+\frac{Z\alpha\hbar cp}{x(-\mathrm{i})\left(E-m_0c^2\right)} \quad . \tag{11}$$

Addition of the equations in (7) followed by division by 2 leads to

Example 9.9.

$$\begin{aligned}
\frac{\mathrm{d}\phi_1}{\mathrm{d}x} &= -\frac{\kappa}{x}\phi_2 + \frac{1}{2}\phi_1 \\
&\quad + \frac{1}{2}\left[\frac{\mathrm{i}Z\alpha\hbar cp}{\left(E+m_0c^2\right)x} + \frac{Z\alpha\hbar cp}{x(-\mathrm{i})\left(E-m_0c^2\right)}\right]\phi_1 \\
&\quad + \frac{1}{2}\left[\frac{-Z\alpha(-\mathrm{i})\hbar cp}{-\left(E+m_0c^2\right)x} + \frac{Z\alpha\hbar cp}{x(-\mathrm{i})\left(E-m_0c^2\right)}\right]\phi_2 \\
&= -\frac{\kappa}{x}\phi_2 + \frac{1}{2}\phi_1 \\
&\quad + \frac{1}{2}\left[\frac{Z\alpha\hbar cp\left(E-m_0c^2\right) + Z\alpha\hbar cp\left(E+m_0c^2\right)}{(-\mathrm{i})x\left(E^2-m_0^2c^4\right)}\right]\phi_1 \\
&\quad + \frac{1}{2}\left[\frac{-Z\alpha\hbar cp\left(E-m_0c^2\right) + Z\alpha\hbar cp\left(E+m_0c^2\right)}{-\mathrm{i}x\left(E^2-m_0^2c^4\right)}\right]\phi_2 \\
&= -\frac{\kappa}{x}\phi_2 + \frac{1}{2}\phi_1 + \frac{\mathrm{i}Z\alpha E}{x\hbar cp}\phi_1 + \frac{\mathrm{i}Z\alpha m_0c^2}{x\hbar cp}\phi_2 \quad ,
\end{aligned} \tag{12}$$

and analogously, the difference of Eqs. (7) yields

$$\frac{\mathrm{d}\phi_2}{\mathrm{d}x} = -\frac{\kappa}{x}\phi_1 - \frac{1}{2}\phi_2 - \frac{\mathrm{i}Z\alpha m_0c^2}{\hbar cpx}\phi_1 - \frac{\mathrm{i}Z\alpha E}{\hbar cpx}\phi_2 \quad . \tag{13}$$

Comparison of (12) and (13) with the system (10) shows that both systems of differential equations are identical, and that for positive and negative energies we have to solve the same system of differential equations.

Now we consider the complex conjugate of (10) (x is purely imaginary):

$$\begin{aligned}
\frac{\mathrm{d}\phi_1^*}{\mathrm{d}x} &= -\left(\frac{1}{2} + \frac{\mathrm{i}Z\alpha E}{\hbar cpx}\right)\phi_1^* - \left(\frac{\kappa}{x} + \frac{\mathrm{i}Z\alpha m_0c^2}{x\hbar cp}\right)\phi_2^* \quad , \\
\frac{\mathrm{d}\phi_2^*}{\mathrm{d}x} &= -\left(\frac{\kappa}{x} - \frac{\mathrm{i}Z\alpha m_0c^2}{\hbar cpx}\right)\phi_1^* + \left(\frac{1}{2} + \frac{\mathrm{i}Z\alpha E}{\hbar cpx}\right)\phi_2^* \quad .
\end{aligned} \tag{14}$$

Equations (14) are identical to (10) if we set

$$\phi_1 = \phi_2^* \quad \text{or} \quad \phi_2 = \phi_1^* \quad . \tag{15}$$

This is the necessary condition for G and F to be real-valued functions and thus to describe standing waves at infinity. We now eliminate ϕ_2 from the system of (10) to give:

$$\begin{aligned}
\phi_2 &= \left(-\frac{\kappa}{x} + \frac{\mathrm{i}Z\alpha m_0c^2}{x\hbar cp}\right)^{-1}\frac{\mathrm{d}\phi_1}{\mathrm{d}x} - \left(\frac{1}{2} + \frac{\mathrm{i}Z\alpha E}{\hbar cpx}\right)\phi_1\left(-\frac{\kappa}{x} + \frac{\mathrm{i}Z\alpha m_0c^2}{x\hbar cp}\right)^{-1} \\
&= \frac{\hbar cpx}{-\kappa\hbar cp + \mathrm{i}Z\alpha m_0c^2}\frac{\mathrm{d}\phi_1}{\mathrm{d}x} - \left(\frac{1}{2} + \frac{\mathrm{i}Z\alpha E}{\hbar cpx}\right)\frac{\hbar cpx}{-\kappa\hbar cp + \mathrm{i}Z\alpha m_0c^2}\phi_1 \\
&= \frac{\hbar cpx}{\mathrm{i}Z\alpha m_0c^2 - \kappa\hbar cp}\frac{\mathrm{d}\phi_1}{\mathrm{d}x} \\
&\quad - \frac{1}{2}\frac{\hbar cpx}{\mathrm{i}Z\alpha m_0c^2 - \kappa\hbar cp}\phi_1 - \frac{\mathrm{i}Z\alpha E}{\mathrm{i}Z\alpha m_0c^2 - \kappa\hbar cp}\phi_1 \quad ,
\end{aligned} \tag{16}$$

Example 9.9.

and calculate its derivative

$$\begin{aligned}\frac{\mathrm{d}\phi_2}{\mathrm{d}x} &= \frac{\hbar cp}{\mathrm{i}Z\alpha m_0c^2-\kappa\hbar cp}\frac{\mathrm{d}\phi_1}{\mathrm{d}x}+\frac{\hbar cpx}{\mathrm{i}Z\alpha m_0c^2-\kappa\hbar cp}\frac{\mathrm{d}^2\phi_1}{\mathrm{d}x^2}\\ &\quad-\frac{1}{2}\frac{\hbar cp}{\mathrm{i}Z\alpha m_0c^2-\kappa\hbar cp}\phi_1-\frac{1}{2}\frac{\hbar cpx}{\mathrm{i}Z\alpha m_0c^2-\kappa\hbar cp}\frac{\mathrm{d}\phi_1}{\mathrm{d}x}\\ &\quad-\frac{\mathrm{i}Z\alpha E}{\mathrm{i}Z\alpha m_0c^2-\kappa\hbar cp}\frac{\mathrm{d}\phi_1}{\mathrm{d}x}\quad\text{[see (13)]}\\ &=-\left(\frac{\kappa\hbar cp+\mathrm{i}Z\alpha m_0c^2}{\hbar cpx}\right)\phi_1-\left(\frac{1}{2}+\frac{\mathrm{i}Z\alpha E}{\hbar cpx}\right)\\ &\quad\times\left(\frac{\hbar cpx}{\mathrm{i}Z\alpha m_0c^2-\kappa\hbar cp}\left(\frac{\mathrm{d}\phi_1}{\mathrm{d}x}-\frac{1}{2}\phi_1\right)-\frac{\mathrm{i}Z\alpha E}{\mathrm{i}Z\alpha m_0c^2-\kappa\hbar cp}\phi_1\right)\quad. \end{aligned}\tag{17}$$

This yields

$$\begin{aligned}&\frac{\hbar cpx}{\mathrm{i}Z\alpha m_0c^2-\kappa\hbar cp}\frac{\mathrm{d}^2\phi_1}{\mathrm{d}x^2}+\frac{\hbar cp-\mathrm{i}Z\alpha E}{\mathrm{i}Z\alpha m_0c^2-\kappa\hbar cp}\frac{\mathrm{d}\phi_1}{\mathrm{d}x}\\ &\quad+\frac{\mathrm{i}Z\alpha E}{\mathrm{i}Z\alpha m_0c^2-\kappa\hbar cp}\frac{\mathrm{d}\phi_1}{\mathrm{d}x}-\frac{1}{2}\frac{\hbar cp}{\mathrm{i}Z\alpha m_0c^2-\kappa\hbar cp}\phi_1+\frac{\kappa}{x}\phi_1\\ &\quad+\frac{\mathrm{i}Z\alpha m_0c^2}{\hbar cpx}\phi_1-\frac{\mathrm{i}Z\alpha E}{2\left(\mathrm{i}Z\alpha m_0c^2-\kappa\hbar cp\right)}\phi_1\\ &\quad-\frac{\mathrm{i}Z\alpha E}{2\left(\mathrm{i}Z\alpha m_0c^2-\kappa\hbar cp\right)}\phi_1+\frac{(Z\alpha E)^2}{\hbar cpx\left(\mathrm{i}Z\alpha m_0c^2-\kappa\hbar cp\right)}\phi_1\\ &\quad-\frac{1}{4}\frac{\hbar cpx}{\mathrm{i}Z\alpha m_0c^2-\kappa\hbar cp}\phi_1=0\quad. \end{aligned}\tag{18}$$

To summarize:

$$\begin{aligned}&\frac{\mathrm{d}^2\phi_1}{\mathrm{d}x^2}+\frac{1}{x}\frac{\mathrm{d}\phi_1}{\mathrm{d}x}-\frac{1}{2x}\phi_1+\frac{\kappa\left(\mathrm{i}Z\alpha m_0c^2-\kappa\hbar cp\right)}{x\hbar cpx}\phi_1\\ &\quad+\frac{\mathrm{i}Z\alpha m_0c^2\left(\mathrm{i}Z\alpha m_0c^2-\kappa\hbar cp\right)}{(\hbar cpx)^2}\phi_1-\frac{\mathrm{i}Z\alpha E}{2\hbar cpx}\phi_1\\ &\quad-\frac{\mathrm{i}Z\alpha E}{2\hbar cpx}\phi_1+\frac{(Z\alpha E)^2}{(\hbar cpx)^2}\phi_1-\frac{1}{4}\phi_1=0 \end{aligned}\tag{19}$$

$$\begin{aligned}&\frac{\mathrm{d}^2\phi_1}{\mathrm{d}x^2}+\frac{1}{x}\frac{\mathrm{d}\phi_1}{\mathrm{d}x}-\frac{1}{2x}\phi_1+\frac{\mathrm{i}Z\alpha m_0c^2\kappa}{x\hbar cpx^2}\phi_1-\frac{\kappa^2}{x^2}\phi_1\\ &\quad-\frac{1}{4}\phi_1+\frac{\left(\mathrm{i}Z\alpha m_0c^2\right)^2}{(\hbar cpx)^2}-\frac{\mathrm{i}Z\alpha m_0c^2\kappa\hbar cp}{(\hbar cpx)^2}\phi_1\\ &\quad-\frac{\mathrm{i}Z\alpha E}{\hbar cpx}\phi_1+\frac{(Z\alpha E)^2}{(\hbar cpx)^2}\phi_1=0\quad, \end{aligned}\tag{20}$$

and finally

$$\frac{\mathrm{d}^2\phi_1}{\mathrm{d}x^2}+\frac{1}{x}\frac{\mathrm{d}\phi_1}{\mathrm{d}x}-\frac{1}{4}\phi_1-\frac{1}{2x}\phi_1-\frac{\kappa^2}{x^2}\phi_1+\frac{(Z\alpha)^2}{x^2}\phi_1-\frac{\mathrm{i}Z\alpha E}{\hbar cpx}\phi_1=0\quad.\tag{21}$$

Rewriting (21) leads to *Example 9.9.*

$$\frac{d^2\phi_1}{dx^2} + \frac{1}{x}\frac{d\phi_1}{dx} - \left[\frac{1}{4} + \left(\frac{1}{2} + \frac{iZ\alpha E}{\hbar cp}\right)\frac{1}{x} + \frac{\gamma^2}{x^2}\right]\phi_1 = 0 \quad , \tag{22}$$

where $\gamma^2 = \kappa^2 - (Z\alpha)^2$. It is sufficient to know $\phi_1(x)$; then $\phi_2(x)$ results from (15). To solve the differential equation (22) we substitute

$$W = x^{1/2}\phi_1 \quad , \tag{23}$$

and a second-order differential equation for ϕ_1 results:

$$\frac{d}{dx}\left[x^{-1/2}\frac{dW}{dx} - \frac{1}{2}x^{-3/2}W\right] + \frac{1}{x}\left[x^{-1/2}\frac{dW}{dx} - \frac{1}{2}x^{-3/2}W\right] - \left[\frac{1}{4} + \left(\frac{1}{2} + \frac{iZ\alpha E}{\hbar cp}\right)\frac{1}{x} + \frac{\gamma^2}{x^2}\right]x^{-1/2}W = 0 \quad .$$

Therefore,

$$\begin{aligned} &-\frac{1}{2}x^{-3/2}\frac{dW}{dx} + x^{-1/2}\frac{d^2W}{dx^2} + \frac{3}{4}x^{-5/2}W - \frac{1}{2}x^{-3/2}\frac{dW}{dx} \\ &+ x^{-3/2}\frac{dW}{dx} - \frac{1}{2}x^{-5/2}W \\ &- \left[\frac{1}{4} + \left(\frac{1}{2} + \frac{iZ\alpha E}{\hbar cp}\right)\frac{1}{x} + \frac{\gamma^2}{x^2}\right]x^{-1/2}W = 0 \quad , \end{aligned}$$

so that the final result reads

$$\frac{d^2W}{dx^2} - \left[\frac{1}{4} + \left(\frac{1}{2} + \frac{iZ\alpha E}{\hbar cp}\right)\frac{1}{x} + \frac{\gamma^2 - 1/4}{x^2}\right]W(x) = 0 \quad . \tag{24}$$

The regular solution (at $x = 0$) of the second-order differential equation, in which there appears now no first order derivative, is the *Whittaker function*[10]

$$W_{-(iy+1/2),\gamma}(x) = x^{\gamma+1/2}\,e^{-x/2}F(\gamma+1+iy, 2\gamma+1; x) \quad , \tag{25}$$

with

$$y = \frac{Z\alpha E}{\hbar cp} \quad . \tag{26}$$

Thus, [and with (23) and (2)], ϕ_1 becomes

$$\begin{aligned} \phi_1 &= N(\gamma+iy)\,e^{i\eta}(2pr)^\gamma\,e^{-ipr}F(\gamma+1+iy, 2\gamma+1; 2ipr) \\ &\equiv N(\gamma+iy)\,e^{i\eta}(2p)^\gamma\phi(r) \quad , \end{aligned} \tag{27}$$

where we have introduced the phase η, which has to be adjusted to make $\phi_2 = \phi_1^*$ valid. Now we rewrite the first equation of the system (10): With $x = 2ipr$ and $d/dr = 2ip\, d/dx$ then

[10] See, e.g. M. Abramowitz, I.A. Stegun: *Handbook of Mathematical Functions* (Dover, New York 1965).

Example 9.9.

$$\frac{\mathrm{d}\phi_1}{\mathrm{d}r} = \left(\mathrm{i}p + \frac{\mathrm{i}Z\alpha E}{\hbar cpr}\right)\phi_1 + \left(-\frac{\kappa}{r} + \frac{\mathrm{i}Z\alpha m_0 c^2}{\hbar cpr}\right)\phi_2 \quad . \tag{28}$$

Inserting ϕ_1 of (27) into (28) yields

$$\begin{aligned} &N(\gamma + \mathrm{i}y)\,\mathrm{e}^{\mathrm{i}\eta}(2p)^\gamma \frac{\mathrm{d}\phi(r)}{\mathrm{d}r} \\ &\quad = \left(\mathrm{i}p + \frac{\mathrm{i}Z\alpha E}{\hbar cpr}\right) N(\gamma + \mathrm{i}y)\,\mathrm{e}^{\mathrm{i}\eta}(2p)^\gamma \phi(r) \\ &\quad\quad + \left(-\frac{\kappa}{r} + \frac{\mathrm{i}Z\alpha m_0 c^2}{\hbar cpr}\right) N(\gamma - \mathrm{i}y)\,\mathrm{e}^{-\mathrm{i}\eta}(2p)^\gamma \phi^*(r) \quad . \end{aligned} \tag{29}$$

Solving the equation with respect to $\mathrm{e}^{\mathrm{i}\eta}$ produces

$$\begin{aligned} \mathrm{e}^{-2\mathrm{i}\eta} &= \frac{\gamma + \mathrm{i}y}{\gamma - \mathrm{i}y}\left(-\frac{\kappa}{r} + \frac{\mathrm{i}Z\alpha m_0 c^2}{\hbar cpr}\right)^{-1} \frac{1}{\phi^*}\frac{\mathrm{d}\phi}{\mathrm{d}x} \\ &\quad - \frac{\gamma + \mathrm{i}y}{\gamma - \mathrm{i}y}\frac{(\mathrm{i}p + \mathrm{i}Z\alpha E/\hbar cpr)}{(-\kappa/r + \mathrm{i}Z\alpha m_0 c^2/\hbar cpr)}\frac{\phi}{\phi^*} \\ &= -\frac{\gamma + \mathrm{i}y}{\gamma - \mathrm{i}y}\left[\frac{-r\hbar cp}{-\kappa\hbar cp + \mathrm{i}Z\alpha m_0 c^2}\frac{1}{\phi^*}\frac{\mathrm{d}\phi}{\mathrm{d}r}\right. \\ &\quad \left. + \frac{\mathrm{i}\hbar cp^2 r + \mathrm{i}Z\alpha E}{-\kappa\hbar cp + \mathrm{i}Z\alpha m_0 c^2}\frac{\phi}{\phi^*}\right] \quad . \end{aligned} \tag{30}$$

Also

$$\begin{aligned} \frac{\mathrm{i}\hbar cp^2 r + \mathrm{i}Z\alpha E}{-\kappa\hbar cp + \mathrm{i}Z\alpha m_0 c^2} &= \frac{\hbar cp(\mathrm{i}pr + \mathrm{i}Z\alpha E/\hbar cp)}{-\hbar cp(\kappa - \mathrm{i}Z\alpha m_0 c^2/\hbar cp)} \\ &= \frac{-\mathrm{i}p(r + Z\alpha E)/\hbar cp^2}{\kappa - \mathrm{i}Z\alpha m_0 c^2/\hbar cp} \\ &= \frac{-\mathrm{i}pr(1 + Z\alpha E/\hbar cp^2 r)}{\kappa - \mathrm{i}Z\alpha m_0 c^2/\hbar cp} \end{aligned} \tag{31}$$

holds. Thus it follows that

$$\mathrm{e}^{-2\mathrm{i}\eta} = -\frac{\gamma + \mathrm{i}y}{\gamma - \mathrm{i}y}\frac{r}{\kappa - \mathrm{i}y m_0 c^2/E}\left[\frac{1}{\phi^*}\frac{\mathrm{d}\phi}{\mathrm{d}r} - \mathrm{i}p\left(1 + \frac{y}{pr}\right)\frac{\phi}{\phi^*}\right] , \tag{32}$$

and in accordance with (27) we have

$$\phi(r) = r^\gamma\,\mathrm{e}^{-\mathrm{i}pr} F(\gamma + 1 + \mathrm{i}y, 2\gamma + 1; 2\mathrm{i}pr) \quad . \tag{33}$$

Hence we may apply the following relations for the confluent hypergeometric function

$$\begin{aligned} \frac{\mathrm{d}}{\mathrm{d}x}F(a,c;x) &= \frac{a}{c}F(a+1,c+1;x) \\ &= \frac{a-c}{c}F(a,c+1;x) + F(a,c;x) \quad , \end{aligned} \tag{34}$$

$$\mathrm{e}^{-x/2}F(\gamma + 1 + \mathrm{i}y, 2\gamma + 1; x) = \mathrm{e}^{x/2}F(\gamma - \mathrm{i}y, 2\gamma + 1; -x) \quad , \tag{35}$$

$$xF(a+1,c+1;x) = c[F(a,+1,c;x) - F(z,c;x)] \tag{36}$$

and obtain

$$\begin{aligned}\phi^* &= r^\gamma \mathrm{e}^{\mathrm{i}pr} F(\gamma+1-\mathrm{i}y, 2\gamma+1; -2\mathrm{i}pr)\\ &= r^\gamma \mathrm{e}^{-\mathrm{i}pr} F(\gamma+\mathrm{i}y, 2\gamma+1; 2\mathrm{i}pr)\end{aligned} \tag{37}$$

and

$$\begin{aligned}\frac{\mathrm{d}\phi}{\mathrm{d}r} &= r^\gamma \mathrm{e}^{-\mathrm{i}pr}\frac{\mathrm{d}}{\mathrm{d}r}F(\gamma+1+\mathrm{i}y, 2\gamma+1; 2\mathrm{i}pr)\\ &\quad + F(\gamma+1+\mathrm{i}y, 2\gamma+1; 2\mathrm{i}pr)\mathrm{e}^{-\mathrm{i}pr}\left(\gamma r^{\gamma-1} - \mathrm{i}pr^\gamma\right)\\ &= r^\gamma \mathrm{e}^{-\mathrm{i}pr}\left[\frac{\gamma+1+\mathrm{i}y-2\gamma-1}{2\gamma+1} 2\mathrm{i}pF(\gamma+1+\mathrm{i}y, 2\gamma+2; 2\mathrm{i}pr)\right.\\ &\quad \left. + 2\mathrm{i}pF(\gamma+1+\mathrm{i}y, 2\gamma+1; 2\mathrm{i}pr)\right]\\ &\quad + F(\gamma+1+\mathrm{i}y, 2\gamma+1; 2\mathrm{i}pr)\mathrm{e}^{-\mathrm{i}pr}\left(\gamma r^{\gamma-1} - \mathrm{i}pr^\gamma\right)\\ &= -\frac{\gamma+\mathrm{i}y}{2\gamma+1} r^\gamma 2\mathrm{i}p\,\mathrm{e}^{-\mathrm{i}pr}F(\gamma+1+\mathrm{i}y, 2\gamma+2; 2\mathrm{i}pr)\\ &\quad + F(\gamma+1+\mathrm{i}y, 2\gamma+1; 2\mathrm{i}pr)\left(\mathrm{i}pr^\gamma \mathrm{e}^{-\mathrm{i}pr} + \gamma r^{\gamma-1}\mathrm{e}^{-\mathrm{i}pr}\right)\\ &= \frac{-\gamma+\mathrm{i}y}{2\gamma+1} r^{\gamma-1}\mathrm{e}^{-\mathrm{i}pr}(2\gamma+1)\\ &\quad \times\left[F(\gamma+1+\mathrm{i}y, 2\gamma+1; 2\mathrm{i}pr) - F(\gamma+\mathrm{i}y, 2\gamma+1; 2\mathrm{i}pr)\right]\\ &\quad + F(\gamma+1+\mathrm{i}y, 2\gamma+1; 2\mathrm{i}pr)\left[\mathrm{i}pr^\gamma \mathrm{e}^{-\mathrm{i}pr} + \gamma r^{\gamma-1}\mathrm{e}^{-\mathrm{i}pr}\right]\\ &= \mathrm{i}yr^{\gamma-1}\mathrm{e}^{-\mathrm{i}pr}F(\gamma+1+\mathrm{i}y, 2\gamma+1; 2\mathrm{i}pr)\\ &\quad - (-\gamma+\mathrm{i}y)r^{\gamma-1}\mathrm{e}^{-\mathrm{i}pr}F(\gamma+\mathrm{i}y, 2\gamma+1; 2\mathrm{i}pr)\\ &\quad + \mathrm{i}pr^\gamma \mathrm{e}^{-\mathrm{i}pr}F(\gamma+1+\mathrm{i}y, 2\gamma+1; 2\mathrm{i}pr)\\ &= \mathrm{e}^{-\mathrm{i}pr}r^{\gamma-1}\left[(\mathrm{i}y+\mathrm{i}pr)F(\gamma+1+\mathrm{i}y, 2\gamma+1; 2\mathrm{i}pr)\right.\\ &\quad \left. - (-\gamma+\mathrm{i}y)F(\gamma+\mathrm{i}y, 2\gamma+1; 2\mathrm{i}pr)\right] \quad .\end{aligned} \tag{38}$$

Now we state that

$$\frac{r}{\gamma-\mathrm{i}y}\,\frac{[\mathrm{d}\phi/\mathrm{d}r - \mathrm{i}p(1+y/pr)\phi]}{\phi^*} = 1 \tag{39}$$

or

$$\left[r\frac{\mathrm{d}\phi}{\mathrm{d}r} - \mathrm{i}pr\phi - \mathrm{i}y\phi\right] = \phi^*(\gamma-\mathrm{i}y) \quad , \tag{40}$$

which can indeed easily be proved:

$$\begin{aligned}&(\mathrm{i}y+\mathrm{i}pr)F(\gamma+1+\mathrm{i}y, 2y+1; 2\mathrm{i}pr) - (-\gamma+\mathrm{i}y)F(\gamma+\mathrm{i}y, 2\gamma+1; 2\mathrm{i}pr)\\ &\qquad - (\mathrm{i}pr+\mathrm{i}y)F(\gamma+1+\mathrm{i}y, 2\gamma+1; 2\mathrm{i}pr)\\ &\quad = (\gamma-\mathrm{i}y)F(\gamma+\mathrm{i}y, 2\gamma+1; 2\mathrm{i}pr) \quad , \quad \text{qed.}\end{aligned} \tag{41}$$

From the statement (39), the phase (32) follows as

$$\mathrm{e}^{2\mathrm{i}\eta} = -\frac{\kappa - \mathrm{i}ym_0c^2/E}{\gamma+\mathrm{i}y} \quad . \tag{42}$$

Example 9.9.

For the radial functions of equations (4) and (5) we may write

$$\begin{aligned} G &= C_1 N (2pr)^\gamma \left\{ (\gamma + \mathrm{i}y)\,\mathrm{e}^{-\mathrm{i}pr+\mathrm{i}\eta} F(\gamma+1+\mathrm{i}y, 2\gamma+1; 2\mathrm{i}pr) + \text{c.c.} \right\} \quad , \\ F &= \mathrm{i} C_2 N (2pr)^\gamma \left\{ (\gamma + \mathrm{i}y)\,\mathrm{e}^{-\mathrm{i}pr+\mathrm{i}\eta} F(\gamma+1+\mathrm{i}y, 2\gamma+1; 2\mathrm{i}pr) - \text{c.c.} \right\} \quad , \end{aligned} \tag{43}$$

taking (27) and (15) into account, (by c.c. we mean here the complex conjugate) where

$$C_1 = \begin{cases} \sqrt{E + m_0 c^2} & \text{for} \quad E > m_0 c^2 \\ \sqrt{-E - m_0 c^2} & \text{for} \quad E < -m_0 c^2 \end{cases} \quad , \tag{44}$$

$$C_2 = \begin{cases} \sqrt{E - m_0 c^2} & \text{for} \quad E > m_0 c^2 \\ -\sqrt{-E + m_0 c^2} & \text{for} \quad E < -m_0 c^2 \end{cases} \quad . \tag{45}$$

Let us now determine the normalization factor N. We normalize the continuum wave functions with respect to the energy axis, i.e. to delta functions of energy:

$$\int \mathrm{d}^3 r \, \psi^\dagger_{E'} \psi_E = \delta(E' - E) \quad . \tag{46}$$

Furthermore, we assume that for $r \to \infty$

$$\begin{aligned} G &= A C_1 \cos(pr + \delta) \quad , \\ F &= -A C_2 \sin(pr + \delta) \end{aligned} \tag{47}$$

holds. We will discuss the proof of (47) later on. According to (4) and (15), for ϕ_1 this means that

$$\phi_1 = \tfrac{1}{2} A \exp\left[\mathrm{i}(pr + \delta)\right] \quad . \tag{48}$$

Now we show that for

$$A = \frac{1}{\sqrt{\pi p}} \tag{49}$$

the normalization condition (46) is fulfilled. Indeed, for $E > m_0 c^2$ it follows that

$$\begin{aligned} \int_0^\infty \mathrm{d}r \Bigg\{ & \sqrt{E + m_0 c^2} \frac{1}{2\sqrt{\pi p}} \left[\mathrm{e}^{\mathrm{i}(pr+\delta)} + \mathrm{e}^{-\mathrm{i}(pr+\delta)}\right] \\ & \times \sqrt{E' + m_0 c^2} \frac{1}{2\sqrt{\pi p'}} \left[\mathrm{e}^{\mathrm{i}(p'r+\delta')} + \mathrm{e}^{-\mathrm{i}(p'r+\delta')}\right] \\ & - \sqrt{E - m_0 c^2} \frac{1}{2\sqrt{\pi p}} \left[\mathrm{e}^{\mathrm{i}(pr+\delta)} - \mathrm{e}^{-\mathrm{i}(pr+\delta)}\right] \\ & \times \sqrt{E' - m_0 c^2} \frac{1}{2\sqrt{\pi p'}} \left[\mathrm{e}^{\mathrm{i}(p'r+\delta')} - \mathrm{e}^{-\mathrm{i}(p'r+\delta')}\right] \Bigg\} \end{aligned}$$

Example 9.9.

$$
\begin{aligned}
&= \frac{1}{2}\int_{-\infty}^{\infty} \mathrm{d}r \Bigg\{ \sqrt{E+m_0c^2}\sqrt{E'+m_0c^2}\frac{1}{4\pi}\frac{1}{(pp')^{1/2}} \\
&\quad \times \left[\mathrm{e}^{\mathrm{i}(p+p')r+\delta+\delta'} + \mathrm{e}^{\mathrm{i}(p-p')r+\delta-\delta'}\,\mathrm{e}^{\mathrm{i}(p'-p)r+\delta'-\delta} + \mathrm{e}^{-\mathrm{i}(p+p')r-\delta-\delta'} \right] \\
&\quad - \sqrt{E-m_0c^2}\sqrt{E'-m_0c^2}\frac{1}{4\pi}\frac{1}{(pp')^{1/2}} \\
&\quad \times \left[\mathrm{e}^{\mathrm{i}(p+p')r+\delta+\delta'} - \mathrm{e}^{\mathrm{i}(p-p')r+\delta-\delta'} - \mathrm{e}^{\mathrm{i}(p'-p)r-\delta+\delta'} + \mathrm{e}^{-\mathrm{i}(p+p')r-\delta-\delta'} \right] \Bigg\} \\
&= \frac{1}{2}\Bigg[\sqrt{E+m_0c^2}\sqrt{E'+m_0c^2}\frac{1}{2}\frac{1}{(pp')^{1/2}}\left[\delta(p-p')+\delta(p-p')\right] \\
&\quad + \sqrt{E-m_0c^2}\sqrt{E'-m_0c^2}\frac{1}{2}\frac{1}{(pp')^{1/2}}\left[\delta(p-p')+\delta(p-p')\right] \Bigg] \\
&= \frac{1}{2}\left[\frac{(E+m_0c^2)}{p}\delta(p-p') + \frac{(E-m_0c^2)}{p}\delta(p-p') \right] \\
&= \frac{E}{p}\delta(p-p') = \frac{E}{p}\frac{\delta(E-E')}{\mathrm{d}p/\mathrm{d}E} = \frac{E}{p}\frac{\delta(E-E')}{E/p} \\
&= \delta(E-E') \quad ,
\end{aligned} \tag{50}
$$

where we have set

$$
p = \sqrt{E^2 - m_0c^2} \quad . \tag{51}
$$

The derivation of relation (50) is only valid in a strict sense if the phase δ does not depend on r. Moreover, the following relations for the δ function were used:

$$
\frac{1}{2\pi}\int_{-\infty}^{\infty} \mathrm{e}^{\mathrm{i}kx}\,\mathrm{d}k = \delta(x) \quad , \tag{52}
$$

$$
\delta(-x) = \delta(x) \quad , \tag{53}
$$

$$
f(x)\delta(x-a) = f(a)\delta(x-a) \quad , \tag{54}
$$

$$
\delta\left[\varphi(x') - \varphi(x)\right] = \frac{1}{|\mathrm{d}\varphi(x)/\mathrm{d}x|}\delta(x'-x) \quad , \tag{55}
$$

$\delta(x) = 0$ for $x > 0$. The result (50) is also valid for $E < -m_0c^2$. In order to describe the asymptotics of the functions (43), we apply the asymptotic behaviour of the confluent hypergeometric functions for $r \to \infty$:

$$
F(a,c;r) \to \frac{\Gamma(c)}{\Gamma(a)} r^{a-c}\,\mathrm{e}^{r} \quad . \tag{56}
$$

Thereby from (43) we obtain for $r \to \infty$ that

$$
\begin{aligned}
G &= C_1 N(2pr)^{\gamma}\Gamma(2\gamma+1)\left[\frac{(\gamma+\mathrm{i}y)\,\mathrm{e}^{\mathrm{i}pr+\mathrm{i}\eta}}{\Gamma(\gamma+1+\mathrm{i}y)}(2\mathrm{i}pr)^{\mathrm{i}y-\gamma} + \text{c.c.} \right] \quad , \\
F &= \mathrm{i}C_2 N(2pr)^{\gamma}\Gamma(2\gamma+1)\left[\frac{(\gamma+\mathrm{i}y)\,\mathrm{e}^{\mathrm{i}pr+\mathrm{i}\eta}}{\Gamma(\gamma+1+\mathrm{i}y)}(2\mathrm{i}pr)^{\mathrm{i}y-\gamma} - \text{c.c.} \right] \quad .
\end{aligned} \tag{57}
$$

Example 9.9.

Now we perform some transformations:

$$\mathrm{i}^{-\gamma} = \mathrm{e}^{z} \Rightarrow z = -\gamma \ln(\mathrm{i}) \quad .$$

For $z = x + \mathrm{i}y = r\,\mathrm{e}^{\mathrm{i}\theta}$, $\ln z = \ln r + \mathrm{i}\theta \Rightarrow \ln \mathrm{i} = \mathrm{i}\frac{\pi}{2}$ and it follows that

$$\mathrm{i}^{-\gamma} = \mathrm{e}^{-\pi \mathrm{i}\gamma/2} \quad . \tag{58}$$

Analogously one gets

$$\begin{aligned}(2\mathrm{i}pr)^{\mathrm{i}y} = \mathrm{e}^{z} \Rightarrow z &= \mathrm{i}y \ln(2\mathrm{i}pr) = \mathrm{i}y\big(\ln \mathrm{i} + \ln(2pr)\big) \\ &= \mathrm{i}y\mathrm{i}\tfrac{\pi}{2} + \mathrm{i}y\ln(2pr) \quad ,\end{aligned}$$

i.e.

$$(2\mathrm{i}pr)^{\mathrm{i}y} = \mathrm{e}^{-\pi y/2}\,\mathrm{e}^{\mathrm{i}y\ln(2pr)} \quad . \tag{59}$$

Furthermore, with $\Gamma(z+1) = z\Gamma(z)$ one calculates

$$\begin{aligned}\frac{\gamma + \mathrm{i}y}{\Gamma(\gamma + 1 + \mathrm{i}y)} &= \frac{1}{\Gamma(\gamma + \mathrm{i}y)} = \frac{1}{|\Gamma(\gamma + \mathrm{i}y)|\mathrm{e}^{i\arg\Gamma(\gamma+\mathrm{i}y)}} \\ &= \frac{\mathrm{e}^{-\mathrm{i}\arg\Gamma(\gamma+\mathrm{i}y)}}{|\Gamma(\gamma + \mathrm{i}y)|} \quad .\end{aligned} \tag{60}$$

With that the radial functions from (57) have the asymptotic form (47). In doing so, one gets

$$A = \frac{1}{\sqrt{\pi p}} = \frac{2N\,\mathrm{e}^{-\pi y/2}\Gamma(2\gamma + 1)}{|\Gamma(\gamma + \mathrm{i}y)|} \quad . \tag{61}$$

The normalization constant N is fixed by this equation. For a phase δ we obtain

$$\delta = y\ln(2pr) - \arg\Gamma(\gamma + \mathrm{i}y) - \frac{\pi\gamma}{2} + \eta \quad . \tag{62}$$

This is the *Coulomb phase*, which is already known to us from the nonrelativistic problem. Thus, the final results for the wave function are

$$\begin{aligned}G_\kappa &= \frac{C_1(2pr)^{\gamma}\,\mathrm{e}^{\pi y/2}|\Gamma(\gamma + \mathrm{i}y)|}{2(\pi p)^{1/2}\Gamma(2\gamma + 1)} \\ &\quad \times \left\{\mathrm{e}^{-\mathrm{i}pr + \mathrm{i}\eta}(\gamma + \mathrm{i}y)F(\gamma + 1 + \mathrm{i}y, 2\gamma + 1; 2\mathrm{i}pr) + \mathrm{c.c.}\right\} \quad , \\ F_\kappa &= \frac{\mathrm{i}C_2(2pr)^{\gamma}\,\mathrm{e}^{\pi y/2}|\Gamma(\gamma + \mathrm{i}y)|}{2(\pi p)^{1/2}\Gamma(2\gamma + 1)} \\ &\quad \times \left\{\mathrm{e}^{-\mathrm{i}pr + \mathrm{i}\eta}(\gamma + \mathrm{i}y)F(\gamma + 1 + \mathrm{i}y, 2\gamma + 1; 2\mathrm{i}pr) - \mathrm{c.c.}\right\} \quad .\end{aligned} \tag{63}$$

We see that the calculation of the continuum wave function is a difficult task. Besides the partial waves we have discussed here, one can also construct travelling waves.[11] Let us merely remark that for the calculation of all scattering processes with heavier nuclei, and also for the calculation of quantum electrodynamic processes, the here-determined continuum waves represent the necessary and appropriate technicals tools. For the study of "new" quantum electrodynamic processes in the strong fields of heavy ion collisions, the continuum waves have even to be calculated for extended nuclei.

[11] For a detailed discussion of this calculation, see M.E. Rose: *Relativistic Electron Theory* (Wiley, New York, London)

EXAMPLE

9.10 Muonic Atoms

A muon is a particle which has the same properties as the electron in almost every attribute, except that it is about 207 times heavier than the electron, i.e.

$$m_\mu = 207\, m_\mathrm{e}$$

($m_\mathrm{e}c^2 = 0.511004\,\mathrm{MeV}$, $m_\mu c^2 = 105.655\,\mathrm{MeV}$).

Since it also has spin $\frac{1}{2}$ and negative charge (the antiparticle of the μ^- is the μ^+, which has positive charge), it also obeys the Dirac equation. Some effects, which play only a minor role for the electron in usual atoms, become important for the μ^- when it "circles" around the nucleus forming a *muonic atom*. These effects are linked to the large mass of the muon, which implies that the Bohr radius of the μ^- is smaller than that of the electron by a factor of

$$\frac{m_\mathrm{e}}{m_\mu} = \frac{1}{207} \quad . \tag{1}$$

Thereby all the effects which are connected with the extension of the nucleus (modified Coulomb potential at smaller distances, quadrupole interaction, etc.) are more important in the case of the muon than for the electron. Some of these effects we will briefly explain in the following. Before that, though, a brief abstract of the formation and the most important physical processes involved in the creation of muonic atoms seems appropriate.

Production of Muonic Atoms. Muons are created by performing inelastic scattering experiments of high-energy protons with protons (hydrogen). There, pions are produced according to the reactions:

$$\mathrm{p} + \mathrm{p} \begin{cases} \rightarrow \mathrm{p} + \mathrm{p} + \pi^- + \pi^+ \dots \\ \rightarrow \mathrm{p} + \mathrm{n} + \pi^+ + \dots \end{cases} \quad . \tag{2}$$

The pions are separated from the beam with magnetic fields. They decay into muons within about 10^{-8} s due to the weak interaction:

$$\pi^- \rightarrow \mu^- + \nu_\mu \quad . \tag{3}$$

The μ^- produced in this way are now decelerated; then, the slow μ^- are bombarded onto ordinary electronic atoms, e.g. Pb. Through the interaction of the μ^- with the electrons, some electrons will be knocked out of the atom, the μ^- will be further decelerated and finally captured by the nucleus. The "capture orbit" of a muon is one of the outer orbits with principal quantum number $n \approx 14$. Such a muonic atom (with one muon and still many electrons) is then in a highly excited state. In the first step, the de-excitation takes place via transitions within the outer shells in which the muon transfers energy to further electrons, which are consequently emitted (Auger emission). The transitions between inner muonic atom shells result in γ emission, which can be observed and measured (see Fig. 9.14).

We now discuss a few remarkable differences between muonic atoms and electronic atoms.

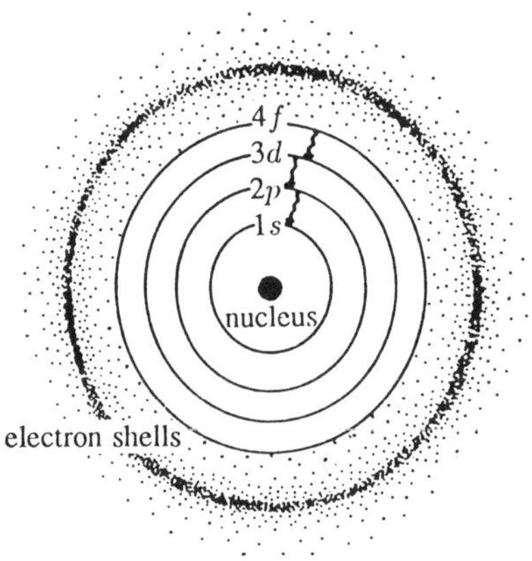

Fig. 9.14. Illustration of the inner μ orbits around an atomic nucleus. The influence of the electrons, which are positioned far out, is schematically indicated by a cloud

(a) *Recoil Effects.* Since, in comparison to the muon, the nucleus does not have infinitely high mass, both particles move around a common centre of mass, which is not in the centre of the nucleus. For the Schrödinger equation the centre-of-mass motion can, as in classical mechanics, easily be separated. Then the relative motion is simply described by a one-particle equation with the reduced mass μ_μ,

$$\mu_\mu = \frac{m_\mu \cdot M_{\text{nucleus}}}{m_\mu + M_{\text{nucleus}}} \quad . \tag{4}$$

Here m_μ is the mass of the muon. In the case of the Dirac equation, this separation does not work, because:

(1) A satisfying two-body Dirac equation[12] does not exist.
(2) The centre-of-mass system can no longer be defined geometrically (as in classical mechanics), but only dynamically; namely as a Lorentz system, in which the sum of the momenta of all the involved particles vanishes, i.e. $\sum_{\nu=1}^{A} \boldsymbol{p}_\nu = 0$. In an approximation one can manage the problem by assigning the above-mentioned reduced mass μ_μ to the muon in the Dirac equation. The binding energies of the muons are considerably influenced by the thus (approximately) considered recoil effect. In view of the experimental precision of $\Delta E/E \leq 10^{-4}$–10^{-5}, there appears no difficulty in verifying these effects.

(b) *Retardation Effects.* If one considers the interaction between the nucleus and the muon in the form of a static Coulomb potential in the Dirac equation, one thereby performs an *external field approximation.* The interaction within this approximation takes place without time delay; the effects of retardation which are due to the recoil motion are neglected. If the velocity of the muon in the inner shells of heavy nuclei becomes comparable to the velocity of light c, the retardational effects should become large and then also the previously mentioned "reduced mass approximation" turns out to be insufficient. In this case it is necessary to consider

[12] A relativistic many-body mechanics has been proposed by F. Rohrlich [see Annals of Physics **117**, 292 (1979)]. It remains, though, to be seen if this theory can be quantized in a satisfactory way. We refer also to the Bethe–Salpeter equation, which is discussed in W. Greiner, J. Reinhardt: *Quantum Electrodynamics*, 2nd ed. (Springer, Berlin, Heidelberg 1994).

Example 9.10.

a two-body equation derived from the quantized theory (field theory), the so-called *Bethe–Salpeter equation*. Since these effects are just near to the edge of the maximal precision of present experiments, we will not discuss them in this context any further.

(c) *Screening by Electrons.* Though the Bohr radius of the many electrons in the atom is much larger than the one of the muon, the tails of the electrons' wave functions reach into the inner region (see Fig. 9.15).

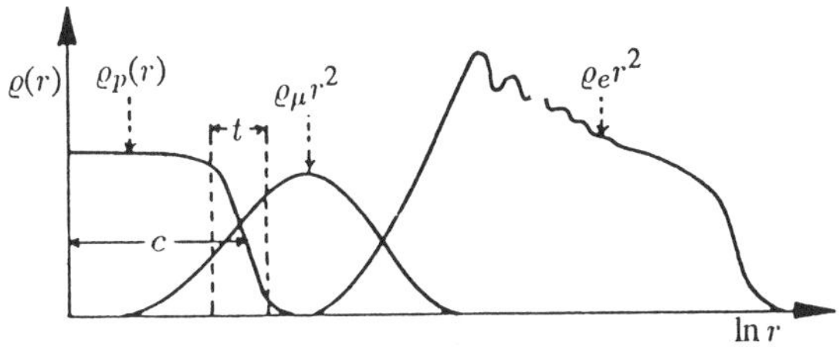

Fig. 9.15. Qualitative picture of the charge distribution of a nucleus $\varrho_p(r)$ with half-value radius c and surface thickness t. The charge distributions of the muon ϱ_μ and the electrons ϱ_e are multiplied by r^2. It can be recognized that the muon stays partly in the interior of the nucleus and the electrons stay partly within the muonic orbit. Note the logarithmic r scale!

Thereby the electrons cause a screening potential for the muons which is not negligible. For symmetric charge distributions $\varrho_e(r)$ it can easily be calculated that

$$V_{\text{screening}}(r) = 4\pi e^2 \left\{ \frac{1}{r} \int_0^r \varrho_e(r') r'^2 \, dr' + \int_r^\infty \varrho_e(r') r' \, dr' \right\} \quad . \tag{5}$$

Here

$$\varrho_e(r) = \sum_{i=1}^{N} \psi_i^* \psi_i(r) \tag{6}$$

is the charge density of the N electrons still bound within the atom, ψ_i being their wave function. For more precise calculations one even has to determine the wave function of the muon and those of the electrons self-consistently. As a typical case for the screening of the muon by the electrons, we consider the $5g - 4f$ transition in muonic lead, whose transition energy amounts to $\Delta E \sim 400\,\text{keV}$. Due to the electron screening, the single muon levels will the less bound by an energy of about 17 keV. Calculating the transition energy, the effect of screening then reduces to a few eV, because both levels are shifted in the same direction (to weaker binding).

(d) *Vacuum Polarization, Self-energy and Anomalous Magnetic Moment.* In Chap. 12 we will learn about field-theoretical effects,[13] which cause small, but measurable, deviations from the Coulomb interaction of the electrons or muons with the nucleus. One has to consider the effect of vacuum polarization, the self-energy and the anomalous magnetic moment. The vacuum polarization describes the creation of

[13] In particular this is discussed (with more precise calculations) in W. Greiner, J. Reinhardt: *Quantum Electrodynamics*, 2nd ed. (Springer, Berlin, Heidelberg 1994).

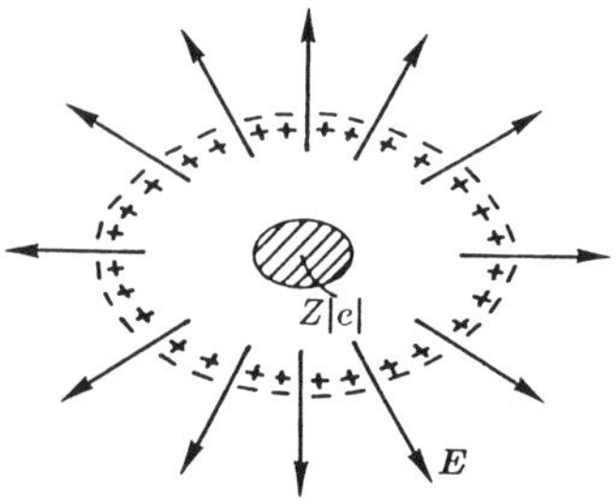

Fig. 9.16. Illustration of the vacuum polarization as a "dipole cloud" in the electric field of the nucleus

virtual electron–positron pairs in strong electromagnetic fields. The virtual electron–positron pairs form a cloud of dipoles, leading to a modification of the Coulomb potential (see Fig. 9.16).

The properties of the vacuum are similar to a dielectric medium. The dominant part of the vacuum polarization potential is the so-called *Uehling potential* given by

$$V_{\mathrm{vp}}(r) = -e^2 \frac{4\alpha}{3} \frac{\lambda\!\!\!^{-}_{\mathrm{e}}}{2} \int_0^\infty \varrho(r') \left[Z_1(|r - r'|) - Z_1(r + r') \right] \mathrm{d}r \quad , \tag{7}$$

where the structure function is

$$Z_1(r) = \int_1^\infty \exp\left\{ -\frac{2}{\lambda\!\!\!^{-}_{\mathrm{e}}} r\xi \right\} \left(1 + \frac{1}{2\xi^2} \right) \frac{(\xi^2 - 1)^{1/2}}{\xi^2} \frac{1}{\xi} \, \mathrm{d}\xi \quad , \tag{8}$$

with $\lambda\!\!\!^{-}_{\mathrm{e}} = 386.1592\,\mathrm{fm}$ being the Compton wavelength of the electron. For $r > \lambda\!\!\!^{-}_{\mathrm{e}}$, i.e. in the region where mainly atomic electrons are present, the vacuum polarization potential falls of exponentially. Therefore very heavy particles, like the muons (and pions), whose probability is large in the region $R_{\mathrm{nucleus}} < r < \lambda\!\!\!^{-}_{\mathrm{e}}$, are particularly suitable for examining the vacuum polarization potential. Nowadays the precision of experiments with muonic atoms is so accurate that processes of higher order than those in (7) and (8) also have to be considered. The self-energy describes the interaction of one particle with itself by emission and reabsorbtion of a photon (see Fig. 9.17). There is a similar cause for the so-called anomalous magnetic moment, whose contribution to the energy is illustrated by the graph in Fig. 9.17. Here we only note that the self-energy and the anomalous magnetic moment are of the order $1/m$. Therefore they are more important in the case of electrons than for muons. In view of the present experimental accuracy the latter only play a minor role.

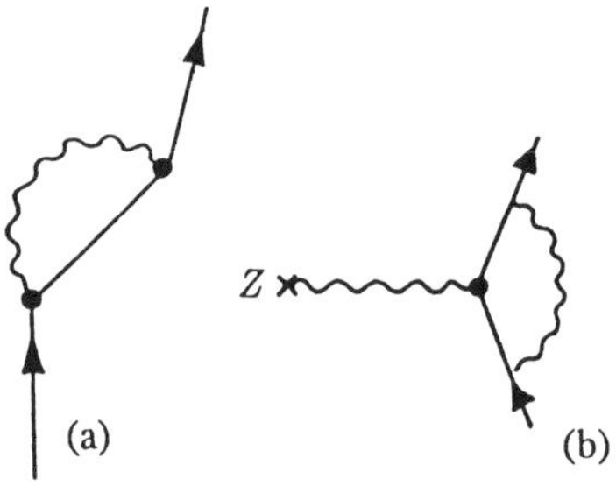

Fig. 9.17. (**a**) Self-energy diagram, (**b**) anomalous magnetic moment

(e) *Nuclear Deformation.* Until now we have assumed that nuclei have a spherical symmetric charge distribution, which may be well described by means of a two-parameter charge distribution with half-density radius c and surface thickness t as parameters. The most common one is the so-called *Fermi distribution*, which is given by

$$\varrho = \frac{\varrho_0}{1 + \mathrm{e}^{(r-c)/t}} \quad . \tag{9}$$

The *half-density radius* c and the *surface thickness* t are free parameters, which are determined from fits to experimental cross-sections of fast electrons. Another

Example 9.10.

method for their determination is the following: When measuring the transition energies of various γ transitions in a muonic atom, accurate values for c and t can be determined by means of so-called $c-t$ diagrams. In other words: The transition energies of the muonic atom depend on the parameters c and t of the nuclear charge distribution (9). However, many nuclei show deviations from the spherical shape. For this reason one has to expand the nuclear charge distribution into multipole moments:

$$\varrho(\boldsymbol{r}) = \varrho_0(r) + \sum_{lm} \varrho_{lm}(r) Y_{lm}(\theta, \varphi) \quad . \tag{10}$$

In most cases nuclei are axially symmetric, but deformed in a cigar-like manner, i.e. they exhibit a prolate deformation. Then (10) reduces to

$$\varrho(\boldsymbol{r}) = \varrho_0(r) + \varrho_2(r) Y_{20} + \dots \quad , \tag{11}$$

where $\varrho_2 \equiv \varrho_{20}$. The second term describes the quadrupole deformation. Accordingly, the quadrupole moment of the nucleus is defined as

$$Q_0 = \int \varrho(\boldsymbol{r}) r^2 Y_{20}^* \, \mathrm{d}V = 2\sqrt{\frac{4\pi}{5}} \int_0^\infty \varrho_2(r) r^4 \, \mathrm{d}r \quad . \tag{12}$$

Not only the absolute values, but also signs of these quadrupole moments can be determined from the pronounced hyperfine structure of the muonic atoms. Hyperfine structure means observable level shifts, additional to the $\boldsymbol{l} \cdot \boldsymbol{s}$ splitting (fine structure).

In order to reproduce transition energies, e.g., in a muonic uranium atom one introduces even four-parametric charge distributions, like

$$\varrho(r, c, t, \beta, \gamma) = \varrho_0 \left[1 + \exp\left(4 \ln 3 \frac{r - c(1 + \beta Y_{20})}{t(1 + \beta\gamma Y_{20})} \right) \right]^{-1} \quad . \tag{13}$$

In comparison to (9) the half-value radius and the surface thickness are now treated as angular-dependent quantities, implying that the density of the "atmosphere" at the tip of the "nuclear cigar" is different from that on its short axis. A typical spectrum of a heavy muonic atom is sketched for uranium in Fig. 9.18. Furthermore it illustrates the difference between the solutions for point-like and extended nuclei, and the different effects discussed above are indicated as far as their order of magnitude allow.

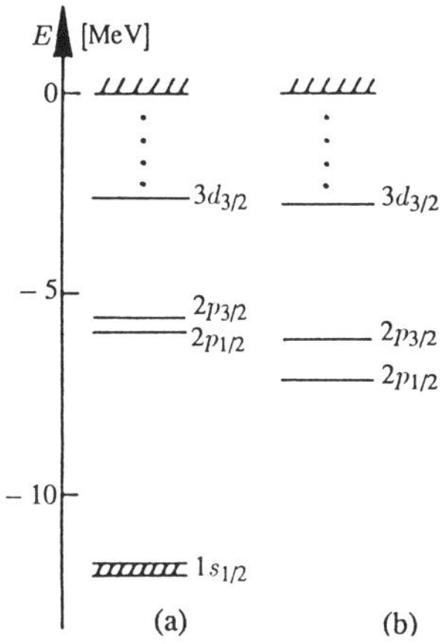

Fig. 9.18a, b. Binding energies of muons in a uranium atom. (**a**) The more realistic spectrum with extended nucleus. The hyperfine splitting due to the quadrupole moments of the uranium nucleus is indicated for the $1s$ level. (**b**) The same spectrum for a point-like U nucleus. The $1s_{1/2}$ state lies at -27.35 MeV, i.e. outside the diagram

(f) *Nuclear Polarization.* Until now the nucleus has been assumed to be static. This implies that the nucleus will not be influenced by the presence of the muon. However, the nucleus behaves like a dielectric medium and thus it is polarizable. It can be excited by the muon and its charge distribution can be slightly changed. We write the total Hamiltonian

$$\hat{H} = \hat{H}_{\text{nucleus}}(\boldsymbol{r}') + \hat{H}_{\text{meson}}(\boldsymbol{r}) + \hat{H}_{\text{int}}(\boldsymbol{r}, \boldsymbol{r}') \quad , \tag{14}$$

where $\hat{H}_{\text{int}} = -Ze^2/|\boldsymbol{r} - \boldsymbol{r}'|$ describes the electromagnetic interaction between nucleus and meson. One performs a multipole expansion

Example 9.10.

$$\hat{H}_{\text{int}} = -\sum_{l,m} \left[\frac{4\pi e^2}{2l+1}\right] \frac{r_<^l}{r_>^{l+1}} Y_{lm}(\Omega') Y_{lm}^*(\Omega_{\text{meson}}) \tag{15}$$

and treats $\hat{H}_{\text{int}}$ in higher-order perturbation theory.[14]

Let us note a typical value for the contribution of nuclear polarization effects. In muonic Pb the transition from $2p_{1/2}$ to $1s_{1/2}$ has a transition energy of $\Delta E \sim 10\,\text{MeV}$, whereas the contribution due to nuclear polarization ΔE_{np} is of the order of 10 keV.

(g) *Isomeric Shift.* The ground and excited states of a nucleus do not need to have equal charge radii. If a nucleus has been excited, then usually the average quadratic nuclear radius changes by Δr^2. Such an isomeric shift gives rise to a level shift in muonic atoms and can be deduced (measured) from studying the level structure. For the first excited nuclear state one finds typical changes of the radius of the order of

$$\frac{\Delta\langle r^2\rangle}{\langle r^2\rangle} \sim 10^{-4} \quad \text{to} \quad 2\times 10^{-3} \quad .$$

Such an excited state may be formed during the cascade of a muon within the atom. The mesonic transition energy is not always emitted by radiation of a real photon, but can also be absorbed directly in the nucleus. Accordingly, the nucleus will be excited into higher states. During the subsequent transitions the meson feels a different charge radius.[15]

(h) *Polarization of Muons and Pions.* In our previous considerations the muon and pion have been considered as structureless particles, disregarding their internal degrees of freedom. However, one knows that the pion has a charge radius of $r_{\text{pion}} \cong 0.8\,\text{fm}$. Moreover, considering that a pion consists of two quarks, it becomes obvious that also the pion can be polarized in the presence of the strong field of a nucleus. However, the corresponding energy shifts turn out to be just too small for an experimental verification at the present time. Concerning the muon, which belongs to the lepton family, an "extension" (a substructure) is still unknown, and up to now leptons (electrons, muons, tauons) also seem to be point-like particles.

EXERCISE

9.11 Dirac Equation for the Interaction Between a Nuclear and an External Field Taking Account of the Anomalous Magnetic Moment

The Dirac equation for a nucleon interacting with an external electromagnetic field contains an additional term describing the interaction between the anomalous magnetic moment of the nucleon with the electromagnetic field:

[14] We refer to the literature. See, e.g., J.M. Eisenberg, W. Greiner: *Nuclear Theory, Vol. II: Excitation Mechanisms of the Nucleus*, 3rd ed. (North-Holland, Amsterdam 1988).

[15] See: J.M. Eisenberg, W. Greiner: *Nuclear Theory, Vol. I: Nuclear Models*, 3rd ed. (North-Holland, Amsterdam 1987) p. 73.

$$\left(\mathrm{i}\hbar c \nabla\!\!\!/ - \frac{e_j}{c} A\!\!\!/ + K_j \frac{e\hbar}{4M_j c} \hat{\sigma}_{\mu\nu} F^{\mu\nu} - M_j c^2 \right) \psi(x) = 0 \quad ,$$

Exercise 9.11.

where $j = \mathrm{p}$ and $e_\mathrm{p} = |e|$ for the proton and $j = \mathrm{n}$ and $e_\mathrm{n} = 0$ for the neutron, respectively.

Problem. (a) Show that the additional term does not violate the relativistic invariance of the Dirac equation.

(b) Show that this term satisfies the hermiticity condition

$$\gamma_0 \left(K_j \frac{e\hbar}{4M_j c} \hat{\sigma}_{\mu\nu} F^{\mu\nu} \right)^\dagger \gamma_0 = K_j \frac{e\hbar}{4M_j c} \hat{\sigma}_{\mu\nu} F^{\mu\nu} \quad .$$

Remembering that when turning from the matrices $\hat{\boldsymbol{\alpha}}$, $\hat{\beta}$ to the gamma matrices in the Dirac equation, a term γ_0 has been factorized out; see (3.8).

(c) Show that the choice $K_\mathrm{p} = 1.79$ and $K_\mathrm{n} = -1.91$ corresponds to the experimentally observed magnetic moment.

Solution. (a) Relativistic invariance of the term $\hat{\sigma}_{\mu\nu} F^{\mu\nu}$ as a contraction of two Lorentz tensors (covariants) is obviously manifest.

(b) Since $\hat{\sigma}_{\mu\nu} F^{\mu\nu}$ represents a product of an operator $\hat{\sigma}_{\mu\nu}$ with a real tensor $F^{\mu\nu} = (F^{\mu\nu})^*$ [see (1.15)], i.e.

$$\hat{\sigma}_{\mu\nu} F^{\mu\nu} = \sum_{\mu,\nu} \hat{\sigma}_{\mu\nu} F^{\mu\nu} \quad ,$$

with the field strength tensor $F^{\mu\nu}$, it is sufficient to show that

$$\gamma_0 \hat{\sigma}^\dagger_{\mu\nu} \gamma_0 = \hat{\sigma}_{\mu\nu}$$

holds. For this purpose we use the relation $\gamma_0^\dagger = \gamma_0$ and $\boldsymbol{\gamma}^\dagger = -\boldsymbol{\gamma}$, and in shorthand notation we have $\gamma^\dagger_\mu = \gamma_0 \gamma_\mu \gamma_0$. We explicitly derive

$$\begin{aligned}
\gamma_0 \hat{\sigma}^\dagger_{\mu\nu} \gamma_0 &= \gamma_0 \left[\frac{\mathrm{i}}{2} \left(\gamma_\mu \gamma_\nu - \gamma_\nu \gamma_\mu \right) \right]^\dagger \gamma_0 \\
&= -\frac{\mathrm{i}}{2} \gamma_0 \left(\gamma^\dagger_\nu \gamma^\dagger_\mu - \gamma^\dagger_\mu \gamma^\dagger_\nu \right) \gamma_0 \\
&= -\frac{\mathrm{i}}{2} \gamma_0 \left(\gamma_0 \gamma_\nu \gamma_0^2 \gamma_\mu \gamma_0 - \gamma_0 \gamma_\mu \gamma_0^2 \gamma_\nu \gamma_0 \right) \gamma_0 \\
&= -\frac{\mathrm{i}}{2} \left(\gamma_\nu \gamma_\mu - \gamma_\mu \gamma_\nu \right) = \hat{\sigma}_{\mu\nu} \quad ,
\end{aligned}$$

where $\gamma_0^2 = \hat{1}\!\!1$ has been used.

(c) First we rewrite the interaction term:

$$\begin{aligned}
\hat{\sigma}_{\mu\nu} F^{\mu\nu} &= \frac{\mathrm{i}}{2} \gamma_\mu \gamma_\nu F^{\mu\nu} - \frac{\mathrm{i}}{2} \gamma_\nu \gamma_\mu F^{\mu\nu} \\
&= \frac{\mathrm{i}}{2} \gamma_\mu \gamma_\nu F^{\mu\nu} - \frac{\mathrm{i}}{2} \gamma_\mu \gamma_\nu F^{\nu\mu} = \mathrm{i} \gamma_\mu \gamma_\nu F^{\mu\nu} \quad ,
\end{aligned}$$

Exercise 9.11.

where we have renamed the dummy indices, and the relation $F^{\mu\nu} = -F^{\nu\mu}$ has been used. By considering that

$$E^i = F^{0i} \quad \text{and} \quad \boldsymbol{B} = \left(F^{23}, F^{31}, F^{12}\right)$$

and $\gamma_k\gamma_l = \mathrm{i}\hat{\Sigma}_j$, with

$$\hat{\Sigma}_j = \begin{pmatrix} \hat{\sigma}_j & 0 \\ 0 & \hat{\sigma}_j \end{pmatrix}$$

as well as $\gamma_k\gamma_0 = \hat{\alpha}_k$, the sum reads explicitly:

$$\sum_{\mu\nu} \gamma_\mu\gamma_\nu F^{\mu\nu} = \sum_{j<k} \gamma_j\gamma_k F^{jk} + \sum_{j<k} \gamma_k\gamma_j F^{kj} - \gamma_0\gamma_k F^{0k} - \gamma_k\gamma_0 F^{k0} + \gamma_0\gamma_0 F^{00} .$$

In view of the relation $F^{00} = 0$, $\gamma_j\gamma_k = -\gamma_k\gamma_j$ and $F^{jk} = -F^{kj}$, the sum becomes

$$\sum_{\mu\nu} \gamma_\mu\gamma_\nu F^{\mu\nu} = 2\left(\sum_{j<k} \gamma_j\gamma_k F^{jk} - \sum_j \gamma_j\gamma_0 F^{j0}\right) .$$

We obtain

$$\gamma_1\gamma_2 F^{12} = \mathrm{i}\hat{\Sigma}_3 B_3 \quad , \quad \gamma_2\gamma_3 F^{23} = \mathrm{i}\hat{\Sigma}_1 B_1 \quad .$$
$$\gamma_3\gamma_1 F^{31} = \mathrm{i}\hat{\Sigma}_2 B_2 = \gamma_1\gamma_3 F^{13} \quad ,$$

and thus

$$\sum_{j<k} \gamma_j\gamma_k F^{jk} = \mathrm{i}\hat{\boldsymbol{\Sigma}}\cdot\boldsymbol{B} \quad \text{and} \quad \sum_j \gamma_j\gamma_0 F^{j0} = -\hat{\boldsymbol{\alpha}}\cdot\boldsymbol{E} \quad ,$$

which yields

$$\sum_{\mu,\nu} \hat{\sigma}_{\mu\nu} F^{\mu\nu} = 2\mathrm{i}\left\{\mathrm{i}\hat{\boldsymbol{\Sigma}}\cdot\boldsymbol{B} + \hat{\boldsymbol{\alpha}}\cdot\boldsymbol{E}\right\} \quad .$$

Instead of the original Dirac equation we obtain

$$\left(\mathrm{i}\hbar c \nabla\!\!\!/ - \frac{e_j}{c} A\!\!\!/ - M_j c^2\right)\psi(x) = K_j \frac{e\hbar}{2M_j c}\left\{\hat{\boldsymbol{\Sigma}}\cdot\boldsymbol{B} - \mathrm{i}\hat{\boldsymbol{\alpha}}\cdot\boldsymbol{E}\right\}\psi(x) \quad . \tag{1}$$

The lhs exclusively describes the coupling with the electromagnetic field. We know that this part of the equation leads to a nonvanishing magnetic moment if $e_j \neq 0$ (i.e. only for proton), namely

$$\mu_\mathrm{p} = \mu_\mathrm{B} = \frac{e\hbar}{2M_\mathrm{p}c} \quad \text{and} \quad \mu_\mathrm{n} = 0 \quad .$$

In the nonrelativistic limit the rhs of (1) takes the form

$$K_j \frac{e\hbar}{2M_j c}\left\{\hat{\boldsymbol{\sigma}}\cdot\boldsymbol{B} - \mathrm{i}\frac{v}{c}\hat{\boldsymbol{\sigma}}\cdot\boldsymbol{E}\right\} \quad ,$$

since

$$\hat{\Sigma} = \begin{pmatrix} \hat{\sigma} & 0 \\ 0 & \hat{\sigma} \end{pmatrix} \quad \text{and} \quad \hat{\alpha} \sim \frac{v}{c}\hat{\sigma} \quad .$$

Exercise 9.11.

We can neglect the second term, because in the nonrelativistic case $v/c =\ll 1$ it is small compared with the first term, which just corresponds to the interaction of the magnetic moment with the magnetic field. Thus, we obtain the total magnetic moment

$$\mu_\mathrm{p} = \left(1 + K_\mathrm{p}\right) \frac{e\hbar}{2M_\mathrm{p}c}$$

for the proton, i.e. $\mu_\mathrm{p} = 2.79\,\mu_\mathrm{B}$, with $K_\mathrm{p} = 1.79$ and

$$\mu_\mathrm{n} = K_\mathrm{n}\frac{e\hbar}{2M_\mathrm{n}c} = -1.91\,\mu_\mathrm{B}$$

for the neutron.

EXAMPLE

9.12 The Impossibility of Additional Solutions for the Dirac–Coulomb Problem Beyond $Z = 118$

Motivation: For an electron coupled to a point particle with charge Z we can rewrite the Dirac equation into the two coupled equations

$$\begin{aligned} &G' + \frac{\kappa G}{r} - \left(\frac{E + m_0c^2}{\hbar c} + \frac{Z\alpha}{r}\right) F = 0 \quad , \\ &F' - \frac{\kappa F}{r} + \left(\frac{E - m_0c^2}{\hbar c} + \frac{Z\alpha}{r}\right) G = 0 \quad , \end{aligned} \tag{1}$$

To study the asymptotic behaviour of the solutions for $r \to 0$ we have to take into account only the derivative terms and the terms containing a factor $1/r$, whereas we can neglect the terms proportional to m and E. The resulting equations can be solved with the ansatz $F(r) = F_0 r^\gamma$, $G(r) = G_0 r^\gamma$, yielding

$$\begin{aligned} &(\kappa + \gamma)G_0 - Z\alpha F_0 = 0 \quad , \\ &Z\alpha G_0 - (\kappa - \gamma)F_0 = 0 \quad . \end{aligned} \tag{2}$$

This leads to the condition

$$\frac{F_0}{G_0} = \frac{Z\alpha}{\kappa - \gamma} = \frac{\kappa + \gamma}{Z\alpha} \quad , \tag{3}$$

which requires

$$\gamma = \pm\sqrt{\kappa^2 - (Z\alpha)^2} \quad . \tag{4}$$

One of the two independent solutions is regular at the origin, while the other diverges like $r^{-|\gamma|}$. If the charge Z is small, this divergence is so strong that the wave function cannot be normalized and thus is to be rejected as a physical

Example 9.12.

solution. However, for a sufficiently weak divergence the normalization integral $\int d^3r(F^2 + G^2)$ is finite. This happens in the case $\gamma > -\frac{1}{2}$, which means that $Z\alpha > \sqrt{\kappa^2 - \frac{1}{4}}$. For $s_{\frac{1}{2}}$ and $p_{\frac{1}{2}}$ states ($\kappa = \pm 1$) the charge has to exceed $Z > \sqrt{\frac{3}{4}} \times 137 \approx 118$. Under this condition it might appear that both types of solution of the Dirac equation are admissible since they both are normalizable. This would have drastic consequences because the rejection of the irregular solution is essential for obtaining discrete energy eigenvalues. If both solutions were acceptable then there would exist a continuum of solutions for the following reason. For any given energy one can find a solution which is well behaved for $r \to \infty$. For $r \to 0$, on the other hand, this solution contains admixtures behaving like $r^{+|\gamma|}$ and $r^{-|\gamma|}$ as well. For $|\gamma| < \frac{1}{2}$ this cannot be accepted since the solution is no longer normalizable. Only for certain values of the energy does the component $r^{-|\gamma|}$ vanish. These are the energy eigenvalues. For $|\gamma| > \frac{1}{2}$, on the other hand, any linear combination of $r^{\pm|\gamma|}$ leads to a solution normalizable at $r \to 0$. For any given solution one only has to make sure that the wave function is normalizable also at $r \to \infty$, which can be done for every energy.

Are these solutions physically relevant? A static point charge represents a very singular object which we actually do not find in nature, since any particle of finite mass cannot be localized at a single point. Furthermore, only nuclei (in fact superheavy nuclei)[16] have charges of the order $Z\alpha \approx \sqrt{\frac{3}{4}}$, and these are clearly extended objects. Even if such a singular point source of such a high charge does not exist, it is still interesting to study the problem of an extended source in the limit that its extension gets smaller and smaller, either by shrinking the nuclear radius artificially or by studying a hypothetical elementary particle of that charge in the limit that its mass goes to infinity and the particle is localized more and more. The following exercise gives some insight into the problem.

Exercise: Study $\kappa = -1$ states of Dirac electrons in the potential of a hollow sphere of charge $Z\alpha > \sqrt{\frac{3}{4}}$ and radius R and show that in the limit $R \to 0$ only the solution behaving like $r^{+\sqrt{1-(Z\alpha)^2}}$ is recovered.[17]

Solution: We proceed in the usually way. We solve the wave function for the inside and outside regions separately and then try to match the ratio F/G. In the outside region the differential equation contains the ordinary Coulomb potential. In the inside of a charged spherical shell the potential takes on the constant value $Z\alpha/R$, so that the differential equation becomes ($\kappa = -1$):

$$
\begin{aligned}
G' - \frac{G}{r} - \left(\frac{E + m_0c^2}{\hbar c} + \frac{Z\alpha}{R}\right) F = 0 \quad , \\
F' - \frac{F}{r} + \left(\frac{E - m_0c^2}{\hbar c} + \frac{Z\alpha}{R}\right) G = 0 \quad .
\end{aligned} \tag{5}
$$

[16] See J.M. Eisenberg, W. Greiner: *Nuclear Theory*, Vol. III: Microscopic theory of the nucleus, 3rd ed. (North-Holland, Amsterdam 1990).

[17] This exercise has been worked out by W. Grabiak.

Example 9.12.

This is just the free Dirac equation for a particle with energy $\varepsilon = E + \hbar c Z\alpha/R$. The general solution is given by

$$F = k\cos(kr+\delta) - \tfrac{1}{2}\sin(kr+\delta)$$
$$G = \left(\frac{m_0c^2}{\hbar c} + \varepsilon\right)\sin(kr+\delta) \tag{6}$$

with $\hbar ck = \sqrt{\varepsilon^2 - (m_0c^2)^2}$. Taking $\delta = 0$ this is the solution found in (33), Exercise 9.5. The validity of the generalization to $\delta \neq 0$ is easily checked by insertion into the differential equation. However, even without performing this calculation explicitly it is clear that the constant phase shift δ can have no influence on the neutral cancellation of the sine or cosine terms.

In contrast to the Coulomb problem with $Z > 118$ the constant potential admits only one normalizable solution, namely $\delta = 0$. Thus we get

$$\left.\frac{F}{G}\right|_{r=R,\,\text{inside}} = -\frac{\hbar c}{m_0c^2+\varepsilon}\left(\frac{1}{R} - k\cot(kR)\right) \quad . \tag{7}$$

Now we are interested in the limit $R \to 0$. In this limit m and E can be neglected against $Z\alpha/R$, so that we can set $\varepsilon/\hbar c = k = Z\alpha/R$, and (7) becomes

$$\frac{F}{G} \longrightarrow \frac{-1}{Z\alpha} + \cot(Z\alpha) \quad . \tag{8}$$

Now we have to study the behaviour of the wave function in the outside region. For a given energy the wave function will in general have admixtures behaving both like $r^{+|\gamma|}$ and $r^{-|\gamma|}$ for $r \to 0$, but the latter will dominate its asymptotic behaviour. In this limit we can also neglect the mass as well as the energy. According to (3) the F/R ratio then reads

$$\left.\frac{F}{G}\right|_{r\to R} \longrightarrow -\frac{1+|\gamma|}{Z\alpha} = \frac{1+\sqrt{1-(Z\alpha)^2}}{Z\alpha} \tag{9}$$

in the limit $R \to 0$. Comparing (7) and (9), we see that the matching condition becomes

$$\frac{1+\sqrt{1-(Z\alpha)^2}}{Z\alpha} = \frac{1}{Z\alpha} - \cot(Z\alpha) \tag{10}$$

in this limit. It is impossible to solve this equation, since for $\alpha < 1$ the lhs is greater than $1/Z\alpha$, whereas the rhs is smaller. We conclude that we cannot find a solution in the limit $R \to 0$, i.e. if the potential approaches the potential of a point charge, if the component $r^{-|\gamma|}$ dominates the behaviour of the wave function for small r. Only in the special-case limit where this component vanishes does (9) no longer hold and is it possible to get a solution. But this happens only if the energy approaches an energy eigenvalue. Thus one retrieves the usual energy eigenvalues of the Dirac equation, whereas no new solutions are found.

The same result is found for a homogeneously charged sphere. The potential in the inside region is too small to produce an F/G which could fulfil the matching condition. We can convince ourselves of this in the following way: we can ask which value of the potential is needed on the inside in order to fulfil the matching

Example 9.12.

condition, assuming that there is a constant potential V_0 on the inside which is not equal to $Z\alpha/R$. The ratio F/G in the outside region is at most 1, and matching is only possible if the potential in the inside is larger than $2.04/R$ (the solution for $F/G = 1$).[18]

The fact that one can find solutions if the potential in the inside region is large enough is not surprising. Even without the $1/r$ part on the outside one can get arbitrarily deeply bound states if only the strength of the potential is big enough. On the other hand, if one decreases the potential in the inside region, the energy of these solutions increases until they are finally turned into ordinary solutions of the $1/r$ potential. Thus no new type of solution is found for the limiting case of a well-behaved potential approaching $1/r$ variation.

[18] This solution also determines the energy eigenvalues for the MIT bag – see W. Greiner, B. Müller: *Gauge Theory of Weak Interactions*, 2nd ed. and W. Greiner, A. Schäfer: *Quantum Chromodynamics* (Springer, Berlin, Heidelberg 1994).

10. The Two-Centre Dirac Equation

The description of one or more electrons in the field of two nuclei is one of the fundamental problems in quantum mechanics. Among other things it contains the theoretical understanding of the phenomenon of chemical binding and thus is connects physics and chemistry. In 1927 W. Heitler and F. London[1] showed for the first time the possibility of the existence of the H_2 molecule. They used an approximation procedure and were able to calculate its physical properties (binding energy, binding length).

Soon afterwards E. Teller[2] (1930) and E.A. Hylleraas[3] (1931) gave, independently of each other, methods for a mathematically exact solution of the nonrelativistic single-electron two-centre problem. Later, G. Jaffé[4] (1934) came up with a third solution. But the numerical evaluation of these procedures is very costly, so that the energies of the higher states can be calculated only with computers. Good nonrelativistic calculations in recent times – also including some deviations from the Coulomb potential – came from K. Helfrich and H. Hartmann[5] (1968).

Calculations for many-electron systems, similar to the Hartree–Fock model, only became available in the early 1970s, e.g. those by F.P. Larkins[6] (1972). Here one uses a finite set of basis functions (one-, two-, or even three-centre functions), which are combined with the intention of finding the lowest total energy. In view of these difficulties, and the fact that experiments up to now were possible only in a small energy range (some eV up to a few keV), an extensive study of the corresponding relativistic equations did not take place. Solely S.K. Luke et al.[7] have treated the relativistic corrections of the single-electron problem in first-order perturbation theory in $Z\alpha$. However, especially from the experimental point of view, one urgently needs a solution of the two-centre Coulomb problem which is exact in all orders of $Z\alpha$. Heavy-ion accelerators allow highly charged nuclei to approach for a short time so closely that even the innermost electron shells belong to both nuclei and form molecular orbitals (cf. Fig. 10.1).

Now we consider collisions of heavy ions with kinetic energies near the Coulomb barrier, so that both nuclei can just touch. An overlap of the involved nuclei and the related inelastic processes shall be excluded here. If both nuclei are very heavy, it is possible to generate for a short period ($\tau \sim 10^{-17}$ s) *superheavy*

[1] W. Heitler, F. London: Z. Phys. **44**, 455 (1927).
[2] E. Teller: Z. Phys. **61**, 458 (1930).
[3] E.A. Hylleraas: Z. Phys. **71**, 739 (1931).
[4] G. Jaffé: Z. Phys. **87**, 535 (1934).
[5] K. Helfrich, H. Hartmann: Theor. Chim. Acta **10**, 406 (1968).
[6] F. Larkins: J. Phys. **B5**, 571 (1972).
[7] S.K. Luke et al.: J. Chem. Phys. **50**, 1644 (1969).

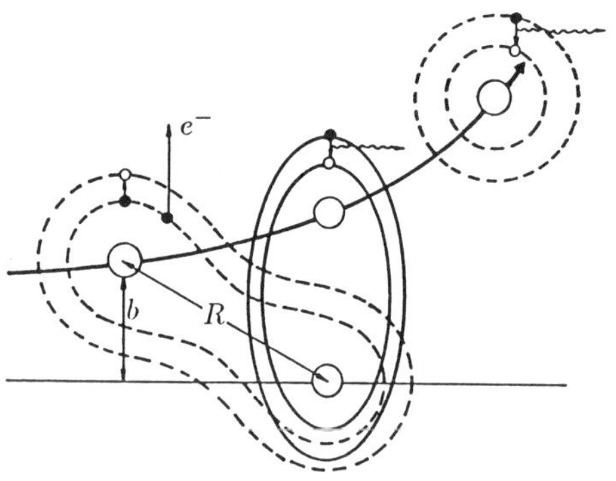

Fig. 10.1. Quasimolecular orbitals. The projectile ion moves along a curved trajectory. The individual atomic orbitals reappear when the colliding ions are separated enough (*upper right*). During the collision electrons (so-called δ electrons – marked by $\uparrow^{e^-}$) and X rays ($\rightsquigarrow$) are emitted. The emission is caused by the varying molecular orbitals, which adjust themselves with respect to the varied internuclear separation

quasimolecules with total charge $Z = Z_1 + Z_2$, which are far above the end of the periodic system: $107 < Z_1 + Z_2 \lesssim 190$. For these systems the velocity of the heavy ions is about 1/10 the velocity of light, while the "velocity" of the electrons in the inner most states is near the velocity of light. This becomes obvious if we remember that for $Z\alpha \sim 1$ the binding energy of the 1*s* electrons becomes comparable with their rest mass. This means that the electrons of the inner shells during a heavy-ion collision have time enough to adjust with respect to the varying distance R between the two Coulomb centres. Thus the properties of the quasimolecular electronic orbitals are determined by the sum of the projectile charge Z_p and the target charge Z_t.

If the distance R between both colliding ions becomes less than the radius of the K shell, we call the system a *superheavy quasiatom*. For a Pb–Pb collision this situation arises for $R \lesssim 500\,\mathrm{fm}$. Induced by the varying Coulomb field during the collision, the possibility is given that inner-shell ionization occurs by excitation of electrons into vacant higher bound states or by direct excitation into the continuum. The highly energetic parts of these final-state continuum electrons are called "delta rays" or "delta electrons", a term originating from when these radioactive rays were discovered: At that time, one bombarded different targets with protons and discovered a highly energetic component of radiation which could not be explained by the classical laws of collisions. Therefore it was assumed that beside the know alpha, beta, and gamma rays, a new kind of radiation had been discovered; hence "delta rays". Nowadays it is well known that the high-energy component of the momentum distribution of the bound electrons allows very much higher energy transfers that can be expected in accordance with the classical laws of collisions. Clearly, the mentioned momentum distribution of the quasimolecularly bound electrons depends on the distance between the colliding ions.

These facts suggest the use of the measurement of delta electrons for the spectroscopy of electronic states in these superheavy quasimolecules. Further possibilities are given by the measurement of the decay of the electron–hole pair which has been created during the collision. Indeed, during the collision the so-called quasimolecular γ radiation will be emitted or – as long as the hole is filled a long time after the collision – the ordinary characteristic radiation of the single atoms can be measured. Some of these processes are illustrated in Fig. 10.1. Such spectroscopy allows the extension of the periodic system by a factor two with respect to the atomic number Z, at least as far as interior electronic shell structure is concerned. Furthermore, the search for the vacuum decay in supercritical fields (see, e.g. Exercises 9.5–9.7 and Chap. 12) necessitates the precise knowledge of the electronic structure in the course of a heavy-ion collision.[8] Fundamental to all such investigations is the solution of the Dirac equation for two charged centres. Therefore the energies and the wave functions of the electrons need to be determined for the given distance R between the nuclei.

In 1973, the two-centre Dirac equation was solved for the first time by Müller and Greiner.[9] The solution was given in terms of prolate elliptic (spheroidal) coordinates ξ, η, ϕ with

$$
\begin{aligned}
x &= R\sqrt{\left(\xi^2-1\right)\left(1-\eta^2\right)}\cos\varphi \quad , \\
y &= R\sqrt{\left(\xi^2-1\right)\left(1-\eta^2\right)}\sin\varphi \quad , \\
z &= R\xi\eta
\end{aligned}
\tag{10.1}
$$

defined in the domain

$$
1 \leq \xi \quad , \quad -1 \leq \eta \leq +1 \quad , \quad 0 \leq \varphi \leq 2\pi \quad .
$$

The problem is rotationally symmetric round the z axis; thus the component J_z of angular momentum is a good quantum number and the eigenfunction in the φ variable is easily separated. It is given by $\exp(im\varphi)$. However, the resulting differential equations in the variables ξ and η do not decouple. The differential equations were diagonalized in the so-called Hylleraas basis, which is essentially composed of Laguerre polynomials in ξ and Legendre polynomials in η.

Meanwhile two centre Hartree–Fock solutions were also obtained, by Fricke et al.[10] (1975). Here, however, we restrict ourselves to the study of a very successful procedure by Müller et al. for the solution of the two-centre Dirac equation, which is essentially based on a multipole expansion of the two-centre potential.

For that purpose let us consider again the general form of the Dirac equation in polar coordinates. Using the relation

$$
\boldsymbol{a} \times (\boldsymbol{b} \times \boldsymbol{c}) = \boldsymbol{b}(\boldsymbol{a} \cdot \boldsymbol{c}) - \boldsymbol{c}(\boldsymbol{a} \cdot \boldsymbol{b})
$$

we obtain

$$
-\boldsymbol{e}_r \times (\boldsymbol{e}_r \times \boldsymbol{\nabla}) = -\boldsymbol{e}_r\,(\boldsymbol{e}_r \cdot \boldsymbol{\nabla}) + \boldsymbol{\nabla}\,(\boldsymbol{e}_r \cdot \boldsymbol{e}_r) \quad , \tag{10.2}
$$

where $\boldsymbol{e}_r$ denotes the unit vector in the $\boldsymbol{r}$ direction. Using this, the $\boldsymbol{\nabla}$ operator may be written as

---

[8] See W. Greiner, J. Reinhardt: *Quantum Electrodynamics*, 2nd ed. (Springer, Berlin, Heidelberg 1994) and especially W. Greiner, B. Müller, J. Rafelski: *Quantum Electrodynamics of Strong Fields* (Springer, Berlin, Heidelberg 1985).

[9] B. Müller, W. Greiner: Phys. Lett. B **47**, 5 (1973); Z. Naturf. **31a**, 1 (1976).

[10] B. Fricke, K. Rashid, P. Bertoncini, A.C. Wahl: Phys. Rev. Lett. **34**, 243 (1975).

$$\begin{aligned}\boldsymbol{\nabla} &= \boldsymbol{e}_r\,(\boldsymbol{e}_r\cdot\boldsymbol{\nabla}) - \boldsymbol{e}_r\times(\boldsymbol{e}_r\times\boldsymbol{\nabla})\\ &= \boldsymbol{e}_r\frac{\partial}{\partial r} - \frac{\mathrm{i}}{\hbar}\left(\frac{\boldsymbol{e}_r}{r}\times\hat{\boldsymbol{L}}\right) \quad ,\end{aligned} \tag{10.3}$$

where $\hat{\boldsymbol{L}} = -\mathrm{i}\hbar(\boldsymbol{r}\times\boldsymbol{\nabla})$ is the orbital angular momentum operator. Hence it follows for the operator of the kinetic energy

$$c\hat{\boldsymbol{\alpha}}\cdot\hat{\boldsymbol{p}} = -\mathrm{i}\hbar c\hat{\alpha}_r\frac{\partial}{\partial r} - \frac{1}{r}c\hat{\boldsymbol{\alpha}}\cdot(\boldsymbol{e}_r\times\hat{\boldsymbol{L}}) \quad , \tag{10.4}$$

with $\hat{\alpha}_r = \hat{\boldsymbol{\alpha}}\cdot\boldsymbol{e}_r$. Now we use

$$(\hat{\boldsymbol{\alpha}}\cdot\boldsymbol{A})(\hat{\boldsymbol{\alpha}}\cdot\boldsymbol{B}) = \boldsymbol{A}\cdot\boldsymbol{B} + \mathrm{i}\hat{\boldsymbol{\Sigma}}\cdot(\boldsymbol{A}\times\boldsymbol{B}) = (\hat{\boldsymbol{\Sigma}}\cdot\boldsymbol{A})(\hat{\boldsymbol{\Sigma}}\cdot\boldsymbol{B}) \tag{10.5}$$

and insert $\boldsymbol{A} = \boldsymbol{e}_r$ and $\boldsymbol{B} = \hat{\boldsymbol{L}}$, which gives

$$\begin{aligned}(\hat{\boldsymbol{\alpha}}\cdot\boldsymbol{e}_r)(\hat{\boldsymbol{\alpha}}\cdot\hat{\boldsymbol{L}}) &= \boldsymbol{e}_r\cdot\hat{\boldsymbol{L}} + \mathrm{i}\hat{\boldsymbol{\Sigma}}\cdot(\boldsymbol{e}_r\times\hat{\boldsymbol{L}})\\ &= (\hat{\boldsymbol{\Sigma}}\cdot\boldsymbol{e}_r)(\hat{\boldsymbol{\Sigma}}\cdot\hat{\boldsymbol{L}}) = \hat{\Sigma}_r(\hat{\boldsymbol{\Sigma}}\cdot\hat{\boldsymbol{L}})\end{aligned} \tag{10.6}$$

because

$$\boldsymbol{e}_r\cdot\hat{\boldsymbol{L}} = 0 \quad ,$$

since $\boldsymbol{e}_r$ is orthogonal to $\hat{\boldsymbol{L}} = \boldsymbol{r}\times\boldsymbol{p}$. We multiply by $-\gamma_5'$ from the left, where

$$\gamma_5' = \begin{pmatrix} 0 & -1\!\!1 \\ -1\!\!1 & 0\end{pmatrix} \quad , \tag{10.7}$$

and remember that

$$\hat{\boldsymbol{\Sigma}} = -\gamma_5'\hat{\boldsymbol{\alpha}} = -\hat{\boldsymbol{\alpha}}\gamma_5' \quad , \quad \hat{\boldsymbol{\alpha}} = -\gamma_5'\hat{\boldsymbol{\Sigma}} = -\hat{\boldsymbol{\Sigma}}\gamma_5' \quad . \tag{10.8}$$

Hence it results that

$$\mathrm{i}\hat{\boldsymbol{\alpha}}\cdot(\boldsymbol{e}_r\times\hat{\boldsymbol{L}}) = \hat{\alpha}_r(\hat{\boldsymbol{\Sigma}}\cdot\hat{\boldsymbol{L}}) \quad , \tag{10.9}$$

and for the operator of kinetic energy

$$c\hat{\boldsymbol{\alpha}}\cdot\hat{\boldsymbol{p}} = -\mathrm{i}\hbar c\hat{\alpha}_r\frac{\partial}{\partial r} + \mathrm{i}c\frac{\hat{\alpha}_r}{r}(\hat{\boldsymbol{\Sigma}}\cdot\hat{\boldsymbol{L}}) \quad . \tag{10.10}$$

Now it is convenient to introduce the operator $\hat{K}$, defined as

$$\hat{K} = \hat{\beta}(\hat{\boldsymbol{\Sigma}}\cdot\hat{\boldsymbol{L}} + \hbar) \quad , \tag{10.11}$$

and one obtains the stationary Dirac equation in polar coordinates:

$$E\psi = \hat{H}\psi = \left[\mathrm{i}c\gamma_5'\hat{\Sigma}_r\left(\hbar\frac{\partial}{\partial r} - \frac{\hat{\beta}}{r}\hat{K} + \frac{\hbar}{r}\right) + V(\boldsymbol{r}) + \hat{\beta}m_0c^2\right]\psi(\boldsymbol{r}) \quad . \tag{10.12}$$

Let us analyze the properties of the spin-orbit operator $\hat{K}$ in more detail. First we show that it commutes with $\hat{\beta}$:

$$\left[\hat{K},\hat{\beta}\right] = \hat{\beta}(\hat{\boldsymbol{\Sigma}}\cdot\hat{\boldsymbol{L}})\hat{\beta} - \hat{\boldsymbol{\Sigma}}\cdot\hat{\boldsymbol{L}} = 0 \quad , \tag{10.13}$$

because

$$\begin{pmatrix} \hat{\sigma} & 0 \\ 0 & \hat{\sigma} \end{pmatrix} \begin{pmatrix} 1\!\!1 & 0 \\ 0 & -1\!\!1 \end{pmatrix} = \begin{pmatrix} \hat{\sigma} & 0 \\ 0 & -\hat{\sigma} \end{pmatrix} \quad \text{and}$$

$$\begin{pmatrix} 1\!\!1 & 0 \\ 0 & -1\!\!1 \end{pmatrix} \begin{pmatrix} \hat{\sigma} & 0 \\ 0 & \hat{\sigma} \end{pmatrix} = \begin{pmatrix} \hat{\sigma} & 0 \\ 0 & -\hat{\sigma} \end{pmatrix} \quad .$$

Furthermore we have

$$\hat{\beta}\,(\hat{\boldsymbol{\alpha}}\cdot\hat{\boldsymbol{p}}) - (\hat{\boldsymbol{\alpha}}\cdot\hat{\boldsymbol{p}})\,\hat{\beta} = 2\hat{\beta}\,(\hat{\boldsymbol{\alpha}}\cdot\hat{\boldsymbol{p}}) \quad , \tag{10.14}$$

and also

$$\begin{aligned} \left[\hat{\beta}\,(\hat{\boldsymbol{\Sigma}}\cdot\hat{\boldsymbol{L}}), \hat{\boldsymbol{\alpha}}\cdot\hat{\boldsymbol{p}}\right]_- &= \hat{\beta}\,(\hat{\boldsymbol{\Sigma}}\cdot\hat{\boldsymbol{L}})\,(\hat{\boldsymbol{\alpha}}\cdot\hat{\boldsymbol{p}}) - (\hat{\boldsymbol{\alpha}}\cdot\hat{\boldsymbol{p}})\hat{\beta}\,(\hat{\boldsymbol{\Sigma}}\cdot\hat{\boldsymbol{L}}) \\ &= \hat{\beta}\,(\hat{\boldsymbol{\Sigma}}\cdot\hat{\boldsymbol{L}})\,(\hat{\boldsymbol{\alpha}}\cdot\hat{\boldsymbol{p}}) + \hat{\beta}(\hat{\boldsymbol{\alpha}}\cdot\hat{\boldsymbol{p}})\,(\hat{\boldsymbol{\Sigma}}\cdot\hat{\boldsymbol{L}}) \\ &= \hat{\beta}\,[\hat{\boldsymbol{\Sigma}}\cdot\hat{\boldsymbol{L}}, \hat{\boldsymbol{\alpha}}\cdot\hat{\boldsymbol{p}}]_+ \quad . \end{aligned} \tag{10.15}$$

Now

$$\begin{aligned} &[\hat{\boldsymbol{\Sigma}}\cdot\hat{\boldsymbol{L}}, \hat{\boldsymbol{\alpha}}\cdot\hat{\boldsymbol{p}}]_+ = -\gamma_5'\,[\hat{\boldsymbol{L}}\cdot\hat{\boldsymbol{p}} + \hat{\boldsymbol{p}}\cdot\hat{\boldsymbol{L}} + \mathrm{i}\hat{\boldsymbol{\Sigma}}\cdot(\hat{\boldsymbol{L}}\times\hat{\boldsymbol{p}} + \hat{\boldsymbol{p}}\times\hat{\boldsymbol{L}})] \quad , \\ &\hat{\boldsymbol{L}}\cdot\hat{\boldsymbol{p}} = (\boldsymbol{r}\times\hat{\boldsymbol{p}})\cdot\hat{\boldsymbol{p}} = 0 \quad , \\ &\hat{\boldsymbol{p}}\cdot\hat{\boldsymbol{L}} = 0 \quad , \end{aligned} \tag{10.16}$$

and we calculate further that

$$\begin{aligned} \hat{\boldsymbol{L}}\times\hat{\boldsymbol{p}} &= \begin{vmatrix} \boldsymbol{e}_x & \boldsymbol{e}_y & \boldsymbol{e}_z \\ \hat{L}_x & \hat{L}_y & \hat{L}_z \\ \hat{p}_x & \hat{p}_y & \hat{p}_z \end{vmatrix} \\ &= \boldsymbol{e}_x\left(\hat{L}_y\hat{p}_z - \hat{L}_z\hat{p}_y\right) - \boldsymbol{e}_y\left(\hat{L}_x\hat{p}_z - \hat{L}_z\hat{p}_x\right) + \boldsymbol{e}_z\left(\hat{L}_x\hat{p}_y - \hat{L}_y\hat{p}_x\right) , \\ \hat{\boldsymbol{p}}\times\hat{\boldsymbol{L}} &= \begin{vmatrix} \boldsymbol{e}_x & \boldsymbol{e}_y & \boldsymbol{e}_z \\ \hat{p}_x & \hat{p}_y & \hat{p}_z \\ \hat{L}_x & \hat{L}_y & \hat{L}_z \end{vmatrix} \\ &= \boldsymbol{e}_x\left(\hat{p}_y\hat{L}_z - \hat{p}_z\hat{L}_y\right) - \boldsymbol{e}_y\left(\hat{p}_x\hat{L}_z - \hat{p}_z\hat{L}_x\right) + \boldsymbol{e}_z\left(\hat{p}_x\hat{L}_y - \hat{p}_y\hat{L}_x\right) . \end{aligned} \tag{10.17}$$

Summing, we obtain

$$\begin{aligned} &\hat{\boldsymbol{L}}\times\hat{\boldsymbol{p}} + \hat{\boldsymbol{p}}\times\hat{\boldsymbol{L}} \\ &\quad = \boldsymbol{e}_x\left(\hat{L}_y\hat{p}_z - \hat{p}_z\hat{L}_y - \hat{L}_z\hat{p}_y + \hat{p}_y\hat{L}_z\right) - \boldsymbol{e}_y\left(\hat{L}_x\hat{p}_z - \hat{p}_z\hat{L}_x - \hat{L}_z\hat{p}_x + \hat{p}_x\hat{L}_z\right) \\ &\quad\quad + \boldsymbol{e}_z\left(\hat{L}_x\hat{p}_y - \hat{p}_y\hat{L}_x - \hat{L}_y\hat{p}_x + \hat{p}_x\hat{L}_y\right) \quad . \end{aligned} \tag{10.18}$$

Now we look at the first commutator of the $\boldsymbol{e}_x$ component:

$$\begin{aligned} \left[\hat{L}_y, \hat{p}_z\right]_- &= (\boldsymbol{r}\times\hat{\boldsymbol{p}})_y\hat{p}_z - \hat{p}_z(\boldsymbol{r}\times\hat{\boldsymbol{p}})_y \\ &= -x\hat{p}_z\hat{p}_z + z\hat{p}_x\hat{p}_z + \hat{p}_z x\hat{p}_z - \hat{p}_z z\hat{p}_x \\ &= z\hat{p}_x\hat{p}_z - \hat{p}_z z\hat{p}_x \quad . \end{aligned} \tag{10.19}$$

By use of

$$\left[x_i, \hat{p}_k\right]_- = \mathrm{i}\hbar\delta_{ik} \quad ,$$

it follows that

$$\left[\hat{L}_y, \hat{p}_z\right]_- = z\hat{p}_x\hat{p}_z + \mathrm{i}\hbar\hat{p}_x - z\hat{p}_z\hat{p}_x = \mathrm{i}\hbar\hat{p}_x \quad .$$

In an analogous way the second commutator of the e_x component is obtained:

$$\begin{aligned}
-\left[\hat{L}_z, \hat{p}_y\right]_- &= -\{(\boldsymbol{r}\times\hat{\boldsymbol{p}})_z\hat{p}_y - \hat{p}_y(\boldsymbol{r}\times\hat{\boldsymbol{p}})_z\} \\
&= -x\hat{p}_y\hat{p}_y + y\hat{p}_x\hat{p}_y + \hat{p}_y x\hat{p}_y - \hat{p}_y y\hat{p}_x \\
&= y\hat{p}_x\hat{p}_y - \hat{p}_y y\hat{p}_x = y\hat{p}_x\hat{p}_y + \mathrm{i}\hbar\hat{p}_x - y\hat{p}_y\hat{p}_x = \mathrm{i}\hbar\hat{p}_x \quad .
\end{aligned}$$

In vector notation we can generally sum up and get

$$\hat{\boldsymbol{L}}\times\hat{\boldsymbol{p}} + \hat{\boldsymbol{p}}\times\hat{\boldsymbol{L}} = 2\mathrm{i}\hbar\hat{\boldsymbol{p}} \quad . \tag{10.20}$$

Finally, for the commutator of $\hat{K}$ with the kinetic energy we obtain

$$\begin{aligned}
\left[\hat{K}, c\hat{\boldsymbol{\alpha}}\cdot\hat{\boldsymbol{p}}\right]_- &= 2\hbar c\hat{\beta}(\hat{\boldsymbol{\alpha}}\cdot\hat{\boldsymbol{p}}) + \hat{\beta}(-\gamma_5')\mathrm{i}\hat{\boldsymbol{\Sigma}}\cdot(2\mathrm{i}\hbar c\hat{\boldsymbol{p}}) \\
&= \hbar c\left(2\hat{\beta}(\hat{\boldsymbol{\alpha}}\cdot\hat{\boldsymbol{p}}) + 2\hat{\beta}\gamma_5'(\hat{\boldsymbol{\Sigma}}\cdot\hat{\boldsymbol{p}})\right) \\
&= \hbar c\left(2\beta(\hat{\boldsymbol{\alpha}}\cdot\hat{\boldsymbol{p}}) - 2\hat{\beta}(\hat{\boldsymbol{\alpha}}\cdot\hat{\boldsymbol{p}})\right) = 0 \quad ,
\end{aligned} \tag{10.21}$$

i.e. $\hat{K}$ commutes with the Hamiltonian for free particles. The commutator of $\hat{K}$ and $\hat{\boldsymbol{j}}$ can be obtained in a similar way, namely

$$\left[\hat{\beta}(\hat{\boldsymbol{\Sigma}}\cdot\hat{\boldsymbol{L}}), \hat{\boldsymbol{L}} + \tfrac{1}{2}\hbar\hat{\boldsymbol{\Sigma}}\right]_- = \hat{\beta}\,[\hat{\boldsymbol{\Sigma}}\cdot\hat{\boldsymbol{L}}, \hat{\boldsymbol{L}}]_- + \tfrac{1}{2}\hbar\hat{\beta}\,[\hat{\boldsymbol{\Sigma}}\cdot\hat{\boldsymbol{L}}, \hat{\boldsymbol{\Sigma}}]_- \quad . \tag{10.22}$$

First we investigate the second commutator, obtaining

$$\begin{aligned}
[\boldsymbol{\Sigma}\cdot\boldsymbol{L}, \boldsymbol{\Sigma}]_- &= (\hat{\boldsymbol{\Sigma}}\cdot\hat{\boldsymbol{L}})\hat{\boldsymbol{\Sigma}} - \hat{\boldsymbol{\Sigma}}(\hat{\boldsymbol{\Sigma}}\cdot\hat{\boldsymbol{L}}) \\
&= \left(\hat{\Sigma}_x\hat{L}_x + \hat{\Sigma}_y\hat{L}_y + \hat{\Sigma}_z\hat{L}_z\right)\left(\hat{\Sigma}_x, \hat{\Sigma}_y, \hat{\Sigma}_z\right) \\
&\quad - \left(\hat{\Sigma}_x, \hat{\Sigma}_y, \hat{\Sigma}_z\right)\left(\hat{\Sigma}_x\hat{L}_x + \hat{\Sigma}_y\hat{L}_y + \hat{\Sigma}_z\hat{L}_z\right) \quad .
\end{aligned}$$

Now we use the assertion

$$[\hat{\boldsymbol{\Sigma}}\cdot\hat{\boldsymbol{L}}, \hat{\boldsymbol{\Sigma}}]_- = 2\mathrm{i}(\hat{\boldsymbol{\Sigma}}\times\hat{\boldsymbol{L}}) \quad , \tag{10.23}$$

verifying this for the x component. (The proof for the other components proceeds analogously). Thus

$$\begin{aligned}
&\left(\hat{\Sigma}_y\hat{L}_y + \hat{\Sigma}_z\hat{L}_z\right)\hat{\Sigma}_x - \hat{\Sigma}_x\left(\hat{\Sigma}_y\hat{L}_y + \hat{\Sigma}_z\hat{L}_z\right) \\
&\quad = -\mathrm{i}\hat{\Sigma}_z\hat{L}_y + \mathrm{i}\hat{\Sigma}_y\hat{L}_z - \mathrm{i}\hat{\Sigma}_z\hat{L}_y + \mathrm{i}\hat{\Sigma}_y\hat{L}_z \,.
\end{aligned}$$

On the other hand

$$\mathrm{i}\,[\hat{\boldsymbol{\Sigma}}\times\hat{\boldsymbol{L}}]_x = \mathrm{i}\hat{\Sigma}_y\hat{L}_z - \mathrm{i}\hat{\Sigma}_z\hat{L}_y \tag{10.24}$$

holds, where we have used the following relations for the $\hat{\Sigma}$-matrices:

$$\hat{\Sigma}_y\hat{\Sigma}_x = -\mathrm{i}\hat{\Sigma}_z \quad , \quad \hat{\Sigma}_z\hat{\Sigma}_x = \mathrm{i}\hat{\Sigma}_y \quad .$$

Hence the relation (10.23) stated above results, and, furthermore, we suspect the validity of

$$[\hat{\boldsymbol{\Sigma}}\cdot\hat{\boldsymbol{L}}, \hat{\boldsymbol{L}}]_- = -\mathrm{i}\hbar(\hat{\boldsymbol{\Sigma}}\times\hat{\boldsymbol{L}}) \quad . \tag{10.25}$$

By use of the commutation relations of angular momentum

$$\hat{L}_y\hat{L}_x - \hat{L}_x\hat{L}_y = -\mathrm{i}\hbar\hat{L}_z \quad , \quad \hat{L}_z\hat{L}_x - \hat{L}_x\hat{L}_z = \mathrm{i}\hbar\hat{L}_y \quad ,$$

the x component of the commutator is found to satisfy

$$\hat{\Sigma}_y\hat{L}_y\hat{L}_x + \hat{\Sigma}_z\hat{L}_z\hat{L}_x - \hat{L}_x\hat{\Sigma}_y\hat{L}_y - \hat{L}_x\hat{\Sigma}_z\hat{L}_z = -\mathrm{i}\hbar\hat{\Sigma}_y\hat{L}_z + \mathrm{i}\hbar\hat{\Sigma}_z\hat{L}_y \quad .$$

The x component of the vector product yields

$$-\mathrm{i}\hbar(\hat{\boldsymbol{\Sigma}} \times \hat{\boldsymbol{L}})_x = -\mathrm{i}\hbar\big(\hat{\Sigma}_y\hat{\Sigma}_z - \hat{\Sigma}_z\hat{L}_y\big) \quad ,$$

so that in total we have

$$\hat{\beta}[\hat{\boldsymbol{\Sigma}} \cdot \hat{\boldsymbol{L}}, \hat{\boldsymbol{L}}]_- = -\mathrm{i}\hbar\hat{\beta}(\hat{\boldsymbol{\Sigma}} \times \hat{\boldsymbol{L}}) \quad , \tag{10.26}$$

and finally we obtain

$$\left[\hat{K}, \hat{\boldsymbol{j}}\right]_- = 0 \quad . \tag{10.27}$$

For $\hat{K}^2$ it follows that

$$\begin{aligned}\hat{K}^2 &= (\hat{\boldsymbol{\Sigma}} \cdot \hat{\boldsymbol{L}} + \hbar)^2 = \hat{\boldsymbol{L}}^2 + \mathrm{i}\hat{\boldsymbol{\Sigma}} \cdot (\hat{\boldsymbol{L}} \times \hat{\boldsymbol{L}}) + 2\hbar\hat{\boldsymbol{\Sigma}} \cdot \hat{\boldsymbol{L}} + \hbar^2 \\ &= \hat{\boldsymbol{L}}^2 + \hbar\hat{\boldsymbol{\Sigma}} \cdot \hat{\boldsymbol{L}} + \hbar^2 \quad .\end{aligned} \tag{10.28}$$

Now we denote the Dirac equation for the Coulomb potential of two centres of charge, $Z_1 e$ and $Z_2 e$, separated by the distance R, as

$$\begin{aligned}&\left[\mathrm{i}c\gamma_5'\hat{\Sigma}_r\left(\hbar\frac{\partial}{\partial r} + \frac{\hbar}{r} - \frac{\hat{\beta}}{r}\hat{K}\right) - \frac{Z_1 e^2}{|\boldsymbol{r} - \boldsymbol{R}/2|} - \frac{Z_2 e^2}{|\boldsymbol{r} + \boldsymbol{R}/2|} + \hat{\beta}m_0c^2\right]\phi_\mu(\boldsymbol{r}) \\ &\quad = E\phi_\mu(\boldsymbol{r}) \quad .\end{aligned} \tag{10.29}$$

The magnetic quantum number μ is the eigenvalue of the projection of the total angular momentum onto the axis connecting the two nuclei (z axis). The nuclei are assumed to be point-like, and at this stage a multipole expansion of the wave function suggests itself:

$$\phi_\mu(\boldsymbol{r}) = \sum_\kappa \phi_{\kappa,\mu}(\boldsymbol{r}) = \sum_\kappa \begin{pmatrix} g_\kappa(r)\chi_{\kappa,\mu} \\ \mathrm{i}f_\kappa(r)\chi_{-\kappa,\mu} \end{pmatrix} \quad , \tag{10.30}$$

where $f_\kappa(r)$ and $g_\kappa(r)$ are the radial wave functions; κ was defined by

$$\kappa = \begin{cases} l & \text{for } j = l - \frac{1}{2} \\ -l - 1 & \text{for } j = l + \frac{1}{2} \end{cases} \quad ; \tag{10.31}$$

and the spinor spherical harmonics are given by

$$\chi_{\kappa,\mu} = \sum_{m=\pm\frac{1}{2}} \left(l\tfrac{1}{2}j|\mu - m, m\right) Y_{l,\mu-m}(\vartheta, \varphi)\chi_m \quad . \tag{10.32}$$

Now we have the unity spinors χ_m

$$\chi_{\frac{1}{2}} = \begin{pmatrix} 1 \\ 0 \end{pmatrix} \quad \text{and} \quad \chi_{-\frac{1}{2}} = \begin{pmatrix} 0 \\ 1 \end{pmatrix} \quad ,$$

and the Clebsch–Gordan coefficients, which explicitly read

$$\left(l\tfrac{1}{2}j\,|\mu-m,m\right) = \begin{array}{|c|c|c|} \hline j \backslash m & \frac{1}{2} & -\frac{1}{2} \\ \hline l+\frac{1}{2} & \sqrt{\frac{l+\mu+\frac{1}{2}}{2l+1}} & \sqrt{\frac{l-\mu+\frac{1}{2}}{2l+1}} \\ \hline l-\frac{1}{2} & -\sqrt{\frac{l-\mu+\frac{1}{2}}{2l+1}} & \sqrt{\frac{l+\mu+\frac{1}{2}}{2l+1}} \\ \hline \end{array} \quad . \tag{10.33}$$

Next we apply various operators to the two spinors $\chi_{\kappa,\mu}$ and get

$$\begin{aligned} \hat{\sigma}_z\chi_{\frac{1}{2}} &= \begin{pmatrix}1 & 0\\ 0 & -1\end{pmatrix}\begin{pmatrix}1\\0\end{pmatrix} = \begin{pmatrix}1\\0\end{pmatrix} \quad , \\ \hat{\sigma}_z\chi_{-\frac{1}{2}} &= \begin{pmatrix}1 & 0\\ 0 & -1\end{pmatrix}\begin{pmatrix}0\\1\end{pmatrix} = -\begin{pmatrix}0\\1\end{pmatrix} \quad , \\ \hat{\boldsymbol{\sigma}}^2\chi_m &= 3\chi_m \quad , \\ \hat{J}_z\chi_{\kappa,\mu} &= \hbar\mu\chi_{\kappa,\mu} \quad , \\ \hat{\boldsymbol{J}}^2\chi_{\kappa,\mu} &= \hbar^2 j(j+1)\chi_{\kappa,\mu} \quad , \\ \hat{S}^2\chi_{\kappa,\mu} &= \hbar^2 s(s+1)\chi_{\kappa,\mu} = \tfrac{3}{4}\hbar^2\chi_{\kappa,\mu} \quad , \\ \hat{\boldsymbol{L}}^2\chi_{\kappa,\mu} &= \hbar^2 l(l+1)\chi_{\kappa,\mu} \quad , \\ 2\hat{\boldsymbol{L}}\cdot\hat{\boldsymbol{S}}\chi_{\kappa,\mu} &= \left(\hat{\boldsymbol{J}}^2 - \hat{\boldsymbol{L}}^2 - \hat{\boldsymbol{S}}^2\right)\chi_{\kappa,\mu} \\ &= \hbar^2\left[j(j+1) - l(l+1) - s(s+1)\right]\chi_{\kappa,\mu} \quad . \end{aligned} \tag{10.34}$$

With $\hbar\hat{\boldsymbol{\sigma}}\cdot\hat{\boldsymbol{L}} = 2\hat{\boldsymbol{S}}\cdot\hat{\boldsymbol{L}}$, it follows that

$$(\hat{\boldsymbol{\sigma}}\cdot\hat{\boldsymbol{L}} + \hbar)\chi_{\kappa,\mu} = \hbar\left[j(j+1) - l(l+1) - s(s+1) + 1\right]\chi_{\kappa,\mu} \quad . \tag{10.35}$$

This relation can be evaluated in both cases of (10.31):

a):

$$l = j + \tfrac{1}{2} \quad , \quad \kappa = l = j + \tfrac{1}{2}$$

and b):

$$\begin{aligned} l' = j - \tfrac{1}{2} \quad , \quad \kappa &= -l' - 1 = -j + \tfrac{1}{2} - 1 \\ &= -\left(j + \tfrac{1}{2}\right) = -|\kappa| \quad , \end{aligned}$$

with the result

a):

$$\begin{aligned} j(j+1) - l(l+1) - s(s+1) + 1 &= \left(l - \tfrac{1}{2}\right)\left(l + \tfrac{1}{2}\right) - l(l+1) - \tfrac{3}{4} + 1 \\ &= l^2 - \tfrac{1}{4} - l^2 - 1 + \tfrac{1}{4} = -l = -\kappa \quad , \end{aligned}$$

and b):

$$\begin{aligned} j(j+1) - l'(l'+1) - s(s+1) + 1 &= \left(l' + \tfrac{1}{2}\right)\left(l' + \tfrac{3}{2}\right) - l'(l'+1) + \tfrac{1}{4} \\ &= 2l' - l' + \tfrac{3}{4} + \tfrac{1}{4} = l' + 1 = -\kappa \quad , \end{aligned} \tag{10.36}$$

so that generally

$$(\hat{\boldsymbol{\sigma}} \cdot \hat{\boldsymbol{L}} + \hbar)\, \chi_{\kappa,\mu} = -\hbar\kappa\chi_{\kappa,\mu} \tag{10.37}$$

holds. We now move on to investigate the multipole decomposition of the two-centre potential. It is generally known that

$$\frac{1}{|\boldsymbol{r} - \boldsymbol{r}'|} = \frac{1}{\sqrt{r^2 + r'^2 - 2rr' \cos\gamma}} \quad ,$$

where γ is the angle between $\boldsymbol{r}$ and $\boldsymbol{r}'$ and

$$\begin{aligned} \frac{1}{|\boldsymbol{r} - \boldsymbol{r}'|} &= \frac{1}{r_> \sqrt{1 - 2\frac{r_<}{r_>} \cos\gamma + \left(\frac{r_<}{r_>}\right)^2}} \quad , \\ \frac{1}{|\boldsymbol{r} + \boldsymbol{r}'|} &= \frac{1}{r_> \sqrt{1 + 2\frac{r_<}{r_>} \cos\gamma + \left(\frac{r_<}{r_>}\right)^2}} \quad , \end{aligned} \tag{10.38}$$

with

$$r_< = \min(|\boldsymbol{r}|, |\boldsymbol{r}'|) \quad , \quad r_> = \max(|\boldsymbol{r}|, |\boldsymbol{r}'|) \quad .$$

Expansion of the root yields

$$\frac{1}{|\boldsymbol{r} - \boldsymbol{r}'|} = \sum_{l=0}^{\infty} \frac{r_<^l}{r_>^{l+1}} P_l(\cos\gamma)$$

and correspondingly

$$\frac{1}{|\boldsymbol{r} + \boldsymbol{r}'|} = \sum_{l=0}^{\infty} (-1)^l \frac{r_<^l}{r_>^{l+1}} P_l(\cos\gamma) \quad , \tag{10.39}$$

with the Legendre polynomials

$$\begin{aligned} &P_0(x) = 1 \quad , \\ &P_1(x) = x \quad , \\ &P_2(x) = \tfrac{1}{2}(3x^2 - 1) \quad , \\ &\text{etc.} \end{aligned}$$

In general, the recursion relation

$$(l+1)P_{l+1}(x) - (2l+1) \times P_l(x) + lP_{l-1}(x) = 0 \tag{10.40}$$

holds. For point-like nuclei the two-centre Coulomb potential reads

$$V = -\frac{Z_1 e^2}{|\boldsymbol{r} - \boldsymbol{R}/2|} - \frac{Z_2 e^2}{|\boldsymbol{r} + \boldsymbol{R}/2|} \quad , \tag{10.41}$$

which, after expansion into multipoles, yields

$$V(r,R,\gamma) = -Z_1 e^2 \sum_{l=0}^{\infty} \frac{r^l}{(R/2)^{l+1}} P_l(\cos\gamma)$$
$$- Z_2 e^2 \sum_{l=0}^{\infty} (-1)^l \frac{r^l}{(R/2)^{l+1}} P_l(\cos\gamma) \quad \text{for } r < \tfrac{R}{2} \quad ,$$
$$V(r,R,\gamma) = -Z_1 e^2 \sum_{l=0}^{\infty} \frac{(R/2)^l}{r^{l+1}} P_l(\cos\gamma)$$
$$- Z_2 e^2 \sum_{l=0}^{\infty} (-1)^l \frac{(R/2)^l}{r^{l+1}} P_l(\cos\gamma) \quad \text{for } r \geq \tfrac{R}{2} \quad . \tag{10.42}$$

The monopole part for point nuclei of the two-centre Coulomb potential can easily be derived from the multipole expansions, and is given by the $l = 0$ term of (10.42), i.e.

$$V_0(r) = \begin{cases} -2\dfrac{(Z_1+Z_2)e^2}{R} & \text{for } r < \dfrac{R}{2} \\ -\dfrac{(Z_1+Z_2)e^2}{r} & \text{for } r \geq \dfrac{R}{2} \end{cases} \quad . \tag{10.43}$$

Remark on multipole expansions of potentials: We write generally

$$V(\boldsymbol{r},R) = \sum_{l=0}^{\infty} V_l(r,R) P_l(\cos\vartheta) = \sum_{l=0}^{\infty} V_l(r,R) P_l(x) \quad , \quad \text{with} \tag{10.44}$$
$$x = \cos\vartheta \quad ,$$

and use the following normalization condition:

$$\int_{-1}^{1} P_{l'}(x) P_l(x)\,\mathrm{d}x = \frac{2}{2l+1}\delta_{l'l} \quad . \tag{10.45}$$

Here ϑ is the polar angle of $\boldsymbol{r} = \{r\sin\vartheta\cos\varphi, r\sin\vartheta\sin\varphi, r\cos\vartheta\}$. Multiplying by $P_l(x)$ and integrating over x from -1 to $+1$ yields

$$\int_{-1}^{+1} V(\boldsymbol{r},R) P_l(x)\,\mathrm{d}x = V_l(r,R)\frac{2}{2l+1} \quad \text{or}$$
$$V_l(r,R) = \frac{2l+1}{2}\int_{-1}^{1} V(\boldsymbol{r},R) P_l(x)\,\mathrm{d}x \quad . \tag{10.46}$$

Now we will show that the operator $\hat{K}$ has the eigenvalue $-\hbar\kappa$:

$$\begin{aligned} \hat{K}\phi_{\kappa,\mu}(\boldsymbol{r}) &= \hat{\beta}(\hat{\boldsymbol{\Sigma}}\cdot\hat{\boldsymbol{L}} + \hbar)\,\phi_{\kappa,\mu}(\boldsymbol{r}) \\ &= \begin{pmatrix} 1\!\!1 & 0 \\ 0 & -1\!\!1 \end{pmatrix} \begin{pmatrix} \hat{\boldsymbol{\sigma}}\cdot\hat{\boldsymbol{L}} + \hbar & 0 \\ 0 & \hat{\boldsymbol{\sigma}}\cdot\hat{\boldsymbol{L}} + \hbar \end{pmatrix} \begin{pmatrix} g_\kappa(r)\chi_{\kappa,\mu}(\vartheta,\varphi) \\ \mathrm{i}f_\kappa(r)\chi_{-\kappa,\mu}(\vartheta,\varphi) \end{pmatrix} \\ &= \begin{pmatrix} \hat{\boldsymbol{\sigma}}\cdot\hat{\boldsymbol{L}} + \hbar & 0 \\ 0 & -(\hat{\boldsymbol{\sigma}}\cdot\hat{\boldsymbol{L}} + \hbar) \end{pmatrix} \begin{pmatrix} g_\kappa(r)\chi_{\kappa,\mu}(\vartheta,\varphi) \\ \mathrm{i}f_\kappa(r)\chi_{-\kappa,\mu}(\vartheta,\varphi) \end{pmatrix} \\ &= \hbar \begin{pmatrix} -\kappa g_\kappa(r)\chi_{\kappa,\mu}(\vartheta,\varphi) \\ -\kappa \mathrm{i}f_\kappa(r)\chi_{-\kappa,\mu}(\vartheta,\varphi) \end{pmatrix} = (-\hbar\kappa)\phi_{\kappa,\mu}(\boldsymbol{r}) \quad . \end{aligned} \tag{10.47}$$

In order to obtain the differential equations for the radial wave functions we write down the stationary two-centre Dirac equation (10.29) once more, though in greater detail:

$$\left[\mathrm{i}c \begin{pmatrix} 0 & -1\!\!1 \\ -1\!\!1 & 0 \end{pmatrix} \begin{pmatrix} \hat{\sigma}_r & 0 \\ 0 & \hat{\sigma}_r \end{pmatrix} \left(\hbar \frac{\partial}{\partial r} + \frac{\hbar}{r} + \frac{\hbar\kappa}{r} \begin{pmatrix} 1\!\!1 & 0 \\ 0 & -1\!\!1 \end{pmatrix} \right) \right.$$
$$\left. + \sum_{l=0}^{\infty} V_l P_l + \begin{pmatrix} 1\!\!1 & 0 \\ 0 & -1\!\!1 \end{pmatrix} m_0 c^2 \right] \sum_{\kappa=\pm 1}^{\pm\infty} \begin{pmatrix} g_\kappa(r)\chi_{\kappa,\mu} \\ \mathrm{i} f_\kappa(r)\chi_{-\kappa,\mu} \end{pmatrix}$$
$$= \sum_{\kappa=\pm 1}^{\pm\infty} \begin{pmatrix} E g_\kappa(r)\chi_{\kappa,\mu} \\ \mathrm{i} E f_\kappa(r)\chi_{-\kappa,\mu} \end{pmatrix} \quad , \tag{10.48}$$

where the two-centre potential in the form (10.44) has been inserted. Equation (10.48) can be further transformed into

$$\left[\mathrm{i}\hbar c \begin{pmatrix} 0 & -\hat{\sigma}_r \\ -\hat{\sigma}_r & 0 \end{pmatrix} \left(\frac{\partial}{\partial r} + \frac{1}{r} + \frac{\kappa}{r} \begin{pmatrix} 1\!\!1 & 0 \\ 0 & -1\!\!1 \end{pmatrix} \right) \right.$$
$$\left. + \sum_{l=0}^{\infty} V_l P_l + \begin{pmatrix} 1\!\!1 & 0 \\ 0 & -1\!\!1 \end{pmatrix} m_0 c^2 - E \right] \sum_{\kappa=\pm 1}^{\pm\infty} \begin{pmatrix} g_\kappa(r)\chi_{\kappa,\mu} \\ \mathrm{i} f_\kappa(r)\chi_{-\kappa,\mu} \end{pmatrix} = 0$$

$$\sum_{\kappa=\pm 1}^{\pm\infty} \begin{pmatrix} \hbar c \frac{\mathrm{d}}{\mathrm{d}r} f_\kappa \hat{\sigma}_r \chi_{-\kappa,\mu} + \hbar c \frac{f_\kappa}{r} \hat{\sigma}_r \chi_{-\kappa,\mu} - \hbar c \frac{\kappa}{r} f_\kappa \hat{\sigma}_r \chi_{-\kappa,\mu} & + \sum_{l=0}^{\infty} V_l P_l g_\kappa \chi_{\kappa,\mu} + \left(m_0 c^2 - E \right) g_\kappa \chi_{\kappa,\mu} \\ -\mathrm{i}\hbar c \frac{\mathrm{d}}{\mathrm{d}r} g_\kappa \hat{\sigma}_r \chi_{\kappa,\mu} - \mathrm{i}\hbar c \frac{g_\kappa}{r} \hat{\sigma}_r \chi_{\kappa,\mu} - \mathrm{i}\hbar c \frac{\kappa}{r} g_\kappa \hat{\sigma}_r \chi_{\kappa,\mu} & + \sum_{l=0}^{\infty} V_l P_l \mathrm{i} f_\kappa \chi_{-\kappa,\mu} - \left(m_0 c^2 + E \right) \mathrm{i} f_\kappa \chi_{-\kappa,\mu} \end{pmatrix} = 0 \tag{10.49}$$

Later on we shall prove the relation

$$\hat{\sigma}_r \chi_{\kappa,\mu} = -\chi_{-\kappa,\mu} \quad , \tag{10.50}$$

from which

$$\sum_{\kappa=\pm 1}^{\pm\infty} \left[-\hbar c \frac{\mathrm{d}}{\mathrm{d}r} f_\kappa \chi_{\kappa,\mu} - \hbar c \frac{f_\kappa}{r} \chi_{\kappa,\mu} + \hbar c \frac{\kappa}{r} f_\kappa \chi_{\kappa,\mu} \right.$$
$$\left. + \sum_{l=0}^{\infty} V_l P_l g_\kappa \chi_{\kappa,\mu} + \left(m_0 c^2 - E \right) g_\kappa \chi_{\kappa,\mu} \right] = 0$$

$$\sum_{\kappa=\pm 1}^{\pm\infty} \left[-\hbar c \frac{\mathrm{d}}{\mathrm{d}r} g_\kappa \chi_{-\kappa,\mu} - \hbar c \frac{g_\kappa}{r} \chi_{-\kappa,\mu} - \hbar c \frac{\kappa}{r} g_\kappa \chi_{-\kappa,\mu} \right.$$
$$\left. - \sum_{l=0}^{\infty} V_l P_l f_\kappa \chi_{-\kappa,\mu} + \left(m_0 c^2 + E \right) f_\kappa \chi_{-\kappa,\mu} \right] = 0 \tag{10.51}$$

follows.

According to (10.32) the spinors $\chi_{\kappa,\mu}$ are orthonormal. Using this property and multiplying the first differential equation of the two-centre Dirac equation by $-\langle \chi_{\overline{\kappa},\mu}|$ and the second one by $-\langle \kappa_{-\overline{\kappa},\mu}|$, we obtain

$$\left\{\hbar c\frac{\mathrm{d}}{\mathrm{d}r}f_{\overline{\kappa}} + \hbar c\frac{f_{\overline{\kappa}}}{r} - \hbar c\frac{\overline{\kappa}}{r}f_{\overline{\kappa}} - \left(m_0c^2 - E\right)g_{\overline{\kappa}} - \sum_{l=0}^{\infty} V_l \sum_{\kappa=\pm1}^{\pm\infty} g_\kappa\langle\chi_{\overline{\kappa},\mu}|P_l|\chi_{\kappa,\mu}\rangle\right\}$$
$$= 0 \quad ,$$
$$\left\{\hbar c\frac{\mathrm{d}}{\mathrm{d}r}g_{\overline{\kappa}} + \hbar c\frac{g_{\overline{\kappa}}}{r} + \hbar c\frac{\overline{\kappa}}{r}g_{\overline{\kappa}} - \left(m_0c^2 + E\right)f_{\overline{\kappa}} + \sum_{l=0}^{\infty} V_l \sum_{\kappa=\pm1}^{\pm\infty} f_\kappa\langle\chi_{-\overline{\kappa},\mu}|P_l|\chi_{-\kappa,\mu}\rangle\right\}$$
$$= 0 \quad . \tag{10.52}$$

This – in principle infinite – system of coupled radial first-order differential equations must be solved numerically in order to determine the energy E.

As a special case we examine these differential equations for a spherically symmetric monopole potential $V(\boldsymbol{r}) = V_0(r)$ for a special κ; therefore

$$\hbar c\frac{\mathrm{d}}{\mathrm{d}r}f + \hbar c\frac{f}{r} - \hbar c\frac{\kappa}{r}f - \left(m_0c^2 - E\right)g - V_0 g = 0 \quad ,$$
$$\hbar c\frac{\mathrm{d}}{\mathrm{d}r}g + \hbar c\frac{g}{r} + \hbar c\frac{\kappa}{r}g - \left(m_0c^2 + E\right)f + V_0 f = 0 \quad . \tag{10.53}$$

These are the already known radial differential equations for spherically symmetric potentials (cf. Example 9.3). The matrix elements $\langle\chi_{\overline{\kappa},\mu}|P_l|\chi_{\kappa,\mu}\rangle$ can be obtained easily by the usual angular momentum algebra, though we will not pursue that here any further.

However, to conclude our formal derivations we will prove the relation

$$\hat{\sigma}_r\chi_{\kappa,\mu} = -\chi_{-\kappa,\mu} \quad \text{with} \quad r\hat{\sigma}_r = \sum_{i=1}^{3} x_i\hat{\sigma}_i \quad , \tag{10.54}$$

where $\hat{\sigma}_r$ is a scalar operator, so $\hat{\sigma}_r\chi_{\kappa,\mu}$ has the same eigenvalues j and μ as $\chi_{\kappa,\mu}$. First we note some properties of the parity of a state. Parity is determined by the transformation properties of the spherical harmonics. For the transformation $(\vartheta,\varphi) \to (\pi-\vartheta,\varphi+\pi)$,

$$Y_{l,m}(\pi-\vartheta,\varphi+\pi) = (-1)^l Y_{l,m}(\vartheta,\varphi) \tag{10.55}$$

follows. Hence the parity of a state is determined by the orbital angular momentum, and is thus given by $(-1)^l$. Therefore we can make the ansatz

$$\hat{\sigma}_r\chi_{\kappa,\mu} = a\chi_{-\kappa,\mu} + b\chi_{\kappa,\mu} \quad . \tag{10.56}$$

As $\hat{\sigma}_r$ changes its sign under parity transformations, b must be zero; thus we can write

$$\pi_l = (-1)^l = (-1)^{(j+1/2S_\kappa)} \quad , \tag{10.57}$$

too, with $S_\kappa = \kappa/|\kappa|$ and

$$\kappa = \begin{cases} l & \text{for } j = l - \frac{1}{2} \\ -l-1 & \text{for } j = l + \frac{1}{2} \end{cases} \tag{10.58}$$

or

$$l = \begin{cases} \kappa & \text{for } \kappa > 0 \\ -\kappa - 1 & \text{for } \kappa < 0 \end{cases} \quad .$$

Furthermore we define the $\bar{l}$ related to $-\kappa$:

$$\bar{l} = \kappa - 1 \quad \text{for} \quad \kappa > 0 \quad , \quad \bar{l} = -\kappa \quad \text{for} \quad \kappa < 0 \quad , \quad \text{and}$$

$$l - \bar{l} = 1 = S_\kappa \qquad \text{for } \kappa > 0 \quad ,$$
$$l - \bar{l} = -1 = S_\kappa \quad \text{for } \kappa < 0 \quad ,$$

holds. Hence we can write the total angular momentum as

$$j = l - \tfrac{1}{2} S_\kappa \quad ,$$

which we have used already in (10.57).

Changing parity yields, for constant j, a change of the orbital angular momentum by one unit. Hence the sign of κ changes under the parity transformation, too.

With

$$(\hat{\boldsymbol{\sigma}} \cdot \boldsymbol{A})(\hat{\boldsymbol{\sigma}} \cdot \boldsymbol{B}) = \boldsymbol{A} \cdot \boldsymbol{B} + \mathrm{i}\hat{\boldsymbol{\sigma}}(\boldsymbol{A} \times \boldsymbol{B})$$

we find

$$\hat{\sigma}_r^2 = \mathbb{1} \tag{10.59}$$

and therefore

$$a^2 = 1 \quad .$$

Consequently, the phase of a remains to be determined. For that purpose we choose $\boldsymbol{e}_r$ along the z axis and set $\vartheta = 0$ in the spherical harmonics. With

$$Y_{l,m}(\vartheta, \varphi) = \sqrt{\frac{2l+1}{4\pi} \frac{(l-m)!}{(l+m)!}} P_{l,m}(\cos\vartheta)\, \mathrm{e}^{\mathrm{i}m\varphi}$$

and

$$P_{l,m}(x) = (-1)^m (1 - x^2)^{m/2} \frac{\mathrm{d}^m}{\mathrm{d}x^m} P_l(x)$$

we obtain for $\vartheta = 0$

$$Y_{l,m}(\vartheta = 0) = \sqrt{\frac{2l + l}{4\pi}} \delta_{m0} \quad .$$

Moreover, we have

$$\hat{\sigma}_r = \hat{\sigma}_z = \begin{pmatrix} 1 & 0 \\ 0 & -1 \end{pmatrix} \quad , \quad \text{with}$$

$$\hat{\sigma}_z \chi_{\frac{1}{2}} = \begin{pmatrix} 1 \\ 0 \end{pmatrix} \quad \text{and} \quad \hat{\sigma}_z \chi_{-\frac{1}{2}} = -\begin{pmatrix} 0 \\ 1 \end{pmatrix} \quad .$$

Thereby we obtain for fixed μ:

$$\chi_{\kappa,\mu} = \sqrt{\frac{2l+1}{4\pi}}\, (l\tfrac{1}{2}j|0\mu)\, \chi_\mu \quad .$$

If we use 2μ as the eigenvalue of $\hat{\sigma}_z$, we arrive at the defining equation for a:

$$a\sqrt{2\bar{l}+1}\,\left(\bar{l}\tfrac{1}{2}j|0\mu\right) = 2\mu\sqrt{2l+1}\,\left(l\tfrac{1}{2}j|0\mu\right) \quad .$$

It is useful to investigate the four cases $j = l \pm \frac{1}{2}$ and $\mu = \pm\frac{1}{2}$ separately:

(1) $j = l + \frac{1}{2},\ \ \mu = \frac{1}{2}$:

$$\bar{l} = l - S_\kappa = l + 1 \quad ;$$

hence it follows that $j = \bar{l} - \frac{1}{2}$, and

$$a\sqrt{2\bar{l}+1}(-1)\sqrt{\frac{\bar{l}}{2\bar{l}+1}} = \sqrt{2l+1}\sqrt{\frac{l+1}{2l+1}} \quad . \tag{10.60}$$

From this we get $a = -1$.

(2) $j = l + \frac{1}{2},\ \ \mu = -\frac{1}{2}$:

$$a\sqrt{\bar{l}} = -\sqrt{l+1} \quad , \quad a = -1 \quad . \tag{10.61}$$

(3) $j = l - \frac{1}{2},\ \ \mu = \frac{1}{2}$:

$$\bar{l} = l - S_\kappa = l - 1 \quad .$$

From this follows $j = \bar{l} + \frac{1}{2}$ and

$$a\sqrt{\bar{l}+1} = -\sqrt{l} \quad , \quad a = -1 \quad . \tag{10.62}$$

(4) $j = l - \frac{1}{2},\ \ \mu = -\frac{1}{2}$:

$$a\sqrt{\bar{l}+1} = \sqrt{-1} \quad , \quad a = -1 \quad . \tag{10.63}$$

This proves the assertion of (10.54). By numerically solving the coupled differential equations (10.52) one finally obtains the R-dependent wave functions $\phi_i(\boldsymbol{r}, R)$ and the energies $E_i(R)$. The latter are usually represented by a so-called *correlation diagram*, where the energies of the states of the separated systems (Z_1, Z_2) are connected to those of the combined system $(Z_1 + Z_2)$. The molecular states are classified according to the good quantum numbers j_z, with the eigenvalues $|\mu| = \frac{1}{2}, \frac{3}{2}, \frac{5}{2}, \ldots$, which are also specified by $\sigma, \pi, \delta, \ldots$. In most cases one assigns to the molecular state, in addition, the quantum numbers of the appertaining state in the combined system $R = 0$, so that altogether we obtain the designation

$$1s_{1/2}\sigma \quad , \quad 2p_{1/2}\sigma \quad , \quad 2p_{3/2}\sigma \quad , \quad 2p_{3/2}\pi \quad \ldots \quad .$$

In the case of identical partners in the molecule, there exists a further constant of motion. Indeed the parity operator commutes with the Hamiltonian and with $\hat{j}_z$, so that additionally one can distinguish between *even* (positive parity) and *odd* (negative parity with respect to the centre of mass of both nuclei) states.

As an example of a relativistic correlation diagram we show the calculated binding energies of some bound states in the Pb–Pb system (Figs. 10.2–10.4). In

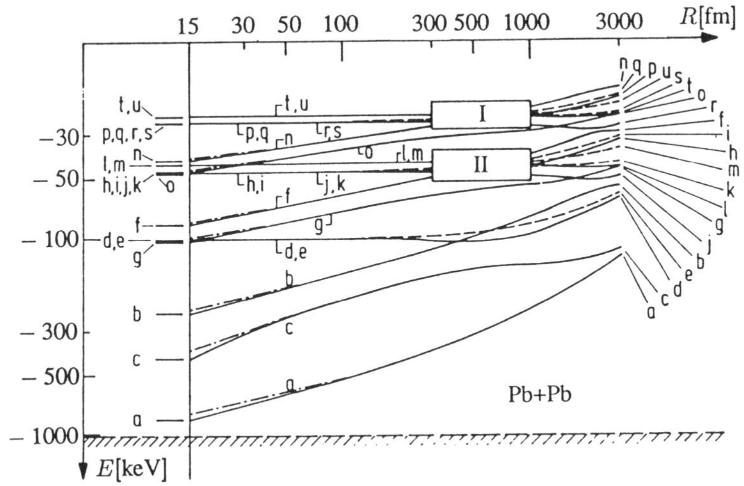

$a = 1s_{1/2}\sigma$, $d = 2p_{3/2}\sigma$, $g = 3p_{1/2}\sigma$, $j = 3d_{3/2}\sigma$, $m = 3d_{5/2}\pi$, $p = 4p_{3/2}\sigma$, $s = 4d_{3/2}\pi$,
$b = 2s_{1/2}\sigma$, $e = 2p_{3/2}\pi$, $h = 3p_{3/2}\sigma$, $k = 3d_{3/2}\pi$, $n = 4s_{1/2}\sigma$, $q = 4p_{3/2}\pi$, $t = 4d_{5/2}\sigma$,
$c = 2p_{1/2}\sigma$, $f = 3s_{1/2}\sigma$, $i = 3p_{3/2}\pi$, $l = 3d_{5/2}\sigma$, $o = 4p_{1/2}\sigma$, $r = 4d_{3/2}\sigma$, $u = 4d_{5/2}\pi$.

Fig. 10.2. Correlation diagram for the (symmetric) Pb–Pb system with double logarithmic scale. The abscissa shows the two-centre distance R, the ordinate shows the binding energy

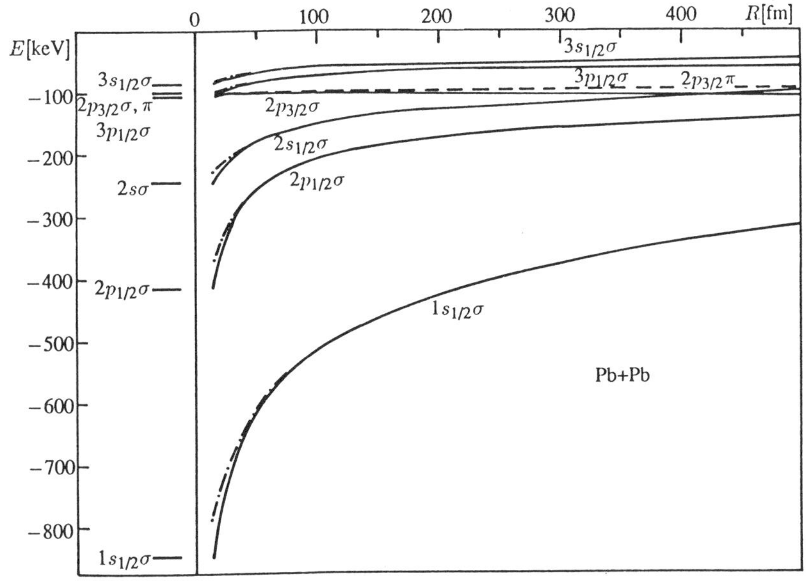

Fig. 10.3. Correlation diagram for the lowest levels in the Pb–Pb system in a linear scale illustrates the strong increase of the binding energies for small distances ($R \to 0$)

order to emphasize the various dependences of the energy eigenvalues on the two-centre distance R we have chosen different representations of the Pb–Pb correlation diagram. First we have displayed in a double logarithmic scale the 21 lowest σ (full lines) and π states (dashed lines) in the range between $R = 15\,\mathrm{fm}$ and $R = 3000\,\mathrm{fm}$ (see Fig. 10.2). The relativistic fine-structure splitting between the states $2p_{3/2}\sigma$ and $2p_{1/2}\sigma$ at $R = 15\,\mathrm{fm}$ with the magnitude of about 316.6 keV is especially noteworthy. The finite extension of the Pb nuclei was considered in these calculations, too; however, the interaction between the electrons was not

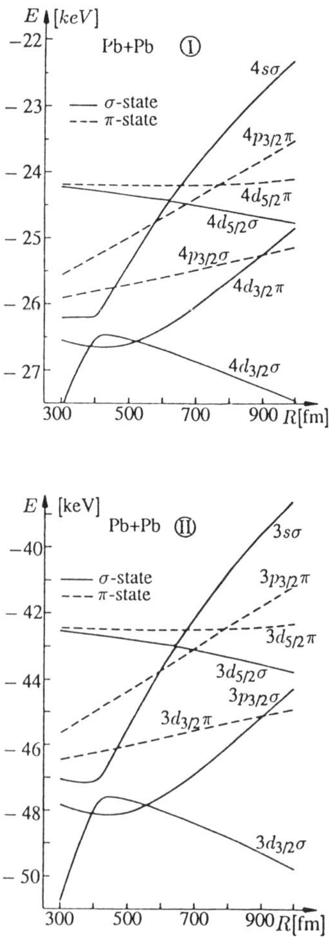

Fig. 10.4. Expanded sections from Fig. 10.3. The noncrossing of levels with at least one different quantum number is clearly noticeable

regarded. Furthermore the delayed crossings between the $3p_{1/2}\sigma$ and the $2p_{3/2}\sigma$ state at $R = 18$ fm and between the $4p_{1/2}\sigma$ and the $3p_{3/2}\sigma$ state at $R = 15.5$ fm are interesting. The energy levels in the squares I and II between 300 fm and 1000 fm are displayed separately with a linear scale (Fig. 10.3) for greater clarity. The rapid change in the energy of the most strongly bound electrons in the range of small two-centre distances is most impressive if one chooses a linear scale for the representation of E as a function of R (see Fig. 10.4). Here the left column (ordinate) represents the binding energies in the combined atom with $Z = 164$. In the range $15\,\text{fm} \leq R \leq 100\,\text{fm}$ we find the following energy changes: 330 keV for the $1s\sigma$ state, 208 keV for the $2p_{1/2}\sigma$ state and 91 keV for the $2s\sigma$ state. For comparison, we mention that the $1s$ binding energy in the element Fermium ($Z = 100$) "only" amounts to 141 keV, while it is close to 900 keV for $Z = 164$. The rapid increase of the binding energies of the interior electron shells with decreasing two-centre distance R reflects the fast approach of overcriticality in the limit $R \to 0$ of these superheavy systems. This is most important with the "decay of the vacuum" in supercritical fields.

11. The Foldy–Wouthuysen Representation for Free Particles

As discussed in the previous chapters, the spinors in the Dirac theory consist of four components. In the nonrelativistic limit, for spinors belonging to positive (negative) energy states, the upper (lower) two components beecome large compared to the lower (uppeer) two components [cf. (2.44), (2.72)]. The question arises, whether there exists a representation that reflects this property as a general feature, i.e. also for large velocities of the particles.

Thus, we will search for a unitary transformation that, when applied to the wave function of a free spin-$\frac{1}{2}$ particle, yields for fixed sign of the energy a wave function completely determined by only two of the four components. However, as mentioned before, in the Dirac representation the wave function is generally determined by four components. In the following we will first discuss two different possibilities of such a unitary transformation and subsequently we will illustrate the transformation of some operators into the new representation, called the *Foldy–Wouthuysen representation* or, alternatively, *Φ representation.*

The change from the original Dirac representation to the Φ representation can be achieved by a unitary transformation

$$\hat{U} = \frac{\hat{\beta}\hat{H}_{\mathrm{f}} + E_p}{\sqrt{2E_p\left(m_0c^2 + E_p\right)}} \quad , \tag{11.1}$$

where $\hat{H}_{\mathrm{f}} = c\hat{\boldsymbol{\alpha}}\cdot\hat{\boldsymbol{p}} + \hat{\beta}m_0c^2$ denotes the Hamiltonian of a free particle, as before. If we use $\hat{\alpha}_k^\dagger = \hat{\alpha}_k$, $\hat{\beta}^\dagger = \hat{\beta}$ and take into account that $\hat{\boldsymbol{p}}$ is a Hermitian operator, we can prove that $\hat{U}$ is indeed a unitary operator:

$$\begin{aligned}
\hat{U}^\dagger\hat{U} &= \frac{1}{2E_p\left(m_0c^2+E_p\right)}\left(c\hat{\beta}\hat{\boldsymbol{\alpha}}\cdot\hat{\boldsymbol{p}} + m_0c^2 + E_p\right)^\dagger\left(c\hat{\beta}\hat{\boldsymbol{\alpha}}\cdot\hat{\boldsymbol{p}} + m_0c^2 + E_p\right)\\
&= \frac{1}{2E_p\left(m_0c^2+E_p\right)}\left(c\hat{\boldsymbol{\alpha}}\cdot\hat{\boldsymbol{p}}\hat{\beta} + m_0c^2 + E_p\right)\left(c\hat{\beta}\hat{\boldsymbol{\alpha}}\cdot\hat{\boldsymbol{p}} + m_0c^2 + E_p\right)\\
&= \frac{1}{2E_p\left(m_0c^2+E_p\right)}\Big(c^2\hat{\boldsymbol{p}}^2 + \left(m_0c^2+E_p\right)c\hat{\beta}\hat{\boldsymbol{\alpha}}\cdot\hat{\boldsymbol{p}}\\
&\quad + c\hat{\boldsymbol{\alpha}}\cdot\hat{\boldsymbol{p}}\hat{\beta}\left(m_0c^2+E_p\right) + \left(m_0c^2+E_p\right)^2\Big)\\
&= \frac{1}{2E_p\left(m_0c^2+E_p\right)}\left(E_p^2 - m_0c^4 + m_0^2c^4 + 2E_pm_0c^2 + E_p^2\right) = 1 \ .
\end{aligned} \tag{11.2}$$

Sometimes $\hat{U}$ is specified in a slightly modified form, which we quote here for completeness:

$$\begin{aligned}\hat{U} &= \frac{\hat{\beta}\hat{H}_\mathrm{f} + E_p}{\sqrt{2E_p\left(m_0c^2 + E_p\right)}} = \frac{\hat{\beta}\hat{\boldsymbol{\alpha}}\cdot\hat{\boldsymbol{p}}}{\sqrt{2E_p\left(m_0c^2 + E_p\right)}} + \sqrt{\frac{m_0c^2 + E_p}{2E_p}} \\ &= \sqrt{\frac{2E_p}{m_0c^2 + E_p}}\frac{1}{2}\left[1 + \hat{\beta}\frac{\hat{H}_\mathrm{f}}{E_p}\right] \\ &= \sqrt{\frac{1}{2}\left(1 + \frac{m_0c^2}{E_p}\right)} + \hat{\beta}\hat{\boldsymbol{\alpha}}\cdot\frac{\hat{\boldsymbol{p}}}{p}\sqrt{\frac{1}{2}\left(1 - \frac{m_0c^2}{E_p}\right)} \quad , \end{aligned} \tag{11.3}$$

where $p = |\boldsymbol{p}|$.

The wave functions in the Φ representation are related to the original representation by

$$\phi = \hat{U}\psi \tag{11.4}$$

Simultaneously, all operators need to be transformed according to

$$\hat{A}_\Phi = \hat{U}\hat{A}\hat{U}^\dagger \quad . \tag{11.5}$$

Since the operator of momentum $\hat{\boldsymbol{p}}$ commutes with $\hat{U}$, it remains invariant under the transformation, i.e.

$$\hat{\boldsymbol{p}}_\Phi = \hat{\boldsymbol{p}} \quad . \tag{11.6}$$

With respect to (11.5) we transform the Hamiltonian $\hat{H}_\mathrm{f}$ into Φ representation:

$$\hat{H}_\Phi = \frac{\hat{\beta}\left(c\hat{\boldsymbol{\alpha}}\cdot\hat{\boldsymbol{p}} + \hat{\beta}m_0c^2\right) + E_p}{\sqrt{2E_p\left(m_0c^2 + E_p\right)}}\left(c\hat{\boldsymbol{\alpha}}\cdot\hat{\boldsymbol{p}} + \hat{\beta}m_0c^2\right)\frac{\left(c\hat{\boldsymbol{\alpha}}\cdot\hat{\boldsymbol{p}} + \hat{\beta}m_0c^2\right)\hat{\beta} + E_p}{\sqrt{2E_p\left(m_0c^2 + E_p\right)}} \; . \tag{11.7}$$

It is useful to evaluate $\hat{H}_\mathrm{f}^2$ first. Using the relation $\hat{\beta}\hat{\alpha}_k = -\hat{\alpha}_k\hat{\beta}$, we find

$$\begin{aligned}\hat{H}_\mathrm{f}^2 &= \left(c\hat{\boldsymbol{\alpha}}\cdot\hat{\boldsymbol{p}} + \hat{\beta}m_0c^2\right)\left(c\hat{\boldsymbol{\alpha}}\cdot\hat{\boldsymbol{p}} + \hat{\beta}m_0c^2\right) \\ &= c^2\hat{\boldsymbol{p}}^2 + c(\hat{\boldsymbol{\alpha}}\cdot\hat{\boldsymbol{p}})\hat{\beta}m_0c^2 + \hat{\beta}m_0c^2c\hat{\boldsymbol{\alpha}}\cdot\hat{\boldsymbol{p}} + m_0^2c^4 = c^2\hat{\boldsymbol{p}}^2 + m_0^2c^4 \\ &\equiv \hat{E}_p^2 = E_p^2 \quad , \end{aligned}$$

where the last equality is justified if the operator $\hat{H}_\mathrm{f}^2$ acts on plane waves. Furthermore we evaluate

$$\hat{\beta}\hat{H}_\mathrm{f}\hat{\beta} = \hat{\beta}\left(c\hat{\boldsymbol{\alpha}}\cdot\hat{\boldsymbol{p}} + \hat{\beta}m_0c^2\right)\hat{\beta} = -c\hat{\boldsymbol{\alpha}}\cdot\hat{\boldsymbol{p}} + \hat{\beta}m_0c^2 = 2m_0c^2\hat{\beta} - \hat{H}_\mathrm{f} \quad ,$$

and hence, we finally have

$$\begin{aligned}\hat{H}_\Phi &= \frac{1}{2E_p\left(m_0c^2 + E_p\right)}\left(\hat{\beta}E_p^2 + E_p\hat{H}_\mathrm{f}\right)\left(\hat{H}_\mathrm{f}\hat{\beta} + E_p\right) \\ &= \frac{1}{2E_p\left(m_0c^2 + E_p\right)}\left(\hat{\beta}E_p^2\hat{H}_\mathrm{f}\hat{\beta} + \hat{\beta}E_p^3 + E_p^3\hat{\beta} + E_p\hat{H}_\mathrm{f}E_p\right) \\ &= \frac{1}{2E_p\left(m_0c^2 + E_p\right)}\left(E_p^2 2m_0c^2\hat{\beta} + 2\hat{\beta}E_p^2\right) \\ &= \frac{2E_p^2\hat{\beta}\left(m_0c^2 + E_p\right)}{2E_p\left(m_0c^2 + E_p\right)} = E_p\hat{\beta} \quad . \end{aligned} \tag{11.8}$$

The Dirac equation

$$\hat{H}_{\mathrm{f}}\psi = \varepsilon\psi \tag{11.9}$$

now reads

$$\hat{\beta}E_p\phi = \varepsilon\phi \quad . \tag{11.10}$$

This concept has already been illustrated in Chap. 1 for the Klein–Gordon equation (cf. the Feshbach–Villars representation). Now we want to introduce a different method to find the Hamiltonian in the Φ representation, originally proposed by *Foldy* and *Wouthuysen*. Again the idea is to search for a unitary transformation $\hat{U}$ which will remove from the Dirac equation all operators of the type $\hat{\boldsymbol{\alpha}}$ that couple the large to the small components. We will term any such operator an *odd operator*, e.g. $\hat{\boldsymbol{\alpha}}$, $\hat{\gamma}_i$, $\hat{\gamma}_5$, whereas an operator that does not couple large and small components, such as $\hat{1\!\!1}$, $\hat{\beta}$, $\hat{\Sigma}$, is called an *even operator*.

We write

$$\phi = \hat{U}\psi = \mathrm{e}^{\mathrm{i}\hat{S}}\psi \quad , \quad \text{where} \tag{11.11}$$

$$\hat{U}^{\dagger}\hat{U} = \mathrm{e}^{-\mathrm{i}\hat{S}}\,\mathrm{e}^{\mathrm{i}\hat{S}} = 1\!\!1 \tag{11.12}$$

and $\hat{S}$ is a yet unknown Hermitian operator. From

$$\hat{H}_{\Phi}\phi = \mathrm{i}\hbar\frac{\partial\phi}{\partial t} \quad , \tag{11.13}$$

it follows that

$$\hat{H}_{\Phi}\,\mathrm{e}^{\mathrm{i}\hat{S}}\psi = \mathrm{i}\hbar\,\mathrm{e}^{\mathrm{i}\hat{S}}\frac{\partial\psi}{\partial t}\,\mathrm{e}^{\mathrm{i}\hat{S}}\psi = \mathrm{e}^{\mathrm{i}\hat{S}}\hat{H}\psi - \hbar\frac{\partial\hat{S}}{\partial t}\,\mathrm{e}^{\mathrm{i}\hat{S}}\psi \quad . \tag{11.14}$$

Multiplication by $\mathrm{e}^{-\mathrm{i}\hat{S}}$ from the rhs results in

$$\hat{H}_{\phi} = \mathrm{e}^{\mathrm{i}\hat{S}}\hat{H}\,\mathrm{e}^{-\mathrm{i}\hat{S}} - \hbar\frac{\partial\hat{S}}{\partial t} \quad . \tag{11.15}$$

If $\hat{S}$ is explicitly known, this relation enables us to determine the Hamiltonian in Φ representation $\hat{H}_{\Phi}$ from the original Hamiltonian $\hat{H}$. In the following we will restrict ourselves to time-independent transformations $\hat{S}$, i.e. we presume $\partial\hat{S}/\partial t = 0$. For $\hat{S}$ we try the following ansatz (note that the field-free case, as considered here, implies that one simply may move over the momentum space representation, where $\hat{\boldsymbol{p}} = \boldsymbol{p}$):

$$\hat{S} = -\left(\frac{\mathrm{i}}{2m_0c}\right)\hat{\beta}\hat{\boldsymbol{\alpha}}\cdot\boldsymbol{p}\,\omega\left(\frac{p}{m_0c}\right) \quad , \tag{11.16}$$

where the function $\omega(p/m_0c)$ will be specified later. It is obvious that $\hat{S}$ is Hermitian. For the Hamiltonian of free Dirac particles in the Φ representation it holds that

$$\begin{aligned}\hat{H}_{\Phi} &= \mathrm{e}^{\mathrm{i}\hat{S}}\left(c\hat{\boldsymbol{\alpha}}\cdot\boldsymbol{p} + \hat{\beta}m_0c^2\right)\mathrm{e}^{-\mathrm{i}\hat{S}} = \mathrm{e}^{\mathrm{i}\hat{S}}\hat{\beta}\left(c\hat{\beta}\hat{\boldsymbol{\alpha}}\cdot\boldsymbol{p} + m_0c^2\right)\mathrm{e}^{-\mathrm{i}\hat{S}}\\ &= \mathrm{e}^{\mathrm{i}\hat{S}}\hat{\beta}\,\mathrm{e}^{-\mathrm{i}\hat{S}}\hat{\beta}\left(c\hat{\boldsymbol{\alpha}}\cdot\boldsymbol{p} + \hat{\beta}m_0c^2\right) \quad ,\end{aligned} \tag{11.17}$$

since $\hat{S}$ commutes with $\hat{\beta}\hat{\boldsymbol{p}}\cdot\hat{\boldsymbol{\alpha}}$. Using the relation

$$\hat{\beta}(\hat{\beta}\hat{\boldsymbol{\alpha}}\cdot\hat{\boldsymbol{p}})^n = (-1)^n(\hat{\beta}\hat{\boldsymbol{\alpha}}\cdot\hat{\boldsymbol{p}})^n\hat{\beta} \quad , \tag{11.18}$$

we can write

$$\begin{aligned}
\hat{\beta}\,\mathrm{e}^{-\mathrm{i}\hat{S}} &= \hat{\beta}\sum_{n=0}^{\infty}\left(\frac{-1}{2m_0c}\right)^n\frac{(\hat{\beta}\hat{\boldsymbol{\alpha}}\cdot\hat{\boldsymbol{p}})^n}{n!}\omega^n \\
&= \sum_{n=0}^{\infty}\left(\frac{1}{2m_0c}\right)^n\frac{(\hat{\beta}\hat{\boldsymbol{\alpha}}\cdot\boldsymbol{p})^2}{n!}\omega^n\hat{\beta} = \mathrm{e}^{\mathrm{i}\hat{S}}\hat{\beta} \quad ,
\end{aligned} \tag{11.19}$$

where we have expanded $\mathrm{e}^{-\mathrm{i}\hat{S}}$ in a power series. Hence (11.17) becomes

$$\hat{H}_\Phi = \mathrm{e}^{2\mathrm{i}\hat{S}}\left(c\hat{\boldsymbol{\alpha}}\cdot\boldsymbol{p} + \hat{\beta}m_0c^2\right) \quad . \tag{11.20}$$

Next, we expand $\mathrm{e}^{2\mathrm{i}\hat{S}}$, i.e.

$$\begin{aligned}
\mathrm{e}^{2\mathrm{i}\hat{S}} &= \exp\left(\frac{\hat{\beta}\hat{\boldsymbol{\alpha}}\cdot\boldsymbol{p}}{m_0c}\omega\left(\frac{p}{m_0c}\right)\right) = \sum_{n=0}^{\infty}\left(\frac{1}{m_0c}\right)^n\frac{(\hat{\beta}\hat{\boldsymbol{\alpha}}\cdot\boldsymbol{p})^n}{n!}\omega^n \\
&= 1 + \frac{\hat{\beta}\hat{\boldsymbol{\alpha}}\cdot\boldsymbol{p}}{m_0c}\omega + \left(\frac{\hat{\beta}\hat{\boldsymbol{\alpha}}\cdot\boldsymbol{p}}{m_0c}\right)^2\frac{\omega^2}{2!} + \left(\frac{\hat{\beta}\hat{\boldsymbol{\alpha}}\cdot\boldsymbol{p}}{m_0c}\right)^3\frac{\omega^3}{3!} \\
&\quad + \left(\frac{\hat{\beta}\hat{\boldsymbol{\alpha}}\cdot\boldsymbol{p}}{m_0c}\right)^4\frac{\omega^4}{4!} + \dots \quad .
\end{aligned} \tag{11.21}$$

With regard to the expansions

$$\begin{aligned}
\cos\left(\frac{p}{m_0c}\omega\right) &= 1 - \frac{(p/m_0c)^2\omega^2}{2!} + \frac{(p/m_0c)^4\omega^4}{4!} - \dots \quad , \\
\frac{\hat{\beta}\hat{\boldsymbol{\alpha}}\cdot\boldsymbol{p}}{p}\sin\left(\frac{p}{m_0c}\omega\right) &= \frac{\hat{\beta}\hat{\boldsymbol{\alpha}}\cdot\boldsymbol{p}}{p}\left[\frac{p}{m_0c}\omega - \frac{[(p/m_0c)\omega]^3}{3!} + \dots\right]
\end{aligned} \tag{11.22}$$

and $(\hat{\beta}\hat{\boldsymbol{\alpha}}\cdot\boldsymbol{p})^2 = -p^2$, we see immediately that $\hat{H}_\Phi$ can also be written as follows:

$$\begin{aligned}
\hat{H}_\Phi &= \left[\cos\left(\frac{p}{m_0c}\omega\right) + \hat{\beta}\frac{\hat{\boldsymbol{\alpha}}\cdot\boldsymbol{p}}{p}\sin\left(\frac{p}{m_0c}\omega\right)\right]\left[c\hat{\boldsymbol{\alpha}}\cdot\boldsymbol{p} + \hat{\beta}m_0c^2\right] \\
&= \hat{\beta}\left[m_0c^2\cos\left(\frac{p}{m_0c}\omega\right) + cp\sin\left(\frac{p}{m_0c}\omega\right)\right] \\
&\quad + \frac{\hat{\boldsymbol{\alpha}}\cdot\boldsymbol{p}}{p}\left[pc\cos\left(\frac{p}{m_0c}\omega\right) - m_0c^2\sin\left(\frac{p}{m_0c}\omega\right)\right] \quad .
\end{aligned} \tag{11.23}$$

We will have reached our aim to remove all odd operators, if we can specify ω such that the bracket $\sim \hat{\boldsymbol{\alpha}}\cdot\boldsymbol{p}/p$ is eliminated. Rearranging this expression yields

$$\begin{aligned}
&\left[pc\cos\left(\frac{p}{m_0c}\omega\right) - m_0c^2\sin\left(\frac{p}{m_0c}\omega\right)\right] \\
&\quad = pc\cos\left(\frac{p}{m_0c}\omega\right)\left[1 - \frac{m_0c}{p}\frac{\sin[(p/m_0c)\omega]}{\cos[(p/m_0c)\omega]}\right] \\
&\quad = pc\cos\left(\frac{p}{m_0c}\omega\right)\left[1 - \frac{m_0c}{p}\tan\left(\frac{p}{m_0c}\omega\right)\right] \quad .
\end{aligned} \tag{11.24}$$

If we now choose

$$\omega = \frac{m_0 c}{p} \arctan\left(\frac{p}{m_0 c}\right) \quad , \tag{11.25}$$

it follows that

$$\begin{aligned} &pc\cos\left(\frac{p}{m_0 c}\omega\right)\left[1 - \frac{m_0 c}{p}\tan\left(\arctan\left(\frac{p}{m_0 c}\right)\right)\right] \\ &\quad = pc\cos\left(\frac{p}{m_0 c}\omega\right)[1-1] = 0 \quad , \end{aligned}$$

i.e. with our choice of ω the odd part of $\hat{H}_\Phi$ vanishes and we get

$$\hat{H}_\Phi = \hat{\beta}\left[m_0 c^2 \cos\left(\frac{p}{m_0 c}\omega\right) + cp\sin\left(\frac{p}{m_0 c}\omega\right)\right] \quad . \tag{11.26}$$

Using the trigonometric relation

$$\arctan x = \arcsin\frac{x}{\sqrt{1+x^2}} = \arccos\frac{1}{\sqrt{1+x^2}} \quad ,$$

which holds for $x > 0$, then

$$\begin{aligned} \hat{H}_\Phi &= \hat{\beta}\left[m_0 c^2 \frac{1}{\sqrt{1+p^2/m_0^2 c^2}} + cp\frac{p/m_0 c}{\sqrt{1+p^2/m_0^2 c^2}}\right] \\ &= \hat{\beta}\left[m_0 c^2 \frac{m_0 c}{\sqrt{p^2+m_0^2 c^2}} + \frac{cp^2}{\sqrt{p^2+m_0^2 c^2}}\right] \\ &= \hat{\beta} c\sqrt{p^2+m_0^2 c^2} = \hat{\beta} E_p \quad . \end{aligned} \tag{11.27}$$

Obviously this result is the same as in our previous calculations [see (11.8)].

Next we calculate the sign operator $\hat{\Lambda}$ in the Φ representation. $\hat{\Lambda}$ was defined as [see (2.48)]:

$$\hat{\Lambda} = \frac{\hat{H}_\mathrm{f}}{\sqrt{\hat{H}_\mathrm{f}^2}} = \frac{c\hat{\boldsymbol{\alpha}}\cdot\hat{\boldsymbol{p}} + \hat{\beta} m_0 c^2}{c\sqrt{\hat{\boldsymbol{p}}^2 + m_0^2 c^2}} \quad , \tag{11.28}$$

or in the momentum representation:

$$\hat{\Lambda} = \frac{c\hat{\boldsymbol{\alpha}}\cdot\boldsymbol{p} + \hat{\beta} m_0 c^2}{E_p} \quad . \tag{11.29}$$

By use of $\hat{H}_\Phi$ we get

$$\hat{\Lambda}_\Phi = \hat{U}\hat{\Lambda}\hat{U}^\dagger = \hat{\beta} \quad . \tag{11.30}$$

For the following calculations it is necessary to know the Φ representation for the operator $\hat{\boldsymbol{\alpha}}$. Therefore we calculate

$$\hat{\alpha}_\Phi = \hat{U}\hat{\alpha}\hat{U}^\dagger = \frac{\hat{\beta}\hat{\alpha}\cdot\boldsymbol{p}c + m_0c^2 + E_p}{2E_p\left(m_0c^2 + E_p\right)}\hat{\alpha}\left(\hat{\alpha}\cdot\boldsymbol{p}c\hat{\beta} + m_0c^2 + E_p\right)$$
$$= \frac{c^2(\hat{\alpha}\cdot\boldsymbol{p})\hat{\alpha}(\hat{\alpha}\cdot\boldsymbol{p})}{2E_p\left(m_0c^2+E_p\right)} + \frac{c\hat{\beta}(\hat{\alpha}\cdot\hat{\boldsymbol{p}})\hat{\alpha} + \hat{\alpha}(\hat{\alpha}\cdot\boldsymbol{p})c\hat{\beta}}{2E_p} + \frac{\hat{\alpha}\left(m_0c^2+E_p\right)}{2E_p} \quad ,$$

and replace

$$c^2\boldsymbol{p}^2 = E_p^2 - m_0^2c^4 = \left(E_p + m_0c^2\right)\left(E_p - m_0c^2\right) \quad .$$

By use of the commutation relation

$$\hat{\alpha}_k\hat{\alpha}_l + \hat{\alpha}_l\hat{\alpha}_k = 2\delta_{kl} \quad ,$$

follows

$$\frac{c^2(\hat{\alpha}\cdot\boldsymbol{p})\hat{\alpha}(\hat{\alpha}\cdot\boldsymbol{p})}{2E_p\left(m_0c^2+E_p\right)} = -\frac{2c^2\boldsymbol{p}(\hat{\alpha}\cdot\boldsymbol{p})}{2E_p\left(m_0c^2+E_p\right)} + \frac{c^2\boldsymbol{p}^2\hat{\alpha}}{2E_p\left(m_0c^2+E_p\right)}$$

and

$$\frac{c\hat{\beta}(\hat{\alpha}\cdot\boldsymbol{p})\hat{\alpha} + \hat{\alpha}(\hat{\alpha}\cdot\boldsymbol{p})c\hat{\beta}}{2E_p} = \frac{c\hat{\beta}(\hat{\alpha}\cdot\boldsymbol{p})\hat{\alpha} + 2c\hat{\beta}\boldsymbol{p} - c\hat{\beta}(\hat{\alpha}\cdot\boldsymbol{p})\hat{\alpha}}{2E_p} = \frac{c\hat{\beta}\boldsymbol{p}}{E_p} \quad .$$

Next we collect terms:

$$\frac{\left(E_p^2 - m_0^2c^4\right)\hat{\alpha} + \hat{\alpha}\left(m_0^2c^4 + E_p^2 + 2E_pm_0c^2\right)}{2E_p\left(m_0c^2+E_p\right)} = \frac{\hat{\alpha}\left(2E_p^2 + 2E_pm_0c^2\right)}{2E_p\left(m_0c^2+E_p\right)} = \hat{\alpha}\,,$$

and finally get

$$\hat{\alpha}_\Phi = \boldsymbol{\alpha} - \frac{c^2\boldsymbol{p}(\hat{\alpha}\cdot\boldsymbol{p})}{E_p\left(E_p + m_0c^2\right)} + \frac{c\hat{\beta}\boldsymbol{p}}{E_p} \quad . \tag{11.31}$$

By a similar calculation it can be shown that the even part of the position operator is given in the Φ representation by

$$[\boldsymbol{r}]_\Phi = \hat{U}[\boldsymbol{r}]\hat{U}^\dagger = \boldsymbol{r} - \frac{\hbar c^2(\hat{\boldsymbol{\Sigma}}\times\boldsymbol{p})}{2E_p\left(E_p + m_0c^2\right)} \quad , \tag{11.32}$$

where

$$[\boldsymbol{r}] = \frac{1}{2}\left(\boldsymbol{r} + \hat{\Lambda}\boldsymbol{r}\hat{\Lambda}\right) = \boldsymbol{r} + \frac{\mathrm{i}\hbar c\hat{\Lambda}}{2E_p}\hat{\alpha} - \frac{\mathrm{i}\hbar c^2\boldsymbol{p}}{2E_p^2} \quad , \tag{11.33}$$

while the odd part reads

$$\{\boldsymbol{r}\}_\Phi = U\{\boldsymbol{r}\}U^\dagger = \frac{\mathrm{i}\hbar c}{2E_p}\left(\hat{\alpha}\hat{\beta} + \frac{c^2\hat{\beta}(\hat{\alpha}\cdot\boldsymbol{p})\boldsymbol{p}}{E_p\left(E_p + m_0c^2\right)}\right) \quad , \tag{11.34}$$

with

$$\{\boldsymbol{r}\} = \tfrac{1}{2}(\boldsymbol{r} - \hat{\Lambda}\boldsymbol{r}\hat{\Lambda}) \quad .$$

For calculation of $\boldsymbol{r}_\Phi$, the relation

$$\boldsymbol{r}_\Phi = \hat{U}\boldsymbol{r}\hat{U}^\dagger = \boldsymbol{r} + \mathrm{i}\hbar\hat{U}\left(\boldsymbol{\nabla}_p\hat{U}^\dagger\right) \qquad \boldsymbol{r} = \mathrm{i}\hbar\boldsymbol{\nabla}_p \tag{11.35}$$

is quite useful (see Exercise 1.18).

Finally we want to transform explicitly the Dirac plane waves into the Φ representation. The projection operators $\hat{\Pi}_{+\Phi}$ and $\hat{\Pi}_{-\Phi}$, which project out from an arbitrary Dirac wave the states with positive or negative energy respectively, read in the Φ representation:

$$\hat{\Pi}_{+\Phi} = \tfrac{1}{2}(1+\hat{\beta}) \quad , \quad \hat{\Pi}_{-\Phi} = \tfrac{1}{2}(1-\hat{\beta}) \quad , \tag{11.36}$$

where $\frac{1}{2}(1\pm\hat{\beta})$ has the explicit form

$$\frac{1+\hat{\beta}}{2} = \begin{pmatrix} 1 & 0 & 0 & 0 \\ 0 & 1 & 0 & 0 \\ 0 & 0 & 0 & 0 \\ 0 & 0 & 0 & 0 \end{pmatrix} \quad , \quad \frac{1-\hat{\beta}}{2} = \begin{pmatrix} 0 & 0 & 0 & 0 \\ 0 & 0 & 0 & 0 \\ 0 & 0 & 1 & 0 \\ 0 & 0 & 0 & 1 \end{pmatrix} \quad . \tag{11.37}$$

With these relations we may show that in the Φ representation the wave function for a given sign of energy is completely described by two components. In the usual representation the orthonormalized wave functions for states with a given momentum in the z direction, given sign $\lambda(=+1$ or $-1)$ of energy and given spin projection $\hat{\boldsymbol{\sigma}}\cdot\hat{\boldsymbol{p}} = \frac{1}{2}$ or $-\frac{1}{2}$, follows from (2.34):

$$\psi_{p,\lambda,1/2} = \sqrt{\frac{m_0c^2+\varepsilon}{2\varepsilon}} \begin{pmatrix} 1 \\ 0 \\ \dfrac{cp}{m_0c^2+\varepsilon} \\ 0 \end{pmatrix} \frac{\mathrm{e}^{\mathrm{i}pz/\hbar}}{(2\pi\hbar)^{3/2}} \quad , \quad \varepsilon = \lambda E_p \quad , \quad \boldsymbol{p}\cdot\hat{\boldsymbol{\sigma}} = p \quad ,$$

$$\psi_{p,\lambda,-1/2} = \sqrt{\frac{m_0c^2+\varepsilon}{2\varepsilon}} \begin{pmatrix} 0 \\ 1 \\ 0 \\ \dfrac{-cp}{m_0c^2+\varepsilon} \end{pmatrix} \frac{\mathrm{e}^{\mathrm{i}pz/\hbar}}{(2\pi\hbar)^{3/2}} \quad , \quad \boldsymbol{p}\cdot\hat{\boldsymbol{\sigma}} = -p \quad . \tag{11.38}$$

Transforming according to $\phi = \hat{U}\psi$, we need

$$\hat{\beta}\hat{\boldsymbol{\alpha}} = \begin{pmatrix} 1\!\!1 & 0 \\ 0 & -1\!\!1 \end{pmatrix} \begin{pmatrix} 0 & \hat{\boldsymbol{\sigma}} \\ \hat{\boldsymbol{\sigma}} & 0 \end{pmatrix} = \begin{pmatrix} 0 & \hat{\boldsymbol{\sigma}} \\ -\hat{\boldsymbol{\sigma}} & 0 \end{pmatrix} \quad \text{and} \quad \hat{\sigma}_z = \begin{pmatrix} 1 & 0 \\ 0 & -1 \end{pmatrix} \quad ;$$

then it follows that

$$\hat{\beta}\hat{\alpha}_z = \begin{pmatrix} 0 & 0 & 1 & 0 \\ 0 & 0 & 0 & -1 \\ -1 & 0 & 0 & 0 \\ 0 & 1 & 0 & 0 \end{pmatrix} \quad ,$$

and, using $c^2p^2 = (E_p + m_0c^2)(E_p - m_0c^2)$ we obtain

$$\phi = \hat{U}\psi = \frac{1}{2E_p}\frac{\mathrm{e}^{\mathrm{i}pz/\hbar}}{(2\pi\hbar)^{3/2}} \begin{pmatrix} \dfrac{c^2p^2}{m_0c^2+\varepsilon} + m_0c^2 + E_p \\ 0 \\ -cp+cp \\ 0 \end{pmatrix}$$

$$= \begin{pmatrix} 1 \\ 0 \\ 0 \\ 0 \end{pmatrix} \frac{\mathrm{e}^{\mathrm{i}pz/\hbar}}{(2\pi\hbar)^{3/2}} \quad , \quad \hat{\boldsymbol{\sigma}}\cdot\boldsymbol{p} = p \quad \text{and} \tag{11.39}$$

$$\phi_{p,1,-1/2} = \frac{1}{2E_p}\frac{\mathrm{e}^{\mathrm{i}pz/\hbar}}{(2\pi\hbar)^{3/2}}\begin{pmatrix} 0 \\ \dfrac{c^2p^2}{m_0c^2+E_p}+m_0c^2+E_p \\ 0 \\ cp-cp \end{pmatrix}$$

$$= \begin{pmatrix} 0 \\ 1 \\ 0 \\ 0 \end{pmatrix}\frac{\mathrm{e}^{\mathrm{i}pz/\hbar}}{(2\pi\hbar)^{3/2}} \quad . \tag{11.40}$$

The states with negative energy read

$$\psi_{p,-1,1/2} = \sqrt{\frac{E_p-m_0c^2}{2E_p}}\begin{pmatrix} 1 \\ 0 \\ \dfrac{-cp}{E_p-m_0c^2} \\ 0 \end{pmatrix}\frac{\mathrm{e}^{\mathrm{i}pz/\hbar}}{(2\pi\hbar)^{3/2}} \quad , \quad \boldsymbol{p}\cdot\hat{\boldsymbol{\sigma}} = p \quad \text{and}$$

$$\psi_{p,-1,-1/2} = \sqrt{\frac{E_p-m_0c^2}{2E_p}}\begin{pmatrix} 0 \\ 1 \\ 0 \\ \dfrac{cp}{E_p-m_0c^2} \end{pmatrix}\frac{\mathrm{e}^{\mathrm{i}pz/\hbar}}{(2\pi\hbar)^{3/2}} \quad , \quad \boldsymbol{p}\cdot\hat{\boldsymbol{\sigma}} = -p \quad .$$

From $\phi = \hat{U}\Psi$ we calculate

$$\phi_{p,-1,1/2} = \frac{1}{2E_p}\frac{\sqrt{E_p-m_0c^2}}{\sqrt{E_p+m_0c^2}}\frac{\mathrm{e}^{\mathrm{i}pz/\hbar}}{(2\pi\hbar)^{3/2}}\begin{pmatrix} \dfrac{-c^2p^2}{E_p-m_0c^2}+m_0c^2+E_p \\ 0 \\ -cp+\dfrac{\left(m_0c^2+E_p\right)(-cp)}{E_p-m_0c^2} \\ 0 \end{pmatrix} . \tag{11.41}$$

The first element of this column matrix is zero, the third element results in $-2cpE_p/(E_p-m_0c^2)$ and we get

$$\frac{\sqrt{E_p-m_0c^2}}{\sqrt{E_p+m_0c^2}} = \frac{E_p-m_0c^2}{\sqrt{E_p+m_0c^2}\sqrt{E_p-m_0c^2}}$$

$$= \frac{E_p-m_0c^2}{\sqrt{E_p^2-\left(m_0c^2\right)^2}} = \frac{E_p-m_0c^2}{cp} \quad .$$

This yields

$$\phi_{p,-1,1/2} = \begin{pmatrix} 0 \\ 0 \\ -1 \\ 0 \end{pmatrix}\frac{\mathrm{e}^{\mathrm{i}pz/\hbar}}{(2\pi\hbar)^{3/2}} \quad , \quad \hat{\boldsymbol{\sigma}}\cdot\boldsymbol{p} = p \quad , \tag{11.42}$$

and, similarly,

$$\phi_{p,-1,1/2} = \begin{pmatrix} 0 \\ 0 \\ 0 \\ -1 \end{pmatrix}\frac{\mathrm{e}^{\mathrm{i}pz/\hbar}}{(2\pi\hbar)^{3/2}} \quad , \quad \hat{\boldsymbol{\sigma}}\cdot\boldsymbol{p} = -p \quad . \tag{11.43}$$

Now, Φ can always we written as

$$\Phi = \Phi_+ + \Phi_- = \begin{pmatrix} w(p) \\ v(p) \end{pmatrix} \quad , \tag{11.44}$$

and using the projection operators one obtains

$$\Phi_+ = \left(\hat{\Pi}_{+\Phi}\right)\Phi = \begin{pmatrix} w(p) \\ 0 \end{pmatrix} \quad , \quad \Phi_- = \left(\hat{\Pi}_{-\Phi}\right)\Phi = \begin{pmatrix} 0 \\ v(p) \end{pmatrix} \quad , \tag{11.45a}$$

where $w(p)$ and $v(p)$ can, in fact, be written as two-component functions

$$w(p) = \begin{pmatrix} w_1 \\ w_2 \end{pmatrix} \quad \text{and} \quad v(p) = \begin{pmatrix} v_1 \\ v_2 \end{pmatrix} \quad . \tag{11.45b}$$

11.1 The Foldy–Wouthuysen Representation in the Presence of External Fields

If external fields are coupled to a Hamiltonian $\hat{H}$, then $\hat{H}$ always contains parts coupling together the free positive and negative energy solutions. In the case of weak fields the role of the odd parts of this operator can be neglected. The Foldy–Wouthuysen transformation, as an approximation of the exact solution, is therefore only applicable to weak fields, where it systematically improves the approach. In performing the transformation we are guided by the example of the case of free fields. Again, we split up the Hamiltonian into

$$\hat{H} = \hat{\beta} m_0 c^2 + \hat{\varepsilon} + \hat{O} \quad , \tag{11.46}$$

where $\hat{\beta}$ corresponds to the remaining even part of $\hat{H}$ ($\hat{\varepsilon} \Leftrightarrow$ even) and $\hat{O}$ to the odd part ($\hat{O} \Leftrightarrow$ odd). Inclusion of external electromagnetic fields yields

$$\hat{O} = c\hat{\boldsymbol{\alpha}}\left(\hat{\boldsymbol{p}} - \frac{e}{c}\boldsymbol{A}\right) \quad \text{and} \quad +\hat{\varepsilon} = eV(\boldsymbol{r}) \quad , \tag{11.47}$$

where $V(\boldsymbol{r})$ is the Coulomb potential. In analogy to the case of free fields (11.16) we introduce a tansformation of the following kind:

$$\hat{H}' = \mathrm{e}^{\mathrm{i}\hat{S}}\hat{H}\,\mathrm{e}^{-\mathrm{i}\hat{S}} \tag{11.48}$$

with the intention of minimizing the odd parts of the Hamiltonian $\hat{H}'$, or even to make them vanish. In analogy to (11.16) we choose

$$\hat{S} = \frac{\mathrm{i}}{2m_0c^2}\hat{\beta}\hat{O} \quad , \tag{11.49}$$

i.e. $\hat{S}$ shall not explicitly depend on time. If the fields occuring in (11.47) depend explicitly on time (and thus also the Hamiltonian) the transformation $\hat{S}$ must generally also be time dependent. Then it is usually not possible to construct $\hat{S}$ in a way that all odd parts of $\hat{H}'$ disappear in any order. Therefore, we restrict ourselves to a nonrelativistic expansion of the Hamiltonian $\hat{H}'$ into an exponential series of $1/m_0c^2$. More precisely, we take into account only terms of order

$$\left(\frac{\text{kinetic energy}}{m_0c^2}\right)^3 \quad \text{and} \quad \frac{(\text{kinetic energy})\cdot(\text{field energy})}{m_0^2c^4} \quad .$$

We expand the exponential function into a power series

$$\begin{aligned}\hat{H}' &= \left(1+\frac{\mathrm{i}\hat{S}}{1!}+\frac{\mathrm{i}^2\hat{S}^2}{2!}+\dots\right)\hat{H}\left(1+\frac{(-\mathrm{i})\hat{S}}{1!}+\frac{(-1)^2\hat{S}^2}{2!}+\dots\right)\\ &= \hat{H}+\mathrm{i}\left[\hat{S},\hat{H}\right]_- + \frac{\mathrm{i}^2}{2!}\left[\hat{S},\left[\hat{S},\hat{H}\right]_-\right]_- + \dots\\ &+\frac{\mathrm{i}^n}{n!}\left[\hat{S},\left[\hat{S},\dots,\left[\hat{S},\hat{H}\right]_-\dots\right]_-\right]_- + \dots \quad . \end{aligned} \tag{11.50}$$

To verify this we write down the second-order terms of $\hat{S}$ separately,

$$\begin{aligned}&\hat{H}\frac{(-\mathrm{i})^2}{2!}\hat{S}^2+\frac{\mathrm{i}^2\hat{S}^2}{2!}\hat{H}+\mathrm{i}\hat{S}\hat{H}(-\mathrm{i})\hat{S}\\ &\quad=\frac{\mathrm{i}^2}{2}\left[\hat{S},\left[\hat{S},\hat{H}\right]_-\right]_- = \frac{\mathrm{i}^2}{2}\left[\hat{S},\hat{S}\hat{H}-\hat{H}\hat{S}\right]_-\\ &\quad=\frac{\mathrm{i}^2}{2}\left(\hat{S}^2\hat{H}-\hat{S}\hat{H}\hat{S}-\hat{S}\hat{H}\hat{S}+\hat{H}\hat{S}^2\right) \quad . \end{aligned} \tag{11.51}$$

We can check the validity of this general commutator expansion (11.52) by considering the operator function

$$\hat{F}(\lambda)=\mathrm{e}^{\mathrm{i}\lambda\hat{S}}\hat{H}\,\mathrm{e}^{-\mathrm{i}\lambda\hat{S}}=\sum_{n=0}^{\infty}\frac{\lambda^n}{n!}\left(\frac{\mathrm{d}^nF}{\mathrm{d}\lambda^n}\right)_{\lambda=0}$$

and by verifying

$$\frac{\mathrm{d}\hat{F}}{\mathrm{d}\lambda}(\lambda)=\mathrm{e}^{\mathrm{i}\lambda\hat{S}}\mathrm{i}\left[\hat{S},\hat{H}\right]_-\ \mathrm{e}^{-\mathrm{i}\lambda\hat{S}} \quad ,$$

as well as

$$\frac{\mathrm{d}^n\hat{F}}{\mathrm{d}\lambda^n}=\mathrm{e}^{\mathrm{i}\lambda\hat{S}}\mathrm{i}^n\left[\hat{S},\left[\hat{S},\dots,\left[\hat{S},\hat{H}\right]_-\right]_-\dots\right]_-\ \mathrm{e}^{-\mathrm{i}\lambda\hat{S}} \quad .$$

Thus the validity of (11.50) easily follows for $\lambda = 1$.

In the following we make use of the relations

$$\hat{\beta}\hat{O}=-\hat{O}\hat{\beta} \quad \text{and} \quad \hat{\beta}\hat{\varepsilon}=\hat{\varepsilon}\hat{\beta} \quad . \tag{11.52}$$

Expanding the Hamiltonian $\hat{H}'$ into powers of $1/m_0c^2$, we restrict ourselves to terms up to order $1/m_0^3c^6$. Then we can write $\hat{H}'$ as

$$\begin{aligned}\hat{H}' &= \hat{H}+\mathrm{i}\left[\hat{S},\hat{H}\right]_- -\frac{1}{2}\left[\hat{S},\left[\hat{S},\hat{H}\right]_-\right]_- -\frac{\mathrm{i}}{6}\left[\hat{S},\left[\hat{S},\left[\hat{S},\hat{H}\right]_-\right]_-\right]_-\\ &+\frac{1}{24}\left[\hat{S},\left[\hat{S},\left[\hat{S},\left[\hat{S},\hat{\beta}m_0c^2\right]_-\right]_-\right]_-\right]_- + \dots \quad . \end{aligned} \tag{11.53}$$

Up to terms of order "one", $\hat{H}'$ is given by

$$\hat{H}'=\hat{\beta}m_0c^2+\hat{\varepsilon}+\hat{O}+\mathrm{i}\left[\hat{S},\hat{\beta}\right]_- m_0c^2 \quad . \tag{11.54}$$

Now we calculate the various commutators of $\hat{S}$ and $\hat{H}$:

$$
\begin{aligned}
\mathrm{i}\left[\hat{S},\hat{H}\right]_- &= \mathrm{i}\left(\left(-\frac{\mathrm{i}}{2m_0c^2}\hat{\beta}\hat{O}\right)(\hat{\beta}m_0c^2+\hat{O}+\hat{\epsilon})-(\hat{\beta}m_0c^2+\hat{O}+\hat{\epsilon})\left(\frac{-\mathrm{i}}{2m_0c^2}\hat{\beta}\hat{O}\right)\right)\\
&= \mathrm{i}\left(\frac{\mathrm{i}}{2}\hat{O}-\frac{\mathrm{i}}{2m_0c^2}\hat{\beta}\hat{O}^2-\frac{\mathrm{i}}{2m_0c^2}\hat{\beta}\hat{O}\hat{\epsilon}+\frac{\mathrm{i}}{2}\hat{O}-\frac{\mathrm{i}}{2m_0c^2}\hat{\beta}\hat{O}^2+\frac{\mathrm{i}}{2m_0c^2}\hat{\epsilon}\hat{\beta}\hat{O}\right)\\
&= -\hat{O}+\frac{\hat{\beta}\hat{O}^2}{m_0c^2}+\frac{\hat{\beta}}{2m_0c^2}\,[\hat{O},\hat{\epsilon}] \quad , \qquad (11.55)
\end{aligned}
$$

$$
\begin{aligned}
\frac{\mathrm{i}^2}{2}\left[\hat{S},\left[\hat{S},\hat{H}\right]_-\right]_- &= \frac{\mathrm{i}}{2}\left\{\left(-\frac{\mathrm{i}}{2m_0c^2}\hat{\beta}\hat{O}\right)\left(-\hat{O}+\frac{\hat{\beta}\hat{O}^2}{m_0c^2}+\frac{\hat{\beta}}{2m_0c^2}\,[\hat{O},\hat{\epsilon}]_-\right)\right.\\
&\quad\left.-\left(-\hat{O}+\frac{\hat{\beta}\hat{O}^2}{m_0c^2}+\frac{\hat{\beta}}{2m_0c^2}\,[\hat{O},\hat{\epsilon}]\right)\left(-\frac{\mathrm{i}}{2m_0c^2}\hat{\beta}\hat{O}\right)\right\}\\
&= \frac{\mathrm{i}}{2}\left(\frac{\mathrm{i}}{2m_0c^2}\hat{\beta}\hat{O}^2+\frac{\mathrm{i}}{2m_0^2c^4}\hat{O}^3+\frac{\mathrm{i}}{4m_0^2c^4}\hat{O}\,[\hat{O},\hat{\epsilon}]_-\right.\\
&\quad\left.+\frac{\mathrm{i}}{2m_0c^2}\hat{\beta}\hat{O}^2+\frac{\mathrm{i}}{2m_0^2c^4}\hat{O}^3-\frac{\mathrm{i}}{4m_0^2c^4}\,[\hat{O},\hat{\epsilon}]_-\,\hat{O}\right)\\
&= -\frac{1}{2m_0c^2}\hat{\beta}\hat{O}^2-\frac{1}{2m_0^2c^4}\hat{O}^3-\frac{1}{8m_0^2c^4}\,\left[\hat{O},[\hat{O},\hat{\epsilon}]_-\right]_- \quad ,
\end{aligned}
$$
(11.56)

$$
\begin{aligned}
&\frac{\mathrm{i}^3}{3!}\left[\hat{S},\left[\hat{S},\left[\hat{S},\hat{H}\right]_-\right]_-\right]_-\\
&= \frac{\mathrm{i}}{3}\left(\frac{-\mathrm{i}}{4m_0^2c^4}\hat{O}^3+\frac{\mathrm{i}}{4m_0^3c^6}\hat{\beta}\hat{O}^4+\frac{\mathrm{i}}{16m_0^3c^6}\hat{\beta}\hat{O}\,\left[\hat{O},[\hat{O},\hat{\epsilon}]_-\right]_-\right.\\
&\quad\left.-\frac{\mathrm{i}}{4m_0^2c^4}\hat{O}^3+\frac{\mathrm{i}}{4m_0^3c^6}\hat{\beta}\hat{O}^4-\frac{\mathrm{i}}{16m_0^3c^6}\,\left[\hat{O},[\hat{O},\hat{\epsilon}]_-\right]_-\hat{\beta}\hat{O}\right)\\
&= \frac{1}{6m_0^2c^4}\hat{O}^3-\frac{1}{6m_0^3c^6}\hat{\beta}\hat{O}^4-\frac{1}{48m_0^3c^6}\hat{\beta}\left[\hat{O},\left[\hat{O},[\hat{O},\hat{\epsilon}]_-\right]_-\right]_- \quad . \qquad (11.57)
\end{aligned}
$$

We take into account only terms of the order of $1/m_0^3c^6$; thus we get

$$
\frac{\mathrm{i}^4}{4!}\left[\hat{S},\left[\hat{S},\left[\hat{S},[\hat{S}\hat{H}]_-\right]_-\right]_-\right]_- \approx \frac{1}{24m_0^3c^6}\hat{\beta}\hat{O}^4 \quad . \qquad (11.58)
$$

Next we collect the terms of (11.54–11.58), which yields

$$
\begin{aligned}
\hat{H}' &= \hat{\beta}m_0c^2+\hat{\epsilon}+\hat{O}-\hat{O}+\frac{1}{2m_0c^2}\hat{\beta}\,[\hat{O},\hat{\epsilon}]_-+\frac{1}{m_0c^2}\hat{\beta}\hat{O}^2-\frac{1}{2m_0c^2}\hat{\beta}\hat{O}^2\\
&\quad-\frac{1}{8m_0^2c^4}\,\left[\hat{O},[\hat{O},\hat{\epsilon}]_-\right]_--\frac{1}{2m_0^2c^4}\hat{O}^3+\frac{1}{6m_0^2c^4}\hat{O}^3-\frac{1}{6m_0^3c^6}\hat{\beta}\hat{O}^4\\
&\quad+\frac{1}{24m_0^3c^6}\hat{\beta}\hat{O}^4-\frac{1}{48m_0^3c^6}\hat{\beta}\left[\hat{O},\left[\hat{O},[\hat{O},\hat{\epsilon}]_-\right]_-\right]_-\\
&= \hat{\beta}\left(m_0c^2+\frac{1}{2m_0c^2}\hat{O}^2-\frac{1}{8m_0^3c^6}\hat{O}^4\right)+\hat{\epsilon}-\frac{1}{8m_0^2c^4}\,\left[\hat{O},[\hat{O},\hat{\epsilon}]_-\right]_-\\
&\quad+\frac{1}{2m_0c^2}\hat{\beta}\,[\hat{O},\hat{\epsilon}]_--\frac{1}{3m_0^2c^4}\hat{O}^3-\frac{1}{48m_0^3c^6}\hat{\beta}\left[\hat{O},\left[\hat{O},[\hat{O},\hat{\epsilon}]_-\right]_-\right]_-\\
&\equiv \hat{\beta}m_0c^2+\hat{\epsilon}'+\hat{O}' \quad . \qquad (11.59)
\end{aligned}
$$

At this step we might simply argue: Let us omit all odd terms of $\hat{H}'$, i.e. the last term of (11.59) with the odd powers of $\hat{O}$. But we may also formally reduce the odd part of $\hat{H}'$ by further Foldy–Wouthuysen transformations. At this point we perform a second transformation

$$\hat{S}' = -\frac{\mathrm{i}}{2m_0c^2}\hat{\beta}\hat{O}' = -\frac{\mathrm{i}}{2m_0c^2}\hat{\beta}\left(\frac{1}{2m_0c^2}\hat{\beta}\,[\hat{O},\hat{\epsilon}]_- - \frac{1}{3m_0^2c^4}\hat{O}^3\right) \tag{11.60}$$

and obtain

$$\hat{H}'' = \mathrm{e}^{\mathrm{i}\hat{S}'}\hat{H}'\,\mathrm{e}^{-\mathrm{i}\hat{S}'} = \hat{\beta}m_0c^2 + \hat{\epsilon}' + \frac{1}{2m_0c^2}\hat{\beta}\left[\hat{O}',\hat{\epsilon}'\right]_- - \frac{1}{3m_0^2c^4}\hat{O}'^3 \quad . \tag{11.61}$$

The term proportional to $\hat{O}'^3$ contains large powers of $1/m_0c^2$ and therefore it can be neglected; thus $\hat{H}''$ is given by

$$\hat{H}'' = \hat{\beta}m_0c^2 + \hat{\epsilon}' + \frac{1}{2m_0c^2}\hat{\beta}\left[\hat{O}',\hat{\epsilon}'\right]_- = \hat{\beta}m_0c^2 + \hat{\epsilon}' + \hat{O}'' \quad . \tag{11.62}$$

$\hat{O}''$ is of the order of $1/m_0^2c^4$. To eliminate $\hat{O}''$ we apply a third transformation

$$\hat{H}''' = \mathrm{e}^{\mathrm{i}\hat{S}''}\hat{H}''\,\mathrm{e}^{-\mathrm{i}\hat{S}''} \quad \text{with} \tag{11.63}$$

$$\hat{S}'' = -\frac{\mathrm{i}}{2m_0c^2}\hat{\beta}\hat{O}'' \quad . \tag{11.64}$$

By neglecting the odd terms proportional to $1/m_0^3c^6$ this yields

$$\begin{aligned}\hat{H}''' &\simeq \hat{\beta}m_0c^2 + \hat{\epsilon}'\\ &= \hat{\beta}\left(m_0c^2 + \frac{1}{2m_0c^2}\hat{O}^2 - \frac{1}{8m_0^3c^6}\hat{O}^4\right) + \hat{\epsilon} - \frac{1}{8m_0^2c^4}\left[\hat{O},[\hat{O},\hat{\epsilon}]_-\right]_-\\ &\equiv \hat{H}_\Phi \quad .\end{aligned} \tag{11.65}$$

To illustrate this procedure we now calculate the various terms of $\hat{H}_\Phi$ explicitly. Here we make use of the following, already known relations for two arbitrary vectors A and B:

$$(\hat{\boldsymbol{\alpha}}\cdot\boldsymbol{A})(\hat{\boldsymbol{\alpha}}\cdot\boldsymbol{B}) = \boldsymbol{A}\cdot\boldsymbol{B} + \mathrm{i}\hat{\boldsymbol{\Sigma}}\cdot(\boldsymbol{A}\times\boldsymbol{B}) \quad , \tag{11.66}$$

with which we obtain

$$\begin{aligned}\frac{1}{2m_0c^2}\hat{O}^2 &\frac{1}{2m_0}\left(\hat{\boldsymbol{\alpha}}\cdot\left(\hat{\boldsymbol{p}}-\frac{e}{c}\boldsymbol{A}\right)\right)^2\\ &= \frac{1}{2m_0}\left(\hat{\boldsymbol{p}}-\frac{e}{c}\boldsymbol{A}\right)^2 + \frac{\mathrm{i}}{2m_0}\hat{\boldsymbol{\Sigma}}\cdot\left(\hat{\boldsymbol{p}}-\frac{e}{c}\boldsymbol{A}\right)\times\left(\hat{\boldsymbol{p}}-\frac{e}{c}\boldsymbol{A}\right)\\ &= \frac{1}{2m_0}\left(\hat{\boldsymbol{p}}-\frac{e}{c}\boldsymbol{A}\right)^2 - \frac{\mathrm{i}e}{2m_0c}\hat{\boldsymbol{\Sigma}}\cdot(\hat{\boldsymbol{p}}\times\boldsymbol{A}+\boldsymbol{A}\times\hat{\boldsymbol{p}}) \quad .\end{aligned}$$

Here $\hat{\boldsymbol{p}} = -\mathrm{i}\hbar\boldsymbol{\nabla}$ and $\mathrm{curl}(f\boldsymbol{A}) = f\,\mathrm{curl}\,\boldsymbol{A} + \mathrm{grad}f\times\boldsymbol{A}$. This expression can be further simplified as

$$\begin{aligned}\frac{1}{2m_0c^2}\hat{O}^2 &= \frac{1}{2m_0}\left(\hat{\boldsymbol{p}}-\frac{e}{c}\boldsymbol{A}\right)^2 - \frac{e\hbar}{2m_0c}\hat{\Sigma}\cdot(\boldsymbol{\nabla}\times\boldsymbol{A})\\ &= \frac{1}{2m_0}\left(\hat{\boldsymbol{p}}-\frac{e}{c}\boldsymbol{A}\right)^2 - \frac{e\hbar}{2m_0c}\hat{\boldsymbol{\Sigma}}\cdot\boldsymbol{B} \quad . \end{aligned} \tag{11.67}$$

Next we look at the commutator $[\hat{O},\hat{\epsilon}]_-$

$$\begin{aligned}\frac{1}{8m_0^2c^4}[\hat{O},\hat{\epsilon}]_- &= \frac{1}{8m_0^2c^4}\left(c\hat{\boldsymbol{\alpha}}\cdot\left(\hat{\boldsymbol{p}}-\frac{e}{c}\boldsymbol{A}\right)eV - eVc\hat{\boldsymbol{\alpha}}\cdot\left(\hat{\boldsymbol{p}}-\frac{e}{c}\boldsymbol{A}\right)\right)\\ &= -\frac{1}{8m_0^2c^4}\mathrm{i}e\hbar c\hat{\boldsymbol{\alpha}}\cdot\boldsymbol{\nabla}V = \frac{\mathrm{i}e\hbar}{8m_0^2c^4}c\hat{\boldsymbol{\alpha}}\cdot\boldsymbol{E} \quad , \end{aligned} \tag{11.68}$$

and, according to (11.65), at the commutator $[\hat{O},\hat{\boldsymbol{\alpha}}\cdot\boldsymbol{E}]_-$

$$\begin{aligned}&\left[\hat{O},\frac{\mathrm{i}e\hbar}{8m_0^2c^4}c\hat{\boldsymbol{\alpha}}\cdot\boldsymbol{E}\right]_-\\ &= \left\{c\hat{\boldsymbol{\alpha}}\cdot\left(\hat{\boldsymbol{p}}-\frac{e}{c}\boldsymbol{A}\right)\frac{\mathrm{i}e\hbar}{8m_0^2c^4}c\hat{\boldsymbol{\alpha}}\cdot\boldsymbol{E} - \frac{\mathrm{i}e\hbar}{8m_0^2c^4}c\hat{\boldsymbol{\alpha}}\cdot\boldsymbol{E}\left(c\hat{\boldsymbol{\alpha}}\cdot\left(\hat{\boldsymbol{p}}-\frac{e}{c}\boldsymbol{A}\right)\right)\right\}\\ &= \left[\hat{\boldsymbol{\alpha}}\cdot\hat{\boldsymbol{p}},\frac{\mathrm{i}e\hbar c^2}{8m_0^2c^4}\hat{\boldsymbol{\alpha}}\cdot\boldsymbol{E}\right]_-\\ &= \frac{\mathrm{i}e\hbar}{8m_0^2c^2}\left(-\mathrm{i}\hbar(\boldsymbol{\nabla}\cdot\boldsymbol{E}) - \mathrm{i}\hbar\boldsymbol{E}\cdot\boldsymbol{\nabla} + \mathrm{i}\hbar\boldsymbol{E}\cdot\boldsymbol{\nabla} + \mathrm{i}\hat{\boldsymbol{\Sigma}}\cdot(\hat{\boldsymbol{p}}\times\boldsymbol{E}) - \mathrm{i}\hat{\boldsymbol{\Sigma}}\cdot(\boldsymbol{E}\times\hat{\boldsymbol{p}})\right)\\ &= \frac{\mathrm{i}e\hbar}{8m_0^2c^2}\left(-\mathrm{i}\hbar(\boldsymbol{\nabla}\cdot\boldsymbol{E}) + \hbar\hat{\boldsymbol{\Sigma}}\cdot(\boldsymbol{\nabla}\times\boldsymbol{E}) - \mathrm{i}\hat{\boldsymbol{\Sigma}}\cdot(\boldsymbol{E}\times\hat{\boldsymbol{p}}) - \mathrm{i}\hat{\boldsymbol{\Sigma}}\cdot(\boldsymbol{E}\times\hat{\boldsymbol{p}})\right)\\ &= \frac{e\hbar^2c^2}{8m_0^2c^4}\operatorname{div}\boldsymbol{E} + \frac{\mathrm{i}c\hbar^2c^2}{8m_0^2c^4}\hat{\boldsymbol{\Sigma}}\cdot\operatorname{curl}\boldsymbol{E} + \frac{e\hbar}{4m_0^2c^2}\hat{\boldsymbol{\Sigma}}\cdot(\boldsymbol{E}\times\hat{\boldsymbol{p}}) \quad . \end{aligned} \tag{11.69}$$

Adding the various contributions, we have

$$\begin{aligned}\hat{H}_\Phi &= \hat{H}'''\\ &= \hat{\beta}\left(m_0c^2 + \frac{1}{2m_0}\left(\hat{\boldsymbol{p}}-\frac{e}{c}\boldsymbol{A}\right)^2 - \frac{1}{8m_0^3c^6}\hat{\boldsymbol{p}}^4\right)\\ &\quad + eV - \frac{1}{2m_0c}\hat{\beta}e\hbar(\hat{\boldsymbol{\Sigma}}\cdot\boldsymbol{B}) - \frac{\mathrm{i}e\hbar^2c^2}{8m_0^2c^4}\hat{\boldsymbol{\Sigma}}\cdot(\operatorname{curl}\boldsymbol{E})\\ &\quad - \frac{e\hbar}{4m_0^2c^2}\hat{\boldsymbol{\Sigma}}\cdot(\boldsymbol{E}\times\hat{\boldsymbol{p}}) - \frac{e\hbar^2c^2}{8m_0^2c^4}\operatorname{div}\boldsymbol{E} \quad . \end{aligned} \tag{11.70}$$

Now it seems advisable to discuss the individual terms of (11.70). The terms in the first parenthesis result from the expansion of $\left[(\hat{\boldsymbol{p}}-e/c\boldsymbol{A})^2+m_0\right]^{1/2}$ and describe the relativistic mass increase. Subsequently follow terms describing the electrostatic energy and the magnetic dipole energy. The next two terms, which actually are Hermitian only if taken together, contain the spin–orbit interaction. This can be seen particularly clearly under the assumption of a spherically symmetric potential with $\operatorname{curl}\boldsymbol{E}=0$, when

$$\hat{\boldsymbol{\Sigma}}\cdot(\boldsymbol{E}\times\hat{\boldsymbol{p}}) = -\frac{1}{r}\frac{\partial V}{\partial r}\hat{\boldsymbol{\Sigma}}\cdot(\boldsymbol{r}\times\hat{\boldsymbol{p}}) = -\frac{1}{r}\frac{\partial V}{\partial r}(\hat{\boldsymbol{\Sigma}}\cdot\hat{\boldsymbol{L}}) \quad . \tag{11.71}$$

Thus we have

$$\hat{H}_{\text{spin–orbit}} = \frac{e\hbar}{4m_0^2c^2}\frac{1}{r}\frac{\partial V}{\partial r}(\hat{\boldsymbol{\Sigma}}\cdot\hat{\boldsymbol{L}}) \quad . \tag{11.72}$$

This *spin–orbit interaction* is responsible for the splitting of states with the same orbital angular momentum l, but with different total angular momentum j. This interaction also is very important in nuclear physics for the classification of single-particle states of nucleons, but there the spin–orbit interaction is not of electromagnetic origin. The last term of (11.70) is the so-called Darwin term. It results from the *Zitterbewegung* of the electron over a region of a magnitude comparable to the Compton wavelength (2.61). For point-like nuclei we can write

$$\operatorname{div} \boldsymbol{E} = -\Delta\phi = 4\pi\varrho_{\mathrm{c}} = 4\pi e\delta(\boldsymbol{r}) \quad .$$

Considering now that for nonrelativistic wave functions only the s state do not vanish at the origin, and thus only $\Psi_{\mathrm{ns}}(0) \neq 0$, one immediately realizes that for light atoms the Darwin term mainly results in an energy shift of the s levels.

Finally we look qualitatively at the level scheme of the hydrogen atom (see Fig. 11.1). The fine-structure splitting due to spin-orbit coupling is quantitatively the largest contribution of the relativistic theory compared to the nonrelativistic Schrödinger description of the hydrogen atom. The previously discussed contributions of the Hamiltonian yield good agreement between the theory and experimental data up to the year 1947, when as an additional effect the hyperfine splitting had to be taken into account. This splitting of the two normally degenerated spin levels. In 1947 Lamb and Retherford discovered a further shift of the $2s_{1/2}$ level against the $2p_{1/2}$ level (indicated in the figure), which should be degenerated if exact solutions of the Dirac equation are considered (refer to Exercise 9.6 and 9.7). The physical origin of this quantum electrodynamical effect is the interaction of the electrons with the fluctuations of the quantized radiation field (self-energy and vacuum polarization), the so-called *Lamb shift.*[1]

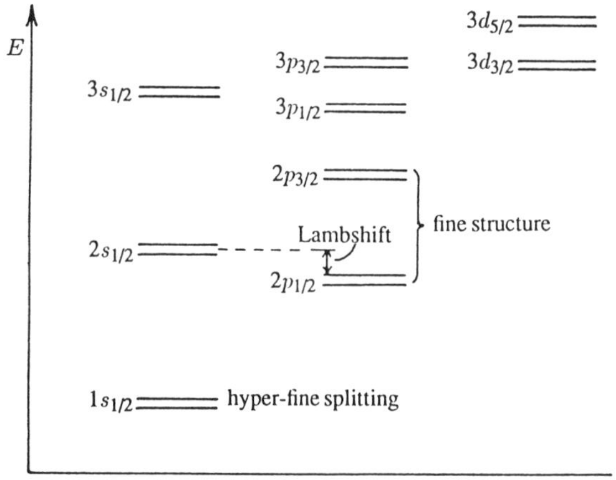

Fig. 11.1. Qualitative scheme of the energy levels of the hydrogren atom

[1] This is discussed in more detail in W. Greiner, J. Reinhardt: *Quantum Electrodynamics*, 2nd ed. (Springer, Berlin, Heidelberg 1994).

12. The Hole Theory

Until now the solutions of the Dirac equation with negative energy have been a puzzle. Attempts similar to those we performed with the solutions of the Klein–Gordon equation, where the energy turned out to be positive (by the Lagrange formalism) for solutions with positive and negative time evolution factors, proved unsuccessful (cf. Exercise 2.3). Solutions with negative energy appear almost everywhere when we are concerned with processes of high energy or with strongly localized wave packets (see Exercises 8.4, 8.5). At this point we have to confront this dilemma and find a proper solution!

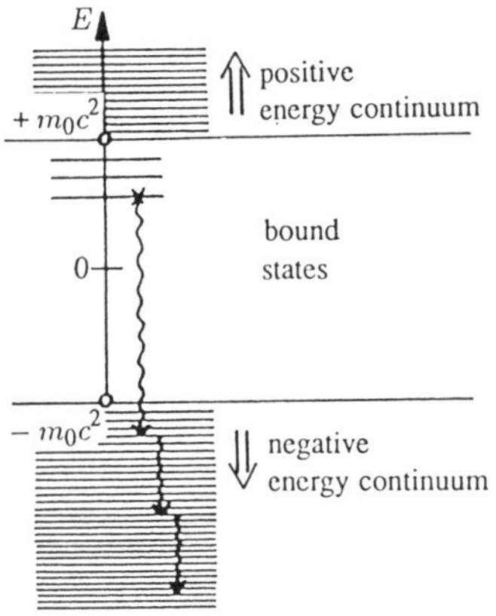

Fig. 12.1. Illustration of the radiation catastrophe of a radiating electron in an atom. It falls deeper and deeper by continued radiative transitions

The existence of solutions with negative energy in the previous interpretation, as single-particle states of the electron, obviously leads to trouble and physical nonsense. Let us consider the electrons in an atom, the spectrum of which is once more given qualitatively in Fig. 12.1.

The bound states directly below the positive energy continuum, with $E < m_0c^2$, are in general in very good agreement with experiments. It is beyond any doubt that these are the bound states of the (one-electron) atom.

An electron in the lowest atomic state ($1s$) could lose more energy by continuous radiative transitions. Thus an atom would be unstable and, because of the continuous emission of light, a *radiation catastrophe* would occur. However, such effects have never been observed! If this decay could happen, our world could not exist. Hence we have both a principle to uphold as well as a practical problem to solve to avoid electrons falling off into the states of negative energy. Neglecting the radiation field, the bound-state electrons would be stationary. By switching on the field (of course it is always "switched on"), and the use of radiation theory and of the wave functions found in Exercise 9.6, an infinite transition probability is obtained particularly if one takes into account the infinitely large number of final states in the lower continuum (see Exercise 12.1). But this is of course sheer nonsense! We must find a new physical idea to remove this dilemma, and, in its original form, this was provided by Dirac.[1] He assumed all states of negative energy to be occupied with electrons (see Fig. 12.2).

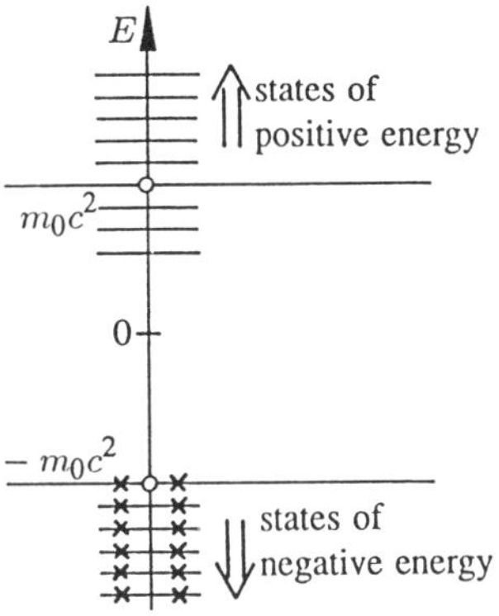

Fig. 12.2. In the hole theory the states of negative energy are occupied with electrons (X). According to the Pauli principle, each state can contain two electrons, namely one with spin up and one with spin down

The *vacuum state* is defined by the absence of real electrons (electrons in states of positive energy), and all the states of negative energy are filled with electrons. *The vacuum state is the energetically deepest stable state, which can be realized under certain conditions (constraints such as, e.g., external fields).* In the absence of the field the vacuum represents the lower (negative) continuum (it is also called the "Dirac sea"), whose states are completely occupied with

[1] P.A.M. Dirac: Proc. R. Soc. (London), **A 126**, 360 (1930); see also J.R. Oppenheimer: Phys. Rev. **35**, 939 (1930).

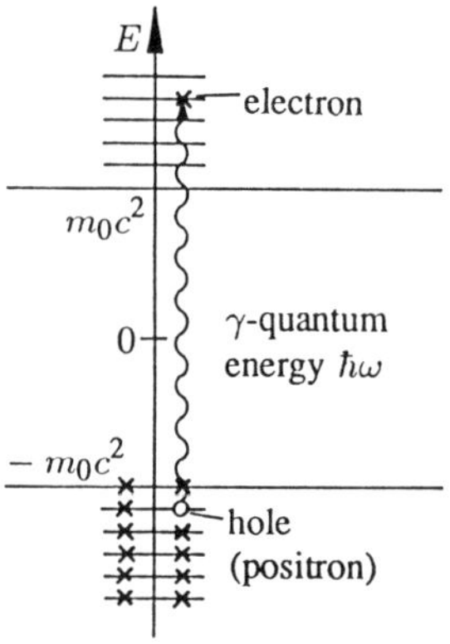

Fig. 12.3. A photon of energy $\hbar\omega > 2m_0c^2$ creates an electron–electron-hole state. The hole is interpreted as a positron; hence the process is just $e^- - e^+$ pair creation

electrons. This physical assumption of the negative energy continuum filled with electrons has very important consequences. We perceive at once that the radiation catastrophe mentioned above is now avoided because of the Pauli principle, which forbids transitions of real electrons into occupied lower states. On the other hand an electron of negative energy can absorb radiation. If the energy $\hbar\omega$ of the absorbed photon is greater than the energy gap ($\hbar\omega > 2m_0c^2$), an electron of negative energy can be excited into a state of positive energy (see Fig. 12.3).

In that case we get a real electron and a hole. The hole behaves like a particle with charge $+|e|$, because it can be annihilated by an electron (e^-) with charge $-|e|$; thus the hole is the *antiparticle of the electron and is named positron* (e^+). Obviously the creation of an electron and electron hole by photons is to be identified with *electron–positron pair creation*, with a threshold energy of

$$\hbar\omega = 2m_0c^2 \quad . \tag{12.1}$$

Alternatively, we call the process where an electron drops into a hole, thereby emitting an appropriate photon, *pair annihilation* (or matter–antimatter annihilation, or e^+e^- annihilation). The energy balance of the pair creation is

$$\begin{aligned}\hbar\omega &= E_{\text{electron with pos. energy}} - E_{\text{electron with neg. energy}}\\ &= \left(+c\sqrt{p^2 + m_0c^2c^2}\right) - \left(-c\sqrt{p'^2 + m_0c^2c^2}\right)\\ &\equiv E_{\text{electron}} + E_{\text{positron}} \quad ,\end{aligned} \tag{12.2}$$

and we can associate the electron with the positive energy

$$E_{\text{electron}} = +c\sqrt{p^2 + m_0^2c^2} \quad . \tag{12.3}$$

So far this is nothing new. What is new, however, is that according to (12.2) we have to give the positron (electron of negative energy) a positive energy, namely

$$E_{\text{positron}} = +c\sqrt{p'^2 + m_0^2c^2} \quad . \tag{12.4}$$

In the special case of vanishing positron momentum $\boldsymbol{p}'$, it follows that the *positron* has the *rest mass*

$$\left(E_{\text{positron}}\right)_{\text{at rest}} = m_0c^2 \quad . \tag{12.5}$$

Therefore positrons (electron holes) have the same rest mass as electrons but opposite charge (as we have shown above). Similarly we obtain the following momentum balance: the photon has momentum $\hbar\boldsymbol{k}$, which is distributed to electrons and positrons. We conclude that initial total momentum = final total momentum, i.e.

$$\hbar\boldsymbol{k} + (\boldsymbol{p}')_{\text{electron with neg. energy}} = (\boldsymbol{p})_{\text{electron with pos. energy}} \quad \text{or} \tag{12.6}$$

$$\hbar\boldsymbol{k} = (\boldsymbol{p})_{\text{electron with pos. energy}} - (\boldsymbol{p}')_{\text{electron with neg. energy}} \tag{12.7}$$

and write

$$\hbar\boldsymbol{k} = (\boldsymbol{p})_{\text{electron}} + (\boldsymbol{p}')_{\text{positron}} \quad . \tag{12.8}$$

Hence, the positron has the opposite momentum to the electron, and negative energy. Indeed, a missing electron (negative energy) with momentum $\boldsymbol{p}'$ should behave like a positively charged particle with the same mass and opposite momentum. In this way we can easily explain the previous paradoxical fact that the mean particle velocity of electrons with negative energy equals

$$-\frac{c^2(\boldsymbol{p})_{\text{electron with neg. energy}}}{E_p} \tag{12.9}$$

[see (2.62) ff.], since we are now dealing with the positron velocity, i.e.

$$\frac{-c^2(\boldsymbol{p})_{\text{electron with neg. energy}}}{E_p} = \frac{+c^2(\boldsymbol{p})_{\text{positron}}}{E_p} \quad . \tag{12.10}$$

The most important result of the hole theory is that it is the first theory which introduces a *model for the vacuum*, i.e. for particle-free space, in a naive sense. The vacuum is here represented by the Dirac sea, which consists of the states of negative energy occupied by electrons. This vacuum should have zero energy (mass) and no charge. However, it is clear that the model in this simple form does not have these properties. The states occupied with electrons of negative energy together have infinitely large negative energy and infinitely large negative charge. Both have to be renormalized to zero, i.e. the zero point of energy and charge is chosen in such a way that the Dirac sea has no mass and no charge. This renormalization procedure is not very satisfactory (aesthetically) but it is feasible (though soon we will discuss a better model, see Sect. 12.4). At this stage we find the qualitatively important fact that the vacuum can be modified, for instance, by the influence of external fields: these can deform the wave functions of the states of negative energy occupied with electrons. Hence they produce a measurable *vacuum polarization* with respect to the state without external fields.

Let us stress the point that the *hole theory is a many-body theory*, describing particles with positive *and* negative charge. Indeed, infinitely many electrons are needed to constitute the Dirac sea. The simple probability interpretation of the wave functions acclaimed in a single-particle theory cannot be true any longer, because the creation and annihilation of electron–positron pairs must be taken into account in the wave functions.

Resume. In early relativistic quantum mechanics, the Klein–Gordon theory was dismissed because it did not seem to allow a proper probability interpretation, and also, the appearance of states of negative energy was problematic. Hence Dirac formulated the Dirac equation with the intention of establishing a true relativistic single-particle theory. As we now know, the difficulties with the negative energy states of the Dirac equation almost of necessity demand a many-body theory (hole theory), and therefore the question arises whether it should also be dropped. On the other hand, though, we have been very successfully applying the Dirac equation to many problems (e.g. prediction of spin, of spin-orbit coupling, of the g factor, of atomic line structure). Additionally, by extending the single-particle Dirac theory to the hole theory, it has new, impressive success:

The prediction of the positron as the antiparticle of the electron was experimentally confirmed in all points, including the correct threshold (12.1). Equally, the vacuum polarization mentioned above, as well as many other effects, were confirmed by experiments. The previously discussed *Zitterbewegung* does not vanish in the hole theory, though because of the filled Dirac sea one might naively suppose that the states of negative energy are not available due to the Pauli principle; however, there still exists an exchange interaction and a related scattering. Accordingly, virtual electron–positron pairs are continuously created in the vacuum. The original electron (in a bound or free state) can fill up the virtual hole in the Dirac sea, and the other electron takes its place, this exchange interaction causing the Zitterbewegung. In its physical content the Dirac theory (extended by the hole theory) is the basis of quantum electrodynamics, which currently is the best-established theory of physics.

Dirac originally formulated his equation with a particular set of motives. The relativistic theory for spin $-\frac{1}{2}$ particles developed in this way had to be reinterpreted (hole theory) to get rid of its contradictions and became even more successful. Although the original motivation was not plausible from our present point of view, it nevertheless apparently showed the right direction. Often in the history of physics, a great success is achieved after a series of more or less erroneous investigations, which finally lead to new concepts and insights. In the following, we retain the Dirac equation and its reinterpretation by hole theory and extend it, also improving on the accuracy; however we drop the one-particle probability interpretation (except for illustrative purposes).

EXAMPLE

12.1 Radiative Transition Probability from the Hydrogen Ground State to the States of Negative Energy

Problem. Estimate the transition probability for a radiative transition from the hydrogen ground state into an electron state of the empty negative continuum with $-m_0c^2 > E > -2m_0c^2$.

Solution. This problem is to show the idea that the electron states with negative energy have to be considered as occupied. The Dirac theory yields for the electromagnetic interaction ($\hbar = c = 1$):

$$\begin{aligned}\hat{H}_{\text{int}} &= -e\hat{\bar{\psi}}(x)\gamma\hat{\psi}(x)\cdot\hat{\boldsymbol{A}}(x)\\ &= -e\hat{\psi}^\dagger(x)\hat{\boldsymbol{\alpha}}\hat{\psi}(x)\cdot\hat{\boldsymbol{A}}(x) \quad . \end{aligned}\tag{1}$$

We quantize the electromagnetic field in a box of volume L^3 and insert the expansion

$$\hat{\boldsymbol{A}} = \sum_{\boldsymbol{k}',\sigma'}\sqrt{\frac{2\pi}{L^3\omega_{k'}}}\boldsymbol{\varepsilon}_{\boldsymbol{k}',\sigma'}\left(\mathrm{e}^{-\mathrm{i}k'\cdot x}\hat{\alpha}_{\boldsymbol{k}'\sigma'} + \hat{\alpha}^\dagger_{\boldsymbol{k}'\sigma'}\mathrm{e}^{\mathrm{i}k'\cdot x}\right) \quad . \tag{2}$$

Here $\hat{\alpha}^+_{\boldsymbol{k}\sigma}$ and $\hat{\alpha}_{\boldsymbol{k}\sigma}$ are the creation and annihilation operators for photons of momentum $\hbar\kappa$ and polarization σ. The initial and the final states are of the following form:

$$\langle 1 \text{ photon with } \boldsymbol{k},\sigma| \cdot \langle\psi_\mathrm{f}| = \langle f| \tag{3}$$

and

Example 12.1.

$$|\mathrm{i}\rangle = |\psi_\mathrm{i}\rangle\,|\text{photon vacuum}\rangle \quad . \tag{4}$$

With this it follows that

$$\langle f|\hat{H}_\mathrm{int}|\mathrm{i}\rangle = -\sqrt{\frac{2\pi}{L^3\omega_k}}\,e\int \mathrm{d}^3x\,\psi_\mathrm{f}^\dagger(x)\hat{\boldsymbol{\alpha}}\psi_\mathrm{i}(x)\cdot\boldsymbol{\varepsilon}_{k\sigma}\,\mathrm{e}^{\mathrm{i}\boldsymbol{k}\cdot\boldsymbol{x}} \quad . \tag{5}$$

We use the dipole approximation $\exp(\mathrm{i}\boldsymbol{k}\cdot\boldsymbol{x})\approx 1$ and the relation

$$c\hat{\boldsymbol{\alpha}} = \frac{\mathrm{i}}{\hbar}\left[\hat{H},\hat{\boldsymbol{x}}\right]_- \quad , \tag{6}$$

which follows from the Dirac equation. Hence

$$\langle f|\hat{H}_\mathrm{int}|\mathrm{i}\rangle = e(E_\mathrm{i}-E_\mathrm{f})\sqrt{\frac{2\pi}{L^3\omega_k}}\left(\int \mathrm{d}^3x\,\psi_\mathrm{f}^\dagger(x)\boldsymbol{x}\psi_\mathrm{i}(x)\right)\cdot\boldsymbol{\varepsilon}_{\boldsymbol{k},\sigma} \quad . \tag{7}$$

Then Fermi's golden rule yields

$$\left(\frac{\text{transitions}}{\text{time}}\right) = \frac{2\pi}{\hbar}\left[\sum_p\sum_{r=3,4}\right]\sum_{\boldsymbol{k},\sigma} e^2\,(E_\mathrm{i}-E_\mathrm{f})^2\,\frac{2\pi}{L^3\omega_k}$$
$$\times\left|\mathrm{d}^3x\,\psi_\mathrm{f}^{r\dagger}(x)\boldsymbol{x}\psi_\mathrm{i}(x)\cdot\boldsymbol{\varepsilon}_{\boldsymbol{k},\sigma}\right|^2\delta(\omega_k - E_\mathrm{i}+E_\mathrm{f}) \tag{8}$$

(summed over all $\boldsymbol{p}$ with $m_0c^2 < E(\boldsymbol{p}) < 2m_0c^2$).

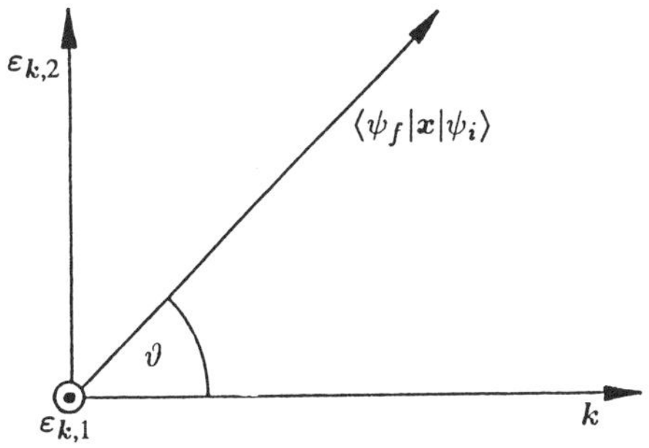

Sketch of the polarization

The first bracket of (8) contains the summation over all permitted final states of the electron: $r = 3,4$ characterizes the two spin directions of the final negative energy states. Now we can chose the polarization vectors $\boldsymbol{\varepsilon}_{\boldsymbol{k},\sigma}$ in such a way that at least one of the polarization vectors, say $\boldsymbol{\varepsilon}_{\boldsymbol{k},1}$, is orthogonal to $\boldsymbol{k}$ (see above figure). From now on we set $\hbar = c = 1$. With

$$\sum_{\boldsymbol{k}}\frac{1}{L^3}\Rightarrow\frac{1}{(2\pi)^3}\int \mathrm{d}^3k \tag{9}$$

one obtains

$$\left(\frac{\text{transitions}}{\text{time}}\right) = \frac{1}{2\pi}\left[\sum_{p,r}\right]\iiint \mathrm{d}\omega_k\,\mathrm{d}\vartheta\,\mathrm{d}\varphi\,\omega_k^2\sin\vartheta\, e^2\,(E_\mathrm{i}-E_\mathrm{f})^2\,\frac{1}{\omega_k}$$
$$\times\left|\int \mathrm{d}^3x\,\psi_\mathrm{f}^{r\dagger}(x)\boldsymbol{x}\psi_\mathrm{i}(x)\right|^2 |\boldsymbol{\varepsilon}_{\boldsymbol{k},2}|^2\sin^2\vartheta\,\delta\,(\omega_{\boldsymbol{k}}-E_\mathrm{i}+E_\mathrm{f}) \; . \tag{10}$$

Example 12.1.

The diagram is useful for a better understanding of (10). With

$$\int_0^{\pi} d\vartheta \sin^3\vartheta = \left[\frac{1}{3}\cos^3\vartheta - \cos\vartheta\right]_0^{\pi} = \frac{4}{3} \quad ,$$

(10) turns out to be

$$\left(\frac{\text{transitions}}{\text{time}}\right) = \frac{4e^3}{3}\sum_{p,r}(E_\mathrm{i} - E_\mathrm{f})^3 \left|\int d^3x\, \psi_\mathrm{f}^{r\dagger}(x)\boldsymbol{x}\psi_\mathrm{i}(x)\right|^2 \quad , \tag{11}$$

and we also normalize the final electron states in a box with volume $L'^3(E = E_\mathrm{f})$:

$$\psi_\mathrm{f}^r = \sqrt{\frac{m_0}{L'^3 E}}\omega^r(p,s)\,\mathrm{e}^{-\mathrm{i}\boldsymbol{p}\cdot\boldsymbol{x}} \quad . \tag{12}$$

These wave functions are normalized to 1:

$$\int_{L'^3} d^3x\, \psi_\mathrm{f}^{r\dagger}(x)\psi_\mathrm{f}^{r'}(x) = \frac{m_0}{L'^3 E}\int_{L'^3} d^3x \frac{E}{m_0}\delta_{rr'} = \delta_{rr'} \tag{13}$$

similarly to the wave function for the hydrogen ground state (we choose the case with spin up):

$$\psi_\mathrm{i} = \frac{(2m_0\alpha)^{3/2}}{\sqrt{4\pi}}\sqrt{\frac{1+\gamma}{2\Gamma(1+\gamma)}}(2m_0\alpha r)^{\gamma-1}\mathrm{e}^{-m_0\alpha r}\begin{pmatrix} 1 \\ 0 \\ \dfrac{\mathrm{i}(1-\gamma)}{\alpha}\cos\vartheta \\ \dfrac{\mathrm{i}(1-\gamma)}{\alpha}\sin\vartheta\,\mathrm{e}^{\mathrm{i}\varphi}\end{pmatrix} \quad ,$$

$$\gamma = \sqrt{1-\alpha^2} \quad , \quad e^2 = 4\pi\alpha \quad . \tag{14}$$

α is the fine-structure constant, i.e. $\alpha = e^2/\hbar c \approx 1/137$. With [cf. (6.30)]

$$\begin{aligned} \omega^{3\dagger}(p,s) &= \sqrt{\frac{E+m_0}{2m_0}}\left(\frac{p_z}{2m_0}, \frac{p_-}{E+m_0}, 1, 0\right) \quad , \\ \omega^{4\dagger}(p,s) &= \sqrt{\frac{E+m_0}{2m_0}}\left(\frac{p_+}{E+m_0}, \frac{-p_z}{E+m_0}, 0, 1\right) \end{aligned} \tag{15}$$

one obtains, with the approximations

$$\frac{1-\gamma}{\alpha} \approx \frac{1-\left(1-\alpha^2/2\right)}{\alpha} = \frac{\alpha}{2} \ll 1$$

and the abbreviation

$$a(r) = \frac{(2m_0\alpha)^{3/2}}{\sqrt{4\pi}}\sqrt{\frac{1+\gamma}{2\Gamma(1+\gamma)}}(2m_0\alpha r)^{\gamma-1}\,\mathrm{e}^{-m_0\alpha r} \quad , \tag{16}$$

Example 12.1.

$$\sum_r \left| \int d^3x\, \psi_f^{r\dagger}(x)\boldsymbol{x}\psi_i(x) \right|^2 \approx \frac{1}{2m_0(E+m_0)} \frac{m_0}{L'^3 E}$$
$$\times \left\{ \left| \iiint r^2\, dr\, d\vartheta\, d\varphi \sin\vartheta \boldsymbol{x} a(r)\, e^{i|\boldsymbol{p}|\,|\boldsymbol{x}|\cos\vartheta} \right|^2 \left(p_z^2 + |p_+|^2\right) \right\}$$
$$= \frac{p^2}{2E(E+m_0)L'^3} \left\{ \left| \iiint r^2\, dr\, d\vartheta\, d\varphi \sin\vartheta\, r\cos\vartheta\, a(r)\, e^{ipr\cos\vartheta} \right|^2 \right.$$
$$+ \left| \iiint r^2\, dr\, d\vartheta\, d\varphi \sin\vartheta\, r \sin\vartheta \sin\varphi\, a(r)\, e^{ipr\cos\vartheta} \right|^2$$
$$\left. + \left| \iiint r^2\, dr\, d\vartheta\, d\varphi \sin\vartheta\, r \sin\vartheta \cos\varphi\, a(r)\, e^{ipr\cos\vartheta} \right|^2 \right\} \quad .$$

Performing the φ integration yields

$$\sum_r \left| \int d^3x\, \psi_f^{r\dagger}(x)\boldsymbol{x}\psi_i(x) \right|^2$$
$$= \frac{p^2 4\pi^2}{2E(E+m_0)L'^3} \left| \int_0^\infty dr \int_0^\pi d\vartheta\, r^3 a(r) \sin\vartheta\cos\vartheta\, e^{ipr\cos\vartheta} \right|^2 \quad , \qquad (17)$$

and with $\gamma \approx 1$ it follows that

$$r^3 a(r) \frac{(2m_0\alpha)^{3/2}}{\sqrt{4\pi}\sqrt{2}} r^3\, e^{-m_0\alpha r} \quad .$$

Insertion of (11) results in

$$\left(\frac{\text{transitions}}{\text{time}}\right) = \frac{8e^2\pi m_0^3\alpha^3}{3} \sum_p \frac{p^2(E_i+E)^3}{L'^3 E(E+m_0)}$$
$$\times \left| \int_0^\infty dr \int_0^\pi d\vartheta\, r^3\, e^{-m_0\alpha r} \sin\vartheta\cos\vartheta\, e^{ipr\cos\vartheta} \right|^2 \quad , \qquad (18)$$

with $e^4 = 4\pi\alpha$, and now the ϑ integration is performed:

$$\int_0^\pi d\vartheta \underbrace{\cos\vartheta}_{u} \underbrace{\sin\vartheta\, e^{ipr\cos\vartheta}}_{v'} = \frac{1}{ipr}\cos\vartheta\, e^{ipr\cos\vartheta}\Big|_0^\pi - \frac{1}{ipr}\int_0^\pi d\vartheta \sin\vartheta\, e^{ipr\cos\vartheta}$$
$$= \frac{1}{ipr}\left(e^{ipr} + e^{-ipr}\right) - \frac{1}{p^2r^2} e^{ipr\cos\vartheta}\Big|_0^\pi$$
$$= \frac{1}{ipr}\left(e^{ipr} + e^{-ipr}\right) + \frac{1}{p^2r^2}\left(e^{ipr} - e^{-ipr}\right) \quad . \qquad (19)$$

With

$$\int_0^\infty r^n\, e^{-\alpha r}\, dr = \frac{n!}{a^{n+1}} \quad ,$$

it follows that

Example 12.1.

$$
\begin{aligned}
&\int_0^\infty \mathrm{d}r \int_0^\pi \mathrm{d}\vartheta\, r^3\, \mathrm{e}^{-m_0\alpha r} \sin\vartheta \cos\vartheta\, \mathrm{e}^{\mathrm{i}pr\cos\vartheta} \\
&= \frac{1}{\mathrm{i}p}\left(\frac{2}{(m_0\alpha-\mathrm{i}p)^3}+\frac{2}{(m_0\alpha+\mathrm{i}p)^3}\right) \\
&\quad + \frac{1}{p^2}\left(\frac{1}{(m_0\alpha-\mathrm{i}p)^2}-\frac{1}{(m_0\alpha+\mathrm{i}p)^2}\right) \\
&= -\frac{\mathrm{i}}{p}\frac{4m_0^3\alpha^3-12m_0\alpha p^2}{\left(m_0^2\alpha^2+p^2\right)^3}+\frac{1}{p}\frac{\mathrm{i}4m_0\alpha}{\left(m_0^2\alpha^2+p^2\right)^2} \\
&= \frac{\mathrm{i}}{p}\frac{16m_0\alpha p^2}{\left(m_0^2\alpha^2+p^2\right)^3} = \mathrm{i}\frac{16m_0\alpha p}{\left(m_0^2\alpha^2+p^2\right)^3} \quad .
\end{aligned} \tag{20}
$$

In (18) we replace $\sum_p 1/L'^3$ by $\int \mathrm{d}^3p\, 1/(2\pi)^3$, which yields

$$
\begin{aligned}
&\left(\frac{\text{transitions}}{\text{time}}\right) \\
&= \frac{4\times 8\times \pi^2}{3}16^2 m_0^5\alpha^6 \int \mathrm{d}^3p \frac{1}{(2\pi)^3}\frac{p^4}{\left(m_0^2\alpha^2+p^2\right)^6}\frac{(E_\mathrm{i}-E)^3}{E(E+m_0)} \\
&= \frac{16^3}{3}m_0^5\alpha^6 \int_0^{\sqrt{3}m_0} \mathrm{d}p \frac{p^6}{\left(m_0^2\alpha^2+p^2\right)^6}\frac{\left(E_\mathrm{i}+\sqrt{p^2+m_0^2}\right)^3}{\sqrt{p^2+m_0^2}\left(\sqrt{p^2+m_0^2}+m_0\right)} \\
&= \frac{16^3}{3}m_0\alpha^6 \int_0^{\sqrt{3}} \mathrm{d}x \frac{x^6}{(\alpha^2+x^2)^6}\frac{\left(E_\mathrm{i}/m_0+\sqrt{x^2+1}\right)^3}{\sqrt{x^2+1}\left(\sqrt{x^2+1}+1\right)} \quad .
\end{aligned} \tag{21}
$$

The hydrogen ground-state energy is approximately $E_\mathrm{i} = m_0 - \alpha^2 m_0/2$. The integral (21) is well defined and yields a finite value, for which we want to estimate the lower bound:

$$
\begin{aligned}
&\int_0^{\sqrt{3}} \mathrm{d}x \frac{x^6}{(\alpha^2+x^2)^6}\frac{\left(E_\mathrm{i}/m_0+\sqrt{x^2+1}\right)^3}{\sqrt{x^2+1}\left(\sqrt{x^2+1}+1\right)} > \int_0^{\sqrt{3}} \mathrm{d}x \frac{x^6}{(\alpha^2+3)^6}\frac{(1+1)^3}{2\cdot 3} \\
&\approx \frac{2^3}{3^6\times 2\times 3}\frac{3^3\sqrt{3}}{7} = \frac{2^2\sqrt{3}}{3^4\times 7} \Rightarrow \left(\frac{\text{transitions}}{\text{time}}\right) > 16m_0\alpha^6 \\
&= \frac{16}{137^6}\frac{0.51\,\mathrm{MeV}}{6.5\times 10^{-22}\,\mathrm{MeV\,s}} \approx 10^9\frac{1}{\mathrm{s}} \quad .
\end{aligned} \tag{22}
$$

If the lower (negative) continuum were empty, then the hydrogen atom would decay in a very short time ($\tau < 10^{-9}$ s). In order to ensure the stability of the hydrogen atom (and of all other atoms) one has to regard the negative energy continuum as fully occupied with electrons, so that the Pauli principle blocks those states from being available for radiative transitions.

12.1 Charge Conjugation

The hole theory leads us also to a *new, fundamental symmetry*: For the electrons there exist antiparticles, the positrons. In general it is true that for every particle in nature there exists an antiparticle. Now we have to formulate this symmetry mathematically in a rigorous way. Thereby we shall see how the wave functions for the positrons follow from the wave functions of the electrons with negative energy, and vice versa.

According to the hole theory a positron is a hole in the filled "sea of electrons with negative energy". According to (12.4–12.7) this positron has the same mass as the electron, positive energy ($E_{\text{positron}} = -E_{\text{electron of neg. energy}}$), and opposite momentum and charge as the electron with negative energy. There exists a one-to-one relation between solutions of the Dirac equation for electrons with negative energy,

$$\left(\mathrm{i}\hbar\nabla\!\!\!/ - \frac{e}{c}A\!\!\!/ - m_0c\right)\psi = 0 \quad , \tag{12.11}$$

and the positron eigenfunctions.

The positron wave function ψ_c has to fulfill the equation

$$\left(\mathrm{i}\hbar\nabla\!\!\!/ + \frac{e}{c}A\!\!\!/ - m_0c\right)\psi_c = 0 \quad , \tag{12.12}$$

because positrons should have all the properties of positively charged electrons. Note that the positron wave function ψ_c should be a solution of (12.12) with *positive energy*!

Remark. If is of no consequences which one of (12.11) or (12.12) is labelled as the first (particle). Historically (12.11) was the starting point for electrons, but (12.12) could be considered as the particle equation, as well. The sign of the charge e of the particles described in the initial equation does not matter either: If we had chosen (12.12) to start with, we would have obtained a spectrum of solutions for the free positrons like that known for electrons from (12.11) (see Fig. 12.4). The negative energy states would then have been filled with positrons (•) in the framework of the hole theory, while electrons would then have been positron holes with wave functions given by the negative energy solutions of (12.12). It is now our aim to find an operator which connects the solutions ψ_c of (12.12) with ψ of (12.11). This operator must change the *relative* sign of $\mathrm{i}\hbar\nabla\!\!\!/$ and $(e/c)A\!\!\!/$. This is simply done by complex conjugation:

$$\left(\mathrm{i}\hbar\frac{\partial}{\partial x^\mu}\right)^* = -\mathrm{i}\hbar\frac{\partial}{\partial x^\mu} \quad , \quad A_\mu^* = A_\mu \quad . \tag{12.13}$$

The electromagnetic four-potential A_μ is always real; thus we take the complex conjugate of (12.11) and after the conversion of all signs we arrive at

$$\left[\left(\mathrm{i}\hbar\frac{\partial}{\partial x^\mu} + \frac{e}{c}A_\mu\right)\gamma^{\mu *} + m_0c\right]\psi^* = 0 \quad . \tag{12.14}$$

Now we wish to find a non-singular matrix $\hat{U} \equiv \hat{C}\gamma^0$ in such a way that

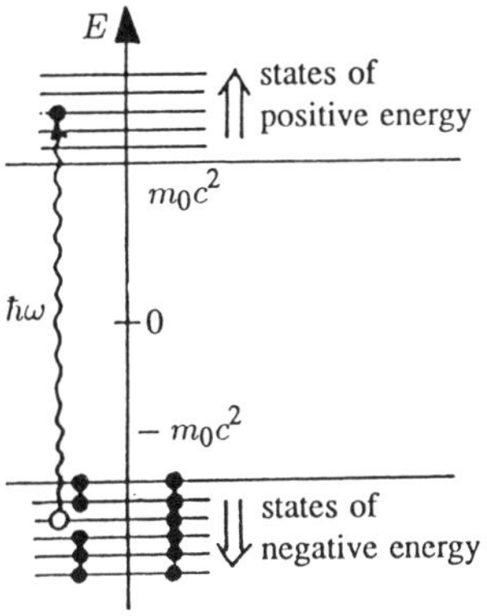

Fig. 12.4. The vacuum of the hole theory based on positrons as initial particles. Here a positron hole is an electron. As marked by the *full circles* (•) the states of negative energy are occupied by positrons

$$\hat{U}\gamma^{\mu *}\hat{U}^{-1} = -\gamma^{\mu} \tag{12.15}$$

holds. Before doing so, however, let us first proceed as if $\hat{U}$ is already known. After multiplication with $\hat{U}$ (12.14) transforms into

$$\left[\left(\mathrm{i}\hbar\frac{\partial}{\partial x^{\mu}} + \frac{e}{c}A_{\mu}\right)\hat{U}\gamma^{\mu *}\hat{U}^{-1} + m_0 c\right]\hat{U}\psi^* = 0$$

or, with equation (12.15),

$$\left[\left(\mathrm{i}\hbar\frac{\partial}{\partial x^{\mu}} + \frac{e}{c}A_{\mu}\right)\gamma^{\mu} - m_0 c\right]\hat{U}\psi^* = 0 \quad \text{or}$$

$$\left[\mathrm{i}\hbar\nabla\!\!\!/ + \frac{e}{c}A\!\!\!/ - m_0 c\right]\hat{U}\psi^* = 0 \quad . \tag{12.16}$$

For the positron wave function a comparison of (12.16) with (12.12) yields

$$\psi_{\mathrm{c}} = \hat{U}\psi^* \equiv \hat{C}\gamma^0\psi^* = \hat{C}\overline{\psi}^{\mathrm{T}} \quad .$$

The superscript "T" indicates "transposition" and means

$$\overline{\psi}^{\mathrm{T}} = \left(\psi^{\dagger}\gamma^0\right)^{\mathrm{T}} = \gamma^{0\mathrm{T}}\psi^{\dagger\mathrm{T}} = \gamma^0\left(\psi^{*\mathrm{T}}\right)^{\mathrm{T}} = \gamma^0\psi^* \quad , \tag{12.17}$$

because of $\gamma^{0\mathrm{T}} = \gamma^0$ and $\psi^{\dagger} = \psi^{*\mathrm{T}}$. To determine $\hat{U}$ explicitly we rewrite (12.15) as

$$\hat{C}\gamma^0\gamma^{\mu *}\left(\hat{C}\gamma^0\right)^{-1} = \hat{C}\gamma^0\gamma^{\mu *}\gamma^0\hat{C}^{-1} = -\gamma^{\mu} \quad . \tag{12.18}$$

Now, according to our explicit representation (3.13) for the γ matrices, i.e.

$$\gamma^i = \begin{pmatrix} 0 & \hat{\sigma}^i \\ -\hat{\sigma}^i & 0 \end{pmatrix} \quad , \quad \gamma^0 = \begin{pmatrix} 1\!\!1 & 0 \\ 0 & -1\!\!1 \end{pmatrix} \quad , \tag{12.19}$$

the identity

$$\gamma^0\gamma^{\mu *}\gamma^0 = \gamma^{\mu\mathrm{T}} \tag{12.20}$$

can easily be derived. Thus for (12.18) we get

$$\left(\hat{C}^{-1}\right)^{\mathrm{T}}\gamma^{\mu}(\hat{C})^{\mathrm{T}} = -\gamma^{\mu\mathrm{T}} \quad , \tag{12.21}$$

and

$$\gamma^{1\mathrm{T}} = -\gamma^1 \quad , \quad \gamma^{2\mathrm{T}} = \gamma^2 \quad , \quad \gamma^{3\mathrm{T}} = -\gamma^3 \quad , \quad \gamma^{0\mathrm{T}} = \gamma^0 \quad , \tag{12.22}$$

$(\hat{C})^{\mathrm{T}}$ must commute with γ^1 and γ^3 and anticommute with γ^2 and γ^0. Therefore

$$\hat{C} = \mathrm{i}\gamma^2\gamma^0 \tag{12.23}$$

is a useful choice for the operator $\hat{C}$, and it holds that

$$\hat{C} = \mathrm{i}\gamma^2\gamma^0 = -\hat{C}^{-1} = -\hat{C}^{\dagger} = -\hat{C}^{\mathrm{T}} \quad . \tag{12.24}$$

This clearly illustrates the non-singularity of $\hat{C}$ (the inverse matrix is explicitly constructed). Of course (12.23) is valid in the special representation (12.19), but

$\hat{C}$ can easily be given in every other representation via a unitary transformation. The phase of the operator $\hat{C}$ is rather arbitrarily fixed by the factor i in (12.24). Indeed, the choice of the phase does not influence at all our current investigations. Thus the charge-conjugate state of $\psi(x)$ is given by

$$\psi_c = \hat{C}\psi \equiv \hat{C}\gamma^0\hat{K}\psi = \hat{C}\gamma^0\psi^* = \hat{C}\overline{\psi}^{\mathrm{T}} = \mathrm{i}\gamma^2\psi^*(x) \quad , \tag{12.25}$$

where $\hat{K}$ is the operator of complex conjugation. According to (12.12) the wave equation for $\psi_c(x)$ differs from the wave equation for $\psi(r)$ (12.11) just by the sign of the charge. Thus it follows: *If $\psi(x)$ describes the motion of a Dirac particle with mass m_0 and charge e in a potential $A_\mu(x)$, then $\psi_c(x)$ represents the motion of a Dirac particle with the same mass m_0 and opposite charge $(-e)$ in the same potential $A_\mu(x)$.* The spinors ψ and ψ_c are charge conjugate to each other. With the relations (12.20), (12.25) and $\gamma^{2\mathrm{T}} = \gamma^2$, it holds that

$$\begin{aligned}(\psi_c)_c &= \left(\mathrm{i}\gamma^2\psi^*(x)\right)_c = \mathrm{i}\gamma^2\left\{\mathrm{i}\gamma^2\psi^*(x)\right\}^* = \gamma^2\gamma^{2*}\psi(x) = \gamma^2\gamma^0\gamma^{2\mathrm{T}}\gamma^0\psi(x)\\ &= \gamma^2\gamma^0\gamma^2\gamma^0\psi(x) = -\gamma^2\gamma^2\gamma^0\gamma^0\psi(x) = \psi(x) \quad , \end{aligned}\tag{12.26}$$

that is

$$(\psi_c)_c = \psi \quad . \tag{12.27}$$

Thus the correspondence between ψ and ψ_c is reciprocal. Besides, we draw the following interesting conclusion about the expectation values of operators. Let

$$\langle\hat{Q}\rangle = \langle\psi|\hat{Q}|\psi\rangle \tag{12.28}$$

be the expectation value of an operator $\hat{Q}$ in the state ψ. Then the expectation value of the *same* operator $\hat{Q}$ in the charge-conjugate state ψ_c is given by

$$\begin{aligned}\langle\hat{Q}\rangle_c &= \langle\psi_c|\hat{Q}|\psi_c\rangle = \int \psi_c^\dagger\hat{Q}\psi_c\,\mathrm{d}^3x = \int \left(\mathrm{i}\gamma^2\psi^*\right)^\dagger \hat{Q}\mathrm{i}\gamma^2\psi^*\,\mathrm{d}^3x\\ &= \int \psi^{*\dagger}\gamma^{2\dagger}\hat{Q}\gamma^2\psi^*\,\mathrm{d}^3x = \int \psi^{*\dagger}\gamma^0\gamma^2\gamma^0\hat{Q}\gamma^2\psi^*\,\mathrm{d}^3x\\ &= -\int \psi^{*\dagger}\gamma^2\gamma^0\gamma^0\hat{Q}\gamma^2\psi^*\,\mathrm{d}^3x = -\int \psi^{*\dagger}\gamma^2\hat{Q}\gamma^2\psi^*\,\mathrm{d}^3x\\ &= -\left(\int \psi^\dagger\left(\gamma^2\hat{Q}\gamma^2\right)^*\psi\,\mathrm{d}^3x\right)^* = -\langle\psi|\gamma^{2*}\hat{Q}^*\gamma^{2*}|\psi\rangle^*\\ &= -\langle\psi|\gamma^2\hat{Q}^*\gamma^2|\psi\rangle^* \quad . \end{aligned}\tag{12.29}$$

In this manner one easily proves the following relations (cf. Exercise 12.2):

$$\begin{aligned}&\text{(a)}\quad \langle\hat{\beta}\rangle_c = -\langle\hat{\beta}\rangle \quad ,\\ &\text{(b)}\quad \langle\boldsymbol{x}\rangle_c = \langle\boldsymbol{x}\rangle \quad ,\\ &\text{(c)}\quad \langle\hat{\alpha}_i\rangle_c = \langle\gamma^0\gamma^i\rangle_c = \langle\hat{\alpha}_i\rangle \quad ,\\ &\text{(d)}\quad \langle\hat{\boldsymbol{p}}\rangle_c = -\langle\hat{\boldsymbol{p}}\rangle \quad ,\\ &\text{(e)}\quad \psi_c^\dagger\psi_c = \psi^\dagger\psi \quad ,\\ &\text{(f)}\quad \psi_c^\dagger\hat{\boldsymbol{\alpha}}\psi_c = \psi^\dagger\hat{\boldsymbol{\alpha}}\psi \quad ,\\ &\text{(g)}\quad \langle\hat{\boldsymbol{\Sigma}}\rangle_c = -\langle\hat{\boldsymbol{\Sigma}}\rangle \quad ,\\ &\text{(h)}\quad \langle\hat{\boldsymbol{L}}\rangle_c = -\langle\hat{\boldsymbol{L}}\rangle \quad , \quad \hat{\boldsymbol{L}} = \boldsymbol{r}\times\hat{\boldsymbol{p}} \quad ,\\ &\text{(i)}\quad \langle\hat{\boldsymbol{J}}\rangle_c = -\langle\hat{\boldsymbol{J}}\rangle \quad , \quad \hat{\boldsymbol{J}} = \boldsymbol{r}\times\hat{\boldsymbol{p}} + \tfrac{1}{2}\hat{\boldsymbol{\Sigma}} \quad . \end{aligned}\tag{12.30}$$

Because the Hamiltonians of both Dirac equations (12.14) and (12.12) are given by

$$\hat{H}(e) \equiv c\hat{\alpha}^i \left(\hat{p}_i + \frac{e}{c}A_i\right) + eA_0 + \hat{\beta}m_0c^2$$
$$= c\hat{\alpha}\cdot\left(\hat{\boldsymbol{p}} - \frac{e}{c}\boldsymbol{A}\right) + eA_0 + \hat{\beta}m_0c^2 \qquad (12.31a)$$

or

$$\hat{H}(-e) \equiv c\hat{\alpha}^i \left(\hat{p}_i - \frac{e}{c}A_i\right) - eA_0 + \hat{\beta}m_0c^2$$
$$= c\hat{\alpha}\cdot\left(\hat{\boldsymbol{p}} + \frac{e}{c}\boldsymbol{A}\right) - eA_0 + \hat{\beta}m_0c^2 \quad , \qquad (12.31b)$$

one also deduces with the help of (12.29) the relation (cf. Example 12.3):

$$\langle\hat{H}(-e)\rangle_{\mathrm{c}} = -\langle\hat{H}(e)\rangle \quad . \qquad (12.32)$$

The results of (12.30) and (12.31) are quite interesting. Accordingly the charge-conjugate solutions ψ_{c} have the same probability density and probability current density in all space-time points (12.30e,f). Therefore the electric charge density and the electric current density for ψ and ψ_{c} are contragredient. Equations (12.30) and (12.32) express the important result that a charge-conjugated state ψ_{c} has the opposite momentum and energy to the state ψ. The relations of the hole theory expressed in (12.4) and (12.7) read in their most precise form: *charge conjugation changes the sign of the momentum and energy.*

EXERCISE

12.2 Expectation Values of Some Operators in Charge-Conjugate States

Problem. Prove the relations (12.30).

Solution. According to (12.9) the expectation value of an operator $\hat{Q}$ with regard to a charge-conjugate state ψ_{c} is given by

$$\langle\hat{Q}\rangle_{\mathrm{c}} = \langle\psi_{\mathrm{c}}|\hat{Q}|\psi_{\mathrm{c}}\rangle = -\left[\int \psi^\dagger(x)\gamma^2\hat{Q}^*\gamma^2\psi(x)\,\mathrm{d}^3x\right]^*$$
$$= -\langle\psi|\gamma^2\hat{Q}^*\gamma^2|\psi\rangle^* \quad . \qquad (1)$$

From this we conclude in particular:

(a) $\hat{Q} = \hat{\beta} = \gamma^0$:

$$\langle\hat{\beta}\rangle_{\mathrm{c}} = -\langle\psi|\gamma^2\gamma^0\gamma^2|\psi\rangle^* = \langle\psi|\gamma^2\gamma^2\gamma^0|\psi\rangle^* = -\langle\psi|\gamma^0|\psi\rangle^* = -\langle\psi|\gamma^0|\psi\rangle \quad ,$$

and therefore

$$\langle\hat{\beta}\rangle_{\mathrm{c}} = -\langle\hat{\beta}\rangle \quad . \qquad (2)$$

(b) $\hat{Q} = \hat{\alpha}_i = \gamma^0\gamma^i$: first case ($i = 2$):

$$\langle\hat{\alpha}_2\rangle_{\mathrm{c}} = -\langle\psi|\gamma^2\gamma^0(-\gamma^2)\gamma^2|\psi\rangle^* = \langle\psi|\gamma^0\gamma^2|\psi\rangle = \langle\hat{\alpha}_2\rangle \quad .$$

Exercise 12.2.

Second case ($i = 1, 3$):

$$\begin{aligned}\langle\hat{\alpha}_i\rangle_c &= -\langle\psi|\gamma^2\gamma^0(+\gamma^i)\gamma^2|\psi\rangle^* = -\langle\psi|\gamma^0\gamma^2\gamma^2\gamma^i|\psi\rangle^* \\ &= -\langle\psi|\gamma^i\gamma^0|\psi\rangle^* = +\langle\psi|\gamma^0\gamma^i|\psi\rangle \\ &= \langle\hat{\alpha}_i\rangle \quad ,\end{aligned}$$

and therefore

$$\langle\hat{\alpha}_i\rangle_c = \langle\hat{\alpha}_i\rangle \quad i = 1,2,3 \quad . \tag{3}$$

(c) $\hat{Q} = \hat{\boldsymbol{x}}$:

$$\begin{aligned}\langle\hat{\boldsymbol{x}}\rangle_c &= -\left[\int \psi^\dagger(\boldsymbol{x})\gamma^2\hat{\boldsymbol{x}}\gamma^2\psi(\boldsymbol{x})\,\mathrm{d}^3x\right]^* = \left[\int \psi^\dagger(\boldsymbol{x})\boldsymbol{x}\psi(\boldsymbol{x})\,\mathrm{d}^3x\right]^* \\ &= \langle\psi|\hat{\boldsymbol{x}}|\psi\rangle^* = \langle\psi|\hat{\boldsymbol{x}}|\psi\rangle\end{aligned}$$

and therefore

$$\langle\hat{\boldsymbol{x}}\rangle_c = \langle\hat{\boldsymbol{x}}\rangle \quad . \tag{4}$$

(d) $\hat{Q} = \hat{\boldsymbol{p}} = (\hbar/\mathrm{i})\boldsymbol{\nabla}$:

$$\begin{aligned}\langle\hat{\boldsymbol{p}}\rangle_c &= -\left[\int \psi^\dagger(\boldsymbol{x})\gamma^2\left(-\frac{\hbar}{\mathrm{i}}\boldsymbol{\nabla}\right)\gamma^2\psi(\boldsymbol{x})\,\mathrm{d}^3x\right]^* = -\left[\int \psi^\dagger(\boldsymbol{x})\frac{\hbar}{\mathrm{i}}\boldsymbol{\nabla}\psi(\boldsymbol{x})\,\mathrm{d}^3x\right]^* \\ &= -\langle\psi|\hat{\boldsymbol{p}}|\psi\rangle^* = -\langle\psi|\hat{\boldsymbol{p}}|\psi\rangle \quad ,\end{aligned}$$

and therefore

$$\langle\hat{\boldsymbol{p}}\rangle_c = -\langle\hat{\boldsymbol{p}}\rangle \quad . \tag{5}$$

(e) Because $\psi^\dagger\psi$ is a real number, the following holds:

$$\begin{aligned}\psi_c^\dagger\psi_c &= \left(\psi^{*\dagger}\gamma^{2\dagger}(-\mathrm{i})\right)\left(\mathrm{i}\gamma^2\psi^*\right) = -\psi^{*\dagger}\gamma^2\gamma^2\psi^* = \psi^{*\dagger}\psi^* \\ &= \psi^{\mathrm{T}}\psi^* = \left(\psi^{*\dagger}\psi\right)^{\mathrm{T}} = \left(\psi^\dagger\psi\right)^{\mathrm{T}} = \psi^\dagger\psi \quad .\end{aligned}$$

From this we conclude that

$$\psi_c^\dagger\psi_c = \psi^\dagger\psi \quad . \tag{6}$$

(f) If we apply the identity

$$\psi_c^\dagger\hat{\alpha}_i\psi_c = \psi^{*\dagger}\gamma^{2\dagger}\hat{\alpha}_i\gamma^2\psi^* = -\psi^{\mathrm{T}}\gamma^2\gamma^0\gamma^i\gamma^2\psi^*$$

to the cases $i = 2$

$$\gamma^2\gamma^0\gamma^2\gamma^2 = -\gamma^2\gamma^0 = \gamma^0\gamma^2 = \hat{\alpha}_2 = -\hat{\alpha}_2^{\mathrm{T}}$$

and $i = 1, 3$

$$\gamma^2\gamma^0\gamma^i\gamma^2 = \gamma^2\gamma^2\gamma^0\gamma^i = -\hat{\alpha}_i = -\hat{\alpha}_i^{\mathrm{T}} \quad ,$$

we get the relation

$$\psi_c^\dagger\hat{\alpha}_i\psi_c = \psi^{\mathrm{T}}\hat{\alpha}_i^{\mathrm{T}}\left(\psi^\dagger\right)^{\mathrm{T}} = \left(\psi^\dagger\hat{\alpha}_i\psi\right)^{\mathrm{T}} = \psi^\dagger\hat{\alpha}_i\psi$$

Exercise 12.2.

because the expression in parentheses is an ordinary number; hence (compare with b):

$$\psi_c^\dagger \hat{\boldsymbol{\alpha}} \psi_c = \psi^\dagger \hat{\boldsymbol{\alpha}} \psi \quad . \tag{7}$$

(g) $\hat{Q} = \hat{\boldsymbol{\Sigma}} = \begin{pmatrix} \hat{\boldsymbol{\sigma}} & 0 \\ 0 & \hat{\boldsymbol{\sigma}} \end{pmatrix}$:

$$\langle \hat{\Sigma}_i \rangle_c = -\langle \psi | \gamma^2 \hat{\Sigma}_i^* \gamma^2 | \psi \rangle^* \quad ,$$

with

$$\begin{aligned} \gamma^2 \hat{\Sigma}_i^* \gamma^2 &= \begin{pmatrix} 0 & \hat{\sigma}_2 \\ -\hat{\sigma}_2 & 0 \end{pmatrix} \begin{pmatrix} \hat{\sigma}_i^* & 0 \\ 0 & \hat{\sigma}_i^* \end{pmatrix} \begin{pmatrix} 0 & \hat{\sigma}_2 \\ -\hat{\sigma}_2 & 0 \end{pmatrix} \\ &= \begin{pmatrix} 0 & \hat{\sigma}_2 \hat{\sigma}_i^* \\ -\hat{\sigma}_2 \hat{\sigma}_i^* & 0 \end{pmatrix} \begin{pmatrix} 0 & \hat{\sigma}_2 \\ -\hat{\sigma}_2 & 0 \end{pmatrix} = - \begin{pmatrix} \hat{\sigma}_2 \hat{\sigma}_i^* \hat{\sigma}_2 & 0 \\ 0 & \hat{\sigma}_2 \hat{\sigma}_i^* \hat{\sigma}_2 \end{pmatrix} \quad . \end{aligned}$$

We get for $i = 2$

$$-\hat{\sigma}_2 \hat{\sigma}_2^* \hat{\sigma}_2 = \hat{\sigma}_2 \hat{\sigma}_2 \hat{\sigma}_2 = \hat{\sigma}_2 = -\hat{\sigma}_2^* = \hat{\sigma}_2^{*\mathrm{T}} \quad ,$$

and $i = 1, 3$

$$-\hat{\sigma}_2 \hat{\sigma}_i^* \hat{\sigma}_2 = -\hat{\sigma}_2 \hat{\sigma}_i \hat{\sigma}_2 = \hat{\sigma}_2 \hat{\sigma}_2 \hat{\sigma}_i = \hat{\sigma}_i = \hat{\sigma}_i^{*\mathrm{T}} \quad ,$$

and therefore

$$\langle \Sigma_i \rangle_c = -\left\langle \psi \left| \begin{pmatrix} \hat{\sigma}_i^{*\mathrm{T}} & 0 \\ 0 & \hat{\sigma}_i^{*\mathrm{T}} \end{pmatrix} \right| \psi \right\rangle^* = -\langle \psi | \Sigma_i^\dagger | \psi \rangle^* = -\langle \psi | \hat{\Sigma}_i | \psi \rangle \quad ,$$

i.e.

$$\langle \hat{\Sigma}_i \rangle_c = -\langle \hat{\Sigma}_i \rangle \quad . \tag{8}$$

(h) $\hat{Q} = \hat{\boldsymbol{L}} = \hat{\boldsymbol{x}} \times \hat{\boldsymbol{p}} = \hat{\boldsymbol{x}} \times \hbar/\mathrm{i}\boldsymbol{\nabla}$:

$$\begin{aligned} \langle \hat{\boldsymbol{L}} \rangle_c &= -\left[\int \psi^\dagger(\boldsymbol{x}) \gamma^2 \left(\hat{\boldsymbol{x}} \times \frac{\hbar}{\mathrm{i}} \boldsymbol{\nabla} \right)^* \gamma^2 \psi(\boldsymbol{x}) \, \mathrm{d}^3 x \right]^* \\ &= -\left[\int \psi^\dagger(\boldsymbol{x}) \left(\boldsymbol{x} \times \frac{\hbar}{\mathrm{i}} \boldsymbol{\nabla} \right) \psi(\boldsymbol{x}) \, \mathrm{d}^3 x \right]^* = -\langle \psi | \hat{\boldsymbol{L}} | \psi \rangle^* = -\langle \psi | \hat{\boldsymbol{L}} | \psi \rangle \quad , \end{aligned}$$

i.e.

$$\langle \hat{\boldsymbol{L}} \rangle_c = -\langle \hat{\boldsymbol{L}} \rangle \tag{9}$$

(i) $\hat{Q} = \hat{\boldsymbol{J}}$:

$$\langle \hat{\boldsymbol{J}} \rangle_c = \langle \hat{\boldsymbol{J}} \rangle_c + \tfrac{1}{2} \langle \hat{\boldsymbol{\Sigma}} \rangle_c = -\langle \hat{\boldsymbol{L}} \rangle - \tfrac{1}{2} (\hat{\boldsymbol{\Sigma}}) \quad ,$$

i.e.

$$\langle \hat{\boldsymbol{J}} \rangle_c = -\langle \hat{\boldsymbol{J}} \rangle \quad . \tag{10}$$

EXERCISE

12.3 Proof of $\langle \hat{H}(-e)\rangle_{\mathrm{c}} = -\langle \hat{H}(e)\rangle$

Problem. Prove the relation $\langle \hat{H}(-e)\rangle_{\mathrm{c}} = -\langle \hat{H}(e)\rangle$ and interpret the result.

Solution. From (12.31a) and (12.31b) we have

$$\hat{H}(e) = c\hat{\boldsymbol{\alpha}}\cdot\left(\hat{\boldsymbol{p}} - \frac{e}{c}\boldsymbol{A}\right) + eA_0 + \hat{\beta}m_0c^2 \quad , \tag{1}$$

$$\hat{H}(-e) = c\hat{\alpha}\cdot\left(\hat{\boldsymbol{p}} + \frac{e}{c}\boldsymbol{A}\right) - eA_0 + \hat{\beta}m_0c^2 \quad . \tag{2}$$

Using the relations derived in Exercise 12.2 we find

$$\langle \hat{\boldsymbol{\alpha}}\cdot\hat{\boldsymbol{p}}\rangle_{\mathrm{c}} = \left[\int \psi^{\dagger}(\boldsymbol{x})\left(-\gamma^2\hat{\boldsymbol{\alpha}}^*\gamma^2\right)\left(-\frac{\hbar}{\mathrm{i}}\boldsymbol{\nabla}\right)\psi(\boldsymbol{x})\,\mathrm{d}^3x\right] \quad .$$

In Exercise (12.2b) we derived the relation $-\gamma^2\hat{\boldsymbol{\alpha}}^2\gamma^2 = \gamma^{\dagger}\gamma^{0\dagger} = \hat{\boldsymbol{\alpha}}^{\dagger}$. Therefore

$$\langle \hat{\boldsymbol{\alpha}}\cdot\hat{\boldsymbol{p}}\rangle_{\mathrm{c}} = -\langle\psi|\hat{\boldsymbol{\alpha}}^{\dagger}\cdot\hat{\boldsymbol{p}}|\psi\rangle = -\langle\psi|\hat{\boldsymbol{p}}\cdot\hat{\boldsymbol{\alpha}}^{\dagger}|\psi\rangle = -\langle\psi|\hat{\boldsymbol{\alpha}}\cdot\hat{\boldsymbol{p}}|\psi\rangle \quad ,$$

$$\langle \hat{\boldsymbol{\alpha}}\cdot\hat{\boldsymbol{p}}\rangle_{\mathrm{c}} = -\langle \hat{\boldsymbol{\alpha}}\cdot\hat{\boldsymbol{p}}\rangle \quad . \tag{3}$$

In addition,

$$\langle\hat{\beta}\rangle_{\mathrm{c}} = -\langle\hat{\beta}\rangle \quad , \tag{4}$$

$$\langle e\hat{\alpha}\cdot\boldsymbol{A}\rangle_{\mathrm{c}} = \langle e\hat{\boldsymbol{\alpha}}\cdot\boldsymbol{A}\rangle \quad , \tag{5}$$

$$\langle eA_0\rangle_{\mathrm{c}} = \langle eA_0\rangle \quad . \tag{6}$$

From (1) to (6) follows

$$\langle\hat{H}(-e)\rangle_{\mathrm{c}} = -c\langle\hat{\boldsymbol{\alpha}}\cdot\hat{\boldsymbol{p}}\rangle + e\langle\hat{\boldsymbol{\alpha}}\cdot\boldsymbol{A}\rangle - \langle eA_0\rangle - \langle\hat{\beta}m_0c^2\rangle = -\langle\hat{H}(e)\rangle \quad ,$$

i.e.

$$\langle\hat{H}(-e)\rangle_{\mathrm{c}} = -\langle\hat{H}(e)\rangle \quad .$$

This result means that the negative energy solutions of the Dirac equation correspond to the charge-conjugate solutions of positive energy (and vice versa). In Exercise 12.2 we derived that these solutions have opposite spin and momentum, properties which allow us to interpret the solutions of negative energy (charge e, spin $\boldsymbol{s}$, momentum $\boldsymbol{p}$) as wave functions of particles with positive energy (charge $-e$, spin $-\boldsymbol{s}$ and momentum $-\boldsymbol{p}$). Compare this with the discussion following somewhat later on charge conjugation of bound states!

EXERCISE

12.4 Effect of Charge Conjugation on an Electron with Negative Energy

Problem. Examine in detail the influence of the transformation

$$\psi_c = \hat{C}\overline{\psi}^{\mathrm{T}} = \mathrm{i}\gamma^2\psi^*$$

on the eigenfunctions of an electron at rest with negative energy.

Solution. A (free) electron at rest is described by the wave function (6.1). For an electron with *negative energy* and spin down (↓) we have

$$\psi^4 = \frac{1}{\sqrt{2\pi\hbar}^3}\begin{pmatrix}0\\0\\0\\1\end{pmatrix}\mathrm{e}^{+\mathrm{i}(m_0c^2/\hbar)t} \quad . \tag{1}$$

The corresponding solution with positive energy and spin up reads

$$\begin{aligned}\left(\psi^4\right)_c = \mathrm{i}\gamma^2\psi^{4*} &= \mathrm{i}\begin{pmatrix}0&0&0&-\mathrm{i}\\0&0&\mathrm{i}&0\\0&\mathrm{i}&0&0\\-\mathrm{i}&0&0&0\end{pmatrix}\begin{pmatrix}0\\0\\0\\1\end{pmatrix}\left(\frac{1}{\sqrt{2\pi\hbar}^3}\mathrm{e}^{+\mathrm{i}(m_0c^2/\hbar)t}\right)^*\\ &= \frac{1}{\sqrt{2\pi\hbar}^3}\begin{pmatrix}1\\0\\0\\0\end{pmatrix}\mathrm{e}^{-\mathrm{i}(m_0c^2/\hbar)t} = \psi^1 \quad .\end{aligned} \tag{2}$$

Similarly, for an electron with negative energy and spin up (↑)

$$\psi^3 = \frac{1}{\sqrt{2\pi\hbar}^3}\begin{pmatrix}0\\0\\1\\0\end{pmatrix}\mathrm{e}^{+\mathrm{i}(m_0c^2/\hbar)t} \quad , \tag{3}$$

we get the positron wavefunction for positive energy and spin down (↓)

$$\begin{aligned}\left(\psi^3\right)_c = \mathrm{i}\gamma^2\psi^{3*} &= \mathrm{i}\begin{pmatrix}0&0&0&-\mathrm{i}\\0&0&\mathrm{i}&0\\0&\mathrm{i}&0&0\\-\mathrm{i}&0&0&0\end{pmatrix}\begin{pmatrix}0\\0\\1\\0\end{pmatrix}\left(\frac{1}{\sqrt{2\pi\hbar}^3}\mathrm{e}^{+\mathrm{i}(m_0c^2/\hbar)t}\right)^*\\ &= (-1)\begin{pmatrix}0\\1\\0\\0\end{pmatrix}\frac{1}{\sqrt{2\pi\hbar}^3}\mathrm{e}^{-\mathrm{i}(m_0c^2/\hbar)t} \quad .\end{aligned} \tag{4}$$

Here an inessential phase factor (-1) appears. This example demonstrates explicitly the influence of charge conjugation: The absence of an electron at rest with negative energy and spin ↑ (↓) is equivalent to the presence of a positron at rest with positive energy and spin ↓ (↑). If there are no fields present, there is no difference between electron and positron: From (2) and (4) one sees that the transformation of charge conjugates leads back to other electron solutions in the field-free case. This example makes the rather strange re-definition of the spinors ω^3 and ω^4 [see (6.56)] conceivable: An electron of negative energy with spin ↑ (↓) corresponds to a positron with spin ↓ (↑). The $v(x,s)$ spinors of (6.56) describe positrons with spin projection s.

Exercise 12.4.

EXERCISE

12.5 Representation of Operators for Charge Conjugation and Time Reversal

Let γ^μ and $\gamma^{\mu'}$ be two representations of the γ matrices connected by a unitary transformation

$$\gamma^\mu = \hat{U}\gamma^{\mu'}\hat{U}^{-1} \quad . \tag{1}$$

Problem. (a) Show that

$$\hat{C}' = \hat{U}^{-1}\hat{C}\left(\hat{U}^{\mathrm{T}}\right)^{-1} \quad , \tag{2}$$

where $\hat{C}$ and $\hat{C}'$ are the respective matrices for the transformation of charge conjugation

(b) Are the relations

$$\hat{C} = -\hat{C}^{-1} = -\hat{C}^\dagger = -\hat{C}^{\mathrm{T}} = \mathrm{i}\gamma^2\gamma^0 \tag{3}$$

also valid for $\hat{C}'$?

(c) Analogously free

$$\hat{T}_0 = \mathrm{i}\gamma^1\gamma^3 \tag{4}$$

from the common representation of γ matrices.

Solution. (a) If ψ solves the Dirac equation

$$\left(\mathrm{i}\hbar\gamma^\mu\partial_\mu - \frac{e}{c}\gamma^\mu A_\mu - m_0c\right)\psi = 0 \quad , \tag{5a}$$

the charge conjugated equation is

$$\left(\mathrm{i}\hbar\gamma^\mu\partial_\mu + \frac{e}{c}\gamma^\mu A_\mu - m_0c\right)\psi_{\mathrm{c}} = 0 \quad . \tag{5b}$$

Inserting the transformation (1) into (5a), one deduces the transformation for the wave functions as $\left(\gamma^{\mu'} = \hat{U}^{-1}\gamma^\mu\hat{U}\right)$

$$\psi' = \hat{U}^{-1}\psi \quad , \tag{6}$$

and analogously

$$\psi'_{\mathrm{c}} = \hat{U}^{-1}\psi_{\mathrm{c}} \quad . \tag{7}$$

Exercise 12.5.

The charge conjugated wave function follows from

$$\psi_c = \hat{C}\overline{\psi}^{\mathrm{T}} \quad . \tag{8}$$

With (7), (8) and the relation $\gamma^{0*} = \gamma^0 = \gamma^{0\mathrm{T}}$

$$\overline{\psi}^{\mathrm{T}} = \left(\psi^\dagger\gamma^0\right)^{\mathrm{T}} = \gamma^{0\mathrm{T}}\psi^* = \gamma^0\psi^* \tag{9}$$

we obtain

$$\psi_c' = \hat{U}^{-1}\hat{C}_{\overline{\psi}}^{\mathrm{T}} = \hat{U}^{-1}\hat{C}\gamma^0\psi^* \quad . \tag{10}$$

Because of

$$\psi'^* = \left(\hat{U}^{-1}\right)^*\psi^* \quad \text{and} \quad \psi^* = \hat{U}^*\psi'^*$$

it is valid that

$$\psi_c' = \hat{U}^{-1}\hat{C}\gamma^0\hat{U}^*\psi'^* \quad . \tag{11}$$

Further we obtain

$$\begin{aligned}\gamma^{0'*} &= \left(\hat{U}^{-1}\right)^*\gamma^{0*}\hat{U}^* = \left(\hat{U}^{-1}\right)^*\gamma^0\hat{U}^* \\ &= \hat{U}^{\mathrm{T}}\gamma^0\left(\hat{U}^{-1}\right)^{\mathrm{T}} = \left(\hat{U}^{-1}\gamma^0\hat{U}\right)^{\mathrm{T}} = \gamma^{0'\mathrm{T}} \quad .\end{aligned} \tag{12}$$

Now we use the unitarity of the transformation $\hat{U}$

$$\hat{U}^{-1} = \hat{U}^\dagger \equiv \left(\hat{U}^{\mathrm{T}}\right)^* \quad , \tag{13}$$

so that (11) leads to

$$\begin{aligned}\psi_c' &= \hat{U}^{-1}\hat{C}\gamma^0\hat{U}^*\psi'^* \\ &= \hat{U}^{-1}\hat{C}\left(\hat{U}^*\hat{U}^{*-1}\right)\gamma^0\hat{U}^*\psi'^* \\ &= \hat{U}^{-1}\hat{C}\hat{U}^*\left(\hat{U}^{*-1}\gamma^0\hat{U}^*\right)\psi'^* \\ &= \hat{U}^{-1}\hat{C}\hat{U}^*\gamma^{0'\mathrm{T}}\psi'^* \\ &= \hat{U}^{-1}\hat{C}\hat{U}^*\left(\psi'^\dagger\gamma^{0'}\right)^{\mathrm{T}} = \hat{U}^{-1}\hat{C}\hat{U}^*\overline{\psi}'^{\mathrm{T}} \quad .\end{aligned} \tag{14}$$

Comparing this equation with the definition of $\hat{C}'$

$$\psi_c' = \hat{C}'\overline{\psi}'^{\mathrm{T}} \tag{15}$$

we immediately get

$$\hat{C}' = \hat{U}^{-1}\hat{C}\left(\hat{U}^{\mathrm{T}}\right)^{-1} \quad . \tag{16}$$

(b) We analyse the various relations (3) one after another

$$\begin{aligned}C'^{\mathrm{T}} &= \left[\hat{U}^{-1}\hat{C}\left(\hat{U}^{\mathrm{T}}\right)^{-1}\right]^{\mathrm{T}} = \hat{U}^{-1}\hat{C}^{\mathrm{T}}\left(\hat{U}^{-1}\right)^{\mathrm{T}} \\ &= -\hat{U}^{-1}\hat{C}\left(\hat{U}^{\mathrm{T}}\right)^{-1} = -\hat{C}' \quad .\end{aligned} \tag{17}$$

This relation is thus conserved. For arbitrary unitary matrices $\hat{U}$, with $\hat{U} = \hat{U}^*$, we have on the other hand:

Exercise 12.5.

$$\hat{C}'^{\dagger} = \left[\hat{U}^{-1}\hat{C}\left(\hat{U}^{\mathrm{T}}\right)^{-1}\right]^{\dagger} = \left[\left(\hat{U}^{\mathrm{T}}\right)^{-1}\right]^{\dagger}\hat{C}^{\dagger}\left(\hat{U}^{-1}\right)^{\dagger}$$
$$= \hat{U}^{\mathrm{T}}\hat{C}^{\dagger}\hat{U} = -\hat{U}^{\mathrm{T}}\hat{C}\hat{U} \neq \hat{C}' \quad , \tag{18}$$

$$\hat{C}'^{*} = \hat{U}^{\mathrm{T}}\hat{C}^{*}\hat{U} = \hat{U}^{\mathrm{T}}\hat{C}\hat{U} \neq \hat{C}' \quad (\hat{C} \text{ is real}) \quad , \tag{19}$$

$$\left(\hat{C}'\right)^{-1} = \hat{U}^{\mathrm{T}}\hat{C}^{-1}\hat{U} = -\hat{U}^{\mathrm{T}}\hat{C}\hat{U} = \hat{C}'^{\dagger} \neq -\hat{C}' \quad , \tag{20}$$

$$\mathrm{i}\gamma^{0'}\gamma_0' = \mathrm{i}\hat{U}^{-1}\gamma^2\hat{U}\hat{U}^{-1}\gamma^0\hat{U} = \mathrm{i}\hat{U}^{-1}\gamma^2\gamma^0\hat{U}$$
$$= \hat{U}^{-1}\hat{C}\hat{U} \neq \hat{U}^{-1}\hat{C}\left(\hat{U}^{\mathrm{T}}\right)^{-1} = \hat{C}' \quad . \tag{21}$$

Remark. For real matrices $\hat{U}\,(\hat{U} = \hat{U}^*)$ all relations are still conserved.

(c) The operator of time inversion is defined by the equation

$$\psi_T(t') = \hat{T}_0\psi^*(t) \quad , \tag{22}$$

where $\psi_T(t')$ is the time inverse transformed spinor with $t' = -t$. We will proceed analogously to (a) and obtain

$$\psi_T(t') = \hat{U}^{-1}\psi_T(t') = \hat{U}^{-1}\hat{T}_0\psi^*(t) = \hat{U}^{-1}\hat{T}_0\hat{U}^*\psi'^*(t) = \hat{T}_0'\psi'^*(t) \quad , \tag{23}$$

i.e.

$$\hat{T}_0' = \hat{U}^{-1}\hat{T}_0\left(\hat{U}^{\mathrm{T}}\right)^{-1} \quad , \tag{24}$$

12.2 Charge Conjugation of Eigenstates with Arbitrary Spin and Momentum

Let $\psi(x)$ be an arbitrary plane wave with momentum $\boldsymbol{p}$. We know how by use of the projection operators [see (7.8), (7.21)] to construct from $\psi(x)$ plane waves of the four-momentum $p^\mu = \{\epsilon p^0, \boldsymbol{p}\}$ with $p^0 = +c\sqrt{\boldsymbol{p}^2 + m_0c^2}$ and spin $s^\mu = \{s^0, \boldsymbol{s}\}$, namely

$$\psi_{\epsilon ps}(x) = \left(\frac{\epsilon \not{p} + m_0c}{2m_0c}\right)\left(\frac{1+\gamma_5\not{s}}{2}\right)\psi(x) \quad . \tag{12.33}$$

These plane waves have positive or negative energy for $\epsilon = \pm 1$, respectively, and spin in the $\boldsymbol{s}$ direction. Because of

$$\left[\hat{C},\gamma_5\right]_- = \left[\mathrm{i}\gamma^2\gamma^0, \mathrm{i}\gamma^0\gamma^1\gamma^2\gamma^3\right]_- = 0 \tag{12.34}$$

and

$$\gamma_5^{\mathrm{T}} = \left(\mathrm{i}\gamma^0\gamma^1\gamma^2\gamma^3\right)^{\mathrm{T}}$$
$$= \mathrm{i}\gamma^{3\mathrm{T}}\gamma^{2\mathrm{T}}\gamma^{1\mathrm{T}}\gamma^{0\mathrm{T}}$$
$$= \mathrm{i}\left(-\gamma^3\right)\left(\gamma^2\right)\left(-\gamma^1\right)\gamma^0 = \mathrm{i}\gamma^0\gamma^1\gamma^2\gamma^3 = \gamma_5$$
$$= \mathrm{i}\gamma^0\gamma^{3*}\gamma^0\gamma^0\gamma^{2*}\gamma^0\gamma^0\gamma^{1*}\gamma^0\gamma^0\gamma^{0*}\gamma^0$$

[using (12.20)]

$$\begin{aligned} &= \mathrm{i}\gamma^0\gamma^{3*}\gamma^{2*}\gamma^{1*}\gamma^{0*}\gamma^0 = -\mathrm{i}\gamma^{3*}\gamma^{2*}\gamma^{1*}\gamma^{0*} \\ &= -\mathrm{i}\gamma^{0*}\gamma^{1*}\gamma^{2*}\gamma^{3*} = \gamma_5^* \quad , \end{aligned} \tag{12.35}$$

after charge conjugation (12.33) yields the state

$$\begin{aligned} (\psi_{\epsilon ps})_\mathrm{c} &= \hat{C}\overline{\psi}^\mathrm{T}_{\epsilon ps} = \hat{C}\gamma^0\psi^*_{\epsilon ps} \\ &= \hat{C}\gamma^0 \left(\frac{\epsilon \not{p} + m_0 c}{2m_0 c}\right)^* \left(\frac{1+\gamma_5 \not{s}}{2}\right)^* \psi^* \\ &= \hat{C}\gamma^0 \left(\frac{\epsilon \not{p} + m_0 c}{2m_0 c}\right)^* \left(\frac{1+\gamma_5 \not{s}}{2}\right)^* \left(\hat{C}\gamma^0\right)^{-1} \hat{C}\gamma^0\psi^* \quad . \end{aligned} \tag{12.36}$$

From (12.18) we get

$$\hat{C}\gamma^0\gamma^{\mu *}\left(\hat{C}\gamma^0\right)^{-1} = -\gamma^\mu \quad ,$$

and, further,

$$p^{\mu *} = +p^\mu \quad , \tag{12.37}$$

because only the momentum (a real number), not the operator of the momentum, appears in the projector [see (7.8)]. Therefore from (12.36), with the aid of (3.1) and with $\gamma^0\gamma_5\gamma^0 = -\gamma_5$, we get

$$\begin{aligned} (\psi_{\epsilon ps})_\mathrm{c} &= \left(\frac{-\epsilon \not{p} + m_0 c}{2m_0 c}\right) \left(\frac{1+\gamma_5 \not{s}}{2}\right) \hat{C}\gamma^0\psi^* \\ &= \left(\frac{-\epsilon \not{p} + m_0 c}{2m_0 c}\right) \left(\frac{1+\gamma_5 \not{s}}{2}\right) \psi_\mathrm{c} \quad . \end{aligned} \tag{12.38}$$

The charge-conjugated solution of $\psi_{\epsilon ps}$ (12.33) *thus has the same polarization* s_μ *but opposite energy and momentum* $(-\epsilon p^\mu)$ *as the original solution.* The latter fact is expressed by $-\epsilon p = -\epsilon\{p^0, \boldsymbol{p}\}$. The inversion of momentum and energy was expected, but (naively) so was the inversion of the polarization s_μ. To understand why the polarization is not changed under charge conjugation, let us remember (6.56), where the spin projection (polarization) of electrons with negative energy is indeed defined with inverted sign. The same is also expressed in (7.19) and (7.20).

12.3 Charge Conjugation of Bound States

We now consider an electron in an attractive Coulomb potential:

$$eA_0(r) = -\frac{Ze^2}{r} \quad . \tag{12.39}$$

An electron state of negative energy corresponds via charge conjugation to a state of positive energy of a particle with the same mass m_0 but opposite charge (i.e. *a positron*) *in the same potential* or – which amounts to the same – a state of *an electron in a repulsive potential.* Due to this correspondence the energy changes

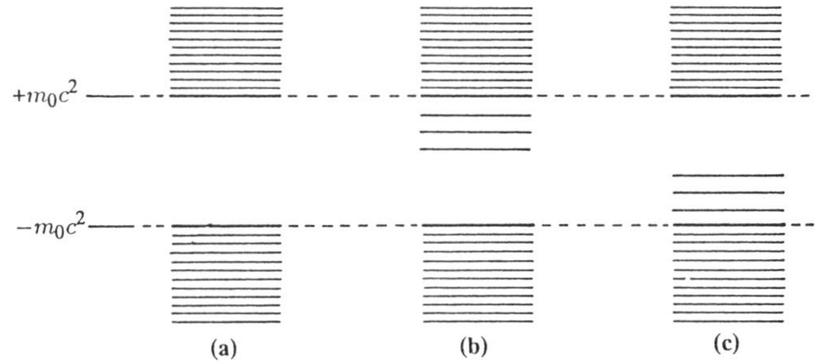

Fig. 12.5. Energy spectrum of a Dirac electron in an attractive Coulomb potential, (**a**) free electron without potential, (**b**) electron in an attractive potential $-Ze^2/r$, (**c**) electron in a repulsive potential $+Ze^2/r$

its sign, and, because of the γ^2 matrix in the operator $\hat{C}\gamma^0 = \mathrm{i}\gamma^2$, the large (upper) and small (lower) components are exchanged under charge conjugation, but the *densities and current densities remain the same.*

The spectrum of an electron in a repulsive potential is shown in Fig. 12.5c. It shows a series of bound states (the same number as for an electron in an attractive potential – case b) which lie near the lower continuum. Switching on the repulsive potential adiabatically, these *bound positron states emerge from the lower continuum*; analogously to the bound electron states, which are pulled down from the upper continuum while switching on the attractive potential (Fig. 12.5b). The positive energy continuum in the repulsive potential corresponds exactly to the negative energy continuum in the attractive potential. This has important consequences for the construction of the vacuum[2] in the hole and field theory. Having defined earlier the vacuum state as a Dirac sea of negative energy states filled up with electrons, we can now, because of the equality of electrons and positrons (charge-conjugation symmetry), consider the *vacuum as a "symmetrized sea of electrons and positrons"* (see Fig. 12.6a below). A one-electron state is visualized in Fig. 12.6b. In this example the electron occupies the second strongest bound level. It is not distinguishable from a corresponding positron hole in *its* second strongest bound level (emerged out of the positron sea of negative energy into the gap $-m_0c^2 \leq E \leq m_0c^2$). Both configurates exist equally beside each other. If they were distinguishable, then no particle–antiparticle symmetry (charge-conjugation symmetry) would exist. The existence of this symmetry allows for the suppression of one half of the figures (usually the positron side) in Figs. 12.6a and b.

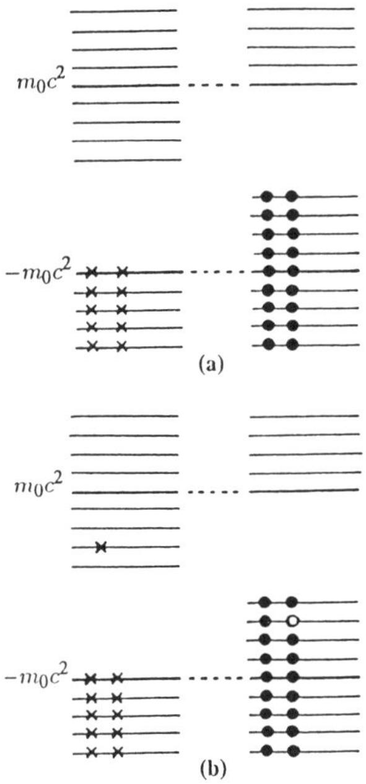

Fig. 12.6. (**a**) Illustration of the "symmetrized" vacuum state, consisting of states of negative energy filled with electrons (×) and positrons (•). (**b**) Illustration of a state with an electron in the second strongest bound level, which is indistinguishable from a positron hole in the second strongest bound positron level

We note that the vacuum state defined in this symmetrized form obviously has total charge zero, but infinitely large energy. The latter must be renormalized to zero, a subject that is tackled in quantum electrodynamics.

From earlier studies (Exercises and Examples 9.6 to 9.9), we know that in this case the spectrum consists of the positive energy continuum $m_0c^2 < E < \infty$, of a number (more precisely an infinite number) of discrete bound energy levels with $-m_0c^2 < E < m_0c^2$, and of the negative energy continuum $-\infty < E < -m_0c^2$. The bound states constitute the discrete energy levels in the energy gap between $-m_0c^2$ and $+m_0c^2$ (see Fig. 12.5). The states of negative energy can be obtained from the states of positive energy by charge conjugation, and vice versa. One should understand the following point very clearly.

[2] Refer also to W. Greiner, J. Reinhardt: *Quantum Electrodynamics*, 2nd ed. (Springer, Berlin, Heidelberg 1994).

Historical comment. It is amusing to read papers on the quantum theory of radiation.[3] Trials were made, e.g., to identify the *electron hole as a proton*. A number of arguments were given to justify that the mass of a hole should be greater than the mass of the electron, even though they were predicted by the hole theory to be equal. On the other hand, it should be possible, due to the hole theory, that an electron and a hole annihilate, with the emission of *two photons*. This probability was calculated by R. Oppenheimer, P.A.M. Dirac and I. Tamm[4] with the result that matter and antimatter annihilate each other in a very short time. However, when the positron was discovered, that which had previously caused the most serious difficulties became the greatest triumph of the theory. Indeed, the prediction of the existence of the antiparticle must be considered as one of the greatest successes of theoretical physics.

12.4 Time Reversal and PCT Symmetry

In this section we want to investigate the time-reversal transformation ($\hat{T}$ transformation) and show its connection with parity transformation and charge conjugation. Similarly to space reflection (parity transformation), time reflection is an *improper Lorentz transformation.* Yet another symmetry is the gauge invariance of the Dirac field ψ, which interacts with the electromagnetic fields A_μ, and which is – as we know – ensured by the minimal coupling $\hat{p}_\mu - (e/c)A_\mu$. However, we do not want to elaborate further on this point, because for the present discussion it will be irrelevant. Later on, however, (when considering the problems of renormalization in quantum electrodynamics) the gauge invariance will be very important.

From (4.9) we already know the parity transformation or spatial reflection, which is represented by

$$\hat{P}\psi(\boldsymbol{x},t) = \psi'(\boldsymbol{x}',t) = \psi'(-\boldsymbol{x},t) = \mathrm{e}^{\mathrm{i}\varphi}\gamma^0\psi(-\boldsymbol{x},t) \quad , \tag{12.40}$$

with

$$\boldsymbol{x}' = -\boldsymbol{x} \quad .$$

The spinor $\psi'(\boldsymbol{x}',t)$ is usually referred to as the *spatially reflected spinor* or *spatially reflected wave function*. In the case of plane waves the momentum, but not the spin is reversed under spatial reflection; exactly as one would expect for classical quantities. The parity transformation has the following effect on the various operators:

$$\hat{P}\boldsymbol{x}\hat{P}^{-1} = \boldsymbol{x}' = -\boldsymbol{x} \quad , \tag{12.41a}$$

$$\hat{P}x_0\hat{P}^{-1} = x_0' = x_0 \quad , \tag{12.41b}$$

---

[3] For example, we recommend Fermi's paper: Rev. Mod. Phys. **4**, 87 (1932), which was written between the formulation of the Dirac equation in 1928 and C.D. Anderson's discovery of the positron in 1933.

[4] J.R. Oppenheimer: Phys. Rev. **35**, 939 (1930); P.A.M. Dirac: Proc. Cambr. Phil. Soc. **26**, 361 (1930); I. Tamm: Zeitschr. für Physik **62**, 7 (1930).

$$\hat{P}\hat{\boldsymbol{p}}\hat{P}^{-1} = \hat{\boldsymbol{p}}' = -\hat{\boldsymbol{p}} \quad , \tag{12.41c}$$

$$\hat{P}\hat{p}_0\hat{P}^{-1} = \hat{p}_0' = \hat{p}_0 \quad , \tag{12.41d}$$

$$\hat{P}A_0(\boldsymbol{x},t)\hat{P}^{-1} = A_0'(\boldsymbol{x}',t) = A_0(\boldsymbol{x},t) \quad , \tag{12.41e}$$

$$\hat{P}\boldsymbol{A}(\boldsymbol{x},t)\hat{P}^{-1} = \boldsymbol{A}'(\boldsymbol{x}',t) = -\boldsymbol{A}(\boldsymbol{x},t) \quad , \tag{12.41f}$$

where

$$\boldsymbol{x}' = -\boldsymbol{x} \quad .$$

The first four relations are immediately understandable, while the last two (12.41e,f) denote the scalar and vectorial nature of the potential $A_0(\boldsymbol{x},t)$ and vector potential $\boldsymbol{a}(\boldsymbol{x},t)$, repsectively. Applying the same arguments leading to (3.30), but now with the special parity transformation (12.40) yields

$$\left(\hat{\not{p}} - \frac{e}{c}\not{A} - m_0c\right)\psi(\boldsymbol{x},t) = 0$$

$$\left(\hat{p}_0\gamma^0 + \hat{p}_i\gamma^i - \frac{e}{c}A_0\gamma^0 - \frac{e}{c}A_i(\boldsymbol{x},t)\gamma^i - m_0c\right)\hat{P}^{-1}\psi'(\boldsymbol{x}',t) = 0$$

and after multiplying with $\hat{P}$ from the left

$$\hat{P}\left(\hat{p}_0\gamma^0 + \hat{p}_i\gamma^i - \frac{e}{c}A_0\gamma^0 - \frac{e}{c}A_i(\boldsymbol{x},t)\gamma^i - m_0c\right)\hat{P}^{-1}\psi'(\boldsymbol{x}',t) = 0$$

$$\left(\hat{p}_0\gamma^0 + (+\hat{p}_i)(-\gamma^i) - \frac{e}{c}A_0\gamma^0 - \frac{e}{c}\big(+A_i(\boldsymbol{x},t)\big)(-\gamma^i) - m_0c\right)\psi'(\boldsymbol{x}',t) = 0$$

$$\left(\hat{p}_0'\gamma^0 + \hat{p}_i'\gamma^i - \frac{e}{c}A_0'(\boldsymbol{x}',t)\gamma^0 - \frac{e}{c}A_i'(\boldsymbol{x}',t)\gamma^i - m_0c\right)\psi'(\boldsymbol{x}',t) = 0$$

$$\left[\not{p}' - \frac{e}{c}\not{A}'(\boldsymbol{x}',t) - m_0c\right]\psi'(\boldsymbol{x}',t) = 0 \quad . \tag{12.42}$$

Since under the parity transformation (4.1, 4.2)

$$\hat{p}_0' = \hat{p}_0 \quad , \quad t' = t \quad ,$$

$$\hat{p}_i' = -\hat{p}_i \quad , \quad \boldsymbol{x}' = -\boldsymbol{x} \quad ,$$

$$A_0(\boldsymbol{x},t) = A_0'(\boldsymbol{x}',t') \quad , \quad \boldsymbol{A}(\boldsymbol{x},t) = -\boldsymbol{A}'(\boldsymbol{x}',t')$$

one finally arrives at

$$\left(\hat{\not{p}}' - \frac{e}{c}\not{A}' - m_0c\right)\psi'(x',t') = 0 \quad .$$

This means: The parity-transformed wave function $\Psi'(\boldsymbol{x}',t) = \hat{P}\Psi(\boldsymbol{x},t) = \Psi'(-\boldsymbol{x},t)$ [see (12.40)] obeys the same Dirac equation as the original wave function $\Psi(\boldsymbol{x},t)$. We say: *The parity transformation $\hat{P}$ leaves the Dirac equation and all physical observables unchanged.*

How does the parity invariance express itself physically? One can express this most simply by a sequence of observations of a state described by a wave function $\Psi(\boldsymbol{x},t)$: These observations are registered on a film, but the pictures are taken via a mirror. In other words: The camera films the observation of a state $\Psi(\boldsymbol{x},t)$ in a plane mirror (see Fig. 12.7). We then call the dynamics described by a state $\Psi(\boldsymbol{x},t)$ parity invariant, if the events registered on the film as reflected images can also

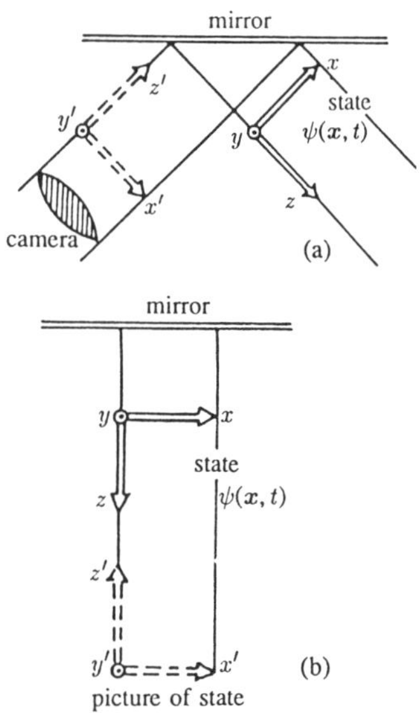

Fig. 12.7a,b. Illustration of parity. The state $\Psi(\boldsymbol{x},t)$ is represented by a Dreibein (three unit vectors of a Cartesian coordinate system), which is filmed via the mirror. On the film the z axis appears reversed (because of reflection). If the camera stood directly beside (in front of or under) the state, one would get the picture shown in (**b**), with opposite z axis

be possible direct observations of the state $\Psi(\boldsymbol{x},t)$; i.e. observations of the state without the mirror. From the observation of the events in the mirror image of the film, we must not be able to say whether we are looking at the picture of the mirror image. Both must be equally possible in parity-invariant dynamics.

We remark that a mirror image is not identical with full spatial reflection, since – as Fig. 12.7b clearly shows – in the mirror only the z axis is reversed (i.e. the axis $\perp$ to the mirror plane). Around the latter a rotation of π must be performed to obtain a full spatial reflection. However, such a rotation is a proper Lorentz transformation. The reflection described is thus a spatial inversion plus a proper Lorentz transformation (rotation). Since the theory (dynamics) is invariant under proper Lorentz transformations, the "mirror movie" yields exactly the information about parity invariance that we require.

We now move on to discuss *time reflection* and *time-reflection invariance*. The physical sense is also explainable with the film example. This time the film registers the observations of the state $\Psi(\boldsymbol{x},t)$ just in its time sequence: we do not need the mirror, instead we run the film backwards. One calls the dynamics *time-reversal invariant* if the observations on the movie running backward could have happened the same way on the forward-running movie. It must not be possible to determine by watching the filmed events, whether the movie is running forwards or backwards. In other words: Both the observations on the forward-running and backward-running movies must be realizable in the state $\Psi(\boldsymbol{x},t)$.

In our case of the Dirac theory the dynamics will be time-reversal invariant if, by performing the transformation

$$t' = -t \quad , \quad \boldsymbol{x}' = \boldsymbol{x} \quad , \tag{12.43}$$

the form of the Dirac equation remains unchanged; the interpretation must not be changed, either (cf. Exercise 12.6). Then, the transformed wave function

$$\psi_n'(\boldsymbol{x},t') = \hat{T}\psi_n(\boldsymbol{x},t) = \psi_n'(\boldsymbol{x},-t) \tag{12.44}$$

describes a Dirac particle which propagates backwards in time. This is physically possible if $\Psi_n'(\boldsymbol{x},t')$ also satisfies the Dirac equation. Equation (12.44) is a special case of the general definition (3.27) and the time inversion (12.43) is a special improper Lorentz transformation. $\hat{T}$ is an operator which acts on the spinor components, but not on space and time coordinates, a fact that should also be made evident in connection with the general scheme (3.27).

Let us now construct the time-inversion transformation explicitly. Therefore we rewrite the Dirac equation in Schrödinger form:

$$\begin{aligned} \mathrm{i}\hbar\frac{\partial\psi_n(\boldsymbol{x},t)}{\partial t} &= \hat{H}(\boldsymbol{x},t)\psi_n(\boldsymbol{x},t) \\ &= \left[c\hat{\boldsymbol{\alpha}}\cdot\left(-\mathrm{i}\hbar\boldsymbol{\nabla}-\frac{e}{c}\boldsymbol{A}\right)+\hat{\beta}m_0c^2+eA_0(\boldsymbol{x},t)\right]\psi_n(\boldsymbol{x},t) \quad , \end{aligned} \tag{12.45}$$

where n characterizes the spinor's quantum numbers. The time inversion of (12.45) is achieved by multiplying from the left by $\hat{T}$:

$$\hat{T}\mathrm{i}\hbar\hat{T}^{-1}\frac{\partial}{\partial t}\hat{T}\psi_n(\boldsymbol{x},t) = \hat{T}\hat{H}(\boldsymbol{x},t)\hat{T}^{-1}\hat{T}\psi_n(\boldsymbol{x},t) \quad . \tag{12.46}$$

The notation $\hat{T}\mathrm{i}\hbar\hat{T}^{-1}$ includes the possibility that the $\hat{T}$ operator may also contain complex conjugation. Since the time inversion as a special Lorentz transformation causes

$$t' = -t \quad ,$$

it follows, together with (12.44), that

$$-\hat{T}\mathrm{i}\hbar\hat{T}^{-1}\frac{\partial}{\partial t'}\psi_n'(\boldsymbol{x},t') = \hat{T}\hat{H}(\boldsymbol{x},t)\hat{T}^{-1}\psi_n'(\boldsymbol{x},t') \quad . \tag{12.47}$$

For $\psi_n'(\boldsymbol{x},t) = \hat{T}\psi_n(\boldsymbol{x},t)$ the same Schrödinger-like Dirac equation should hold as for $\psi_n(\boldsymbol{x},t)$ in (12.45); hence

$$\mathrm{i}\hbar\frac{\partial}{\partial t'}\psi_n'(\boldsymbol{x},t') = \hat{H}(\boldsymbol{x},t')\psi_n'(\boldsymbol{x},t') \quad . \tag{12.48}$$

In case of time-inversion symmetry one obviously has to demand that $\hat{H}(\boldsymbol{x},t') = \hat{H}(\boldsymbol{x},-t) = \hat{H}(\boldsymbol{x},t)$, and in this case an observer can, in principle, not distinguish between a forward and a backward running "movie".

Now the comparison of (12.47) with (12.48) yields a priori two possibilities of procedure: either we demand

$$\hat{T}\mathrm{i}\hat{T}^{-1} = \mathrm{i} \quad \text{and} \quad \hat{T}\hat{H}(\boldsymbol{x},t)\hat{T}^{-1} \equiv \hat{H}(\boldsymbol{x},-t) = -\hat{H}(\boldsymbol{x},t) \tag{12.49}$$

or

$$\hat{T}\mathrm{i}\hat{T}^{-1} = -\mathrm{i} \quad \text{and} \quad \hat{T}\hat{H}(\boldsymbol{x},t)\hat{T}^{-1} \equiv \hat{H}(\boldsymbol{x},t') = \hat{H}(\boldsymbol{x},-t) \stackrel{!}{=} \hat{H}(\boldsymbol{x},t) \quad . \tag{12.50}$$

The latter possibility is the only possible one, because the operator equation $\hat{T}\hat{H}(\boldsymbol{x},t)\hat{T}^{-1} = \hat{H}(\boldsymbol{x},t)$ can be included into the general scheme (3.27) of symmetry transformations. In addition, the condition (12.49) $\hat{T}\hat{H}(\boldsymbol{x},t)\hat{T}^{-1} = -\hat{H}(\boldsymbol{x},t)$ would alter the spectrum of $\hat{H}$ by the time inversion in the special case of time-independent Hamiltonians, which can physically not be accepted. We also notice this fact if we look at the explicit form of $\hat{H}(\boldsymbol{x},t)$:

$$\hat{H}(\boldsymbol{x},t) = c\hat{\boldsymbol{\alpha}}\cdot\left(-\mathrm{i}\hbar\boldsymbol{\nabla} - \frac{e}{c}\boldsymbol{A}(\boldsymbol{x},t)\right) + \hat{\beta}m_0c^2 + eA_0(\boldsymbol{x},t) \quad . \tag{12.51}$$

The conditions (12.49) cannot be satisfied: The vector potential $\hat{A}(\boldsymbol{x},t)$ is created by electric currents $\boldsymbol{j}(\boldsymbol{x},t)$, which change sign under the transformation $t \to -t$; therefore it holds that

$$\hat{T}\boldsymbol{A}(\boldsymbol{x},t)\hat{T}^{-1} = \boldsymbol{A}(\boldsymbol{x},-t) = -\boldsymbol{A}(\boldsymbol{x},t) \quad . \tag{12.52}$$

On the contrary the Coulomb potential is created by the electric charge density $\varrho(\boldsymbol{x},t)$, which remains unchanged under time inversion, and hence

$$\hat{T}A_0(\boldsymbol{x},t)\hat{T}^{-1} = A_0(\boldsymbol{x},-t) = A_0(\boldsymbol{x},t) \quad . \tag{12.53}$$

Furthermore we have

$$\hat{T}\boldsymbol{\nabla}\hat{T}^{-1} = \boldsymbol{\nabla} \quad \text{and} \quad \hat{T}\boldsymbol{x}\hat{T}^{-1} = \boldsymbol{x} \quad , \tag{12.54}$$

because the application of $\hat{T}$ does not effect the space coordinates. Therefore we see that, for example, the choice (12.49) would yield

$$\hat{T}\hat{\alpha}\hat{T}^{-1} = \hat{\alpha}$$

due to the term $\hat{\alpha} \cdot \boldsymbol{A}(\boldsymbol{x}, t)$. On the other hand the sign change of the term $\hat{\alpha} \cdot \mathrm{i}\boldsymbol{\nabla}$ would imply

$$\hat{T}\hat{\alpha}\hat{T}^{-1} = -\hat{\alpha} \quad .$$

Obviously both conditions are contradictory and, as a consequencc, only (12.50) remains as a valid choice: It can be satisfied consistently, because with $\hat{T}\mathrm{i}\hat{T}^{-1} = -\mathrm{i}$ it fillows that

$$\begin{aligned}\hat{T}\hat{H}(\boldsymbol{x}, t)\hat{T}^{-1} &= \hat{T}c\hat{\alpha}\hat{T}^{-1} \cdot \left[\left(\hat{T}(-\mathrm{i}\hbar)\hat{T}^{-1}\right)\hat{T}\boldsymbol{\nabla}\hat{T}^{-1} - \frac{e}{c}\hat{T}\boldsymbol{A}(\boldsymbol{x}, t)\hat{T}^{-1}\right] \\ &\quad + m_0c^2\hat{T}\hat{\beta}\hat{T}^{-1} + e\hat{T}A_0(\boldsymbol{x}, t)\hat{T}^{-1} \\ &= -\hat{T}c\hat{\alpha}\hat{T}^{-1} \cdot \left[-\mathrm{i}\hbar\boldsymbol{\nabla} - \frac{e}{c}\boldsymbol{A}(\boldsymbol{x}, +t)\right] + m_0c^2\hat{T}\hat{\beta}\hat{T}^{-1} + eA_0(\boldsymbol{x}, -t) \\ &= +\hat{H}(\boldsymbol{x}, t) \end{aligned} \tag{12.55}$$

[due to (12.50), (12.52) and (12.53)], if

$$\hat{T}\hat{\alpha}\hat{T}^{-1} = -\hat{\alpha} \quad , \quad \hat{T}\hat{\beta}\hat{T}^{-1} = +\hat{\beta} \tag{12.56}$$

holds. Because $\hat{T}\mathrm{i}\hat{T}^{-1} = -\mathrm{i}$ the operator $\hat{T}$ must contain the complex conjugation $\hat{K}$, and therefore we set

$$\hat{T} = \hat{T}_0\hat{K} \quad , \tag{12.57}$$

where the matrix $\hat{T}_0$ must still be determined. If we insert (12.57) into (12.56), then

$$\begin{aligned} &\hat{T}_0\hat{\alpha}^*\hat{T}_0^{-1} = -\hat{\alpha} \quad \text{or} \quad \hat{T}_0\gamma^{i*}\hat{T}_0^{-1} = -\gamma^i \\ &\hat{T}_0\hat{\beta}\hat{T}_0^{-1} = \hat{\beta} \qquad \text{or} \quad \hat{T}_0\gamma^0\hat{T}_0^{-1} = \gamma^0 \quad . \end{aligned} \tag{12.58}$$

Since only the matrix $\hat{\alpha}_2$ is purely imaginary, all other matrices being real [see (2.13)], these conditions explicitly read

$$\begin{aligned} &\hat{T}_0\hat{\alpha}_1\hat{T}_0^{-1} = -\hat{\alpha}_1 \quad , \quad \hat{T}_0\hat{\alpha}_3\hat{T}_0^{-1} = -\hat{\alpha}_3 \quad , \\ &\hat{T}_0\hat{\alpha}_2\hat{T}^{-1} = \hat{\alpha}_2 \quad , \quad \hat{T}_0\hat{\beta}\hat{T}_0^{-1} = \hat{\beta} \quad . \end{aligned} \tag{12.59}$$

With the help of the commutation relations (2.8) it can be shown that

$$\hat{T}_0 = -\mathrm{i}\hat{\alpha}_1\hat{\alpha}_3 \quad , \quad \hat{T}_0^{-1} = \mathrm{i}\hat{\alpha}_3\hat{\alpha}_1 \tag{12.60}$$

satisfies these conditions, the factor i guaranteeing the unitarity of $\hat{T}$. With $\gamma^0 = \hat{\beta}$ and $\gamma^i = \hat{\beta}\hat{\alpha}_i$ [see (3.8)] the complete *time-inversion operator* can then be written as

$$\hat{T} = -\mathrm{i}\hat{\alpha}_1\hat{\alpha}_3\hat{K} = \mathrm{i}\gamma^1\gamma^3\hat{K} = \hat{T}_0\hat{K} \quad . \tag{12.61}$$

In addition, we mention that the time-inverted Dirac equation (12.48) is form invariant, too. This was our intention. Indeed it was formulated in that way, and the comparison of (12.48) with (12.45) shows this immediately.

Next we want to prove that the time inversion transformation $\hat{T}$ accords to the classical concept of time inversion. To that end we apply $\hat{T}$ to a free-particle solution (a plane Dirac wave) with positive energy. We know how a plane wave with four-momentum $p^\mu = \{\epsilon p^0, \boldsymbol{p}\}$ [p^0 is given by $p^0 = +c(p^2 + m_0^2c^2)^{1/2}$] and spin $s^\mu = \{s^0, \boldsymbol{s}\}$ can be constructed (projected) from an arbitrary plane wave $\psi_\alpha(x)$, with the help of the projection operators, namely [see (12.33)]

$$\psi_{\epsilon ps} = \left(\frac{\epsilon \not{p} + m_0 c}{2m_0 c}\right)\left(\frac{1+\gamma_5 \not{s}}{2}\right)\psi_\alpha(\boldsymbol{x},t) \quad . \tag{12.62}$$

For particle with positive energy one has to set $\epsilon = 1$, and the $\hat{T}$ operator can act on this equation. With $\hat{T}\psi_\alpha(\boldsymbol{x},t) = \psi'_\alpha(\boldsymbol{x},t') = \hat{T}_0\psi^*_\alpha(\boldsymbol{x},t)$ we get

$$\begin{aligned}
\hat{T}\psi_{\epsilon ps} &= \hat{T}\left(\frac{\epsilon \not{p} + m_0 c}{2m_0 c}\right)\left(\frac{1+\gamma_5 \not{s}}{2}\right)\hat{T}^{-1}\hat{T}\psi_\alpha(\boldsymbol{x},t) \\
&= \hat{T}_0\left(\frac{\epsilon \not{p}^* + m_0 c}{2m_0 c}\right)\left(\frac{1+\gamma_5 \not{s}^*}{2}\right)\hat{T}_0^{-1}\psi'_\alpha(\boldsymbol{x},t') \\
&= \hat{T}_0\left(\frac{\epsilon \not{p}^* + m_0 c}{2m_0 c}\right)\hat{T}_0^{-1}\hat{T}_0\left(\frac{1+\gamma_5 \not{s}^*}{2}\right)\hat{T}_0^{-1}\psi'_\alpha(\boldsymbol{x},t') \\
&= \left(\frac{\epsilon \not{p}' + m_0 c}{2m_0 c}\right)\left(\frac{1+\gamma_5 \not{s}'}{2}\right)\psi'_\alpha(\boldsymbol{x},t')
\end{aligned} \tag{12.63}$$

(because γ_5 is real). In the last step, equation (12.58) was used and the four-vectors

$$p' = \{p^0, -\boldsymbol{p}\} \quad \text{and} \quad s' = \{s^0, -\boldsymbol{s}\} \tag{12.64}$$

were introduced. These projectors yield a free solution with opposite direction of the spatial momentum $\boldsymbol{p}$ and the spin $\boldsymbol{s}$, if applied to the time-inverted plane wave $\psi'_\alpha(\boldsymbol{x},t')$. This is the so-called *Wigner time inversion.* It was introduced for the first time by Eugene *Wigner* in 1932.[5]

Now we want to see how the electron and the positron wave functions can easily be connected with the help of the time-inversion transformation. From our earlier considerations we know that the charge conjugate state ψ_c, which is obtained from the state ψ by

$$\psi_c = \hat{U}\psi^* = \hat{C}\gamma^0\psi^* = \hat{C}\gamma^0\hat{K}\psi \equiv \hat{\mathbb{C}}\psi \tag{12.65}$$

($\hat{C} = \mathrm{i}\gamma^2\gamma^0$), describes a particle with the same mass m_0 and the same spin direction (polarization), but with opposite charge, opposite sign of energy and opposite momentum [see, for example, (12.38)]. If $\psi(\boldsymbol{x},t)$ describes an electron with momentum $\boldsymbol{p}$ and negative energy, then $\psi_c(\boldsymbol{x},t)$ describes a positron with momentum $-\boldsymbol{p}$ and positive energy, and both particles propagate forward in time.

Now we combine the parity operation $\hat{P}$ [see (12.40)], the charge conjugation $\hat{C}$ [see (12.25)] and the time inversion $\hat{T}$ (12.61) and construct the wave function

[5] E.P. Wigner: Göttinger Nachrichten **31**, 546 (1932).

$$\begin{aligned}\psi_{PCT}(x') &= \hat{P}\hat{\mathbb{C}}\hat{T}\psi(x) = \hat{P}\hat{C}\gamma^0\left(\hat{T}\psi(\boldsymbol{x},t)\right)^* \\ &= \hat{P}\hat{C}\gamma^0\left(\mathrm{i}\gamma^1\gamma^3\hat{K}\psi(\boldsymbol{x},-t')\right)^* = \hat{P}\hat{C}\gamma^0\left(\mathrm{i}\gamma^1\gamma^3\psi^*(\boldsymbol{x},-t')\right)^* \\ &= -\mathrm{i}\hat{P}\hat{C}\gamma^0\gamma^1\gamma^3\psi(\boldsymbol{x},-t') = -\mathrm{i}\hat{P}\mathrm{i}\gamma^2\gamma^0\gamma^0\gamma^1\gamma^3\psi(\boldsymbol{x},-t') \\ &= \mathrm{e}^{\mathrm{i}\varphi}\gamma^0\gamma^2\gamma^1\gamma^3\psi(-\boldsymbol{x}',-t') = \mathrm{i}\mathrm{e}^{\mathrm{i}\varphi}\mathrm{i}\gamma^0\gamma^1\gamma^2\gamma^3\psi(-\boldsymbol{x}',-t') \\ &= \mathrm{i}\,\mathrm{e}^{\mathrm{i}\varphi}\gamma_5\psi(-x') \end{aligned} \tag{12.66}$$

(here $\gamma_5 = \mathrm{i}\gamma^0\gamma^1\gamma^2\gamma^3$ was used.) If $\psi(\boldsymbol{x},t)$ is an electron wave function of negative energy, then ψ_{PCT} is a positron wave function of positive energy, and this is effected by the charge conjugation $\hat{C}$. Therefore, we can also read (12.66) as $\psi_{PCT}(\boldsymbol{x},t)$ bcing a positron wave function with positive energy moving forward in time and space (positive t, positive $\boldsymbol{x}$). It is – up to a factor $\mathrm{i}\,\mathrm{e}^{\mathrm{i}\varphi}\gamma_5$ – identical with an electron wave function of negative energy, moving backwards in time and space [negative t and negative $\boldsymbol{x}$ in the argument of $\psi(-\boldsymbol{x},-t)$ on the rhs in (12.66)]. For a plane wave with definite spin s^μ and momentum p^μ one can deduce this result explicitly. The plane electron wave with negativee energy, momentum $-\boldsymbol{p}$ and spin $-\boldsymbol{s}$ is given by

$$\left(\frac{-\not{p}+m_0c}{2m_0c}\right)\left(\frac{1-\gamma_5\not{s}}{2}\right)\psi(-\boldsymbol{x},-t)$$

and moves backwards in space and time. If we apply the $\hat{P}\hat{C}\hat{T}$ transformation (12.66), we obtain

$$\begin{aligned}\psi_{PCT}(\boldsymbol{x},t) &= \mathrm{i}\,\mathrm{e}^{\mathrm{i}\varphi}\gamma_5\left(\frac{-\not{p}+m_0c}{2m_0c}\right)\left(\frac{1-\gamma_5\not{s}}{2}\right)\psi(-\boldsymbol{x},-t) \\ &= \left(\frac{\not{p}+m_0c}{2m_0c}\right)\left(\frac{1+\gamma_5\not{s}}{2}\right)\mathrm{i}\,\mathrm{e}^{\mathrm{i}\varphi}\gamma_5\psi(+\boldsymbol{x},+t) \quad . \end{aligned} \tag{12.67}$$

This is evidently a positron wave function (because of the charge conjugation) with positive energy p_0, positive momentum $\boldsymbol{p}$ and positive spin $\boldsymbol{s}$, moving forward in space and time.

EXERCISE

12.6 Behaviour of the Current with Time Reversal and Charge Conjugation

If the time inversion $\hat{T}$ is a given symmetry operation of the Dirac theory, then the rules for the interpretation of the wave function

$$\psi'_\alpha(\boldsymbol{x},t') = \hat{T}\psi_\alpha(\boldsymbol{x},t) \quad , \quad t'=-t \tag{1}$$

must remain the same for $\psi'_\alpha(\boldsymbol{x},t')$. This means that observables consisting of bilinear forms of $\psi'_\alpha(\boldsymbol{x},t')$ and $\psi'^+_\alpha(\boldsymbol{x},t')$ must be interpreted (i.e. physically explained) in the same way as those with $\psi_\alpha(\boldsymbol{x},t)$ and $\psi^+_\alpha(\boldsymbol{x},t)$. Naturally this is valid only up to the expected behaviour under time reversal of the special observable. The following examples will illustrate this.

Problem. (a) Prove that the following relation is valid for the current:

$$j'_\mu(x'^\nu) = j^\mu(x^\nu) \quad . \tag{2}$$

Exercise 12.6.

Note: The indices on the lhs are lower indices (covariant indices) and on the rhs upper ones (contravariant indices). Similarly, show that

$$\langle \boldsymbol{r}\rangle' = \langle \boldsymbol{r}\rangle \quad , \quad \langle \boldsymbol{p}\rangle' = -\langle \boldsymbol{p}\rangle \quad . \tag{3}$$

(b) Show also the behaviour of these observables under charge conjugation $\hat{C}$. Demonstrate especially that

$$\overline{\psi}_{\mathrm{c}}(x)\gamma_\mu\psi_{\mathrm{c}}(x) = +\overline{\psi}(x)\gamma_\mu\psi(x)$$

and interpret this result.

Solution. (a) The operation of time reversal transforms a spinor $\psi(t)$ into a spinor

$$\psi_T(t') = \hat{T}_0\psi^*(t) \quad t' = -t \tag{4}$$

or

$$\psi_T^\dagger(t') = \psi^{\mathrm{T}}(t)\hat{T}_0^\dagger \quad , \tag{5}$$

where the upper index "T" means transposition. In the usual representation of the γ matrices [see (12.60, 12.61)] one finds that

$$\hat{T}_0 = \mathrm{i}\gamma^1\gamma^3 \quad . \tag{6}$$

Since

$$\gamma^{\mu\dagger} = \gamma^0\gamma^\mu\gamma^0 \tag{7}$$

is valid, it may easily be seen that

$$\hat{T}_0^\dagger = -\mathrm{i}\gamma^{3\dagger}\gamma^{1\dagger} = -\mathrm{i}\gamma^0\gamma^3\gamma^1\gamma^0 = -\mathrm{i}\gamma^0\gamma^0\gamma^3\gamma^1 = -\mathrm{i}(-\gamma^1\gamma^3) = \hat{T}_0 \quad , \tag{8}$$

where we have used the commutation relations of the γ matrices and

$$\hat{T}_0 \cdot \hat{T}_0 = \mathrm{i}\gamma^1\gamma^3\gamma^1\gamma^3 = -\gamma^1\gamma^3\gamma^1\gamma^3 = \gamma^1\gamma^3\gamma^3\gamma^1 = \mathbb{1} \quad , \tag{9}$$

i.e. $\hat{T}_0$ is unitary and Hermitian. From (6) it follows that

$$\begin{aligned}
\hat{T}_0\gamma^1\hat{T}_0 &= -\gamma^1 = -\gamma^{1*} \quad , \\
\hat{T}_0\gamma^2\hat{T}_0 &= \gamma^2 = -\gamma^{2*} \quad , \\
\hat{T}_0\gamma^3\hat{T}_0 &= -\gamma^3 = -\gamma^{3*} \quad , \\
\hat{T}_0\gamma^0\hat{T}_0 &= \gamma^0 = \gamma^{0*} \quad ,
\end{aligned} \tag{10}$$

because γ^1, γ^3, γ^0 are real and γ^2 is purely imaginary, and we can simply write these four equations as

$$\hat{T}\gamma^\mu\hat{T} = \gamma_\mu^* \quad . \tag{11}$$

Our task is to investigate for the current $j_T^\mu(x')$ the expression

$$j_T^\mu(x') = \overline{\psi}_T(t')\gamma^\mu\psi_T(t') \quad , \tag{12}$$

so we obtain from (4), (5) and (11):

Exercise 12.6.

$$
\begin{aligned}
j_T^\mu(t') &= \left(\psi^\dagger(t')\gamma_0\right)_T \gamma^\mu \psi_T(t') \\
&= \psi^\dagger(t)\gamma_0 \hat{T}_0 \gamma^\mu \hat{T}_0 \psi^*(t) \\
&= \psi^{\mathrm{T}}(t)\gamma^0\gamma_\mu^*\psi^*(t) = \psi_\alpha(t)\left[\gamma^0\gamma_\mu^*\right]_{\alpha\beta}\psi_\beta^*(t) \\
&= \psi_\beta^*(t)\left[\gamma_0\gamma_\mu^*\right]_{\beta\alpha}^{\mathrm{T}}\psi_\alpha(t) = \psi^{\mathrm{T}*}(t)\gamma_\mu^{\mathrm{T}*}\gamma^0\psi(t) \\
&= \psi^\dagger(t)\gamma_\mu^\dagger\gamma^0\psi(t) = \psi^\dagger(t)\gamma^0\gamma_\mu\gamma^0\gamma^0\psi(t) \\
&= \overline{\psi}(t)\gamma_\mu\psi(t) = j_\mu(t)
\end{aligned}
\tag{13}
$$

[in the last but one transformation we used (7)]. The time argument alone was noted in $\psi(\boldsymbol{x},t)$, $\psi(\boldsymbol{x}',t')$, because the position vector ($\boldsymbol{x}=\boldsymbol{x}'$) remains unchanged under time inversion. The operators $\boldsymbol{r}$ and $\boldsymbol{p}$ commute with $\hat{T}_0$, because they do not carry spinor indices (they are proportional to the unit matrix). For the position operator it holds, in particular, that

$$
\begin{aligned}
\psi_T^\dagger(t')\boldsymbol{r}\psi_T(t') &= \psi^{\mathrm{T}}(t)\hat{T}_0\hat{T}_0\boldsymbol{r}\psi^*(t) \\
&= \boldsymbol{r}\psi^{\mathrm{T}}(t)\psi^*(t) = \boldsymbol{r}\psi^{\mathrm{T}*}(t)\psi(t) \\
&= \psi^\dagger(t)\boldsymbol{r}\psi(t)
\end{aligned}
\tag{14}
$$

[$\psi^{\mathrm{T}}(t)\psi^*(t)$ is purely real], wherefrom $\langle\boldsymbol{r}\rangle' = \langle\boldsymbol{r}\rangle$ follows immediately by integration. In the case of the momentum operator $\hat{\boldsymbol{p}} = -\mathrm{i}\hbar\boldsymbol{\nabla}$, we must explicitly write down the expectation value as an integral, in order to express clearly that $\hat{\boldsymbol{p}}$ is Hermitian. Because of $\hat{T}_0\hat{\boldsymbol{p}}\hat{T}_0 = \hat{\boldsymbol{p}}$, one gets

$$
\begin{aligned}
\langle\hat{\boldsymbol{p}}\rangle' &= \int \psi_T^\dagger(\boldsymbol{x},t')\hat{\boldsymbol{p}}\psi_T(\boldsymbol{x},t')\,\mathrm{d}^3x \\
&= \int \psi^{\mathrm{T}}(\boldsymbol{x},t)\hat{\boldsymbol{p}}\psi^*(\boldsymbol{x},t)\,\mathrm{d}^3x \\
&= \int \left(-\mathrm{i}\hbar\boldsymbol{\nabla}\psi^\dagger(x)\right)\psi(x)\,\mathrm{d}^3x \\
&= -\int \psi^\dagger(x)(-\mathrm{i}\hbar\boldsymbol{\nabla})\psi(x)\,\mathrm{d}^3x - \mathrm{i}h \int\limits_{\text{surface of volume } V} \psi^\dagger(x)\psi(x)\,\mathrm{d}\boldsymbol{F} \\
&= -\int \psi^\dagger(x)\hat{\boldsymbol{p}}\psi(x)\,\mathrm{d}^3x + 0 = -\langle\boldsymbol{p}\rangle \quad .
\end{aligned}
\tag{15}
$$

In the last but one line a partial integration has been performed, the surface term vanishing due to the general properties of wave functions at infinity.

(b) For $\hat{C} = \mathrm{i}\gamma^2\gamma^0$, then, in analogy to (10) and (11):

$$
\begin{aligned}
\hat{C}\gamma^0\hat{C} &= \gamma^0 = \gamma^{0*} \quad , \\
\hat{C}\gamma^2\hat{C} &= \gamma^2 = -\gamma^{2*} \quad , \\
\hat{C}\gamma^1\hat{C} &= -\gamma^1 = -\gamma^{1*} \quad , \\
\hat{C}\gamma^3\hat{C} &= -\gamma^3 = -\gamma^{3*} \quad , \\
\hat{C}\gamma^\mu\hat{C} &= \gamma_\mu^* \quad ,
\end{aligned}
\tag{16}
$$

which may be easily checked, so that

$$
\begin{aligned}
\psi_{\mathrm{c}}(x) &= \hat{C}\gamma^0\psi^*(x) \quad , \\
\psi_{\mathrm{c}}^\dagger(x) &= \psi^{\mathrm{T}}(x)\gamma^0\hat{C}^\dagger \quad ,
\end{aligned}
\tag{17}
$$

Exercise 12.6.

and, further,

$$\hat{C}^\dagger = -\hat{C} \quad , \quad \hat{C}^2 = -1 \quad . \tag{18}$$

Now we can write

$$\begin{aligned}\overline{\psi}_c\gamma^\mu\psi_c &= \psi_c^\dagger\gamma^0\gamma^\mu\psi_c \\ &= \psi^T\gamma^0\hat{C}^\dagger\gamma^0\gamma^\mu\hat{C}\gamma^0\psi^* \\ &= \psi^T\gamma^0\hat{C}\gamma^0\hat{C}\hat{C}\gamma^\mu\hat{C}\gamma^0\psi^* \\ &= \psi^T\gamma^0\gamma_0^*\gamma_\mu^*\gamma^0\psi^* = \psi^T\gamma_\mu^*\gamma^0\psi^* \\ &= \psi_\alpha\left(\gamma_\mu^*\gamma^0\right)_{\alpha\alpha'}\psi_{\alpha'}^* = \psi^{T*}\gamma^0\gamma_\mu^\dagger\psi \\ &= \overline{\psi}\gamma_\mu^\dagger\psi = \overline{\psi}\gamma^\mu\psi \quad ,\end{aligned} \tag{19}$$

which shows everything that is required. In the penultimate line, care is needed since we have to interchange the order of ψ^* and ψ. This is possible if one regards ψ as the *wave function* of an electron or positron; however, if ψ is regarded as a quantized *field operator* $\hat{\psi}$, then an additional minus sign appears, due to the fact that $\hat{\psi}$ and $\hat{\psi}^*$ are fermion operators, so that the charge conjugated *current operator* reads

$$:\hat{\overline{\psi}}_c\gamma^\mu\hat{\psi}_c: = -:\hat{\overline{\psi}}\gamma^\mu\hat{\psi}: \quad , \tag{20}$$

where $:\ldots:$ implies normal ordering.[6] We now have to interpret the result (19) in the hole picture. If ψ is the wave function of a state with positive energy, then ψ_c is a state of negative energy which is already occupied in the vacuum, so that the corresponding current must not be counted, since the current density of the vacuum is defined to be zero. On the other hand, if we create a positron in that state, then this state will be empty and the current $\overline{\psi}_c\gamma^\mu\psi_c$ is missing, so that we have to ascribe the current density $-\overline{\psi}_c\gamma^\mu\psi_c$ to the positron, which has the opposite sign as the corresponding charge density of the electron. From that we can conclude that electrons and positrons have opposite charges.

The correctness of the interpretation given at the end of Exercise 12.6 can be explicitly verified for the Dirac equation involving electromagnetic interaction. To do this we write the *Dirac eigenvalue equation for states with negative energy* in the presence of the electromagnetic four-potential $A_\mu = (A_0, \boldsymbol{A})$ as

$$\left[\hat{\alpha}\cdot\left(-\mathrm{i}\hbar\boldsymbol{\nabla} - \frac{e}{c}\boldsymbol{A}\right) + \hat{\beta}m_0c^2 + eA_0(x)\right]\psi(x) = -E\psi(x) \tag{12.68}$$

and operate with the $\hat{P}\hat{C}\hat{T}$ transformation (12.66). Since under space-time inversion $[x'_\mu = -x_\mu$, see (12.52) and (12.40)]

$$\hat{P}\hat{T}\hat{A}_\mu(x)\left(\hat{P}\hat{T}\right)^{-1} = A'_\mu(x') = +A_\mu(x) \tag{12.69}$$

[6] See W. Greiner, J. Reinhardt: *Quantum Electrodynamics*, 2nd ed. (Springer, Berlin, Heidelberg 1994).

is valid, then (12.68) transforms under $\hat{P}\hat{C}\hat{T}$ into

$$\hat{P}\hat{\mathbb{C}}\hat{T}\left[\hat{\boldsymbol{\alpha}}\cdot\left(-\mathrm{i}\hbar\boldsymbol{\nabla}-\frac{e}{c}\boldsymbol{A}(\boldsymbol{x},t)\right)+\hat{\beta}m_0c^2+eA_0(\boldsymbol{x},t)\right](\hat{P}\hat{\mathbb{C}}\hat{T})^{-1}(\hat{P}\hat{\mathbb{C}}\hat{T})\psi(\boldsymbol{x},t)$$
$$=-E(\hat{P}\hat{\mathbb{C}}\hat{T})\psi(\boldsymbol{x},t) \quad . \tag{12.70}$$

This results in

$$\begin{aligned}\psi_{PCT}(x') &= \psi_{PCT}(-\boldsymbol{x},-t)\\ &= \hat{P}\hat{\mathbb{C}}\hat{T}\psi(\boldsymbol{x},t) \quad ,\end{aligned}$$

$$\begin{aligned}\hat{P}\hat{\mathbb{C}}\hat{T}\hat{\alpha}_i(\hat{P}\hat{\mathbb{C}}\hat{T})^{-1} &= \hat{P}\hat{\mathbb{C}}\hat{T}\hat{\alpha}_i\hat{T}^{-1}\hat{\mathbb{C}}^{-1}\hat{P}^{-1}\\ &= -\hat{P}\hat{\mathbb{C}}\hat{\alpha}_i\hat{\mathbb{C}}^{-1}\hat{P}^{-1}\\ &= -\hat{P}\hat{\alpha}_i\hat{P}^{-1}\\ &= \hat{\alpha}_i \quad ,\end{aligned} \tag{12.71}$$

$$\hat{P}\hat{\mathbb{C}}\hat{T}\boldsymbol{\nabla}(\hat{P}\hat{\mathbb{C}}\hat{T})^{-1} = -\boldsymbol{\nabla}\equiv\boldsymbol{\nabla}' \quad , \tag{12.72}$$

$$\hat{P}\hat{\mathbb{C}}\hat{T}\hat{\beta}(\hat{P}\hat{\mathbb{C}}\hat{T})^{-1} = -\hat{\beta} \quad , \tag{12.73}$$

and therefore (12.70) becomes

$$\left[c\hat{\boldsymbol{\alpha}}\cdot\left(-\mathrm{i}\hbar\boldsymbol{\nabla}'+\frac{e}{c}\boldsymbol{A}'(x')\right)+\hat{\beta}m_0c^2-eA_0'(x')\right]\psi_{PCT}(x')$$
$$=E\psi_{PCT}(x') \quad , \tag{12.74}$$

with $x'=-x$.

If (12.68) was the Dirac equation for particles with charge e, rest mass m_0 and negative energy $(-E)$ moving forward in space and time $(\boldsymbol{x},t)$, then (12.74) is the Dirac equation for particles with charge $-e$, rest mass m_0 and positive energy $(+E)$ moving backwards in space and time [in the argument of $\psi_{PCT}(x')$, $x'=-x=-\{ct,-\boldsymbol{x}\}$]. We can therefore interpret positrons as electrons of negative energy moving backwards in space and time. This important result serves as one of the fundamental concepts of positron theory,[7] which was founded by Stückelberg and Feynman. In quantum-electrodynamical perturbation theory (which is based on it) we shall extensively both make use of, and recognize the great advantages of, this formulation.[8]

Remark. The form of the interaction between the electron–positron field and the electromagnetic field has been assumed to be

$$j_\mu(x)A^\mu(x)=\frac{e}{c}\overline{\psi}\gamma_\mu\psi(x)A^\mu(x) \quad . \tag{12.75}$$

This followed as the simplest interaction which is gauge invariant, and it is equivalent to the interaction between electrons and the electromagnetic field which we know from the nonrelativistic limiting case. This interaction is $\hat{C}$, $\hat{P}$ and $\hat{T}$ invariant (see Exercise 12.6). Whether or not these symmetries are realized depends

[7] E.C.G. Stückelberg: Helv. Phys. Acta **14.32L**, 588 (1941); R.P. Feyman: Phys. Rev. **76**, 749 (1949); ibid. 769.

[8] See W. Greiner, J. Reinhardt: *Quantum Electrodynamics*, 2nd ed. (Springer, Berlin, Heidelberg 1994).

on the actual interaction. As we know, an additional interaction with the magnetic moment of the form

$$\overline{\psi}\hat{\sigma}_{\mu\nu}\psi(x)F^{\mu\nu}(x) \tag{12.76}$$

for spin-$\frac{1}{2}$ particles with anomalous magnetic moment (g factor $\neq 2$ as for example for protons and neutrons), is in general necessary. This additional interaction also shows all the symmetries mentioned above, as can be easily verified. Extending the Dirac theory to other spin-$\frac{1}{2}$ particles, for example μ mesons, and to other kinds of interaction (e.g. the weak interaction), the assumption that the $\hat{C}$, $\hat{P}$, and $\hat{T}$ symmetries are also valid suggests itself.

This is an assumption which has to be verified by experiment, i.e. success or failure of phenomena predicted by it. Lee and Yang showed[9] that the symmetry of parity is no longer valid in the weak interaction; however, the much weaker assumption of Lorentz invariance and the connection between spin and statistics (spin-$\frac{1}{2}$ particles obey Fermi statistics, spin-0 particles obey Bose statistics)[10] always guarantees the invariance of the interaction under the product $\hat{P}\hat{C}\hat{T}$. This is the famous $\hat{P}\hat{C}\hat{T}$ theorem, which was derived by R. Lüders.[11]

12.5 Biographical Notes

WIGNER, Eugene Paul, Hungarian-American physicist, * 17.11.1902 in Budapest, professor at Princeton University and afterwards at Louisiana State University in Baton Rouge/Louisiana. He studied at the Universities of Berlin and received his Ph.D. in electrial engineering from the Technical University in Berlin in 1925. He made important contributions to theoretical physics, above all by introducing systematically group theoretical methods into physics. His book "Group Theory and its Application to the Quantum Mechanics of the Atomic Spectra" became a classic. But also his contributions to the theory of nuclear forces, neutron absorption and quantum mechanics, and parity conservation are widely known. W. decisively participated in the development of the American atom bomb and the construction of the first nuclear reactor. In 1963 W. received the Nobel Prize in physics, together with Maria Goeppert-Mayer and J.H.D. Jensen. In 1958 he received the Enrico Fermi prize and in 1961 the Max Planck Medaille [BR].

[9] T.D. Lee, C.N. Yang: Phys. Rev. **105**, 167 (1957).
This is extensively discussed in W. Greiner, B. Müller: *Gauge Theory of Weak Interactions*, 2nd ed. (Springer, Berlin, Heidelberg 1996).

[10] This is discussed in W. Greiner, J. Reinhardt: *Quantum Electrodynamics* 2nd ed. (Springer, Berlin, Heidelberg 1994).

[11] R. Lüders: Kgl. Dansk. Vid. Sels. Mat.-Fys. Medd **28**, no. 5 (1954).

13. Klein's Paradox

In the following, we want to concern ourselves with the scattering of an electron with energy E and momentum $p = p_z$ at an infinitely extended potential step (Fig. 13.1). First we shall study this problem from the point of view of the one-particle interpretation of the Dirac equation and then, in Example 13.1, we shall look at the same problem using the framework of hole theory, understanding better the resulting situation, which looks paradoxical at first sight.[1]

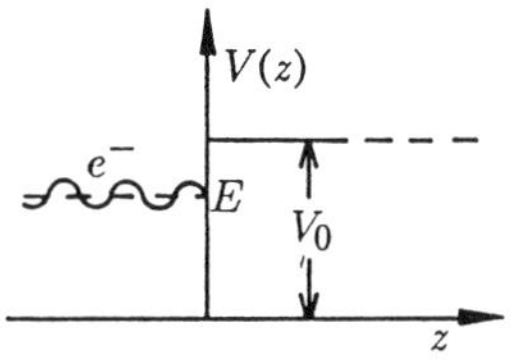

Fig. 13.1. An electron wave propagates along the z axis and hits a potential step of strength V_0

For the free electron we have $(E/c)^2 = p^2 + m_0^2c^2$, whereas in the presence of the constant potential,

$$\left(\frac{E - V_0}{c}\right)^2 = \overline{p}^2 + m_0^2c^2 \tag{13.1}$$

is valid, where $\overline{p}$ denotes the momentum of the electron inside the potential. The Dirac equation and its adjoint then read

$$\left\{\frac{E - eV}{c} - \hat{\beta}m_0c\right\}\psi + \mathrm{i}\hbar\sum_{k=1}^{3}\hat{\alpha}_k\frac{\partial\psi}{\partial x_k} = 0 \quad , \tag{13.2a}$$

$$\overline{\psi}\left\{\frac{E - eV}{c} - \hat{\beta}m_0c\right\} + \mathrm{i}\hbar\sum_{k=1}^{3}\frac{\partial\overline{\psi}}{\partial x_k}\hat{\alpha}_k = 0 \quad . \tag{13.2b}$$

We now assume that

$$\begin{aligned} eV &= V_0 \quad \text{for} \quad z > 0 \quad , \\ eV &= 0 \quad \text{for} \quad z < 0 \quad , \end{aligned}$$

and that the incoming wave is given by

$$\psi_{\mathrm{i}} = u_{\mathrm{i}}\exp\left\{\frac{\mathrm{i}}{\hbar}(pz - Et)\right\} \quad , \tag{13.3}$$

so that, inserting (13.3) into (13.2a) and using $\hat{\alpha} = \hat{\alpha}_3$ it follows that

$$\left\{\frac{E}{c} - \hat{\alpha}p - \hat{\beta}m_0c\right\}u_{\mathrm{i}} = 0 \quad . \tag{13.4}$$

[1] O. Klein: Z. Phys. **53**, 157 (1929).

Since we require $u_\mathrm{i} \neq 0$, then because of $\hat{\alpha}\hat{\beta} + \hat{\beta}\hat{\alpha} = 0$ we conclude that

$$\frac{E^2}{c^2} = p^2 + m_0^2 c^2 \quad , \tag{13.5}$$

and, moreover, due to our interest in the incoming electrons we choose $E > 0$. The momentum of the reflected wave must be $-p$, whereas the momentum $\overline{p}$ of the transmitted wave is given by (13.1). For small V_0, $\overline{p}$ is positive, so that in the first instance we can set

$$\psi_\mathrm{r} = u_\mathrm{r} \exp\left\{\frac{\mathrm{i}}{\hbar}(-pz - Et)\right\} \quad , \quad \psi_\mathrm{t} = u_\mathrm{t} \exp\left\{\frac{\mathrm{i}}{\hbar}(\overline{p}z - Et)\right\} \tag{13.6}$$

and therefore, due to (13.2a),

$$\left\{\frac{E}{c} + \hat{\alpha}p - \hat{\beta}m_0 c\right\} u_\mathrm{r} = 0 \quad \text{and} \quad \left\{\frac{E - V_0}{c} - \hat{\alpha}\overline{p} - \hat{\beta}m_0 c\right\} u_\mathrm{t} = 0 \quad . \tag{13.7}$$

The total wave function must be continuous at the boundary, i.e. for $z = 0$

$$u_\mathrm{i} + u_\mathrm{r} = u_\mathrm{t} \tag{13.8}$$

must be valid. From (13.4) and (13.8) therefore follows

$$\left(\frac{E}{c} - \hat{\beta}m_0 c\right)(u_\mathrm{t} + u_\mathrm{r}) = +\hat{\alpha}p(u_\mathrm{i} - u_\mathrm{r}) \quad , \tag{13.9}$$

and with (13.7) and (13.8) we get

$$\left(\frac{E}{c} - \hat{\beta}m_0 c\right)(u_\mathrm{i} + u_\mathrm{r}) = \left(\frac{V_0}{c} + \hat{\alpha}\overline{p}\right)(u_\mathrm{i} + u_\mathrm{r}) \quad . \tag{13.10}$$

Thus we have

$$\left\{\frac{V_0}{c} + \hat{\alpha}\overline{p}\right)(u_\mathrm{i} + u_\mathrm{r}) = +\hat{\alpha}p(u_\mathrm{i} - u_\mathrm{r}) \tag{13.11}$$

or

$$\left\{\frac{V_0}{c} + \hat{\alpha}(p + \overline{p})\right\} u_\mathrm{r} = -\left\{\frac{V_0}{c} - \hat{\alpha}(p - \overline{p})\right\} u_\mathrm{i} \quad . \tag{13.12}$$

We multiply both sides by $V_0/c - \hat{\alpha}(p + \overline{p})$, which, because of $\hat{\alpha}^2 = 1$ and with (13.1) and (13.5), leads to

$$u_\mathrm{r} = \frac{(2V_0/c)(-E/c + \hat{\alpha}p)}{V_0^2/c^2 - (p + \overline{p})^2} u_\mathrm{i} \equiv r u_\mathrm{i} \quad . \tag{13.13}$$

Analogously we find for the adjoint amplitude

$$u_\mathrm{r}^\dagger = r u_\mathrm{i}^\dagger \quad , \tag{13.14}$$

i.e.

$$u_\mathrm{r}^\dagger u_\mathrm{r} = \left(\frac{2V_0/c}{V_0^2/c^2 - (p + \overline{p})^2}\right)^2 u_\mathrm{i}^\dagger \left(-\frac{E}{c} + \hat{\alpha}p\right)^2 u_\mathrm{i} \quad , \tag{13.15}$$

so that using the identity

$$c u_i^\dagger \hat{\alpha} u_i = \frac{pc^2}{E} u_i^\dagger u_i \tag{13.16}$$

(which can be easily derived from the equations of motion for $u_i^\dagger$ and u_i) it follows from (13.15) that

$$\begin{aligned} u_r^\dagger u_r &= \left(\frac{2V_0/c}{V_0^2/c^2 - (p+\overline{p})^2}\right)^2 \left\{\left(\frac{E^2}{c^2} + p^2\right) u_i^\dagger u_i - \frac{2Ep}{c} u_i^\dagger \hat{\alpha} u_i\right\} \\ &= \left(\frac{2V_0 m_0}{V_0^2/c^2 - (p+\overline{p})^2}\right)^2 u_i^\dagger u_i \equiv R u_i^\dagger u_i \quad . \end{aligned} \tag{13.17}$$

Thus the quantity R is the fraction of electrons which are reflected: For $V_0 = 0$, $R = 0$, whereas for $V_0 = E - m_0c^2$ [i.e. $\overline{p} = 0$, from (13.1)], $R = 1$, and all electrons are reflected. If V_0 increases still further ($V_0 > E - m_0c^2$), then $\overline{p}$ becomes imaginary. Then we set

$$\psi_t = u_t \exp\left\{-\mu z - \mathrm{i}\frac{E}{\hbar}t\right\} \quad , \tag{13.18}$$

where μ is now a real quantity. μ must be greater than zero, since otherwise the density on the rhs of the barrier would be infinitely large for $z \to \infty$. On the other hand, because of (13.6), $\overline{p} = +\mathrm{i}\hbar\mu$, i.e. for (13.13) we get

$$u_r = -\frac{(2V_0/c)(E/c - \hat{\alpha}p)}{V_0^2/c^2 - (p + \mathrm{i}\hbar\mu)^2} u_i \quad , \quad u_r^\dagger = -\frac{(2V_0/c)(E/c - \hat{\alpha}p)}{V_0^2/c^2 - (p - \mathrm{i}\hbar\mu)^2} u_i^\dagger \quad , \tag{13.19}$$

and therefore

$$u_r^\dagger u_r = \frac{(2V_0/c)^2(E^2/c^2 - p^2)}{\left[(V_0/c + p)^2 + \mu^2\hbar^2\right]\left[(V_0/c - p)^2 + \mu^2\hbar^2\right]} u_i^\dagger u_i \quad . \tag{13.20}$$

Using (13.1) and (13.5) we conclude that

$$\overline{p}^2 = p^2 - \frac{V_0(2E - V_0)}{c^2} \quad , \tag{13.21}$$

i.e. since $\overline{p}^2 = -\mu^2\hbar^2$,

$$\left(\frac{V_0}{c} \pm p\right)^2 + \mu^2\hbar^2 = 2\frac{V_0}{c}\left(\frac{E}{c} \pm p\right) \tag{13.22}$$

and therefore $u_r^\dagger u_r = u_i^\dagger u_i$.

This means that the reflected current is equal to the incoming one. Behind the boundary there is an exponentially decreasing solution for the wave. Due to (13.20) the condition for that case is $p^2 < V_0(2E - V_0)/c^2$, and for increasing V_0 this condition is fulfilled as soon as V_0 exceeds the value $E - c(E^2/c^2 - p^2)^{1/2} = E - m_0c^2$. If V_0 increases further, then, due to (13.21), μ first increases, reaches its maximal value for $E = V_0$ and then again decreases, becoming zero for $V_0 = E + m_0c^2$. For still greater values $V_0 > E + m_0c^2$, $\overline{p}$ again assumes real values, so that (13.13) and (13.17) are again solutions of the problem. However, in this region the kinetic energy $E - V_0$ is negative, so that this is a classically forbidden situation. The group velocity, which due to (13.16) and (13.1) is given by

$$v_{\text{gr}} = \frac{c^2}{E - V_0}\overline{p} \quad , \tag{13.23}$$

therefore has the opposite direction as the momentum $\overline{p}$ in this region if we choose $\overline{p} > 0$. Since the group velocity is the velocity of the moving wave packet, it looks as if the transmitted wave packet came in from $z = +\infty$; on the other hand this contradicts the condition which only allows an incoming wave packet from $z = -\infty$. We thus have to choose $\overline{p} < 0$ [i.e. the negative sign of the root in (13.1)]. However, this condition is not included in the Dirac equation, but is forced upon us by the physical boundary conditions.[2] In the discussion given by Bjorken and Drell[3] this has not been taken into account, the reflection coefficient reading

$$R = \frac{(1-r)^2}{(1+r)^2} \quad , \quad \text{with} \tag{13.24}$$

$$r = \frac{\overline{p}}{p}\frac{E + m_0c^2}{E - V_0 + m_0c^2} \quad . \tag{13.25}$$

For $V_0 > E + m_0c^2$ the fraction indeed becomes negative, but r always remains greater than zero, because we have to choose $\overline{p} < 0$ due to the boundary conditions. Hence for the reflection coefficient we always have $R \leq 1$ and not, as given by Bjorken and Drell, $R > 1$.

We thus have seen that for $V_0 > E + m_0c^2$ a fraction of the electrons can traverse the potential barrier by transforming the original positive value of the kinetic energy to a negative one. The group velocity of the tunneling electrons is, due to (13.1), given by

$$\frac{c^2}{V_0 - E}|\overline{p}| = c\sqrt{1 - \left(\frac{m_0c^2}{V_0 - E}\right)^2} \quad , \tag{13.26}$$

which for $V_0 = E + m_0c^2$ is just zero and for $V_0 \to \infty$ approaches the velocity of light. When $V_0 = E + m_0c^2$, the reflection coefficient from (13.17) is just $R = 1$ (total reflection); it decreases for increasing V_0 down to the value

$$\alpha = R_{\text{min}} = \lim_{V_0 \to \infty} R(V_0) = \frac{(E/c - p)}{(E/c + p)} \tag{13.27}$$

for $V_0 \to \infty$. The corresponding fraction of electrons travelling through the boundary surface is thus

$$\beta = \frac{2p}{E/c + p} \quad , \tag{13.28}$$

where β is called the *transmission coefficient* and $\alpha + \beta = 1$! For $p = m_0c$ (i.e. electrons with a velocity 80 % that of light) we get with (13.5): $\beta \approx 2/(2^{1/2} +$

---

[2] H.G. Dosch, J.H.D. Jensen, V.L. Mueller: Phys. Norv. **5**, 151 (1971).

[3] J.D. Bjorken, S.D. Drell: *Relativistic Quantum Mechanics*, ed. by L. Schiff, International Series in Pure and Applied Physics (McGraw-Hill, New York 1964).

1) $\sim$ 0.83, i.e. 83 % of the incoming electrons penetrate the potential barrier. This large transmission coefficient also remains for V_0 not approaching infinity but only several rest masses. Calculations by F. Sauter[4] using a smoothened potential edge have shown that this large transmission coefficient, which is classically not understandable at all, does not occur if the width of the increase from $V = 0$ to $V = V_0$ is of the order of the Compton wavelength, i.e.

$$d \lesssim \frac{\hbar}{m_0 c} \quad . \tag{13.29}$$

This unexpected largeness of the transmission coefficient is known as *Klein's paradox*, and its interpretation given here is completely that of a single particle. In the framework of this representation it is not necessary at all to consider pair production (as has already been done by Bjorken and Drell at this level), though we shall see in connection with the hole theory that the boundary conditions change (no wave coming in from $z = +\infty$), that is we do not have to demand $\overline{p} < 0$ any longer (see the following Example 13.1).

EXAMPLE

13.1 Klein's Paradox and the Hole Theory

The hole theory becomes important if one wants to describe the behaviour of a particle wave striking a potential barrier with $V_0 > m_0c^2 + E$.

The Dirac equation for a plane wave moving in the z direction with spin up is:

(a) for region I (see the following figure)

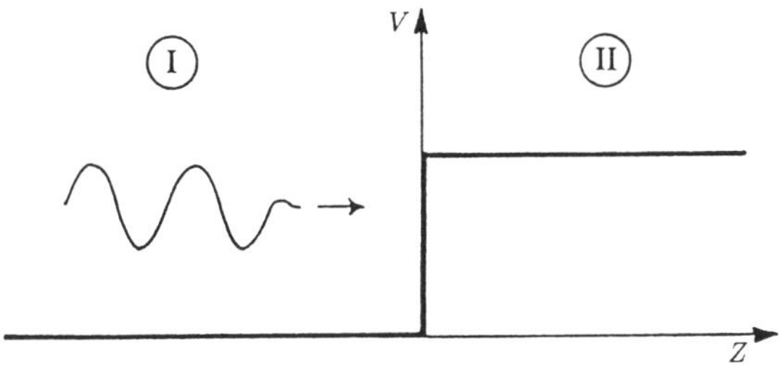

Electron wave and potential barrier

$$\left(c\hat{\alpha}_3\hat{p}_z + \hat{\beta}m_0c^2\right)\psi = E\psi \quad ; \tag{1}$$

(b) for region II

$$\left(c\hat{\alpha}_3\hat{p}_z + \hat{\beta}m_0c^2\right)\psi = (E - V_0)\psi \quad , \tag{2}$$

with solutions

[4] F. Sauter: Z. Physik **73**, 547 (1931).

Example 13.1.

$$\psi_{\rm I} = A \begin{pmatrix} 1 \\ 0 \\ \dfrac{p_1 c}{E + m_0 c^2} \\ 0 \end{pmatrix} e^{ip_1 z/\hbar} \quad ,$$

$$p_1 c = \sqrt{E^2 - m_0^2 c^4} \quad , \tag{3}$$

$$\psi_{\rm II} = B \begin{pmatrix} 1 \\ 0 \\ \dfrac{-p_2 c}{V_0 - E - m_0 c^2} \\ 0 \end{pmatrix} e^{ip_2 z/\hbar} \quad ,$$

$$p_2 c = \sqrt{(V_0 - E)^2 - m_0^2 c^4} \quad . \tag{4}$$

Decisive for the explanation is the fact that for $V_0 > E + m_0 c^2$ the momentum p_2 becomes real again, allowing for free plane waves to propagate in region II. This can only be understood by the existence of a second energy continuum corresponding to the solutions of the Dirac equation with negative energy (see below). From the impacting wave (3) one part is reflected (maintaining energy and momentum conservation):

$$\psi_{\rm I}^{\rm r} = C \begin{pmatrix} 1 \\ 0 \\ \dfrac{-p_1 c}{E + m_0 c^2} \\ 0 \end{pmatrix} e^{-ip_1 z/\hbar} \tag{5}$$

and the other part propagates further (4). We must require at $z = 0$ that the wave functions be equal inside and outside the potential (this does not mean continuity!):

$$\psi_{\rm I}(z = 0) + \psi_{\rm I}^{\rm r}(z = 0) = \psi_{\rm II}(z = 0) \quad . \tag{6}$$

From this the equations determining the various coefficients follow:

$$A + C = B \quad \text{and} \tag{7}$$

$$\begin{aligned} A - C &= -B \frac{p_2}{p_1} \frac{E + m_0 c^2}{V_0 - E - m_0 c^2} \\ &= -B \sqrt{\frac{\left(V_0 - E + m_0 c^2\right)\left(E + m_0 c^2\right)}{\left(V_0 - E - m_0 c^2\right)\left(E - m_0 c^2\right)}} =: -B\gamma \quad . \end{aligned} \tag{8}$$

Thus we have

$$\left.\begin{aligned} (7) + (8) &\Rightarrow A = \tfrac{B}{2}(1 - \gamma) \\ (7) - (8) &\Rightarrow C = \tfrac{B}{2}(1 + \gamma) \end{aligned}\right\} \Rightarrow \frac{C}{A} = \frac{1 + \gamma}{1 - \gamma} \tag{9}$$

and

$$\frac{B}{A} = \frac{2}{1 - \gamma} \quad . \tag{10}$$

With the expression for the particle current

Example 13.1.

$$j(x) = c\psi^\dagger(x)\hat{\boldsymbol{\alpha}}\psi(x) \tag{11}$$

and

$$\psi_\mathrm{I}^\dagger\hat{\alpha}_1 = A^*\left(0, \frac{p_1 c}{E+m_0c^2}, 0, 1\right)\mathrm{e}^{-\mathrm{i}p_1 z/\hbar} \quad , \tag{12}$$

$$\psi_\mathrm{I}^\dagger\hat{\alpha}_2 = A^*\left(0, -\mathrm{i}\frac{p_1 c}{E+m_0c^2}, 0, -\mathrm{i}\right)\mathrm{e}^{-\mathrm{i}p_1 z/\hbar} \quad , \tag{13}$$

$$\psi_\mathrm{I}^\dagger\hat{\alpha}_3 = A^*\left(\frac{p_1 c}{E+m_0c^2}, 0, 1, 0\right)\mathrm{e}^{-\mathrm{i}p_1 z/\hbar} \quad , \tag{14}$$

it follows that

$$j_\mathrm{I} = AA^*\frac{2p_1c^2}{E+m_0c^2}\,e_z \quad . \tag{15}$$

Correspondingly

$$j_\mathrm{I}^\mathrm{r} = -CC^*\frac{2p_1c^2}{E+m_0c^2}\,e_z \quad , \tag{16}$$

$$j_\mathrm{II} = -BB^*\frac{2p_2c^2}{V_0-\left(E+m_0c^2\right)}\,e_z \quad , \tag{17}$$

Thus the ratios of the currents [γ in (9) is real] are

$$\frac{|j_\mathrm{I}^\mathrm{r}|}{|j_\mathrm{I}|} = \frac{(1+\gamma)^2}{(1-\gamma)^2} \quad ,$$
$$\frac{|j_\mathrm{II}|}{|j_\mathrm{I}|} = \frac{4}{(1-\gamma)^2}|-\gamma| = \frac{4\gamma}{(1-\gamma)^2} \quad . \tag{18}$$

It can be seen from (8) and $\gamma > 1$, leading to

$$|j_\mathrm{I}^\mathrm{r}| > |j_\mathrm{I}| \quad . \tag{19}$$

This result corresponds to the fact that the flow of j_II is in the $(-z)$ direction, i.e. electrons leave region II, but according to our assumptions up to now, there are no electrons in there anyway. A reinterpretation is thus necessary, and in doing this the solutions of negative energy formally derived previously are treated seriously; thus there exist two electron continua:

To prevent the transition of all electrons to states of negative energy one has to require that all electron states with $E < -m_0c^2$ are occupied with electrons (see Chap. 12). This hypothesis permits the following explanation: The potential $V_0 > m_0c^2 + E$ raises the electron energy in region II sufficiently for there to be an overlap between the negative continuum for $z > 0$ and the positive continuum for $z < 0$. In the case of $V_0 > m_0c^2 + E$ the electrons striking the potential barrier from the left are able to knock additional electrons out of the vacuum on the right, leading to positron current flowing from left to right in the potential region.

From this notion it is understandable that there occur free plane wave solutions in region II given by (4) (see following figure), called positron waves. Furthermore,

Energy levels of the free Dirac equation

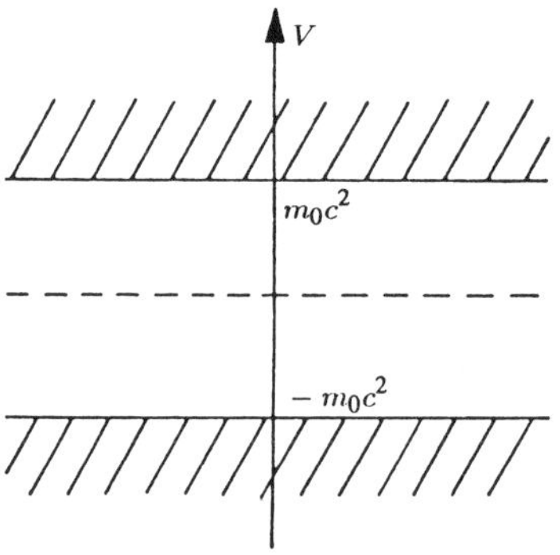

according to the hole theory, the negative continuum states are occupied, so that it is now possible to understand the sign of $j_{\rm II}$ in (17) by assuming that the electrons entering region I are coming from the negative continuum: Correspondingly the relation

$$j_{\rm I}+j_{\rm I}^{\rm r}=j_{\rm I}\left(1-\frac{|C|^2}{|A|^2}\right)=j_{\rm I}\frac{-4\gamma}{(1-\gamma)^2}=j_{\rm II} \tag{20}$$

holds. Since the holes remaining in region II are interpreted as positrons, it is possible to describe this effect in an alternative manner: The phenomena described above can be understood as electron–positron pair creation at the potential barrier (as shown in the figure below) and is related to the decay of the vacuum in the presence of supercritical fields.[5]

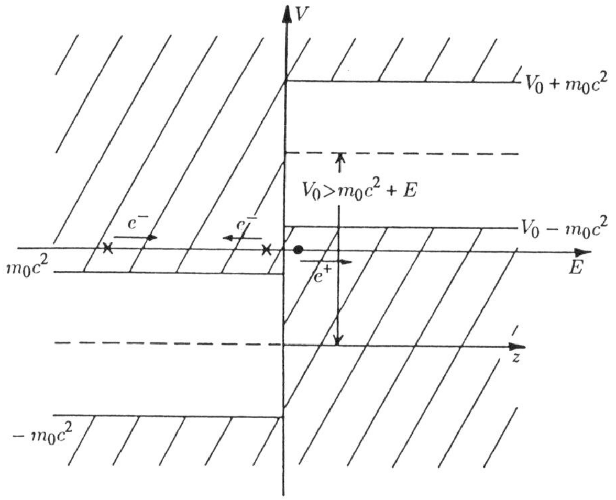

Energy continua of the Dirac equation at a potential barrier

[5] This is discussed by J. Reinhardt, W. Greiner: Rep. Prog. Phys. **40**, 219 (1977); and is covered in more detail in W. Greiner, W. Müller, J. Rafelski: *Quantum Electrodynamics of Strong Fields* (Springer, Berlin, Heidelberg 1985).

14. The Weyl Equation – The Neutrino

In 1930 W. Pauli postulated the existence of the neutrino in order to guarantee the energy and momentum conservation for the weak interaction, which at that time seemed to be violated in the β-decay experiments. Since the energy of the neutrino could not be determined even in the most sensitive measurements of the β decay of nuclei, the interaction of this postulated particle with matter must be extremely small. For example, it must not have electric charge, and accordingly, mass and magnetic moment must be assumed to be nearly vanishing, or even zero. The particle was named "neutrino" and abbreviated by "ν". Because of the relativistic mass–energy relation, a particle with rest mass $m_\nu = 0$ moves with the velocity of light. The experimental upper bound for the rest mass of the electronic neutrino is about a few electron volts and thus less than a thousandth of the electron's rest mass. Therefore the assumption $m_\nu = 0$ seems to be reasonable. Further experimental observations of the angular momentum balance during β decay showed that the neutrino has spin $\frac{1}{2}$. Consequently the Dirac equation for $m_0 = 0$ should be the fundamental equation of motion for the neutrino.

Neutrinos and photons can be regarded as equal as far as charge, magnetic moment, mass and velocity are concerned, the main difference between the particles being, respectively, the half-integer and integer half-integer spin. There is also a difference between the neutrino ν that occurs in β^+ decays and the antineutrino $\overline{\nu}$ that is emitted during the β^- decay. Historically the existence of the neutrino can be experimentally proved as an outcome of the β decay (β^- decay) through the recoil of the atomic nucleus in the reaction

$$\mathrm{n} \rightarrow \mathrm{p} + \mathrm{e}^- + \overline{\nu} \quad .$$

Further, the inverse β decay (β^+ decay) can be initiated by the neutrino and observed in experiments:

$$\mathrm{p} + \overline{\nu} \rightarrow \mathrm{n} + \mathrm{e}^+ \quad .$$

Another empirical property which has been found by accurate studies of the inverse β decays is that the antineutrino always has the same distinct spin orientation compared to its momentum direction. Assuming in the calculations that the spin of the antineutrino can be oriented parallel as well as antiparallel to its momentum results in theoretical cross-sections half as large as the experimental values. Precise experimental analyses have shown that the spin of the neutrino is antiparallel and the spin of the antineutrino is parallel to its momentum direction (see Fig. 14.1). This is the basic phenomenon of parity violation: If parity was conserved, neutrinos as well as antineutrinos must exist in nature with both spin directions. Finally,

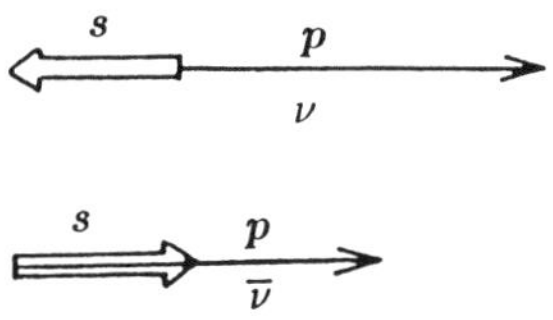

Fig. 14.1. The spin of the neutrino is always directed antiparallel to its direction of motion; the opposite is true for the antineutrino

in this short introduction we mention that there are different kinds of neutrinos, which can be related to electrons, muons and tauons by their specific behaviour in weak-interaction processes.[1]

In 1929 Hermann Weyl proposed a two-component equation to describe massless spin-$\frac{1}{2}$ particles,[2] but the Weyl equation violates parity invariance and therefore it was at first rejected. Then, after parity violation of the weak interaction was experimentally proved in 1957, Landau, Salam as well as Lee and Yang took this proposal[3] and regarded the Weyl equation as the basic equation of motion of the neutrino.

We first consider the Dirac equation of a particle with mass $m_0 = 0$:

$$\mathrm{i}\hbar\frac{\partial\Psi}{\partial t} = c\hat{\boldsymbol{\alpha}}\cdot\hat{\boldsymbol{p}}\Psi(x) \quad , \tag{14.1}$$

and this equation no longer contains the $\hat{\beta}$ matrix. The anticommutation relations for the three matrices $\hat{\alpha}_1$, $\hat{\alpha}_2$ and $\hat{\alpha}_3$,

$$\{\hat{\alpha}_i, \hat{\alpha}_j\} = 2\delta_{ij} \quad , \tag{14.2}$$

can be satisfied by the 2×2 Pauli matrices $\hat{\sigma}_i$. Merely the necessity of constructing $\hat{\beta}$ as the fourth anticommuting matrix requires the introduction of 4×4 matrices, and the necessity of describing particles with spin-$\frac{1}{2}$ by two spinors is obviously connected with the particles' masses. Therefore this reason disappears if the mass is zero: The wave equation of such a particle can be set up with only one spinor.

The wave equation for the *two-component amplitude* $\Psi^{(+)}(x)$ describing the neutrino reads

$$\mathrm{i}\hbar\frac{\partial\Phi^{(+)}}{\partial t} = c\hat{\boldsymbol{\sigma}}\cdot\hat{\boldsymbol{p}}\Phi^{(+)}(x) \tag{14.3}$$

or, after dividing by $\mathrm{i}\hbar$,

$$\frac{\partial\Phi^{(+)}}{\partial t} = -c\hat{\boldsymbol{\sigma}}\cdot\boldsymbol{\nabla}\Phi^{(+)}(x) \quad , \tag{14.4}$$

where $\hat{\sigma}_i$ are the 2×2 Pauli matrices. The plane-wave solutions of the Weyl equation are given by

$$\Phi^{(+)}(x) = \frac{1}{\sqrt{2E(2\pi)^3}}\,\mathrm{e}^{-\mathrm{i}p\cdot x/\hbar}u^{(+)}(p) \tag{14.5}$$

with

$$p = \{p_0, \boldsymbol{p}\} = \left\{\frac{E}{c}, \boldsymbol{p}\right\} \quad , \quad x = \{x_0, \boldsymbol{x}\} \quad \text{and}$$
$$p\cdot x = p_0x_0 - \boldsymbol{p}\cdot\boldsymbol{x} \quad . \tag{14.6}$$

---

[1] The theory of weak interaction is extensively discussed in W. Greiner, B. Müller: *Gauge Theory of Weak Interactions* 2nd ed. (Springer, Berlin, Heidelberg 1996).

[2] H. Weyl: Z. Physik **56**, 330 (1929).

[3] L. Landau: Nucl. Phys. **3**, 127 (1957); T.D. Lee, C.N. Yang: Phys. Rev. **105** 1671 (1957); A. Salam: Nuovo Cimento **5**, 299 (1957).

Following the plane-wave solutions for electrons, the neutrino wave functions are normalized in such a way that the norm remains invariant under Lorentz transformation (see Chap. 6). $u^{(+)}(p)$ is a two-component spinor which satisfies the equation

$$p_0 u^{(+)} = \hat{\boldsymbol{\sigma}} \cdot \boldsymbol{p} u^{(+)} \quad . \tag{14.7}$$

This relation is of special interest, since we note that due to (14.7) the solution for a given sign of energy p_0 corresponds to a certain orientation of the spin $\boldsymbol{\sigma}$ with respect to the direction of motion $\boldsymbol{p}$. By applying the *helicity operator* $\hat{\boldsymbol{\sigma}} \cdot \hat{\mathbf{p}}/|\boldsymbol{p}|$ to both sides of the equation and using the relation

$$(\hat{\boldsymbol{\sigma}} \cdot \boldsymbol{A})(\hat{\boldsymbol{\sigma}} \cdot \boldsymbol{B}) = \boldsymbol{A} \cdot \boldsymbol{B} + \mathrm{i}\hat{\boldsymbol{\sigma}} \cdot (\boldsymbol{A} \times \boldsymbol{B}) \quad ,$$

we get $(\hat{\boldsymbol{\sigma}} \cdot \boldsymbol{p})^2$, and therefore

$$\left(p_0^2 - \boldsymbol{p}^2\right) u^{(+)} = 0 \quad , \tag{14.8}$$

so that only non-vanishing solutions for u exist if

$$p_0 = \pm|\boldsymbol{p}| \tag{14.9}$$

is valid. This is naturally the relativistic energy of massless particles, and here it again follows that neutrinos move with the velocity of light. But, of course, this result was already implied in our initial demand that the massless particle has non-zero energy. In the common representation of the Pauli matrices, with the z axis in the $\boldsymbol{p}$ direction, the solution of (14.7) reads

$$u^{(+)} = \begin{pmatrix} 1 \\ 0 \end{pmatrix} \quad . \tag{14.10}$$

This solution describes right-handed massless particles with spin in the direction of motion. By this we mean

$$\frac{\hat{\boldsymbol{\sigma}} \cdot \boldsymbol{p}}{|\boldsymbol{p}|} u^{(+)} = \hat{\sigma}_z \begin{pmatrix} 1 \\ 0 \end{pmatrix} = + \begin{pmatrix} 1 \\ 0 \end{pmatrix} \quad , \tag{14.11}$$

or, literally: The helicity operator has a positive eigenvalue, the spin is directed parallel to $\boldsymbol{p}$, which corresponds to a right-handed screw if we look in the direction of motion (i.e. in the direction of $\boldsymbol{p}/|\boldsymbol{p}|$), and which is illustrated in Fig. 14.2.

For states with positive energy ($p_0 = +|\boldsymbol{p}|$) the wave equation (14.3) only has waves of positive helicity as solutions. With (14.7) and the aid of (14.9) we see immediately that we have the reverse result for states with negative energy ($p_0 = -|\boldsymbol{p}|$); there, the wave equation (14.3) only contains waves of negative helicity as solutions. This remains valid even in the hole theory, where a wave function with negative energy, negative momentum, and negative spin direction is interpreted as an antiparticle with positive energy, positive momentum, and positive spin direction. This assignment of $\pm p_0 \Leftrightarrow$ particles, antiparticles to the helicity is contrary to the requirement of the experiments dealing with weak interactions (see Fig. 14.1).

To describe left-handed massless particles, we must obviously start with the equation

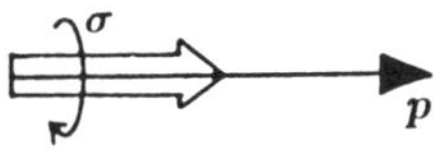

Fig. 14.2. The spin of the massless particle described by the wave equations (14.3) or (14.7) points in the direction of $\boldsymbol{p}$, corresponding to a right-handed screw. By convention the helicity of the particle $\hat{\boldsymbol{\sigma}} \cdot \boldsymbol{p}/|\boldsymbol{p}| = +1$. A neutrino cannot be such a particle, because it is known from experiments that the neutrino has helicity -1

$$\mathrm{i}\hbar\frac{\partial\Phi^{(-)}}{\partial t} = -c\hat{\boldsymbol{\sigma}}\cdot\hat{\boldsymbol{p}}\Phi^{(-)}(x) \quad . \tag{14.12}$$

Replacing $\hat{\boldsymbol{\sigma}}$ by $-\hat{\boldsymbol{\sigma}}$ also yields a realization of the commutation relation (14.2), so that (14.12) is then a possible Dirac equation for massless spin-$\frac{1}{2}$ particles, in the same way as (14.3). With the ansatz

$$\Phi^{(-)}(x) = \frac{1}{\sqrt{2E(2\pi)^3}}\,\mathrm{e}^{-\mathrm{i}p\cdot x/\hbar}u^{(-)}(p) \tag{14.13}$$

we find

$$p_0 u^{(-)} = -\hat{\boldsymbol{\sigma}}\cdot\boldsymbol{p}u^{(-)} \quad , \quad \text{with} \tag{14.14}$$

$$u^{(-)} = \begin{pmatrix}0\\1\end{pmatrix} \quad . \tag{14.15}$$

In (14.14), $p_0 = E/c$ can again be either positive or negative. We call solution (14.13) a *neutrino state*, because, as already mentioned repeatedly, the experiments show that neutrinos appear only as left-handed particles (with negative helicity). In this state the spin is antiparallel to $\boldsymbol{p}$ and can be representend by a left-handed screw (see Fig. 14.1). For the antineutrino state ($p_0 < 0$) we have the opposite case, the particle is right-handed. Results achieved in β-decay experiments show that the neutrino always moves antiparallel to its spin direction. This means that the helicity or longitudinal polarization of a neutrino with positive energy is negative, while the helicity of a neutrino with negative energy is positive. The solution with $p_0 = -|\boldsymbol{p}|$ yields a spin parallel to $\boldsymbol{p}$ and can be represented by a right-handed screw.

On the other hand the helicity of a particle with positive energy, described by (14.4), is also positive. Therefore, this particle of (14.4) may be identified with the antineutrino of (14.12). As already mentioned, according to the interpretation of the negative energy states within the hole theory, the antineutrino has a momentum which is opposite to the empty negative energy state, and also the spin flips. Therefore, the relation between spin direction and momentum of the antineutrino is represented by a right-handed screw according to (14.14). The massless antiparticle which belongs to (14.3) or (14.7), respectively, has negative helicity; it obviously behaves exactly like the neutrino described by (14.14). Nevertheless, it should be mentioned that stating that the neutrino (antineutrino) is always left handed (right handed), only makes sense if the rest-mass is exactly zero. Otherwise a Lorentz transformation, which transforms a left-handed particle into a right-handed one, can always be found.

To derive the neutrino current we rewrite (14.12) as

$$\frac{1}{c}\frac{\partial\Phi^{(-)}}{\partial t} - \hat{\boldsymbol{\sigma}}\cdot\boldsymbol{\nabla}\Phi^{(-)} = 0 \quad . \tag{14.16}$$

Combining the unit matrix and $\hat{\boldsymbol{\sigma}}$ to form a four-vector $\hat{\sigma}_\mu = \{\mathbb{1}, +\hat{\boldsymbol{\sigma}}\}$, we can write the Weyl equation in a more compact way:

$$\hat{\sigma}_\mu\boldsymbol{\nabla}^\mu\Phi^{(-)} = 0 \quad , \tag{14.17}$$

and the Hermitian conjugate equation reads

$$\boldsymbol{\nabla}^\mu\left(\Phi^{(-)}\right)^\dagger\hat{\sigma}_\mu = 0 \quad . \tag{14.18}$$

We multiply the first equation from the left by $\left(\Phi^{(-)}\right)^\dagger$, the latter one from the right by $\Phi^{(-)}$ and then add both equations. As usual this yields

$$\nabla^\mu \left(\Phi^{(-)}\right)^\dagger \hat{\sigma}_\mu \Phi^{(-)} = 0 \quad , \tag{14.19}$$

which is a continuity equation with the four-current

$$j_\mu = \left(\Phi^{(-)}\right)^\dagger \hat{\sigma}_\mu \Phi^{(-)} \quad . \tag{14.20}$$

Its space and time components are

$$\boldsymbol{j} = -\left(\Phi^{(-)}\right)^\dagger \hat{\boldsymbol{\sigma}} \Phi^{(-)} \quad , \tag{14.21}$$

$$\varrho = \left(\Phi^{(-)}\right)^\dagger \Phi^{(-)} \quad . \tag{14.22}$$

The normalization constant of the neutrino wave function can be derived from the integral with positive definite density ϱ, if we require that

$$\int \varrho \, \mathrm{d}^3 x = 1 \quad . \tag{14.23}$$

By considering various types of interactions, neutrinos can appear together with other spin-$\frac{1}{2}$ particles which have a finite mass and therefore are described by four-component wave functions. To have a unified description in such cases, it is appropriate to also introduce a bispinor wave function for the neutrino. To provide the connection between the two-component solutions of the Weyl equation with the already known four-component electron spinors, we go back to the Dirac equation for a particle with rest mass m_0. However we choose a different representation of the Dirac matrices $\hat{\alpha}$ and $\hat{\beta}$, namely

$$\hat{\alpha}_i = \begin{pmatrix} \hat{\sigma}_i & 0 \\ 0 & -\hat{\sigma}_i \end{pmatrix} \quad , \tag{14.24}$$

$$\hat{\beta} = \begin{pmatrix} 0 & -1\!\!1 \\ -1\!\!1 & 0 \end{pmatrix} \quad . \tag{14.25}$$

One proves immediately the anticommutator relations of the Dirac matrices:

$$\begin{aligned} \hat{\alpha}_i \hat{\alpha}_j + \hat{\alpha}_j \hat{\alpha}_i &= \begin{pmatrix} \hat{\sigma}_i & 0 \\ 0 & -\hat{\sigma}_i \end{pmatrix} \begin{pmatrix} \hat{\sigma}_j & 0 \\ 0 & -\hat{\sigma}_j \end{pmatrix} + \begin{pmatrix} \hat{\sigma}_j & 0 \\ 0 & -\hat{\sigma}_j \end{pmatrix} \begin{pmatrix} \hat{\sigma}_i & 0 \\ 0 & -\hat{\sigma}_i \end{pmatrix} \\ &= \begin{pmatrix} \hat{\sigma}_i \hat{\sigma}_j + \hat{\sigma}_j \hat{\sigma}_i & 0 \\ 0 & \hat{\sigma}_i \hat{\sigma}_j + \hat{\sigma}_j \hat{\sigma}_i \end{pmatrix} = 2\delta_{ij} \quad , \end{aligned} \tag{14.26}$$

$$\begin{aligned} \hat{\alpha}_i \hat{\beta} + \hat{\beta} \hat{\alpha}_i &= \begin{pmatrix} \hat{\alpha}_i & 0 \\ 0 & -\hat{\sigma}_i \end{pmatrix} \begin{pmatrix} 0 & -1\!\!1 \\ -1\!\!1 & 0 \end{pmatrix} + \begin{pmatrix} 0 & -1\!\!1 \\ -1\!\!1 & 0 \end{pmatrix} \begin{pmatrix} \hat{\sigma}_i & 0 \\ 0 & -\hat{\sigma}_i \end{pmatrix} \\ &= \begin{pmatrix} 0 & -\hat{\sigma}_i \\ \hat{\sigma}_i & 0 \end{pmatrix} + \begin{pmatrix} 0 & \hat{\sigma}_i \\ -\hat{\sigma}_i & 0 \end{pmatrix} = 0 \quad . \end{aligned} \tag{14.27}$$

This representation has the disadvantage that the four components of the bispinor do not split up into small and large components in the non-relativistic limit. However, as neutrinos have at least approximately zero mass and are therefore relativistic particles, this disadvantage is in practice insignificant. In this representation, with the help of the two-component notation

$$\Phi = \begin{pmatrix} \Phi^{(+)} \\ \Phi^{(-)} \end{pmatrix} \tag{14.28}$$

the Dirac equation can be split up into

$$\begin{aligned} \mathrm{i}\hbar\frac{\partial \Phi^{(+)}}{\partial t} &= -\mathrm{i}\hbar c\hat{\boldsymbol{\sigma}} \cdot \boldsymbol{\nabla}\Phi^{(+)} - m_0c^2\Phi^{(-)} \\ &= c\hat{\boldsymbol{\sigma}} \cdot \hat{\boldsymbol{p}}\Phi^{(+)} - m_0c^2\Phi^{(-)} \quad , \end{aligned} \tag{14.29}$$

$$\begin{aligned} \mathrm{i}\hbar\frac{\partial \Phi^{(-)}}{\partial t} &= \mathrm{i}\hbar c\hat{\boldsymbol{\sigma}} \cdot \boldsymbol{\nabla}\Phi^{(-)} - m_0c^2\Phi^{(+)} \\ &= -c\hat{\boldsymbol{\sigma}} \cdot \hat{\boldsymbol{p}}\Phi^{(-)} - m_0c^2\Phi^{(+)} \quad . \end{aligned} \tag{14.30}$$

Note that the upper and the lower components of Ψ are coupled by the mass term. In the limiting case $m_0 \to 0$ two decoupled two-component equations are obtained, corresponding to the respective Weyl equations. Then $\Phi^{(+)}$ describes right-handed and $\Phi^{(-)}$ left-handed massless particles. Because of the fact that in nature neutrinos or antineutrinos appear only with a definite helicity, we must require that in the four-component description two of the components vanish. This is achieved by applying the projection operators

$$\hat{P}_\pm = \tfrac{1}{2}(1\!\!1 \pm \gamma_5) \tag{14.31}$$

to the spinor Ψ of (14.28). In the representation used here (14.24) and (14.25) γ_5 is diagonal, and with $\gamma^5 = \mathrm{i}\gamma^0\gamma^1\gamma^2\gamma^3 = \gamma_5$ it follows that

$$\gamma_5 = \begin{pmatrix} 1\!\!1 & 0 \\ 0 & -1\!\!1 \end{pmatrix} \quad . \tag{14.32}$$

Thus we get

$$\Psi^{(-)} = \frac{1}{2}(1\!\!1 - \gamma_5)\Psi = \begin{pmatrix} 0 \\ \Phi^{(-)} \end{pmatrix} \quad , \tag{14.33}$$

$$\Psi^{(+)} = \frac{1}{2}(1\!\!1 + \gamma_5)\Psi = \begin{pmatrix} \Phi^{(+)} \\ 0 \end{pmatrix} \quad . \tag{14.34}$$

Also, by application of $\gamma_5^2 = 1\!\!1$, one evidently gets

$$\gamma_5\Psi^{(-)} = -\Psi^{(-)} \quad , \tag{14.35}$$

$$\gamma_5\Psi^{(+)} = \Psi^{(+)} \quad . \tag{14.36}$$

Equations (14.33) and (14.34) are in agreement with the Dirac equation only in the case that the particle mass is exactly zero, i.e. only then are $\Psi^{(+)}$ and $\Psi^{(-)}$ solutions of the Dirac equation. Indeed γ^5 anticommutes with all γ matrices and it obviously commutes with the mass term. Hence, for vanishing rest mass m_0, $\Psi^{(+)}$ and $\Psi^{(-)}$ are eigenfunctions of the Hamiltonian, the helicity operator and of γ_5.

The two-component Weyl theory is thus equivalent to a four-component Dirac representation. However, in the framework of the Weyl equation, the distinction between particles and antiparticles is superfluous. The two possible states of the neutrino are only characterized by parallel and antiparallel orientation of the spin with respect to its momentum.

Now, we want to analyse the non-invariance of the two-component theory under parity transformations. Let us have a look at the two-component equations (14.3) and (14.7). By performing a space inversion ($\boldsymbol{p} \to -\boldsymbol{p}$, $\boldsymbol{x} \to -\boldsymbol{x}$, $\hat{\boldsymbol{\sigma}} \to \hat{\boldsymbol{\sigma}}$) a state with the energy $p_0 = |\boldsymbol{p}|$, momentum $\boldsymbol{p}$ and helicity $\hat{\boldsymbol{\sigma}} \cdot \boldsymbol{p}/|\boldsymbol{p}| = 1$ will be transformed into a state with $p_0 = |\boldsymbol{p}|$, momentum $-\boldsymbol{p}$ and helicity -1. Such a state, though, does not exist in this two-component theory of massless spin-$\frac{1}{2}$ particles.

There is another way to recognize the non-invariance of a two-component theory for space inversion. $\boldsymbol{p}$ is a polar vector, $\hat{\boldsymbol{\sigma}}$ is an axial vector and thus $\hat{\boldsymbol{\sigma}} \cdot \boldsymbol{p}$ is a pseudoscalar under inversion. This is also obvious from (14.35) and (14.36) which give rise to

$$\overline{\Psi^{(\pm)}}\gamma_5\Psi^{(\pm)} = \pm\overline{\Psi^{(\pm)}}\Psi^{(\pm)} \quad . \tag{14.37}$$

The left-hand side is a pseudoscalar density while the right-hand side is a scalar density. Thus helicity eigenstates are no parity eigenstates. The wave equations (14.29) and (14.30), which contain the mass m_0, are symmetric with respect to reflection. In the description of a particle by only *one* spinor this symmetry gets lost. This symmetry, though, is not essential, because the reflection symmetry need not be a universal property of nature. Reflection symmetry only exists if the particle is replaced by the antiparticle simultaneously. Equations (14.29) and (14.30) are in fact invariant under $\boldsymbol{x} \to -\boldsymbol{x}$ and $\Psi^{(+)} \to \Psi^{(-)}$. The situation might be further clarified by observing that the mass term in the Dirac equation mixes the two helicity states while the kinetic terms conserve it

$$\begin{aligned}
&\overline{\Psi}\left(\not{p} - e\not{A} - m_0 c\right)\frac{1\!\!1 \pm \gamma_5}{2}\Psi^{(\pm)} \\
&= \Psi^\dagger\gamma_0\left[\frac{1\!\!1 \mp \gamma_5}{2}(\not{p} - e\not{A}) - \frac{1\!\!1 \pm \gamma_5}{2}m_0 c\right]\Psi^{(\pm)} \\
&= \left[\left(\frac{1\!\!1 \pm \gamma_5}{2}\right)\Psi\right]^\dagger\gamma_0(\not{p} - e\not{A})\Psi^{(\pm)} - \left[\frac{1\!\!1 \mp \gamma_5}{2}\Psi\right]^\dagger\gamma_0 m_0 c\Psi^{(\pm)} \\
&= \overline{\Psi}^{(\pm)}(\not{p} - e\not{A})\Psi^{(\pm)} - \overline{\Psi}^{(\mp)}m_0 c^2\Psi^{(\pm)} \quad .
\end{aligned} \tag{14.38}$$

Left- and right-handed fermions are thus strongly coupled if they have non-vanishing mass.

Finally we consider the angular-momentum representation of the Weyl equation. With

$$\Phi^{(+)} = \phi^{(+)}\,\mathrm{e}^{-\mathrm{i}Et/\hbar} \quad , \tag{14.39}$$

then from (14.4) we get

$$\hbar\hat{\boldsymbol{\sigma}} \cdot \boldsymbol{\nabla}\phi^{(+)} = \mathrm{i}p_0\phi^{(+)} \quad . \tag{14.40}$$

The operators $\hat{\boldsymbol{J}}^2$ and $\hat{J}_z$ commute with $\hat{\boldsymbol{\sigma}} \cdot \boldsymbol{\nabla}$, while $\hat{\boldsymbol{L}}^2$ does not. Accordingly, the angular momentum representation of the neutrino and antineutrino wave function, respectively, contain spherical spinors $\chi_{\pm\kappa,\mu}$. As we have already learned from the discussion of electron wave function including the coupling to external fields (see Chap. 9), the spin-orbit operator $\hat{\boldsymbol{\sigma}} \cdot \hat{\boldsymbol{L}}$ also does not commute with $\hat{\boldsymbol{\sigma}} \cdot \boldsymbol{\nabla}$. The same holds for the parity operator multiplied by any arbitrary 2×2 matrix. Furthermore, there exists no 2×2 matrix which anticommutes with $\hat{\boldsymbol{\sigma}}$. As a consequence of these commutation properties the solutions will not have a definite parity and thus they will become a linear combination of both spherical spinors $\chi_{\pm\kappa,\mu}$. We try the ansatz (see also Exercise 14.3)

$$\phi^{(+)}_{\kappa,\mu} = g(r)\chi_{\kappa,\mu} + \mathrm{i}f(r)\chi_{-\kappa,\mu} \tag{14.41}$$

and then proceed to make use of the following relations, which have been derived in the context of the two-centre Dirac equation [see (10.2)ff., (10.37), (10.54)ff.]:

$$\hat{\boldsymbol{\sigma}} \cdot \boldsymbol{\nabla} = \hat{\sigma}_r \left(\frac{\partial}{\partial r} - \frac{1}{\hbar r} \hat{\boldsymbol{\sigma}} \cdot \hat{\boldsymbol{L}} \right) \quad , \tag{14.42}$$

$$\hat{\boldsymbol{\sigma}} \cdot \hat{\boldsymbol{L}} \chi_{\kappa,\mu} = -\hbar(\kappa + 1)\chi_{\kappa,\mu} \quad , \tag{14.43}$$

$$\hat{\sigma}_r \chi_{\kappa,\mu} = -\chi_{-\kappa,\mu} \quad . \tag{14.44}$$

This finally leads to the differential equations for the radial functions $g(r)$ and $f(r)$:

$$\frac{\mathrm{d}g}{\mathrm{d}r} = \frac{p_0}{\hbar} f - \frac{\kappa + 1}{r} g \quad , \tag{14.45}$$

$$\frac{\mathrm{d}f}{\mathrm{d}r} = \frac{\kappa - 1}{r} f - \frac{p_0}{\hbar} g \quad . \tag{14.46}$$

These are the same differential equations as those we derived previously for the case of electrons in the presence of a constant, spherical-symmetric potential V_0, assuming that we set $m_0 = 0$ and $V_0 = 0$. The solutions which are regular at $r = 0$ can be directly taken from Exercise 9.5, yielding the results:

$$g(r) = j_l \left(\frac{p_0 r}{\hbar} \right) \quad , \tag{14.47}$$

$$f(r) = \frac{\kappa}{|\kappa|} j_l \left(\frac{p_0 r}{\hbar} \right) \quad . \tag{14.48}$$

Note that the wave functions with $\kappa = |\kappa|$ and $\kappa = -|\kappa|$ are not linearly independent. Instead, one has

$$\phi^{(+)}_{-\kappa,\mu} = \mathrm{i}\phi^{(+)}_{\kappa,\mu} \quad . \tag{14.49}$$

As a consequence of the unique helicity of the massless, two-component particle there exist only half as many states in a four-component description. If we start with (14.11) for left-handed neutrinos, the angular momentum representation would read (see Exercise 14.3):

$$\phi^{(-)}_{\kappa,\mu} = g(r)\chi_{\kappa,\mu} - \mathrm{i}f(r)\chi_{-\kappa,\mu} \quad . \tag{14.50}$$

EXERCISE

14.1 Dirac Equation for Neutrinos

Problem. Solve the Dirac equation for neutrinos and determine the eigenvalues of the helicity operator and those of γ_5 for both energy solutions. Make use of the standard representation of the Dirac matrices

$$\hat{\beta} = \begin{pmatrix} \mathbb{1} & 0 \\ 0 & -\mathbb{1} \end{pmatrix} \quad , \quad \hat{\boldsymbol{\alpha}} = \begin{pmatrix} 0 & \hat{\boldsymbol{\sigma}} \\ \hat{\boldsymbol{\sigma}} & 0 \end{pmatrix} \quad .$$

Solution. Consider the Dirac equation for massless particles:

$$\mathrm{i}\hbar \frac{\partial \Psi}{\partial t} = -\mathrm{i}\hbar c \hat{\boldsymbol{\alpha}} \cdot \boldsymbol{\nabla} \Psi(x) = \hat{H} \Psi(x) \quad . \tag{1}$$

By means of the usual ansatz for the time evolution

$$\Psi = \psi \, \mathrm{e}^{-\mathrm{i}Et/\hbar} \tag{2}$$

we obtain

$$E\psi = -\mathrm{i}\hbar c \hat{\boldsymbol{\alpha}} \cdot \boldsymbol{\nabla} \psi \quad . \tag{3}$$

The solution of this equation can be represented in terms of plane waves. Accordingly, the ansatz

$$\psi = \mathrm{e}^{\mathrm{i}\boldsymbol{p}\cdot\boldsymbol{x}/\hbar} u(p) \tag{4}$$

yields

$$Eu(p) = c\hat{\boldsymbol{\alpha}} \cdot \boldsymbol{p} u(p) \quad , \tag{5}$$

where

$$E = \pm E_p = \pm |\boldsymbol{p}| c \quad . \tag{6}$$

Using the standard representation of the Dirac matrices, one has $\gamma^5 = \gamma^0\gamma^1\gamma^2\gamma^3$,

$$\gamma_5 = \begin{pmatrix} 0 & \mathbb{1} \\ \mathbb{1} & 0 \end{pmatrix} \quad , \tag{7}$$

and obviously

$$\gamma_5 \hat{\boldsymbol{\Sigma}} = \begin{pmatrix} 0 & \mathbb{1} \\ \mathbb{1} & 0 \end{pmatrix} \begin{pmatrix} \hat{\boldsymbol{\sigma}} & 0 \\ 0 & \hat{\boldsymbol{\sigma}} \end{pmatrix} = \begin{pmatrix} 0 & \hat{\boldsymbol{\sigma}} \\ \hat{\boldsymbol{\sigma}} & 0 \end{pmatrix} = \hat{\boldsymbol{\alpha}} \tag{8}$$

holds. Thus, the Hamiltonian in (1) can also be written as

$$\hat{H} = c\hat{\boldsymbol{\alpha}} \cdot \hat{\boldsymbol{p}} = c\gamma_5 \hat{\boldsymbol{\Sigma}} \cdot \hat{\boldsymbol{p}} \quad . \tag{9}$$

The eigenfunctions of the Hamiltionian $\hat{H}$ are simultaneously eigenfunctions of the helicity operator $\hat{\boldsymbol{\Sigma}} \cdot \hat{\boldsymbol{p}}/|\boldsymbol{p}|$ and γ_5. The solutions of the eigenvalue equation can be directly taken from the plane-wave solutions of the free Dirac equation (see

Exercise 14.1.

Chap. 2). We set $m_0 = 0$ and obtain four linearly independent solutions for the spinor u. Hereby the z axis is chosen as the direction of the momentum $\boldsymbol{p}$. For u it follows that

helicity:

$$\underbrace{\overset{+1}{\begin{pmatrix}1\\0\\1\\0\end{pmatrix}}\quad\overset{-1}{\begin{pmatrix}0\\1\\0\\-1\end{pmatrix}}}_{\text{positive energy}}\qquad\underbrace{\overset{+1}{\begin{pmatrix}1\\0\\-1\\0\end{pmatrix}}\quad\overset{-1}{\begin{pmatrix}0\\1\\0\\1\end{pmatrix}}}_{\text{negative energy}} ,$$

whereby normalization factors have been suppressed. The eigenvalues of γ_5 are found to be

E	Helicity	Eigenvalue of γ_5
$+E_p$	$+1$	$+1$
$+E_p$	-1	-1
$-E_p$	$+1$	-1
$-E_p$	-1	$+1$

Obviously γ_5 and the helicity operator have equal eigenvalues in the case of positive energy solutions. Opposite signs result for solutions with negative energy, and the eigenvalues of γ_5 are just the negative of the helicity eigenvalues.

EXERCISE

14.2 CP as a Symmetry for the Dirac Neutrino

Problem. Show that the product of charge conjugation and parity transformation represents a symmetry transformation of the Dirac neutrino.

Solution. In the standard representation of the Dirac matrices γ_S^μ the charge-conjugated spinor is found via

$$\psi_c(x) = \hat{C}\gamma^0\psi^* \quad . \tag{1}$$

Hereby the charge-conjugation operator $\hat{C}$ satisfies the condition

$$\left(\hat{C}\gamma^0\right)\gamma^{\mu *}\left(\hat{C}\gamma^0\right)^{-1} = -\gamma^\mu \tag{2}$$

together with

$$\hat{C} = \mathrm{i}\gamma^2\gamma^0 = -\hat{C}^{-1} = -\hat{C}^\dagger = -\hat{C}^\mathrm{T} \quad . \tag{3}$$

By means of a unitary transformation $\hat{U}$, which transforms the standard representation of the γ matrices into the one we have used for describing the neutrinos [see (14.24) and (14.25)], the corresponding representation of the charge-conjugation operator can be found. The unitary operator $\hat{U}$ reads

Exercise 14.2.

$$\hat{U} = \frac{1}{\sqrt{2}}\left(1\!\!1 + \gamma_S^0\gamma_S^5\right) \quad . \tag{4}$$

Using

$$\gamma_S^0\gamma_S^5 = \begin{pmatrix} 1\!\!1 & 0 \\ 0 & -1\!\!1 \end{pmatrix}\begin{pmatrix} 0 & 1\!\!1 \\ 1\!\!1 & 0 \end{pmatrix} = \begin{pmatrix} 0 & 1\!\!1 \\ -1\!\!1 & 0 \end{pmatrix} \quad . \tag{5}$$

it follows that

$$\hat{U} = \frac{1}{\sqrt{2}}\begin{pmatrix} 1\!\!1 & 1\!\!1 \\ -1\!\!1 & 1\!\!1 \end{pmatrix} \quad . \tag{6}$$

Indeed, we obtain

$$\begin{aligned} \hat{U}\gamma_S^i\hat{U}^{-1} = \hat{U}\gamma_S^i\hat{U}^\dagger &= \frac{1}{2}\begin{pmatrix} 1\!\!1 & 1\!\!1 \\ -1\!\!1 & 1\!\!1 \end{pmatrix}\begin{pmatrix} 0 & \hat{\sigma}_i \\ -\hat{\sigma}_i & 0 \end{pmatrix}\begin{pmatrix} 1\!\!1 & -1\!\!1 \\ 1\!\!1 & 1\!\!1 \end{pmatrix} \\ &= \frac{1}{2}\begin{pmatrix} 1\!\!1 & 1\!\!1 \\ -1\!\!1 & 1\!\!1 \end{pmatrix}\begin{pmatrix} \hat{\sigma}_i & \hat{\sigma}_i \\ -\hat{\sigma}_i & \hat{\sigma}_i \end{pmatrix} = \begin{pmatrix} 0 & \hat{\sigma}_i \\ -\hat{\sigma}_i & 0 \end{pmatrix} = \gamma^i \quad , \end{aligned} \tag{7}$$

$$\begin{aligned} \hat{U}\gamma_S^0\hat{U}^{-1} &= \frac{1}{2}\begin{pmatrix} 1\!\!1 & 1\!\!1 \\ -1\!\!1 & 1\!\!1 \end{pmatrix}\begin{pmatrix} 1\!\!1 & 0 \\ 0 & -1\!\!1 \end{pmatrix}\begin{pmatrix} 1\!\!1 & -1\!\!1 \\ 1\!\!1 & 1\!\!1 \end{pmatrix} \\ &= \frac{1}{2}\begin{pmatrix} 1\!\!1 & 1\!\!1 \\ -1\!\!1 & 1\!\!1 \end{pmatrix}\begin{pmatrix} 1\!\!1 & -1\!\!1 \\ -1\!\!1 & -1\!\!1 \end{pmatrix} = \begin{pmatrix} 0 & -1\!\!1 \\ -1\!\!1 & 0 \end{pmatrix} = \gamma^0 \quad . \end{aligned} \tag{8}$$

Now we transform the charge-conjugation operator $\hat{C}$ into the new representation:

$$\hat{C} = \hat{U}\hat{C}_S\hat{U}^{-1} \quad . \tag{9}$$

Accordingly, from

$$\hat{C}_S = \mathrm{i}\gamma_S^2\gamma_S^0 = \mathrm{i}\begin{pmatrix} 0 & \hat{\sigma}_2 \\ -\hat{\sigma}_2 & 0 \end{pmatrix}\begin{pmatrix} 1\!\!1 & 0 \\ 0 & -1\!\!1 \end{pmatrix} = \mathrm{i}\begin{pmatrix} 0 & -\hat{\sigma}_2 \\ -\hat{\sigma}_2 & 0 \end{pmatrix} \tag{10}$$

we derive

$$\begin{aligned} \hat{C} &= \frac{\mathrm{i}}{2}\begin{pmatrix} 1\!\!1 & 1\!\!1 \\ -1\!\!1 & 1\!\!1 \end{pmatrix}\begin{pmatrix} 0 & -\hat{\sigma}_2 \\ -\hat{\sigma}_2 & 0 \end{pmatrix}\begin{pmatrix} 1\!\!1 & -1\!\!1 \\ 1\!\!1 & 1\!\!1 \end{pmatrix} \\ &= \frac{\mathrm{i}}{2}\begin{pmatrix} 1\!\!1 & 1\!\!1 \\ -1\!\!1 & 1\!\!1 \end{pmatrix}\begin{pmatrix} -\hat{\sigma}_2 & -\hat{\sigma}_2 \\ -\hat{\sigma}_2 & \hat{\sigma}_2 \end{pmatrix} = \mathrm{i}\begin{pmatrix} -\hat{\sigma}_2 & 0 \\ 0 & \hat{\sigma}_2 \end{pmatrix} \quad . \end{aligned} \tag{11}$$

Performing a parity transformation of the Dirac spinors we obtain

$$\psi p(x') = \mathrm{e}^{\mathrm{i}\varphi}\gamma_S^0\psi(-\boldsymbol{x}, t) \quad . \tag{12}$$

In the following considerations the phase factor $\mathrm{e}^{\mathrm{i}\varphi}$ does not play any role and will be omitted. Combining charge conjugation and parity transformation it clearly follows that

$$\begin{aligned} \psi_{CP}(\boldsymbol{x}, t) &= \hat{C}\gamma_S^0\gamma_S^0\psi^*(-\boldsymbol{x}, t) = \hat{C}\psi^*(-\boldsymbol{x}, t) \\ &= \mathrm{i}\begin{pmatrix} -\hat{\sigma}_2 & 0 \\ 0 & \hat{\sigma}_2 \end{pmatrix}\psi^*(-\boldsymbol{x}, t) \quad . \end{aligned} \tag{13}$$

Finally we check whether the spinor ψ_{CP} fulfills the Dirac equation. For this purpose we use the relation

Exercise 14.2.

$$\hat{\alpha}_i \begin{pmatrix} -\hat{\sigma}_2 & 0 \\ 0 & \hat{\sigma}_2 \end{pmatrix} = \begin{pmatrix} \hat{\sigma}_i & 0 \\ 0 & -\hat{\sigma}_i \end{pmatrix} \begin{pmatrix} -\hat{\sigma}_2 & 0 \\ 0 & \hat{\sigma}_2 \end{pmatrix}$$
$$= -\begin{pmatrix} -\hat{\sigma}_2 & 0 \\ 0 & \hat{\sigma}_2 \end{pmatrix} \begin{pmatrix} \hat{\sigma}_i^* & 0 \\ 0 & -\hat{\sigma}_i^* \end{pmatrix}$$
$$= -\begin{pmatrix} -\hat{\sigma}_2 & 0 \\ 0 & \hat{\sigma}_2 \end{pmatrix} \hat{\alpha}_i^* \quad . \tag{14}$$

With $\boldsymbol{x}' = -\boldsymbol{x}$ we derive

$$\left(\mathrm{i}\hbar\frac{\partial}{\partial t} + \mathrm{i}\hbar\hat{\boldsymbol{\alpha}} \cdot \boldsymbol{\nabla}_x\right) \psi_{CP}(\boldsymbol{x}, t)$$
$$= \left(\mathrm{i}\hbar\frac{\partial}{\partial t} + \mathrm{i}\hbar\hat{\boldsymbol{\alpha}} \cdot \boldsymbol{\nabla}_x\right) \mathrm{i} \begin{pmatrix} -\hat{\sigma}_2 & 0 \\ 0 & \hat{\sigma}_2 \end{pmatrix} \psi^*(-\boldsymbol{x}, t)$$
$$= \mathrm{i} \begin{pmatrix} -\hat{\sigma}_2 & 0 \\ 0 & \hat{\sigma}_2 \end{pmatrix} \left(\mathrm{i}\hbar\frac{\partial}{\partial t} - \mathrm{i}\hbar\hat{\boldsymbol{\alpha}}^* \cdot \boldsymbol{\nabla}_x\right) \psi^*(-\boldsymbol{x}, t)$$
$$= \mathrm{i} \begin{pmatrix} -\hat{\sigma}_2 & 0 \\ 0 & \hat{\sigma}_2 \end{pmatrix} \left(\mathrm{i}\hbar\frac{\partial}{\partial t} + \mathrm{i}\hbar\hat{\boldsymbol{\alpha}}^* \cdot \boldsymbol{\nabla}'_x\right) \psi^*(\boldsymbol{x}', t)$$
$$= -\mathrm{i} \begin{pmatrix} -\hat{\sigma}_2 & 0 \\ 0 & \hat{\sigma}_2 \end{pmatrix} \left[\left(\mathrm{i}\hbar\frac{\partial}{\partial t} + \mathrm{i}\hbar\hat{\boldsymbol{\alpha}} \cdot \boldsymbol{\nabla}'_x\right) \psi(\boldsymbol{x}', t)\right]^*$$
$$= 0 \quad . \tag{15}$$

In (15) the factor in brackets vanishes; thus the combination of charge conjugation and parity transformation is indeed a symmetry transformation.

EXERCISE

14.3 Solutions of the Weyl Equation with Good Angular Momentum

Problem. Derive once more the angular momentum representation of the solutions of the Weyl equation (14.41) and (14.50).

Solution. We start from the Dirac spinor in standard representation:

$$\phi_{\kappa,\mu} = \begin{pmatrix} g(r) & \chi_{\kappa,\mu} \\ \mathrm{i}f(r) & \chi_{-\kappa,\mu} \end{pmatrix} \quad . \tag{1}$$

The transformation to the new representation is achieved by the unitary matrix

$$\hat{U} = \frac{1}{\sqrt{2}} \begin{pmatrix} 1 & 1 \\ -1 & 1 \end{pmatrix} \quad . \tag{2}$$

Furthermore, the neutrino states are generated from $\Psi = \psi\,\mathrm{e}^{-\mathrm{i}Et/\hbar}$ by

$$\psi_{\mathrm{L}} = \tfrac{1}{2}(\mathbb{1} - \gamma_5)\,\psi = \tfrac{1}{2}(\mathbb{1} - \gamma_5)\,\hat{U}\,\psi_{\kappa,\mu} \quad ,$$
$$\psi_{\mathrm{R}} = \tfrac{1}{2}(\mathbb{1} + \gamma_5)\,\psi = \tfrac{1}{2}(\mathbb{1} + \gamma_5)\,\hat{U}\,\psi_{\kappa,\mu} \quad . \tag{3}$$

Exercise 14.3.

In view of

$$(\mathbb{1}-\gamma_5)\hat{U} = \frac{1}{\sqrt{2}}\begin{pmatrix}0 & 0\\ 0 & 2\end{pmatrix}\begin{pmatrix}1 & 1\\ -1 & 1\end{pmatrix} = \frac{1}{\sqrt{2}}\begin{pmatrix}0 & 0\\ -2 & 2\end{pmatrix} \tag{4}$$

one obtains

$$\begin{aligned}\psi_{\mathrm{L}} &= \frac{1}{\sqrt{2}}\begin{pmatrix}0 & 0\\ -1 & 1\end{pmatrix}\begin{pmatrix}g(r) & \chi_{\kappa,\mu}\\ \mathrm{i}f(r) & \chi_{-\kappa,\mu}\end{pmatrix}\\ &= \frac{1}{\sqrt{2}}\left(-g(r)\chi_{\kappa,\mu}+\mathrm{i}f(r)\chi_{-\kappa,\mu}\right) \quad ,\end{aligned} \tag{5}$$

and correspondingly it holds that

$$(\mathbb{1}+\gamma_5)\,\hat{U} = \frac{1}{\sqrt{2}}\begin{pmatrix}2 & 0\\ 0 & 0\end{pmatrix}\begin{pmatrix}1 & 1\\ -1 & 1\end{pmatrix} = \frac{1}{\sqrt{2}}\begin{pmatrix}2 & 2\\ 0 & 0\end{pmatrix} \tag{6}$$

and thus

$$\psi_{\mathrm{R}} = \frac{1}{\sqrt{2}}\begin{pmatrix}1 & 1\\ 0 & 0\end{pmatrix}\begin{pmatrix}g(r) & \chi_{\kappa,\mu}\\ \mathrm{i}f(r) & \chi_{-\kappa,\mu}\end{pmatrix} = \frac{1}{\sqrt{2}}\left(g(r)\chi_{\kappa,\mu}+\mathrm{i}f(r)\chi_{-\kappa,\mu}\right) \quad . \tag{7}$$

Since the factors $\sqrt{2}$ and global signs can be absorbed within normalization or phase factors, they can be omitted.

15. Wave Equations for Particles with Arbitrary Spins

15.1 Particles with Finite Mass

Here we want to outline briefly how to construct wave functions which describe particles with spin $s = 1, \frac{3}{2}, \dots$ out of solutions of the Dirac equation and also to study by what kind of wave equation they are generated. As already seen in Chap. 6, the lower components of free solutions of the Dirac equation with positive energy vanish in the case $m_0 \neq 0$ in the rest system of the particles [cf. (6.13)]. Thus, for $E_p = m_0c^2$ (which means $p_i = 0$ when we are in the rest system) the spinor components are given by $\omega_\alpha^{(r)}(0) = \delta_{r\alpha}$ and thus

$$\omega_\alpha^{(+)} = 0 \quad , \quad \alpha = 3, 4 \quad .$$

The superscript $(+)$ denotes $r = 1, 2$, which characterizes solutions of positive energy. The tensor product

$$\tilde{\omega}^{(+)}_{\alpha\beta\dots\tau} =: \underbrace{\omega_\alpha^{(+)}(0)\omega_\beta^{(+)}(0)\dots\omega_\tau^{(+)}(0)}_{2s} \tag{15.1}$$

has $2s$ indices which are, at first, independent of each other. But considering the *total symmetric part of this multispinor*

$$\omega^{(+)}_{\alpha\beta\dots\tau} =: \sum_p \tilde{\omega}^{(+)}_{\alpha\beta\dots\tau}(0) = \tilde{\omega}^{(+)}_{\{\alpha\beta\dots r\}}(0) \tag{15.2}$$

(the summation index p denotes permutations of the indices), we find the following linearly independent combinations in the rest system:

$$\begin{aligned}
\omega^{(+)}_{\alpha\beta\dots\tau}(0, i = 0) &= \delta_{\alpha 1}\delta_{\beta 1}\dots\delta_{\nu 1}\delta_{\tau 1} \quad , \\
\omega^{(+)}_{\alpha\beta\dots\tau}(0, i = 1) &= \delta_{\alpha 2}\delta_{\beta 1}\delta_{\gamma 1}\dots\delta_{\nu 1}\delta_{\tau 1} \\
&\quad + \delta_{\alpha 1}\delta_{\beta 2}\delta_{\gamma 1}\dots\delta_{\nu 1}\delta_{\tau 1} \\
&\quad + \delta_{\alpha 1}\delta_{\beta 1}\delta_{\gamma 2}\dots\delta_{\nu 1}\delta_{\tau 1} + \dots \quad , \\
&\vdots \\
\omega^{(+)}_{\alpha\beta\dots\tau}(0, i = 2s) &= \delta_{\alpha 2}\delta_{\beta 2}\dots\delta_{\nu 2}\delta_{\tau 2} \quad .
\end{aligned} \tag{15.3}$$

The ith multispinor has i indices equal to 2 and $2s - i$ indices equal to 1. Each of these multispinors represents an eigenvector of the operator of total spin, with in the rest system is defined by

$$\begin{aligned}\tfrac{1}{2}\hbar\hat{\Sigma}^3_{\alpha\alpha'\beta\beta'\ldots\nu\nu'\tau\tau'} =: &\ \tfrac{1}{2}\hbar\hat{\Sigma}^3_{\alpha\alpha'}\delta_{\beta\beta'}\ldots\delta_{\tau\tau'}\\ &\ \vdots\\ &+\tfrac{1}{2}\hbar\hat{\Sigma}^3_{\tau\tau'}\delta_{\alpha\alpha'}\delta_{\beta\beta'}\ldots\delta_{\nu\nu'}\\ &(\alpha,\alpha';\beta,\beta';\ldots\nu,\nu';\tau,\tau'=1,\ldots,4)\quad ,\end{aligned} \tag{15.4}$$

where

$$\hat{\Sigma}^3_{\alpha\alpha'} = \begin{pmatrix}\hat{\sigma}_3 & 0\\ 0 & \hat{\sigma}_3\end{pmatrix} = \begin{pmatrix}1 & 0 & 0 & 0\\ 0 & -1 & 0 & 0\\ 0 & 0 & 1 & 0\\ 0 & 0 & 0 & -1\end{pmatrix}$$

is the well-known four-dimensional Pauli matrix. Indeed, it is easily proven that

$$\tfrac{1}{2}\hbar\hat{\Sigma}^3\omega^{(+)}(0,i) = \hbar(s-i)\omega^{(+)}(0,i) \quad . \tag{15.5}$$

This is demonstrated in detail in Exercise 15.1. Since, apparently, $i = 0,\ldots,2s$ is valid, the number of eigenvectors is just $2s+1$, and according to (15.5) the eigenvalues of $(\hbar/2)\hat{\Sigma}^3$ are $s, s-1, \ldots, -s+1, -s$. This directly demonstrates that the symmetric multispinor (15.2) may indeed be interpreted as the wave function of a particle with spin s, where the z component can obviously assume $2s+1$ different values. An analogous consideration allows the construction of solutions of negative energy. In this case the upper instead of the lower components have to vanish in the rest system. This means that we only have to replace the indices 1, 2 in the Kronecker deltas in (15.3) by 3, 4 and the superscript (+) by (−), and this will be verified in Exercise 15.3.

EXERCISE

15.1 Eigenvalue Equation for Multispinors

Problem. Verify that the multispinors $\omega^{(+)}(0,i)$ fulfill the eigenvalue equation

$$\tfrac{1}{2}\hbar\hat{\Sigma}^3\omega^{(+)}(0,i) = \hbar(s-i)\omega^{(+)}(0,i) \qquad i = 0,1,\ldots,2s \quad .$$

Solution. Using

$$\left(\hat{\Sigma}_3\right)_{\alpha\alpha'} = \begin{pmatrix}1 & 0 & 0 & 0\\ 0 & -1 & 0 & 0\\ 0 & 0 & 1 & 0\\ 0 & 0 & 0 & -1\end{pmatrix} \quad ,$$

$$\left(\hat{\Sigma}_3\right)_{\beta\beta'} = \begin{pmatrix}1 & 0 & 0 & 0\\ 0 & -1 & 0 & 0\\ 0 & 0 & 1 & 0\\ 0 & 0 & 0 & -1\end{pmatrix} \quad , \quad \text{etc.}$$

and (15.4) we get

Exercise 15.1.

$$
\begin{aligned}
&\left(\frac{1}{2}\hbar\hat{\Sigma}^3\omega^{(+)}(0,i=0)\right)_{\alpha\beta\ldots\tau}\\
&\quad=\frac{\hbar}{2}\hat{\Sigma}^3_{\alpha\alpha'\beta\beta'\ldots\nu\nu'\tau\tau'}\omega^{(+)}_{\alpha'\beta'\ldots\nu'\tau'}(0,i=0)\\
&\quad=\frac{\hbar}{2}\hat{\Sigma}^3_{\alpha\alpha'}\delta_{\alpha'1}\delta_{\beta1}\ldots\delta_{\nu1}\delta_{\tau1}\\
&\qquad+\delta_{\alpha1}\frac{\hbar}{2}\hat{\Sigma}^3_{\beta\beta'}\delta_{\beta'1}\delta_{\gamma1}\ldots\delta_{\nu1}\delta_{\tau1}+\ldots+\delta_{\alpha1}\delta_{\beta1}\ldots\delta_{\nu1}\frac{\hbar}{2}\hat{\Sigma}^3_{\tau\tau'}\delta_{\tau'1}\\
&\quad=\frac{\hbar}{2}2s\left(\delta_{\alpha1}\delta_{\beta1}\ldots\delta_{\tau1}\right)=\hbar s\omega^{(+)}_{\alpha\beta\ldots\tau}(0,i=0)
\end{aligned}\tag{1}
$$

and analogously

$$
\begin{aligned}
&\left(\frac{1}{2}\hbar\hat{\Sigma}^3\omega^{(+)}(0,i=1)\right)_{\alpha\beta\ldots\tau}\\
&\quad=\frac{\hbar}{2}\hat{\Sigma}^3_{\alpha\alpha'}\delta_{\alpha'2}\delta_{\beta1}\ldots\delta_{\tau1}+\delta_{\alpha2}\frac{\hbar}{2}\hat{\Sigma}^3_{\beta\beta'}\delta_{\beta'1}\ldots\delta_{\tau1}\\
&\qquad+\ldots+\delta_{\alpha2}\delta_{\beta1}\ldots\frac{\hbar}{2}\hat{\Sigma}^3_{\tau\tau'}\delta_{\tau'1}+\frac{\hbar}{2}\hat{\Sigma}^3_{\alpha\alpha'}\delta_{\alpha'1}\delta_{\beta2}\ldots\delta_{\tau1}\\
&\qquad+\delta_{\alpha1}\frac{\hbar}{2}\hat{\Sigma}^3_{\beta\beta'}\delta_{\beta'2}\ldots\delta_{\tau1}+\ldots+\delta_{\alpha1}\delta_{\beta2}\ldots\frac{\hbar}{2}\hat{\Sigma}^3_{\tau\tau'}\delta_{\tau'1}+\ldots\\
&\quad=\frac{\hbar}{2}\Bigg[-\delta_{\alpha2}\delta_{\beta1}\ldots\delta_{\tau1}+\overbrace{\delta_{\alpha2}\delta_{\beta1}\ldots\delta_{\tau1}+\ldots+\delta_{\alpha2}\delta_{\beta1}\ldots\delta_{\tau1}}^{2s-1}\\
&\qquad+\delta_{\alpha1}\delta_{\beta2}\ldots\delta_{\tau1}-\delta_{\alpha1}\delta_{\beta2}\ldots\delta_{\tau1}+\ldots+\delta_{\alpha1}\delta_{\beta2}\ldots\delta_{\tau1}+\ldots\Bigg]\\
&\quad=\hbar(s-1)\left[\delta_{\alpha2}\delta_{\beta1}\ldots\delta_{\tau1}+\delta_{\alpha1}\delta_{\beta2}\ldots\delta_{\tau1}+\ldots\right]\\
&\quad=\hbar(s-1)\omega^{(+)}_{\alpha\beta\ldots\tau}(0,i=1)\quad,
\end{aligned}\tag{2}
$$

$$\vdots$$

$$
\begin{aligned}
&\left(\frac{\hbar}{2}\hat{\Sigma}^3\omega^{(+)}(0,i=2s)\right)_{\alpha\beta\ldots\tau}\\
&\quad=\frac{\hbar}{2}\hat{\Sigma}^3_{\alpha\alpha'}\delta_{\alpha'2}\delta_{\beta2}\ldots\delta_{\tau2}+\delta_{\alpha2}\frac{\hbar}{2}\hat{\Sigma}^3_{\beta\beta'}\delta_{\beta'2}\ldots\delta_{\tau2}\\
&\qquad+\ldots+\delta_{\alpha2}\delta_{\beta2}\ldots\frac{\hbar}{2}\hat{\Sigma}^3_{\tau\tau'}\delta_{\tau'2}\\
&\quad=-\hbar s\omega^{(+)}_{\alpha\beta\ldots\tau}(0,i=2s)\quad.
\end{aligned}\tag{3}
$$

Hence the action of $(\hat{\Sigma}^3)_{\alpha\alpha'\ldots\tau\tau'}$ on the multispinors $\omega^{(+)}(0,i)$ is clarified, and the validity of the eigenvalue equation (15.5) is proven.

EXERCISE

15.2 Multispinor $\omega^{(+)}$ as Eigenvector of $\hat{\Sigma}^2$

Problem. Show by using the example of the multispinor $\omega^{(+)}(0, i=0)$ that it is an eigenvector of $(\hbar^2/4)\hat{\boldsymbol{\Sigma}}^2_{\alpha\alpha'\beta\beta'\dots\tau\tau'}$ with the eigenvalue $\hbar^2 s(s+1)$.

Solution. The vector operator $\hat{\Sigma}$ is defined in a similar manner to $\hat{\Sigma}^3$ in (15.4); then, we have

$$\begin{aligned}\left(\tfrac{1}{4}\hat{\boldsymbol{\Sigma}}^2\right)_{\alpha\alpha'\beta\beta'\dots\tau\tau'} &= \tfrac{1}{4}\hat{\boldsymbol{\Sigma}}^2_{\alpha\alpha'} + \tfrac{1}{4}\hat{\boldsymbol{\Sigma}}_{\alpha\alpha'}\cdot\hat{\boldsymbol{\Sigma}}_{\beta\beta'} + \dots + \tfrac{1}{4}\hat{\boldsymbol{\Sigma}}_{\alpha\alpha'}\cdot\hat{\boldsymbol{\Sigma}}_{\tau\tau'}\\ &\quad + \tfrac{1}{4}\hat{\boldsymbol{\Sigma}}_{\beta\beta'}\cdot\hat{\boldsymbol{\Sigma}}_{\alpha\alpha'} + \tfrac{1}{4}\hat{\boldsymbol{\Sigma}}^2_{\beta\beta'} + \dots + \tfrac{1}{4}\hat{\boldsymbol{\Sigma}}_{\beta\beta'}\cdot\hat{\boldsymbol{\Sigma}}_{\tau\tau'}\\ &\quad\;\vdots\\ &\quad + \tfrac{1}{4}\hat{\boldsymbol{\Sigma}}_{\tau\tau'}\cdot\hat{\boldsymbol{\Sigma}}_{\alpha\alpha'} + \tfrac{1}{4}\hat{\boldsymbol{\Sigma}}_{\tau\tau'}\cdot\hat{\boldsymbol{\Sigma}}_{\beta\beta'} + \dots + \tfrac{1}{4}\hat{\boldsymbol{\Sigma}}^2_{\tau\tau'} \quad . \qquad (1)\end{aligned}$$

Here, the notation has been simplified by omitting the delta functions with respect to the not explicitly listed indices in each term of the sum [cf. (15.4)]. On the rhs of (1) we see $2s$ rows with $2s$ factors in each individual term. The quadratic terms can be easily calculated, e.g.

$$\hat{\boldsymbol{\Sigma}}^2_{\alpha\alpha'} = (\hat{\Sigma}^1)^2_{\alpha\alpha'} + (\hat{\Sigma}^2)^2_{\alpha\alpha'} + (\hat{\Sigma}^3)^2_{\alpha\alpha'} = 3\delta_{\alpha\alpha'} \quad . \qquad (2)$$

A similar expression is valid for every pair of indices $\beta\beta', \gamma\gamma', \dots, \tau\tau'$. Since each row of (1) contains just one of these quadratic terms, the first contribution to the expectation value of (1) is given by $(3/4)2s = (3/2)s$. Just $(2s-1)$ mixed terms of the form

$$\tfrac{1}{4}\hat{\boldsymbol{\Sigma}}\cdot\hat{\boldsymbol{\Sigma}}' \quad , \qquad (3)$$

remain in each of the $2s$ rows, where $\hat{\boldsymbol{\Sigma}}$ and $\hat{\boldsymbol{\Sigma}}'$ act onto different indices because of the total symmetry of $\omega^{(+)}_{\alpha\beta\dots\tau}(0,i)$ in $\alpha, \beta, \dots, \tau$. By considering the example $\omega^{(+)}(0, i=0)$ it is most easily demonstrated that

$$\tfrac{1}{4}\hat{\boldsymbol{\Sigma}}\cdot\hat{\boldsymbol{\Sigma}}'\omega^{(+)}(0, i=0) = \tfrac{1}{4}\omega^{(+)}(0, i=0) \quad . \qquad (4)$$

Inserting $\omega^{(+)}(0, i=0)$ from the first equation of (15.3) and remembering that

$$\hat{\boldsymbol{\Sigma}} = \begin{pmatrix}\hat{\boldsymbol{\sigma}} & 0\\ 0 & \hat{\boldsymbol{\sigma}}\end{pmatrix} \quad ,$$

it follows in the representation of the Pauli matrices [see (1.65)] that

$$\begin{aligned}&\left(\tfrac{1}{4}\hat{\boldsymbol{\Sigma}}_{\alpha\alpha'}\cdot\hat{\boldsymbol{\Sigma}}_{\beta\beta'}\right)\delta_{\alpha'1}\delta_{\beta'1}\dots\\ &\quad = \tfrac{1}{4}\left(\hat{\Sigma}^1_{\alpha\alpha'}\hat{\Sigma}^1_{\beta\beta'} + \hat{\Sigma}^2_{\alpha\alpha'}\hat{\Sigma}^2_{\beta\beta'} + \hat{\Sigma}^3_{\alpha\alpha'}\hat{\Sigma}^3_{\beta\beta'}\right)\delta_{\alpha'1}\delta_{\beta'1}\dots\\ &\quad = \tfrac{1}{4}\left(\hat{\Sigma}^1_{21}\hat{\Sigma}^1_{21}\delta_{\alpha2}\delta_{\beta2} + \hat{\Sigma}^2_{21}\hat{\Sigma}^2_{21}\delta_{\alpha2}\delta_{\beta2} + \hat{\Sigma}^3_{11}\hat{\Sigma}^3_{11}\delta_{\alpha1}\delta_{\beta1}\right)\dots\\ &\quad = \tfrac{1}{4}\left((+1)\delta_{\alpha2}\delta_{\beta2} + (-1)\delta_{\alpha2}\delta_{\beta2} + (+1)\delta_{\alpha1}\delta_{\beta1}\right)\dots\\ &\quad = \tfrac{1}{4}\delta_{\alpha1}\delta_{\beta1}\dots \quad . \qquad (5)\end{aligned}$$

In (1), each term of the form (3) yields the contribution 1/4 to the eigenvalue, and altogether, we obtain

Exercise 15.2.

$$\begin{aligned}\left(\tfrac{1}{4}\hat{\boldsymbol{\Sigma}}^2\right)\omega^{(+)}(0,i=0) &= \left[\tfrac{3}{2}s+\tfrac{1}{4}(2s-1)2s\right]\omega^{(+)}(0,i=0)\\ &= s(s+1)\omega^{(+)}(0,i=0) \quad . \qquad (6)\end{aligned}$$

Multiplication by $\hbar^2$ yields the desired result. Similar, but more involved, is the recalculation that the remaining $\omega^{(+)}(0,i)$ of (15.3) satisfy the same eigenvalue equation. Together with (15.5), this proves that the $2s+1$ multispinors $\omega^{(+)}(0,i)$ of (15.3) describe particles with spin s.

EXERCISE

15.3 Multispinor of Negative Energy

Problem. Construct – analogously to (15.3) – the multispinor of negative energy in the rest system and, based on this, the eigenvalue equation analogous to (15.5).

Solution. It follows immediately from (6.3) that, analogously to (15.3),

$$\begin{aligned}\omega^{(-)}_{\alpha\beta\ldots\tau}(0,i=0) &= \delta_{\alpha 3}\delta_{\beta 3}\ldots\delta_{\tau 3} \quad ,\\ \omega^{(-)}_{\alpha\beta\ldots\tau}(0,i=0) &= \delta_{\alpha 4}\delta_{\beta 3}\ldots\delta_{\tau 3}+\delta_{\alpha 3}\delta_{\beta 4}\delta_{\gamma 3}\ldots\delta_{\tau 3}+\ldots+\delta_{\alpha 3}\delta_{\beta 3}\ldots\delta_{\tau 4} \quad ,\\ &\ \ \vdots\\ \omega^{(-)}_{\alpha\beta\ldots\tau}(0,i=2s) &= \delta_{\alpha 4}\delta_{\beta 4}\ldots\delta_{\tau 4} \quad . \qquad (1)\end{aligned}$$

Since

$$\hat{\Sigma}^3_{\alpha\alpha'} = \begin{pmatrix}1&0&0&0\\0&-1&0&0\\0&0&1&0\\0&0&0&-1\end{pmatrix} \quad ,$$

the action of $\hat{\Sigma}^3$ on $\omega^{(-)}_{\alpha\beta\ldots\tau}(0,i)$ is the same as the action of $\hat{\Sigma}^3$ on $\omega^{(+)}_{\alpha\beta\ldots\tau}(0,i)$. For $i=0$ this means that

$$\begin{aligned}\left(\tfrac{1}{2}\hbar\hat{\Sigma}^3\omega^{(-)}(0,i=0)\right)_{\alpha\beta\ldots\tau} &= \tfrac{1}{2}\hbar\left[\hat{\Sigma}^3_{\alpha\alpha'}\delta_{\alpha' 3}\delta_{\beta 3}\ldots\delta_{\tau 3}+\delta_{\sigma 3}\hat{\Sigma}^3_{\beta\beta'}\delta_{\beta' 3}\ldots\delta_{\tau 3}\right.\\ &\qquad \left.+\ldots+\delta_{\alpha 3}\delta_{\beta 3}\ldots\hat{\Sigma}^3_{\tau\tau'}\delta_{\tau' 3}\right]\\ &= \hbar s\,\delta_{\alpha 3}\delta_{\beta 3}\ldots\delta_{\tau 3} = \hbar s\,\omega^{(-)}_{\alpha\beta\ldots\tau}(0,i=0) \quad .\end{aligned}$$

The further calculation is analogous to that presented in Exercise 15.1.

Now we can transform these multispinors into an arbitrary frame of reference, bearing in mind the work of Chap. 6, particularly (6.30) and Exercise 6.1. By "boosting" all factors of (15.2) at the same time, i.e. applying the operator $\hat{S}(\boldsymbol{p})$

$$\hat{S}_{\alpha\alpha'\beta\beta'\ldots\tau\tau'}\left(\frac{\boldsymbol{p}}{E}\right) =: \hat{S}_{\alpha\alpha'}\left(\frac{\boldsymbol{p}}{E}\right)\hat{S}_{\beta\beta'}\left(\frac{\boldsymbol{p}}{E}\right)\ldots\hat{S}_{\tau\tau'}\left(\frac{\boldsymbol{p}}{E}\right) \quad , \qquad (15.6)$$

we get

$$\omega^{(+)}(\boldsymbol{p},i) = \hat{S}\left(\frac{\boldsymbol{p}}{E}\right)\omega^{(+)}(0,i) \quad , \qquad \omega^{(-)}(\boldsymbol{p},i) = \hat{S}\left(\frac{\boldsymbol{p}}{E}\right)\omega^{(-)}(0,i) \quad . \tag{15.7}$$

(In Exercise 15.4 these spinors $\omega^{(+)}(\boldsymbol{p},i)$ will be explicitly calculated.) Now every wave function may be written as a superposition of plane waves:

$$\begin{aligned} \Psi^{(+)}_{\alpha\ldots\tau}(x;p,i) &= \omega^{(+)}_{\alpha\ldots\tau}(\boldsymbol{p},i)\,\mathrm{e}^{-\mathrm{i}p\cdot x/\hbar} \quad , \\ \Psi^{(-)}_{\alpha\ldots\tau}(x;p,i) &= \omega^{(-)}_{\alpha\ldots\tau}(\boldsymbol{p},i)\,\mathrm{e}^{+\mathrm{i}p\cdot x/\hbar} \end{aligned} \tag{15.8}$$

and thus

$$\begin{aligned} \Psi_{\alpha\beta\ldots\tau}(x) = {} & \sum_i \int c^{(+)}(\boldsymbol{p},i)\Psi^{(+)}_{\alpha\beta\ldots\tau}(x;p,i)\,\mathrm{d}^3p \\ & + \sum_i \int c^{(-)}(\boldsymbol{p},i)\Psi^{(-)}_{\alpha\beta\ldots\tau}(x;p,i)\,\mathrm{d}^3p \quad . \end{aligned} \tag{15.9}$$

Hence these multispinors fulfill in coordinate space the following Dirac equation for each index, separately:

$$\begin{aligned} (\mathrm{i}\hbar\gamma\cdot\partial - m_0c)_{\alpha\alpha'}\,\Psi_{\alpha'\beta\ldots\tau}(x) &= 0 \quad , \\ &\;\vdots \\ (\mathrm{i}\hbar\gamma\cdot\partial - m_0c)_{\tau\tau'}\,\Psi_{\alpha\beta\ldots\tau'}(x) &= 0 \quad . \end{aligned} \tag{15.10}$$

They are named the "***Bargmann–Wigner*** equations", after their inventors.[1] Of course each component is, in accordance with the properties of solutions of the Dirac equation, also a solution of the Klein–Gordon equation

$$\left(\Box + \frac{m_0^2c^2}{\hbar^2}\right)\Psi_{\alpha\ldots\tau}(x) = 0 \quad . \tag{15.11}$$

The quantities $\Psi_{\alpha\beta\ldots\tau}(x)$ can obviously be regarded as the components of the wave function of a particle with mass m_0 and spin s, that is composed of elementary identical spin-$\frac{1}{2}$ fields, since there exist exactly $(2s+1)$ linearly independent components, each of which obeys (15.10) and (15.11). According to (15.5), each of these components is an eigenstate of $\hat{\Sigma}^3$.

EXERCISE

15.4 Construction of the Spinor $\omega^{(+)}_{\alpha\beta\ldots\tau}(\boldsymbol{p},i)$

Problem. Determine from (15.6) and the results of Exercise 6.1 the spinor $\omega^{(+)}_{\alpha\beta\ldots\tau}(\boldsymbol{p},i)$, using the operator

$$\hat{S}_{\alpha\alpha'\beta\beta'\ldots\tau\tau'}(\boldsymbol{p}) \quad ;$$

hence deduce the result

$$\begin{aligned} (\not{p} - m_0c)_{\alpha\alpha'}\omega^{(+)}_{\alpha'\beta\ldots\tau}(\boldsymbol{p},i) &= (\not{p} - m_0c)_{\beta\beta'}\omega^{(+)}_{\alpha\beta'\ldots\tau}(\boldsymbol{p},i) \\ &\;\vdots \\ &= (\not{p} - m_0c)_{\tau\tau'}\omega^{(+)}_{\alpha\beta\ldots\tau'}(\boldsymbol{p},i) = 0 \quad . \end{aligned} \tag{1}$$

[1] V. Bargmann, E. Wigner: Proc. Nat. Sci. (USA) **34**, 211 (1948).

Solution. In the standard representation (3.13) we obtain according to (6.32) the four-component Dirac spinors of momentum $\boldsymbol{p}$ by using the Lorentz transformation

Exercise 15.4.

$$\omega^{\tau}(\boldsymbol{p}) = \hat{S}\left(-\frac{\boldsymbol{p}}{E}\right)\omega^{\tau}(0) \qquad \tau = 1,2 \quad , \tag{2}$$

where, according to Exercise 6.1,

$$\hat{S}\left(-\frac{\boldsymbol{p}}{E}\right) = \sqrt{\frac{E+m_0c^2}{2m_0c^2}} \times \begin{pmatrix} 1 & 0 & \dfrac{p_z c}{E+m_0c^2} & \dfrac{p_- c}{E+m_0c^2} \\ 0 & 1 & \dfrac{p_+ c}{E+m_0c^2} & \dfrac{-p_z c}{E+m_0c^2} \\ \dfrac{p_z c}{E+m_0c^2} & \dfrac{p_- c}{E+m_0c^2} & 1 & 0 \\ \dfrac{p_+ c}{E+m_0c^2} & \dfrac{-p_z c}{E+m_0c^2} & 0 & 1 \end{pmatrix} \tag{3}$$

Hence the two solutions of positive energy with spin up and down are the first two columns of matrix (3). However, in contrast to (15.1) it holds that

$$\omega^{(+)}_{\alpha\beta\ldots\tau}(\boldsymbol{p},i) \neq 0 \quad \text{for} \quad \alpha,\beta\ldots,\tau = 3,4 \quad . \tag{4}$$

Using (15.6), (15.7) and (15.1) we obtain

$$\begin{aligned}
\omega^{(+)}_{\alpha\beta\ldots\tau}(\boldsymbol{p},i=0) &= \hat{S}\left(-\frac{\boldsymbol{p}}{E}\right)_{\alpha 1}\hat{S}\left(-\frac{\boldsymbol{p}}{E}\right)_{\beta 1}\ldots\hat{S}\left(-\frac{\boldsymbol{p}}{E}\right)_{\tau 1} \quad , \\
\omega^{(+)}_{\alpha\beta\ldots\tau}(\boldsymbol{p},i=1) &= \hat{S}\left(-\frac{\boldsymbol{p}}{E}\right)_{\alpha 2}\hat{S}\left(-\frac{\boldsymbol{p}}{E}\right)_{\beta 1}\ldots\hat{S}\left(-\frac{\boldsymbol{p}}{E}\right)_{\tau 1} \\
&\quad +\hat{S}\left(-\frac{\boldsymbol{p}}{E}\right)_{\alpha 1}\hat{S}\left(-\frac{\boldsymbol{p}}{E}\right)_{\beta 2}\ldots\hat{S}\left(-\frac{\boldsymbol{p}}{E}\right)_{\tau 1} \\
&\quad +\ldots \\
&\quad +\hat{S}\left(-\frac{\boldsymbol{p}}{E}\right)_{\alpha 1}\hat{S}\left(-\frac{\boldsymbol{p}}{E}\right)_{\beta 1}\ldots\hat{S}\left(-\frac{\boldsymbol{p}}{E}\right)_{\tau 2} \\
&\;\;\vdots \\
\omega^{(+)}_{\alpha\beta\ldots\tau}(\boldsymbol{p},i=2s) &= \hat{S}\left(-\frac{\boldsymbol{p}}{E}\right)_{\alpha 2}\ldots\hat{S}\left(-\frac{\boldsymbol{p}}{E}\right)_{\tau 2} \quad .
\end{aligned} \tag{5}$$

Each of the indices $\alpha,\beta,\ldots,\tau$ "scans" the columns of matrix (3), which are composed of the solutions $\omega^{\tau}(\boldsymbol{p})$ of (6.33a), i.e.

$$(\not{p} - m_0c)\omega^{\tau}(\boldsymbol{p}) = 0 \qquad \tau = 1,2 \quad . \tag{6}$$

Hence (1) holds.

EXERCISE

15.5 The Bargmann–Wigner Equations

Problem. Show that the spinors (15.9) obey the Dirac equation in the form (15.10).

Solution. We use the Dirac equation (6.33a) in the momentum representation for the four-component spinors $\omega^r(\boldsymbol{p})$, that is

$$(\not{p} - \varepsilon_r m_0 c)\omega^r(\boldsymbol{p}) = 0 \quad \text{for} \quad r = 1,2,3,4 \tag{1}$$

and the "plane waves" (6.31)

$$\psi_{\boldsymbol{p}}^r(x) = \omega^r(\boldsymbol{p})\,\mathrm{e}^{-\mathrm{i}\varepsilon_r p\cdot x/\hbar} \quad . \tag{2}$$

Now the Dirac equation in coordinate representation reads

$$\left(\mathrm{i}\hbar\gamma_\mu\partial^\mu - m_0c\right)\psi_{\boldsymbol{p}}^r(x) = \left(\varepsilon_r\gamma_\mu p^\mu - m_0c\right)\omega^r(\boldsymbol{p})\,\mathrm{e}^{-\mathrm{i}\varepsilon_r p\cdot x/\hbar} = 0 \quad . \tag{3}$$

Each linear combination of these solutions

$$\Psi(x) = \sum_r \int \frac{\mathrm{d}^3p}{(2\pi\hbar)^{3/2}}\sqrt{\frac{m_0c^2}{E_p}}\,c_r(\boldsymbol{p})\psi_{\boldsymbol{p}}^r(x) \tag{4}$$

with arbitrary functions $c_r(\boldsymbol{p})$ obeys the Dirac equation, too:

$$\begin{aligned}
&\left(\mathrm{i}\hbar\gamma_\mu\partial^\mu - m_0c\right)\Psi(x)\\
&\quad = \sum_r \int \frac{\mathrm{d}^3p}{(2\pi\hbar)^{3/2}}c_r(\boldsymbol{p})\left(\mathrm{i}\hbar\gamma_\mu\partial^\mu - m_0c\right)\psi_{\boldsymbol{p}}^r(x)\\
&\quad = \sum_r \int \frac{\mathrm{d}^3p}{(2\pi\hbar)^{3/2}}c_r(\boldsymbol{p})\left(\varepsilon_r\gamma_\mu p^\mu - m_0c\right)\omega^r(\boldsymbol{p})\,\mathrm{e}^{-\mathrm{i}\varepsilon_r p\cdot x/\hbar}\\
&\quad = 0 \quad .
\end{aligned} \tag{5}$$

For an arbitrary Bargmann–Wigner multispinor

$$\begin{aligned}
\Psi_{\alpha\beta\ldots\tau}(x) = \sum_i \int \frac{\mathrm{d}^3p}{(2\pi\hbar)^{3/2}} &\left[\omega^{(+)}_{\alpha\beta\ldots\tau}(\boldsymbol{p},i)\,\mathrm{e}^{-\mathrm{i}p\cdot x/\hbar}c^{(+)}(\boldsymbol{p},i)\right.\\
&\left. + \,\omega^{(-)}_{\alpha\beta\ldots\tau}(\boldsymbol{p},i)\,\mathrm{e}^{\mathrm{i}p\cdot x/\hbar}c^{(-)}(\boldsymbol{p},i)\right]
\end{aligned} \tag{6}$$

we obtain

$$\begin{aligned}
&\left(\mathrm{i}\hbar\gamma_\mu\partial^\mu - m_0c\right)_{\alpha\alpha'}\Psi_{\alpha'\beta\ldots\tau}(x)\\
&\quad = \sum_i \int \frac{\mathrm{d}^3p}{(2\pi\hbar)^{3/2}}\left[\left(\gamma_\mu p^\mu - m_0c\right)_{\alpha\alpha'}\omega^{(+)}_{\alpha'\beta\ldots\tau}(\boldsymbol{p},i)\,\mathrm{e}^{-\mathrm{i}p\cdot x/\hbar}c^{(+)}(\boldsymbol{p},i)\right.\\
&\qquad \left. - \left(\gamma_\mu p^\mu + m_0c\right)_{\alpha\alpha'}\omega^{(-)}_{\alpha'\beta\ldots\tau}(\boldsymbol{p},i)\,\mathrm{e}^{\mathrm{i}p\cdot x/\hbar}c^{(-)}(\boldsymbol{p},i)\right] \quad ,
\end{aligned} \tag{7}$$

etc. for all the other indices. However according to the construction of $\omega^{(+)}_{\alpha\beta\ldots\tau}(\boldsymbol{p},i)$ (cf. Exercises 15.2 and 15.3), it holds that

Exercise 15.5.

$$\begin{aligned}
&\left(\gamma_\mu p^\mu - m_0 c\right)_{\alpha\alpha'} \omega^{(+)}_{\alpha'\beta\ldots\tau}(\boldsymbol{p}, i) \\
&\quad = (\not{p} - m_0 c)_{\alpha\alpha'} \hat{S}_{\alpha'\alpha''\beta\beta'\ldots\tau\tau'} \left(\frac{\boldsymbol{p}}{E}\right) \omega^{(+)}_{\alpha''\beta'\ldots\tau'}(0, i) \\
&\quad = 0 \quad .
\end{aligned} \tag{8}$$

Taking into consideration the relation

$$\omega^r(\varepsilon_r \boldsymbol{p}) = \hat{S}\left(-\frac{\varepsilon_r \boldsymbol{p}}{E}\right)\omega^r(0) \tag{9}$$

[cf. (2) of Exercise 6.6] one gets, furthermore, that

$$\begin{aligned}
&\left(\gamma_\mu p^\mu + m_0 c\right)_{\alpha\alpha'} \omega^{(-)}_{\alpha'\beta\ldots\tau}(\boldsymbol{p}, i) \\
&\quad = (\not{p} - m_0 c)_{\alpha\alpha'} \hat{S}_{\alpha'\alpha''\beta\beta'\ldots\tau\tau'} \left(\frac{\boldsymbol{p}}{E}\right) \omega^{(-)}_{\alpha''\beta'\ldots\tau'}(0, i) \\
&\quad = 0 \quad .
\end{aligned} \tag{10}$$

Of course, we obtain analogous results for the other indices $\beta, \ldots, \tau$, and hence we obtain

$$\left(\mathrm{i}\hbar\gamma_\mu \partial^\mu - m_0 c\right)_{\alpha\alpha'} \Psi_{\alpha'\beta\gamma\ldots\tau}(x) = 0 \tag{11}$$

and, analogously, also

$$\begin{aligned}
&\left(\mathrm{i}\hbar\gamma_\mu \partial^\mu - m_0 c\right)_{\beta\beta'} \Psi_{\alpha\beta'\gamma\ldots\tau}(x) = 0 \quad , \\
&\qquad\vdots \\
&\left(\mathrm{i}\hbar\gamma_\mu \partial^\mu - m_0 c\right)_{\tau\tau'} \Psi_{\alpha\beta\gamma\ldots\tau'}(x) = 0 \quad .
\end{aligned} \tag{12}$$

These are the so-called Bargmann-Wigner equations.

15.2 Massless Particles

We have to modify the derivation of the Bargmann–Wigner equations for massless particles, because we cannot find a rest system for $m_0 = 0$. However, we can choose the z axis colinear to the direction of momentum:

$$p^\mu = \left(p^0, 0, 0, p\right) \quad . \tag{15.12}$$

The Bargmann–Wigner equations (i.e. for each index the Dirac equations) in momentum representation read

$$\begin{aligned}
&p\left(\pm\gamma^0 - \gamma^3\right)_{\alpha\alpha'} \omega^{(\pm)}_{\alpha'\beta\ldots\tau}(p) = 0 \quad , \\
&\qquad\vdots \\
&p\left(\pm\gamma^0 - \gamma^3\right)_{\tau\tau'} \omega^{(\pm)}_{\alpha\beta\ldots\tau'}(p) = 0 \quad ,
\end{aligned} \tag{15.13}$$

where $\pm$ again denotes the sign of the energy. Multiplying these equations by γ^0 yields

$$\left(\gamma^0\gamma^3\right)_{\alpha\alpha'} \omega^{(\pm)}_{\alpha'\beta\ldots\tau}(p) \equiv \hat{\alpha}^3_{\alpha\alpha'}\omega^{(\pm)}_{\alpha'\beta\ldots\tau}(p) = \pm\omega^{(\pm)}_{\alpha\beta\ldots\tau}(p) \quad ,$$
$$\vdots$$
$$\left(\gamma^0\gamma^3\right)_{\tau\tau'} \omega^{(\pm)}_{\alpha\beta\ldots\tau'}(p) \equiv \hat{\alpha}^3_{\tau\tau'}\omega^{(\pm)}_{\alpha\beta\ldots\tau'}(p) = \pm\omega^{(\pm)}_{\alpha\beta\ldots\tau}(p) \quad , \tag{15.14}$$

because $\gamma^0\gamma^0 = \mathbb{1}$. Neither in the standard representation (3.13) nor in the Majorana representation (cf. Exercise 5.2) of the Clifford algebra is the matrix $\gamma^0\gamma^3$ diagonal. But, using a unitary transformation, we can find a new basis so that

$$\gamma^0 = \begin{pmatrix} 0 & -\mathbb{1} \\ -\mathbb{1} & 0 \end{pmatrix} \quad , \quad \gamma^i \begin{pmatrix} 0 & \hat{\sigma}_i \\ -\hat{\sigma}_i & 0 \end{pmatrix} \quad , \quad \hat{\alpha}^i =: \gamma^0\gamma^i = \begin{pmatrix} \hat{\sigma}_i & 0 \\ 0 & -\hat{\sigma}_i \end{pmatrix} \quad . \tag{15.15}$$

This representation is called the *chiral or* ***Weyl*** *representation*, because the chirality operator γ^5 is diagonal in this particular representation (which will be shown in the next exercise). The name "chirality" will be justified in the following text.

EXERCISE

15.6 γ Matrices in the Weyl Representation

Problem. Show that the γ matrices in the Weyl representation obey the common commutation relations.

Solution. In standard representation [cf. (3.13)]

$$\gamma_0^{\mathrm{S}} = \begin{pmatrix} \mathbb{1} & 0 \\ 0 & \mathbb{1} \end{pmatrix} \quad , \quad \gamma_5^{\mathrm{S}} = \mathrm{i}\gamma^0\gamma^1\gamma^2\gamma^3 = \begin{pmatrix} 0 & \mathbb{1} \\ \mathbb{1} & 0 \end{pmatrix} \quad , \tag{1}$$

where the upper index S stands for "standard representation", (5.9) yields

$$\gamma_\mu^{\mathrm{S}}\gamma_5^{\mathrm{S}} + \gamma_5^{\mathrm{S}}\gamma_\mu^{\mathrm{S}} = 0 \quad , \tag{2}$$

and (3.11) gives

$$\gamma_\mu^{\mathrm{S}}\gamma_0^{\mathrm{S}} + \gamma_0^{\mathrm{S}}\gamma_\mu^{\mathrm{S}} = 2g_{\mu 0}\mathbb{1} \quad . \tag{3}$$

In the Weyl representation the matrices γ_i remain unchanged, only γ_5 and γ_0 are exchanged:

$$\gamma_i^{\mathrm{W}} = \gamma_i^{\mathrm{S}} \quad , \quad \gamma_0^{\mathrm{W}} = -\gamma_5^{\mathrm{S}} \quad , \quad \gamma_5^{\mathrm{W}} = \gamma_0^{\mathrm{S}} \quad . \tag{4}$$

Because of (2) and (3) the anticommutation relations (3.11) are not changed by this transformation. The transformation is explicitly given by

$$\gamma_\mu^{\mathrm{W}} \mathrel{\hat{=}} S\gamma_\mu^{\mathrm{S}}\hat{S}^\dagger \quad , \tag{5}$$

with

$$\hat{S} = \frac{1}{\sqrt{2}}\begin{pmatrix} \mathbb{1} & -\mathbb{1} \\ \mathbb{1} & \mathbb{1} \end{pmatrix} \quad .$$

In the Weyl representation the operator on the lhs of the Bargmann–Wigner equations becomes diagonal so that the eigenfunctions may be determined conveniently. Before proceeding to do so, we want to emphasize a remarkable property of the zero-mass Dirac equation. Due to the absence of the mass term an additional symmetry arises: for any solution $\omega(\boldsymbol{p})$ there is always the spinor $\gamma^5\omega(\boldsymbol{p})$ simultaneously eigenfunction of the equation, since the matrix γ^5 commutes with the operator $\hat{\alpha}^3 = \gamma^0\gamma^3$ appearing in the equation of motion (15.14) and two commuting operators can simultaneously be diagonalized. Therefore the solutions of the Bargmann-Wigner equations may be classified by the quantum numbers of γ^5 (*chirality*): it is convenient to employ the chiral representation, where

$$\gamma_5 = \mathrm{i}\gamma_0\gamma_1\gamma_2\gamma_3 = \begin{pmatrix} \mathbb{1} & 0 \\ 0 & -\mathbb{1} \end{pmatrix} \quad . \tag{15.16}$$

Obviously the *chirality operator* is already in diagonal form (which explains the name of the representation). With respect to a chosen inertial system, we define the *chirality of the solution* to be positive if the eigenfunctions satisfy

$$\gamma_5\omega_+^{(\pm)}(p) = (+1)\omega_+^{(\pm)}(p) \quad , \tag{15.17}$$

where the subscript "+" designates positive chirality. With this notation, the solution $\omega_+^{\pm}(p)$, for example, refers to a state with positive energy [superscript (+)] and positive chirality (subscript +). Hence, the spinor $\omega_+^{(\pm)}(p)$ must have the form

$$\omega_+^{(\pm)}(p) = \begin{pmatrix} w_{+1}^{(\pm)}(p) \\ \omega_{+2}^{(\pm)}(p) \\ 0 \\ 0 \end{pmatrix} := \begin{pmatrix} u^{(\pm)}(p) \\ 0 \end{pmatrix} \quad , \tag{15.18}$$

and, on the other hand, solutions with negative chirality must satisfy

$$\gamma_5\omega_-^{(\pm)}(p) = (-1)\omega_-^{(\pm)}(p) \quad , \tag{15.19}$$

i.e. they will be of the form

$$\omega_-^{(\pm)}(p) = \begin{pmatrix} 0 \\ 0 \\ \omega_{-3}^{(\pm)}(p) \\ \omega_{-4}^{(\pm)}(p) \end{pmatrix} := \begin{pmatrix} 0 \\ v^{(\pm)}(p) \end{pmatrix} \quad . \tag{15.20}$$

The functions

$$u^{(\pm)} =: \begin{pmatrix} \omega_{+1}^{(\pm)} \\ \omega_{+2}^{(\pm)} \end{pmatrix} \quad , \quad v^{(\pm)} =: \begin{pmatrix} \omega_{-3}^{(\pm)} \\ \omega_{-4}^{(\pm)} \end{pmatrix} \tag{15.21}$$

just defined are now two-component spinors. Consequently, the Dirac equation (15.14) splits into two-component equations, namely

$$\hat{\sigma}^3 u^{(\pm)} = \pm u^{(\pm)} \quad , \quad \hat{\sigma}^3 v^{(\pm)} = \mp v^{(\pm)} \quad . \tag{15.22}$$

Since there exists only two eigenfunctions of $\hat{\sigma}_3$, we conclude that $v^{(\pm)}(\boldsymbol{p}) = u^{(\mp)}(\boldsymbol{p})$, an identification that implies significant consequences, as we shall see. However, the eigenfunctions of

$$\hat{\sigma}^3 = \begin{pmatrix} 1 & 0 \\ 0 & -1 \end{pmatrix}$$

correspond to the eigenvalues ± 1, i.e. for positive (negative) signs of the energy only the upper (lower) components of the two-spinor $u^{(\pm)}$ do not vanish. That means, in this case the spin orientation must be parallel (antiparallel) to the direction of momentum. In contrast, the spinor $v^{(\pm)}$ has exactly the opposite property: for positive (negative) energy only the lower (upper) component is non-vanishing and the spin orientation is antiparallel (parallel) to the momentum, which, by definition, is assumed to point along the z axis [see (15.12)]. In other words: zero-mass fermions (e.g. neutrinos) with definite chirality have the property that the helicity $\hat{\boldsymbol{\sigma}} \cdot \boldsymbol{p}/|\boldsymbol{p}|$ of the particle depends on the sign of its energy. In the case (15.17), zero-mass particles with positive energy have positive helicity, whereas zero-mass particles with negative energy carry negative helicity, and this partly explains the name chirality ("screw sense"). If we interpret the properties just explained in terms of the hole theory (cf. Chap. 12), where a wave function of negative energy and momentum $-\boldsymbol{p}$ corresponds to an antifermion with positive energy and momentum $+\boldsymbol{p}$, one observes that zero-mass fermions *and* anti-zero-mass fermions with positive chirality both also have positive helicity (they are "right-handed"). Similarly, fermions and antifermions with negative chirality both carry negative helicity (they are "left-handed"). With this insight, the term "chirality" now becomes evident: *chirality for zero-mass particles is equivalent to their helicity.*

In analogy to the wave functions for particles with finite mass, one may now proceed to construct a zero-mass, symmetric Bargmann–Wigner multispinor from the zero-mass Dirac solution in the chiral representation [cf. (15.1–15.3)]. However, since we employed a unitary transformation to pass to the chiral representation, now all the components with indices 1 and 4 in (15.3) contribute to wave functions with positive energy, which is also reflected by (15.22). Due to the chiral symmetry of the particular components, we can now search for eigenfunctions of the Bargmann–Wigner equations for each component that is simultaneously an eigenfunction of γ^5, i.e. with $\omega_{\alpha\beta\ldots\tau}(p)$ being a solution, $\gamma^5_{\alpha\alpha'}\omega_{\alpha'\beta\ldots\tau}(p)$, $\gamma^5_{\beta\beta'}\omega_{\alpha\beta'\ldots\tau}(p)$, etc. are solutions too. Taking into considerations the symmetry of the Bargmann–Wigner multispinors, it follows that the number of possible solutions reduces from $2s+1$ to only two, since for positive energy and positive chirality there are merely

$$\omega^{(+)}_{+\alpha\beta\ldots\tau}(p) \sim \delta_{\alpha 1}\delta_{\beta 1}\ldots\delta_{\tau 1} \quad . \tag{15.23a}$$

Similarly, for positive energy and negative chirality,

$$\omega^{(+)}_{-\alpha\beta\ldots\tau}(p) \sim \delta_{\alpha 4}\delta_{\beta 4}\ldots\delta_{\tau 4} \quad , \tag{15.23b}$$

and, as before, the subscripts "+", "−" designate the chirality, and the superscript (+) denotes the positive energy.

These are the two eigenfunctions of the operator for the total helicity that correspond to the two extreme eigenvalues $\pm s$. For negative energy solutions the conclusions are completely analogous. Hence, the wave functions may be transformed to any Lorentz frame and subsequently be re-expressed in terms of plane waves to construct arbitrary wave packets in configuration space as linear combinations which satisfy the Bargmann–Wigner equations in configurations space.

Unfortunately, the outlined procedure for constructing the Bargmann–Wigner fields for particles with arbitrary masses does not allow us to find a Lagrange formulation, which is of fundamental importance for the quantization of the theory. To infer a Lagrangian density, one needs to transform the general Bargmann–Wigner equations for each particular spin orientation separately in a skillful way. In the following we will illustrate this concept for the case of spin-1 fields.

15.3 Spin-1 Fields for Particles with Finite Mass: Proca Equations

In this case the Bargmann-Wigner field is labelled by two indices. The two Dirac equations for the symmetric matrix $\Psi_{\alpha\beta}(x)$ may be written as follows:

$$\left(\mathrm{i}\hbar\gamma_\mu\partial^\mu - m_0c\right)\Psi(x) = 0 \quad , \quad \Psi(x)\left(\mathrm{i}\hbar\gamma_\mu^{\mathrm{T}}\overleftarrow{\partial}^\mu - m_0c\right) = 0 \tag{15.24}$$

or, in detail

$$\left(\mathrm{i}\hbar\partial_\mu\gamma^\mu_{\alpha\alpha'} - m_0c\delta_{\alpha\alpha'}\right)\Psi_{\alpha'\beta}(x) = 0 \quad , \tag{15.25a}$$

$$\Psi_{\alpha\beta'}(x)\left(\mathrm{i}\hbar\gamma^\mu_{\beta\beta'}\overleftarrow{\partial}_\mu - m_0c\delta_{\beta\beta'}\right) = 0 \quad . \tag{15.25b}$$

Since the 4×4 spinor is symmetric, it may be expanded in terms of a complete set of symmetrical elements of the Clifford algebra standard representation, the latter consisting of ten symmetrical matrices

$$\gamma^\mu\hat{C} \quad , \quad \hat{\sigma}^{\mu\nu}\hat{C} \quad , \tag{15.26}$$

where $\hat{C} = \mathrm{i}\gamma^2\gamma^0$ is the charge conjugation matrix that was defined in (12.23) and satisfies (12.24). The remaining six matrices

$$\gamma^\mu\gamma_5\hat{C} \quad , \quad \mathrm{i}\gamma_5\hat{C} \quad , \quad \hat{C} \tag{15.27}$$

are antisymmetric; thus it holds on the one hand that

$$\begin{aligned}\left(\gamma^\mu\hat{C}\right)^{\mathrm{T}} &= \hat{C}^{\mathrm{T}}\gamma^{\mu\mathrm{T}} = -\hat{C}\left(\gamma^0\gamma^\mu\gamma^0\right)^* \\ &= -\mathrm{i}\gamma^2\gamma^{\mu*}\gamma^0 = \gamma^\mu\mathrm{i}\gamma^2\gamma^0 = \gamma^\mu\hat{C} \quad ,\end{aligned} \tag{15.28a}$$

and on the other hand, we have

$$\left(\mathrm{i}\gamma_5\hat{C}\right)^{\mathrm{T}} = \hat{C}^{\mathrm{T}}\mathrm{i}\gamma_5^{\mathrm{T}} = -\mathrm{i}\gamma^2\gamma^0(+\mathrm{i}\gamma_5) = -\mathrm{i}\gamma_5\hat{C} \quad . \tag{15.28b}$$

Therefore, we define

$$\Psi(x) = \frac{m_0c}{\hbar}\gamma_\mu\hat{C}\varphi^\mu(x) + \frac{1}{2}\hat{\sigma}^{\mu\nu}\hat{C}G_{\mu\nu}(x) \quad , \tag{15.29}$$

where the coefficients $\varphi^\mu(x)$ and $G^{\mu\nu}(x)$ are generally complex and transform under Lorentz transformations like a vector and an antisymmetrical tensor, respectively. The Bargmann–Wigner equations (15.24) now become

$$
\begin{aligned}
&(\mathrm{i}\hbar\gamma\cdot\partial - m_0c)\left(\frac{1}{\hbar}m_0c\gamma_\mu\varphi^\mu(x) + \frac{1}{2}\hat{\sigma}_{\mu\nu}G^{\mu\nu}(x)\right)\hat{C} = 0 \quad , \\
&\left(\frac{1}{\hbar}m_0c\gamma_\mu\varphi^\mu(x) + \frac{1}{2}\hat{\sigma}_{\mu\nu}G^{\mu\nu}(x)\right)\hat{C}\left(\mathrm{i}\hbar\gamma^{\mathrm{T}}\cdot\overleftarrow{\partial} - m_0c\right) = 0 \quad .
\end{aligned}
\tag{15.30a}
$$

With regard to all indices, we explicitly rewrite the first of the above equations, which, for example, reads

$$
\begin{aligned}
&(\mathrm{i}\hbar\gamma\cdot\partial - m_0c)_{\alpha\alpha'}\left(\frac{1}{\hbar}m_0c(\gamma_\mu\hat{C})_{\alpha'\beta}\varphi^\mu(x) + \frac{1}{2}(\hat{\sigma}_{\mu\nu}\hat{C})_{\alpha'\beta}G^{\mu\nu}(x)\right) = 0 \quad , \\
&(\mathrm{i}\hbar\gamma\cdot\partial - m_0c)_{\beta\beta'}\left(\frac{1}{h}m_0c(\gamma_\mu\hat{C})_{\alpha\beta'}\varphi^\mu(x) + \frac{1}{2}(\hat{\sigma}_{\mu\nu}\hat{C})_{\alpha\beta'}G^{\mu\nu}(x)\right) = 0 \quad .
\end{aligned}
\tag{15.30b}
$$

It now becomes obvious that the coefficients $\varphi^\mu(x)$ and $G^{\mu\nu}(x)$ do not take part in the matrix multiplication, and therefore they may equally be placed between the matrices γ^μ and $\hat{C}$ or between $\hat{\sigma}^{\mu\nu}$ and $\hat{C}$, as has been done in (15.30a,b).

Now, in the above representation of $\hat{C}$ it holds that $\hat{C}\gamma_\mu^{\mathrm{T}} = -\gamma_\mu\hat{C}$. Thus, we can factorize $\hat{C}$ out of the equations:

$$
\begin{aligned}
&\big(\mathrm{i}m_0c\partial_\alpha\varphi_\mu(x)\gamma^\alpha\gamma^\mu - m_0^2c^2\gamma_\mu\varphi^\mu(x) + \tfrac{1}{2}\mathrm{i}\hbar\gamma_\alpha\hat{\sigma}^{\mu\nu}\partial^\alpha G_{\mu\nu}(x) \\
&\quad -\tfrac{1}{2}m_0c\hat{\sigma}^{\mu\nu}G_{\mu\nu}(x)\big)\hat{C} = 0 \quad , \\
&\big(\mathrm{i}m_0c\partial_\mu\varphi_\mu(x)\gamma^\alpha\gamma^\mu + m_0^2c^2\gamma_\mu\varphi^\mu(x) + \tfrac{1}{2}\mathrm{i}\hbar\sigma^{\mu\nu}\gamma_\alpha\partial^\alpha G_{\mu\nu}(x) \\
&\quad +\tfrac{1}{2}m_0c\hat{\sigma}^{\mu\nu}G_{\mu\nu}(x)\big)\hat{C} = 0 \quad .
\end{aligned}
\tag{15.31}
$$

Using the relation (cf. Exercise 3.2)

$$
\left[\gamma^\alpha, \hat{\sigma}^{\mu\nu}\right] = 2\mathrm{i}\left(g^{\alpha\mu}\gamma^\nu - g^{\alpha\nu}\gamma^\mu\right) \quad ,
\tag{15.32}
$$

it follows for the difference of the two equations (15.31) that

$$
m_0c\left(\partial^\alpha\varphi^\mu - \partial^\mu\varphi^\alpha - G^{\alpha\mu}\right)\hat{\sigma}_{\alpha\mu}\hat{C} - 2\gamma_\mu\hat{C}\left(\hbar\partial_\alpha G^{\alpha\mu} + \frac{m_0^2c^2}{\hbar}\varphi^\mu\right) = 0 \quad .
\tag{15.33}
$$

The coefficients of the linearly independent matrices $\hat{C}\hat{\sigma}_{\alpha\beta}$ and $\gamma^\mu\hat{C}$ must vanish separately. Hence, for $m_0 \neq 0$, this implies that

$$
G^{\mu\nu} = \partial^\mu\varphi^\nu - \partial^\nu\varphi^\mu \quad ,
\tag{15.34}
$$

$$
\partial_\mu G^{\mu\nu} = -\frac{m_0^2c^2}{\hbar^2}\varphi^\nu \quad .
\tag{15.35}
$$

These are the so-called *Proca equations*.[2] Expressed in terms of the vector field φ^μ they have the form

$$
\Box\varphi^\mu - \partial^\mu\left(\partial_\nu\varphi^\nu\right) + \frac{m_0^2c^2}{\hbar^2}\partial_\mu\varphi^\mu = 0
\tag{15.36}
$$

from which the subsidiary condition

[2] A. Proca: Le Journal de Physique et le Radium **7**, 347 (1936).

$$\frac{m_0^2c^2}{\hbar^2}\partial_\mu\varphi^\mu = 0 \qquad (15.37)$$

follows directly. Thereby we then obtain

$$\left(\Box + \frac{m_0^2c^2}{\hbar^2}\right)\varphi^\mu = 0 \quad , \qquad (15.37a)$$

$$\partial_\mu\varphi^\mu = 0 \quad . \qquad (15.37b)$$

Just as in (15.34) and (15.35), these are again the Proca equations, but now solely obtained using the potentials φ^μ. The appertaining Lagrangian density is

$$\begin{aligned}\mathcal{L}_{\text{Proca}} = &\frac{1}{2}G^*_{\mu\nu}G^{\mu\nu} - \frac{1}{2}G^*_{\mu\nu}\left(\partial^\mu\varphi^\nu - \partial^\nu\varphi^\mu\right)\\ &- \frac{1}{2}\left(\partial_\mu\varphi^*_\mu - \partial_\nu\varphi^{*\nu}\right)G^{\mu\nu} + \frac{m_0^2c^2}{\hbar^2}\varphi^*_\mu\varphi^\mu \quad ,\end{aligned} \qquad (15.38)$$

and we shall further comment on this in Exercise 15.10.

Note that from (15.34) an identity follows in form of the homogeneous equation

$$\partial^\varrho G^{\mu\nu} + \partial^\mu G^{\nu\varrho} + \partial^\nu G^{\varrho\mu} = 0 \quad , \qquad (15.39)$$

which can be formulated by means of the *"dual" tensor*

$$\tilde{G}^{\mu\nu} =: \tfrac{1}{2}\varepsilon^{\mu\nu\varrho\sigma}G_{\varrho\sigma} \quad , \qquad (15.40)$$

also as

$$\partial_\mu\tilde{G}^{\mu\nu} = 0 \quad . \qquad (15.41)$$

15.4 Kemmer Equaton

The Proca equations can also be cast in another linearized form. For that purpose we introduce the ten-dimensional "spinor"

$$\chi = \begin{pmatrix}\chi_1\\ \vdots\\ \chi_{10}\end{pmatrix} \quad . \qquad (15.42)$$

Its components are connected with the Proca fields as follows:

$$\begin{aligned}
\chi_1 &= -\frac{\mathrm{i}}{\sqrt{m_0c}}G^{01} \quad , & \chi_2 &= -\frac{\mathrm{i}}{\sqrt{m_0c}}G^{02} \quad ,\\
\chi_3 &= -\frac{\mathrm{i}}{\sqrt{m_0c}}G^{03} \quad , & \chi_4 &= -\frac{1}{\sqrt{m_0c}}G^{23} \quad ,\\
\chi_5 &= \frac{1}{\sqrt{m_0c}}G^{13} \quad , & \chi_6 &= -\frac{1}{\sqrt{m_0c}}G^{12} \quad ,\\
\chi_7 &= -\frac{\sqrt{m_0c}}{\hbar}\varphi^1 \quad , & \chi_8 &= -\frac{\sqrt{m_0c}}{\hbar}\varphi^2 \quad ,\\
\chi_9 &= -\frac{\sqrt{m_0c}}{\hbar}\varphi^3 \quad , & \chi_{10} &= -\mathrm{i}\frac{\sqrt{m_0c}}{\hbar}\varphi^0 \quad .
\end{aligned} \qquad (15.43)$$

Then the equations of the Proca theory (15.34) and (15.35) can be written as

$$\left(\mathrm{i}\hbar\beta_\mu\partial^\mu - m_0c\right)\chi(x) = 0 \quad , \tag{15.44}$$

where the 10×10 matrices can be chosen as follows:

$$\beta^0 = \begin{pmatrix} \emptyset & \emptyset & -1\!\!1 & \bar{0}^\dagger \\ \emptyset & \emptyset & \emptyset & \bar{0}^\dagger \\ -1\!\!1 & \emptyset & \emptyset & \bar{0}^\dagger \\ \bar{0} & \bar{0} & \bar{0} & 0 \end{pmatrix} \quad , \quad \mathrm{i}\beta^k = \begin{pmatrix} \emptyset & \emptyset & \emptyset & \mathrm{i}K^{k\dagger} \\ \emptyset & \emptyset & S^k & \bar{0}^\dagger \\ \emptyset & -S^k & \emptyset & \bar{0}^\dagger \\ \mathrm{i}K^k & \bar{0} & \bar{0} & 0 \end{pmatrix} \quad , \tag{15.45}$$

with $k = 1, 2, 3$. The relation (15.44) represents a set of 10 differential equations, which are known as the ***Kemmer*** *equations*.[3] The elements of the 10×10 matrices β^μ are given by the matrices

$$\emptyset = \begin{pmatrix} 0 & 0 & 0 \\ 0 & 0 & 0 \\ 0 & 0 & 0 \end{pmatrix} \quad , \quad 1\!\!1 = \begin{pmatrix} 1 & 0 & 0 \\ 0 & 1 & 0 \\ 0 & 0 & 1 \end{pmatrix} \quad ,$$

$$S^1 = \mathrm{i}\begin{pmatrix} 0 & 0 & 0 \\ 0 & 0 & -1 \\ 0 & 1 & 0 \end{pmatrix} \quad , \quad S^2 = \mathrm{i}\begin{pmatrix} 0 & 0 & 1 \\ 0 & 0 & 0 \\ -1 & 0 & 0 \end{pmatrix} \quad , \quad S^3 = \mathrm{i}\begin{pmatrix} 0 & -1 & 0 \\ 1 & 0 & 0 \\ 0 & 0 & 0 \end{pmatrix} \quad ,$$

$$K^1 = (1\ 0\ 0) \quad , \quad K^2 = (0\ 1\ 0) \quad , \quad K^3 = (0\ 0\ 1) \quad ,$$

$$\bar{0} = (0\ 0\ 0) \quad . \tag{15.46}$$

We convince ourselves of the equivalence of (15.44) with (15.34) and (15.35) by showing that the ten equations do indeed coincide. For example, the first line of (15.44) yields

$$\mathrm{i}\sqrt{m_0c}\,\left(\partial^0\varphi^1 - \partial^1\varphi^0 - G^{01}\right) = 0 \quad ,$$

i.e. just the $\mu = 0$, $\nu = 1$ component of the Proca equation (15.34). As a second example we consider the tenth line of (15.44):

$$\mathrm{i}\hbar(\partial_1 G_{01} + \partial_2 G_{02} + \partial_3 G_{03}) + \frac{\mathrm{i}m_0c^2}{\hbar}\varphi^0 = 0 \quad .$$

Because of the antisymmetry of $G^{\mu\nu}$ it can be also written as

$$\partial_1 G^{10} + \partial_2 G^{20} + \partial_3 G^{30} + \frac{m_0^2c^2}{\hbar^2}\varphi^0 = 0 \quad ,$$

which is just the $\nu = 0$ component of the inhomogeneous Proca equation (15.35). The remaining components of (15.34) and (15.35) can be verified in an analogous manner. Now the above defined β matrices fulfill the "commutation relations"

$$\beta^\mu\beta^\lambda\beta^\nu + \beta^\nu\beta^\lambda\beta^\mu = g^{\mu\lambda}\beta^\nu + g^{\nu\lambda}\beta^\mu \tag{15.47}$$

(as can be shown easily by calculation, cf. Exercise 15.7), and these define the so-called *Kemmer algebra*. If one passes to another representation $\beta' = S\beta S^\dagger$

[3] N. Kemmer: Proc. Roy. Soc. **A 177**, 9 (1939). Sometimes the name Duffin-Kemmer equation or Duffin–Kemmer–Petiau equation is used in the literature. As is often the case in the history of physics, several researchers almost simultaneously studied the same topic, here the spin-1 fields.

of the β matrices by a unitary transformation S, then of course, formally neither the commutation relations (15.47) nor the equation of motion (15.44) for the field $\chi' = S\chi$ (called the *free Kemmer equation*) change. Obviously it is fully equivalent to the Proca equation, and this is also the case with regard to the coupling of spin-1 mesons to the electromagnetic field.[4] Therefore in the chosen standard representation (15.46) the reverse transformation of the Kemmer theory into the Proca form can be constructed. If we define with

$$\begin{aligned} U^{\mu} &:- \quad -(\beta_1)^2(\beta_2)^2(\beta_3)^2\left(\beta^{\mu}\beta^0 - g^{\mu 0}\right) \quad , \\ U^{\mu\nu} &:- \quad U^{\mu}\beta^{\nu} = -U^{\nu\mu} \end{aligned} \tag{15.48}$$

ten additional 10×10 matrices, and with

$$E = \begin{pmatrix} 0 \\ \vdots \\ 0 \\ 1 \end{pmatrix} \tag{15.49}$$

a 10-component spinor, we obtain

$$\varphi^{\nu}(x) = -\frac{\mathrm{i}\hbar}{\sqrt{m_0 c}} E^{\dagger} U^{\nu} \chi(x) \quad , \tag{15.50a}$$

$$G^{\mu\nu}(x) = \sqrt{m_0 c}\, E^{\dagger} U^{\mu\nu} \chi(x) \quad . \tag{15.50b}$$

If we pass over to another representation $\beta' = S\beta S^{\dagger}$, then

$$U'_{\nu} = S U_{\nu} S^{\dagger} \quad , \quad U'_{\mu\nu} = S U_{\mu\nu} S^{\dagger} \quad , \quad E' = SE \tag{15.51}$$

holds accordingly. In other words: the Kemmer equations cannot only be represented in the form (15.44), (15.46) but also in many other equivalent representations. (In the following we shall follow these ideas further.) Also in the Kemmer form the spin-1 fields posses a Lagrangian density of the form

$$\mathcal{L}_{\text{Kemmer}} = \tfrac{1}{2}\mathrm{i}\hbar\overline{\chi}\beta_{\mu}\overleftrightarrow{\partial}^{\mu}\chi - m_0 c\overline{\chi}\chi \quad , \tag{15.52}$$

where one defines with $\eta = 2\beta_0^2 - 1\!\!1$

$$\overline{\chi} \equiv \chi^{\dagger}\eta \tag{15.53a}$$

and

$$\overleftrightarrow{\partial}^{\mu} = \overrightarrow{\partial}^{\mu} - \overleftarrow{\partial}^{\mu} \quad . \tag{15.53b}$$

[4] This was shown by Max Riedel in his diploma thesis (Frankfurt University, 1979) who also discussed other spin-1 theories and their relations to the Proca/Kemmer theory. Tragically, he died soon after its completion.

15.5 The Maxwell Equations

If the mass of the Bargmann–Wigner particles vanishes, we have to take into consideration the chiral symmetry if we want to deduce a theory which is analogous to the Proca equation; therefore we have to decompose the fields into eigensolutions of the chirality operator γ_5 for each index (remember the explanations in Sect. 15.2). This can be realized formally by requiring that for the corresponding symmetric 4×4 matrix Ψ it holds that

$$\begin{aligned} &\mathrm{i}\hbar\gamma \cdot \partial\,(\alpha + \beta\gamma_5)\,\Psi(x) = 0 \quad , \\ &\Psi(x)\,(\alpha + \beta\gamma_5)\,\mathrm{i}\hbar\gamma^{\mathrm{T}} \cdot \overleftarrow{\partial} = 0 \quad , \end{aligned} \tag{15.54}$$

and that for arbitrary numbers α and β, if $\Psi(x)$ is additionally an eigenfunction to γ_5 with the eigenvalues $+1$, then $(\alpha + \beta\gamma_5)$ reduces to the number $(\alpha \pm \beta)$ and (15.54) reduces to the Bargmann–Wigner equations with vanishing mass (15.14). Now the ansatz for the solution, in contrast to (15.29), can only read

$$\Psi(x) = \tfrac{1}{2}\hat{\sigma}_{\mu\nu}\hat{C}\,G^{\mu\nu}(x) \quad , \tag{15.55}$$

since $\gamma^5\hat{\sigma}^{\mu\nu}\hat{C}$ is certainly symmetric, though not so, however, $\gamma^5\gamma^\mu\hat{C}$. From both equations for Ψ it then follows that

$$\begin{aligned} &\tfrac{1}{2}\mathrm{i}\hbar\gamma_\alpha\,(\alpha + \beta\gamma_5)\,\hat{\sigma}_{\mu\nu}\partial^\alpha G^{\mu\nu}(x)\hat{C} = 0 \quad , \\ &\tfrac{1}{2}\mathrm{i}\hbar\hat{\sigma}_{\mu\nu}\,(\alpha + \beta\gamma_5)\,\gamma_\alpha\partial^\alpha G^{\mu\nu}(x)\hat{C} = 0 \quad . \end{aligned} \tag{15.56}$$

Because of

$$\gamma_5\hat{\sigma}^{\mu\nu} = \tfrac{1}{2}\mathrm{i}\varepsilon^{\mu\nu\sigma\varrho}\hat{\sigma}_{\sigma\varrho} \tag{15.57}$$

(see Exercise 15.20) and (15.40), this can be written as

$$\begin{aligned} &\frac{1}{2}\mathrm{i}\hbar\gamma_\delta\hat{\sigma}^{\mu\nu}\left(\alpha\partial^\delta G_{\mu\nu} + \frac{\mathrm{i}}{2}\beta\partial^\delta\varepsilon^{\sigma\varrho}{}_{\mu\nu}G_{\sigma\varrho}\right)\hat{C} \\ &\quad = \frac{1}{2}\mathrm{i}\hbar\gamma_\delta\hat{\sigma}^{\mu\nu}\left(\alpha\partial^\delta G_{\mu\nu} + \mathrm{i}\beta\partial^\delta\tilde{G}_{\mu\nu}\right)\hat{C} = 0 \quad . \end{aligned} \tag{15.58a}$$

Analogously the transposed equation reads

$$\tfrac{1}{2}\mathrm{i}\hbar\hat{\sigma}^{\mu\nu}\gamma_\delta\left(\alpha\partial^\delta G_{\mu\nu} + \mathrm{i}\beta\partial^\delta\tilde{G}_{\mu\nu}\right)\hat{C} = 0 \quad , \tag{15.58b}$$

and the difference reduces to

$$-2\hbar\left(\alpha\partial_\mu G^{\mu\nu} + \mathrm{i}\beta\partial_\mu\tilde{G}^{\mu\nu}\right)\gamma_\nu\hat{C} = 0 \quad . \tag{15.59}$$

The coefficients of the linearly independent matrices $\gamma^\mu\hat{C}$ must all vanish, in this case, additionally, for arbitrary numbers α, β. If we set $\alpha = 1$, $\beta = 0$ or $\alpha = 0$, $\beta = 1$, eight Maxwell equations result:

$$\partial_\mu G^{\mu\nu} = 0 \quad , \tag{15.60a}$$

$$\partial_\mu\tilde{G}^{\mu\nu} = 0 \quad . \tag{15.60b}$$

The homogeneous Maxwell equation (15.60b), which according to (15.39) reads

$$\partial^{\varrho} G^{\mu\nu} + \partial^{\mu} G^{\nu\varrho} + \partial^{\nu} G^{\varrho\mu} = 0 \tag{15.61}$$

can be fulfilled identically with the ansatz

$$G^{\mu\nu} = \partial^{\mu}\varphi^{\nu} - \partial^{\nu}\varphi^{\mu} \quad ; \tag{15.62}$$

however, we find no way to fulfill the auxiliary condition (15.37b) as we did for the Proca field. The field φ^{μ} has been introduced here just as a supplementary quantity and has no physical meaning, this supplementary character being emphasized by the fact that its choice is not unique. The so called re-gauging

$$\varphi_{\mu}(x) \rightarrow \varphi'_{\mu}(x) = \varphi_{\mu}(x) - \partial_{\mu}\Lambda(x) \tag{15.63}$$

with an arbitrary scalar field $\Lambda(x)$ leaves the physical field $G^{\mu\nu}$ unchanged, as long as the very general condition

$$\left(\partial_{\mu}\partial_{\nu} - \partial_{\nu}\partial_{\mu}\right)\Lambda(x) = 0$$

holds. The variety of fields $\varphi^{\mu}(x)$ can be reduced if we define additional conditions like (15.37b), which select a *specific gauge*. Equation (15.37b), which plays an important role in electrodynamics, is called the *Lorentz gauge*. But using this gauge, the field $\varphi^{\mu}(x)$ is still not uniquely characterized, because we can define a new field φ'^{μ} such that

$$\partial_{\mu}\varphi^{\mu} = \partial_{\mu}\varphi'^{\mu} = 0 \tag{15.64}$$

for the case $\Box\hat{\Lambda} = 0$! A different choice of gauge would be to demand $\varphi^{0} = 0$ in addition to (15.37b) (*radiation gauge*), but obviously this gauge does not lead to a unique choice of the vector field φ^{μ}, either.

In the following exercise and examples we will further deepen our understanding concerning the various ideas presented here.

EXERCISE

15.7 Commutation Relation of Kemmer Matrices

Problem. Verify the commutation relation (15.47) for the Kemmer matrices (15.45).

Solution. We present the solution for three examples:

(a) All indices are equal ($\mu = \nu = \lambda = 0$).
(b) Two indices are equal ($\mu = \nu = 1$, $\lambda = 2$).
(c) All indices are different ($\mu = 0$, $\nu = 2$, $\lambda = 3$).

Exercise 15.7.

(a) ($\mu = \nu = \lambda = 0$): First we calculate

$$\beta^0\beta^0 = \left(\beta^0\right)^2 = \begin{pmatrix} \emptyset & \emptyset & -1\!\!1 & \bar{0}^\dagger \\ \emptyset & \emptyset & \emptyset & \bar{0}^\dagger \\ -1\!\!1 & \emptyset & \emptyset & \bar{0}^\dagger \\ \bar{0} & \bar{0} & \bar{0} & 0 \end{pmatrix} \begin{pmatrix} \emptyset & \emptyset & -1\!\!1 & \bar{0}^\dagger \\ \emptyset & \emptyset & \emptyset & \bar{0}^\dagger \\ -1\!\!1 & \emptyset & \emptyset & \bar{0}^\dagger \\ \bar{0} & \bar{0} & \bar{0} & 0 \end{pmatrix}$$

$$= \begin{pmatrix} 1\!\!1 & \emptyset & \emptyset & \bar{0}^\dagger \\ \emptyset & \emptyset & \emptyset & \bar{0}^\dagger \\ \emptyset & \emptyset & 1\!\!1 & \bar{0}^\dagger \\ \bar{0} & \bar{0} & \bar{0} & 0 \end{pmatrix} \tag{1}$$

and then

$$\beta^0\left(\beta^0\right)^2 = \begin{pmatrix} \emptyset & \emptyset & -1\!\!1 & \bar{0}^\dagger \\ \emptyset & \emptyset & \emptyset & \bar{0}^\dagger \\ -1\!\!1 & \emptyset & \emptyset & \bar{0}^\dagger \\ \bar{0} & \bar{0} & \bar{0} & 0 \end{pmatrix} \begin{pmatrix} 1\!\!1 & \emptyset & \emptyset & \bar{0}^\dagger \\ \emptyset & \emptyset & \emptyset & \bar{0}^\dagger \\ \emptyset & \emptyset & 1\!\!1 & \bar{0}^\dagger \\ \bar{0} & \bar{0} & \bar{0} & 0 \end{pmatrix}$$

$$= \begin{pmatrix} \emptyset & \emptyset & -1\!\!1 & \bar{0}^\dagger \\ \emptyset & \emptyset & \emptyset & \bar{0}^\dagger \\ -1\!\!1 & \emptyset & \emptyset & \bar{0}^\dagger \\ \bar{0} & \bar{0} & \bar{0} & 0 \end{pmatrix} = \beta^0 \quad . \tag{2}$$

Thus we get

$$\beta^0\beta^0\beta^0 + \beta^0\beta^0\beta^0 = \beta^0 + \beta^0 = g^{00}\beta^0 + g^{00}\beta^0 \quad . \tag{3}$$

(b) ($\mu = \nu = 1$, $\lambda = 2$): Once again we begin with the product

$$\beta^2\beta^1 = \begin{pmatrix} \emptyset & \emptyset & \emptyset & K^{2\dagger} \\ \emptyset & \emptyset & -\mathrm{i}S^2 & \bar{0}^\dagger \\ \emptyset & \mathrm{i}S^2 & \emptyset & \bar{0}^\dagger \\ K^2 & \bar{0} & \bar{0} & 0 \end{pmatrix} \begin{pmatrix} \emptyset & \emptyset & \emptyset & K^{1\dagger} \\ \emptyset & \emptyset & -\mathrm{i}S^1 & \bar{0}^\dagger \\ \emptyset & \mathrm{i}S^1 & \emptyset & \bar{0}^\dagger \\ K^1 & \bar{0} & \bar{0} & 0 \end{pmatrix}$$

$$= \begin{pmatrix} K^{2\dagger}\otimes K^1 & \emptyset & \emptyset & \bar{0}^\dagger \\ \emptyset & -\mathrm{i}S^2\mathrm{i}S^1 & \emptyset & \bar{0}^\dagger \\ \emptyset & \emptyset & -\mathrm{i}S^2\mathrm{i}S^1 & \bar{0}^\dagger \\ \bar{0} & \bar{0} & \bar{0} & K^2K^{1\dagger} \end{pmatrix} \tag{4}$$

Here products of the matrices S^i from (15.46) occur, especially

$$\mathrm{i}S^2\mathrm{i}S^1 = \begin{pmatrix} 0 & 0 & 1 \\ 0 & 0 & 0 \\ -1 & 0 & 0 \end{pmatrix} \begin{pmatrix} 0 & 0 & 0 \\ 0 & 0 & -1 \\ 0 & 1 & 0 \end{pmatrix} = \begin{pmatrix} 0 & 1 & 0 \\ 0 & 0 & 0 \\ 0 & 0 & 0 \end{pmatrix}$$

$$\equiv M^{(12)} \equiv K^{1\dagger}\otimes K^2 \quad , \tag{5}$$

where "$\otimes$" denotes the tensor product of the vectors $K^{1\dagger}$ and K^2. With the thus defined 3×3 matrices $M^{(ij)}$, we now have

$$\beta^1\left(\beta^2\beta^1\right) = \begin{pmatrix} \emptyset & \emptyset & \emptyset & K^{1\dagger} \\ \emptyset & \emptyset & -\mathrm{i}S^1 & \bar{0}^\dagger \\ \emptyset & \mathrm{i}S^1 & \emptyset & \bar{0}^\dagger \\ K^1 & \bar{0} & \bar{0} & 0 \end{pmatrix} \begin{pmatrix} M^{(21)} & \emptyset & \emptyset & \bar{0}^\dagger \\ \emptyset & -M^{(12)} & \emptyset & \bar{0}^\dagger \\ \emptyset & \emptyset & -M^{(12)} & \bar{0}^\dagger \\ \bar{0} & \bar{0} & \bar{0} & 0 \end{pmatrix}$$

$$= \begin{pmatrix} \emptyset & \emptyset & \emptyset & \bar{0}^\dagger \\ \emptyset & \emptyset & \mathrm{i}S^1M^{(12)} & \bar{0}^\dagger \\ \emptyset & -\mathrm{i}S^1M^{(12)} & \emptyset & \bar{0}^\dagger \\ K^1M^{(21)} & \bar{0} & \bar{0} & 0 \end{pmatrix} \quad , \tag{6}$$

Exercise 15.7.

and hence

$$K^1 M^{(21)} = (1\ 0\ 0)\begin{pmatrix} 0 & 0 & 0 \\ 1 & 0 & 0 \\ 0 & 0 & 0 \end{pmatrix} = \bar{0} \tag{7}$$

and

$$\mathrm{i}S^1 M^{(12)} = -\begin{pmatrix} 0 & 0 & 0 \\ 0 & 0 & -1 \\ 0 & 1 & 0 \end{pmatrix}\begin{pmatrix} 0 & 1 & 0 \\ 0 & 0 & 0 \\ 0 & 0 & 0 \end{pmatrix} = \emptyset \quad . \tag{8}$$

The rhs of (6) contains the 10×10 zero matrix $O_{10\times 10}$; hence the final results reads

$$\beta^1\beta^2\beta^1 + \beta^1\beta^2\beta^1 = O_{10\times 10} = g^{12}\beta^1 + g^{12}\beta^1 \quad . \tag{9}$$

(c) $\mu = 0$, $\nu = 2$, $\lambda = 3$: Evaluating the product of the β matrices

$$\beta^3\beta^2 = \begin{pmatrix} K^{3\dagger}\otimes K^2 & \emptyset & \emptyset & \bar{0}^\dagger \\ \emptyset & -\mathrm{i}S^3\mathrm{i}S^2 & \emptyset & \bar{0}^\dagger \\ \emptyset & \emptyset & -\mathrm{i}S^3\mathrm{i}S^2 & \bar{0}^\dagger \\ \bar{0} & \bar{0} & \bar{0} & 0 \end{pmatrix} \tag{10}$$

the products of the 3×3 matrices

$$\begin{aligned}\mathrm{i}S^3\mathrm{i}S^2 &= \begin{pmatrix} 0 & -1 & 0 \\ 1 & 0 & 0 \\ 0 & 0 & 0 \end{pmatrix}\begin{pmatrix} 0 & 0 & 1 \\ 0 & 0 & 0 \\ -1 & 0 & 0 \end{pmatrix} = \begin{pmatrix} 0 & 0 & 0 \\ 0 & 0 & 1 \\ 0 & 0 & 0 \end{pmatrix} \\ &\equiv M^{(23)} \equiv K^{2\dagger}\otimes K^3 \end{aligned} \tag{11}$$

occur. Because

$$\begin{aligned}\mathrm{i}S^2\mathrm{i}S^3 &= \begin{pmatrix} 0 & 0 & 1 \\ 0 & 0 & 0 \\ -1 & 0 & 0 \end{pmatrix}\begin{pmatrix} 0 & -1 & 0 \\ 1 & 0 & 0 \\ 0 & 0 & 0 \end{pmatrix} = \begin{pmatrix} 0 & 0 & 0 \\ 0 & 0 & 0 \\ 0 & 1 & 0 \end{pmatrix} \\ &\equiv M^{(32)} \equiv K^{3\dagger}\otimes K^2 = M^{(23)\dagger} \end{aligned} \tag{12}$$

holds, we obtain

$$\begin{aligned}&\beta^0\beta^3\beta^2 + \beta^2\beta^3\beta^0 \\ &= \begin{pmatrix} \emptyset & \emptyset & -1\!\!1 & \bar{0}^\dagger \\ \emptyset & \emptyset & \emptyset & \bar{0}^\dagger \\ -1\!\!1 & \emptyset & \emptyset & \bar{0}^\dagger \\ \bar{0} & \bar{0} & \bar{0} & 0 \end{pmatrix}\begin{pmatrix} M^{(32)} & \emptyset & \emptyset & \bar{0}^\dagger \\ \emptyset & -M^{(23)} & \emptyset & \bar{0}^\dagger \\ \emptyset & \emptyset & -M^{(23)} & \bar{0}^\dagger \\ \bar{0} & \bar{0} & \bar{0} & 0 \end{pmatrix} \\ &\quad + \begin{pmatrix} M^{(23)} & \emptyset & \emptyset & \bar{0}^\dagger \\ \emptyset & -M^{(32)} & \emptyset & \bar{0}^\dagger \\ \emptyset & \emptyset & -M^{(32)} & \bar{0}^\dagger \\ \bar{0} & \bar{0} & \bar{0} & 0 \end{pmatrix}\begin{pmatrix} \emptyset & \emptyset & -1\!\!1 & \bar{0}^\dagger \\ \emptyset & \emptyset & \emptyset & \bar{0}^\dagger \\ -1\!\!1 & \emptyset & \emptyset & \bar{0}^\dagger \\ \bar{0} & \bar{0} & \bar{0} & 0 \end{pmatrix} \\ &= \begin{pmatrix} \emptyset & \emptyset & M^{(23)} & \bar{0}^\dagger \\ \emptyset & \emptyset & \emptyset & \bar{0}^\dagger \\ -M^{(32)} & \emptyset & \emptyset & \bar{0}^\dagger \\ \bar{0} & \bar{0} & \bar{0} & 0 \end{pmatrix} + \begin{pmatrix} \emptyset & \emptyset & -M^{(23)} & \bar{0}^\dagger \\ \emptyset & \emptyset & \emptyset & \bar{0}^\dagger \\ M^{(32)} & \emptyset & \emptyset & \bar{0}^\dagger \\ \bar{0} & \bar{0} & \bar{0} & 0 \end{pmatrix} \\ &= O_{10\times 10} = g^{03}\beta^2 + g^{20}\beta^3 \quad . \end{aligned} \tag{13}$$

EXERCISE

15.8 Properties of the Kemmer Equation Under Lorentz Transformation

Problem. Discuss the properties of the Kemmer equation under Lorentz transformations.

Solution. From our discussion in Chap. 3, we know that there has to be an explicit prescription which allows for an observer A to recalculate a wave function $\chi(x)$ of the Kemmer field into a wave function $\chi'(x')$ which is the wave function observed by B, who is at rest in another inertial frame. Following the relativity principle, $\chi'(x')$ is the solution of an equation, which is of the form

$$\left(\mathrm{i}\hbar\beta'_\mu\partial'^\mu - m_0c\right)\chi'(x') = 0 \tag{1}$$

in the system of observer B, if $\chi(x)$ is a solution of the Kemmer equation (15.44). Because the Kemmer matrices β'_μ must also obey the commutation relations (15.47), from which they are defined up to a unitary transformation, we can write without loss of generality:

$$\beta'_\mu = \beta_\mu \quad . \tag{2}$$

Covariance of the Kemmer equation means that simultaneously (1) and (15.44) hold, where primed quantities refer to the inertial system of observer B, which is connected to the inertial system of A via the Lorentz transformation

$$x'_\mu = a_\mu{}^\nu x_\nu \quad , \quad \partial'_\mu = a_\mu{}^\nu \partial'_\nu \quad , \tag{3}$$

[see (3.1), (3.3)] where $a^{\mu\nu}$ is an orthogonal matrix [see (3.4)]. As an ansatz for connecting χ with χ' we write

$$\chi'(x') = \hat{S}(\hat{a})\chi(x) \quad , \tag{4}$$

and vice versa [see (3.27–28)]:

$$\chi(x) = \hat{S}^{-1}(\hat{a})\chi'(x') \quad . \tag{5}$$

Inserting (5) into (15.44) and after multiplication with $\hat{S}(\hat{a})$ one gets

$$\left(\mathrm{i}\hbar\hat{S}(\hat{a})\beta_\mu\hat{S}^{-1}(\hat{a})\partial^\mu - m_0c\right)\chi'(x') = 0 \quad , \tag{6}$$

and then, because of (3), for the primed system B it holds that

$$\left(\mathrm{i}\hbar\hat{S}(\hat{a})\beta_\mu\hat{S}^{-1}(\hat{a})a^\mu{}_\nu\partial'^\nu - m_0c\right)\chi'(x') = 0 \quad . \tag{7}$$

In order that (7) is identical with (1),

$$\hat{S}^{-1}(\hat{a})\beta_\mu\hat{S}(\hat{a}) = a_\mu{}^\nu\beta_\nu \tag{8}$$

must hold, and from this equation $\hat{S}(\hat{a})$ can be determined. To reach that goal we start, similary to the procedure described for the Dirac field in (3.36–3.68), with the infinitesimal proper Lorentz transformation

Exercise 15.8.

$$a_{\mu\nu} = g_{\mu\nu} + \Delta\omega_{\mu\nu} \quad , \tag{9}$$

where

$$\Delta\omega_{\mu\nu} \ll 1 \quad .$$

Because of the orthogonality (3.36a) of the Lorentz transformation,

$$\Delta\omega_{\mu\nu} = -\Delta\omega_{\nu\mu} + \mathrm{O}\left(\Delta\omega_{\mu\nu}^2\right) \tag{10}$$

holds. Expanding $\hat{S}(\hat{a})$ up to first order in $\Delta\omega_{\mu\nu}$ gives

$$\hat{S}(\hat{a}) = \mathbb{1} - \frac{\mathrm{i}}{4}\Delta\omega_{\mu\nu}\hat{I}^{\mu\nu} \tag{11a}$$

or

$$\hat{S}^{-1}(\hat{a}) = \hat{S}(\hat{a}^{-1}) = \mathbb{1} + \frac{\mathrm{i}}{4}\Delta\omega_{\mu\nu}\hat{I}^{\mu\nu} \tag{11b}$$

with the unknown 10×10 matrices $\hat{I}_{\mu\nu}$ ($\mathbb{1}$ is the 10×10 unit matrix). As in the case of the Dirac field a formally similar calculation yields

$$\left[\beta^\lambda, \hat{I}^{\mu\nu}\right]_- = -2\mathrm{i}\left(g^{\lambda\mu}\gamma^\nu - g^{\lambda\nu}\gamma^\mu\right) \quad . \tag{12}$$

An additional determining equation for the generators $\hat{I}_{\mu\nu}$ follows from the relation

$$\hat{S}(\hat{a})\hat{S}(\hat{a}')\hat{S}^{-1}(\hat{a}) = \hat{S}(\hat{a}\hat{a}'\hat{a}^{-1}) \quad , \tag{13}$$

which states that $\hat{S}(\hat{a})$ is a representation of the Lorentz group on the vector space spanned by the Kemmer algebra. For infinitesimal Lorentz transformations it follows that

$$\left[\hat{I}^{\mu\nu}, \hat{I}^{\alpha\beta}\right]_- = 2\mathrm{i}\left(g^{\nu\alpha}\hat{I}^{\mu\beta} - g^{\nu\beta}\hat{I}^{\mu\alpha} + g^{\mu\beta}\hat{I}^{\nu\alpha} - g^{\mu\alpha}\hat{I}^{\nu\beta}\right) \quad . \tag{14}$$

The relation (12) and (14) are fulfilled by the antisymmetric matrices

$$\hat{I}^{\mu\nu} = 2\mathrm{i}\left(\beta^\mu\beta^\nu - \beta^\nu\beta^\mu\right) \quad , \tag{15}$$

which can be easily verified by use of the commutation relations (15.47). This result occurs in direct analogy to the matrix $\sigma^{\mu\nu}$ in the Dirac theory [see (3.42)]. The factor "2" in (15) gives rise to the fact that the Kemmer field transforms to itself under a spatial rotation of 2π, and not 4π, as is the case for the Dirac field (c. f. Chap. 3). The problem being solved for infinitesimal proper Lorentz transformations, we now look for the finite transformations which can be constructed by repeated application of the infinitesimal Lorentz transformations. With $\Delta\omega^{\mu\nu} = \omega^{\mu\nu}/N$ it follows that

$$\begin{aligned}
\chi'(x') &= \lim_{N\to\infty}\left(\mathbb{1} - \frac{\mathrm{i}}{4}\frac{\omega^{\mu\nu}}{N}\hat{I}_{\mu\nu}\right)^N \chi(x) \\
&= \exp\left(-\frac{\mathrm{i}}{4}\omega^{\mu\nu}\hat{I}_{\mu\nu}\right)\chi(x) = \hat{S}(\hat{a})\chi(x) \quad .
\end{aligned} \tag{16}$$

With this the transformation properties of the Kemmer field $\chi(x)$ under proper Lorentz transformations $a^{\mu\nu}$ are known. Now we concentrate upon the improper Lorentz transformations $\tilde{a}^{\mu\nu}$, which can always be written as the product

Exercise 15.8.

$$\tilde{a}^{\mu\nu} = b^{\mu}{}_{\alpha} a^{\alpha\nu} \quad , \tag{17}$$

where the matrix $b^{\mu\nu}$ describes either space or time inversion, or both, i. e.

$$\hat{b} = \hat{p} \quad , \quad \hat{t} \quad , \quad \hat{p}\hat{t} \quad . \tag{18}$$

In this connection

$$p^{\mu}{}_{\nu} = \begin{pmatrix} 1 & 0 & 0 & 0 \\ 0 & -1 & 0 & 0 \\ 0 & 0 & -1 & 0 \\ 0 & 0 & 0 & -1 \end{pmatrix} = g^{\mu\nu} \tag{19}$$

[cf. (4.2)] and

$$t^{\mu}{}_{\nu} = \begin{pmatrix} -1 & 0 & 0 & 0 \\ 0 & 1 & 0 & 0 \\ 0 & 0 & 1 & 0 \\ 0 & 0 & 0 & 1 \end{pmatrix} = -g^{\mu\nu} \quad . \tag{20}$$

Since the spinors $\chi(x)$ lie within the representation space of the Lorentz group, then

$$\hat{S}(\hat{\tilde{a}}) = \hat{S}(\hat{b}\hat{a}) = \hat{S}(\hat{b})\hat{S}(\hat{a}) \tag{21}$$

must hold again, i. e. it is sufficient for us to find $\hat{S}(\hat{p})$ and $\hat{S}(\hat{t})$. All other transformations are then calculable as products. Since $\hat{p}$ and $\hat{t}$ are discrete transformations, they cannot be obtained from infinitesimal transformations and we must directly start with (8):

$$\hat{S}^{-1}(\hat{p})\beta_{\mu}\hat{S}(\hat{p}) = p_{\mu}{}^{\nu}\beta_{\nu} = \sum_{\nu} g^{\mu\nu}\beta^{\nu} \quad . \tag{22}$$

By means of the Kemmer algebra (15.47) and η defined in (15.53a), one verifies that

$$\eta\beta^{\mu}\eta = \sum_{\nu} g^{\mu\nu}\beta^{\nu} \tag{23}$$

is true and with that we obtain the parity operator of the Kemmer theory:

$$\hat{P} \equiv \hat{S}(\hat{p}) = \eta \mathrm{e}^{\mathrm{i}\varphi} \quad , \tag{24}$$

with arbitrary $\varphi \in \mathbb{R}$.

In the case of time inversion (cf. Chap. 12) we must start with the Kemmer equation with minimal coupling, thus with

$$\left[\mathrm{i}\hbar\beta^{\mu}\left(\partial_{\mu} + \mathrm{i}\frac{e}{c}A_{\mu}\right) - m_0 c\right]\chi(x) = 0 \quad . \tag{25}$$

We designate the time-inverted wave function $\chi_T(x')$ and require it to obey (25) in the transformed coordinate system; thus

$$\left[\mathrm{i}\hbar\beta^{\mu}\left(\partial'_{\mu} + \mathrm{i}\frac{e}{c}A'_{\mu}\right) - m_0 c\right]\chi_T(x') = 0 \quad . \tag{26}$$

Exercise 15.8.

In this connection, because of (20) and (12.41a) and (12.52, 53),

$$\partial'_\mu = t_\mu{}^\nu \partial_\nu \quad , \tag{27}$$

$$A'_\mu = -t_\mu{}^\nu A_\nu \quad , \tag{28}$$

The additional sign in (28) can be cancelled by complex conjugation. Thus (25) implies that

$$\left[-\mathrm{i}\hbar\left(\partial'_\mu + \frac{\mathrm{i}e}{c}A'_\mu\right) t^{\mu\nu}\beta^*_\nu - m_0 c\right]\chi^*(x) = 0 \quad . \tag{29}$$

If we define the matrix $\hat{T}$ by

$$\chi_T(x') = \hat{T}\chi^*(x) \quad , \tag{30}$$

this implies that

$$\left[-\mathrm{i}\hbar\left(\partial'_\mu + \mathrm{i}eA'_\mu\right) t^{\mu\nu}\hat{T}\beta^*_\nu\hat{T}^{-1} - m_0 c\right]\chi_T(x') = 0 \quad . \tag{31}$$

which coincides with (26) if

$$\hat{T}\beta^*_\nu\hat{T}^{-1} = -t_\nu{}^\mu\beta_\mu \tag{32}$$

is true [in complete analogy to (12.56)] for the Dirac matrices. Now in the representation (15.47)

$$\beta^*_\mu = -t^\nu_\mu\beta_\nu \quad , \tag{33}$$

holds, which can easily be checked, i. e. we can choose

$$\hat{T} = 1\!\mathrm{l}\,\mathrm{e}^{\mathrm{i}\varphi'} \quad , \tag{34}$$

with arbitrary $\varphi' \in \mathbb{R}$. The time-reversed wave function of the Kemmer theory is thus (up to an arbitrary phase) given by

$$\chi_T(x') = \chi^*(x) \quad . \tag{35}$$

EXERCISE

15.9 Verification of the Kemmer Algebra

Problem. Show that the Kemmer algebra (15.47) can be realized by the (16×16) matrices

$$(\beta^\mu)_{\alpha\beta\gamma\delta} = \tfrac{1}{2}\left(\gamma^\mu_{\alpha\beta}\delta_{\gamma\delta} + \gamma^\mu_{\gamma\delta}\delta_{\alpha\beta}\right) \quad . \tag{1}$$

Exercise 15.9.

Solution. Instead of (1) one often writes in an offhand manner

$$\beta_\mu = \tfrac{1}{2}\left(\gamma_\mu + \gamma'_\mu\right) \quad , \tag{2}$$

where besides

$$\{\gamma_\mu, \gamma_\nu\} = 2g_{\mu\nu} = \{\gamma'_\mu, \gamma'_\nu\} \quad , \tag{3}$$

also

$$\left[\gamma_\mu, \gamma'_\nu\right]_- = 0$$

must be required, too. [Strictly speaking $\gamma_\mu = \gamma_\mu \otimes 1\!\!1$ and $\gamma'_\mu = \gamma_\mu \otimes 1\!\!1$, where $1\!\!1$ is the unit matrix of the Clifford algebra (3) in four dimensions.] Now, because of (2), we have

$$\begin{aligned}\beta_\lambda\beta_\mu\beta_\nu = \tfrac{1}{8}\big(&\gamma_\lambda\gamma_\mu\gamma_\nu + \gamma_\lambda\gamma'_\mu\gamma'_\nu + \gamma_\lambda\gamma_\mu\gamma'_\nu + \gamma_\lambda\gamma'_\mu\gamma_\nu \\ &+\gamma'_\lambda\gamma_\mu\gamma_\nu + \gamma'_\lambda\gamma'_\mu\gamma'_\nu + \gamma'_\lambda\gamma_\mu\gamma'_\nu + \gamma'_\lambda\gamma'_\mu\gamma_\nu\big) \quad .\end{aligned} \tag{4}$$

Adding $\beta_\nu\beta_\mu\beta_\lambda$ to this product results in

$$\begin{aligned}\beta_\lambda\beta_\mu\beta_\nu + \beta_\nu\beta_\mu\beta_\lambda = \tfrac{1}{8}\big(&\gamma_\lambda\gamma_\mu\gamma_\nu + \gamma_\nu\gamma_\mu\gamma_\lambda + \gamma'_\lambda\gamma'_\mu\gamma'_\nu + \gamma'_\nu\gamma'_\mu\gamma'_\lambda \\ &+ \gamma_\lambda\left\{\gamma'_\mu, \gamma'_\nu\right\} + \gamma'_\lambda\left\{\gamma_\mu, \gamma_\nu\right\} + \gamma'_\nu\left\{\gamma_\lambda, \gamma_\mu\right\} \\ &+\gamma_\nu\left\{\gamma'_\lambda, \gamma'_\mu\right\} + \gamma'_\mu\left\{\gamma_\lambda, \gamma_\nu\right\} + \gamma_\mu\left\{\gamma'_\lambda, \gamma'_\nu\right\}\big) \quad ,\end{aligned} \tag{5}$$

and since

$$\gamma_\lambda\gamma_\mu\gamma_\nu = \{\gamma_\lambda, \gamma_\mu\}\gamma_\nu - \gamma_\mu\{\gamma_\lambda, \gamma_\nu\} + \{\gamma_\mu, \gamma_\nu\}\gamma_\lambda - \gamma_\nu\gamma_\mu\gamma_\lambda \tag{6}$$

is valid, as well as the corresponding terms for the product $\gamma'_\lambda\gamma'_\mu\gamma'_\nu$, too, we obtain [because of (3)]

$$\begin{aligned}\beta_\lambda\beta_\mu\beta_\nu + \beta_\nu\beta_\mu\beta_\lambda &= \tfrac{1}{8}\left(4g_{\lambda\mu}\gamma_\nu + 4g_{\mu\nu}\gamma_\lambda + 4g_{\lambda\mu}\gamma'_\nu + 4g_{\mu\nu}\gamma'_\lambda\right) \\ &= g_{\lambda\mu}\beta_\nu + g_{\mu\nu}\beta_\lambda \quad ,\end{aligned} \tag{7}$$

and thus (15.47).

EXERCISE

15.10 Verification of the Proca Equations from the Lagrange Density

Problem. Show that the Lagrange density $\mathcal{L}_{\text{Proca}}$ from (15.38) leads to the Proca equations (15.34) and (15.35) and their complex conjugates.

Solution. In this notation of the Lagrange density the fields $G^{\mu\nu}$, $G^{\mu\nu *}$, φ^μ and $\varphi^{\mu *}$ are conceived as *independent*. A variation with respect to $G^*_{\mu\nu}$ yields then because of the definition (15.34) of $G^{\mu\nu}$

$$\frac{\partial\mathcal{L}}{\partial G^*_{\mu\nu}} = \frac{1}{2}G^{\mu\nu} - \frac{1}{2}\partial^\mu\varphi^\nu - \partial^\nu\varphi^\mu = 0 \quad .$$

Variation with respect to φ^*_μ results in

$$\frac{\partial\mathcal{L}}{\partial\varphi^*_\mu} - \partial_\nu\frac{\partial\mathcal{L}}{\partial(\partial_\nu\varphi^*_\mu)} = \frac{m_0^2c^2}{\hbar^2}\varphi^\mu + \frac{1}{2}\partial_\nu(G^{\mu\nu} - G^{\nu\mu}) = 0 \quad ,$$

i. e. the equation of motion (12.35). The complex-conjugate equations are obtained analogously by variation with respect to $G_{\mu\nu}$ and φ_μ.

EXERCISE

15.11 Conserved Current of Vector Fields

Problem. Determine the conserved current for the vector fields.

Solution. The Proca equation (15.35) holds for the field $G_{\mu\nu}$, the corresponding complex conjugate equation

$$\partial^\mu G^*_{\mu\nu} = -\frac{m_0^2 c^2}{\hbar^2}\varphi^*_\nu \tag{1}$$

holding for the field $G^*_{\mu\nu}$. If we multiply (15.35) by φ^*_ν and (1) by φ_ν, then the result is

$$\varphi^*_\nu \partial_\mu G^{\mu\nu} = -\frac{m_0^2 c^2}{\hbar^2}\varphi^*_\nu \varphi^\nu \quad , \tag{2a}$$

$$\varphi_\nu \partial^\mu G^*_{\mu\nu} = -\frac{m_0^2 c^2}{\hbar^2}\varphi^\nu \varphi^*_\nu \quad . \tag{2b}$$

Subtracting (2a) from (2b), we obviously get

$$-\varphi^*_\nu \partial_\mu G^{\mu\nu} + \varphi^\nu \partial^\mu G^*_{\mu\nu} = 0 \quad . \tag{3}$$

This can be written in the form

$$\partial_\mu J^\mu = 0 \quad ,$$

where the conserved current is given by

$$J_\mu = \varphi^\nu G^*_{\mu\nu} - G_\mu{}^\nu \varphi^*_\nu \quad . \tag{4}$$

The reason is that the terms in braces in

$$\partial^\mu J_\mu = \varphi^\nu \partial^\mu G^*_{\mu\nu} - \left(\partial^\mu G_\mu{}^\nu\right)\varphi^*_\nu + \left\{(\partial^\mu \varphi^\nu) G^*_{\mu\nu} - G_{\mu\nu}\partial^\mu \varphi^{\nu *}\right\} \tag{5}$$

equal zero because of the antisymmetry of $G_{\mu\nu}$ and $G^*_{\mu\nu}$, which follows from the definition (15.34), i. e.

$$\begin{aligned}\partial^\mu \varphi^\nu G^*_{\mu\nu} - G_{\mu\nu}\partial^\mu \varphi^{\nu *} &= \tfrac{1}{2}\left[(\partial^\mu \varphi^\nu - \partial^\nu \varphi^\mu) G^*_{\mu\nu} - G_{\mu\nu}(\partial^\mu \varphi^{\nu *} - \partial^\nu \varphi^{\mu *})\right] \\ &= \tfrac{1}{2}\left[G^{\mu\nu} G^*_{\mu\nu} - G_{\mu\nu} G^{\mu\nu *}\right] = 0 \quad .\end{aligned} \tag{6}$$

EXERCISE

15.12 Lorentz Covariance of Vector Field Theory

Problem. Investigate the Lorentz covariance of the vector field theory.

Solution. In order for the Proca equations to transform covariantly under Lorentz transformations, the field φ_μ must transform like a vector, i. e. under the infinitesimal Lorentz transformation [cf. (3.1) and (3.53)]

$$x'_\mu = a_\mu{}^\nu x_\nu = x_\mu + \Delta\omega_\mu{}^\nu x_\nu \quad , \tag{1}$$

Exercise 15.12.

it follows that

$$A'_\mu(x') = A_\mu(x) + \Delta\omega_\mu{}^\nu A_\nu(x) \tag{2}$$

must hold, too. On the other hand the Lorentz transformation of a field can generally be written [in analogy to (3.27) of (3.39)] in the infinitesimal case as

$$A'_\mu(x') = \left[g_\mu{}^\nu + \tfrac{1}{2}\left(\hat{I}_\mu{}^\nu\right)^{\alpha\beta}\Delta\omega_{\alpha\beta}\right]A_\nu(x) \quad . \tag{3}$$

Each of the matrices $\hat{I}_{\mu\nu}$ must be antisymmetric, because the $\Delta\omega^{\mu\nu}$ are antisymmetric too [cf. (3.36)]. The comparison of (2) and (3) yields at once that

$$\left(\hat{I}^{\mu\nu}\right)_{\alpha\beta} = +\left(g^\mu{}_\alpha g^\nu{}_\beta - g^\mu{}_\beta g^\nu{}_\alpha\right) \quad . \tag{4}$$

For pure rotations, for instance, one obtains

$$\left(\hat{I}^{12}\right)_{kl} = +\left(g^1{}_k g^2{}_l - g^1{}_l g^2{}_k\right) = \begin{pmatrix} 0 & +1 & 0 \\ -1 & 0 & 0 \\ 0 & 0 & 0 \end{pmatrix} \quad ,$$

$$\left(\hat{I}^{23}\right)_{kl} = -\left(g^2{}_k g^3{}_l - g^2{}_l g^3{}_k\right) = \begin{pmatrix} 0 & 0 & 0 \\ 0 & 0 & -1 \\ 0 & +1 & 0 \end{pmatrix} \quad ,$$

$$\left(\hat{I}^{31}\right)_{kl} = -\left(g^3{}_k g^1{}_l - g^3{}_l g^1{}_k\right) = \begin{pmatrix} 0 & 0 & -1 \\ 0 & 0 & 0 \\ +1 & 0 & 0 \end{pmatrix} \quad . \tag{5}$$

The indices k and l here take only the values $1, 2, 3$ and these matrices can be conceived as the spin-1 analogues of the Pauli matrices [see (1.65)].

EXAMPLE

15.13 Maxwell-Similar Form of the Vector Fields

The additional condition (15.37) of the Proca theory implies that the four-components of φ^μ are not independent dynamic variables. To isolate the independent quantities, we introduce the three-dimensional fields:

$$B^i = -\tfrac{1}{2}\varepsilon^{ijk}G_{jk} \quad i,j,k = 1,2,3 \tag{1a}$$

$$E^i = -G^{0i} = g^{i0} \tag{1b}$$

so that we can write

$$G^{\mu\nu} = \begin{pmatrix} 0 & -E^1 & -E^2 & -E^3 \\ E^1 & 0 & -B^3 & B^2 \\ E^2 & B^3 & 0 & -B^1 \\ E^3 & -B^2 & B^1 & 0 \end{pmatrix} = -G^{\nu\mu} \quad . \tag{2}$$

In this notation the Proca equations (15.34) read:

$$\begin{aligned} B^i &= -\tfrac{1}{2}\varepsilon^{ijk}G_{jk} \\ &= -\tfrac{1}{2}\varepsilon^{ijk}(\partial_j\varphi_k - \partial_k\varphi_j) = (\boldsymbol{\nabla}\times\boldsymbol{\varphi})^i \quad , \end{aligned} \tag{3a}$$

Example 15.13.

$$E^i = -G^{0i} = -\partial^0\varphi_i + \partial^i\varphi^0 = -\left(\frac{1}{c}\frac{\partial}{\partial t}\boldsymbol{\varphi} + \boldsymbol{\nabla}\varphi^0\right)^i \quad . \tag{3b}$$

Using the inversion of (1a),

$$\begin{aligned} -\varepsilon^{jkl}B_l &= \tfrac{1}{2}\varepsilon^{jkl}\varepsilon_{lmn}G^{mn} = \tfrac{1}{2}\left(g^j{}_m g^k{}_n - g^k{}_m g^j{}_n\right)G^{mn} \\ &= \tfrac{1}{2}(G^{jk} - G^{kj}) = G^{jk} \quad , \end{aligned} \tag{4}$$

one gets the following relation:

$$\partial_i G^{ij} = \partial_i \varepsilon^{ijk}B_k = -\varepsilon^{jik}\partial_i B_k = (\boldsymbol{\nabla}\times\boldsymbol{B})^j \quad , \tag{5a}$$

$$\partial_0 G^{0j} = -\frac{1}{c}\frac{\partial}{\partial t}E^j \quad . \tag{5b}$$

Therefore the spatial part of (15.35), with $\nu = j$, is given by

$$\boldsymbol{\nabla}\times\boldsymbol{B} - \frac{1}{c}\frac{\partial}{\partial t}\boldsymbol{E} = -\frac{m_0^2c^2}{\hbar^2}\boldsymbol{\varphi} \quad . \tag{6}$$

The temporal part, however, ($\nu = 0$) reads as

$$\partial_i G^{i0} = \boldsymbol{\nabla}\times\boldsymbol{E} = -\frac{m_0^2c^2}{\hbar^2}\varphi^0 \quad . \tag{7}$$

Equations (3b) and (6) are the true equations of motion, because they contain the time derivations of $\partial_0\boldsymbol{\varphi}$ and $\partial_0\boldsymbol{E}$, whereas (3a) and (7) can be seen as *definitions* of the fields $\boldsymbol{B}$ and φ_0, respectively, expressed by the *independent* fields $\boldsymbol{\varphi}$ and $\boldsymbol{E}$. Substituting the dependent fields into the equations of motion by their definitions, one obtains from (3b) and (6) that

$$\frac{1}{c}\frac{\partial}{\partial t}\boldsymbol{\varphi} = -\boldsymbol{E} + \frac{\hbar^2}{m_0^2c^2}\boldsymbol{\nabla}(\boldsymbol{\nabla}\cdot\boldsymbol{E}) \quad , \tag{8a}$$

$$\begin{aligned} \frac{1}{c}\frac{\partial}{\partial t}\boldsymbol{E} &= +\frac{m_0^2c^2}{\hbar^2}\boldsymbol{\varphi} + \boldsymbol{\nabla}\times(\boldsymbol{\nabla}\times\boldsymbol{\varphi}) \\ &= \left(-\boldsymbol{\nabla}^2 + \frac{m_0^2c^2}{\hbar^2}\right)\boldsymbol{\varphi} + \boldsymbol{\nabla}(\boldsymbol{\nabla}\times\boldsymbol{\varphi}) \end{aligned} \tag{8b}$$

In the last conversion we used the relation

$$\begin{aligned} [\boldsymbol{\nabla}\times(\boldsymbol{\nabla}\times\boldsymbol{\varphi})]_i &= \varepsilon_{ijk}\partial_j\varepsilon_{klm}\partial_l\varphi_m = (\delta_{li}\delta_{mj} - \delta_{lj}\delta_{mi})\partial_j\partial_l\varphi_m \\ &= \partial_i\partial_m\varphi_m - \partial_l\partial_l\varphi_i = [\boldsymbol{\nabla}(\boldsymbol{\nabla}\cdot\boldsymbol{\varphi}) - \boldsymbol{\nabla}^2\boldsymbol{\varphi}]_i \quad . \end{aligned} \tag{9}$$

EXERCISE

15.14 Plane Waves for the Proca Equation

Problem. Find the solution of the Proca equation (15.36), with the additional conditions (15.35) and (15.37) of the form

$$\varphi_p^\mu(x) = N_p\varepsilon_p^\mu \mathrm{e}^{-\mathrm{i}p\cdot x/\hbar} \quad , \tag{1}$$

(i.e. plane waves) where N_p is a normalizing factor.

Exercise 15.14.

Solution. First of all, from (15.36) and (15.37) it follows that

$$\left(p^2 - m_0^2 c^2\right) \varepsilon_p^\mu = 0 \quad . \tag{2}$$

This means the functions ε_p^μ can only be different from zero if

$$p^0 = \pm \frac{1}{c} E_p \quad , \tag{3a}$$

$$E_p = +c\sqrt{\boldsymbol{p}^2 + m_0^2 c^2} \quad . \tag{3b}$$

Additionally, condition (15.35) implies that

$$p_\mu \varepsilon_p^\mu = 0 \quad . \tag{4}$$

Now for every three-vector $\boldsymbol{p}$ and each sign of p^0, a set of three linearly independent four-vectors $\varepsilon_{p\lambda}^\mu$ ($\lambda = 1, 2, 3$) can be constructed which fulfil (4). Let $\boldsymbol{\varepsilon}_\lambda$ ($\lambda = 1, 2, 3$) be an arbitrary "tripod" with

$$\boldsymbol{\varepsilon}_\lambda \cdot \boldsymbol{\varepsilon}_{\lambda'} = \delta_{\lambda\lambda'} \quad , \tag{5}$$

which, in the rest system, is obviously valid. Now we Lorentz transform to a system which moves with the velocity $-\boldsymbol{p}/p^0 = -\boldsymbol{v}/c$. Then the three-vectors $\boldsymbol{\varepsilon}_\lambda$ convert to the vectors $\varepsilon_{p\lambda}^\mu$ with the components

$$\begin{aligned}
\varepsilon_{p\lambda}^0 &= \frac{1}{\sqrt{1 - v^2/c^2}} \frac{\boldsymbol{v}}{c} \cdot \boldsymbol{\varepsilon}_\lambda = \frac{\boldsymbol{p} \cdot \boldsymbol{\varepsilon}_\lambda}{m_0 c} \quad , \\
\boldsymbol{\varepsilon}_{p\lambda} &= \boldsymbol{\varepsilon}_\lambda + \frac{\sqrt{1 - v^2/c^2} - 1}{v^2/c^2} \left(\frac{\boldsymbol{v}}{c} \cdot \boldsymbol{\varepsilon}_\lambda\right) \frac{\boldsymbol{v}}{c} \\
&= \boldsymbol{\varepsilon}_\lambda + \frac{(\boldsymbol{p} \cdot \boldsymbol{\varepsilon}_\lambda)\boldsymbol{p}}{m_0 c(p^0 + m_0 c)} \quad .
\end{aligned} \tag{6}$$

These of course obey the covariance condition

$$p \cdot \varepsilon_{p\lambda} = p_\mu \varepsilon_{p\lambda}^\mu = 0 \quad , \tag{7}$$

because the latter does not depend on the reference system and in the rest system $\boldsymbol{p} = 0$ and $\varepsilon_{\boldsymbol{p}=0,\lambda}^0 = 0$. Furthermore the following orthogonality is valid:

$$\begin{aligned}
&\varepsilon_{p\lambda} \cdot \varepsilon_{p\lambda'} \\
&= \frac{(\boldsymbol{p} \cdot \boldsymbol{\varepsilon}_\lambda)(\boldsymbol{p} \cdot \boldsymbol{\varepsilon}_{\lambda'})}{m_0^2 c^2} - \left[\boldsymbol{\varepsilon}_\lambda + \frac{\boldsymbol{p}(\boldsymbol{p} \cdot \boldsymbol{\varepsilon}_\lambda)}{m_0 c(p^0 + m_0 c)}\right] \left[\boldsymbol{\varepsilon}_{\lambda'} + \frac{\boldsymbol{p}(\boldsymbol{p} \cdot \boldsymbol{\varepsilon}_{\lambda'})}{m_0 c(p^0 + m_0 c)}\right] \\
&= \frac{(\boldsymbol{p} \cdot \boldsymbol{\varepsilon}_\lambda)(\boldsymbol{p} \cdot \boldsymbol{\varepsilon}_{\lambda'})}{m_0^2 c^2} - \left[\boldsymbol{\varepsilon}_\lambda \cdot \boldsymbol{\varepsilon}_{\lambda'} + \frac{2(\boldsymbol{\varepsilon}_\lambda \cdot \boldsymbol{p})(\boldsymbol{\varepsilon}_{\lambda'} \cdot \boldsymbol{p})}{m_0 c(p^0 + m_0 c)} + \frac{\boldsymbol{p}^2(\boldsymbol{p} \cdot \boldsymbol{\varepsilon}_\lambda)(\boldsymbol{p} \cdot \boldsymbol{\varepsilon}_{\lambda'})}{m_0^2 c^2 (p^0 + m_0 c)^2}\right] \\
&= -\boldsymbol{\varepsilon}_\lambda \cdot \boldsymbol{\varepsilon}_{\lambda'} + \frac{(\boldsymbol{p} \cdot \boldsymbol{\varepsilon}_\lambda)(\boldsymbol{p} \cdot \boldsymbol{\varepsilon}_{\lambda'})}{m_0^2 c^2} \left[1 - \frac{2 m_0 c}{p^0 + m_0 c} - \frac{p^0 - m_0 c}{p^0 + m_0 c}\right] \\
&= -\delta_{\lambda\lambda'} \quad ,
\end{aligned} \tag{8}$$

because the bracket yields

$$\frac{p^0 + m_0 c - 2 m_0 c - p^0 + m_0 c}{p^0 + m_0 c} = 0 \quad .$$

Equation (8) is evident in the rest system, and, due to the invariance of the four-scalar product under Lorentz transformation, the explicit calculation of (8) was not actually necessary. Together with the time-like vector p^μ, the three space-like vectors $\varepsilon^\mu_{p\lambda}$ constitute an orthonormal quadrupod in Minkowski space.

Exercise 15.14.

EXERCISE

15.15 Transformation from the Kemmer to the Proca Representation

Problem. Verify the reverse transformation (15.50) from the Kemmer to the Proca representation.

Solution. The matrices U^μ and $U^{\mu\nu}$ in our representation are given by 10×10 matrices:

$$U^0 = \begin{pmatrix} \emptyset & \emptyset & \emptyset & \bar{0}^\dagger \\ \emptyset & \emptyset & \emptyset & \bar{0}^\dagger \\ \emptyset & \emptyset & \emptyset & \bar{0}^\dagger \\ \bar{0} & \bar{0} & \bar{0} & -1 \end{pmatrix} \quad , \tag{1a}$$

$$U^k = \begin{pmatrix} \emptyset & \emptyset & \emptyset & \bar{0}^\dagger \\ \emptyset & \emptyset & \emptyset & \bar{0}^\dagger \\ \emptyset & \emptyset & \emptyset & \bar{0}^\dagger \\ \bar{0} & \bar{0} & -\mathrm{i}K^k & 0 \end{pmatrix} \quad , \tag{1b}$$

$$U^{k0} = \begin{pmatrix} \emptyset & \emptyset & \emptyset & \bar{0}^\dagger \\ \emptyset & \emptyset & \emptyset & \bar{0}^\dagger \\ \emptyset & \emptyset & \emptyset & \bar{0}^\dagger \\ \mathrm{i}K^k & \bar{0} & \bar{0} & 0 \end{pmatrix} \quad , \tag{1c}$$

$$U^{kl} = \begin{pmatrix} \emptyset & \emptyset & \emptyset & \bar{0}^\dagger \\ \emptyset & \emptyset & \emptyset & \bar{0}^\dagger \\ \emptyset & \emptyset & \emptyset & \bar{0}^\dagger \\ \bar{0} & \mathrm{i}K^k S^l & \bar{0} & 0 \end{pmatrix} \quad . \tag{1d}$$

Now, for example, we have

$$-\frac{\mathrm{i}\hbar}{\sqrt{m_0 c}} E^\dagger U^0 \chi = +\frac{\mathrm{i}\hbar}{\sqrt{m_0 c}} \chi_{10} = \varphi^0 \quad , \tag{2}$$

because the spinor $E^\dagger$ of (15.49) picks out exactly the lowest component of $U^0\chi$ [using (15.42) and (15.43) in the process]. In the same way

$$-\frac{\mathrm{i}\hbar}{\sqrt{m_0 c}} E^\dagger U^1 \chi = -\frac{\hbar}{\sqrt{m_0 c}} \chi_7 = \varphi^1 \quad , \tag{3}$$

etc. are valid. On the other hand, for example,

$$\sqrt{m_0 c} E^\dagger U^{10} \chi = \mathrm{i}\sqrt{m_0 c}\, \chi_1 = G^{01} \tag{4}$$

or

$$\sqrt{m_0 c} E^\dagger U^{23} \chi = \mathrm{i}\sqrt{m_0 c}\, (\mathrm{i}\chi_4) = G^{23} \tag{5}$$

Exercise 15.15.

holds, because

$$K^2S^3 = \mathrm{i}(0\ 1\ 0)\begin{pmatrix} 0 & -1 & 0 \\ 1 & 0 & 0 \\ 0 & 0 & 0 \end{pmatrix} = \mathrm{i}(1\ 0\ 0) = \mathrm{i}K^1 \quad , \tag{6}$$

etc. Now it is clear that the transformation (15.50) does give exactly the components (15.43), so that, indeed, (15.50) is the reverse transformation.

EXERCISE

15.16 Lagrange Density for Kemmer Theory

Problem. Convince yourself of the correctness of the Lagrange density (15.52) of the Kemmer theory.

Solution. It has to be verified that the Euler–Lagrange equations

$$\frac{\partial \mathcal{L}}{\partial \overline{\chi}} - \partial_\mu \frac{\partial \mathcal{L}}{\partial(\partial_\mu \overline{\chi})} = 0 \tag{1}$$

are equvalent to the Kemmer equations (15.44). Now with

$$\frac{\partial \mathcal{L}}{\partial \overline{\chi}} = \frac{1}{2}\mathrm{i}\hbar\beta_\mu \partial^\mu \chi - m_0 c\chi \tag{2}$$

and

$$\frac{\partial \mathcal{L}}{\partial(\partial_\mu \overline{\chi})} = -\frac{1}{2}\mathrm{i}\hbar\beta_\mu \chi \quad , \tag{3}$$

it follows immediately from (1) that

$$\left(\mathrm{i}\hbar\beta_\mu\partial^\mu - m_0 c\right)\chi = 0 \quad ,$$

i. e. (15.44). Alternatively we may write:

$$\begin{aligned} 0 &= \frac{\partial \mathcal{L}}{\partial \chi} - \partial_\mu \frac{\partial \mathcal{L}}{\partial(\partial_\mu \chi)} \\ &= -m_0 c\overline{\chi} - \frac{1}{2}\mathrm{i}\hbar\overline{\chi}\beta_\mu \overleftarrow{\partial}^\mu - \partial_\mu \frac{1}{2}\mathrm{i}\hbar\overline{\chi}\beta_\mu \\ &= -\overline{\chi}\left(m_0 c + \mathrm{i}\hbar\beta_\mu \overleftarrow{\partial}^\mu\right) = 0 \quad , \end{aligned} \tag{4}$$

the equation for the adjoint spinor $\overline{\chi}$.

EXAMPLE

15.17 The Weinberg–Shay–Good Equations

There is yet another six-dimensional spinor representation of the Proca theory, first given by Weinberg in 1964 and later investigated further by Shay and Good.[5] To

[5] D. Shay, R.H. Good: Phys. Rev. **179**, 141 (1969) and S. Weinberg: Phys. Rev. **133**, B1318 (1964).

Example 15.17.

substantiate it one can proceed in a completely analogous way to that of the derivation of the Kemmer equation. In order to do this we define the six-dimensional "spinor"

$$\chi = \begin{pmatrix} \chi_1 \\ \vdots \\ \chi_6 \end{pmatrix} \quad , \tag{1}$$

which is connected to the Proca field by

$$\begin{aligned}
\chi_1 &= \frac{1}{4m_0c}\left(G^{23} - \mathrm{i}G^{01}\right) \quad , \\
\chi_2 &= -\frac{1}{4m_0c}\left(G^{13} + \mathrm{i}G^{02}\right) \quad , \\
\chi_3 &= \frac{1}{4m_0c}\left(G^{12} - \mathrm{i}G^{03}\right) \quad , \\
\chi_4 &= -\frac{1}{4m_0c}\left(G^{23} + \mathrm{i}G^{01}\right) \quad , \\
\chi_5 &= \frac{1}{4m_0c}\left(G^{13} - \mathrm{i}G^{02}\right) \quad , \\
\chi_6 &= -\frac{1}{4m_0c}\left(G^{12} + \mathrm{i}G^{03}\right) \quad .
\end{aligned} \tag{2}$$

With the notation (cf. Exercise 15.13) of the field strengths

$$\begin{aligned}
E^i &= G^{i0} \quad , \\
B_i &= \varepsilon^{ijk} G^{jk}
\end{aligned} \tag{3}$$

equation (2) reads

$$\begin{aligned}
\chi_1 &= \frac{1}{4m_0c}\left(B^1 - \mathrm{i}E^1\right) \quad , \quad \chi_4 = -\frac{1}{4m_0c}\left(B^1 + \mathrm{i}E^1\right) \quad , \\
\chi_2 &= \frac{1}{4m_0c}\left(B^2 - \mathrm{i}E^2\right) \quad , \quad \chi_5 = -\frac{1}{4m_0c}\left(B^2 + \mathrm{i}E^2\right) \quad , \\
\chi_3 &= \frac{1}{4m_0c}\left(B^3 - \mathrm{i}E^3\right) \quad , \quad \chi_6 = -\frac{1}{4m_0c}\left(B^3 + \mathrm{i}E^3\right)
\end{aligned} \tag{4}$$

(cf. also Exercise 2.1, where the components χ_4, χ_5, χ_6 are unnessary because the Maxwell field is real). The field equations are given by

$$\begin{aligned}
&\left[\gamma^{\mu\nu}(\mathrm{i}\hbar\partial_\mu)(\mathrm{i}\hbar\partial_\nu) - (\mathrm{i}\hbar\partial_\mu)(\mathrm{i}\hbar\partial^\mu) + 2m_0^2c^2\right]\chi(x) \\
&\quad = \left[\mathrm{i}\hbar\partial_\mu(\gamma^{\mu\nu} - g^{\mu\nu})\mathrm{i}\hbar\partial_\nu + 2m_0^2c^2\right]\chi = 0 \quad ,
\end{aligned} \tag{5}$$

where the 6×6 matrices are given by

$$\begin{aligned}
\gamma^{ij} &= \begin{pmatrix} \emptyset & \delta_{ij}\mathbb{1} + M^{(ij)} + M^{(ji)} \\ \delta_{ij}\mathbb{1} + M^{(ij)} + M^{(ji)} & \emptyset \end{pmatrix} = \gamma^{ji} \quad , \\
\gamma^{0i} &= \gamma^{i0} = \begin{pmatrix} \emptyset & S^i \\ -S^i & \emptyset \end{pmatrix} \quad , \quad \gamma^{00} = -\begin{pmatrix} \emptyset & \mathbb{1} \\ \mathbb{1} & \emptyset \end{pmatrix} \quad .
\end{aligned} \tag{6}$$

Example 15.17.

The definitions of the 3×3 matrices S^i, $M^{(ij)}$, 0 and 1 can be taken from Exercise 15.7 (commutation relation of the Kemmer matrices) and (15.46). That the Weinberg–Shay–Good equations are equivalent to the Proca equations cannot be shown as easily as in the case of the Kemmer equations, because first one has to give a suitable linear combination of the six Shay–Good equations, which then lead to the Proca form. This algebra problem can be most easily presented by first finding the inverse transformation, which is then used to reconstruct the Proca field $G^{\mu\nu}$ from the spinor χ.

Here we proceed in close analogy to the Kemmer theory. In the same manner as in (15.48) and (15.49), the antisymmetric 6×6 matrix $U^{\mu\nu}$ is defined by

$$\begin{aligned} U^{ij} &= \tfrac{1}{2}\varepsilon^{ij}{}_k B^k \quad , \\ U^{0j} &= \tfrac{1}{2}\mathrm{i}A^j \quad , \end{aligned} \tag{7}$$

where

$$\begin{aligned} A^i &= \begin{pmatrix} \emptyset & \emptyset \\ M^{(3i)} & M^{(3i)} \end{pmatrix} \quad , \\ B^i &= \begin{pmatrix} \emptyset & \emptyset \\ M^{(3i)} & -M^{(3i)} \end{pmatrix} \quad , \end{aligned} \tag{8}$$

are 6×6 matrices, too. The six-component spinor E is defined as

$$E = \begin{pmatrix} 0 \\ \vdots \\ 0 \\ 1 \end{pmatrix} \quad . \tag{9}$$

Hence, analogously to (15.50), the following relation holds:

$$G^{\mu\nu}(x) = \frac{1}{2m_0c} E^\dagger U^{\mu\nu} \chi(x) \quad . \tag{10}$$

Applying this matrix from the left to the Weinberg–Shay–Good equation (5) one has to take note of the fact that

$$U^{\mu\nu}\gamma^{\alpha\beta} = g^{\alpha\beta}U^{\mu\nu} - \left(g^{\alpha\mu}U^{\beta\nu} - g^{\alpha\nu}U^{\beta\mu} + g^{\beta\mu}U^{\alpha\nu} - g^{\beta\nu}U^{\alpha\mu}\right) \tag{11}$$

holds. Consequently we find that

$$-E^\dagger U^{\mu\nu}\left[\mathrm{i}\hbar\partial_\alpha\left(\gamma^{\alpha\beta} - g^{\alpha\beta}\right)\mathrm{i}\hbar\partial_\beta + 2m_0^2c^2\right]\chi = 0 \quad , \tag{12}$$

or

$$\begin{aligned} &\frac{1}{2m_0c}E^\dagger\left[\mathrm{i}\hbar\partial_\alpha\left(g^{\alpha\mu}U^{\beta\nu} - g^{\alpha\nu}U^{\beta\mu} + g^{\beta\mu}U^{\alpha\nu} - g^{\beta\nu}U^{\alpha\mu}\right)\mathrm{i}\hbar\partial_\beta\right]\chi \\ &\quad = 2m_0^2c^2G^{\mu\nu} \quad . \end{aligned} \tag{13a}$$

The lhs of this equation reads as

$$\begin{aligned} &\mathrm{i}\hbar\partial^\mu\mathrm{i}\hbar\partial_\beta G^{\beta\nu} - \mathrm{i}\hbar\partial^\nu\mathrm{i}\hbar\partial_\beta G^{\beta\mu} + \mathrm{i}\hbar\partial^\mu\mathrm{i}\hbar\partial_\beta G^{\beta\nu} - \mathrm{i}\hbar\partial^\nu\mathrm{i}\hbar\partial^\beta G^{\beta\mu} \\ &\quad = 2m_0^2c^2\left(\partial^\mu\varphi^\nu - \partial^\nu\varphi^\mu\right) \quad , \end{aligned} \tag{13b}$$

defining the field φ^μ by

Example 15.17.

$$\varphi^\mu = -\frac{\hbar^2}{m_0^2 c^2}\partial_\beta G^{\beta\mu} \quad . \tag{14}$$

Equations (13) and (14) agree with both of the Proca equations (15.33) and (15.34). The Lagrangian of the Weinberg–Shay–Good theory is written as

$$\mathcal{L}_{\mathrm{WSG}} = \hbar^2\,(\partial_\nu\overline{\chi})\,(\gamma^{\mu\nu} - g^{\mu\nu})\,\partial_\mu\chi + 2m_0c^2\overline{\chi}\chi \quad , \tag{15}$$

where

$$\overline{\chi} =: \overline{\chi}^\dagger\gamma^{00} \quad , \tag{16}$$

and this is examined in more detail in Exercise 15.18.

Besides the representation used here there is another interesting representation, in which the spinor χ has a simple form. Defining the unitary matrix

$$S = \frac{1}{\sqrt{2}}\begin{pmatrix} 1\!\!1 & 1\!\!1 \\ 1\!\!1 & -1\!\!1 \end{pmatrix} \quad , \tag{17}$$

we get

$$\chi' = S\chi = \frac{\sqrt{2}}{4m_0c}\begin{pmatrix} \mathrm{i}\boldsymbol{E} \\ \boldsymbol{B} \end{pmatrix} \quad . \tag{18}$$

This spinor satisfies (5) if one inserts the transformed matrices

$$\gamma'^{\mu\nu} = S\gamma^{\mu\nu}S^\dagger \quad . \tag{19}$$

In (10), naturally $U^{\mu\nu}$ and E have to be substituted by $U'^{\mu\nu}$ and E':

$$U'^{\mu\nu} = SU^{\mu\nu}S^\dagger \quad , \quad E' = SE \quad .$$

EXERCISE

15.18 Lagrangian Density for the Weinberg–Shay–Good Theory

Problem. Prove that the Lagrangian of the Weinberg–Shay–Good theory [Exercise 15.17, (15)] yields the proper equation of motion.

Solution. With the help of the Euler–Lagrange equation we obtain

$$0 = \frac{\partial\mathcal{L}}{\partial\overline{\chi}} - \partial_\mu\frac{\partial\mathcal{L}}{\partial(\partial_\mu\overline{\chi})} = 2m_0c^2\chi - \hbar^2\partial_\mu(\gamma^{\mu\nu} - g^{\mu\nu})\partial_\nu\chi \tag{1}$$

[cf. Exercise 15.17, (5)]. The adjoint equation is obtained accordingly as

$$0 = \frac{\partial\mathcal{L}}{\partial\chi} - \partial_\mu\frac{\partial\mathcal{L}}{\partial(\partial_\mu\chi)} = 2m_0c^2\overline{\chi} - \hbar^2\partial_\mu\partial_\nu\overline{\chi}(\gamma^{\mu\nu} - g^{\mu\nu}) \quad . \tag{2}$$

EXAMPLE

15.19 Coupling of Charged Vector Mesons to the Electromagnetic Field

The complex spin-1 fields appearing in the Proca, Kemmer, Weinberg–Shay–Good equations describe charged vector mesons. To let the equations of motion remain invariant, if we apply a local gauge transformation to the fields, the derivation ∂_μ must be substituted by the "gauge-invariant derivation"

$$D_\mu = \partial_\mu + \mathrm{i}\frac{e}{c}A_\mu \quad ,$$

where e is the charge of the vector field [cf. (1.132–136)]. Therefore the field equations of the three different theories read:

$$D_\mu G^{\mu\nu} + \frac{m_0^2c^2}{\hbar^2}\varphi^\nu = 0 \quad , \tag{1a}$$

$$G^{\mu\nu} = D^\mu\varphi^\nu - D^\nu\varphi^\mu \quad , \quad \text{Proca} \tag{1b}$$

$$\left(\mathrm{i}\hbar D_\mu\beta^\mu - m_0c\right)\chi(x) = 0 \quad , \quad \text{Kemmer} \tag{2}$$

$$\left[\hbar^2 D_\mu(\gamma^{\mu\nu} - g^{\mu\nu})D_\nu - 2m_0^2c^2\right]\chi(x) = 0 \quad , \quad \text{Weinberg–Shay–Good} \quad . \tag{3}$$

Since observable massive charged mesons are not elementary particles, but composed of quarks, as are the nucleons, we expect an anomalous coupling [in the case of the proton (cf. Exercise 9.11)]. In the simplest cases it can appear in form of a dipole or quadrupole coupling like

$$\mathcal{L}_{\text{dipole}} \sim b^{\mu\nu}F_{\mu\nu} \quad , \quad \mathcal{L}_{\text{quadrupole}} \sim c^{\lambda\mu\nu}\partial_\lambda F_{\mu\nu} \quad . \tag{4}$$

The tensors $b^{\mu\nu}$ and $c^{\lambda\mu\nu}$ are bilinear forms in the vector fields formed with matrices of the algebra of the correspondling theory.

In the case of free vector mesons we saw that the Kemmer, as well as the Weinberg–Shay-Good equations can be brought into Proca form. In doing this with minimal and anomalous couplings, the algebra becomes more involved. In these calculations one finds that the anomalous couplings in the Proca equations transform into the corresponding anomalous terms of the Kemmer equations and that the minimal coupling of the Proca theory is equivalent to that of the Kemmer theory. The Weinberg–Shay–Good theory exhibits different behaviour: after transformation into the Proca form the minimal coupling in $\mathcal{L}_{\text{WSG}}$ creates an anomalous dipole moment. However, since the Proca theory cannot exclude an anomalous dipole moment, a measurable difference of the three theories cannot be found (at least in the first order of the coupling constant $\alpha = e^2/\hbar c$).

EXERCISE

15.20 A Useful Relation

Problem. Prove the relation

$$\gamma_5\hat{\sigma}^{\mu\nu} = \tfrac{1}{2}\mathrm{i}\varepsilon^{\mu\nu\sigma\varrho}\hat{\sigma}_{\sigma\varrho} \quad .$$

Solution. To prove this relation we take the standard representation, for which it holds [cf. (2.13) as well as (3.57)] that

$$\begin{aligned}
\gamma_5 &= \begin{pmatrix} 0 & \mathbb{1} \\ \mathbb{1} & 0 \end{pmatrix} \quad , \\
\hat{\sigma}^{0i} &= \mathrm{i}\begin{pmatrix} 0 & \hat{\sigma}^i \\ \hat{\sigma}^i & 0 \end{pmatrix} \quad , \\
\hat{\sigma}^{ij} &= \varepsilon^{ij}{}_k \begin{pmatrix} \hat{\sigma}^k & 0 \\ 0 & \hat{\sigma}^k \end{pmatrix} \quad , \\
\hat{\Sigma}^i &= \begin{pmatrix} \hat{\sigma}^i & 0 \\ 0 & \hat{\sigma}^i \end{pmatrix} = \tfrac{1}{2}\varepsilon^{ijk}\hat{\sigma}_{jk} \quad .
\end{aligned}$$

Hence we write

$$\begin{aligned}
\gamma_5\hat{\sigma}^{0i} &= \mathrm{i}\begin{pmatrix} \hat{\sigma}^i & 0 \\ 0 & \hat{\sigma}^i \end{pmatrix} = \frac{\mathrm{i}}{2}\varepsilon^{ijk}\hat{\sigma}_{jk} = \frac{\mathrm{i}}{2}\varepsilon^{0ijk}\hat{\sigma}_{jk} = \frac{\mathrm{i}}{2}\varepsilon^{0i\mu\nu}\hat{\sigma}_{\mu\nu} \quad , \\
\gamma_5\hat{\sigma}^{ij} &= \varepsilon^{ijk}\begin{pmatrix} 0 & \hat{\sigma}_k \\ \hat{\sigma}_k & 0 \end{pmatrix} = -\mathrm{i}\varepsilon^{ijk}\hat{\sigma}^{0k} = +\mathrm{i}\varepsilon^{ijk0}\hat{\sigma}^{k0} = \frac{\mathrm{i}}{2}\varepsilon^{ij\mu\nu}\hat{\sigma}_{\mu\nu} \quad .
\end{aligned}$$

Since the transition to other representations does not change anything, the relation is thus proven in general.

15.6 Spin-$\frac{3}{2}$ Fields

For simplicity we restrict ourselves to fields with finite mass. For $s = \frac{3}{2}$ the Bargmann–Wigner multispinor has three indices, and it is totally symmetric with respect to these. The equations of motion (15.10) read as

$$(\mathrm{i}\hbar\gamma\cdot\partial - m_0c)_{\alpha\alpha'}\,\Psi_{\alpha\beta\gamma}(x) = 0 \quad , \tag{15.65a}$$

$$(\mathrm{i}\hbar\gamma\cdot\partial - m_0c)_{\beta\beta'}\,\Psi_{\alpha\beta'\gamma}(x) = 0 \quad , \tag{15.65b}$$

$$(\mathrm{i}\hbar\gamma\cdot\partial - m_0c)_{\gamma\gamma'}\,\Psi_{\alpha\beta\gamma'}(x) = 0 \quad . \tag{15.65c}$$

We now try to expand the field in totally symmetric matrices similar to (15.29). Because of the total symmetry in the first two indices we make the ansatz [cf. (15.29)]

$$\Psi_{\alpha\beta\gamma}(x) = \frac{m_0c}{\hbar}(\gamma_\mu\hat{C})_{\alpha\beta}\psi^\mu{}_\gamma(x) + \frac{1}{2}(\hat{\sigma}_{\mu\nu}\hat{C})_{\alpha\beta}\psi^{\mu\nu}_\gamma(x) \quad . \tag{15.66}$$

Here $\psi^{\mu}{}_{\gamma}(x)$ is a "vector spinor" ($\mu = 0, 1, 2, 3$ is a Lorentz index, whereas $\gamma = 1, 2, 3, 4$ is a spinor index), $\psi^{\mu\nu}{}_{\gamma}(x)$ transforms like the product of a covariant antisymmetric tensor and a spinor. Symmetries in the indices β and γ, and consequently also the symmetry with respect to α, β, and γ, is guaranteed if the coefficients (15.26) of the expansion of the matrix $(\Psi_\alpha)_{\beta\gamma}$ in the basis of a Clifford algebra also vanish, as was the case in the expansion of $(\Psi_\gamma)_{\alpha\beta}$. One gets these coefficients by contracting (15.66), with respect to the indices β and γ, with the matrices $\hat{C}^{-1}_{\beta\gamma}$, $(\hat{C}^{-1}\mathrm{i}\gamma_5)_{\beta\gamma}$ and $(\hat{C}^{-1}\gamma_5\gamma^\mu)_{\beta\gamma}$, which yields, for example,

$$\begin{aligned}\Psi_{\alpha\beta\gamma}\hat{C}^{-1}_{\beta\gamma} &= \frac{m_0c}{\hbar}(\gamma_\mu\hat{C})_{\alpha\beta}\hat{C}^{-1}_{\beta\gamma}\psi^{\mu}{}_{\gamma}(x) + \frac{1}{2}(\hat{\sigma}_{\mu\nu}\hat{C})_{\alpha\beta}\hat{C}^{-1}_{\beta\gamma}\psi^{\mu\nu}{}_{\gamma}(x) \\ &= \frac{m_0c}{\hbar}(\gamma_\mu)_{\alpha\gamma}\psi^{\mu}{}_{\gamma}(x) + \frac{1}{2}(\hat{\sigma}_{\mu\nu})_{\alpha\gamma}\psi^{\mu\nu}{}_{\gamma}(x) = 0 \quad .\end{aligned}$$

In this way one gets the three constraint equations

$$\frac{m_0c}{\hbar}\gamma_\mu\psi^\mu(x) + \frac{1}{2}\hat{\sigma}_{\mu\nu}\psi^{\mu\nu}(x) = 0 \quad , \tag{15.67a}$$

$$\frac{m_0c}{\hbar}\gamma_\mu\gamma_5\psi^\mu(x) + \frac{1}{2}\hat{\sigma}_{\mu\nu}\gamma_5\psi^{\mu\nu}(x) = 0 \quad , \tag{15.67b}$$

$$\frac{m_0c}{\hbar}\gamma_\mu\gamma_5\gamma_\lambda\psi^\mu(x) + \frac{1}{2}\hat{\sigma}_{\mu\nu}\gamma_5\gamma_\lambda\psi^{\mu\nu}(x) = 0 \quad , \tag{15.67c}$$

where we have not written out the spinor indices explicitly any more. Multiplying (15.67b) by γ_5 and adding and subtracting (15.67a), respectively, leads to two equivalent conditions:

$$\gamma_\mu\psi^\mu(x) = 0 \quad , \tag{15.68a}$$

$$\hat{\sigma}_{\mu\nu}\psi^{\mu\nu}(x) = 0 \quad . \tag{15.68b}$$

Furthermore we multiply (15.67c) by γ_5 on the lhs and make use of the commutation relations (3.11) and the relation (15.32), which yields

$$\begin{aligned}&-\frac{m_0c}{\hbar}(2g_{\mu\lambda} - \gamma_\lambda\gamma_\mu)\psi^\mu(x) + \frac{1}{2}\gamma_\lambda\hat{\sigma}_{\mu\nu}\psi^{\mu\nu}(x) \\ &\quad - \mathrm{i}(g_{\mu\lambda}\gamma_\nu - g_{\nu\lambda}\gamma_\mu)\psi^{\mu\nu}(x) = 0 \quad .\end{aligned} \tag{15.69}$$

Because of (15.68) and the antisymmetry of $\psi^{\mu\nu}(x)$, (15.69) reduces to

$$\frac{m_0c}{\hbar}\psi_\lambda(x) - \mathrm{i}\gamma^\mu\psi_{\mu\lambda}(x) = 0 \quad . \tag{15.70}$$

Since the condition (15.68b) follows from (15.70) and (15.68a), namely

$$\begin{aligned}\frac{m_0c}{\hbar}\gamma_\lambda\psi^\lambda(x) - \mathrm{i}\gamma_\lambda\gamma_\mu\psi^{\mu\lambda}(x) &= 0 - \frac{\mathrm{i}}{2}(\gamma_\lambda\gamma_\mu - \gamma_\mu\gamma_\lambda)\psi^{\mu\lambda}(x) \\ &= \hat{\sigma}_{\mu\lambda}\psi^{\mu\lambda}(x) = 0 \quad ,\end{aligned}$$

we are left with the two independent conditions:

$$\gamma_\mu\psi^\mu(x) = 0 \quad , \tag{15.71a}$$

$$\frac{m_0c}{\hbar}\psi^\mu(x) = \mathrm{i}\gamma_\lambda\psi^{\lambda\mu}(x) \quad . \tag{15.71b}$$

This corresponds to 4 + 16 = 20 linear conditions between the 16 + 24 = 40 components of ψ^μ_α and $\psi^{\mu\nu}_\alpha$, so that the number of independent conditions reduces to 20, which is the correct number of components of totally symmetric tensor of the third rank in a four-dimensional vector space. From the Bargmann–Wigner equations (15.65a) and (15.65b) then follows:

$$\begin{aligned}&\frac{m_0c}{\hbar}(\mathrm{i}\hbar\gamma\cdot\partial - m_0c)_{\alpha\alpha'}(\gamma_\mu\hat{C})_{\alpha'\beta}\psi^\mu{}_\gamma(x)\\&\quad+\frac{1}{2}(\mathrm{i}\hbar\gamma\cdot\partial - m_0c)_{\alpha\alpha'}(\hat{\sigma}_{\mu\nu}\hat{C})_{\alpha'\beta}\psi^{\mu\nu}{}_\gamma(x)=0 \quad ,\end{aligned} \tag{15.72a}$$

$$\begin{aligned}&\frac{m_0c}{\hbar}\gamma_\mu\hat{C}_{\alpha\beta'}(\mathrm{i}\hbar\gamma\cdot\partial - m_0c)_{\beta\beta'}\psi^\mu{}_\gamma(x)\\&\quad+\frac{1}{2}(\hat{\sigma}_{\mu\nu}\hat{C})_{\alpha\beta'}(\mathrm{i}\hbar\gamma\cdot\partial - m_0c)_{\beta\beta'}\psi^{\mu\nu}{}_\gamma(x)=0 \quad .\end{aligned} \tag{15.72b}$$

In (15.72b), again the products $\gamma_\mu\hat{C}\gamma_\nu^{\mathrm{T}}$ and $\sigma_{\mu\nu}\hat{C}\gamma_\lambda^{\mathrm{T}}$ appear [cf. (15.30)]. Using the properties of $\hat{C}$ we rewrite (15.72) in the form

$$\begin{aligned}&\frac{m_0c}{\hbar}\left[(\mathrm{i}\hbar\gamma_\nu\partial^\nu - m_0c)\gamma_\mu\hat{C}\right]_{\alpha\beta}\psi^\mu{}_\gamma(x)\\&\quad+\frac{1}{2}\left[(\mathrm{i}\hbar\gamma_\lambda\partial^\lambda - m_0c)\hat{\sigma}_{\mu\nu}\hat{C}\right]_{\alpha\beta}\psi^{\mu\nu}{}_\gamma(x)=0 \quad ,\end{aligned} \tag{15.73a}$$

$$\begin{aligned}&\frac{m_0c}{\hbar}\left[\gamma_\mu(\mathrm{i}\hbar\gamma_\nu\partial^\nu - m_0c)\hat{C}\right]_{\alpha\beta}\psi^\mu{}_\gamma(x)\\&\quad+\frac{1}{2}\left[\hat{\sigma}_{\mu\nu}(\mathrm{i}\hbar\gamma_\lambda\partial^\lambda + m_0c)\hat{C}\right]_{\alpha\beta}\psi^{\mu\nu}{}_\gamma(x)=0 \quad .\end{aligned} \tag{15.73b}$$

The difference of (15.73a) and (15.73b) yields

$$\begin{aligned}&2m_0c(\hat{\sigma}_{\nu\mu}\hat{C})_{\alpha\beta}\partial^\nu\psi^\mu{}_\gamma(x) - \frac{2m_0^2c^2}{\hbar}(\gamma_\mu\hat{C})_{\alpha\beta}\psi^\mu{}_\gamma(x)\\&\quad - m_0c(\hat{\sigma}_{\mu\nu}\hat{C})_{\alpha\beta}\psi^{\mu\nu}{}_\gamma(x)+\frac{1}{2}\mathrm{i}\hbar\left([\gamma_\lambda,\hat{\sigma}_{\mu\nu}]\hat{C}\right)_{\alpha\beta}\partial^\lambda\psi^{\mu\nu}{}_\gamma(x)=0 \quad .\end{aligned}$$

Because of (15.32), this corresponds to

$$\begin{aligned}&\frac{m_0}{\hbar}(\hat{\sigma}_{\nu\mu}\hat{C})_{\alpha\beta}\left[\partial^\nu\psi^\mu{}_\gamma(x) - \partial^\mu\psi^\nu{}_\gamma(x) + \psi\mu\nu_\gamma(x)\right]\\&\quad - (\gamma_\mu\hat{C})_{\alpha\beta}\left[\frac{2m_0^2c^2}{\hbar^2}\psi^\mu{}_\gamma(x) + 2\partial_\mu\psi^{\mu\nu}{}_\gamma(x)\right]=0 \quad .\end{aligned} \tag{15.74}$$

The coefficients of the linear basis matrices $\hat{\sigma}_{\mu\nu}\hat{C}$ and $\gamma_\mu\hat{C}$ must vanish. For $m_0 \neq 0$ this yields similar relations as were found in (15.34) and (15.35) for the Proca fields:

$$\psi^{\mu\nu}(x) = \partial^\mu\psi^\nu(x) - \partial^\nu\psi^\mu(x) \quad , \tag{15.75a}$$

$$\partial_\mu\psi^{\mu\nu}(x) = -\frac{m_0^2c^2}{\hbar^2}\psi^\mu(x) \quad . \tag{15.75b}$$

Because of the antisymmetry of $\psi^{\mu\nu}(x)$, then from (15.75b) it again follows that

$$\partial_\mu\psi^\mu{}_\gamma(x) = 0 \quad , \tag{15.76}$$

and therefore also that

$$\Box\psi^\mu{}_\gamma(x) + \frac{m_0^2c^2}{\hbar^2}\psi^\mu{}_\gamma(x) = 0 \quad . \tag{15.77}$$

Thus each spinor component of $\psi^\mu(x)$ fulfils the Proca equation, and the tensor field $\psi^{\mu\nu}(x)$ is uniquely determined by ψ^μ. Correspondingly the third Bargmann–Wigner equation yields

$$\begin{aligned}\frac{m_0c}{\hbar}(\gamma_\mu\hat{C})_{\alpha\beta}\left[(\mathrm{i}\hbar\gamma\cdot\partial - m_0c)\psi^\mu(x)\right]_\gamma \\ + \frac{1}{2}(\hat{\sigma}_{\mu\nu}\hat{C})_{\alpha\beta}\left[(\mathrm{i}\hbar\gamma\cdot\partial - m_0c)\psi^{\mu\nu}(x)\right]_\gamma = 0 \quad .\end{aligned} \tag{15.78}$$

However, this is automatically fulfilled since [because of (15.71a)], (15.71b) and (15.75a) combine to form

$$m_0c\psi^\mu = \hbar\mathrm{i}\gamma_\lambda\psi^{\lambda\mu} = \hbar\mathrm{i}\gamma_\lambda\left(\partial^\lambda\psi^\mu - \partial^\mu\psi^\lambda\right) = \mathrm{i}\hbar\gamma\cdot\partial\psi^\mu \quad ,$$

which is just the Dirac equation for each Lorentz component of $\psi^\mu(x)$. Furthermore from (15.75) it also follows that

$$(\mathrm{i}\hbar\gamma\cdot\partial - m_0c)\,\psi^{\mu\nu}(x) = 0 \quad , \tag{15.79}$$

so that the validity of (15.78) has been proven. Thus the equations of motion for the spin-$\frac{3}{2}$ field reduce to the Dirac equation for the independent vector-spinor field

$$(\mathrm{i}\hbar\gamma\cdot\partial - m_0c)\psi^\mu(x) = 0 \quad , \tag{15.80a}$$

together with the four constraints

$$\gamma_\mu\psi^\mu(x) = 0 \quad . \tag{15.80b}$$

The "gauge" condition (15.76) can be derived by multiplying (15.80a) with γ_μ to give

$$\begin{aligned}0 &= \left(\mathrm{i}\hbar\gamma_\mu\gamma_\nu\partial^\nu - m_0c\gamma_\mu\right)\psi^\mu(x) = 2\mathrm{i}\hbar\left(g_{\mu\nu} - \tfrac{1}{2}\gamma_\nu\gamma_\mu\right)\partial^\nu\psi^\mu(x) \\ &= 2\mathrm{i}\hbar\partial_\mu\psi^\mu(x) \quad ,\end{aligned}$$

so that this is not an independent relation. The equations (15.80) are known as the ***Rarita–Schwinger** equation*.[6] They can also be combined into one single equation, namely

$$\begin{aligned}&\left[(\mathrm{i}\hbar\gamma\cdot\partial - m_0c)g_{\mu\lambda} - \tfrac{1}{3}\mathrm{i}\hbar(\gamma_\mu\partial_\lambda + \partial_\mu\gamma_\lambda) + \tfrac{1}{3}\gamma_\mu(\mathrm{i}\hbar\gamma\cdot\partial + m_0c)\gamma_\lambda\right] \\ &\quad\times\psi^\lambda(x) = 0 \quad ,\end{aligned} \tag{15.81}$$

since multiplying this equation by γ^μ yields

$$\left(\tfrac{2}{3}\mathrm{i}\hbar\partial_\lambda + \tfrac{1}{3}m_0c\gamma_\lambda\right)\psi^\lambda(x) = 0 \quad , \tag{15.82a}$$

whereas the application of ∂^μ results in

$$\left[\gamma\cdot\partial\left(\tfrac{2}{3}\mathrm{i}\hbar\partial_\lambda + \tfrac{1}{3}m_0c\gamma_\lambda\right) - m_0\partial_\lambda\right]\psi^\lambda(x) = 0 \quad . \tag{15.82b}$$

Inserting (15.82a) into (15.82b), then because of $m_0 \neq 0$, it follows that

$$\partial_\lambda\psi^\lambda(x) = 0 \quad ,$$

---

[6] W. Rarita, J. Schwinger: Phys. Rev. **60**, 61 (1941).

and thus (15.76), and (15.82a) are equivalent to (15.71a), i.e.

$$\gamma_\lambda \psi^\lambda(x) = 0 \quad .$$

With these two conditions (15.81) immediately reduces to (15.80a). Combining the four four-spinors $\psi^\lambda(x)$ into a 16-component spinor

$$\chi = \begin{pmatrix} \psi^0 \\ \psi^1 \\ \psi^2 \\ \psi^3 \end{pmatrix} \quad ,$$

one can write (15.81) as

$$(\mathrm{i}\hbar\alpha_\nu \partial^\nu - \beta m_0 c)\,\chi(x) = 0 \quad , \tag{15.83}$$

where the 16×16 matrices β and α_μ have a Lorentz index as well as a spinor index:

$$(\alpha_\nu)_{\mu\lambda} =: \gamma_\nu g_{\mu\lambda} - \tfrac{1}{3}\gamma_\mu(g_{\nu\lambda} - \gamma_\nu\gamma_\lambda) - \tfrac{1}{3}\gamma_\lambda g_{\mu\nu} \quad , \tag{15.84a}$$

$$\beta_{\mu\lambda} =: g_{\mu\lambda} - \tfrac{1}{3}\gamma_\mu\gamma_\lambda \quad . \tag{15.84b}$$

Each element of these matrices with the indices μ, λ is a 4×4 matrix in the spinor indices. Now one has the equation

$$\beta_{\mu\lambda}\left(g^\lambda{}_\nu - \gamma^\lambda\gamma_\nu\right) = \left(g_\mu{}^\lambda - \gamma_\mu\gamma^\lambda\right)\beta_{\lambda\nu} = g_{\mu\nu} \quad , \tag{15.85}$$

from which it follows that the matrix β has an inverse,

$$\left(\beta^{-1}\right)_{\mu\nu} = g_{\mu\nu} - \gamma_\mu\gamma_\nu \quad . \tag{15.86}$$

Using

$$\begin{aligned}
(\Gamma_\nu)_{\mu\sigma} &= \left(\beta^{-1}\alpha_\nu\right)_{\mu\sigma} = \left(\beta^{-1}\right)_{\mu\lambda}(\alpha_\nu)^\lambda{}_\sigma \\
&= (g_{\mu\lambda} - \gamma_\mu\gamma_\lambda)\left[\gamma_\nu g^\lambda{}_\sigma - \tfrac{1}{3}\gamma^\lambda(g_{\nu\sigma} - \gamma_\nu\gamma_\sigma) - \tfrac{1}{3}\gamma_\sigma g^\lambda{}_\nu\right] \\
&= \gamma_\nu g_{\mu\sigma} - \tfrac{1}{3}\gamma_\mu(g_{\nu\sigma} - \gamma_\nu\gamma_\sigma) - \tfrac{1}{3}\gamma_\sigma g_{\mu\nu} \\
&\quad - \gamma_\mu\gamma_\sigma\gamma_\nu + \tfrac{4}{3}\gamma_\nu(g_{\nu\sigma} - \gamma_\nu\gamma_\sigma) + \tfrac{1}{3}\gamma_\mu\gamma_\nu\gamma_\sigma \\
&= \gamma_\nu g_{\mu\sigma} + \gamma_\mu g_{\nu\sigma} - \gamma_\mu\{\gamma_\nu, \gamma_\sigma\} - \tfrac{1}{3}\gamma_\sigma g_{\mu\nu} + \tfrac{1}{3}\gamma_\mu\gamma_\nu\gamma_\sigma \\
&= \gamma_\nu g_{\mu\sigma} - \gamma_\mu g_{\nu\sigma} - \tfrac{1}{3}\gamma_\sigma g_{\nu\mu} + \tfrac{1}{3}\gamma_\mu\gamma_\nu\gamma_\sigma \quad ,
\end{aligned} \tag{15.87}$$

Equation (15.83) takes the so-called ***Fierz–Pauli–Gupta*** form

$$\left(\mathrm{i}\hbar\Gamma_\mu\partial^\mu - m_0 c\right)\chi(x) \quad . \tag{15.88}$$

Here the matrices Γ_μ obey the commutation relations

$$\sum_{(P)}\left(\Gamma_\mu\Gamma_\nu - g_{\mu\nu}\right)\Gamma_\lambda\Gamma_\varrho = 0 \quad , \tag{15.89}$$

where $\sum_{(P)}$ implies a sum over all possible permutations of μ, ν, λ and ϱ.

As supplementary literature to this chapter we recommend:
Y. Takahashi: *Introduction to Field Quantization* (Pergamon Press, Oxford 1969)
A.S. Wightman: *Invariant Wave Equations; General Theory and Applications to the External Field Problem*, Lecture Notes in Physics, Vol. 73 (Springer, Berlin, Heidelberg 1978)
D. Lurie: *Particles and Fields* (Interscience, New York 1968)
M. Riedel: *Relativistische Wellengleichung für Spin-1-Teilchen*, Diploma thesis, Institut für Theoretische Physik der Johann Wolfgang Goethe-Universität, Frankfurt am Main (1979)
G. Labonte: Nuovo Cimento **80**, 77 (1984)

15.7 Biographical Notes

BARGMANN, Valentine, * April 6, 1908 in Berlin, wrote phD thesis in Zürich (1936), Associate Professor of Mathematics in Pittsburgh (1948) and, since 1957, Professor in Princeton. Main fields of activity: quantum theory, theory of relativity and group theory.

WEYL, Claus Hugo Hermann, mathematician, * 9.11.1885 Elmshorn (Germany), † 9.12.1955, Zürich. W. was appointed as professor in 1913 at ETH Zürich, in 1930 at Göttingen and then in 1933 at Princeton. W. worked on the theory of differential and integral equations and later connected topological considerations with the conceptions of Riemannian surfaces. Meeting with A. Einstein inspired him to his fundamental publication Raum, Zeit, Materie [*Space-Time-Matter* (Dover 1950)], which contains a chapter where he tries to unify gravitation and electromagnetism. This is considered by many as the first approach to what we call nowadays gauge theories. For the representation of mathematical groups used in quantum mechanics he developed an integral method, contrary to the infinitesimal methods of S. Lie and E. Cartan. W. stood for intuitionism (a method for a constructive foundation of mathematics) and tried to maintain a close connection between mathematics, physics and philosophy in his own work.

KEMMER, Nicholas, British theoretical physicist, * 7.12.1911, St. Petersburg, trained at the Bismarckschule in Hannover as well as at the universities in Göttingen and Zürich; he took up the appointment of professor of mathematical physics in Edinburgh in 1953, in 1979 becoming emeritus professor. 1983 he as awarded the Max Planck medal of the German Physical Society.

SCHWINGER, Julian Seymour, * 12.03.1918, New York, Professor at Harvard University and California University, contributed fundamentals to QED, discovered charge and mass renormalization, with the aid of which he calculated the Lamb shift. For this he got, together with R. Feynman and S. Tomonaga the Nobel Prize in Physics in 1965. Moreover he worked on quantum field theory, many-body problems, etc.

FIERZ, Markus, * 1912, Basel, assistent to W. Pauli, 1944–60 professor at the university of Basel, from 1960 professor at ETH Zürich, 1959 director of the Theoretical Division at CERN. He made various fundamental contributions to theoretical physics, e.g. the "Fierz transformations" [see W. Greiner, B. Müeller: *Gauge Theory of Weak Interactions*, 2nd ed. (Springer, Berlin, Heidelberg 1996)].

GUPTA, Suraj Narayan, * 1.12.1924, Haryana (India), professor at the Wayne state university (since 1956). Main fields of research: theory of relativity, gravitation, quantum electrodynamics, nuclear physics, high-energy physics.

16. Lorentz Invariance and Relativistic Symmetry Principles

16.1 Orthogonal Transformations in Four Dimensions

We consider the four-dimensional space with coordinates x_μ ($\mu = 0, 1, 2, 3$), which – assuming the most general case – may be complex numbers. The absolute value of the position vector is given by (summation convention)

$$s = \sqrt{g_{\mu\nu} x^\mu x^\nu} = \sqrt{x_\mu x^\mu} \quad , \tag{16.1}$$

and the length s can also take complex values. Examining an orthogonal transformation $a_{\mu\nu}$, which relates each point with coordinates x_μ to new ones x'^μ:

$$x'^\nu = a^\nu{}_\mu x^\mu \quad , \tag{16.2}$$

the absolute value should remain unchanged by this transformation (this is the fundamental, defining condition for orthogonal transformations), i. e.

$$s' = \sqrt{x'_\mu x'^\mu} = \sqrt{x_\mu x^\mu} = s \quad . \tag{16.3}$$

From this relation $x'^\mu x'_\mu = a^\mu{}_\sigma a_\mu{}^\tau x_\sigma x_\tau = x^\sigma x_\sigma$ it directly follows that

$$a^\mu{}_\sigma a_\mu{}^\tau = \delta^\tau \sigma \quad (\sigma, \tau = 0, 1, 2, 3) \quad , \tag{16.4}$$

implying that the $a_{\mu\nu}$ generate a linear orthogonal transformation. Equivalently to (16.2) and (16.3), (16.4) can be envisaged as a defining equation for orthogonal transformations: Sequential application of two transformations gives

$$x''^\nu = a'^\nu{}_\mu x'^\mu = a'^\nu{}_\mu a^\mu{}_\sigma x^\sigma = \beta^\nu{}_\sigma x^\sigma \quad , \tag{16.5}$$

where the transformation $\beta^\nu{}_\sigma$ again represents an orthogonal transformation, since

$$\beta_\sigma{}^\nu \beta^\sigma{}_\varrho = a'_\sigma{}^\mu a_\mu{}^\nu a'^\sigma{}_\varepsilon a^\varepsilon \varrho = \delta^\mu{}_\varepsilon a_\mu{}^\nu a^\varepsilon{}_\varrho = \delta^\nu{}_\varrho \quad . \tag{16.6}$$

Thus the orthogonal transformations $a^\nu{}_\mu$ form a group. The unit element of this group is given by $a_\mu{}^\nu = \delta_\mu{}^\nu$ (the identity transformation) and the existence of the inverse elements follows from (16.4) ($a \cdot a^{-1} = 1\!\!1$). Accordingly all conditions characterizing the group structure are fulfilled (the associative law is also valid because the transformations $a^\nu{}_\mu$ are linear and orthogonal).

This group of homogeneous, linear transformations that keep the distance between two points invariant is called a *group of the four-dimensional, complex, orthogonal transformations.* In analogy to the three-dimensional group O(3) we use the compact notation O(4) [more precisely, $O(4, C)$ because this group is defined over the field C of the complex numbers].

At first $a_{\mu\nu}$ has sixteen components; however, due to the ten orthogonality relations (16.1) only six independent (complex) parameters remain. Additional restrictions for the $a_{\mu\nu}$ yield the subgroup of $0(4, C)$: Let us assume all x_μ to be real, then the same must hold for all $a_{\mu\nu}$. We are led to the *real four-dimensional group* $O(4, R)$, whose group properties follow from those of $O(4, C)$.

Furthermore, we may consider the case that the coordinates entering the scalar product of two vectors may occur with different sign, eg.

$$ds^2 = dt^2 - dx^2 = -dx^2 + dt^2 \quad .$$

The orthogonal group keeping this line element invariant is called $O(3, 1)$. (In general the group $O(p, q)$ leaves a bilinear form $-x_1^2 - x_2^2 - \ldots - x_p^2 + x_{p+1}^2 + \ldots + x_{p+q}^2$ invariant.) This non-symmetric expression for the line element can also be treated in the framework of $O(4, C)$, if one does choose the coordinates x_1, x_2, x_3 as real and the coordinate $x_0 = it$ as purely imaginary – the corresponding space being called *Minkowski space*. In that case the transformed coordinates x'_μ must also have the same properties, which yields for the elements $a_{\mu\nu}$:

$$\left.\begin{array}{ll} a_{ik} & (i, k = 1, 2, 3) \\ a_{00} & \end{array}\right\} \text{real} \quad ,$$
$$a_{0i}, a_{i0} \quad (i = 1, 2, 3) \quad \text{imaginary} \quad . \tag{16.7}$$

The group, whose elements are restricted by the conditions (16.7), is called the *homogeneous Lorentz group* L of the Minkowski space, and the elements are given by the Lorentz transformations.

Further subgroups already known to us can be obtained by the restrictions $a_{i0} = a_{0i} = 0$ and $a_{00} = 1$. The group with these properties represents the complex, three-dimensional group $O(3, C)$ [O(3) and O(4) always means $O(3, C)$ and $O(4, C)$, respectively]. For real x_μ we are led to the real, three-dimensional group $O(3, R)$. In the following we shall discuss the groups O(4) and L in more detail.

16.2 Infinitesimal Transformations and the Proper Subgroup of O(4)

First we write down an infinitesimal element of O(4):

$$a_\mu{}^\nu = \delta_\mu{}^\nu + \varepsilon_\mu{}^\nu \quad \left(|\varepsilon_\mu{}^\nu| \ll 1\right) \quad . \tag{16.8}$$

The transformation of x^ν then reads

$$x'^\nu = a^\nu{}_\mu = \left(\delta^\nu{}_\mu + \varepsilon^\nu{}_\mu\right) x^\mu \quad , \tag{16.9}$$

from which, according to the orthogonality relations for a_μ^ν, it follows that

$$\begin{aligned} a_\mu{}^\nu a^\mu{}_\sigma = \delta^\nu{}_\sigma &= \left(\delta_\mu{}^\nu + \varepsilon_\mu{}^\nu\right)\left(\delta^\mu{}_\sigma + \varepsilon^\mu{}_\sigma\right) \\ &= \delta_\mu{}^\nu \delta^\mu{}_\sigma + \delta_\mu{}^\nu \varepsilon^\mu{}_\sigma + \varepsilon_\mu{}^\nu \delta^\mu{}_\sigma + O(\varepsilon^2) \\ &= \delta^\nu{}_\sigma + \varepsilon^\nu{}_\sigma + \varepsilon_\sigma{}^\nu + O(\varepsilon^2) \quad . \end{aligned}$$

In order to satisfy the orthogonality relations at least up to terms of the order $O(\varepsilon^2)$,

$$\varepsilon^\nu{}_\sigma = -\varepsilon_\sigma{}^\nu \tag{16.10}$$

must be valid. Since

$$\varepsilon_\sigma{}^\nu = g^{\nu\varrho}\varepsilon_{\sigma\varrho} \quad , \quad \varepsilon^\nu{}_\sigma = g^{\nu\varrho}\varepsilon_{\varrho\sigma} \quad ,$$

we also obtain [in view of (16.3)] $\varepsilon_{\varrho\sigma} = \varepsilon_{\sigma\varrho}$, i. e. the infinitesimal quantities $\varepsilon_{\mu\nu}$ have to be antisymmetric. In accordance with our result obtained previously, this implies that there exist six independent parameters of an infinitesimal transformation. In the case of O(4) the $\varepsilon_{\mu\nu}$ are arbitrary complex numbers, while they are all real for $O(4,R)$. As already mentioned before, the group parameters of L ε_{ik} $(i,k = 1,2,3)$ are real and $\varepsilon_{i0} = \varepsilon_{0i}$ are imaginary.

In order to construct generators of O(4) group transformations we rewrite (16.9) as

$$x'^\nu = \left(\delta^\nu\mu + \alpha_s(\hat{I})^{\nu(s)}_{\ \ \mu}\right) x^\mu \quad , \tag{16.11}$$

where α_s denote the parameters of the transformation. In order to understand the meaning of the $(\hat{I})^{\nu(s)}_{\ \ \mu}$ let us have a look at the transformation of a vector with components A^σ:

$$A'^\varrho = D^\varrho{}_\sigma(\alpha_i)A^\sigma \quad .$$

$D^\varrho{}_\sigma$ stands for the transformation matrix dependent on the parameter α_i. Expansion of the matrix elements for small α_i yields

$$A'^\varrho = \left(\delta^\varrho{}_\sigma + \sum_i \left(\frac{\partial D^\varrho{}_\sigma}{\partial\alpha_i}\right)_{\alpha=0} \alpha_i + O(\alpha^2)\right) A^\sigma \quad , \tag{16.12}$$

and comparison with (16.11) shows that

$$(\hat{I})^{\nu(s)}_{\ \ \mu} = \left(\frac{\partial D^\nu{}_\mu}{\partial\alpha_s}\right)_{\alpha=0} \tag{16.13}$$

is valid. Such a Taylor expansion about the identity transformation ($\alpha_i = 0$) is only possible if the $D^\nu{}_\mu$ are continuously differential functions of the parameters α_i; however, this is the essential condition for a group to be a Lie group, and since all groups considered in the following are Lie groups [Lorentz group, O(4), $O(4,R)$, ...], the improper Lorentz transformations will be excluded from this discussion. In our case the parameters α_s, stand for the infinitesimal quantities $\varepsilon^\varrho{}_\sigma$, which means that s is the short-hand notation for two indices ($s \to \varrho\sigma$). Accordingly the infinitesimal operators (generators) $(\hat{I})^\nu{}_\mu$ have to be characterized not only by s, but by two indices ϱ, σ. In order to avoid confusion and to distinguish between four-dimensional matrix indices and generator indices, we write the latter in brackets[1] and rewrite (16.11) as

[1] Note, this notation appears somewhat different from the one used in Chap. 3 [e. g. (3.44)ff]. It is assumed that the reader will be able to establish the connection without major problems.

$$x'^{\nu} = \left(\delta^{\nu}{}_{\mu} + \frac{1}{2} \sum_{\sigma\varrho} \varepsilon_{\sigma\varrho} \left(\hat{I}^{\nu}{}_{\mu} \right)^{(\sigma\varrho)} \right) x^{\mu} \quad , \tag{16.14}$$

Since we sum over ϱ and σ, each term is counted twice. For this reason we have to multiply the sum by the factor 1/2. In view of the antisymmetry of $\varepsilon_{\varrho\sigma}$, we also choose $\left(\hat{I}^{\nu}{}_{\mu}\right)^{(\sigma\varrho)}$ to be antisymmetric with respect to the indices ϱ, σ. Parts symmetric in ϱ, σ would not contribute when contracting with the antisymmetric $\varepsilon_{\varrho\sigma}$. Thus we obtain six infinitesimal operators $\left(\hat{I}^{\nu}{}_{\mu}\right)^{(\sigma\varrho)}$, the *generators of the orthogonal group* O(4). Comparing (16.14) with (16.9) we are led to a defining equation for $\hat{I}^{\nu}{}_{\mu}$:

$$\frac{1}{2} \sum_{\sigma\varrho} \varepsilon_{\sigma\varrho} \left(\hat{I}^{\nu}{}_{\mu} \right)^{(\sigma\varrho)} = \varepsilon^{\nu}{}_{\mu} \quad . \tag{16.15}$$

First we lower the tensor index ν:

$$\sum_{\sigma,\varrho} \frac{1}{2} \varepsilon_{\sigma\varrho} g_{\tau\nu} \left(\hat{I}^{\nu}{}_{\mu} \right)^{(\sigma\varrho)} = g_{\tau\nu} \varepsilon^{\nu}{}_{\mu} = \varepsilon_{\tau\mu} \quad ,$$

or

$$\sum_{\sigma,\varrho} \frac{1}{2} \varepsilon_{\sigma\varrho} \left(\hat{I}_{\tau\mu} \right)^{(\sigma\varrho)} = \varepsilon_{\tau\mu} \quad ,$$

respectively. Since the sum over ϱ and σ only contributes for $\varrho = \mu$ and $\sigma = \tau$, then it must terminate. Accordingly an ansatz for a solution of (16.15) must look similar to $(\hat{I}_{\mu\nu})^{(\sigma\varrho)} = g^{\sigma}{}_{\mu} g^{\varrho}{}_{\nu}$. Taking into account the antisymmetry with respect to μ, ν, it follows that

$$\left(\hat{I}_{\mu\nu} \right)^{(\sigma\varrho)} = g^{\sigma}{}_{\mu} g^{\varrho}{}_{\nu} - g^{\sigma}{}_{\nu} g^{\varrho}{}_{\mu} = \left(-\hat{I}_{\nu\mu} \right)^{(\sigma\varrho)} \quad , \tag{16.16}$$

and the antisymmetry with respect to σ, ϱ also becomes obvious:

$$\left(\hat{I}_{\mu\nu} \right)^{(\sigma\varrho)} = - \left(\hat{I}_{\mu\nu} \right)^{(\varrho\sigma)} \quad . \tag{16.17}$$

With the aid of the relation $g^{\mu}{}_{\varrho} = g^{\mu\nu} g_{\varrho\nu} = \delta^{\mu}{}_{\varrho}$ we can directly denote the *matrix representation of the generators*, e. g.

$$\left(\hat{I}_{\mu\nu} \right)^{(10)} = g^{1}{}_{\mu} g^{0}{}_{\nu} - g^{1}{}_{\nu} g^{0}{}_{\mu} = \delta^{1}{}_{\mu} \delta^{0}{}_{\nu} - \delta^{1}{}_{\nu} \delta^{0}{}_{\mu} = \begin{pmatrix} 0 & 1 & 0 & 0 \\ -1 & 0 & 0 & 0 \\ 0 & 0 & 0 & 0 \\ 0 & 0 & 0 & 0 \end{pmatrix} \quad ,$$

and for mixed tensors

$$\left(\hat{I}^{\mu}{}_{\nu} \right)^{(10)} = g^{\mu\tau} \left(\hat{I}_{\tau\nu} \right)^{(10)} = \begin{pmatrix} 0 & -1 & 0 & 0 \\ -1 & 0 & 0 & 0 \\ 0 & 0 & 0 & 0 \\ 0 & 0 & 0 & 0 \end{pmatrix} \quad .$$

Furthermore,

$$\left(\hat{I}_{\mu\nu}\right)^{(13)} = g^1{}_\mu g^3{}_\nu - g^1{}_\nu g^3{}_\mu = \begin{pmatrix} 0 & 0 & 0 & 0 \\ 0 & 0 & 0 & -1 \\ 0 & 0 & 0 & 0 \\ 0 & 1 & 0 & 0 \end{pmatrix} \quad ,$$

$$\left(\hat{I}_{\mu\nu}\right)^{(21)} = g^2{}_\mu g^1{}_\nu - g^2{}_\nu g^1{}_\mu = \begin{pmatrix} 0 & 0 & 0 & 0 \\ 0 & 0 & 1 & 0 \\ 0 & -1 & 0 & 0 \\ 0 & 0 & 0 & 0 \end{pmatrix} \quad ,$$

$$\left(\hat{I}_{\mu\nu}\right)^{(32)} = g^3{}_\mu g^2{}_\nu - g^3{}_\nu g^2{}_\nu = \begin{pmatrix} 0 & 0 & 0 & 0 \\ 0 & 0 & 0 & 0 \\ 0 & 0 & 0 & 1 \\ 0 & 0 & -1 & 0 \end{pmatrix} \quad ,$$

$$\left(\hat{I}_{\mu\nu}\right)^{(20)} = g^2{}_\mu g^0{}_\nu - g^2{}_\nu g^0{}_\mu = \begin{pmatrix} 0 & 0 & 1 & 0 \\ 0 & 0 & 0 & 0 \\ -1 & 0 & 0 & 0 \\ 0 & 0 & 0 & 0 \end{pmatrix} \quad ,$$

$$\left(\hat{I}_{\mu\nu}\right)^{(30)} = g^3{}_\mu g^0{}_\nu - g^3{}_\nu g^0{}_\mu = \begin{pmatrix} 0 & 0 & 0 & 1 \\ 0 & 0 & 0 & 0 \\ 0 & 0 & 0 & 0 \\ -1 & 1 & 0 & 0 \end{pmatrix} \quad .$$

Now the commutation relations of $\left(\hat{I}_{\mu\nu}\right)^{(\sigma\varrho)}$ can be easily verified: for example

$$\begin{aligned}
&\left[\left(\hat{I}_{\mu\nu}\right)^{(21)}, \left(\hat{I}_{\mu\nu}\right)^{(32)}\right]_- \\
&= \begin{pmatrix} 0 & 0 & 0 & 0 \\ 0 & 0 & 1 & 0 \\ 0 & -1 & 0 & 0 \\ 0 & 0 & 0 & 0 \end{pmatrix} \begin{pmatrix} 0 & 0 & 0 & 0 \\ 0 & 0 & 0 & 0 \\ 0 & 0 & 0 & 1 \\ 0 & 0 & -1 & 0 \end{pmatrix} - \begin{pmatrix} 0 & 0 & 0 & 0 \\ 0 & 0 & 0 & 0 \\ 0 & 0 & 0 & 1 \\ 0 & 0 & -1 & 0 \end{pmatrix} \\
&\quad \times \begin{pmatrix} 0 & 0 & 0 & 0 \\ 0 & 0 & 1 & 0 \\ 0 & -1 & 0 & 0 \\ 0 & 0 & 0 & 0 \end{pmatrix} \\
&= \begin{pmatrix} 0 & 0 & 0 & 0 \\ 0 & 0 & 0 & 1 \\ 0 & 0 & 0 & 0 \\ 0 & 0 & 0 & 0 \end{pmatrix} - \begin{pmatrix} 0 & 0 & 0 & 0 \\ 0 & 0 & 0 & 0 \\ 0 & 0 & 0 & 0 \\ 0 & 1 & 0 & 0 \end{pmatrix} = \begin{pmatrix} 0 & 0 & 0 & 0 \\ 0 & 0 & 0 & 1 \\ 0 & 0 & 0 & 0 \\ 0 & -1 & 0 & 0 \end{pmatrix} \\
&= -\left(\hat{I}_{\mu\nu}\right)^{(13)} \quad .
\end{aligned}$$

Analogously we get

$$\begin{aligned}
\left[\left(\hat{I}_{\mu\nu}\right)^{(32)}, \left(\hat{I}_{\mu\nu}\right)^{(13)}\right]_- &= -\left(\hat{I}_{\mu\nu}\right)^{(21)} \quad , \\
\left[\left(\hat{I}_{\mu\nu}\right)^{(13)}, \left(\hat{I}_{\mu\nu}\right)^{(21)}\right]_- &= -\left(\hat{I}_{\mu\nu}\right)^{(32)} \quad , \\
\left[\left(\hat{I}_{\mu\nu}\right)^{(10)}, \left(\hat{I}_{\mu\nu}\right)^{(20)}\right]_- &= -\left(\hat{I}_{\mu\nu}\right)^{(21)} \quad , \qquad (16.18)
\end{aligned}$$

$$\begin{aligned}\left[(\hat{I}_{\mu\nu})^{(20)},(\hat{I}_{\mu\nu})^{(30)}\right]_- &= -\left(\hat{I}_{\mu\nu}\right)^{(32)} \quad ,\\ \left[(\hat{I}_{\mu\nu})^{(30)},(\hat{I}_{\mu\nu})^{(10)}\right]_- &= -\left(\hat{I}_{\mu\nu}\right)^{(13)} \quad .\end{aligned} \tag{16.19}$$

(There are nine further commutators left, but they appear rather seldom in applications.) All fifteen commutation relations can be summarized within a single one (for the sake of more clarity we omit the tensor indices μ and ν and note only the parameter indices σ and ϱ; we also drop the brackets ( ) on the parameter indices)

$$\left[\hat{I}^{\alpha\beta},\hat{I}^{\gamma\delta}\right]_- = -\delta^{\alpha\gamma}\hat{I}^{\beta\delta} + \delta^{\alpha\delta}\hat{I}^{\beta\gamma} + \delta^{\beta\gamma}\hat{I}^{\alpha\delta} - \delta^{\beta\delta}\hat{I}^{\alpha\gamma} \tag{16.20a}$$

or, if the parameter indices are lowered,

$$\left[\hat{I}_{\alpha\beta},\hat{I}_{\gamma\delta}\right]_- = -g_{\alpha\gamma}\hat{I}_{\beta\delta} + g_{\alpha\delta}\hat{I}_{\beta\gamma} + g_{\beta\gamma}\hat{I}_{\alpha\delta} - g_{\beta\delta}\hat{I}_{\alpha\gamma} \quad . \tag{16.20b}$$

These equations represent the Lie algebra of the group O(4). It is convenient to introduce six linearly independent combinations instead of $\hat{I}^{\sigma\varrho}$ $(k,l=1,2,3)$

$$\begin{aligned}\hat{I}^{i(+)} &= \tfrac{1}{2}\left(\hat{I}^{kl}+\hat{I}^{i0}\right) \quad ,\\ \hat{I}^{i(-)} &= \tfrac{1}{2}\left(\hat{I}^{kl}-\hat{I}^{i0}\right) \quad ,\end{aligned} \tag{16.21}$$

where i,k,l are cyclic permutations of $1,2,3$. From (16.20a) the commutation relations of $\hat{I}^{i(+)},\hat{I}^{i(-)}$ are derived as

$$\begin{aligned}\left[\hat{I}^{i(+)},\hat{I}^{k(+)}\right]_- &= -\hat{I}^{l(+)} \quad ,\\ \left[\hat{I}^{i(-)},\hat{I}^{k(-)}\right]_- &= -\hat{I}^{l(-)} \quad ,\\ \left[\hat{I}^{i(+)},\hat{I}^{k(-)}\right]_- &= 0 \quad ,\\ i,k,l \quad &= 1,2,3 \text{ cyclic} \quad .\end{aligned} \tag{16.22}$$

Since $\hat{I}^{i(+)}$ and $\hat{I}^{i(-)}$ are not hermitian, we introduce the hermitian operators

$$\hat{J}^{i(+)} = -\mathrm{i}\hat{I}^{i(+)} \quad , \quad \hat{J}^{i(-)} = -\mathrm{i}\hat{I}^{i(-)} \quad . \tag{16.23}$$

Then, it follows from (16.22) that

$$\begin{aligned}\left[\hat{J}^{i(+)},\hat{J}^{k(+)}\right]_- &= \mathrm{i}\hat{J}^{l(+)} \quad ,\\ \left[\hat{J}^{i(-)},\hat{I}^{k(-)}\right]_- &= \mathrm{i}\hat{J}^{l(-)} \quad ,\\ \left[\hat{J}^{i(+)},\hat{J}^{k(-)}\right]_- &= 0 \quad ,\\ i,k,l \quad &= 1,2,3 \text{ cyclic} \quad .\end{aligned} \tag{16.24}$$

By introducing $\hat{I}^{i(\pm)}$ we have constructed two 3×3 matrices from the 4×4 matrices $\hat{I}^{\sigma\varrho}$. Consequently we have left four-dimensional space and moved to a three-dimensional subspace, where $\hat{I}^{i(\pm)}$ are the infinitesimal generators. Therefore, each four-dimensional transformation can be performed by combining appropriate transformations on the subspaces. Because of (16.21), one has

$$\begin{aligned}&\hat{I}^{kl} = \hat{I}^{i(+)} + \hat{I}^{i(-)} \quad \text{and} \quad \hat{I}^{i0} = \hat{I}^{i(+)} - \hat{I}^{i(-)}\\ &i,k,l = 1,2,3 \quad \text{cyclic} \quad ,\end{aligned} \tag{16.25}$$

and the $\hat{J}^{i(\pm)}$ are the *generators of the three-dimensional subgroup of* O(4). As (16.24) shows, they obey the same commutation relations as the angular momentum components $\hat{L}_x$, $\hat{L}_y$ and $\hat{L}_z$; hence they are generators of an O(3) subgroup. Since we have selected *two* three-dimensional subspaces of the four-dimensional space with the operators $\hat{I}^{i(+)}$ and $\hat{I}^{i(-)}$, the reduction from O(4) to O(3) yields six operators $\hat{J}^{i(\pm)}$ $(i = 1, 2, 3)$, instead of three. This point will later on be studied more rigorously in connection with the spin, but for the moment we want to get a more intuitive understanding of the meaning of the parameters $\varepsilon^{\nu}{}_{\mu}$ in the case of the real rotation group $O(4, R)$. To do this we choose $\varepsilon^1{}_2 = -\varepsilon^2{}_1 = \varepsilon$ and all other $\varepsilon^{\nu}{}_{\mu} = 0$. The components of the transformation (16.9) then read:

$$\begin{aligned} x'^0 &= x^0 \quad , \\ x'^1 &= x^1 + \varepsilon x^2 \quad , \\ x'^2 &= x^2 - \varepsilon x^1 \quad , \\ x'^3 &= x^3 \quad . \end{aligned} \tag{16.26}$$

In fact this describes a rotation about the infinitesimal angle ε within the $(x^1 - x^2)$ plane, i. e. $\varepsilon^1{}_2$ represents a rotation angle; hence $(\hat{I}^{\nu}{}_{\mu})^{(12)}$ is the operator which effects infinitesimal rotations in the $(x^1 - x^2)$ plane. Therefore we can say, in general: $\varepsilon_{\mu\nu}$ is the rotation angle about two axes[2] x_σ and x_ϱ which are perpendicular to the $\mu\nu$ plane and, in addition, the $\mu\nu$ plane remains unchanged by this rotation. All infinitesimal rotations of the group O(4) are rotations in the six planes:

$$x_1x_2, \; x_2x_3, \; x_3x_1, \; x_1x_0, \; x_2x_0, \; x_3x_0, \quad .$$

The situation is different for the *Lorentz group* L of the Minkowski space $(x_0 = ict)$. Here ε_{12}, ε_{23} and ε_{31} are real and have the same meaning as in the case of $O(4, R)$; but ε_{10} and ε_{01} are imaginary $(\varepsilon_{10} = -\varepsilon_{01} = \mathrm{i}\varepsilon)$: if all other $\varepsilon_{\mu\nu}$ vanish, we have I

$$\begin{aligned} x'^0 &= -\mathrm{i}\varepsilon x^1 + x^0 \quad , \\ x'^1 &= x^1 + \mathrm{i}\varepsilon x^0 \quad , \\ x'^2 &= x^2 \quad , \\ x'^3 &= x^3 \quad . \end{aligned} \tag{16.27}$$

This is no rotation, but an infinitesimal *special* Lorentz transformation along the x^1 axis, which becomes clear if we look at this Lorentz transformation in more detail:

$$\begin{aligned} x'^1 &= \gamma\left(x^1 + \mathrm{i}\frac{v}{c}x^0\right) \quad , \\ x'^2 &= x^2 \quad , \\ x'^3 &= x^3 \quad , \\ x'^0 &= \gamma\left(-\mathrm{i}\frac{v}{c}x^1 + x^0\right) \quad , \end{aligned}$$

where

$$\gamma = \left(1 - \frac{v^2}{c^2}\right)^{-1/2} \quad .$$

[2] This is just a manner of speaking analogous to that describing rotations of a plane in $\mathbb{R}^3$.Mathematically it is not correct only that $\varepsilon_{\mu\nu}$ causes a rotation of the plane spanned by the μ and the ν axes into itself.

In the case of the two coordinate systems $v/c = \varepsilon \ll 1$ this yields the result (16.27); hence ε_{10} represents the ratio of the relative velocity and the velocity of light, and $(\hat{I}^{\nu}{}_{\mu})^{(10)}$ is the operator which effects a special Lorentz transformation along the x^1 axis (*Lorentz boost*). The parameters ε_{20} and ε_{30} and the operators $(\hat{I}^{\nu}{}_{\mu})^{(20)}$ and $(\hat{I}^{\nu}{}_{\mu})^{(30)}$ have analogous meanings.

By multiple application of the infinitesimal rotation we can perform a rotation about the finite angle ε_{12}:

$$x'^{\nu} = \lim_{N\to\infty}\left(\delta^{\nu}{}_{\mu} + \frac{1}{N}\left(\frac{1}{2}\sum_{\sigma\varrho}\varepsilon_{\sigma\varrho}\left(\hat{I}^{\nu}{}_{\mu}\right)^{(\sigma\varrho)}\right)\right)^{N} x^{\mu}$$

$$x'^{\nu} = \exp\left(\frac{1}{2}\sum_{\sigma\varrho}\varepsilon_{\sigma\varrho}\left(\hat{I}^{\nu}{}_{\mu}\right)^{(\sigma\varrho)}\right) x^{\mu} \quad , \tag{16.28}$$

i. e. in the case $\varepsilon_{12} \neq 0$ and all other $\varepsilon_{\mu\nu} = 0$, we have

$$x'^{\nu} = \exp\left\{\varepsilon_{12}\left(\hat{I}^{\nu}{}_{\mu}\right)^{(12)}\right\} x^{\mu} \quad . \tag{16.29}$$

Here we should note that there exist transformations in O(4) which cannot be achieved by infinitesimal transformations (i. e., the space and time inversions). Such examples will be discussed later on.

Remark. If all group elements can be created from the unit element by infinitesimal continuous variation of the parameters, then the group is called *connected* [O(4) is not connected!].

16.3 Classification of the Subgroups of O(4)

Now we want to study the properties of the group O(4) more rigorously. To that end we consider an arbitrary transformation $\hat{a} = [a^{\mu}{}_{\nu}]$. The orthogonality relations read

$$a_{\mu}{}^{\nu} a^{\mu}{}_{\sigma} = \delta^{\nu}{}_{\sigma} \tag{16.30}$$

or, in abbreviated notation,

$$\tilde{\hat{a}}\hat{a} = \mathbb{1} \quad , \tag{16.31}$$

where r denotes the transposed matrix. Therefore, for the determinants it holds that

$$\det(\tilde{\hat{a}}\hat{a}) = \det\tilde{\hat{a}} \cdot \det\hat{a} = (\det\hat{a})^2 = 1 \quad ,$$

which implies

$$\det\hat{a} = \pm 1 \tag{16.32}$$

Transformations with determinants ± 1 are called *unimodular*: the group O(4) and its subgroups are therefore unimodular. By classifying the elements into those with $\det\hat{a} = +1$ and $\det\hat{a} = -1$, respectively, O(4) is divided into two parts. First we consider the group $O(4,R)$: The part with $\det\hat{a} = +1$ is denoted by $S0(4,R)$ (SO for *special orthogonal transformations*). Besides the identity transformation, it contains all infinitesimal transformations ($\det(\delta^\nu{}_\mu + \varepsilon^\nu{}_\mu) = +1$), i. e. all finite transformations which can be assembled by these infinitesimal transformations belong to the group $SO(4,R)$ (proper four-dimensional rotations). A typical member of the other part of $O(4,R)$ with $\det\hat{a} = -1$ is the *coordinate inversion*

$$x'^k = -x^k \quad , \quad x'^0 = x^0 \quad (k = 1,2,3) \quad . \tag{16.33}$$

Since

$$a^\nu{}_\mu = \begin{pmatrix} 1 & 0 & 0 & 0 \\ 0 & -1 & 0 & 0 \\ 0 & 0 & -1 & 0 \\ 0 & 0 & 0 & -1 \end{pmatrix} \quad ,$$

this transformation belongs to the group $O(4,R)$ (all $a^\nu{}_\mu$ are real) and it cannot be decomposed into infinitesimal transformations. Furthermore this second part of $O(4,R)$ is no group, because it does not contain the identity transformation; but, together with the $SO(4,R)$, it does make up the whole group $O(4,R)$! [These statements are also valid for the $O(3,R)$.] To study the properties of the Lorentz group L in more detail, the situation is more complicated because, in addition to $\det\hat{a} = \pm 1$, we find with the orthogonality relations that

$$a_\mu{}^0 a^\mu{}_0 = a_k{}^0 a^k{}_0 + a_0{}^0 a^0{}_0 = 1 \quad .$$

Since $a^k{}_0$ is imaginary, it holds that $a_k{}^0 a^k{}_0 = -|a_k^0|^2$ and consequently

$$\left(a_0{}^0\right)^2 = 1 + |a_k{}^0|^2 \geq 1 \quad ,$$

i. e.

$$a_0{}^0 \geq 1 \quad \text{or} \quad a_0{}^0 \leq -1 \quad . \tag{16.34}$$

Thus the Lorentz group can be split up into four parts, namely:

$$\begin{aligned}
&\text{Part I} && \text{or} \quad L_+^\uparrow : \det\hat{a} = +1,\ a_0{}^0 \geq 1 \quad \text{(orthochronous LT)} \quad , \\
&\text{Part II} && \text{or} \quad L_-^\uparrow : \det\hat{a} = -1,\ a_0{}^0 \geq 1 \quad \text{(orthochronous LT)} \quad , \\
&\text{Part III} && \text{or} \quad L_-^\downarrow : \det\hat{a} = -1,\ a_0{}^0 \leq -1 \quad \text{(antichronous LT)} \quad , \\
&\text{Part IV} && \text{or} \quad L_+^\downarrow : \det\hat{a} = +1,\ a_0{}^0 \leq -1 \quad \text{(antichronous LT)} \quad ,
\end{aligned} \tag{16.35}$$

Part I is named the group of *proper Lorentz transformations*. It includes the unit transformation, infinitesimal transformations and their iterations, i. e. all spatial rotations and the special Lorentz transformations of the Minkowski space.

In the following we shall denote Part I by L_p (p = proper). L_p is a subgroup of L. On the contrary Part II or $L_-^\uparrow$ is no group, and a typical representative member is again space inversion. [Thus everything mentioned in connection with $O(4,R)$ is valid.] Together with L_p, $L_-^\uparrow$ forms the so-called *full Lorentz group* L_f, which is a subgroup of L. It includes the unit transformation, space rotations, special Lorentz transformations, space inversion as well as an combination of all members.

The basic element of Part III is *time reversal*

$$x'^i = x^i \quad (i = 1, 2, 3) \quad ,$$
$$x'^0 = -x^0 \tag{16.36}$$

that is

$$a^\nu{}_\mu = \begin{pmatrix} -1 & 0 & 0 & 0 \\ 0 & 1 & 0 & 0 \\ 0 & 0 & 1 & 0 \\ 0 & 0 & 0 & 1 \end{pmatrix} \quad ,$$
$$\det a^\nu{}_\mu = -1 \quad , \quad a^0{}_0 = -1 \quad .$$

However, $L_-^\downarrow$ does not form a group because this part does not include the unit transformation. The same is true for Part IV, one example being the total inversion of Minkowski space:

$$x'^\mu = -x^\mu \quad (\mu = 0, 1, 2, 3) \quad . \tag{16.37}$$

$L_+^\downarrow$ is not a group, either.

Together with L_f, $L_-^\downarrow$ and $L_+^\downarrow$ form the so-called *extended Lorentz group*. Because L_f (full Lorentz group) does not change the sign of the zero component of a time-like vector, it is frequently denoted *orthochronous Lorentz group*.

16.4 The Inhomogeneous Lorentz Group

The inhomogeneous Lorentz group L keeps the distance between two points of the Minkowski space invariant, and the transformation of points x^μ of the Minkowski space by L has been given by $x'^\nu = a^\nu{}_\mu x^\mu$. We now drop the requirement of homogeneity and use the transformation

$$x'^\nu = a^\nu{}_\mu x^\mu + \beta^\nu \quad . \tag{16.38}$$

It can be seen that the term β^ν cancels in the expression for the distance between two points, i. e. (16.38) also keeps the distance between two points invariant, whereby the orthogonality relation $a_\mu{}^\nu a^\mu{}_\sigma = \delta_\sigma{}^\nu$ holds once more. [Of course, the transformations (16.38) do *not* keep the length of x^μ invariant.] Furthermore β^0 has to be imaginary and the β^i $(i = 1, 2, 3)$ have to be real and obviously the β^ν characterize space-time translations. The transformations (16.38) form a group, the so-called *inhomogeneous Lorentz group* or ***Poincaré*** group P.

Like the homogeneous Lorentz group, P also disintegrates into four parts, e. g. the proper inhomogeneous Lorentz group [the group theoretical expression: SO(4) corresponds to ISO(4); I for "inhomogeneous"].

P has ten parameters (six independent $a^\nu{}_\mu$ and the four components of β^ν). The subgroups of **P** are the homogeneous Lorentz group **L** and the *four-dimensional group of translations* **S**, whose elements are given by $(x^\nu)' = x^\nu + \beta^\nu$. Because the transformations do not commute with the homogeneous Lorentz transformations, the group **P** is not the direct product of **L** and **S**, i. e.

$$a^\nu{}_\mu(x^\mu + \beta^\nu) \neq a^\nu{}_\mu x^\mu + \beta^\nu \quad . \tag{16.39}$$

We now consider the translational group and first look for the infinitesimal *generators of the transformation* $\hat{P}_\varrho$ ($\varrho = 0, 1, 2, 3$). The infinitesimal transformation is analogous to (11):

$$\psi \to \psi' = \left(\mathbb{1} + \frac{\mathrm{i}}{\hbar} \varepsilon^\varrho \hat{P}_\varrho \right) \psi \quad . \tag{16.40}$$

The ε^ϱ are the infinitesimal transformation, i.e. $x'^\varrho = x^\varrho + \varepsilon^\varrho$ ($|\varepsilon^\varrho| \ll 1$), though because of (16.12) we have

$$\psi x'^\varrho = \psi(x^\varrho + \varepsilon^\varrho) = \psi(x^\varrho) + \varepsilon^\varrho \frac{\partial \psi}{\partial x^\varrho} \quad . \tag{16.41}$$

Comparison of (16.40) and (16.41) yields

$$\hat{P}_\varrho = -\mathrm{i}\hbar \frac{\partial}{\partial x^\varrho} \quad , \tag{16.42}$$

which are the *generators of the infinitesimal translations*. As in the case of non-relativistic quantum mechanics, the momentum operator is connected with space (time) translations. Each finite translation can be generated by iterative application of (16.42), i.e. according to (16.28) and (16.29) the unitary operator

$$\hat{U} = \exp\left[\frac{\mathrm{i}}{\hbar} \beta^\varrho \hat{P}_\varrho \right] \tag{16.43}$$

is the *relativistic translation operator.*

The translational group **S** is an *abelian group*, because

$$\left[\hat{P}_\mu, \hat{P}_\nu \right] = 0 \quad . \tag{16.44}$$

However, $\hat{P}_\varrho$ does not commute with the generators $\left(\hat{I}^\nu{}_\mu \right)^{(\varrho\sigma)}$ (16.13) of the homogeneous Lorentz group and therefore does not commute with the invariants constructed with $\hat{J}^{i(+)}$ and $\hat{J}^{i(-)}$, though it is possible to show that the operators

$$\hat{P}_\mu^2 = \hat{P}_\mu \hat{P}^\mu \quad \text{and} \quad \hat{W}_\lambda = \frac{1}{2} \sum \varepsilon_{\lambda\mu\nu\varrho} \hat{P}^\mu \hat{I}^{(\nu\varrho)} \tag{16.45}$$

commute with all $\hat{I}^{(\varrho\sigma)}$ and $\hat{P}_\varrho$, i.e. *$\hat{P}_\mu^2$ and $\hat{W}_\lambda$ are the invariants (Casimir operators) of the Poincaré group.* For a free particle we have $P^\mu P_\mu = m_0^2 c^2$, the particle's mass, and the operators $\hat{W}_\lambda$ are connected with the spin of the particle, as will be seen later. Obviously the Casimir operators provide all quantities (mass, spin) necessary for the description of the free particle, which is the reason for the great importance of this group in modern elementary particle physics.

16.5 The Conformal Group

In the preceding section we have seen that the generators of the physically important Poincaré group are given by the six generators of the homogeneous Lorentz group $(\hat{I}^{\nu}{}_{\mu})^{(\varrho\sigma)}$ and the four generators $\hat{P}_{\mu}$, of the space-time translations. The Poincaré group can be extended to the 15-parametric *conformal group* of the four-dimensional Minkowski space, by considering transformations that keep the light-like line element $\mathrm{d}s^2 = (\mathrm{d}x^0)^2 - (\mathrm{d}x^1)^2 - (\mathrm{d}x^2)^2 - (\mathrm{d}x^3)^2 = 0$ invariant. That is, the conformal transformations include changes of length, but keep the angle between two vectors invariant (hence "conformal mapping"); thus we have to add five more generators to the ten generators of the Poincaré group. The first operator is that of scale transformations or *dilatation* $\hat{D}$ and it yields the transformation

$$x'^{\mu} = \varrho x^{\mu} \quad (\varrho > 0) \quad . \tag{16.46}$$

The four remaining generators $\hat{K}_{\mu}$ create the so-called *proper (special) conformal transformations*, which change the length scale point by point, thus forming position-dependent dilatations. These transformations can be written as a product of an *inversion* $\hat{I}_1$:

$$x'^{\mu} = \frac{k^2}{(x_{\nu}x^{\nu})} x^{\mu} \quad ,$$

a *translation* $\hat{T}$:

$$x''^{\mu} = x'^{\mu} - a^{\mu} \quad , \tag{16.47}$$

and a further *inversion* $\hat{I}_2$:

$$x'''^{\mu} = \frac{k^2}{\left(x''_{\nu}x^{\nu''}\right)} x''^{\mu} \quad ,$$

All together, this yields the special conformal transformation $\hat{K} = \hat{I}_2\hat{T}\hat{I}_1$:

$$x'^{\mu} = \frac{x^{\mu} - a^{\mu}x^2}{\sigma(x)} \quad , \quad \text{with} \tag{16.48}$$

$$\sigma(x) = 1 - 2a^{\nu}x_{\nu} + a^2x^2 \quad \text{and}$$

$$a^2 = a_{\nu}a^{\nu} \quad , \quad x^2 = x_{\nu}x^{\nu} \quad .$$

For the conformal group we get a total of fifteen infinitesimal generators:

$(\hat{I}^{\nu}{}_{\mu})^{(\varrho\sigma)}$: six generators of the Lorentz group } generators of the
$\hat{P}_{\mu}$: four generators of the translational group } Poincaré group
$\hat{D}$: one generator of dilatations
$\hat{K}^{\mu}$: four generators of special conformal transformations.

It is possible to give an explicit representation of $\hat{D}$ and $\hat{K}_\mu$, by

$$\begin{aligned}\hat{D} &= \mathrm{i}x^\nu \frac{\partial}{\partial x^\nu}\\ \hat{K}_\mu &= \mathrm{i}\left(2x_\mu x^\nu \frac{\partial}{\partial x^\nu} - x^2 \frac{\partial}{\partial x^\mu}\right) \quad .\end{aligned} \tag{16.49}$$

Furthermore, one may calculate the commutators of $\hat{D}$ and $\hat{K}_\mu$ with the remaining generators of the conformal group by help of the explicit representations (16.16), (16.20), (16.42) and (16.49): We find

$$\begin{aligned}\left[\hat{P}_\mu, \hat{D}\right] &= \mathrm{i}\hat{P}_\mu \quad ,\\ \left[\hat{P}_\mu, \hat{K}_\nu\right] &= 2\mathrm{i}\left(g_{\mu\nu}\hat{D} - \left(\hat{I}_{\mu\nu}\right)^{\varrho\sigma}\right) \quad ,\\ \left[\hat{D}, \left(\hat{I}_{\mu\nu}\right)^{\varrho\sigma}\right] &= 0 \quad ,\\ \left[\hat{D}, \hat{K}_\mu\right] &= \mathrm{i}\hat{K}_\mu \quad ,\\ \left[\hat{K}_\mu, \hat{K}_\nu\right] &= 0 \quad ,\\ \left[\hat{K}_\sigma, \left(\hat{I}_{\mu\nu}\right)^{(\varrho\lambda)}\right] &= \mathrm{i}\left(g^{\varrho\lambda}\hat{K}_\sigma - g^{\lambda\sigma}\hat{K}_\varrho\right) \quad .\end{aligned} \tag{16.50}$$

The importance of the conformal group is based on the fact that, e. g., the Maxwell equations without sources are not only Lorentz invariant but also conformal invariant! However problems do occur in the discussion of the Maxwell equations with source terms or of equations of motion for massive particles: Because of $m_0^2 c^2 = \hat{P}_\mu \hat{P}^\mu$ and

$$\mathrm{e}^{\mathrm{i}\theta\hat{D}}\hat{P}_\mu\hat{P}^\mu \mathrm{e}^{-\mathrm{i}\theta\hat{D}} = \mathrm{e}^{2\theta}\hat{P}_\mu\hat{P}^\mu \quad , \tag{16.51}$$

the rest mass is not invariant with respect to dilatations, i. e. equations of motion like the Dirac equation including a mass term are not conformal invariant. An exact symmetry of dilatation is thus only possible for a continuous mass spectrum or vanishing mass.

EXERCISE

16.1 Transformation Relations of the Rest Mass Under Dilatations

Problem. Show the validity of (16.51) with the help of (16.50)!

Solution. Use $[\hat{P}_\mu, \hat{D}] = \mathrm{i}\hat{P}_\mu$. For any given $\hat{A}$ and $\hat{B}$ the Champbell–Hausdorf relation holds:

$$\mathrm{e}^{-\hat{A}}\hat{B}\mathrm{e}^{\hat{A}} = \hat{B} + \frac{1}{1!}[\hat{B},\hat{A}] + \frac{1}{2!}\left[[\hat{B},\hat{A}],\hat{A}\right] + \dots \tag{1}$$

In our case we have $\hat{A} = -\mathrm{i}\theta\hat{D}$ and $\hat{B} = \hat{P}_\mu\hat{P}^\mu$, so that

$$\begin{aligned}[\hat{B},\hat{A}] &= -\mathrm{i}\theta\left[\hat{P}_\mu\hat{P}^\mu, \hat{D}\right] = -\mathrm{i}\theta\left\{\left[\hat{P}_\mu\hat{D}\right]\hat{P}^\mu + \hat{P}_\mu\left[\hat{P}^\mu, \hat{D}\right]\right\}\\ &= -\mathrm{i}\theta\left\{\mathrm{i}\hat{P}_\mu\hat{P}^\mu + \mathrm{i}\hat{P}_\mu\hat{P}^\mu\right\} = 2\theta\hat{P}_\mu\hat{P}^\mu = 2\theta\hat{B} \quad ,\end{aligned} \tag{2}$$

Exercise 16.1.

i.e. the multiple commutators in (1) always reduce to $[\hat{B},\hat{A}] = 2\theta\hat{B}$. Thus it is possible to write down the series in a closed form and perform the summation

$$\begin{aligned} \mathrm{e}^{\mathrm{i}\theta\hat{D}}\hat{P}_\mu\hat{P}^\mu\mathrm{e}^{-\mathrm{i}\theta\hat{D}} &= \hat{P}_\mu\hat{P}^\mu + 2\theta\hat{P}_\mu\hat{P}^\mu + \frac{1}{2!}(2\theta)^2\hat{P}_\mu\hat{P}^\mu + \dots \\ &= \hat{P}_\mu\hat{P}^\mu\left(1 + \frac{2\theta}{1!} + \frac{1}{2!}(2\theta)^2 + \dots\right) \\ &= \mathrm{e}^{2\theta}\hat{P}_\mu\hat{P}^\mu \quad . \end{aligned}$$

16.6 Representations of the Four-Dimensional Orthogonal Group and Its Subgroups

In this section we consider only the homogeneous proper groups.

16.6.1 Tensor Representation of the Proper Groups

We consider a linear space of dimension 4^n. One element or "vector" of this space has 4^n components, which is denoted by $\psi_{\mu_1,\mu_2,\dots,\mu_n}$ with indices μ_ν equal to $0, 1, 2, 3$. In the case of the group SO$(4,R)$ all components are real; however for $\mathrm{L_p}$ all components including an odd number of indices "0" are imaginary, all the others being real. [This will become clear at once by taking into account the law of transformation (16.52).] First we look at a proper transformation $x^\nu \to x'^\nu = a^\nu{}_\mu x^\mu$ of SO$(4,R)$ or $\mathrm{L_p}$. Here ψ should transform like

$$\psi_{\mu_1\dots\mu_n} \to \psi'_{\mu_1\dots\mu_n} = a_{\mu_1}{}^{\nu_1}a_{\mu_2}{}^{\nu_2}\dots a_{\mu_n}{}^{\nu_n}\psi_{\nu_1\dots\nu_n} \quad , \tag{16.52}$$

and the identity transformation reads

$$\psi'_{\mu_1\dots\mu_n} = \delta_{\mu_1}{}^{\nu_1}\delta_{\mu_2}{}^{\nu_2}\dots\delta_{\mu_n}{}^{\nu_n}\psi_{\nu_1\dots\nu_n} \quad . \tag{16.53}$$

Because of the orthogonality relations for $a_\mu{}^\nu$ the transformations (16.52) form a linear transformation group in 4^n dimensions; thus we have found a connection between the group SO$(4,R)$ (of $\mathrm{L_p}$) and the linear transformation group in 4^n dimensions (mathematically: an isomorphism).

Hence one says: The transformations (16.52) form a 4^n-dimensional *representation* of the group SO$(4,R)$ (or $\mathrm{L_p}$). The elements of the 4^n-dimensional representation space we shall name *tensors of rank n*, and the tensor transformation has the form

$$\psi_{\mu\nu\dots\varepsilon} \to \psi'_{\mu\nu\dots\varepsilon} = a_\mu{}^\varrho a_\nu{}^\sigma \dots a_\varepsilon{}^\omega \psi_{\varrho\sigma\dots\omega} \quad . \tag{16.54}$$

A tensor of rank 2 accordingly transforms like

$$\psi'_{\mu\nu} = a_\mu{}^\varrho a_\nu{}^\sigma \psi_{\varrho\sigma} \quad . \tag{16.55}$$

The most simple representation is the one without indices, the *scalar representation*, where the transformation obeys

$$\psi \to \psi' = \psi \quad . \tag{16.56}$$

The next possibility is for $n = 1$, i. e. in 4 dimensions:

$$\psi_\mu \to \psi'_\mu = \sum_{\nu=0}^{3} a_\mu{}^\nu \psi_\nu \quad . \tag{16.57}$$

This is the *vector representation*. Since the group SO$(4, R)$ is four dimensional, the ψ_μ are just the self-representation of the group: we denote the quantities ψ_μ as *four-vectors*. The next representation is the *tensor representation* of rank 2, which is $4^2 = 16$ dimensional. The behaviour of this transformation follows

$$\psi_{\mu\nu} \to \psi'_{\mu\nu} = a_\mu{}^\varrho a_\nu{}^\sigma \psi_{\varrho\sigma} \quad . \tag{16.58}$$

We can also interpret this as follows: The tensors $\psi_{\mu\nu}$ can be written as a 16-component vector in representation space, using the transformation law (16.58), as

$$\psi'_N = A_N{}^M \psi_M \quad (M, N = 1, \dots, 16) \quad , \tag{16.59}$$

whereby the 16×16 matrix $A_N{}^M$ is built up of the corresponding $a_\mu{}^\nu$. Symbolically we can write this as $A = a \otimes a$. The components of a tensor of rank 2 thus transform like the product of two vector components. In general, tensors of rank n transform with the $4^n \times 4^n$ matrix:

$$\hat{A} = \underbrace{\hat{a} \otimes \hat{a} \otimes \hat{a} \otimes \dots \otimes \hat{a}}_{n \text{ factors}} \quad ,$$

i. e. the components transform like the product of n vector components. Hence we conclude:

All tensor representations of the groups SO$(4, R)$ and $\mathrm{L_p}$ can be reduced to the vector representation (they are reducible). Only the scalar and the vector representation are irreducible.

16.6.2 Spinor Representations

First we consider the representations of the group SU$(2, C)$, i. e. the group of two-dimensional unitary, unimodular transformations:

$$u'_1 = au_1 + bu_2 \quad , \quad u'_2 = cu_1 + du_2 \quad . \tag{16.60}$$

Because of unimodularity

$$\det \hat{M} = \begin{vmatrix} a & b \\ c & d \end{vmatrix} = ad - bc = 1 \quad , \tag{16.61}$$

and because of unitary

$$\hat{M}^\dagger = \begin{pmatrix} a^* & c^* \\ b^* & d^* \end{pmatrix} = \hat{M}^{-1} = \begin{pmatrix} d & -b \\ -c & a \end{pmatrix} \tag{16.62}$$

must hold; therefore $d = a^*$ and $c = -b^*$, i. e.

$$\begin{pmatrix} u_1' \\ u_2' \end{pmatrix} = \begin{pmatrix} a & b \\ -b^* & a^* \end{pmatrix} \begin{pmatrix} u_1 \\ u_2 \end{pmatrix} \tag{16.63}$$

and (16.61) becomes

$$aa^* + bb^* = |a|^2 + |b|^2 = 1 \quad . \tag{16.64}$$

The group elements are the unitary, unimodular matrices

$$\hat{M} = \begin{pmatrix} a & b \\ -b^* & a^* \end{pmatrix} \quad , \tag{16.65}$$

and the group has three real parameters: the complex numbers a and b minus the condition (16.64). These are called the ***Cayley–Klein*** parameters.

The elements of the complex linear space which have the transformation behaviour (16.63) are called elementary (or two-component) *spinors* of the three-dimensional space [SU(2) is an isomorph to $SO(3, R)$].

We consider the linear space of *monomials* (a monomial of degree v is an expression of the form: $x^a y^b \dots z^c$, where $a + b + \dots c = v$) of degree v:

$$P_k = u_1^{v-k} u_2^k \quad , \tag{16.66}$$

(e. g. $P_0 = u_1^v$, $P_1 = u_1^{v-1} u_2$, etc.), where v and k are integers and $0 \leq k \leq v$. For a given v there are $v + 1$ monomials of the form (16.66), i. e. the space is $(v + 1)$ dimensional. It is also the desired representation space of the group SU(2). Applying the transformation $\hat{M}$ on P_k, one notes the following:

1. Because of the linearity of the transformations, application of the $\hat{M}$ results once more in P_k's, or linear combinations of P_k's.
2. The product of two SU(2) transformations has the same effect on P_k as the single transformations applied to P_k one after another.
3. The identity transformation of $\hat{M}$ with $a = 1$ and $b = 0$ corresponds to the identity transformation $P_k \to P_k$. in representation space. To the matrix $\hat{M}$ corresponds a $(v + 1) \times (v + 1)$ matrix in representation space. To find this matrix, we apply $\hat{M}$ on P_k:

$$P_k' = (au_1 + bu_2)^{v-k} \left(-b^* u_1 + a^* u_2\right)^k \quad . \tag{16.67}$$

Obviously it transforms according to

$$P_k' = \sum_{l=0}^{v} D_{kl} u_1^{v-l} u_2^l = \sum_{l=0}^{v} D_{kl} P_l \tag{16.68}$$

after ordering according to *monomials* of P_k in (16.67). The matrix elements D_{kl} of the matrix $\hat{D}$ [a $(v + 1) \times (v + 1)$ matrix] depend on the parameters a and b of the transformation in SU(2).

It is advantageous to introduce the abbreviation $j = v/2$, whereby $j = 0$,
It is advantagous to introduce the abbreviation $j = v/2$, whereby $j = 0, 1/2, 1, \dots$ can assume all integer or half-integer values. We denote the representation which we have constructed this way by D^j, where $D_j = \{\dots D^j(\hat{M})\dots\}$ is the set of $(2j+1)\times(2j+1)$-dimensional matrices which depend on the parameters of the SU(2) transformations $\hat{M}$. The corresponding elements of the $(2j+1)$-dimensional representation space are called *spinors of order* $2j$ of the three-dimensional real space.

The most simple representation is D^0: All elements $\hat{M}$ of SU(2) have as their image the number 1, and the representation $D^{1/2}$ is SU(2) itself: $P_1 = u_1$, $P_2 = u_2$ are the two components of the elements of representation space, higher representations being obtained analogously.

EXERCISE

16.2 D^1 Representation of SU(2)

Problem. What does the $D^1(\hat{M})$ representation of SU(2) look like? Give the matrix $D^1(\hat{M})$ in the case of

$$\hat{M} = \begin{pmatrix} a & b \\ -b^* & a^* \end{pmatrix} \quad .$$

Solution. D^1 operates on three-dimensional space since there are three components of representation space:

$$P_0 = u_1^2 \quad , \quad P_1 = u_1 u_2 \quad , \quad P_2 = u_2^2 \quad ,$$

and, additionally, $v = 2j = 2$. With (16.67) and (16.68), it follows that

$$P_k' = \sum_{l=0}^{2} D_{kl} P_l = (au_1 + bu_2)^{v-k}\left(-b^* u_1 + a^* u_2\right)^k \quad ,$$

i. e.

$$\begin{aligned} P_0' &= D_{00}u_1^2 + D_{01}u_1u_2 + D_{02}u_2^2 \\ &= (au_1 + bu_2)^2 = a^2u_1^2 + 2abu_1u_2 + b^2u_2^2 \quad , \end{aligned}$$

$$\begin{aligned} P_1' &= D_{10}u_1^2 + D_{11}u_1u_2 + D_{12}u_2^2 \\ &= (au_1 + bu_2)\left(-b^*u_1 + a^*u_2\right) \\ &= -ab^*u_1^2 + \left(aa^* - bb^*\right)u_1u_2 + a^*bu_2^2 \quad , \end{aligned}$$

$$\begin{aligned} P_2' &= D_{20}u_1^2 + D_{21}u_1u_2 + D_{22}u_2^2 \\ &= \left(-b^*u_1 + a^*u_2\right)^2 = b^{*2}u_1^2 - 2a^*b^*u_1u_2 + a^{*2}u_2^2 \quad . \end{aligned}$$

Exercise 16.2.

Comparison of coefficients with

$$\begin{pmatrix} P_0' \\ P_1' \\ P_2' \end{pmatrix} = \begin{pmatrix} D_{00} & D_{01} & D_{02} \\ D_{10} & D_{11} & D_{12} \\ D_{20} & D_{21} & D_{22} \end{pmatrix} \begin{pmatrix} P_0 \\ P_1 \\ P_2 \end{pmatrix} \quad \text{yields}$$

$$D^1(\hat{M}) = \begin{pmatrix} a^2 & 2ab & b^2 \\ -ab^* & (aa^* - bb^*) & a^*b \\ b^{*2} & -2a^*b^* & a^{*2} \end{pmatrix}$$

One can now show in a lengthly algebraic proof that the representations D^j:

(a) are irreducible, i. e. there is no representation D^{j_1} which could be built up from D^{j_2}, with $j_2 < j_1$; and

(b) the D^{j_2} represent all possible representations of $U(2)$.

16.7 Representation of SL(2, C)

The group SL$(2, C)$ is the group of the linear complex 2×2 matrices with the determinant +1. We consider now the four-dimensional unimodular group SL$(2, C)$. The transformations are, analogously to SU(2), given by

$$u_1' = au_1 + bu_2 \quad , \quad u_2' = cu_1 + du_2 \quad . \tag{16.69a}$$

The only restriction now is the unimodularity

$$\det \hat{M} = \begin{vmatrix} a & b \\ c & d \end{vmatrix} = ad - bc = 1 \quad . \tag{16.69b}$$

[The unitarity drops in the case of SL$(2, C)$.]

Since SL$(2, C)$ is isomorphic to $\mathrm{L_p}$, we denote the elements $u = (u_1, u_2)$ of SL$(2, C)$ as *two-component spinors of Minkowski space.*

The complex conjugate transformation to (16.69a) reads

$$u_{\dot{1}}' = a_1^* u_{\dot{1}} + b^* u_{\dot{2}} \quad , \quad u_{\dot{2}}' = c^* u_{\dot{1}} + d^* u_{\dot{2}} \quad , \tag{16.70a}$$

with

$$\det \hat{\overline{M}} = \begin{vmatrix} a^* & b^* \\ c^* & d^* \end{vmatrix} = a^*d^* - b^*c^* = 1 \tag{16.70b}$$

The dot on the components (e. g. $u_{\dot{1}}$, $u_{\dot{2}}$ etc.) denotes complex conjugation. Because of the independence of a and a^* etc., the elements $\dot{u} = (u_{\dot{1}}, u_{\dot{2}})$ behave differently under transformations to the elements u, i. e. the group has six real parameters [between the four complex coefficients a, b, c, d there exists only the one complex relation (16.69b)]. Thus the representations split into two groups: The spinors u, to which belong all matrices $\hat{M}$, and the dotted spinors $\dot{u}$, to which belong the matrices $\hat{\overline{M}}$. Correspondingly we construct the monomials analogous to P_k in the spinors of degree v and in the dotted spinors of degree v' as

$$P_{kk'} = u_1^{v-k} u_{\dot{1}}^{v'-k'} u_2^k u_{\dot{2}}^{k'} \quad . \tag{16.71}$$

Altogether the monomial (16.71) has the degree $v + v'$, e. g. one has

$$P_{00} = u_1^v u_{\dot 1}^{v'} \quad , \qquad P_{01} = u_1^v u_{\dot 1}^{v'-1} u_{\dot 2} \quad ,$$
$$P_{10} = u_1^{v-1} u_{\dot 1}^{v'} u_2 \quad , \quad P_{11} = u_1^{v-1} u_{\dot 1}^{v'-1} u_2 u_{\dot 2} \quad ,$$

and since k, k' are integers, then $0 \le k \le v$ and $0 \le k' \le v'$. For fixed v and v' there are altogether $(v+1)(v'+1)$ monomials, i. e. the representation space has the dimension $(v+1)(v'+1)$ and the monomials (16.71) span this representation space. Now we again introduce $j = v/2$ and $j = v'/2$ and denote the representations by $D^{jj'}$, where $j, j' = 0, 1/2, 1, 3/2, 2\ldots$. The representation has the dimension $(2j+1)(2j'+1)$ and its elements are called *spinors of rank* $(2j+1)(2j'+1)$ *of Minkowski space*. Of course, the representations are not unitary. Now we consider the matrix form of the representation $D^{jj'} = \{\ldots D^{jj'}(\hat{M}, \hat{\overline{M}})\ldots\}$, whose most simple representation is D^{00}, which is one-dimensional and yields the image 1 for every element. The next representation is $D^{\frac{1}{2}0}$, which is the self-representation of SL(2, C) by the spinors u_1 and u_2. The four-dimensional representation is based on the elements

$$P_{00} = u_1 u_{\dot 1} \quad , \quad P_{01} = u_1 u_{\dot 2} \quad , \quad P_{10} = u_2 u_{\dot 1} \quad , \quad P_{11} = u_2 u_{\dot 2} \quad .$$

With the aid of the unimodularity relation (16.69b) one obtains analogously, for example for SU(2), the image of the transformation $\hat{M}$ of SL(2, C):

$$\hat{D}^{\frac{1}{2}\frac{1}{2}}(\hat{M}, \hat{\overline{M}}) = \begin{pmatrix} aa^* & ab^* & ba^* & bb^* \\ ac^* & ad^* & bc^* & bd^* \\ ca^* & cb^* & da^* & db^* \\ cc^* & cd^* & dc^* & dd^* \end{pmatrix} \quad . \tag{16.72}$$

The $D^{jj'}$ represent *all* irreducible spinor representations of SL(2, C) and thus also of L_p [because of the isomorphism of SL(2, C) and L_p].

16.8 Representations of SO(3, R)

The three-dimensional rotation group SO(3, R) is isomorphous to the group SU(2). We have seen that all irreducible representations of SU(2) are spinor representations, and therefore this holds for SO(3, R), too. We have already constructed the tensor representation to SO(3, R): we now show that the tensor representations are included in the representations D^j, starting with D^1. Therefore we transform the basis in representation space (P_0, P_1, P_2) with the help of the matrix

$$\hat{T} = \begin{pmatrix} -1 & 0 & 1 \\ -i & 0 & -i \\ 0 & 2 & 0 \end{pmatrix}$$

into

$$\psi = \hat{T}P = \begin{cases} \psi_1 = -P_0 + P_2 \\ \psi_2 = -\mathrm{i}P_0 - \mathrm{i}P_2 \\ \psi_3 = -2P_1 \end{cases} \quad . \tag{16.73}$$

Because of $P' = \hat{D}^1(\hat{M})P$ [we have calculated $\hat{D}^1(\hat{M})$ in Exercise 16.2], it holds that

$$\psi' = \hat{T}\hat{D}^1(\hat{M})\hat{T}^{-1}\psi \quad ,$$

where

$$\hat{D}^1(\hat{M})\hat{T}^{-1} = \frac{1}{2}\begin{pmatrix} a^2 - b^2 - b^{*2} + a^{*2} & \mathrm{i}\left(a^{*2} + b^{*2} - a^2 - b^2\right) & -2\,(ab + a^*b^*) \\ \mathrm{i}\left(a^2 - b^2 - a^{*2} + b^{*2}\right) & a^2 + b^2 - a^{*2} - b^{*2} & -2\mathrm{i}\,(ab - a^*b^*) \\ 2\,(ab^* + a^*b) & 2\mathrm{i}\,(a^*b - ab^*) & 2\,(aa^* - bb^*) \end{pmatrix} .$$

Because of $aa^* + bb^* = 1$, one directky obtains

$$\psi_i'\psi_i' = \psi_i\psi_i = \text{invariant} \quad .$$

Furthermore $\det(\hat{T}\hat{D}^1\hat{T}^{-1}) = \det\hat{D}^1 = +1$ and all matrix elements are real. The transformation is therefore identical to the vector transformation $\psi_i \to \psi_i' = \alpha_i{}^k\psi_k$. Hence it follows that D^1 is identical to the vector representation. So one can show that (we don't want to give the proof here):

All representations D^j, where j is integer, are tensor representations, while all representations with half-integer j are proper spinor representations.

Now we turn to the Lorentz group.

16.9 Representations of the Lorentz Group L_p

The proper Lorentz group L_p is isomorphic to the group $SL(2, C)$; therefore the spinor representations $D^{jj'}$ are all the irreducible representations of L_p. First we consider the representation $D^{\frac{1}{2}\frac{1}{2}}$, which is based on: $P_{00} = u_1 u_{\dot{1}}$, $P_{o1} = u_1 u_{\dot{2}}$ and $P_{11} = u_2 u_{\dot{2}}$, these transforming with the matrix $\hat{D}^{\frac{1}{2}\frac{1}{2}}(\hat{M}, \hat{\overline{M}})$. Because of $ad - bc = 1$ and $a^*d^* - b^*c^* = 1$, it follows that

$$-P_{00}'P_{11}' + P_{01}'P_{10}' = -P_{00}P_{11} + P_{01}P_{10} \quad . \tag{16.74}$$

With the coordinate transformation $\psi = 1/\sqrt{2}\hat{T}P$ or $P = \sqrt{2}\hat{T}^{-1}\psi$, where

$$\hat{T} = \frac{1}{\sqrt{2}}\begin{pmatrix} 0 & 1 & 1 & 0 \\ 0 & \mathrm{i} & -\mathrm{i} & 0 \\ 1 & 0 & 0 & -1 \\ \mathrm{i} & 0 & 0 & \mathrm{i} \end{pmatrix} \quad , \tag{16.75}$$

we now introduce new axes in P space. This transformation reads explicitly:

$$\begin{aligned} \psi_1 &= \tfrac{1}{2}(P_{10} + P_{01}) \quad , \quad \text{or} \quad & P_{00} &= \psi_3 - \mathrm{i}\psi_4 \quad , \\ \psi_2 &= \tfrac{1}{2\mathrm{i}}(P_{10} - P_{01}) \quad , & P_{01} &= \psi_1 - \mathrm{i}\psi_2 \quad , \\ \psi_3 &= \tfrac{1}{2}(P_{00} - P_{11}) \quad , & P_{10} &= \psi_1 + \mathrm{i}\psi_2 \quad , \\ \psi_4 &= -\tfrac{1}{2\mathrm{i}}(P_{00} + P_{11}) \quad , & P_{11} &= -\psi_3 - \mathrm{i}\psi_4 \quad , \end{aligned} \tag{16.76}$$

and one obtains

$$\psi' = \frac{1}{\sqrt{2}}\hat{T}P' = \frac{1}{\sqrt{2}}\hat{T}\hat{D}^{\frac{1}{2}\frac{1}{2}}(\hat{M}\hat{\overline{M}})P = \hat{T}\hat{D}^{\frac{1}{2}\frac{1}{2}}(\hat{M}\hat{\overline{M}})\hat{T}^{-1}\psi \quad . \tag{16.77}$$

If we insert the P_{ik} $(i,k=0,1)$ from (16.76) in (16.74), we see that the transformation (16.77) keeps

$$\psi'_\mu\psi'_\mu = \psi_\mu\psi_\mu$$

invariant.

Because the dot in $P_{00} = u_1 u_{\dot{1}}$ etc. denotes complex conjugation, oer recognizes from (16.77) that ψ_1, ψ_2, ψ_3 are real and ψ_4 is complex. Furthermore

$$\det\left(\hat{T}\hat{D}^{\frac{1}{2}\frac{1}{2}}\hat{T}^{-1}\right) = \det\hat{D}^{\frac{1}{2}\frac{1}{2}} = \det\hat{M} \times \det\hat{\overline{M}} = +1 \quad \text{and}$$

$$\left(\hat{T}\hat{D}^{\frac{1}{2}\frac{1}{2}}\hat{T}^{-1}\right)_{44} = \tfrac{1}{2}\left(|a|^2+|b|^2+|c|^2+|d|^2\right) \geq 1 \quad ,$$

because of unimodularity. The transformation $\hat{T}\hat{D}^{\frac{1}{2}\frac{1}{2}}\hat{T}^{-1}$ transforms the ψ_μ in a manner similar to a four-vector in a proper Lorentz transformation, i. e. the representation $\hat{D}^{\frac{1}{2}\frac{1}{2}}$ is just the vector representation. The representation of the vector ψ_μ by the spinor components P is given in (16.76).

We now show a relation between $\hat{M}$ from SL$(2,C)$ and an element of L_p, taking as an example a rotation around the z axis:

$$\begin{aligned} x'_1 &= x_1\cos\varphi + x_2\sin\varphi \quad , \\ x'_2 &= -x_1\sin\varphi + x_2\cos\varphi \quad , \\ x'_3 &= x_3 \quad , \\ x'_0 &= x_0 \end{aligned} \tag{16.78}$$

or, equivalently,

$$\begin{aligned} x'_3 - \mathrm{i}x'_0 &= x_3 - \mathrm{i}x_0 \quad , \\ x'_1 - \mathrm{i}x'_2 &= \mathrm{e}^{\mathrm{i}\varphi}(x_1 - \mathrm{i}x_2) \quad , \\ x'_1 + \mathrm{i}x'_2 &= \mathrm{e}^{-\mathrm{i}\varphi}(x_1 + \mathrm{i}x_2) \quad , \\ -x'_3 - \mathrm{i}x'_0 &= -x_3 - \mathrm{i}x_0 \quad . \end{aligned} \tag{16.79}$$

If we compare this with (16.76), it follows for the spinor components that

$$\begin{aligned} P'_{00} &= P_{00} \quad , \qquad P'_{10} = \mathrm{e}^{\mathrm{i}\varphi}P_{10} \quad , \\ P'_{01} &= \mathrm{e}^{\mathrm{i}\varphi}P_{01} \quad , \qquad P'_{11} = P_{11} \quad . \end{aligned} \tag{16.80}$$

Comparing this with the general transformation (16.72) we find:

$$\begin{aligned} &aa^* = 1 \quad , \quad ad^* = \mathrm{e}^{\mathrm{i}\varphi} \quad , \quad da^* = \mathrm{e}^{-\mathrm{i}\varphi} \quad , \quad dd^* = 1 \\ &bb^* = 0 \quad , \quad cc^* = 0 \quad , \quad \text{etc.} \end{aligned}$$

The solution of these equations is

$$a = \pm\mathrm{e}^{\mathrm{i}\varphi/2} \quad , \quad b = 0 \quad , \quad c = 0 \quad , \quad d = \pm\mathrm{e}^{-\mathrm{i}\varphi/2} \quad .$$

The rotation around the z axis by the angle φ corresponds in SL(2, C) or in $\boldsymbol{D}^{\frac{1}{2}0}$ to the transformation

$$\hat{M} = \hat{D}^{\frac{1}{2}0}(\hat{M}, \hat{\overline{M}}) = \pm \begin{pmatrix} \mathrm{e}^{\mathrm{i}\varphi/2} & 0 \\ 0 & \mathrm{e}^{-\mathrm{i}\varphi/2} \end{pmatrix} \quad \text{or} \tag{16.81a}$$

$$\hat{\overline{M}} = \hat{D}^{0\frac{1}{2}}(\hat{M}, \hat{\overline{M}}) = \pm \begin{pmatrix} \mathrm{e}^{-\mathrm{i}\varphi/2} & 0 \\ 0 & \mathrm{e}^{\mathrm{i}\varphi/2} \end{pmatrix} \quad , \tag{16.81b}$$

respectively, for the dotted spinors.

In analogy to this one can show, e. g., that the transformation

$$\hat{M} = \hat{D}^{\frac{1}{2}0}(\hat{M}, \hat{\overline{M}}) = \pm \begin{pmatrix} \gamma & 0 \\ 0 & 1/\gamma \end{pmatrix}$$

corresponds to a special Lorentz transformation of the z axis, where

$$\gamma^2 = \sqrt{\frac{1-\beta}{1+\beta}} \quad \text{and} \quad \beta = \frac{v}{c}$$

is valid. Therefore the representations $\boldsymbol{D}^{\frac{1}{2}0}$ and $\boldsymbol{D}^{0\frac{1}{2}}$ are not tensor representations, but proper spinor representations. In general one finds for $\boldsymbol{D}^{jj'}$ that

$(j + j')$ integer $\Rightarrow \boldsymbol{D}^{jj'}$ is a tensor representation,
$(j + j')$ half integer $\Rightarrow \boldsymbol{D}^{jj'}$ is a proper spinor representation.

16.10 Spin and the Rotation Group

We consider a field $\psi_\alpha(x_\mu)$ whose components should have a distinct property of transformation, i. e. $\psi_\alpha(x_\mu)$ is a member of a representation space $\boldsymbol{D}^{jj'}$ of the proper Lorentz group $\mathrm{L_p}$. If we perform an infinitesimal Lorentz transformation, then the change in $\psi_\alpha(x_\mu)$ is given with the help of the generators $\left(\hat{I}^\nu{}_\mu\right)^{(\sigma\varrho)}$ of the transformation (16.16) and the infinitesimal transformation $\varepsilon_{\sigma\varrho}$ by

$$\delta\psi_\alpha(x_\mu) = \psi'_\alpha(x'_\tau) - \psi_\alpha(x_\tau) = \tfrac{1}{2}\varepsilon_{\sigma\varrho}\left(\hat{I}^\beta{}_\alpha\right)^{(\sigma\varrho)}\psi_\beta(x_\tau) \tag{16.82}$$

or, in matrix form,

$$\delta\psi = \psi'(x') - \psi(x) = \tfrac{1}{2}\varepsilon_{\sigma\varrho}\hat{I}^{(\sigma\varrho)}\psi(x) \quad . \tag{16.83}$$

On the other hand we can only consider the local variation of ψ, not of x; then, we have

$$\delta^*\psi = \psi'(x) - \psi(x) \quad .$$

The quantities $\delta\psi$ and $\delta^*\psi$ are not independent, because $x'^\tau = x^\tau + \varepsilon^\tau{}_\sigma x^\sigma$ and therefore

$$\delta^*\psi = \psi'\left(x'^\tau - \varepsilon^\tau{}_\sigma x^\sigma\right) - \psi(x) \quad . \tag{16.84}$$

Now we evaluate this expression in a Taylor series and neglect terms of second order, which yields

$$\begin{aligned}\delta^*\psi &= \psi'\left(x'^{\tau} - \psi(x^{\tau}) - \varepsilon^{\tau}{}_{\sigma}x^{\sigma}\partial_{\tau}\right)\psi \\ &= \delta\psi - \varepsilon^{\tau}{}_{\sigma}x^{\sigma}\partial_{\tau}\psi \end{aligned} \qquad (16.85)$$

and therefore

$$\delta^*\psi = \tfrac{1}{2}\varepsilon_{\sigma\varrho}\hat{I}^{(\sigma\varrho)}\psi - \varepsilon^{\nu}{}_{\mu}x^{\mu}\partial_{\nu}\psi \quad . \qquad (16.86)$$

As we sum over σ, ϱ we relabel the indices:

$$\delta^*\psi = \left(\tfrac{1}{2}\varepsilon_{\mu\nu}\hat{I}^{(\mu\nu)} - \varepsilon^{\nu}{}_{\mu}x^{\mu}\partial_{\nu}\right)\psi \quad . \qquad (16.87)$$

Rewriting the second term:

$$\begin{aligned}\varepsilon^{n}u_{\mu}x^{\mu}\partial_{\nu}\psi &= \tfrac{1}{2}\varepsilon^{\nu}{}_{\mu}\left[(x^{\mu}\partial_{\nu} + x_{\nu}\partial^{\mu}) + (x^{\mu}\partial_{\nu} - x_{\nu}\partial^{\mu})\right]\psi \\ &= \tfrac{1}{2}\varepsilon^{\nu}{}_{\mu}\,(x^{\mu}\partial_{\nu} - x_{\nu}\partial^{\mu})\,\psi \quad , \end{aligned} \qquad (16.88)$$

then, because of the antisymmetry of $\varepsilon^{\nu}{}_{\mu}$, it results that

$$\delta^*\psi = \tfrac{1}{2}\left[-\varepsilon^{\nu}{}_{\mu}\,(x^{\mu}\partial_{\nu} - x_{\nu}\partial^{\mu}) + \varepsilon_{\mu\nu}\hat{I}^{(\mu\nu)}\right]\psi \quad .$$

Hence, for the space-like components ($i, k = 1, 2, 3$) one gets

$$\delta^*\psi = \tfrac{1}{2}\left[-\varepsilon^{k}{}_{i}\left(x^{i}\partial_{k} - x_{k}\partial^{i}\right) + \varepsilon_{ik}\hat{I}^{(ik)}\right]\psi \quad . \qquad (16.89)$$

Up to the factor $\mathrm{i}/\hbar$ the first expression in brackets is just the angular momentum operator

$$\hat{L}^{i}{}_{k} = \frac{\mathrm{i}}{\hbar}\left(x^{i}\partial_{k} - x_{k}\partial^{i}\right) \quad . \qquad (16.90)$$

The second term, i. e. $(\mathrm{i}/\hbar)\hat{I}^{(ik)}$, is independent of the coordinates and therefore also of the coordinate system; thus it is straightforward to interpret this term as an " inner" angular momentum, i. e. as a spin. With the help of the generators we see that [see (16.21–16.23)]

$$\begin{aligned}\hat{S}_1 &= \frac{\hbar}{\mathrm{i}}\hat{I}^{(32)} = \hbar\left(\hat{J}^{1(+)} + \hat{J}^{1(-)}\right) \quad , \\ \hat{S}_2 &= \frac{\hbar}{\mathrm{i}}\hat{I}^{(13)} = \hbar\left(\hat{J}^{2(+)} + \hat{J}^{2(-)}\right) \quad , \\ \hat{S}_3 &= \frac{\hbar}{\mathrm{i}}\hat{I}^{(21)} = \hbar\left(\hat{J}^{3(+)} + \hat{J}^{3(-)}\right) \quad , \end{aligned} \qquad (16.91)$$

Now, the commutation relations of the $\hat{S}_i$ are a direct consequence of the commutation relations of the $\hat{J}^{i(\pm)}$ and thus of the generators of the infinitesimal Lorentz transformation, i. e.

$$\hat{S}_1\hat{S}_2 - \hat{S}_2\hat{S}_1 = \mathrm{i}\hbar\hat{S}_3 \qquad (16.92)$$

and cyclic commutations therefrom.

Consequently the operators $\hat{S}_i$ have all the properties of an angular momentum operator: thus for a given integer or half-integer s, $\hat{S}_3$ has the $(2s+1)$ eigenvalues

$(-s, -s+1, \ldots, s-1, s)$. Furthermore there exists a Casimir operator $\hat{S}^2 = \sum_i \hat{S}_i^2$, with the eigenvalues $\hbar^2 s(s+1)$.

Using these considerations, we define the operator of the total angular momentum to be

$$\hat{M}^{ik} = \frac{\hbar}{\mathrm{i}}\left(x^i\partial^k - x^k\partial^i\right) + \frac{\hbar}{\mathrm{i}}\hat{I}^{(ik)} = \hat{\boldsymbol{L}} + \hat{\boldsymbol{S}} \quad , \tag{16.93}$$

and hence we get from (16.89) that

$$\delta^*\psi = \tfrac{1}{2}\varepsilon_{ik}\frac{\mathrm{i}}{\hbar}\hat{M}^{ik}\psi \quad . \tag{16.94}$$

Thus the coefficient of the local variation of the field defines the operator of the total angular momentum, and we conclude: To determine the spin of the field $\psi(x)$ we must find infinitesimal operators generating space-like rotations in the representation space $D^{jj'}$ which is determined by the transformation properties of ψ. Let us now only consider the subgroup of the Lorentz group $\mathrm{L_p}$ which characterizes the space-like rotations. Furthermore we can then conclude: In this case also, the matrices $\hat{M}$ from SL(2, C), which belong to space-like rotations, are unitary [not only unimodular, and thus they do not belong to SU(2)]. To understand the consequences we define the *direct product* of two matrices $\hat{A}$ and $\hat{B}$ as

$$\hat{C} = \hat{A} \otimes \hat{B} \tag{16.95}$$

or, in matrix notation,

$$\hat{C}_{\alpha\beta,\gamma\delta} = \hat{A}_{\alpha\gamma}\hat{B}_{\beta\delta} \quad . \tag{16.96}$$

This definition becomes clearer in the following example:

$$\begin{aligned}\hat{A}\otimes\hat{B} &= \begin{pmatrix} a_{11} & a_{12} \\ a_{21} & a_{22}\end{pmatrix} \otimes \begin{pmatrix} b_{11} & b_{12} \\ b_{21} & b_{22}\end{pmatrix} \\ &= \begin{pmatrix} a_{11}b_{11} & a_{11}b_{12} & a_{12}b_{11} & a_{12}b_{12} \\ a_{11}b_{21} & a_{11}b_{22} & a_{12}b_{21} & a_{12}b_{22} \\ a_{21}b_{11} & a_{21}b_{12} & a_{22}b_{11} & a_{22}b_{12} \\ a_{21}b_{21} & a_{21}b_{22} & a_{22}b_{21} & a_{22}b_{22}\end{pmatrix} \\ &= \begin{pmatrix} a_{11}\hat{B} & a_{12}\hat{B} \\ a_{21}\hat{B} & a_{22}\hat{B}\end{pmatrix} \quad .\end{aligned} \tag{16.97}$$

Of course this operation is not commutative, i. e.

$$\hat{B}\otimes\hat{A} = \begin{pmatrix} b_{11}\hat{A} & b_{12}\hat{A} \\ b_{21}\hat{A} & b_{22}\hat{A}\end{pmatrix} \neq \hat{A}\otimes\hat{B} \quad .$$

With the help of definition (16.97), one can easily derive the following properties:

1. The direct product of two diagonal matrices is diagonal.
2. $(\hat{A}+\hat{B})\otimes\hat{C} + (\hat{A}\otimes\hat{C}) + (\hat{B}\otimes\hat{C})$.
3. If $\hat{A}$, $\hat{A}'$ are $n\times n$ matrices and $\hat{B}$, $\hat{B}$ $m\times m$ matrices, then it holds that $(\hat{A}\otimes\hat{B})(\hat{A}'\otimes\hat{B}') = \hat{A}\hat{A}'\otimes\hat{B}\hat{B}'$.
4. If $\hat{A}$ and $\hat{B}$ are unitary, then $\hat{A}\otimes\hat{B}$ is also unitary
5. $\mathrm{tr}(\hat{A}\otimes\hat{B}) = \mathrm{tr}\,\hat{A}\,\mathrm{tr}\,\hat{B}$.

With the help of these definitions we are now able to rewrite the representation (16.72) as

$$\hat{D}^{\frac{1}{2}\frac{1}{2}}(\hat{M},\hat{\overline{M}}) = \begin{pmatrix} aa^* & ab^* & ba^* & bb^* \\ ac^* & ad^* & bc^* & bd^* \\ ca^* & cb^* & da^* & db^* \\ cc^* & cd^* & dc^* & dd^* \end{pmatrix} \overset{!}{=} \begin{pmatrix} a & b \\ c & d \end{pmatrix} \otimes \begin{pmatrix} a^* & b^* \\ c^* & d^* \end{pmatrix} \quad . \tag{16.98}$$

From this example one can conclude that this representation can be described as a direct product, i. e. for the matrices we have

$$\hat{D}^{jj'}(\hat{M},\hat{\overline{M}}) = \hat{D}^{j}(\hat{M}) \otimes \hat{D}^{j'}(\hat{\overline{M}}) \quad . \tag{16.99}$$

The matrices $\hat{D}^{j}(\hat{M})$ and $\hat{D}^{j'}(\hat{\overline{M}})$ are not identical with the matrices of the representation $\boldsymbol{D}^{j}$, $\boldsymbol{D}^{j'}$ of the SU(2)! Of course this is only the case if the matrix $\hat{M}$ is unitary. But this is just the case for space-like rotations. Thus for the total representation space we can write:

$$\boldsymbol{D}^{jj'} \mathrel{\widehat{=}} \boldsymbol{D}^{j} \otimes \boldsymbol{D}^{j'} \quad , \tag{16.100}$$

i. e. for space-like rotations the representation $\boldsymbol{D}^{jj'}$ of the Lorentz group L_p splits up into a direct product of the representations $\boldsymbol{D}^{j}$ and $\boldsymbol{D}^{j'}$ of the three-dimensional rotation group, each $\boldsymbol{D}^{j}$ representing a different spin. Therefore, in general a covariant Lorentz field $\psi_\alpha(x_\mu)$ does not possess unique spin representations, but splits up into different spin representations $\boldsymbol{D}^{j}$, $\boldsymbol{D}^{j'}$.

First we consider the representation $\boldsymbol{D}^{j}$. As we have seen it is $(2j + 1)$ dimensional. In this space the infinitesimal operators which determine the spin are $(2j+1)\times(2j+1)$ matrices and consequently have $(2j+1)$ eigenvalues. Degeneracy does not appear, because the quantities spanning up the representation space are linearly independent. But we have seen that these are the $2s + 1$ eigenvalues of the spin, i.e. *the spin is given by the index j of the representation* $\boldsymbol{D}^{j}$. For example, $j = 0$ belongs to the scalar representation $\boldsymbol{D}^{0}$ describing spinless fields. For the vector representation $\boldsymbol{D}^{1}$ the spin has the value 1. This becomes clear with help of the infinitesimal operators: For space-like rotations we have:

$$\hat{I}^3 = \hat{I}^{(21)} = \begin{pmatrix} 0 & 1 & 0 \\ -1 & 0 & 0 \\ 0 & 0 & 0 \end{pmatrix} \quad .$$

EXERCISE

16.3 Vector Representation and Spin

Problem. Show that spin 1 belongs to the vector representation $\boldsymbol{D}^{1}$.

Solution. The vector representation $\boldsymbol{D}^{1}$ is just the self-representation of the space of three-dimensional rotations, i. e. the spin values are given by the eigenvalues of the corresponding generators of the Lorentz group. For three-dimensional rotations this is

Exercise 16.3.

$$\hat{I}^3 = \hat{I}^{(21)} = \begin{pmatrix} 0 & 1 & 0 \\ -1 & 0 & 0 \\ 0 & 0 & 0 \end{pmatrix} \quad .$$

The characteristic equation reads

$$\det(\hat{I}^3 - \lambda 1\!\!1) = 0 \quad , \quad \text{i.e.} \quad \begin{vmatrix} -\lambda & 1 & 0 \\ -1 & -\lambda & 0 \\ 0 & 0 & -\lambda \end{vmatrix} = 0 = -\lambda(1+\lambda^2) \quad ,$$

hence $\lambda = 0, \pm\text{i}$. Because $s = (\hbar/\text{i})\lambda$, one gets $s = (-1, 0, 1)\hbar$.

For the representation $\boldsymbol{D}^{\frac{1}{2}}$, one finds analogously, $s = \hbar/2$, etc. Without proof we note that (16.100) can be written as a sum with the help of a generalized Clebsch–Gordan theorem:

$$\boldsymbol{D}^{jj'} \mathrel{\hat{=}} \boldsymbol{D}^{j} \otimes \boldsymbol{D}^{j'} = \boldsymbol{D}^{j+j'} \otimes \boldsymbol{D}^{j+j'-1} \otimes \ \ldots \ \otimes \boldsymbol{D}^{j-j'} \quad . \tag{16.101}$$

Therefore we have, for example,

$$\boldsymbol{D}^{\frac{1}{2}0} \mathrel{\hat{=}} \boldsymbol{D}^{\frac{1}{2}} \otimes \boldsymbol{D}^{0} = \boldsymbol{D}^{\frac{1}{2}} \quad ,$$

because

$$D^{\frac{1}{2}} = \pm \begin{pmatrix} \text{e}^{-\text{i}\varphi/2} & 0 \\ 0 & \text{e}^{\text{i}\varphi/2} \end{pmatrix} \quad .$$

$\boldsymbol{D}^{\frac{1}{2}}$ is a self-representation of SU(2) (i.e. of all unitary, unimodular matrices), and indeed, $\boldsymbol{D}^{0\frac{1}{2}}$ is a matrix of SU(2) ($\boldsymbol{D}^{\frac{1}{2}}$).

From (16.101) it follows that: the spin which belongs to the representation $\boldsymbol{D}^{jj'}$ is

half integer, if $j + j'$ is half-integer
integer, if $j + j'$ is integer .

However, this means: *the tensor representation belongs to the integer spins, the spinor representation to the half-integer spins.*

The described representation is not unique, which is illustrated in the following example: One has

$$\boldsymbol{D}^{\frac{1}{2}\frac{1}{2}} \mathrel{\hat{=}} \boldsymbol{D}^{\frac{1}{2}} \otimes \boldsymbol{D}^{\frac{1}{2}} = \boldsymbol{D}^{1} \otimes \boldsymbol{D}^{0} \quad . \tag{16.102}$$

According to our rules $\boldsymbol{D}^{\frac{1}{2}\frac{1}{2}}$ describes spin-1 fields. However, we see that (16.102) also contains scalar parts, which do not lead to spin values $s = 1$. To get a unique representation one has to introduce auxiliary fields, which just eliminate this scalar part ($\boldsymbol{D}^{0}$). Such auxiliary conditions play an important role in the quantum theory of the Maxwell field.

16.11 Biographical Notes

POINCARÉ, Henri Jules, French mathematician and philosopher, * 29.4.1854 Nancy, † 17.7.1912 Paris, was a cousin of Raymond Poincaré, President of the French Republic during World War I. Between 1879 and 1881 at the University of Caen, and from 1881 at the University of Paris, P. worked in the field of pure mathematics (automorpheous functions), made important contributions to the theory of equilibrium properties of rotating fluids and achieved – independently of Einstein – a number of results of the special theory of relativity in his famous paper on the dynamics of the electron, published in 1906.

CAYLEY, Arthur, British mathematician, * 16.08.1821 Richmond, † 26.01.1895 Cambridge. C. was first a lawyer in London and from 1863 a professor at Cambridge. With J. J. Sylvester he founded the "theory of invariants" and algebraic geometry. By using complex coordinates C. showed that metric geometry is contained in projective geometry: with his projective measure (1859) he gave a new foundation to geometry which allowed the treatment of euclidian and noneuclidian geometry from a common point of view. He invented matrix calculus and was the first to formulate group theory (the representation of finite groups by multiplication tables or permutations) in an abstract way. C. also worked on conformal mappings, elliptic and hyperelliptic functions, the theory of differential equations, theoretical mechanics, the motion of the moon, and spherical astronomy. [BR]

KLEIN, Felix, German mathematician, * 25.04.1849 Düsseldorf, † 22.06.1925 Göttingen. K. studied from 1865 to 1870 in Bonn. During an educational stay 1870 in Paris he came into contact with the rapidly developing group theory. From 1871 K. taught at Göttingen and became professor at Erlangen in 1872, at München in 1875, at Leipzig in 1880 and at Göttingen in 1886. He made fundamental contributions to function theory, geometry, and algebra. He was especially interested in group theory and its applications. In 1872 K. published the Erlanger program. When he was older he occupied himself more intensely with pedagogical and historical problems.

Subject Index

Printing: Saladruck, Berlin
Binding: Buchbinderei Lüderitz & Bauer, Berlin